Practical Design of Ships and Other Floating Structures

Elsevier Science Internet Homepage

http://www.elsevier.nl (Europe)
http://www.elsevier.com (America)
http://www.elsevier.co.jp (Asia)

Consult the Elsevier homepage for full catalogue information on all books, journals and electronic products and services

Elsevier Titles of Related Interest

BAI
Pipelines and Risers
ISBN: 008-043712-5

BJORHOVDE, COLSON & ZANDONINI
Connections in Steel Structures III.
ISBN: 008-042821-5

CHAN & TENG
ICASS '99, Advances in Steel Structures. (2 Volume Set).
ISBN: 008-043015-5

DUBINA
SDSS '99 – Stability and Ductility of Steel Structures.
ISBN: 008-043016-3

FRANGOPOL, COROTIS & RACKWITZ
Reliability and Optimization of Structural Systems.
ISBN: 008-042826-6

FUKUMOTO
Structural Stability Design.
ISBN: 008-042263-2

GODOY
Thin-Walled Structures with Structural Imperfections: Analysis and Behavior.
ISBN: 008-042266-7

GUEDES-SOARES
Advances in Safety and Reliability (3 Volume Set).
ISBN: 008-042835-5

MOAN & BERGE
13th Int Ship & Offshore Structures Congress (ISSC 1997).
ISBN: 008-042829-0

OHTSUBO & SUMI
Proceedings of the 14th International Ship and Offshore Structures Congress.
ISBN: 008-043602-1

OWENS
Steel in Construction (CD-ROM Proceedings with Printed Abstracts Volume, 268 papers).
ISBN: 008-042997-1

SRIVASTAVA
Structural Engineering World Wide 1998 (CD-ROM Proceedings with Printed Abstracts Volume 702 papers).
ISBN: 008-042845-2

USAMI & ITOH
Stability and Ductility of Steel Structures.
ISBN: 008-043320-0

VASSALOS
Contemporary Ideas on Ship Stability.
ISBN: 008-043652-8

VUGTS
BOSS '97 Behaviour of Offshore Structures (3 Volume Set).
ISBN: 008-042834-7

WATSON
Practical Ship Design.
ISBN: 008-042999-8

YOUNG
Wind Generated Ocean Waves.
ISBN: 008-043317-0

Related Journals

Free specimen copy gladly sent on request. Elsevier Science Ltd, The Boulevard, Langford Lane, Kidlington, Oxford, OX5 1GB, UK

Applied Ocean Research
Advances in Engineering Software
CAD
Coastal Engineering
Composite Structures
Computers and Structures
Construction and Building Materials
Engineering Failure Analysis
Engineering Fracture Mechanics
Engineering Structures
Finite Elements in Analysis and Design
International Journal of Solids and Structures
Journal of Constructional Steel Research
Marine Structures
NDT & E International
Ocean Engineering
Structural Safety
Thin-Walled Structures

To Contact the Publisher

Elsevier Science welcomes enquiries concerning publishing proposals: books, journal special issues, conference proceedings, etc. All formats and media can be considered. Should you have a publishing proposal you wish to discuss, please contact, without obligation, the publisher responsible for Elsevier's ocean engineering publishing programme:

Ian Salusbury
Senior Publishing Editor
Elsevier Science Ltd
The Boulevard, Langford Lane
Kidlington, Oxford
OX5 1GB, UK

Phone: +44 1865 843425
Fax: +44 1865 843920
E.mail: i.salusbury@elsevier.co.uk

General enquiries, including placing orders, should be directed to Elsevier's Regional Sales Offices – please access the Elsevier homepage for full contact details (homepage details at the top of this page).

Practical Design of Ships and Other Floating Structures

Proceedings of the Eighth International Symposium on Practical Design of Ships and Other Floating Structures

16 - 21 September 2001
Shanghai, China

Edited by

You-Sheng Wu
China Ship Scientific Research Center,
Wuxi, Jiangsu, China

Wei-Cheng Cui
School of Naval Architecture & Ocean Engineering,
Shanghai Jiao Tong University,
Shanghai, China

and

Guo-Jun Zhou
China Ship Scientific Research Center,
Wuxi, Jiangsu, China

Volume II

2001
ELSEVIER
AMSTERDAM – LONDON – NEW YORK – OXFORD – PARIS – SHANNON – TOKYO

ELSEVIER SCIENCE Ltd
The Boulevard, Langford Lane
Kidlington, Oxford OX5 1GB, UK

First edition 2001

British Library Cataloguing in Publication Data
International Symposium on Practical Design of Ships and other Floating Structures (8th : 2001 : Shanghai, China)
Practical design of ships and other floating structures : proceedings of the Eighth International Symposium on Practical Design of Ships and other Floating Structures 16-21 September 2001, Shanghai, China
1.Naval architecture - Congresses 2.Shipbuilding - Congresses
I.Title II.Wu, You-Sheng III.Cui, Wei-Cheng IV.Zhou, Guo-Jun
623.8

ISBN 0080439500

Library of Congress Cataloging in Publication Data
A catalog record from the Library of Congress has been applied for.

ISBN: 0-08-043950-0 (2 Volume set)

∞ The paper used in this publication meets the requirements of ANSI/NISO Z39.48-1992 (Permanence of Paper).
Printed in The Netherlands.

PREFACE

During the last century the science and technology of ships and marine structures experienced extremely great progress, and thus created the modern shipbuilding, shipping and ocean industries. The relevant achievements were a part of the driving sources, which changed the whole world and the society. Among the efforts towards these achievements was the creation of "The First International Symposium on Practical Design in Shipbuilding" in 1977 in Tokyo. Later on it became a series of symposia, PRADS as the abbreviation. Last century seven PRADS symposia were held in Tokyo ('77 and '83), Seoul ('83 and '95), Trondheim ('87), Varna ('89), Newcastle ('92) and The Hague ('98).

This proceedings contains the papers presented at "The 8th International Symposium on Practical Design of Ships and Other Floating Structures" held at Shanghai Everbright Convention & Exhibition Center, China on 16-21 September 2001. This is the first of the PRADS Symposia in the 21st century.

The overall aim of PRADS symposia is to advance the design of ships and other floating structures as a professional discipline and science by exchanging knowledge and promoting discussion of relevant topics in the fields of naval architecture and marine and offshore engineering. Inline with the aim, in welcoming the new era this Symposium is particularly for an increase in international cooperation and giving a momentum for the new development of design and production technology of ships and other floating structures for efficiency, economy, safety, and environmental production.

The main themes of this Symposium are Design Synthesis, Production, Hydrodynamics, Structures and Materials of Ships and Floating Systems. Proposals for over 270 papers from 26 countries and regions within the themes were received for PRADS'2001, and about 170 papers were accepted for presentation at the symposium. With the high quality of the proposed papers the Local Organizing Committee had a difficult task to make a balanced selection and to control the total number of papers for fitting into the allocated time schedule approved by the Standing Committee of PRADS.

Volume I of the proceedings covers the subjects about design synthesis, production and part of hydrodynamics. Volume II contains the subjects for the rest of hydrodynamics, structures and materials.

On behalf of the Standing Committee of PRADS and the Local Organizing Committee of PRADS'2001, we would like to thank all the participants for their great contributions to the successful symposium. The full support from the sponsors, Mechanical and Vehicle Engineering Division of Chinese Academy of Engineering, Chinese Society of Naval Architects and Marine Engineers, and Chinese Institute of Navigation are greatly acknowledged. Sincere gratitude is also extended to China Ship Scientific Research Center, Shanghai Jiao Tong University and other institutes and shipyards in China, who have helped the preparation of this Symposium.

You-Sheng Wu
Wei-Cheng Cui
Guo-Jun Zhou

These Proceedings of Volumes I and II contain papers presented at the 8th International Symposium on Practical Design of Ships and Other Floating Structures. The Symposium was held at the Shanghai Everbright Convention & Exhibition Center in Shanghai, China, on 16-21 September 2001, and organized by:

CAE	**Chinese Academy of Engineering, Mechanical and Vehicle Engineering Division**
CSNAME	**Chinese Society of Naval Architects and Marine Engineers**
CIN	**China Institution of Navigation**

These organizations are represented in the Local Organizing Committee.

The Local Organizing Committee organized the Symposium under supervision of the PRADS's Standing Committee. The Symposium gained the generous support of many sponsors. They are listed together with the membership of the committees in the following.

HONORARY ADVISORY COMMITTEE

Mr. Rong-Sheng Wang, *President, Chinese Society of Naval Architects and Marine Engineers*
Mr. Xiao-Jin Chen, *President, China State Shipbuilding Corporation*
Mr. Shan-Xiang Hong, *Vice Minister, Ministry of Communications*
Mr. Ping-Tao Huang, *President, China Shipbuilding Industrial Corporation*
Mr. Zai-Kuan Jin, *Vice President, China State Shipbuilding Corporation*
Mr. Ke-Jun Li, *President, China Classification Society*
Mr. Zu-Yi Lin, *President, China Institute of Navigation*
Prof. Dian-Zuo Wang, *Vice President, Chinese Academy of Engineering*
Mr. Hui Wang, *Vice President, China Shipbuilding Industrial Corporation*
Mr. Guang-Qin, Zhang, *Vice President of Commission of Science, Technology and Industry for National Defence*

PRADS STANDING COMMITTEE

Prof. S. Motora (Honorary Chairman), *Previously, Ship and Ocean Foundation, Japan*
Prof. You-Sheng Wu (Chairman), *China Ship Scientific Research Center, China*
Prof. T. Borzecki, *Technical University of Gdansk, Poland*
Dr. L.L. Buxton, *University of Newcastle, UK*
Prof. O.M. Faltinsen, *The Norwegian Institute of Technology, Norway*
Dr. R. Porcari, *Italian Ship Research Center, Italy*
Prof. J. J. Jensen, *Technical University of Denmark, Denmark*
Prof. H. Kim, *Seoul National University, Korea*
Dr. D. Liu, *American Bureau of Shipping, U.S.A.*
Prof. H. Ohtsubo, *University of Tokyo, Japan*
Dr. M.W.C. Osterveld (Ex-officio), *MARIN, The Netherlands*
Prof. H. Petershagen, *University of Hamburg, Germany*
Dr. S. G. Tan, *MARIN, The Netherlands*
Prof. Wei-Cheng, Cui (Secretary), *Shanghai Jiao Tong University, China*

CONTENTS

VOLUME I

VOLUME II

3. HYDROMECHANICS (*continued*)

PROPULSOR AND PROPULSION

PODDED DRIVES

HULL-PROPULSOR-APPENDAGES INTERACTION

3. HYDROMECHANICS (*continued*)

Practical Design of Ships and Other Floating Structures
You-Sheng Wu, Wei-Cheng Cui and Guo-Jun Zhou (Eds)

PROPELLER DESIGN AND ANALYSIS SYSTEM USING AN OBJECT-ORIENTED DATABASE IN WINDOWS ENVIRONMENT

Chang-Sup Lee[1] and Chung-Ho Cho[1]

[1] Department of Naval Architecture and Ocean Engineering,
Chungnam National University, Taejon, Korea

ABSTRACT

An integrated propeller design and analysis system using an objected-oriented database in Windows environment is developed. The system consists of various computational modules such as the basic geometry definition, optimization, lifting-line and lifting-surface design and analysis, finite-element steady and unsteady analysis, and cavity analysis. The whole system is controlled by a user-friendly graphic user interface (GUI), which enables the user to perform the design swiftly and accurately. A new database-drive class is designed and implemented to enable the exchange of information between the object-oriented database and the computational modules. The dynamic link library (DLL) technique enables the whole system disintegrated into modules. Selected graphical outputs are demonstrated very helpful in performing design and analysis.

KEYWORDS

Propeller design, Propeller analysis, Database, Interface design, Graphic user interface, Windows

1 INTRODUCTION

An integrated program package for the design and analysis usually consists of many elemental modules covering various computing needs. It is not an easy task even for the trained expert to follow correctly the input/output procedure of various modules avoiding human errors. It gets even more difficult if various modules are interrelated interchanging information among them. In order to use this kind of package effectively, the user has to be familiar with the design flow, which must be simple, and also can control the information flow between various modules.

The easiness in carrying out the design can be achieved by adopting the graphic user interface (GUI) to control the computational flow. The GUI is now very popular in many applications working in Windows environment and facilitates the use of the complicated programs through a series of simple mouse clicks. To maximize the visual effect in judging the input parameters and the computational results, the graphic package commercially available may be linked to the design package as a dynamically linked library (DLL) file. The techniques are well known in the computer software technology and have been successfully applied to the propeller design and analysis (ProDAS) package by Jung et al(1996, 1997).

In order to provide the input data to each computational module and also to exchange the information between modules, the data has to be stored and managed systematically, and to be easily accessible by the user and by each program module. This paper will describe the procedure to build the systematic relation between modular programs and the database (DB) system together with the sample application of the package for the design and analysis of a propeller.

We note first that most of the existing programs in the ship design and marine hydrodynamics does not use the database. Even when the database is in use, most programs only accept the ASCII file to communicate between the database and the modular programs. Use of ASCII file is traditional as an input/output form. With the increase of the package size, the code will generate unwanted junk files and produce and save duplicate data files deteriorating the efficiency and accuracy of the design procedure. The ASCII file is very inefficient in re-processing the information and is very difficult in providing the proper data to the other modules. The number of the ASCII files increases with the number of the modular programs, whereas only one database file is necessary to support the whole package. If separate treatment of the data among different modules is necessary, the data set may be grouped into separate tables to match the modular programs.

To secure the linkage between the modules and the database, we selected a relational database package, Access of Microsoft, which is known easy to use and reliable. Figure 1 shows the linkage between the database system and the computational modules under the control of the design system. If the user asks, the modules can also be connected to the traditional file system.

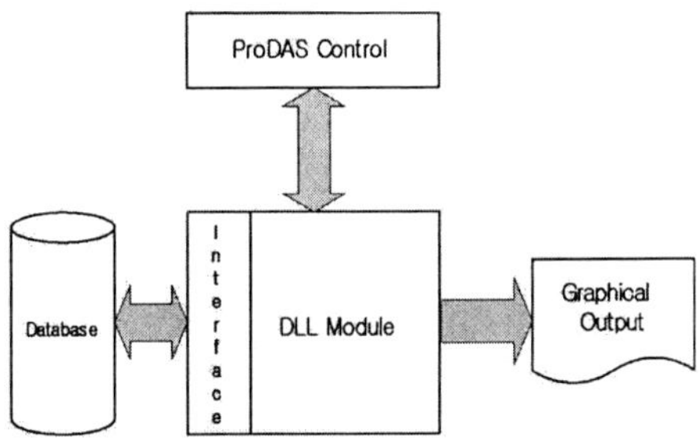

Figure 1: A Dll module communicating with DB

2 DATABASE INTERFACE DESIGN

We will now describe the procedure developing the interface to enable the data exchange between the application and the database, using the driver, e.g., DAO (Data Access Objects) as adopted in the present work, supplied by the developer of the database system. Figure 2 shows the relations between the database interface classes, which may be grouped into three regions (Allison 1996). The classes CDaoDatabase and CDaoRecordset, supplied by the database developer, connect the database and the application, and provide the information passage between them. The programmer needs only to know the meaning of the member variables and the usage of the member functions of these classes. The classes in the upper-left region of Figure 2, CProdasDB, …, CUpdate, provide the application modules with the basic functions which locate the proper record set and enable the read/write operation. These classes are very general and not related to the content of each module or to the specific tables or fields in the database. The classes in the lower-left region of Figure 2, CDBMopti, …, CDBMoptiRecSet, provide the connectivity to exchange information between the module and database record set. This portion has to be implemented into each module within the application package.

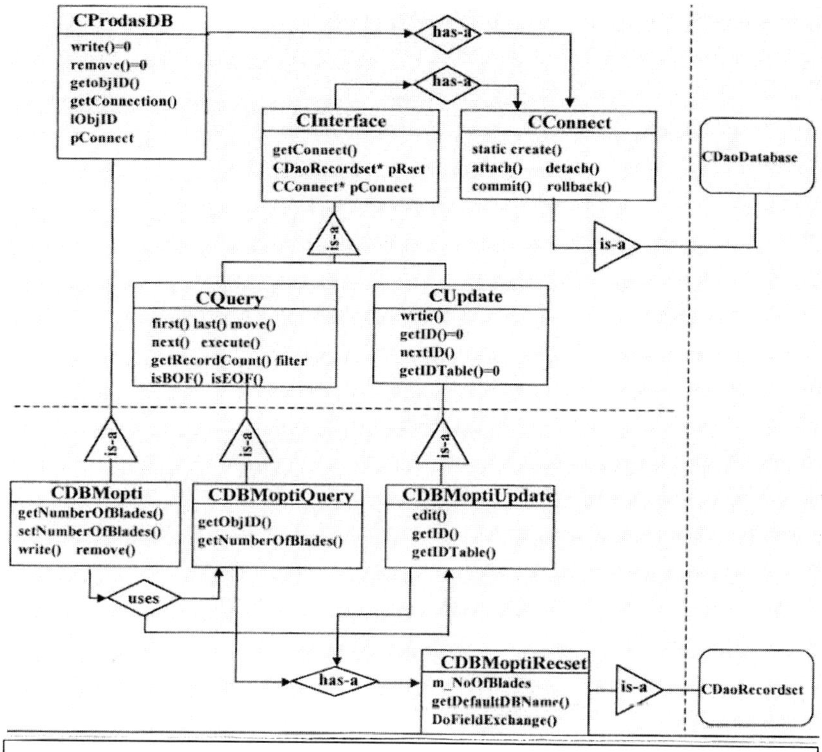

Figure 2 Database interface class structure

2.1 *Function and Interrelations of Interface Classes*

The CProdasDB class is the top-level class in the application containing the member variables, pConnect and lObjID, that connect the database and point the record, respectively. The class CConnect creates an object to connect, to open, to attach or to detach from the database and also has the functions to commit the data-writing into the disc or to roll-back the works that are not yet stored permanently into the disc. The class CInterface contains the CConnect pointer pConnect and CDaoRecordset pointer pRset, which are to be inherited into two derived classes CQuery and CUpdate and used to point the database and the proper record. The CQuery class is used to move the pointer into the requested record according to the query conditions. The class CUpdate is an abstract class which updates the data modified by the application modules. The five classes described above, which are the basic classes in ProDAS, provides the functions to create and to control the data flow management.

The classes shown in the lower-left region in Figure 2 are dependent upon each computational module. The class CDBMopti is derived from CProdasDB; CDBMoptiQuery and CDBMoptiUpdate from CQuery and CUpdate, respectively; CDBMoptiRecSet from CDaoRecordset. The most basic class among them, CDBMopti, has the right to use two friends CDBMoptiQuery and CDBMoptiUpdate and hence can query and modify the data, upon creation of the CDBMopti object. This object may also use its own member functions (e.g. getNumberOfBlades() and setNumberOfBlade()) to read and write the data. The class CDBMoptiRecSet initializes the data in the record and provides gateways to do data exchange between the database record set and the application.

```
// DBMoptiQuery.cpp
... ... ...
char* CDBMoptiQuery::queryTable[CDBMoptiQuery::NFIELDS+1] =
        {"Mopti", ... , "[NoOfBlades]", ...};
... ... ...
int CDBMoptiQuery::getNumberOfBlades() const
{
 return ((CDBMoptiRecSet *)pRset)->m_nblade;
}

... ... ...

// DBMoptiRecSet.cpp: implementation file
... ... ...
CDBMoptiRecSet::CDBMoptiRecSet(CDaoDatabase* pdb)
: CDaoRecordset(pdb)
{
 //{{AFX_FIELD_INIT(CDBMoptiRecSet)
 m_NoOfBlades = 0;
 ... ... ...
 //}}AFX_FIELD_INIT
 m_nDefaultType = dbOpenDynaset;
 //m_bCheckCacheForDirtyFields = FALSE;
}

void CDBMoptiRecSet::DoFieldExchange(CDaoFieldExchange* pFX)
{
 ... ... ...
 DFX_Long(pFX, _T("[NoOfBlades]"), m_NoOfBlades);
 ... ... ...
}
```

Figure 3 Segments of Query and Recordset

```
// Optimization Module in ProDAS
extern "C" int WINAPI MOpti(CListBox FAR* mDlg, long id)
{
 CConnect *pc = CConnect::create("..\\ProDBase.mdb");
 CDBMopti mopti(pc, id);
 InputFromDBFile(mopti);
 ... code segments ...
 OutputToDBFile(mopti);
 pc->detach();
}

void InputFromDBFile(CDBMopti &mopti)
{
 ident   = mopti.getIdent();
 nvel    = mopti.getNumberOfVelocities();
 igivend = mopti.getGivenDiameter();
  ... ... ...
 for (n=0; n<nvel; n++) {
   svs[n] = mopti.getShipVs(n);  // ship speed
   ... ... ...
 }
}

void OutputToDBFile(CDMopti &mopti)
{
 mopti.setPropDiameter(ddiam);
 mopti.write();
}
```

Figure 4 Segments of a sample DLL code

2.2 *Exchange of Information between Application and Database*

Once the interface classes are completed, they may be implemented into the source code as shown in Figure 3. The upper portion shows the query table array of the CDBMoptiQuery class corresponding to the fields of the database. For simplicity, shown only here are the table name "Mopti" and one particular field [NoOfBlades] in the optimization module in ProDAS. Figure also shows how the function getNumberOfBlades() returns the value of m_NoOfBlades, which is the member variable of the object pRset pointing the recordset. The lower portion shows how the data m_NoOfBlades gets the value [NoOfBlades] through DoFieldExchange() function implemented in DBMoptiRecSet code.

2.3 *Creating Objects in Applications*

If the interface classes and the implementation programs are completed, the remaining task is now to create the object in the application program. Figure 4 shows a code segment extracting or storing the data between the database and a DLL module in ProDAS. The function MOpti first connects to the database by creating the CConnect object *pc, and then creates a CDBMopti object mopti. Two functions InputFromDBFile() and OutputToDBFile() are then shown. The objective and contents of the functions are obvious. The global variables belonging to the particular module, e.g. nvel or svs[n], are assigned by calling the member functions of the CDBMopti class object mopti. All other variables required by computation modules are extracted from the database similarly. Values determined by the modules can also be stored by calling the member functions of the CDBMopti class as shown in OutputToDBFile() function. As demonstrated so far, the interface class and implemented functions are very simple to use and easily applicable to the computing modules.

3 DESIGN IN WINDOWS ENVIRONMENT

Existing design programs require high level of knowledge in using the codes, making decision and evaluating designs. When various computation modules are involved, it is important to select proper programs and to exchange information between them. In practice, any computations can be repeated as many times as needed with the updated input data when passing through the design spiral. An easiest way to control the design flow in a complicated package is to introduce the menu-driven method in Windows environment as shown in Figure 5. Visual display of various design steps in the pull-down menu enables selections of proper "buttons" representing pre-combined design stages. Figure 5 shows the package consists of six major design steps including the data preparation, the preliminary design, the lifting surface design, the strength analysis, the cavity analysis and the diagnosis.

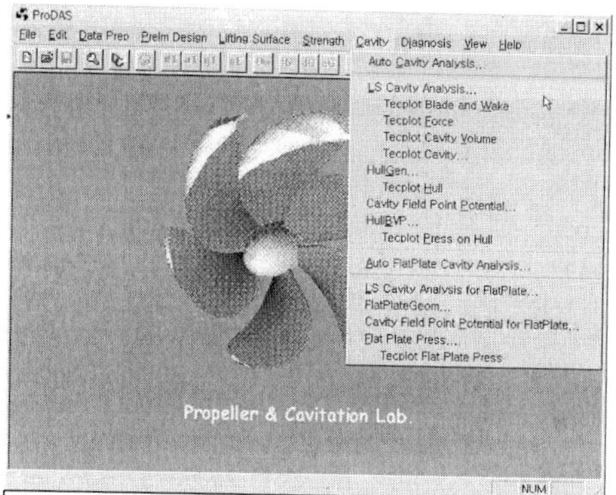

Figure 5 Main window of ProDAS showing the menus

3.1 *Data Preparation via Database System*

The most significant advantage of adopting the database is to use SQL to locate the record, to filter and manipulate the data easily. Existing data may be used as the stock propeller. Each module in ProDAS has the corresponding data tables in the database as shown in Figure 6. Each button in the figure, when clicked by the mouse, shows the data input/out windows as typically shown in Figure 7. Each box in this figure contains the data for computing module. This set of data may be manually typed, corrected or read from the ASCII files, which might be generated by other computations or experiments. The computed output may also be displayed.

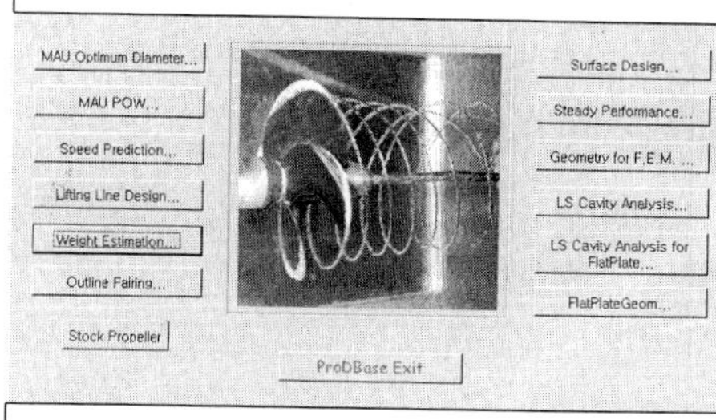

Figure 6 Initial window of database for ProDAS

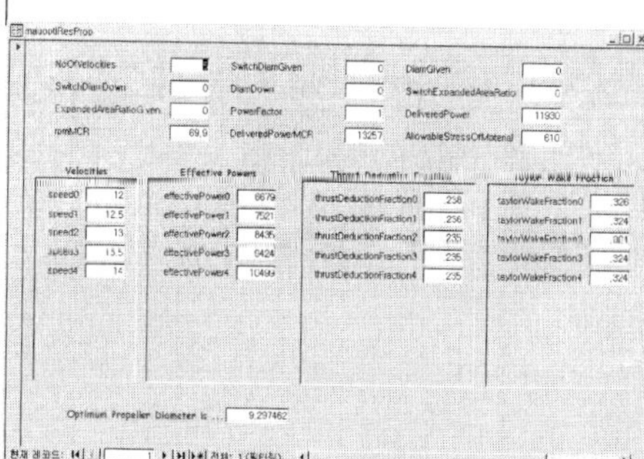

Figure 7 Typical view of the database showing the data updated

3.2 *Visual Display of Outputs*

By selecting the computation steps one by one as suggested in the design menu in Figure 5, all the design works can be performed without much difficulty. The DLL technique also allows us to link the commercially available graphic packages as the library dynamically. The computed output may be directed into the database and at the same time to the graphic routines. Figure 8 shows the sample plotting of the computed hull pressure fluctuation induced by the cavitating propeller. The color contour maximizes the visual effect in locating the region of high loads.

3.3 *Onsite Guidelines for Decision Making*

Since the design criteria can also be stored in the database, the computed output may be directly compared with the criteria and displayed as shown in Figure 9. The computed unsteady force amplitudes (in red) are compared with the guideline (in green). This online guideline feature is especially useful to alert the designer, minimizing the likelihood of erroneous judgement and is useful to the beginners who have limited amount of experiences.

Online Help Message

When the designer is not experienced or when the design procedure is very complicated, the online help system is very helpful. Figure 10 shows that methods using various steps in design are fully supported. Help files are written in HTML (Hypertext Markup Language), which can be accessed with a simple mouse click over the underlined keywords. The help files remains open on top of all other windows and thus the user is assisted by the detailed online information from the Help file.

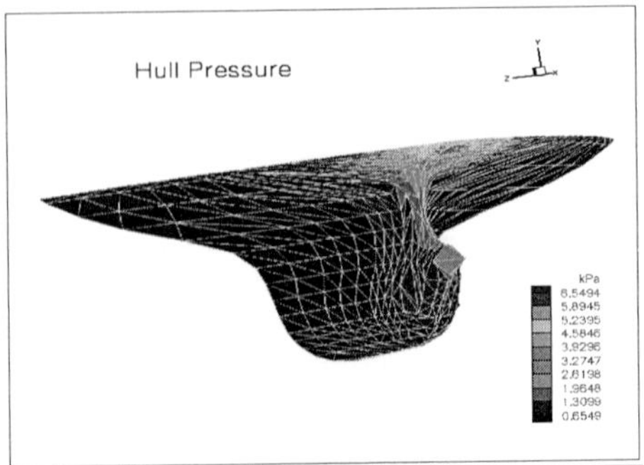

Figure 8 Pressure due to a cavitating propeller

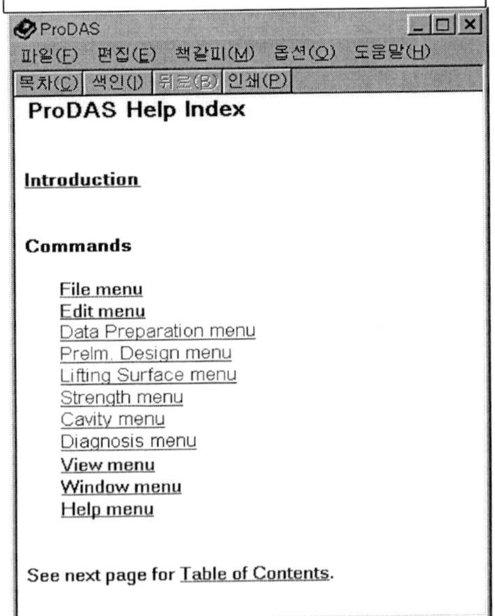

Figure 10 Help topics showing the design guideline

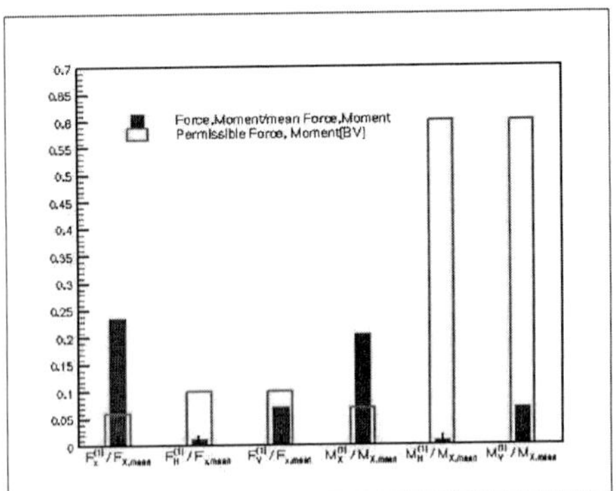

Figure 9 Onsite guideline for decision making

4 CONCLUSIONS

1. The database interface classes for easy exchange of information between the database and computing modules are designed, illustrated in detail, so that the same technique can be applied to other applications.

2. Each elemental program is modularized for dynamic linkage with the main program when needed, reducing the size of the execution file, and ensuring easy maintenance in the future.

3. Storage of all design data into a single database can play the role of the stock propellers, and provide information for other designs.
4. The menu-driven windows system is shown very easy to use, providing the maximum accuracy and

swiftness. Visual effect to demonstrate the output of the computations helps the designer to correctly interpret the results and to make decisions for the subsequent design stages.

References

Allison, C. (1996). Object Persistence with Relational Database. C/C++ Users Journal, **14:5**, 37-46.

Jung, I.-S., Lee, C.-S. and Cho, C.-H. (1997). Design and implementation of interface between the database and the application program. *J. SNAK*, **34:4**, 139-147 (in Korean).

Jung, I.-S., Lee, C.-S. and Lim, H.-K. (1996). Propeller design using GUI technique. *Proc. Spring Conference*, Society of Naval Architects of Korea (SNAK), 220-223 (in Korean).

Practical Design of Ships and Other Floating Structures
You-Sheng Wu, Wei-Cheng Cui and Guo-Jun Zhou (Eds)

A PROPELLER DESIGN METHOD WITH NEW BLADE SECTION FOR IMPROVING CAVITATION INCEPTION UNDER UNSTEADY CONDITION

Wei-Xin Zhou, You-Hua Wu and Shi-Tang Dong

China Ship Scientific Research Center, Wuxi, Jiangsu 214082, China

ABSTRACT

A propeller design procedure for improving cavitation inception in non-uniform inflow is presented. Special attention is paid to the design of blade sections with Eppler Method, in order to behave higher cavitation inception speed. The two-dimensional section directly obtained by Eppler method is converted into a new blade section in three-dimensional case since they are significantly different from each other. Calculations show that the cavitation-free bucket diagram of new blade sections is wider than the conventional NACA sections.

The lift-line and lift-surface method are used for the initial propeller design with the circumferential averaged inflow. Using steady/unsteady panel method to check and to modify the design for meeting the thrust requirement and selecting a key section suffering from cavitation. New blade sections are adopted to improve the cavitation bucket. To investigate the influence of the chordwise loading distribution forms on the propeller performance, two propellers with different chordwise loading are designed. Their model test results confirm the effectiveness of the design method.

KEYWORDS

Propeller design, New blade section, Cavitation inception, Lifting-surface method, Surface Panel method, Eppler section design

1 INTRODUCTION

It is well known that the object of wake adopted marine propeller design is to improve efficiency or cavitation performance including reducing the propeller-induced vibration and noise. Therefore, theoretical design methods, which take into consideration the radially varied but circumferentially averaged inflow and adopt NACA66 thickness form and a=0.8 mean line for section design, have been commonly employed.

For higher-speed marine propellers operating in non-uniform wake fields, some extent of cavity is inevitable. It is not good enough to use usual design method with choosing the conventional blade sections. However, use of the Eppler's method[2] makes it possible to design new blade sections for prescribed pressure distributions around the blade section to delay cavitation inception.

During recent decade, numerous researchers have made contributions to developing propeller design methods with new blade sections based on Eppler method for improving cavitation performance. It has been theoretically and experimentally verified that the cavitation-free buckets of the new sections are much wider than those from NACA series. Among them, the methods by Yamaguchi et.al. [3-5], Kuiper [6], Nakazaki et.al.[7], Jin-Tae Lee et.al. [8], Dang et al. [9] and C. Kawakita and T. Hoshino[10] have gotten satisfactory results.

This paper presents a design procedure of marine propeller with new blade sections. The existing lifting-line and lifting-surface design method are employed for an initial design, by which only radially varied but circumferential averaged inflow can be involved.As a check, the steady and unsteady panel method are used to predict the hydrodynamics performance of the designed propeller at steady and unsteady conditions. From unsteady calculations, in which both radial and circumferential non-uniformity of inflow are taken into account, the sections suffering from cavitation can be found and a key section is chosen. New blade section based on Eppler method is designed to raise cavitation inception speed. An approach to convert a 2-D Eppler section into the one suitable for 3-D propeller case is applied. The thickness form and the chordwise load distribution form of the key section are employed for all radii to avoid the difficulty of surface fairing and smoothing.

To investigate the influence of the chordwise loading distribution form on the propeller performance, two propellers with different chordwise loading forms are designed, one with NACA series sections and the other with new sections. The model test results verify that the cavitation inception speed with the application of new blade sections is significantly higher and its induced fluctuating pressure and noise are much better than those of the conventional one.

2 DESIGN METHOD OF MARINE PROPELLERS BY USING NEW BLADE SECTIONS

Our new procedure of propeller design consists of six parts. Figure 1 shows its flow chart. Every part needs iteration until the desired propeller performance is reached. The brief descriptions of each part are described below.

2.1 Principle dimensions

The principle dimensions, such as diameter, tentative blade area ratio and radial chord-length distribution, number of blades, skew and thickness are designed as usual way after examination on wake harmonic analysis and strength. Those parameters except the blade area ratio and radial chord-length distribution, which may be modified in lifting line design, are not changed in the further optimization.

2.2 Propeller Lifting-line and lifting-surface design

In order to achieve the required thrust and better cavitation performance, the existing propeller lifting-line and lifting-surface design methods are used to design detailed propeller geometry. At this stage, only circumferentially averaged, radial wake distribution can be considered.

In the lifting-line calculation, discretized vortex model is proposed instead of continuous one, including each free vortex represented by segments of vortex line. The induced velocities are calculated by numerical integration directly instead of Lerb's methods. By this way, the deformation of the slipstream can be assigned to be closing to the real case in the near wake. The lifting-line design is mainly utilized to obtain the initial hydrodynamic pitch angle and the chord-length distribution for the lifting-surface code.

The lifting-surface design is to get the geometrical pitch angle distribution and the geometry of each blade section. The blade shapes is usually chosen from the existing NACA a=0.8 camber lines and the modified NACA66 thickness form, but now is from new blade sections if necessary. It will be described in details in section 3.

2.3 Analysis of blade strength

The blade strength is checked according to the rules of ship classification societies Then the static stress distribution over the blade surface is analyzed by utilizing Finite Element Method (FEM) under the static hydrodynamic loading distribution from the steady panel method calculation. Also the dynamic blade stress can be examined by unsteady panel method and FEM.

2.4 Prediction of hydrodynamic performance

The propeller's thrust and torque are predicted by the steady panel method in circumferentially averaged wake field. Using the unsteady panel method checks its time-averaged thrust and torque. Both the steady and unsteady panel methods used were developed in CSSRC. Once the thrust is not satisfied, modification of the design should be made until it is fulfilled.

2.5 Prediction of cavitation inception

For checking the cavitation inception on the blade sections at each radius, the pressure distributions of each blade section are calculated over one revolution in both radial and circumferential non-uniform wake field by the unsteady panel method.The operating curves for each section that is lift coefficient C_L versus $-c_{pmin}$, where c_{pmin} is minimum pressure coefficient, can be drawn. Among the sections, a representative one suffering from cavitation is selected as a key section, usually 0.7R or 0.8R section to be the candidate. Meanwhile, an equivalent 2-D section of the key blade section is defined based on the same chordwise loading distribution and thickness distribution as in 3-D case, referring to the concept of [6]. .

3 DEVELOPMENT OF A NEW BLADE SECTION

3.1 Design of two-dimensional sections

Taking into account the cavitation number, Reynolds number and lifting coefficient of the key section, Eppler's method is used to design a new two-dimensional section, which is tried to widen and deepen the cavitation–free bucket as much as possible. During the design process, many parameters have to be adjusted and meanwhile, it should be kept in mind that as much as possible to avoid separation occurring within the wide range of angles of attack, i.e., enough region for pressure recovery on suction side to be required. In the present design procedure, a boundary layer calculation code is provided for checking separation possibility. No separation occurring implies low drag of the section.

Fig.2 shows the cavitation-free bucket diagram and the operating curve of the key section of our example. The cavitation bucket of conventional NACA blade section can't envelop the operating curve but the new blade section can.

3.2 Design of three-dimensional blade sections of propeller

After gaining the two-dimensional new section design, a 2-D panel method is used to analyze the chordwise loading distribution of that section. Fig.3 is the comparison between it and that of NACA a=0.8 section.

The chordwise load distribution form and the chordwise thickness distribution form of that new section is adopted for all blade sections of the new blade design by using lifting surface method again.

The camber lines from 2-D design and 3-D design are shown in figure 4, in which NACA a=0.8 camber line from 3-D design also shown. Fig.5 shows the difference between the sections from two-dimensional and three-dimensional design.

3.3 Verification of hydrodynamic performance of the new design

Utilizing the steady and unsteady panel method, its hydrodynamic performance and the pressure distributions on the blade surface are checked in order to verify if the new design satisfies thrust requirement and the cavitation inception speed is improved. Figures 6-8 are the comparisons of chord-wise pressure distribution of 0.8R section between the new design and the conventional one. They also show that the propeller with new blade section has more even pressure distribution than that of conventional one, even if at the unsteady condition. That implies that the new design will have better cavitation performance, such as higher inception speed, lower exciting forces and noise.

It should be mentioned that all the steps described above have been assembled into a package code.

4 DESIGN EXAMPLE AND ITS MODEL TEST RESULTS

In order to verify the effectiveness of our method, two propellers were designed for a high speed ship. One propeller is designed with sections formed with $NACA66_{(mod)}$ and NACA a=0.8 meanline. The other was designed with new blade sections by using present method. Fig2.-Fig.8 show some of the design results and their test verification is presented and discussed below.

4.1 Propeller open-water characteristics

The present design procedure uses panel method to check and to ensure the design accuracy. In order to verify the accuracy of the panel method, the designed example's open water hydrodynamic forces are compared between the test results and calculated ones, and shown in figure 9. It shows in good agreement with each other.

4.2 Cavitation observation and inception speed

Cavitation observation and inception tests were carried out in the simulated wake inside the cavitation tunnel. The variations of cavitation pattern on the propeller blade during one revolution are shown in Fig.10-12 with angle interval 10 degree. Fig.10 shows that there exists bubble cavitation on the backside from 0.5R to 0.88R for the propeller with conventional sections when the blade location at 80°. It is in good correspondence with the pressure prediction by the panel method (see Fig.7). Fig.11 shows that there is face cavitation from the root to 0.55R along the leading edge within the range of blade locations 280^{0}-320^{0} for that propeller. It is also in correspondence with the results shown in Fig.8 . Fig.12 shows that neither bubble cavitation on the backside nor face cavitation was observed for the propeller with new blade sections.

Fig.13 is the cavitation inception curve in the uniform inflow and Fig.14 is one in the non-uniform inflow. It is proved that the cavitation bucket of the propeller with new blade sections is wider than the one with the conventional blade section. And Cavitation inception tests in non-uniform inflow also show that the inception speed of the propeller with new blade section is about 1.5knots higher than that of the conventional propeller.

4.3 Fluctuating pressure and noise

Induced fluctuating pressures on a flat plate above the propeller models were measured and the harmonic components of the maximum measured pressure among the transducers are shown in Table 1. The fluctuating pressure level of the propeller with the new blade sections is lower than that with conventional sections.

Table 1 Comparison of the fluctuating pressure level for harmonic components

	1st order	2nd order	3rd order	4th order
The conventional propeller	3.16kPa	1.23kPa	0.32kPa	0.31kPa
The new propeller	1.85kPa	1.05kPa	0.35kPa	0.32 kPa

Table 1 Comparison of the fluctuating pressure level for harmonic components

The noise tests were also carried out in the cavitation tunnel and its results are shown in Table 2. The noise level of the propeller with the new blade sections is about 4dB lower at the maximum speed than that of with conventional section. The overall noise level of the propeller with the new blade sections lower over different ship speeds.

	Max. speed	29knots	26knots	18knots
Conventional sections	144.0dB	144.0 dB	140.6 dB	136.4 dB
New blade sections	140.3 dB	139.1 dB	138.0 dB	135.5 dB

Table 2 Comparison of noise level for different ship speeds

5 CONCLUSIONS

From the design process and the model test results, following conclusions can be drawn:

(1) Due to marine propellers that operate in circumferential non-uniform wake fields, the propeller with new blade sections can provide with better cavitation performance at wide range of angle positions over one revolution than that of one with the NACA blade sections.

(2) From the open-water test, it shows that the panel method used in present design procedure provides good accuracy.

(3) Model test results also show that all blade harmonic amplitudes as well as the noise levels of the propeller with the new blade sections are significantly lower than those of the propeller with conventional sections, and the cavitation inception speed of the new propeller is higher than the conventional one.

References

[1] Abbott I.H. and Von Doenhoff A.E., *Theory of Wing Sections*. Dover Publications.

[2] Eppler R. And Somers D.M. (1980). A Computer Program for the Design and Analysis of low Speed Aifoils, *NASA Technical Memorandum 80210*.

[3] Yamaguchi H.etal. (1985). Development of Marine Propellers with Better Cavitation Performance(1st Report). *Journal of SNAJ*. **158**.

[4] Yamaguchi H. et. al. (1986). Development of New Marine Propellers with Improved Cavitation Performance. *Procedings of ISPC*, Wuxi, China, CSNAME.

[5] Yamaguchi H.etal. (1988). Development of Marine Propellers with Better Cavitation Performance (2nd Report)， *Journal of SNAJ*. **163**.

[6] Kuiper G. et. al. (1993). A Propeller Design Method for Unsteady Conditions. *Technical Session of the Centennial Meeting of the Society of Naval Architects and Marine Engineers*.

[8] Lee J.T., Kim M.C., Ahn, J.W.Kim, K.S. and Kim H.C. (1991). Development of Marine propellers with New Blade Sections for Container Ships. *Proceedings of Propellers Shafting 91 Symposium*. Virginia Beach, Virginia.

[9] Dang J., Chen J. and Tang D. (1992). A Design Method of Highly Skewed Propellers with New Blade Sections in Circumferentially Non-Uniform Ship Wake. *China Ship Scientific Research Center, Report English version 92004*.

[10] Kawakita C., Hoshino T. (1998). Design System of marine Propellers with new Blade Sections. *23rd International Symposium on Cavitation*. Mistubishi Heavy Industries.

Practical Design of Ships and Other Floating Structures
You-Sheng Wu, Wei-Cheng Cui and Guo-Jun Zhou (Eds)

AN OPTIMISATION METHOD BASED ON HILBERT SPACE THEORY FOR DESIGN OF MARINE PROPELLERS AND HULL FORM

Taek S. Jang [1], Takeshi Kinoshita [1] and Takanori Hino [2]

[1] Institute of Industrial Science, University of Tokyo, Tokyo, Japan
[2] Ship Research Institute 6-38-1, Shinkawa, Mitaka, Tokyo, 181-0004, Japan

ABSTRACT

By using the infinite dimensional optimization[Jang and Kinoshita(2000)], which is based on the Hilbert space theory, optimal marine propellers and optimal hull forms are studied. In the first study, Koyama's code[Koyama(1980)], based on potential flow with simple viscous correction, are applied to numerical calculation of hydrodynamic forces on blades of marine propellers. As a numerical example, the MAU type propeller is considered and used as the initial guess for the optimization method. In the second study, the Navier-Stokes solver[Hino(1996)] is used as a practical tool for hull design. For the first step, two dimensional hull form is studied to minimize its drag. The numerical results for an optimal marine propeller as well as hull form are illustrated and it is shown that practically reasonable optimizations could be found.

KEYWORDS

Infinite dimensional optimization method, Optimal marine propellers, Optimal hull forms

1 OPTIMAL MARINE PROPELLERS

Forward speed of a propeller U, a diameter of a propeller d and revolution per time n being fixed constant, the advance ratio $J = U/nd$ is constant, so that thrust T and torque Q of the propeller are functionals of the geometry of the propeller.(Figure 1) Because the geometry can be described by a distribution of pitch $p(r)$, it is possible to assume that both the thrust T and the torque Q are functionals of the pitch distribution $p(r)$;

$$T = T(p) \tag{1}$$

$$Q = Q(p) \tag{2}$$

According to Eqn. 1. and Eqn. 2., the propeller efficiency η_p is also a functional of the pitch distribution $p(r)$; i.e.

$$\eta_p = \frac{JK_T(p)}{2\pi K_Q(p)} = \eta_p(p) \tag{3}$$

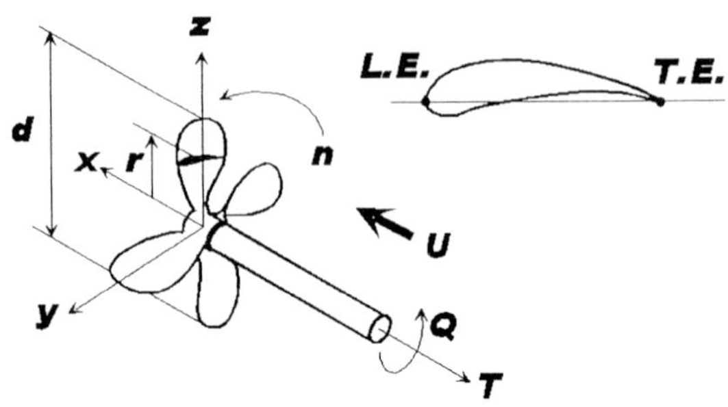

Figure1: A propeller sketch for terminology

Let $h = h(r)$ be a small variation in the sectional pitch $p(r)$ with a regularity such that

$$\|h\| = \left(\int_{RB}^{RO} |h(r)|^2 dr \right)^{1/2} = O(\varepsilon) \tag{4}$$

Where $\varepsilon << 1$ indicates a positive small parameter and the symbol $\|\cdot\|_X$ is a norm[Roman(1975)]. In Eqn. 4., RB and RO stand for hub (radius of boss) and tip(radius of propeller) respectively. Jang and Kinoshita(2000) derived the following infinite linear matrix equation so that components g_i of Eqn. 7. can be solved;

$$\eta_p{}^k = \sum_{i=1}^{\infty} A_{k,i} g_i \tag{5}$$

where

$$\eta_p{}^k = \eta_p(p + h^k) - \eta_p(p) \quad , \quad A_{k,i} = < h^k, e_i > = \int_{RB}^{RO} h^k(r) e_i(r) dr$$

To generate the variations suitable for the present optimization, the test function [Jang, Kinoshita and Yamaguchi(2000)] in the theory of generalized function is introduced ; i.e. a function which is infinitely differentiable and which vanishes outside some finite interval. Then the variations in Eqn. 5 has the following form

$$h^k(r) = \begin{cases} \varepsilon \cdot \exp[\dfrac{\delta^2}{(r-b_k)^2 - \delta^2}], & |r - b_k| < \delta \\ 0 & |r - b_k| > \delta \end{cases}$$

where δ and $b_k \in (RB, RO)$ are positive reals such that $RB < b_k \pm \delta < RO$. The desired distribution of pitch distribution, which increases the propeller efficiency, is

$$p_{new}(r) = p_0(r) + h_{\max}(r) \tag{6}$$

$$\text{where} \qquad h_{\max}(r) = a \sum_{i=0}^{\infty} g_i e_i(r) \tag{7}$$

Here a is positive real. Here, $p_0(r)$ denotes an initial guess of pitch distribution. With $p_{new}(r)$ being substituted into Eqn. 6. instead of $p_0(r)$, the above optimal procedure can be repeated while a constraint is satisfied. The constraint that local change of pitch must be bounded is imposed in Eqn. 8.; i.e. in this study, maximal local change of pitch is chosen as 20%.

$$\left|\frac{p_{opti}(r)}{p_0(r)}\right| < 1.2, \qquad RB < r < RO \tag{8}$$

The fluid surrounding a propeller is assumed to be inviscid, incompressible, free of cavitation and infinite in extent so that the velocity potential function can exist and fluid velocity is expressed by the gradient of the potential function. Assuming that the blades of the propeller is thin, the mean surface of them can be replaced with bound vortex sheets. Theoretical evaluation of hydrodynamic forces acting on the propeller requires the blade loading distribution obtained from lifting surface theory. To evaluate the blade loading distribution of the propeller with a given geometry is reduced to a two-dimensional singular integral equation. In this study, the Kernel function expansion method or series expansion method and Koyama's code[Koyama(1980)] are used for numerical calculation of hydrodynamic forces. Frozen wake model is adopted to calculate fluid flow and a simple viscous correction is made by taking into account the section lift slope reduction and viscous drag. As a numerical example, the MAU type propeller is considered and used as the initial guess; the two constant pitch distributions of $p_0 = 171.44(mm)$ and $p_1 = 182.315(mm)$ are used in the optimization process. By the way, for the propeller of constant pitch distribution p_0, numerical calculation indicates that the maximal value of the efficiency is 0.6635 at the advance ratio;

$$J_{\max} = 0.7494\,. \tag{9}$$

It is reasonable to optimize the MAU type propellers of constant pitch distributions of p_0 and p_1 at the advance ratio $J_{\max}$. As shown in solid line in Figure 2, in the case of the p_0 propeller, the present optimization is stopped when iteration number reaches 14,

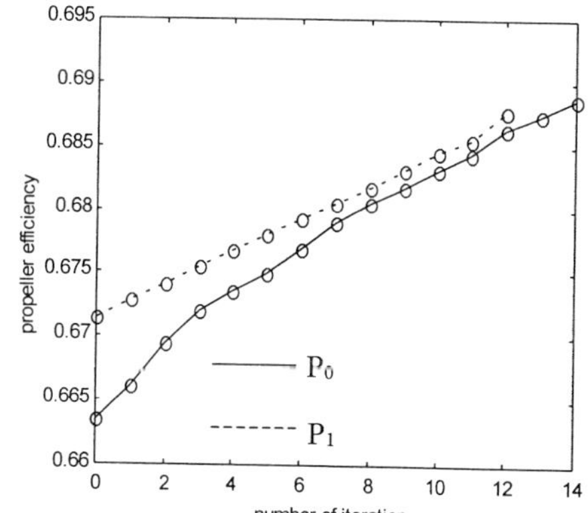

Figure 2: Optimization process for the MAU type propeller

because of the constraint of Inequality 8. From Figure 3, it is easy to find that the two pitch distributions are in reasonably good agreement with each other.

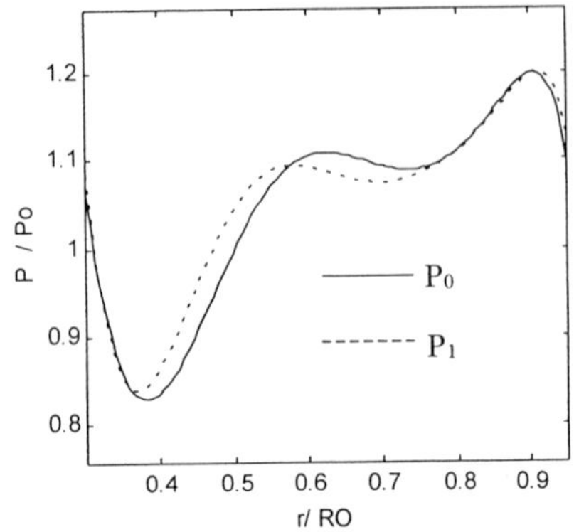

Figure 3: The comparison of the two optimized pitch distributions

It is interesting that two slightly different pitch distributions in Figure 3 gives nearly coincided open-water characteristics[Jang, Kinoshita and Yamaguchi(2000)]. In fact, mathematically, Jang, Kinoshita and Yamaguchi(2000) showed the uniqueness of the optimized pitch distribution i.e. the present optimization method gives only one solution in a Hilbert space, which is independent of an initial guess of pitch distribution. Thereby propeller open- characteristic curve becomes unique.

2 OPTIMAL HULL FORM

As first step, a two dimensional body is considered without free surface flow in this study. The initial body for optimization is given as(Figure 4)

$$B(x) = 0.6x(1-x) \qquad 0 \le x \le 1,\ y \ge 0 \tag{10}$$

where the leading edge is $x = 0$. Provided that a body moves with a constant speed U, then the drag D of the body is a functional of the geometry of the body B;

$$D = D(B) \tag{11}$$

In the present study, the Reynolds number is taken as 10^6 and the constraint is imposed as follows;

$$B_{opt}(0.5) = B(0.5) \tag{12}$$

The governing equations are the incompressible Navier-Sotkes equations with introduction of artificial compressibility. Turbulence models implemented are the algebraic model by Baldwin and Lomax(1978) and its variant[Kodama(1998)]. The Navier-Stokes solver[Hino(1996)] is used to design optimal hull form. Efficiency of the optimization procedure is strongly dependent on CPU time of a flow solver, since the Navier-Stokes analysis is most time-consuming among the elements of the optimization system. Using Eqn. 5. 6. 7., the initial body of Eqn. 10. can be optimized and the numerical result for the optimal shape is seen in Figure 4. The same class of variations in the previous section are generated and applied to optimize the hull form of Equation 10. Here the coefficient of drag is defined as follows;

$$C_d = \frac{D}{\frac{1}{2}\rho U^2 L} \tag{13}$$

Figure 5 shows the history of convergence of Cd in the optimization process; 1000 iteration is required to obtain numerical convergence. And the histories for pressure drag and frictional drag are shown in Figure 6.

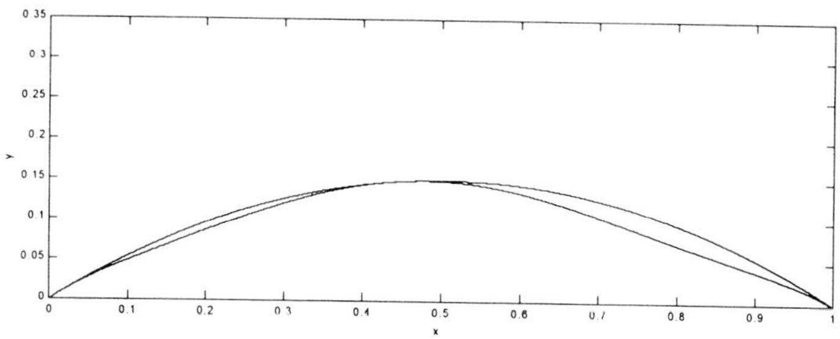

Figure 4: The comparison of optimized body(solid line) and the initial body(dotted line)

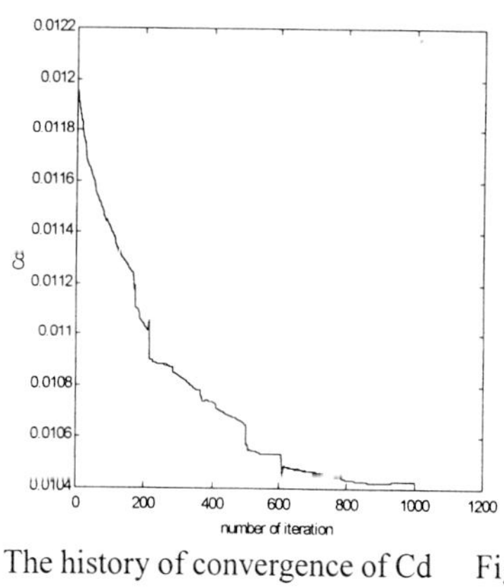

Figure 5: The history of convergence of Cd

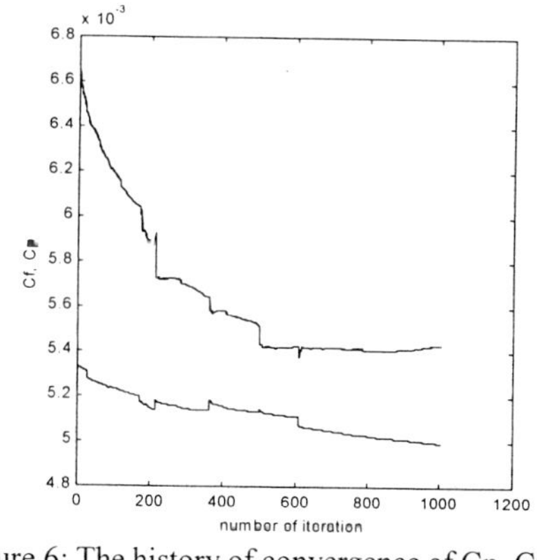

Figure 6: The history of convergence of Cp, Cf (solid: Cp, dotted: Cf)

3 CONCLUSIONS

In this paper, the infinite dimensional optimization method based on the Hilbert space theory is applied to find the optimal pitch distribution for marine propellers and the optimal shape for the two dimensional hull form. Thereby desired performances are significantly improved compared to original ones. Especially, it is interesting that pressure drag can be significantly reduced compared to frictional one for the optimal hull. Unfortunately, in the case of the optimal hull form, there are some changes in displacement, which looks like weak point of the present method; it is necessary to develop the method further so that it can overcome the problem of change in displacement.

References

Baldwin B.S. and Lomax H. (1978). Thin layer approximation and algebraic model for separated turbulent flows. *AIAA* , 78-257

Hino T. (1996). Towards Drag Reduction by Hull Form Optimization. *Proc. 68th General Meeting of Ship Research Institute*, December [in Japanese]

Jang T.S. and Kinoshita T. (2000). A minimization theory in a Hilbert space and its application to two dimensional cavity flow with a numerical study. *Journal of Marine Science and Technology* (submitted).

Jang T.S., Kinoshita T. and Yamaguchi H.(2000). Application of an infinite dimensional optimization method to marine propellers; unique optimal pitch distribution. *Journal of Marine Science and Technology* (submitted)

Kodama Y. (1998). Scope of CFD for computing ship flow , *Proc. The third Osaka Colloquium on advanced CFD application to ship flow and Hull form design*. 395-405

Koyama K. (1980). On application of the lifting surface theory to marine propellers. *Proc. 13th Symposium on naval hydrodynamics*, Tokyo, 13-40.

Roman P.(1975). Some *Mordern Mathematics For Physicists and Other Outsiders*. **1,2**, Pergamon Press Inc.

Practical Design of Ships and Other Floating Structures
You-Sheng Wu, Wei-Cheng Cui and Guo-Jun Zhou (Eds)

NUMERICAL ANALYSIS OF CAVITATING PROPELLERS INCLUDING VISCOUS FLOW EFFECTS

F. Salvatore[1] and P. G. Esposito[1]

[1]INSEAN - Istituto Nazionale per Studi ed Esperienze di Architettura Navale, Rome, Italy

ABSTRACT

Theoretical modeling of sheet cavitation by using a viscous/inviscid coupling technique is addressed. The inviscid cavitating flow is studied by a boundary element approach including a sheet cavitation model. The viscous-flow correction is given by 2D boundary-layer equations used at strip theory. Viscous/inviscid coupling is obtained via Lighthill's transpiration velocity concept. Comparisons of present numerical results with experiments in the literature show the importance of viscosity correction and show the capability of the present methodology to give reliable cavity pattern predictions.

KEYWORDS

Theoretical Hydrodynamics, Sheet Cavitation, Inviscid Flows, BEM, Boundary Layers, Viscous/Inviscid Coupling, Hydrofoils, Marine Propellers.

1 INTRODUCTION

Cavitation phenomena affect performance of high speed propellers and cause vibrations, noise and blade surface erosion. The demand for large fast vessels motivates the current interest for developing reliable prediction tools for the analysis of cavitating propellers.

Experimental investigations highlight the influence of viscosity effects on cavitation. In particular, in the case of sheet cavitation, the location of the cavity detachment is typically associated with the occurrence of laminar separation of the boundary layer. Hence, an accurate cavitation modelling should include viscous flow effects.

Existing literature related to theoretical modelling of viscous cavitating flows is limited to two-dimensional or three-dimensional flows around hydrofoils. Two-dimensional boundary layer analysis coupled with inviscid cavitating flow solution by BEM is proposed in Kinnas *et al.* (1994). Recently, commercial Reynolds-Averaged Navier-Stokes (RANS) solvers have been coupled with cavitation models, Hirschi *et al.* (1998) and Chahine & Hsiao (2000), with applications to 3D hydrofoils in steady motion.

In the present paper, the analysis of cavitation in high-Reynolds number flows by means of a viscous/inviscid coupling technique is proposed. This methodology extends a formulation presented by

Esposito & Salvatore (2000) that is limited to steady two-dimensional cavitating viscous flows. The inviscid cavitating flow is studied by a boundary integral formulation for the velocity potential with a sheet cavitation. The numerical solution is obtained by a zero-th order BEM.
The boundary layer surrounding the blade surface and the vortical wake is studied by using a boundary layer approximation that is valid in the case of attached or slightly separated high-Reynolds number flows. The boundary layer equations are solved in integral form. A viscosity correction to the potential flow solution is obtained in terms of Lighthill's transpiration velocity approach (Lighthill, 1958), adapted here to the case of cavitating flows. The boundary layer and the potential flow solution including the viscosity correction are matched via an iterative process.

2 THEORETICAL MODEL

Under the assumption that the perturbation induced by the propeller to the velocity field is irrotational, it is possible to introduce a scalar potential ϕ for the perturbation velocity. The total velocity field in a frame of reference $(Oxyz)$ that is fixed with the propeller (hereafter referred to as the rotating frame of reference) is given by

$$\mathbf{q} = \mathbf{v}_I + \nabla\phi \tag{1}$$

where $\mathbf{v}_I$ is the relative incoming flow. The velocity potential ϕ satisfies the Laplace equation $\nabla^2\phi = 0$. The pressure field is given by the Bernoulli equation that, in the rotating frame of reference, reads

$$\frac{\partial\phi}{\partial t} + \frac{1}{2}q^2 + \frac{p}{\rho} + gz_0 = \frac{1}{2}v_I^2 + \frac{p_0}{\rho} \tag{2}$$

where $q = \|\mathbf{q}\|$, $v_I = \|\mathbf{v}_I\|$, p_0 is the undisturbed flow pressure, ρ is the fluid density, g is the gravity acceleration and z_0 is the local depth referred to the free surface.
In order to solve the Laplace equation, boundary conditions must be imposed. On the wetted portion of the propeller surface S_{WP} the impermeability condition yields $\mathbf{q}\cdot\mathbf{n} = 0$, or, recalling Eq. (1),

$$\frac{\partial\phi}{\partial n} = -\mathbf{v}_I \cdot \mathbf{n} \qquad \text{on } S_{WP} \tag{3}$$

where $\mathbf{n}$ is the outward unit normal to S_{WP}.
Boundary conditions for ϕ on the wake surface S_W are determined by applying mass and momentum conservation laws across the surface. As a result, pressure and normal component of the perturbation velocity must be continuous across S_W. Thus, applying the Bernoulli equation (2) on the two sides of the wake, the potential discontinuity $\Delta\phi$ at an arbitrary wake point $\mathbf{x}_W$ and time t may be expressed as

$$\Delta\phi(\mathbf{x}_W, t) = \Delta\phi(\mathbf{x}_{WTE}, t - \tau_W) \tag{4}$$

where $\mathbf{x}_{WTE}$ is the trailing edge wake point lying on the same streamline as $\mathbf{x}_W$, whereas τ_W denotes the convection time from $\mathbf{x}_{WTE}$ to $\mathbf{x}_W$.
A further condition on ϕ is required in order to assure that no finite pressure jump may exist at the blade trailing edge (Kutta condition). Following Morino *et al.* (1975), this is equivalent to impose

$$\Delta\phi(\mathbf{x}_{WTE}, t) = \phi(\mathbf{x}_{TE}^+, t) - \phi(\mathbf{x}_{TE}^-, t) \tag{5}$$

where $\mathbf{x}_{TE}^+$ and $\mathbf{x}_{TE}^-$ denote, respectively, blade trailing edge points on suction and pressure sides.
On the external boundary of the cavity, S_C, a constant pressure condition $p = p_v$, is assumed, with p_v denoting the vapor pressure. Hence, by Bernoulli's theorem (2) and by introducing the cavitation number $\sigma = (p_0 - p_v)/0.5\rho n^2 D^2$, where n and D are respectively the number of propeller revolutions per time unit and the propeller diameter, one obtains

$$q = \sqrt{(nD)^2\sigma - 2\left(\frac{\partial\phi}{\partial t} + gz_0\right) + \|\mathbf{v}_I\|^2} \qquad \text{on } S_C \tag{6}$$

Hereafter, it is assumed that the cavity extension is limited to the blade surface (partial cavitation). On the blade surface, a curvilinear coordinate system (s,ν) is introduced, with constant s lines directed along the span and constant ν lines directed along the chord; observing that, by Eq. (1), $\partial\phi/\partial s = (\mathbf{q}-\mathbf{v}_I)\cdot\hat{\mathbf{s}}$, Eq. (6) provides a nonlinear differential equation for ϕ.
Integrating along s and after some manipulations, one obtains

$$\phi(s,u) = \phi(s_{CLE},u) + \int_{s_{CLE}}^{s}\left[q_\nu\cos\theta - v_{I_s} + |\sin\theta|\sqrt{(nD)^2\sigma - 2\left(\frac{\partial\phi}{\partial t} + gz_0\right) + \|\mathbf{v}_I\|^2 - q_\nu^2 - q_n^2}\,\right]ds' \tag{7}$$

where s_{CLE} is the abscissa s of the cavity detachment line, $q_\nu = \mathbf{q}\cdot\hat{\mathbf{v}}$ and $v_{I_s} = \mathbf{q}\cdot\hat{\mathbf{s}}$.
A cavity termination model is introduced in order to prescribe the transition between the constant pressure condition $p = p_v$, valid on the cavity, and the variable pressure condition valid on the wetted portion of the blade surface. In the present approach a pressure recovery model based on Lemonnier & Rowe (1988) is used.
A relationship for the cavity thickness h_c is obtained by assuming that the cavity surface is a material surface, *i.e.*, $D[\hat{S}_C(s,\nu,\eta,\tau)]/Dt = 0$. Observing that $\eta = h_c(s,\nu,t)$ represents the cavity surface S_C, one has

$$\frac{1}{\sin^2\theta}(q_\nu\cos\theta - q_s)\frac{\partial h_c}{\partial s} + \frac{1}{\sin^2\theta}(q_s\cos\theta - q_\nu)\frac{\partial h_c}{\partial \nu} - \frac{\partial h_c}{\partial t} = -q_n \tag{8}$$

where $q_s = \mathbf{q}\cdot\hat{\mathbf{s}}$. The expression above represents a partial differential equation for h_c that may be solved once the potential field is known.
The viscous flow model considered here is valid in the case of attached high–Reynolds number flows. Under these assumptions, a classical boundary–layer/potential–flow coupling technique is used. In order to limit the model complexity, a steady flow integral boundary layer solver has been used. In fact, boundary layer equations in integral form provide a consistent reduction of computational costs in comparison with differential methods, with no lack of accuracy in the prediction of the global viscosity effects. In addition, a strip–theory approach is used in which two–dimensional boundary layer equations are coupled with the three–dimensional potential flow solver described above.
In the laminar portion of the boundary layer the classical von Kármán equation is solved by using the collocation method described in Thwaites (1949). The turbulent portion of the boundary layer and the wake are studied by the "lag-entrainment" method by Green *et al.* (1973). Michel's method (1952) is used to predict the transition from laminar to turbulent flow.
The boundary layer equations are solved by using the velocity field given by the potential flow method as an input. Once the boundary layer equations are solved, the viscous flow correction for the potential flow model is evaluated by means of Lighthill's transpiration velocity given by $\chi_\nu = \partial(q_s^e\delta_s^*)/\partial s + \partial(q_\nu^e\delta_\nu^*)/\partial\nu$, where δ_s^*, δ_ν^* are, respectively, displacement thickness of the vortical layer in s and ν directions, whereas the superscript e denotes velocity given by the potential flow model with viscosity correction. Due to the strip—theory approach, here it is $\delta_\nu^* = 0$.
As a consequence, the boundary condition over the wetted body surface, Eq.(3) is modified as follows

$$\frac{\partial\phi}{\partial n} = -\mathbf{v}_I\cdot\mathbf{n} + \chi_\nu \qquad \mathbf{x}\in S_{WP} \tag{9}$$

whereas, on the cavitating portion of the blade surface

$$\frac{\partial\phi}{\partial n} = -\mathbf{v}_I\cdot\mathbf{n} + \chi_\nu + \chi_c \qquad \mathbf{x}\in S_C \tag{10}$$

where χ_c is obtained by Eq. (8) by recalling that $q_n = \partial\phi/\partial n + \mathbf{v}_I\cdot\mathbf{n}$.

Finally, on the wake surface S_W, one has

$$\Delta\left(\frac{\partial\phi}{\partial n}\right)=\chi_v^+ + \chi_v^- \qquad (11)$$

Boundary layer and potential flow solutions are matched via an iterative process.

3 SOLUTION PROCEDURE

In the present approach, the Laplace equation for ϕ is solved by means of a boundary integral formulation. Under the assumption that the cavity thickness is small as compared to the thickness of the propeller blade, the boundary integral equation is written by replacing the external boundary of the cavity with its projection onto the blade surface. This is done to avoid variability of the boundary with consequent increase of computational effort. A numerical solution of the boundary integral formulation for the velocity potential is then obtained by using a boundary element method. The propeller surface and the wake are divided into hyperboloidal quadrilateral elements.
The distributions of ϕ over S_{WP} and of $\partial\phi/\partial n$ on S_C are evaluated in discretized form at each centroid on the propeller surface (collocation points). Flow quantities are assumed to be constant on each element. Denoting by N_P the number of elements on the propeller surface, and by N_W the number of elements on the wake, one obtains

$$\sum_{j=1}^{N_P}\left(\delta_{ij}+D_{ij}\right)=\sum_{j=1}^{N_P}S_{ij}\left(\frac{\partial\phi}{\partial n}\right)_j+\sum_{j=1}^{N_W}S_{ij}\Delta\left(\frac{\partial\phi}{\partial n}\right)_j-\sum_{j=1}^{N_W}D_{ij}\Delta\phi_j \qquad (12)$$

where $\delta_{ij}=1$ if $i=j$ and $\delta_{ij}=0$ otherwise, whereas S_{ij}, D_{ij} denote, respectively, the influence at $\mathbf{x}_i$ of, respectively, sources and doublets distributed on the j–th surface element on S_P or S_W. The evaluation of coefficients S_{ij}, D_{ij} in Eq. (12) is performed analytically as in Morino *et al.* (1975).
At each time step, a coupled boundary-layer/non cavitating potential flow solution is performed in order to provide the location of the cavity detachment line from the laminar separation condition as described in Kinnas *et al* (1994). Next, the iterative cavitating-flow solution is performed until convergence of the cavity shape is reached.

4 NUMERICAL RESULTS

First, the analysis of marine propellers in non uniform flow by the inviscid formulation described above is addressed. Propeller computations have been performed by using 36x10 panels on each blade surface, 38x30 panels on the hub surface and 60x10 panels on the wake. A sensitivity analysis (see Fig. 4) shows that grid refinement should be used to provide grid-independency of results.
The test case chosen is the three-bladed DTRC 4148 propeller tested in a screen-generated wake, with J=0.91, σ_n = 2.76 and F_n = 9.16. Figures 1 to 3 show the predicted cavitation pattern at three angular positions, ψ = -30°, 6, and 30° (For clarity, cavity thickness is multiplied by a factor 3). Numerical results are compared to experimental observations available in Mishima *et al* (1995).
The radial extent of the cavity is accurately predicted, whereas the chordwise extension is overestimated by present results at ψ = 6° and ψ = 30°. This is confirmed by Fig. 5, where predicted and measured chordwise extensions of the cavity at r/R = 0.8 are depicted. In the calculations above, cavity detachment at leading edge has been enforced. Unfortunately, no experimental measure of the actual location of the cavity detachment line is available.
Application of the viscous/inviscid technique to the propeller flow is underway. Here, the methodology has been applied to the case of three-dimensional hydrofoils in uniform flow. As a test case, a windsurfer fin geometry has been considered. The fin has an elliptic distribution of chord and small positive sweep angle (see Kinnas & Fine, 1993 for details) and has been tested at angle of attack α=6.5°, σ=1.148 and

$Re = 1\times10^6$. Numerical results have been obtained by using 80x40 panels on the fin surface and 40x40 panels on the wake. First, the influence of the cavity detachment location x_d on the cavity extent is addressed. Figures 6 and 7 show computations by using the viscous/inviscid model with x_d forced at, respectively, 0 and 2.2% of chord. Figure 6 shows predicted locations of the cavity trailing edge, whereas Fig. 7 shows cavity thickness distributions at some spanwise locations (in Figs. 6 and 9, cavity thickness is multiplied by a factor 5).

It is apparent that cavity detachment location strongly affects the accuracy of cavity pattern prediction. For the same flow conditions, Fig. 8 shows the location of the cavity trailing edge obtained by using the viscous/inviscid model with x_d computed by the laminar transition/separation condition. Present results are compared to experimental ones given in Kinnas & Fine (1993). and to numerical results in Pelleat & Pellone (1996) obtained by an inviscid-flow model with $x_d/c = 0.006$ imposed.

It may be seen that the laminar transition/separation condition provides results that are in good agreement with experimental observations. The result above proves that viscosity effects are fundamental for determining the exact location of the cavity detachment line. On the contrary, if cavitation detachment is enforced, viscous-flow correction is of minor importance in influencing cavity extension and thickness. This is shown in Fig. 9, where numerical results obtained by including and neglecting viscous-flow effects are compared. In order to have fully comparable results, $x_d/c = 0.016$ has been forced in both cases.

5 CONCLUDING REMARKS

A viscous/inviscid technique that allows to take into account viscosity effects in sheet cavitation modeling via BEM has been presented. Numerical results by using inviscid-flow modeling and viscous/inviscid modeling are discussed. It is shown that viscous flow correction is fundamental in order to determine an accurate prediction of the cavity detachment line. It is also shown that small errors in predicting the location of this line may produce large errors in the estimation of the cavity pattern. On the other hand, if cavity detachment is prescribed, viscosity slightly affects both cavity extension and thickness.

Here reliable results have been obtained by coupling a two-dimensional boundary-layer model with a potential based BEM solver. A validation test has been performed in the case of three dimensional hydrofoils, whereas the application to cavitating propellers in unsteady flow is underway.

Acknowledgments

This work was supported by the *Ministero dei Trasporti e della Navigazione* in the frame of INSEAN Research Program 2000-2002.

References

Chahine G.L. & Hsiao C.T. (2000). Modeling 3D Unsteady Sheet Cavities Using a Coupled UnRANS-BEM Code. *Proceedings of the 23rd ONR Symposium on Naval Hydrodynamics*, Val de Reuil (France).

Esposito P.G. and Salvatore F. (2000). Optimal Design of Cavitating Blade Sections by Genetic Algorithms. *Proc. of NCT 50*, 275-288.

Green J.E., Weeks D.J., and Brooman W.F. (1973). Prediction of Turbulent Boundary Layers and Wakes in Compressible Flow by a Lag-Entrainment Method. *A.R.C.R. & M., n. 3791.*

Hirschi R., Dupont P., Avellan, F. (1998). Partial Sheet Cavities Prediction on a Twisted Elliptical Planform Hydrofoil Using a Fully 3-D Approach. *Proceedings of the 3rd Int. Symp. on Cavitation,* Grenoble (France).

Kinnas S.A. and Fine N.E. (1993). A Numerical Non-linear Analysis of the Flow Around Two and Three-Dimensional Partially Cavitating Hydrofoils. *Journal of Fluid Mechanics* **254**, 151-181.
Kinnas S.A, Mishima S. and Brewer W. (1994). Nonlinear analysis of Viscous Flow around Cavitating Hydrofoils. *Proc. of the Twentieth Symposium on Naval Hydrodynamics*, 446-465.
Lemonnier H. and Rowe A. (1988). Another Approach in Modelling Cavitating Flows. *J. Fluid Mech.* **195**, 557-580.
Michel R. (1952). Etude de la Transition sur les Profils d'Aile - Establissment d'un Point de Transition et Calcul de la Trainée de Profil en Incompressible. *ONERA Report n. 1/1578A*.
Mishima S., Kinnas S.A., Egnor D. (1995). The Cavitating Propeller Experiment (CAPREX), Phases I & II. *Technical Report, Dept. Ocean Engineering, MIT.*
Morino L., Chen L.-T. and Suciu E. O. (1975). Steady and Oscillatory Subsonic and Supersonic Aerodynamics around Complex Configurations. *AIAA J.* **13**, 368-374.
Pelleat J.M. and Pellone C. (1996). Experimental Validation of Two- and Three-Dimensional Numerical Analysis of Partially Cavitating Hydrofoils. *Journal of Ship Research* **40:3**, 211-223.
Thwaites B. (1949). Approximate Calculation of the Laminar Boundary Layer. *Aeronaut. Quart.* **1**, 245-280.

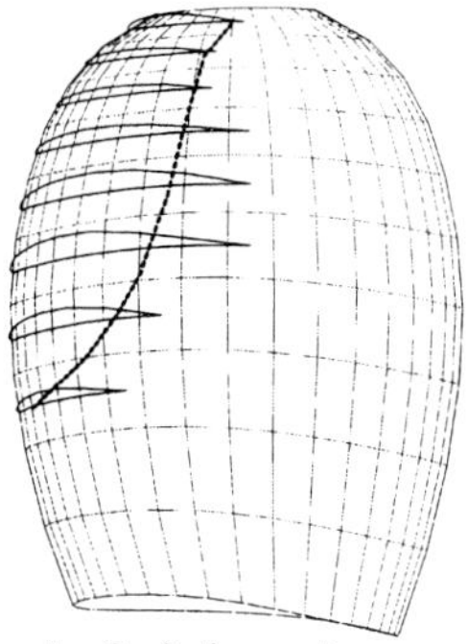

Figure 1: Cavitation pattern on the propeller DTRC4148 at ψ=-30°; thick dashed line: experiments; thin solid line: present method.

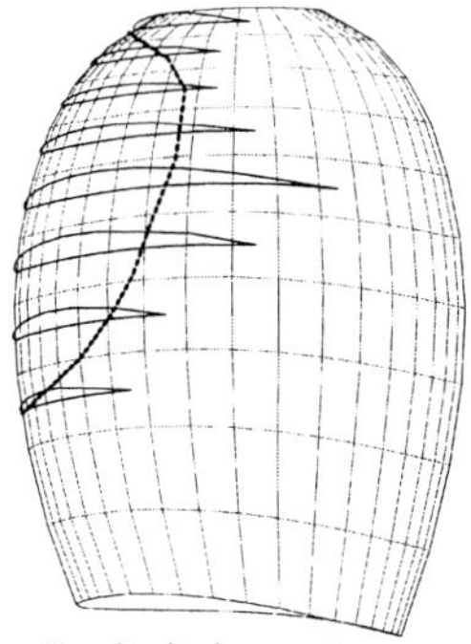

Figure 2: Cavitation pattern on the propeller DTRC4148 at ψ=6°; thick dashed line: experiments; thin solid line: present method.

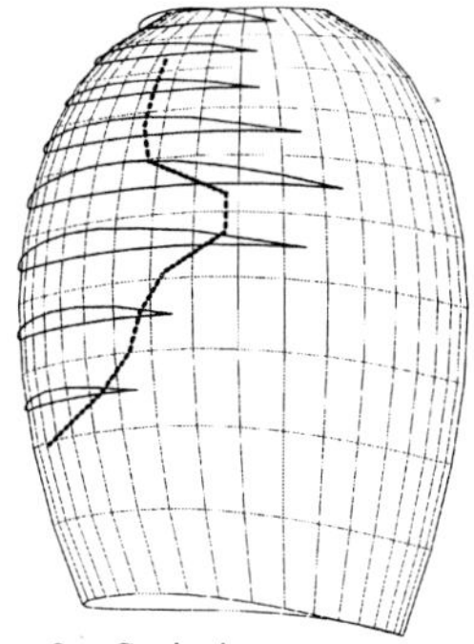

Figure 3: Cavitation pattern on the propeller DTRC4148 at ψ=30°; thick dashed line: experiments; thin solid line: present method.

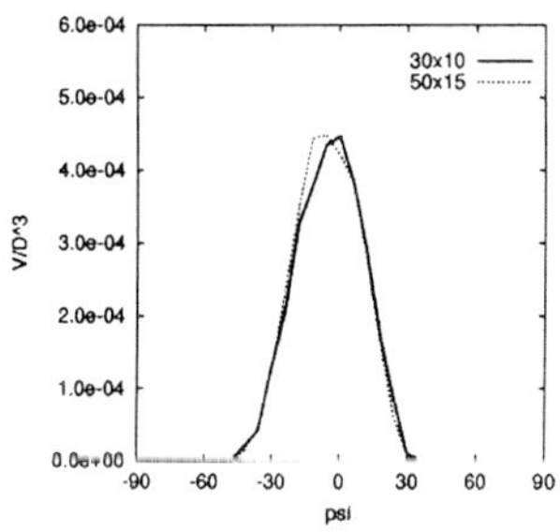

Figure 4: Effect of blade grid refinement on cavity volume

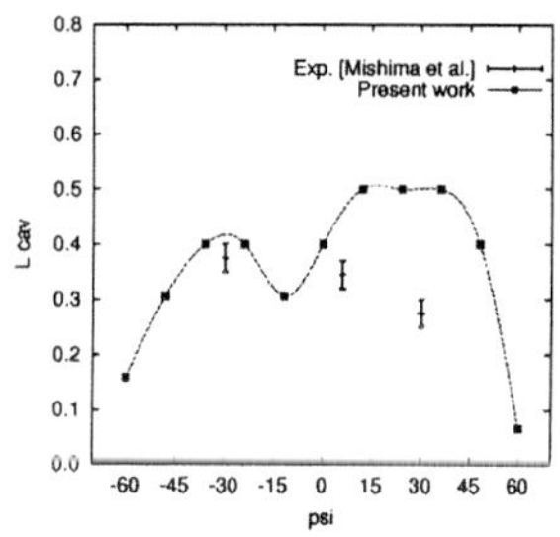

Figure 5: Chordwise cavity extension at r/R=0.8 on the DTRC4148 propeller.

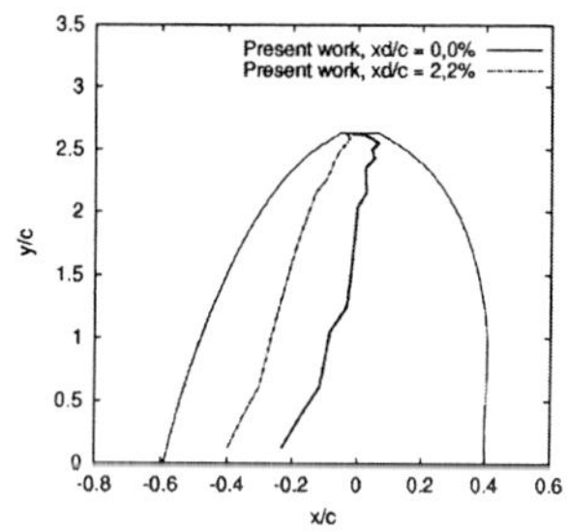

Figure 6: Influence of cavity detachment line on cavitation pattern of windsurfer fin (inviscid model).

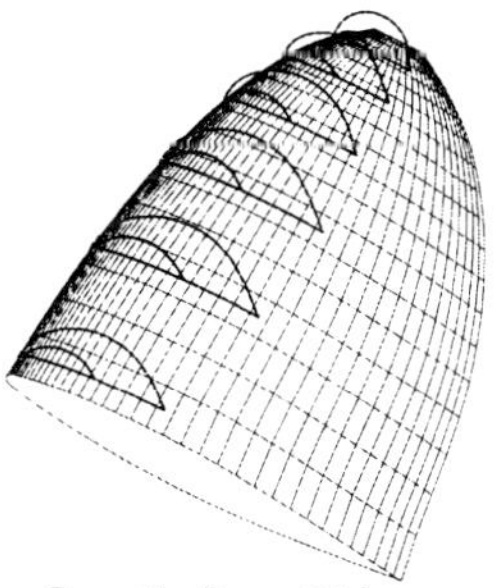

Figure 7: Cavity thickness on windsurfer fin: solid: x_d/c=0%; dashed: x_d/c=2.2%.

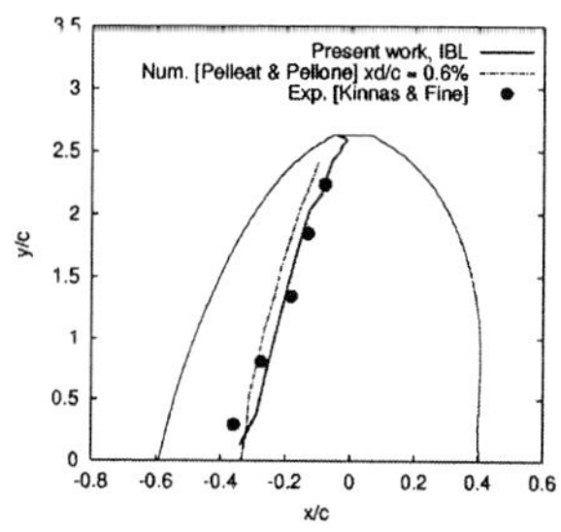

Figure 8: Comparison of cavitation patterns on windsurfer fin among numerical and experimental results.

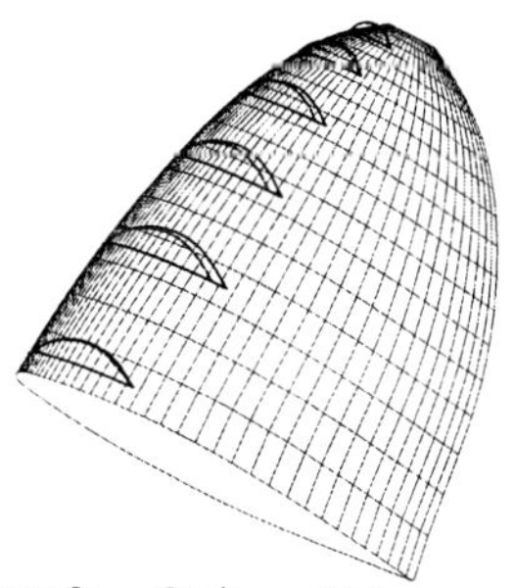

Figure 9: Cavity thickness on windsurfer fin: solid: inviscid model; dashed: viscous/inviscid model.

Practical Design of Ships and Other Floating Structures
You-Sheng Wu, Wei-Cheng Cui and Guo-Jun Zhou (Eds)

PROPELLER DESIGN BASED ON SURFACE PANEL METHOD BY PRESCRIBED PRESSURE DISTRIBUTION

Ting-Shou Tan

College of Naval Architecture and Civil Engineering, Wuhan University of Technology
Yu Jia Tou, Wuhan 430063, China

ABSTRACT

Since the speed of ship is increasing or the propeller operates in non-uniform flow, the cavitation on the propeller blades is often unavoidable. The cavitation on the propeller blades, giving rise to some trouble such as blade cavitation erosion or severe vibration and highly radiated noise, is tightly related to the distribution of pressure on the propeller blades. In order to improve the propeller cavitation performance, this paper presented the technique of the propeller design based on the surface panel method by prescribed three-dimensional pressure distribution that can be decided by the propeller designer. It may also be obtained from the result of propeller lifting surface design or modifying the pressure distribution of the existing propeller. The blade surface geometry is generated by the B-spline method that enables it to be completely specified using only a small number of parameters. The square summation of the differences between the pressure distribution calculated by the surface panel method and prescribed at the control points on the blade surface is regarded as the objective function. The parameters can be obtained by minimizing the objective function.

KEYWORDS

Propeller design, Surface panel method, B-spline, Pressure distribution

1 INTRODUCTION

Marine propeller design based on methodical series tests still use for typical commercial ship. In order to improve the propeller cavitation performance, such as reducing the propeller induced vibration and noise, or to improve the efficiency of the propeller, the theoretical design method perhaps is better tools than the design charts method. The most familiar propeller design method is the lifting-surface theory design. In this method, a part of the geometry is given; they are the radial distribution of chord length, rake, skew, thickness and the type of the blade section. For given radial circulation distribution and chordwise circulation shape under the condition of circumferential average inflow, the radial pitch and camber surface are determined by the lifting-surface theory.

In present design method, for given thickness distribution and the desired three- dimensional pressure distribution that can be obtained from the result of lifting-surface theory design or by modifying the existing propeller. The propeller geometrical parameters to be determined still are the radial pitch distribution and camber surface. They are generated by the B-spline curve and surface that ensure the blade surface is fair by a small number of parameters. Therefore, an objective function, the square summation of the differences between the prescribed pressure distribution and that calculated by surface panel method is minimized by modifying the geometry parameters.

2 PROPELLER BLADE SURFACE BY B-SPLINE

Propeller blade are generally defined by the radial distribution of geometrical quantities, such as chord length, pitch, rake, skew, and chordwise distribution of camber and thickness. At each step in the design process, the smooth blade surface is required in order to perform correctly the hydrodynamic analysis of the design propeller. It is therefore desirable that geometrical parameters designed, such as radial distribution of pitch and chordwise distribution of camber, are sufficiently smooth. The B-spline not only have this property, but can define the blade surface with a relatively few number of parameters, this is particularly convenient for design blades by numerical method. A B-spline curve or surface is described by means of a control polygon net. The number of control polygon vertices reflects the computational effort of the design process. For the present application, we choose the cubic integral B-spline. The B-spline curve and surface are applied to the radial pitch distribution and the camber surface, respectively. The radial pitch distribution is uniquely defined by following B-spline curve.

$$P(w)=\sum_{j=1}^{N_w} P_j N_j(w) \tag{1}$$

In Eqn.1, $P(w)$ is the pitch distribution, P_j are the vertices of the control polygon that set the shape of the curve, $N_j(w)$ are the B-spline basis function. The parameter w is a monotonically increasing function along the curve, the curve begin at the first control polygon vertex with the parameter value $w=0$, and end at the final vertex with a parameter value $w=N_W-3$. Secondly, consider the parametric B-spline surface given by

$$f(u,w)=\sum_{i=1}^{N_U}\sum_{j=1}^{N_W} X_{i,j} N_i(u) N_j(w) \tag{2}$$

The points on the camber surface $f(u,w)$ are now expressed in term of two parameters, (u,w), N_U and N_W are the number of B-spline surface control polygon vertices in u (chordwise) and w (radial) direction, respectively, $X_{i,j}$ are control polygon vertices. For the blade camber surface, $f=0$ for the leading and trailing edge, therefore it is such as above that the chordwise first and final vertices, $X_{1,j}$ and $X_{N_U,j}$, is equal to zero.

3 SURFACE PANEL METHOD FOR PROPELLER

It is assumed that the propeller with a finite number of blades and a constant angular velocity Ω rotates in an inviscid, incompressible, irrotational fluid. Since the flow is potential, there is a perturbation velocity potential ϕ satisfying Laplace's equation $\nabla^2\phi=0$. The ϕ must satisfy the boundary conditions that are the kinematics boundary condition that fluid can not flow across body surface, the Kutta condition at the blade trailing edge and other conditions. Applying Green's identity and boundary condition, the perturbation potential ϕ at the point P on the propeller surface can be expressed as follows

$$\left.\begin{aligned} & 2\pi\phi(P) - \iint_{S_b} \phi(q)\frac{\partial}{\partial n_q}\left(\frac{1}{R(P,q)}\right)dS - \iint_{S_W} \Delta\phi(q)\frac{\partial}{\partial n_q}\left(\frac{1}{R(P,q)}\right)dS \\ & \quad = \iint_{S_b} \frac{\partial \phi(q)}{\partial n_q}\frac{1}{R(P,q)}dS \end{aligned}\right\} \tag{3}$$

The boundary surfaces S_b and S_W are the body surface and wake surface, respectively, $R(P,q)$ is the distance from the point P to the point q on the boundary surface. The source strength can be determined by kinematics boundary condition. The doublet strength $\Delta\phi$ on wake is determined by Kutta condition. The unknown perturbation velocity potential ϕ on the body surface can be uniquely determined by Eqn.3. Taking the derivative of the perturbation velocity potential, we can get the total velocity and the pressure distribution. [1]

Generally, the Eqn.3 can be only solved numerically, so the propeller and its wake surface are approximated by a number of small quadrilateral hyperboloidal panels. The parts of hub intersected with blades are considered a cylinder of the equal radius. The downstream and upstream parts are regarded as parabolic solid of revolution. At the intersections between the blade and the hub, the axis position of the hub panels meet with those of the blade panels. The shape of wake surface is assumed previously. The trailing vortex leaves the trailing edge in the direction coincided with the geometric pitch angle of the blade section, and extends downstream with various radius and pitch until the prescribed location and then maintains constant. The discretization of the Propeller is shown in Fig 1. The constant doublet on each panel will be solved by satisfying the Eqn.3 at the centroid of each panel. Thus, a system of N linear algebraic equations is obtained.

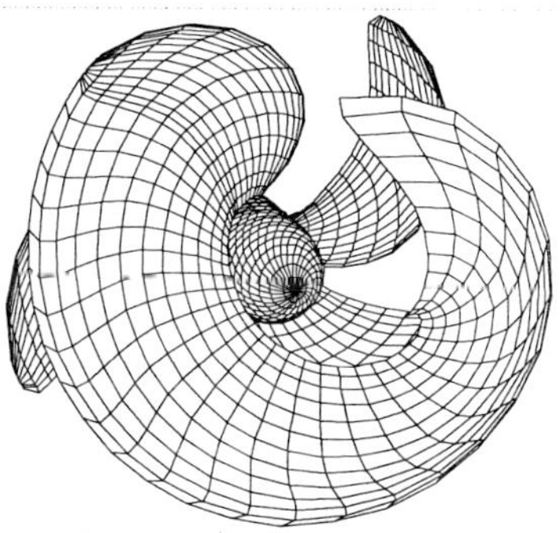

Figure1: The discretization of propeller

$$\sum_{j=1}^{N}\left(2\pi\delta_{ij} - D_{ij}\right)\phi_j - \sum_{m=1}^{N_R} W_{im}\Delta\phi_m = \sum_{j=1}^{N} S_{ij}\left(V_I \cdot \boldsymbol{n}\right)_j \qquad i = 1,2,3,\cdots,N \tag{4}$$

where the N is the number of the total panels on the propeller surface, the D_{ij}, W_{im} and S_{ij} are the influence coefficients at the control point of $i-th$ panel by unit strength doublet and source distribution over $j-th$ or $ml-th$ panel, respectively. The ϕ_j can be obtained by solving the equation (4).

In order to obtain a close-form of equation (4), Kutta condition must be applied. This paper introduce the equal pressure Kutta condition that the pressure is equal at the control points of the upper and lower panels adjacent to the trailing edge. The pressure difference Δp_{TE} on the trailing edge is non-linearly dependent of the unknown $\Delta\phi_m$; therefore, an iterative process is necessary to meet the equal pressure Kutta condition. Namely, using the values of $\{\Delta\phi\}^n$ and $\{\Delta p\}^n$ at the $n-th$ iteration, an iterative equation is constructed to determine the next step solution $\{\Delta\phi\}^{n+1}$.

$$\Delta\phi_m^{(n+1)} = \Delta\phi_m^{(n)} - \sum_{j=1}^{N_R} P_{mj}^{-1} \Delta p_j^{(n)} \tag{5}$$

where P_{ij} is the element of the iterative matrix.[3] the above process is repeated until the value $\left|\Delta p_m^{(n)}\right| < \varepsilon$ (ε is a desired small parameter) . Then the pressure distribution on blade surface can be determined by applying the Bernoulli's equation

4 PROPELLER BLADE DESIGN

The propeller design process used in present paper consists of the following steps

1. The desired pressure distribution on the blade surface, from the initial design, such as the lifting surface design, can directly be obtained by the result of lifting surface design. It may also be prescribed by the propeller designer or by modifying the existing propeller.
2. Calculate the pressure distribution at a set of control points on the blade surface by the surface panel method.
3. Adjust the blade surface shape (pitch and camber) until the differences between prescribed and calculated pressure distribution at the control points are minimized in a least square sense.
4. calculate the propeller hydrodynamic by the pressure integral on the blade surface

For present design problem, the radial pitch distribution and the blade camber surface are regarded as unknown and controlled by a set of the B-spline control polygon vertices, or design variables. In addition, the prescribed pressure distributions on the blade surface should be reasonable. In order to minimize the differences between the desired and calculated pressure distribution at each control points, the adjustment of the blade surface is carried out by direct modification of the control polygon vertices defining the blade shape. Let the desired pressure distribution (here is pressure coefficient) on the i-th control point of the blade surface be denoted by C_{Pi}, and $\hat{C}_{Pi}$ stand for that calculated by the surface panel method. The solution of the present design problem is to minimize the following objective function

$$F(X) = \sum_{i=1}^{M} \Delta C_{Pi}^2(X) = \sum_{i=1}^{M} \left[\hat{C}_{Pi}(X) - C_{Pi}(X)\right]^2 \Rightarrow min \tag{6}$$

where $X = (x_1, x_2, x_3, x_4, x_5, \cdots, x_N)^T$ is the design variable which number is N, the M represents the number of the control points on the blade surface. Since the calculated pressure coefficient $\hat{C}_{Pi}$ in Eqn.6 is nonlinear with respect to the design variables X, the Eqn.6 can be solved by solving a series of the linear least square problem. Let X_n is the solution of the $n-th$ iteration approximation; we expand the pressure differences $\hat{C}_{Pi}$ in the first order Taylor series at design space point X_n. The minimization of the objective function can be realized by following equation with an iterative process

$$\boldsymbol{X}_{n+1} = \boldsymbol{X}_n - \left(\boldsymbol{A}^T \boldsymbol{A} + u_n \boldsymbol{I}\right)^{-1} \boldsymbol{A}^T \left(\hat{C}_P - C_P\right) \tag{7}$$

where A denotes the Jacobian matrix which elements are the changes in pressure coefficient calculated by surface panel method due to a small change in each design variables, the u_n is damping parameter of the Levenberg-Marquardt method to improve convergence, the I is the identity matrix. However, the new blade surface controlled by the design variables will not necessarily be the desired surface since the elements of the Jocobian matrix depend in a nonlinear way on the shape of the blade surface. A new Jocobian matrix must therefore be formed and process repeated until the blade surface has converged.

5 DESIGN RESULTS AND DISCUSSION

In order to show the validity of present design method, Two design examples are given, the first is an existing propeller. The propeller DTMB4118 as mother propeller which principal data come from reference [1], another is a cargo ship propeller with new blade section, this propeller was designed by the lifting surface theory in a non-uniform flow.

In this first example, the parent propeller is DTMB 4118 propeller; the design advanced is 0.833. Now we change the propeller blade surface by reducing its section camber by $25\,\%$, and increasing its pitch ratio 5% to obtain a desired propeller and meet the thrust coefficient. Then the pressure distribution calculated based on surface panel method is regarded as the desired pressure distribution C_P. The initial guesses of design variables (i.e. the B-spline control polygon vertices) are obtained by using B-spline fitting for the parent propeller. The B-spline control polygon vertices by using $N_W = 5$ (the radial) and $N_U = 9$ (the chordwise), the number of the vertices $N = N_W \times (N_U - 1) = 40$. The pressure distribution and the chordwise distribution of the camber are shown in Fig.2 and Fig.3, respectively. From those figures, we can conclude that the present design method has been applied successfully in designing a new propeller.
In the second example, the propeller was designed by the lifting surface theory method, the desired pressure distribution come from the results of the lifting-surface design. However, the pressure distributions calculated by the lifting surface method and the actual pressure distribution probably have some discrepancy. The propeller is redesigned by the present design method to obtain correct blade geometry. The pressure coefficient on blade surface and the camber distribution are shown as in Fig.4 and Fig.5, respectively. From Fig.4, the pressure distributions prescribed and calculated based on panel method are to agree well. But the camber distributions designed by lifting-surface and panel method have a great discrepancy, this reflect that lifting force from pressure distribution of lifting-surface is too large so that a larger camber is required to match the larger lifting force.

6 CONCLUSIONS

The propeller design based on surface panel method from the knowledge of the desired pressure distribution by the B-spline Method has been applied successfully. Design results show that the present design method needs only a relatively few number of the design variables to obtain the desired blade surface. The present design method by prescribed pressure distribution can more effectively control and improve the blade cavitation performance than the lifting surface design method which prescribed the circulation distribution.

References

1 Kerwin, J. E. and Lee, C. S. (1978) · Prediction of Steady and Unsteady Marine Propeller Performance by Numerical Lifting-Surface Theory · Trans. SNAME, 86, 218~253

2 Tan Tingshou.(1997) · Prediction of Hydrodynamic Performance on Propeller by Surface Panel Method · Journal of Wuhan Transportation University, China, 21:2, 534~541

3 Tan Tingshou.(2000) · Improvement on Numerical Kutta Condition for Propeller Panel Method ·

Journal of Ship Mechanics, China, 4:2, 12~18

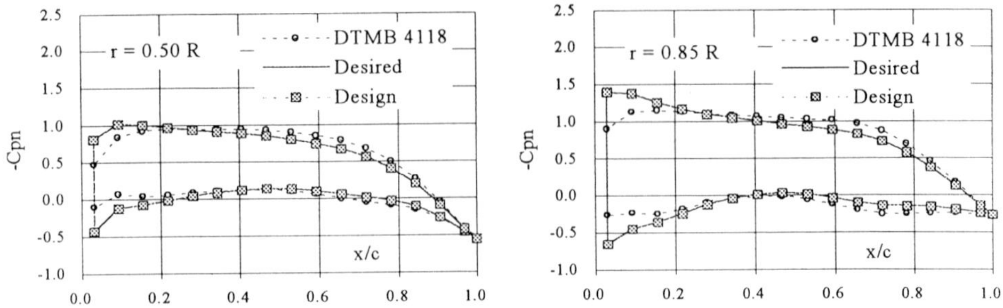

Figure 2: The pressure distribution

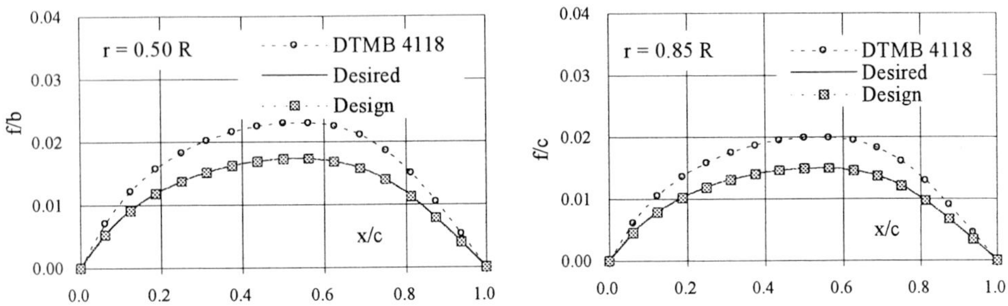

Figure 3: The chordwise distribution of the camber

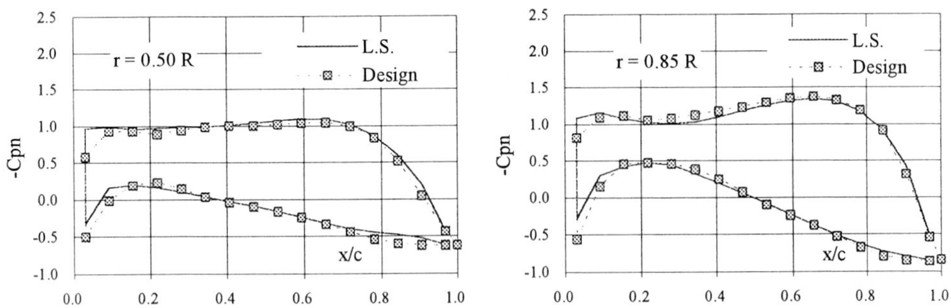

Figure 4: The pressure distribution of the cargo propeller

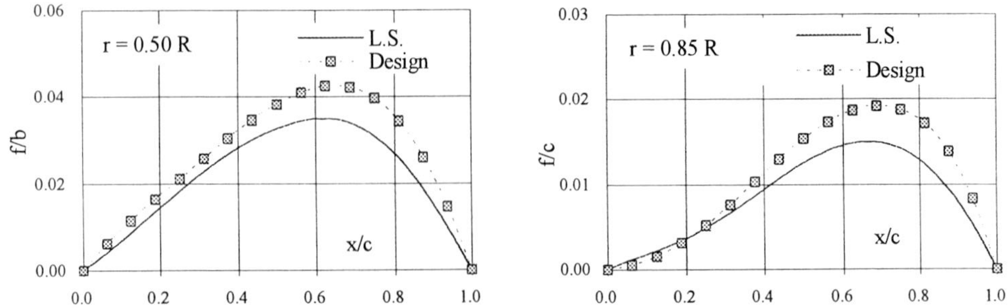

Figure 5: The chordwise distribution of the camber of the cargo propeller

CFD-BASED OPTIMIZATION OF TANKER STERN FORM - MINIMIZATION OF DELIVERED HORSEPOWER USING SELF-PROPULSION SIMULATOR -

Y. Tahara[1], J. Ando[2], and Y. Himeno[1]

([1]Osaka Prefecture University, Japan, [2]Kyushu University, Japan)

ABSTRACT

The present work concerns tanker stern form optimization using CFD-based optimization method. The main focus is placed on minimization of delivered horsepower (*DHP*) in conjunction with application of self-propulsion simulator, which is based on extension of a Reynolds-averaged Navier-Stokes (RaNS) equation solver including propeller effects evaluated by the infinite-bladed propeller theory. The overall numerical method follows that developed in earlier work of the authors, i.e., coupling of RaNS equation solver and successive quadratic programming. Constraints considered in the present study are geometrical requirements of optimized hull form for displacement, profile, maximum beam and depth, and stern frame lines to secure enough space for engine room. In the present paper, an overview of the numerical method is given, and discussion of the results includes identification of salient differences of forms and integral parameters between for the original and optimized hull forms. In conclusion, the present method appeared to successfully optimize the given tanker stern form for minimum *DHP*, and the modification trends automatically demonstrated in the present work agree well with those commonly in use in traditional tanker hull form design.

KEYWORDS

Tanker Stern Form Optimization, RaNS Equations, Nonlinear Programming, Self-Propulsion Simulator

INTRODUCTION

Determination of tanker stern form is one of the most important and time-consuming tasks for ship designers. The designers must find the best compromise between the so-called V- and U-shaped stern cross sections. V-shaped sterns induce weaker stern bilge vortices and lower viscous resistance. In contrast, U-shaped sterns result in generation of stronger vortices and higher viscous resistance; however, the propeller inflow is more uniform with better cavitation characteristics and lower noise. In addition to conventional tank tests, hull form designers have demanded for reliable design tools to support their decision making, in which an optimized hull form is determined so that various design requirements are simultaneously satisfied (Tahara et al., 1998, 1999a, 1999b, 2000).

The present work is related to development of a simulation-based design tool, i.e., it concerns tanker stern form optimization using CFD-based optimization method. The main focus is placed on minimization of delivered horsepower (*DHP*) in conjunction with application of self-propulsion simulator, which is based on extension of a Reynolds-averaged Navier-Stokes (RaNS) equation solver including propeller effects evaluated by the infinite-bladed propeller theory (Nakatake, 1981). The overall numerical method follows that developed in earlier work of the authors, i.e., coupling of RaNS equation solver and successive quadratic programming (SQP). In the following, an overview of the numerical method is given, and discussion of the results includes identification of salient differences of forms and integral parameters between for the original and optimized hull forms.

COMPUTATIONAL METHOD

The present numerical method is based on coupling of the recently developed self-propulsion simulator and SQP module implemented in the authors' earlier work (Tahara et al., 1998, 1999a, 1999b, 2000). The self-propulsion simulator consists of three parts, i.e., a RaNS solver, a propeller performance program, and a root finder module to determine propeller rotational speed n_p so that the hull resistance R balances with the propeller thrust T. The present RaNS solver solves the unsteady RaNS and continuity equations (written in Cartesian coordinates defined in Fig.1) for mean velocity, pressure, and eddy viscosity, by using the algebraic turbulence model, finite-analytic discretization, PISO-type velocity-pressure coupling algorithm, and a numerically generated body conforming grid. In the present work, free-surface effects are not considered. Propeller effects are included in RaNS equations by the body force approach. The body force distribution is interactively and iteratively determined by propeller performance calculation based on infinitely-bladed propeller theory (Nakatake, 1981), by using bound vortex distribution on the propeller disk, which is divided into 36×5=180 panels in the present work. Fig.2 illustrates the overall computational procedure of the present self-propulsion simulator. On the other hand, the present initial computational grid for the original hull form was generated by an elliptic-algebraic method developed by the authors. During the optimization, the grid is updated at every optimization cycle as the hull form is modified, that is accomplished by the use of algebraic scheme to increase the computational efficiency.

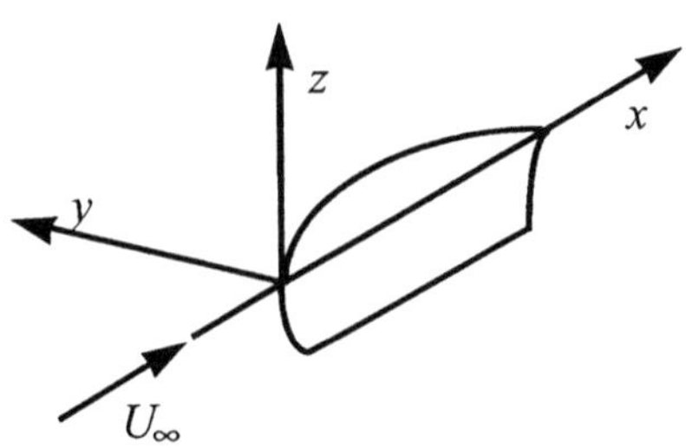

Figure 1: Definition sketch of coordinate system

In the present optimization, the design parameters are used to express modified body geometry through the use of a 6-design-parameter modification function developed by the authors (Tahara et al., 1998, 1999a, 1999b, 2000), where the values of the parameters are solutions of the present optimization problem. Constraints considered are geometrical requirements of optimized hull form, such that displacement, profile, maximum beam and depth of the optimized hull form are same as those of the original hull form, and stern frame line at a control section (S.S.7/8) secures enough space for a-priori designed engine room. The objective function to be minimized is *DHP* directly evaluated by the present self-propulsion simulator. Hence, the present optimization problem is nonlinear, since the constraint functions as well as the objective function are nonlinear functions of the design parameters. As the earlier-mentioned, a SQP algorithm was used to solve the nonlinear optimization problem.

RESULTS

Prior to application of the present method to tanker stern form optimization, accuracy of the present self-propulsion simulator was evaluated through comparison of the results with experimental data for Case-A tanker hull form. The computational condition simulates the measurements, i.e., $Rn=8\times10^6$. The computational grid is C-O-type, whose number is 200×40×60 (in longitudinal × radial × girthwise directions, respectively)=480,000, where the domain includes both port and starboard sides of the hull to compute asymmetric flow fields due to influences of propeller action. The first grid points off the body surface are located in the range $y^+<1$, in order to correctly resolve the flow in near wall region. Fig.3 provides comparison of computed surface pressure contours between for the towing and self propulsion conditions. The pressure contours for the latter clearly show expected decrease due to the action of the propeller, which is especially evident in the region the boundary layer is thick, i.e., the region near the stern bulb and near the propeller. It is also shown that the influences of the propeller are restricted to the stern region. The trends agree well with those displayed in the measurements. Fig.4 shows comparison of effective wake coefficient $1-w_t$, propeller efficiency ratio η_r, and thrust deduction coefficient $1-t$ between for the computational and experimental results, where satisfactory agreement between the two results is indicated.

Next, the present optimization method was applied to tanker stern form optimization for minimum *DHP*. For comparison, precursory results for minimum viscous resistance (*Rv*) (Tahara et al. 1999a) are also included in the following discussion. Hereafter, those results are referred to as DHP-Min. and VR-Min., respectively. The original hull form is Case-X, which is one of the modern tanker hull forms selected by the domestic Japan research project. The computational condition is $Rn=1.9\times10^6$, which follows that used in the authors' precursory work (Tahara et al. 1999a), and number of the computational grid (C-O-type) is 170×40×60 (in longitudinal × radial × girthwise directions, respectively) = 408,000. The modification region is between S.S.4 and the stern end; and only transverse modification is considered.

The optimized solution was obtained in 8 optimization cycles. Fig.5 shows comparison of geometry (bodyplan) between for the original and DHP-Min. hull forms, and Fig.6 for the original and VR-Min. hull forms. In addition, Figs.7 and 8 compare cross-sectional area distribution (CP curve) and the computed integral variables for those hull forms, respectively. Differences in stern frame lines are obvious between for the original and optimized hull forms, in which clear differences between for DHP-Min. and VR-Min. are also indicated, i.e., frame line modification from U to V types is significant for the latter, in contrast, the trend is somewhat reduced for the former. From U to V frame line modification of VR-Min. results in lower viscous resistance; however, also larger effective wake $1\text{-}w_t$ coefficient which causes lower propeller efficiency and larger *DHP*. This implies an important fact that hull form optimization for minimum viscous resistance does not yield minimum delivered-horsepower hull form. The volume distribution is moved backward for both DHP-Min. and VR-Min.; however, the trend of the former is slightly reduced, which leads to better thrust deduction coefficient $1\text{-}t$ for the former. The present DHP-Min. indicates the lowest *DHP* among the three hull forms, which supports a conclusion that the present method successfully optimized the original hull form for minimum *DHP*. All of the above-mentioned modification trends agree well with those commonly in use in traditional tanker hull form design.

CONCLUSIONS

The present work concerns tanker stern form optimization using CFD-based optimization method. The main focus is placed on minimization of delivered horsepower in conjunction with application of self-propulsion simulator, which is based on extension of a RaNS equation solver including propeller effects evaluated by the infinite-bladed propeller theory. In conclusion, the present method appeared to successfully optimize the given tanker stern form, and the modification trends automatically demonstrated in the present work agree well with those commonly in use in traditional tanker hull form design. Further modification and extension of the present self-propulsion simulator are currently in progress, in association with inclusion of free surface effects and extension of the RaNS solver for full-scale *Rn*.

REFERENCES

Nakatake, K. (1981). A Practical Method to Calculate Propulsive Performance of Ships. Memoirs of the Faculty of Engineering, Kyushu University, Vol. 41, No.1, pp.87-122.

Tahara, Y., Himeno, Y., and Tsukahara, T. (1998). An Application of Computational Fluid Dynamics to Tanker Hull Form Optimization Problem. Proceedings 3rd Osaka Colloquium on Advanced CFD Applications to Ship Flow and Hull Form Design, Osaka, Japan, May 25-27, pp. 515-531.

Tahara, Y., Saitoh, Y. and Himeno, Y. (1999a). CFD-Aided Optimization of Tanker Stern Form (1st Report) - Minimization of Viscous Resistance -. J. Kansai Society of Naval Architects, No. 231, pp. 29-36.

Tahara, Y., Saitoh, Y. and Himeno, Y. (1999b). CFD-Aided Optimization of Tanker Stern Form (2nd Report) - Minimization of Delivered Horsepower -. J. Kansai Society of Naval Architects, No. 232, pp. 9-18.

Tahara, Y., Nishida, R., Ando, J. and Himeno Y. (2000). CFD-Aided Optimization of Tanker Stern Form (3rd Report) - Minimization of Delivered Horsepower Using Self-Propulsion Simulator -. J. Kansai Society of Naval Architects, No. 234, pp.41-50.

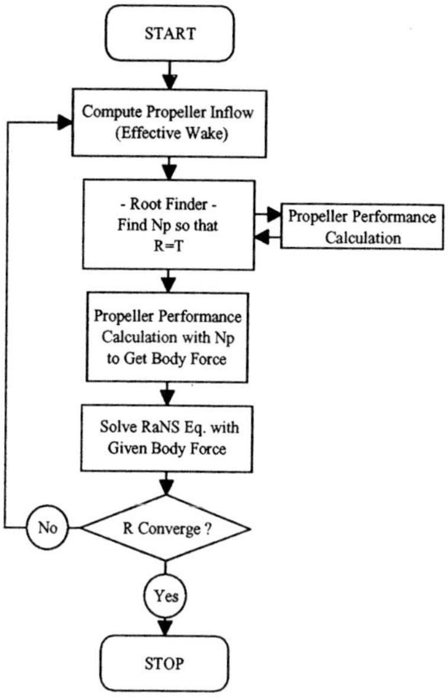

Figure 2: Overall computational procedure of the present self-propulsion simulator

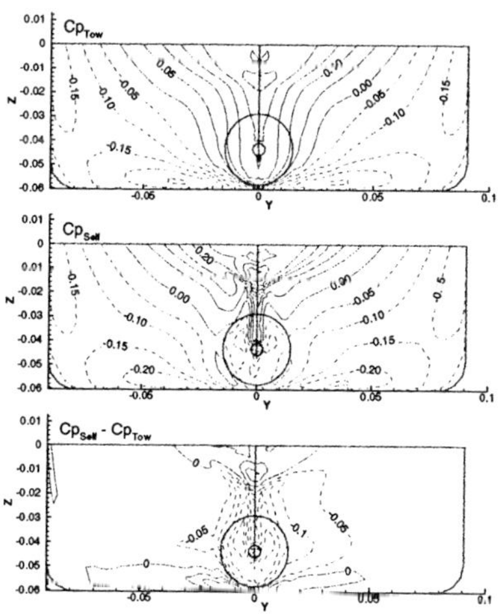

Figure 3: Comparison of surface pressure distributions between for towing and self-propulsion conditions (Case-A hull form)

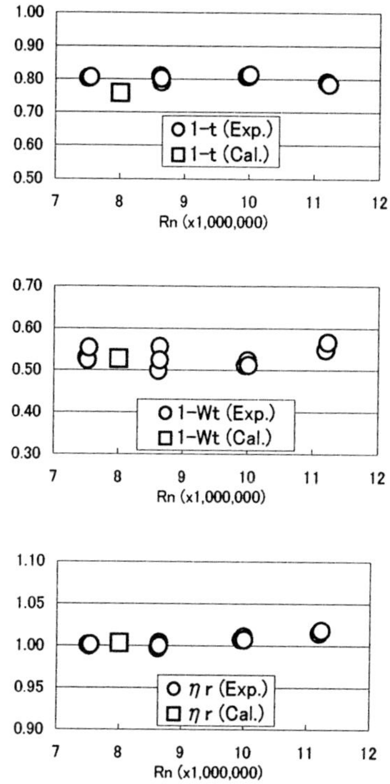

Figure 4: Comparison of $1\text{-}t$, $1\text{-}w_t$, and η_r between for computational and experimental results for self-propulsion condition (Case-A hull form)

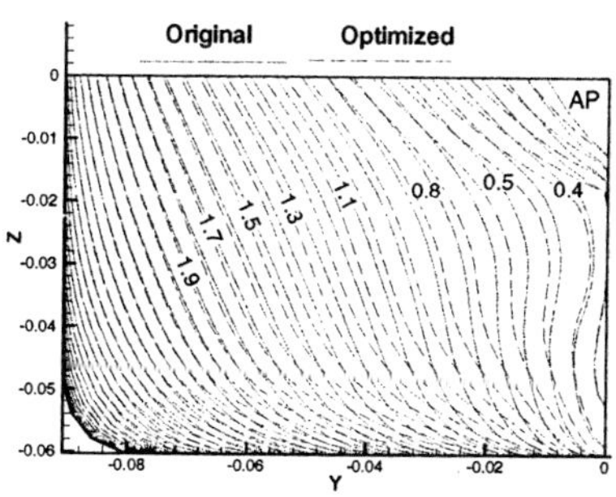

Figure 5: Comparison of bodyplan between for the original (Case-X) and optimized hull forms (DHP-Min. as the modified)

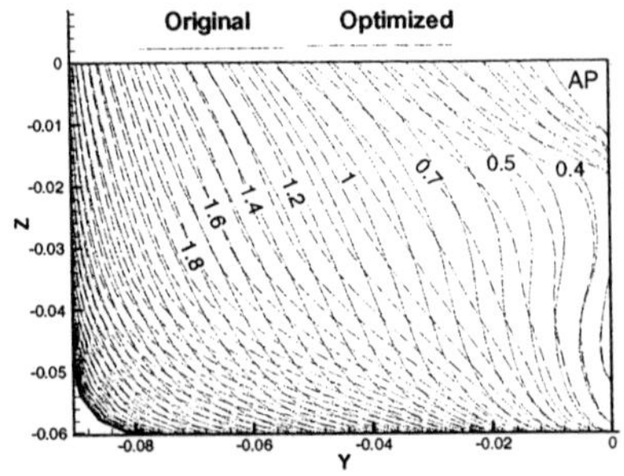

Figure 6: Comparison of bodyplan between for the original (Case-X) and optimized hull forms (VR-Min. as the modified)

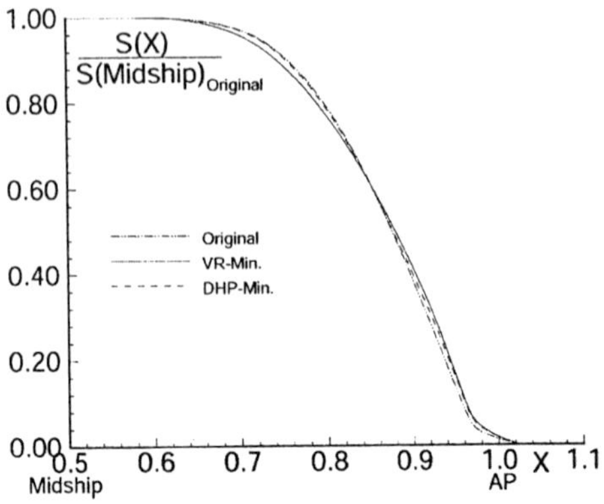

Figure 7: Cross-sectional area distribution (CP curve) for the original (Case-X) and optimized hull forms (VR-Min. and DHP-Min. as *Rv* and *DHP* minimized hull forms, respectively)

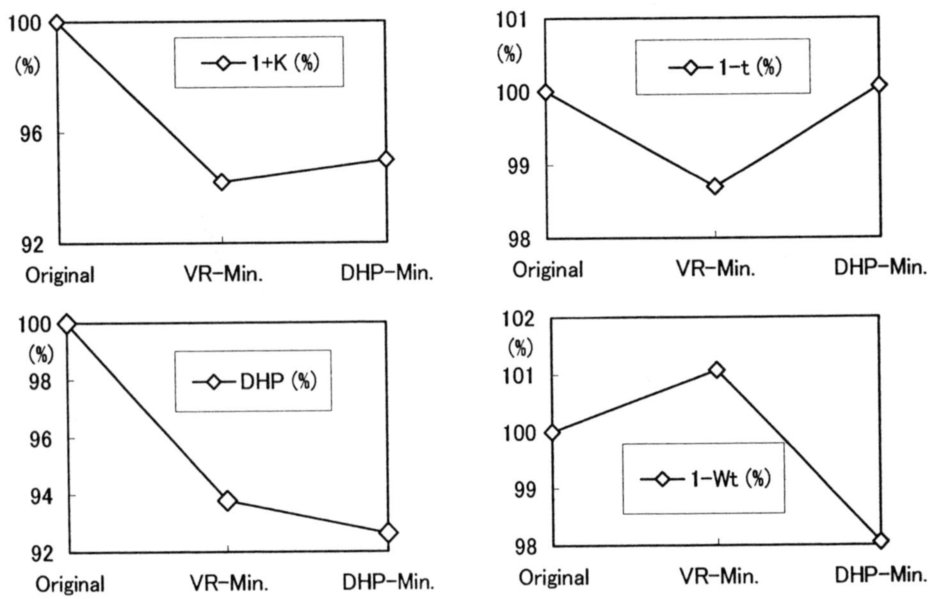

Figure 8: Comparison of computed 1+K, 1-t, *DHP*, and 1-w_t between for the original (Case-X) and optimized hull forms (VR-Min. and DHP-Min. as *Rv* and *DHP* minimized hull forms, respectively)

Practical Design of Ships and Other Floating Structures
You-Sheng Wu, Wei-Cheng Cui and Guo-Jun Zhou (Eds)

NUMERICAL AND EXPERIMENTAL STUDIES OF DUCTED PROPELLER

Rong Zhao

MARINTEK, P.O.Box 4125 Valentinlyst, Trondheim, Norway

ABSTRACT

Numerical approaches for steady and unsteady analysis of ducted propeller in combination of an experimental study of two duct propellers are carried out in this paper.

KEYWORDS

Ducted propeller, Propulsion, Model test, Thrust, Torque, Numerical analysis

1 INTRODUCTION

A numerical method to predict steady and unsteady performance of ducted propeller is developed. A vortex lattice method is used for the propeller and a surface panel method is used for the duct. In the lifting surface theory, the lifting effect is simulated by a bound vortex system, free vortex system, and the influence of the blade thickness by source distribution. The hydrodynamic interaction between the propeller and the duct is done by an iterative method in the steady analysis. The generalized image method is used for unsteady analysis of ducted propeller. The flow around the duct is calculated by using a three dimensional boundary element method. The surface of the duct is approximated with quadrilateral panels, and the boundary condition is satisfied at the center of each panel. An equal-pressure Kutta condition is used at trailing edge of the duct. The velocities induced by the propeller are circumferentially averaged in the steady analysis, so that the physically periodical flow around the duct is simplified as a steady, axis-symmetrical flow. The total thrust and torque coefficients of ducted propeller and the thrust coefficient of duct are predicted by the program. In addition the pressure distribution on the propeller blade and duct is calculated.
In order to validate the theory, model tests has been carried out at MARINTEK. One duct and two propellers have been tested. For each condition two to four tests are carried out. The tip clearance between the duct and propeller is investigated by model test for the propeller model P1287 with tip clearance 0.5%, 1.0% and 1.5% of the propeller diameter. It is found that the tip clearance is not very important for the total thrust and torque. The theoretical results are compared with experimental data. Both steady and unsteady analysis are used. The comparison are presented in this paper. Good agreement is obtained.

2 THEORY

Theory to predict steady and unsteady performance of ducted propeller is based on vortex lattice method for the propeller and a boundary element method for the duct. The velocities induced by propeller is taken into account in predicting the flow around the duct and the velocity induced by the duct is included in the inflow to the propeller. The hydrodynamic interaction between the propeller and the duct is solved by an iterative method in the steady analysis. In the unsteady analysis, a generalized image method is used which is based on an approach of Kinnas et al. (1990).

2.1 Theory to Predict Steady Performance of Propeller

A numerical method based on a lifting-surface theory of Wang and Hu(1988) is used to predict the steady and unsteady performance of propeller. A brief description about the theory will be given here.

A vortex lattice method was developed by Wang and Hu(1988). The propeller loading and blade thickness are represented by vortex and source distribution respectively. The method provide suitable results for moderately loaded propellers.

A trailing vortex wake model based on a modification of the methods by Kerwin and Lee(1978) and Greely and Kerwin(1982) is used. The numerical results show that this wake model gives better agreement with experiments for highly skewed propeller.

The duct is calculated by using a potential-based panel method. A detail description of theory is given here.

2.2 Boundary Element Method to Predict Flow around a Duct

A potential theory is used to predict flow around a duct. Here one assumes that the duct is axissymmetric about the x-axis. The x-axis is pointed to the down stream of the ducted propeller. The inflow to the duct is given. The velocities induced by the propeller and the inflow velocity are circumferentially averaged in the steady analysis. In the unsteady analysis, the exact inflow and propeller induced velocity are used.

It is assumed that the flow around the duct is described by a velocity potential ϕ and that ϕ satisfies the Laplace equation $\nabla^2\phi = 0$ in the fluid domain.

In order to solve the ϕ one must know the inflow velocity on the duct surface. The body boundary condition satisfied on the exact duct surface can be written as $\frac{\partial\phi}{\partial n} = -V_n$. Where V_n is the total velocity of inflow and propeller induced velocity in the normal direction $\vec{n} = (n_1, n_2, n_3)$ on the duct surface. Positive direction of $\vec{n}$ is pointed into the fluid domain.

The velocity potential can be represented by Green's second identity, i.e.

$$4\pi\phi(x,y,z) = \iint_S \left[-\frac{1}{R}\frac{\partial\phi(x_1,y_1,z_1)}{\partial n} + \phi(x_1,y_1,z_1)\frac{\partial(\frac{1}{R})}{\partial n} \right] dS(x_1,y_1,z_1) \tag{1}$$

where $R = [(x-x_1)^2 + (y-y_1)^2 + (z-z_1)^2]^{0.5}$. The surface S consists of S_{inf} and S_B. Here S_B is the duct surface and S_{inf} is the surface from the trailing edge of the duct to far down stream.

A integral equation based on equation (1) is set up by letting (x,y,z) approach points on the duct surface. On S_{inf} surface the velocity potential has a jump and the velocities are continuous. On the body surface ϕ is unknown and $\partial\phi/\partial n$ is known. On S_{inf} surface there is no contribution from $\partial\phi/\partial n$. A system of linear equations is then obtained. The jump of velocity can be obtained by introducing a Kutta condition and solved from a system of linear equations.

A Kutta condition is applied at the trailing edge of duct. It is assumed that the pressure is continuous at the trailing edge. The linear equations is solved by a standard subroutine. The velocity potential is the solution from the linear equation system. After one knows the velocity potential, the velocity on the

duct surface can be predicted by $\vec{v} = \nabla\phi$. The dynamic pressure p on the duct surface is calculated by using Bernoulli's equation.
The equation (1) is used to evaluate the duct induced velocity potential on the propeller. Then the velocity can be obtained by using $\vec{v} = \nabla\phi$.

2.3 Details on Numerical Method

In the steady analysis an iteration method is used to calculate interaction between the propeller and the duct. It is found out 6 iterations are always enough to get convergence results. The generalized image method is used in the unsteady analysis. The total inflow velocity to the duct includes two parts. One is from inflow without propeller induced velocity. The other is from propeller induced velocity. The velocities induced from the propeller are solved by using a generalized image method. For each unit strength vortex loop on the propeller blade and its trailing wake, a generalized image flow which satisfies the body boundary condition on the duct is calculated. This flow will introduce an additional velocity on the propeller blade. This has been taken in the total inflow velocity to the propeller. An unsteady analysis method of propeller in combination of an generalized image approach is used to solved unsteady problem of ducted propeller.

3 VALIDATION

In order to validate the theory, model test of ducted propellers has been carried out at MARINTEK for "open water" condition. One duct(D20) which is similar to 19A is used. Two propellers(P790 and P1287) with diameter 300 mm have been tested. For the propeller model P790, three pitch values (P/D=0.7, 0.86 and 1.0) are tested. For the propeller model P1287 four pitch values(P/D=0.9, 1.10, 1.30 and 1.50) are tested. Both propellers have four blades. The duct thrust, propeller thrust and torque are measured. For each conditions two rotational speed of propeller were tested. For some conditions two tests were done for each rotational 3 speed. The main propeller geometry data for P790 and P1287 is given in table 1. The geometry of duct is given in table 2.
The comparison between the theoretical predictions(steady analysis) and experimental data are presented in figure (1) and (2). Figure (1) presents the results for the propeller model P790. Figure (2) is the results for the propeller model P1287. The experimental data is for tip clearance of 1%.
Generally speaking the steady program predicted well the total thrust and torque, as well as efficiency. The program underestimates the duct thrust for high pitch ratio and overestimates the duct thrust for low pitch ratio. The program can predict the steady performance when the advance coefficient J going to zero. The model test has some difficulties at low advance coefficient J. At small J-values the duct has large vibration, that means the propeller may touch the duct. It seems that the theory predict less good results for small values of advance coefficient J. This may be due to nonlinear effects of propeller characteristics and the unsteady effect of the duct.
An unsteady analysis of ducted propeller is used to study the effect of unsteady flow on the duct. The results have been compared with experimental data and the results of steady analysis. Figure (3) shows the comparison of the results of steady and unsteady analysis, and the experimental results. It seems that the unsteady analysis give improvement of numerical results compared with experimental data, specially for the duct thrust.
The importance of tip clearance between the propeller blade and duct is studied by choosing the values of 0.5%, 1.0% and 1.5% of the diameter of the propeller model P1287. This is achieved by chopping off the propeller tip. In figure (4) the results from experimental data are compared with the theoretical results. The pitch ratio(P/D) is 1.3. It seems that there is almost no difference between the experimental results of tip clearance of 1.0% and 1.5%.
But there is some difference between tip clearance of 0.5% and 1.0% (or 1.5%). The experimental results of tip clearance of 0.5% are based on two rotational speeds of propeller. The difference in experimental results is large for these two rotational speeds. One may expect that the largest difference should be between the tip clearances 1.0% and 1.5%, since the area of propeller blade changes most.

Based on the experimental data one may conclude that the tip clearance may be not so important for the total thrust and torque of duct propeller.

4 CONCLUSIONS

A numerical program to predict steady and unsteady performance of ducted propeller is developed. The theories are validated by model test which has been carried out at MARINTEK.
The agreement between the theories and experimental results are satisfactory for duct thrust, propeller thrust and torque, specially for the results of unsteady analysis.

TABLE 1

MAIN GEOMETRY DATA OF P790 AND P1287

```
a) Number of chord length, thickness and ...
b) Nondimensional radius(r/R), where chord length and thickness are input
c) Pitch ratio(P/D) (D-propeller diameter)
d) Skew angle of the blade sections along the radial direction(degree)
e) Rake distribution in radial direction (x_m/D)
f) Chord length distribution (c/D)
g) Max. thickness of blade section in percentage of propeller diameter(t/D)
h) Camber distribution along radial direction (f/c)
Main geometry data of P790
a) 10
b) 0.28 0.30 0.40 0.50 0.60 0.70 0.80 0.90 0.95 1.00
c) 0.603 0.633 0.743 0.803 0.840 0.860 0.867 0.857 0.85 0.84
d) 0.000 0.000 0.000 0.000 0.000 0.000 0.000 0.000 0.000 0.000
e) 0.000 0.000 0.000 0.000 0.000 0.000 0.000 0.000 0.000 0.000
f) 0.190 0.203 0.250 0.293 0.330 0.347 0.347 0.307 0.253 0.000
g) 0.040 0.040 0.033 0.030 0.023 0.017 0.013 0.010 0.007 0.005
h) 0.0 0.0148 0.0267 0.0238 0.0222 0.0202 0.0192 0.0196 0.0197 0.000
Main geometry data of P1287
14
0.269 0.30 0.325 0.35 0.40 0.50 0.60 0.70 0.80 0.90 0.95 0.975 0.99 1.00
.827 .884 .927 .966 1.038 1.148 1.214 1.236 1.220 1.152 1.092 1.055 1.031 1.013
.0 -2.15 -3.57 -4.75 -6.48 -6.79 -5.66 -2.83 1.93 8.50 12.31 14.10 15.35 16.48
.0 .0043 .0057 .0064 .0076 .0097 .0117 .0132 .0135 .0118 .0092 .0068 .0044 0.0
.157 .182 .203 .223 .261 .320 .374 .416 .436 .417 .373 .326 .261 .051
.064 .042 .034 .031 .027 .0207 .0154 .0112 .0081 .0060 .0054 .0053 .0052 0.0051
0.0 .0236 .0287 .0286 .0292 .0302 .0313 .0317 .0310 .0283 .0247 .0210 .0170 0.0
```

TABLE 2

GEOMETRY OF THE DUCT

```
! length of duct(mm) = 300.0 mm
! coordinates of propeller center in the coordinate system of duct(mm)
! (x1 , r1)= (150.0 mm, -302.0 mm)
```

! xr-coordinate on the inside of ! an axis-symmetric duct. ! The length of the duct is unit(1.0 mm) ! The points start from trailing ! edge of the duct.		xr-coordinate on the outside side of an axis-symmetric duct The length of the duct is unit(1.0mm) The points start from leading edge of the duct	
! x-coor.	r-coor.	x-coor.	r-coor.
1.000000	0.0400	0.000000	0.1825
0.950000	0.0186	0.012500	0.2072
0.900000	0.0145	0.025000	0.2107
0.800000	0.0082	0.050000	0.2080
0.700000	0.0029	0.075000	0.2042
0.600000	0.0033	0.100000	0.2004
0.500000	0.0038	0.150000	0.1928
0.400000	0.0043	0.200000	0.1852
0.300000	0.0048	0.250000	0.1776
0.250000	0.0110	0.300000	0.1700
0.200000	0.0217	0.400000	0.1548
0.150000	0.0387	0.500000	0.1396
0.100000	0.0634	0.600000	0.1244
0.075000	0.0800	0.700000	0.1092
0.050000	0.1014	0.800000	0.0940
0.025000	0.1285	0.900000	0.0788
0.012500	0.1465	0.950000	0.0712
0.000000	0.1825	1.000000	0.0400

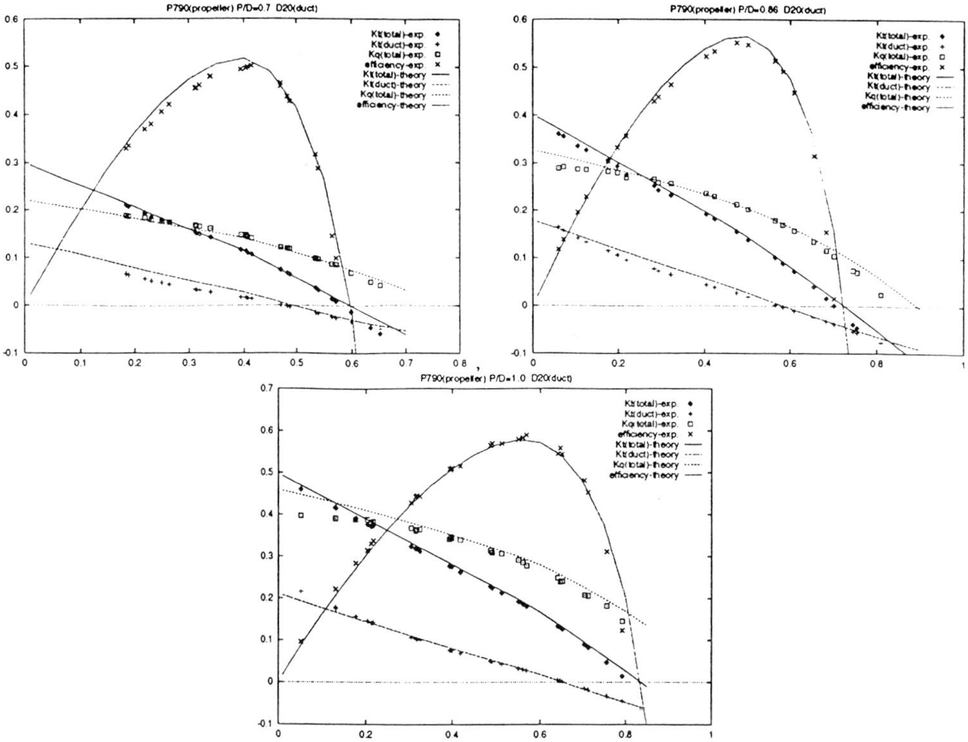

Figure 1: Comparison of numerical and experimental results of duct thrust, total thrust, torque and efficiency coefficients in "open water" condition. P790(propeller) and D20(duct) are used. a) Pitch ratio P/D is 0.70. b) Pitch ratio P/D is 0.86 c) Pitch ratio P/D is 1.0. The horizontal axis is J=V/(nD). Steady analysis is used in the theoretical prediction. The Kq values presented in the figure are multiplied by a factor 10.

References

Greely,D.S. and Kerwin, J.E.(1982). Numerical methods for propeller design and analysis in steady flow, *Trans. SMAME*, **90**.

Kerwin, J.E. and Lee,C.S.(1978). Prediction of steady and unsteady marine propeller performance by numerical lifting-surface theory, **Trans. SNAME, 86**.

Kinnas, S.A., Hsin, C.Y., and Keenan, D.P.(1990). A potential based method for the unsteady flow around open and ducted propeller, *18th ONR*.

Minsaas, K. (1967). *Propellteori*, Skipsmodelltankens Meddelelse no.96.

Wang, G.Q. and Hu, S.G.(1988). Improvement of Prediction Method for Propeller Characteristics and Blade Pressure Distridution, *Shipbuilding of China*, Tot. No.100.

Zhao, R.(1999). Ducted Propeller Computer Program Development, Verification Tests in Open Water Condition, *MARINTEK REPORT* MT60 F99-421, 603526.00.01

Practical Design of Ships and Other Floating Structures
You-Sheng Wu, Wei-Cheng Cui and Guo-Jun Zhou (Eds)

DESIGN OF CAVITATING PROPELLERS BY LIFTING SURFACE THEORY

Guo-Qiang Wang, Chen-Jun Yang

Shanghai Jiao Tong University, Shanghai, 200030 China

ABSTRACT

The paper presents a theoretical method for the design of supercavitating propellers or trans-cavitating propellers (a type of hybrid of the supercavitating propeller) by using the lifting surface theory. Based on performances of the 2-D supercavitating hydrofoil obtained by linear vortex panel method, the Lerbs' induction factor method is adopted in the preliminary stage. The geometry of the propeller, such as blade outline, section profiles and circulation and pitch radial distribution etc., is determined. The final pitch radial distribution and detailed off-sets of section profiles of the blades are then obtained by lifting surface theory for the cavitating propeller to meet with design requirements.

KEYWORDS

Panel method, Lifting surface theory, Propeller, Cavitation, Supercavitating propeller

1 INTRODUCTION

Choosing a supercavitating propeller as a propulsive device for high speed ships is a wise decision to meet successfully a demand for high thrust while simultaneously lessening the erosion effects of cavitation. SSPA of Sweden published a series of supercavitating propellers in 1979.[1] Corresponding charts were provided, which can be easily employed by designers. Early in 1962, DTMB design method was developed finally by Venning, in which the Goldstein's lifting line theory and Tulin's linearized cavity flow theory together with some sophisticated corrections were applied.[4] Systematic investigations upon the design method of supercavitating propeller have been made by Ukon of SRI in Japan. Several papers related to this topic have been published since 1994.[2,3] In this paper, a linear vortex panel method is adopted for the calculation of hydrodynamic characteristics of 2-D supercavitating hydrofoils. Based on the cavity geometry and hydrodynamic characteristics of 2-D supercavitating hydrofoils, the lifting line theory together with the lifting surface corrections is employed for the design of supercavitating propellers or the trans-cavitating propellers. The preliminary blade outline, section profiles, and load and pitch radial distribution can be determined in this stage. Then a vortex lattice method for the performance prediction of partial cavitating and supercavitating propeller is applied to compute the cavity geometry and hydrodynamic performance of the designed propeller. Section profiles and pitch distribution of the blade of the designed cavitating propeller are modified until all the design requirements are fulfilled.

2 LINEAR VORTEX PANEL METHOD FOR THE HYDRODYNAMIC CHARACTERISTICS OF 2-D SUPERCAVITATING HYDROFOILS

In this method the supercavitating hydrofoil is represented by a set of singularities. Vortices with the strength of linear distribution are arranged on the wetted surface of the hydrofoil. Vortices of equal density with opposite sign are distributed on the upper and lower surfaces of the cavity respectively. At the leading edge of the foil and in the small region in front of the cavity, vortices with singularity are specified. Upon arranging all the singularities, the strength of vortex distribution and cavity geometry can be evaluated by using the dynamic and kinematic conditions on the wetted surface of the hydrofoil and cavity surface. The hydrodynamic characteristics are calculated finally.

Fig.1 shows a 2-D supercavitating hydrofoil. The lower wetted surface of the hydrofoil is composed of three parts. A small region *sle2* with *nf12* elements and a small region *ste* with *nf3* elements are chosen near the leading and trailing edges respectively, on which vortices with singularity are distributed. The middle large part of the lower surface is divided into *nf2* elements, on which vortices with the strength of linear distribution are arranged. The upper surface is consists of two part. A small region *sle1* with *nf11* elements is near the leading edge, on which vortices with singularity are distributed. Other large part of the upper surface is covered by cavity totally. The cavity starts from the end points of *sle1* on the upper side and the end point of *ste* on the lower side respectively. As mentioned above, vortices of equal density with opposite sign are distributed on the upper and lower surfaces of the cavity respectively. Upper surface of the cavity is divided into *nc2* elements and lower surface *nc1*. The control point is located in the middle of each element. It should be noticed that at the end of the cavity the Riabouchinsky model is used for the closure condition of the cavity. A small circular arc of length *send* with *ne* elements is put at the end of cavity. Vortices with the strength of linear distribution are arranged on this circular arc. The strength of the vortex is specified to be zero at the middle point of the circular arc due to the opposite sign of the vortices on the upper and lower cavity surfaces.

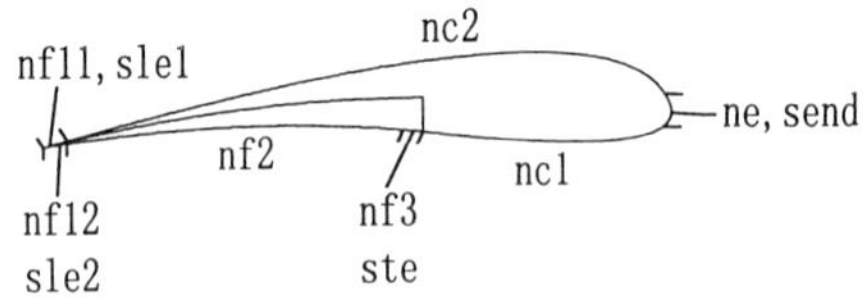

Figure 1: 2-D supercavitating hydrofoil

This approach is validated for the 2-term supercavitating section shape. Fig.2 shows the comparison of the lift coefficient at different cavitation number for the 2-term supercavitating section shape obtained from present calculation and experiments. It can be seen from Fig.2 that under conditions of fully developed cavitation or supercavitation, calculated results have good agreement with experimental data. Only in the case of small attack angle and large cavitation number results obtained by the present calculation are a little higher than ones from experiments due to that the cavity length is short so that the cavity surface intersects the section profile near the trailing edge.

Sections of the supercavitating propeller are usually operating near the condition that the geometric angle of attack measured from the nose-tail line $\theta = 2.5^0$, that is the angle of attack measured from the so-called $A_0 = 0$ reference line $\alpha = 2^0$. Thus, systematical calculations for relations among total

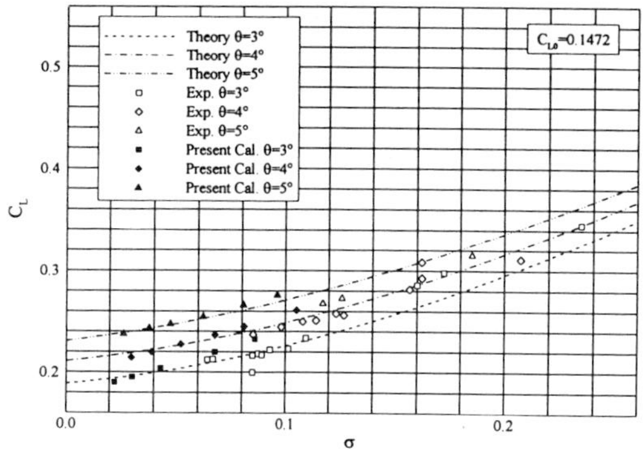

Figure 2: Comparison of experimental and theoretical lift coefficient for a 2-term supercavitating section

lift coefficient C_L, drag coefficient C_D and cavity length L_C are carried out at different cavitation number σ and lift due to camber C_{L0} from 0.01 to 0.12 for 2-term section under $\alpha = 2^0$, namely $\theta = 2.5^0$. Fig.3 gives an example of those calculated results, which will be used for the supercavitating propeller design.

3 DESIGN OF CAVITATING PROPELLER BY LIFTING LINE THEORY

At first, conventional lifting line theory together with lifting surface corrections is applied. The only difference from the method for the sub-cavitating propeller is that the supercavitating sections are used for the blade instead of conventional airfoil sections. Hydrodynamic characteristics of supercavitating sections are obtained from the linear vortex panel method as mention above. However, conventional airfoil sections are still used for sections not operating under the condition of supercavitation. The design procedure is similar to what given in reference [4].

3.1 Design Procedure

The well known induction factor method of Lerbs is employed to determine the radial distribution of nondimensional circulation, hydrodynamic pitch angle and ($C_L \cdot L/D$) ratio under the design condition of given power or thrust first. Then the lifting coefficient C_L and cavitation number can be evaluated by known radial distribution of blade chord length. A set of lift coefficient C_L, drag coefficient C_D and cavity length L_C at different C_{L0} and obtained section cavitation number can be determined from Fig.4 through 6, which are derived from figures similar to Fig.3. The lift coefficient C_{L0} due to camber, drag coefficient C_D and cavity length L_C for each section are determined by interpolation from the section lift coefficient C_L obtained above.

Upon having C_{L0}, face ordinates and thickness distribution of the 2-term supercavitating section can be computed.[4]

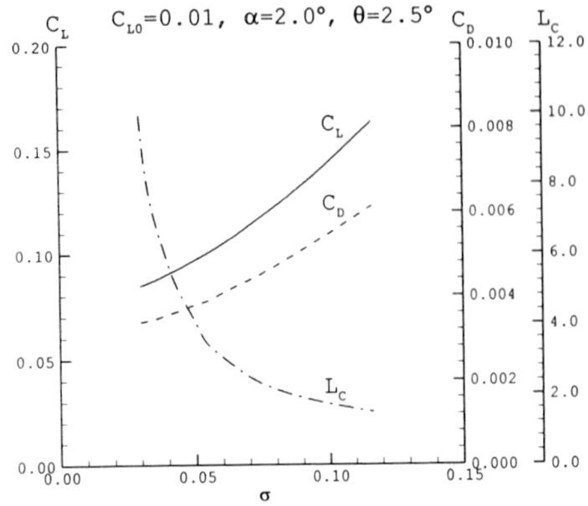

Figure3: Relations among total lift coefficient C_L drag coefficient C_D and cavity length L_C

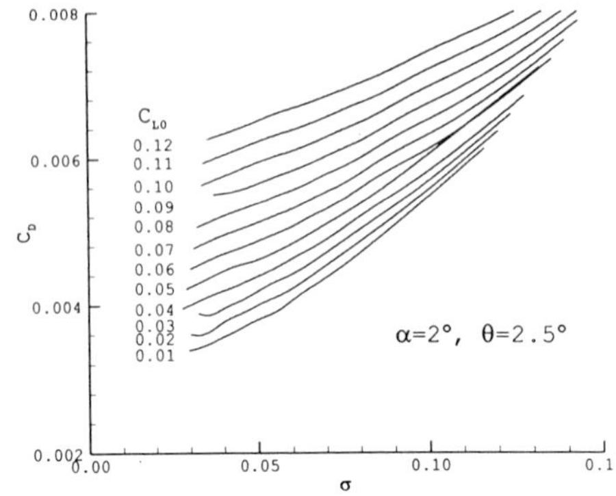

Figure 4: Drag coefficient C_D at different C_{L0}

If the condition of supercavitation is not fulfilled in the inner part of the blade radius, conventional airfoil profiles are employed, for example NACA66-mod, and determined by the design method for sub-cavitating propeller. Final geometric parameters of the cavitating propeller are obtained after lifting surface corrections. Propeller thrust coefficient K_T and torque coefficient K_Q are estimated.

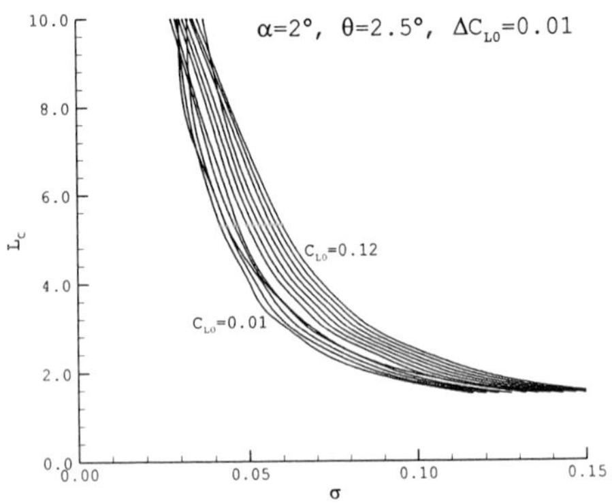

Figure 5: Cavity length L_C at different C_{L0}

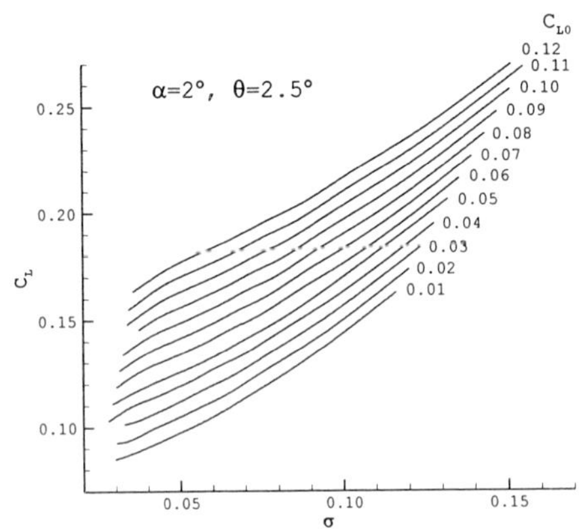

Figure 6: Lift coefficient C_L at different C_{L0}

It should be mentioned here that the following formula is used to estimated the drag-lift ratio in the present design method:

$$\frac{C_D}{C_L}=\left[\left(2.14-0.05866\frac{\alpha^o}{C_L}\right)+\frac{1.25}{C_L}\right]\times\left[C_L^2\left(0.3192+0.01316\frac{\alpha^0}{C_L}\right)^2+0.0042\right] \qquad (1)$$

3.2 Design Sample

A design sample of a supercavitating propeller in reference [2] is redesigned in this paper. Design conditions are as follows:

design speed: $V = 50.0$ knots, required thrust: $T = 100000.0$ kgf, propeller diameter: $D = 3.318$ m,

hub ratio $r_h = 0.2$, rate of revolution: $n = 423.0$ rpm, horse power: $P = 53200.0$ hp, immersion depth of propeller shaft: 4.0m, rake angle 10^0, thrust deduction factor $t = 0.0$, blade expanded ratio: $EAR = 0.55$.
According to the required thrust, namely $K_{T0} = 0.1587$, a cavitating propeller is designed by using above mentioned design method. Design results are as follows:

$$K_T = 0.1571, \quad K_Q = 0.0404, \quad \eta_0 = 0.6811$$

Fig.7 describes the propeller blade outline and section profiles.

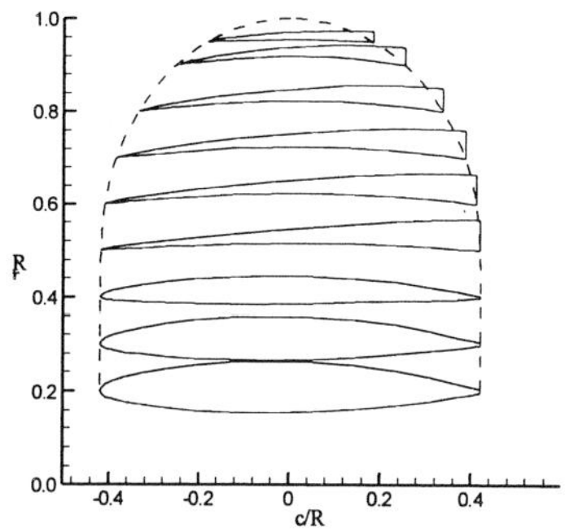

Figure 7: Propeller blade outline and section profiles

Three supercavitating propellers were designed in reference [2] according to the same design conditions. One among those is designed by a similar method to the present one. Results from [2] are $K_T = 0.160$, $K_Q = 0.0356$ and $\eta_0 = 0.786$. A model for the designed propeller was made and corresponding experimental data at design conditions are $K_T = 0.183$, $K_Q = 0.0474$ and $\eta_0 = 0.676$. Compared with the results shown above, cavitating propeller obtained by present method is quite acceptable from view point of the demand accuracy.

4 PREDICTION OF HYDRODYNAMIC PERFORMANCE OF PROPELLER CAVITY

A theoretical method for prediction of cavity geometry and hydrodynamic forces of 3-D cavitating hydrofoil and cavitating propeller was developed.[5, 6, 7]

References [5, 6] presented an effective unified numerical method for the calculations of steady and unsteady cavity extent and shape on 3-D partial cavitating, supercavitating hydrofoils or those with mixed planforms consisting of partial and supercavitation, without the need of separate algorithms or separate discretizations. The problem of partial cavitation and supercavitation on 3-D hydrofoils were investigated systematically. This method was extended to the case of cavitating propeller.[7] The singularities representing loading, section thickness and cavity on the propeller blades, and control points, where boundary conditions should be fulfilled, are all located on the camber surface and the wake vortex sheets of blades. In unsteady flow, the wake vortex sheet contains trailing vortices and shed vortices. A conical helicoid to describe the deformed trailing vortex sheet in the transitional wake region and a semi-closure cavity model at the cavity end are employed in the calculation. The cavity extent and shape and hydrodynamic performances of the propeller operating in the uniform or non-unioform flow field were studied.

For the cavitating propeller in an unbound, incompressible fluid, the solution of the problem must fulfil the law of fluid mass conservation, quiescence condition at infinity upstream, Kutta condition at the trailing edge and Kelvin's theorem for conservation of circulation. They will all be satisfied if a suitable discrete singularity arrangement and grid generation are applied. Besides, boundary conditions, such as the tangency condition on the wetted surface, dynamic condition, kinematic condition on the cavity surface and cavity semi-closure condition, must be satisfied.

A set of algebraic equations fulfilling the boundary conditions on the control points of propeller blade operating in the non-uniform flow filed can be derived as follows:

$$\sum_{m=1}^{M}\sum_{n=1}^{N}\left[(K_{ij1}^{G})_{nm1}^{n}-(K_{ij1}^{sG})_{nm1}^{n}\right]\cdot\Gamma_{nm1}+\sum_{m=1}^{M}\sum_{n=1}^{L(m)}(K_{ij1}^{Q})_{nm1}^{n}\cdot Q_{nm1}^{c}$$
$$=-\mathbf{U}_r\cdot\mathbf{n}_{ij1}-(V^b)_{ij1}^{n}-(V^o)_{ij1}^{n}-\sum_{m=1}^{M}(\sum_{n=1}^{N}\Gamma_{nm1}^{(t-\Delta t)})(K_{ij1}^{sG})_{nm1}^{n}-\sum_{m=1}^{M}\sum_{n=2}^{NW}(K_{ij1}^{sG})_{nm1}^{n}\cdot\Gamma_{nm1}^{s}$$
$$i=1,2,\cdots,N;\qquad j=1,2,\cdots,M \tag{2}$$

$$\sum_{m=1}^{M}\sum_{n=1}^{N}\left[(\boldsymbol{K}_{ij1}^{G})_{nm1}^{t}-(\boldsymbol{K}_{ij1}^{sG})_{1m1}^{t}\right]\cdot\Gamma_{nm1}+\sum_{m=1}^{M}\sum_{n=1}^{L(m)}(\boldsymbol{K}_{ij1}^{Q})_{nm1}^{t}\cdot\boldsymbol{Q}_{nm1}^{c}-\frac{1}{2U_r}\cdot\sigma_{j1}$$
$$=-\phi_t-\frac{1}{2}\tilde{u}^2-\frac{1}{U_r}gh-(V^b)_{ij1}^{t}-(V^o)_{ij1}^{t}$$
$$-\sum_{m=1}^{M}(\sum_{n=1}^{N}\Gamma_{ij1}^{t-\Delta t})(\boldsymbol{K}_{ij1}^{sG})_{1m1}^{t}-\sum_{m=1}^{M}\sum_{n=2}^{NW}(\boldsymbol{K}_{ij1}^{sG})_{nm1}^{t}$$
$$i=1,2,\cdots,L(j);\qquad j=1,2,\cdots,M \tag{3}$$

$$\sum_{i=1}^{L(j)}Q_{ij1}^{c}=2\beta\alpha_r L(j)U_r+\frac{\partial V(r,t)}{\partial t}\qquad j=1,2,\cdots,M \tag{4}$$

where treatments for terms of $\tilde{u}^2$、ϕ_t、$\dfrac{\partial V(r,t)}{\partial t}$ refer to reference [6]. $(K_{ijk}^{G})_{nml}^{n}$、$(K_{ijk}^{G})_{nml}^{t}$ respectively denote the normal and tangential induced velocities at ijk-th control point due to the lmn-th spanwise vortex element with unit strength and the corresponding chordwise and trailing vortex lines. The value with superscript s stands for that due to the sheded vortex. (k=1, l=1 in equations mean the corresponding values on the key blade). $(K_{ijk}^{Q})_{nml}^{n}$、$(K_{ijk}^{Q})_{nml}^{t}$ respectively denote the normal and tangential induced velocities at ijk-th control point due to the lmn-th source line element representing the cavity with unit strength. $\Gamma_{ijk}^{(t-\Delta t)}$ denotes the ijk-th spanwise vortex strength at time $(t-\Delta t)$ and Γ_{ijk} at time *t*. $\boldsymbol{Q}_{nml}^{C}$ is nml-th source strength representing cavity. α_r is angle of attack of inflow at radius r. β is a coefficient equal to 0.35. *L(j)* cavity length at r. g is gravity acceleration. h is the height of the point to the shaft. $U_r=\sqrt{V^2(1-w_x)^2+(2\pi nr-Vw_t)^2}$ incoming flow at the section. $\boldsymbol{V}^b$ induced velocity due to blade thickness. $\boldsymbol{V}^o$ induced velocity due to singularity system of other blades.

The cavitating propeller calculation becomes the solution of the set of linear algebraic equations (2) to (4). As cavity dynamic condition should be fulfiled on the cavity region, which is unkown prior, iteration procedure is necessary. Initial value of $\boldsymbol{L(m)}$ is pressumed, (2) to (4) become a set of linear algebraic equations for unknown quantities Γ_{ijk}、Q_{ijk}^{c}、σ_{j1}. The solution of σ_{jk} is compared with given cavitation number σ. The cavity length $\boldsymbol{L(m)}$ is increased or decreased when σ_{jk} is greater or less than σ. The corrected $\boldsymbol{L(m)}$ is then substituted into (2) to (4), equations are solved again till $\sigma_{jk}=\sigma$ on every section. The values of Γ_{ijk}、Q_{ijk}^{c}、σ_{jk} obtained are the solution for the cavitating propeller.

5 DESIGN OF CAVITATING PROPELLER BY VORTEX LATTICE METHOD

5.1 Design Procedure

As mentioned above, conventional lifting line theory together with lifting surface correction for the design of cavitating propellers can be accepted for the engineering practical application, but it is not enough, especially so-called lifting surface correction is obtained for sub-cavitating propeller. Thus a lifting surface theory for the cavitating propeller is employed in this paper.

First, the design method described in paragraph 3 is used as the first step of cavitating propeller design. Then the vortex lattice method is applied to estimate the hydrodynamic performance, K_T and K_Q, of the preliminary designed propeller. Compared with the given K_{Q0} obtained from design condition, change the geometry of propeller, for example the camber of the section in the present method, and repeat the lifting surface part until $|K_{Q0} - K_Q| \le \varepsilon_Q$, a specified allowance error.

Second, K_T is compared with demand K_{T0}, if $|K_{T0} - K_T| \ge \varepsilon_T$, change the ship speed and redesign from the lifting line part util $|K_{T0} - K_T| \le \varepsilon_T$.

If the condition of supercavitation is not fulfilled in the inner part of the blade radius, conventional airfoil profiles are employed, for example NACA66-mod, and determined by the design method for subcavitating propeller.

5.2 Design Sample

5.2.1 A Supercavitating Propeller Designed by SSPA Chart

A supercavitating propeller was deigned by SSPA chart under the design condition as: delivered horse power 11214.0hp, rate of revolution 1000.0rpm, shaft efficiency 0.97, wake factor 0.35, thrust deduction coefficient 0.21, relative rotation efficiency 1.0, immersion depth of propeller shaft 2.95m.

The geometry of the supercavitating propeller designed by SSPA chart is used as input for the present method to calculate the hydrodynamic performance. Tab.1 gives the comparison of the calculated results obtained by SSPA chart and present method respectively.

TABLE 1

COMPARISON OF CALCULATED RESULTS BY SSPA CHART AND PRESENT METHOD

	SSPA Chart	Present Method
K_T	0.1357	0.1386
$10K_Q$	0.2744	0.2783
η_0	0.5727	0.5762

It can be seen that hydrodynamic performances are correctly be evaluated by the present method compared with the experimental results from SSPA supercavitating propeller series.

5.2.2 Propeller designed in section 3

The cavitating propeller designed in section 3 is checked by the present method. The results are compared in Tab.2.

TABLE 2

COMPARISON OF RESULTS BY LIFTING LINE AND PRESENT METHOD

	Lifting Line	Present Method
K_T	0.1571	0.1417
$10K_Q$	0.4040	0.3729
η_0	0.6811	0.6651

It is found that both the thrust and torque coefficients obtained by the present method are lower than those from lifting line theory. It might be that the lifting surface corrections for the sub-cavitating propeller seem not enough to consider the 3-D effect. Of course, further study is necessary.

6 CONCLUSIONS

The following conclusions can be drawn:

1. The method for the design of supercavitating propellers or trans-cavitating propellers (a type of hybrid of the supercavitating propeller) by using the vortex lattice method proposed in this paper is validated. This method will be a useful tool for the design of cavitating propellers for the purpose of practical application.
2. The flow-curvature corrections used in the design of conventional propellers are not completely applicable to the design of supercavitating propellers.
3. Future study is necessary to improve the present method. In order to accumulate experience, more practice is needed.

References

[1] Rutgersson O.(1979). Supercavitating Propeller Performance, Influence of Propeller Geometry and Interaction between Propeller, Rudder and Hull. *SSPA* **No.82**

[2] Ukon Y., Kudo T., Kurobe Y., Hoshino T.(1994). Design and Evaluation of New Supercavitating Propellers. *The Second International Symposium on Cavitation*, 395-402.

[3] Ukon Y., Kudo T., Kurobe Y., Matsuda N., Kato H.(1995). Design of High Performance Supercavitating Propellers Based on a Vortex Lattice Method. *PROPCAV'95*, 151-162.

[4] Venning E., Haberman W. L.(1962). Supercavitating Propeller Performance. *Trans. SNAME*, 354-417.

[5] Wang G.-Q., Huang Y., Yang C.-J., Yang J.-M.(1996). Investigations on the Stability of Cavity on 3-D Hydrofoil. *Proceedings of Sixth Symposium of Propulsor and Cavitation, CSNAME*, 145-152.

[6] Wang, G.-Q., Huang, Y., Yang C.-J.(1997). Analysis of the Flow around 3-D Cavitating Hydrofoil by a Vortex Lattice Method. *Proceedings of China-Korea Marine Hydrodynamics Meeting*, 286-294.

[7] Wang G.-Q., Huang Y., Yang J.-M., Yang C.-J.(1996). Prediction of Propeller Unsteady Cavitation. *Proceedings of Sixth Symposium of Propulsor and Cavitation, CSNAME*, 137-144.

You-Sheng Wu, Wei-Cheng Cui and Guo-Jun Zhou (Eds)

PREDICTION OF TRANSIENT LOADING ON A PROPELLER FROM AN APPROACHING ICE BLOCK

P. Liu[1], B. Colbourne[2] and Chin Shin[3]

[1,2,3]National Research Council Canada, Institute for Marine Dynamics, Box 12093, Station "A"
1 Kerwin Place, St. John's, Newfoundland, A1B 3T5, Canada,

ABSTRACT

An unsteady 3D surface panel method has been developed to predict hydrodynamic load fluctuations on an ice class propeller induced by continuous variation of proximity to an ice block. The low order, time domain, combined doublet and source panel method approximates the doublet and source distribution uniformly over each panel on the propeller blades. For non-lifting bodies, i.e., the hub and ice block, only sources are distributed over the body surfaces. The simulation model is contrived in such a manner that the ice block and surrounding fluid remain stationary; and at each time step, the propeller rotates and advances forward in the inertial reference frame. This numerical model is validated with previous fixed-proximity experimental measurements and good agreement is obtained. Prediction of the fluctuating hydrodynamic load is carried out as a full dynamic interaction between the ice block and the propeller. Results for this study are compared with previous fixed-proximity numerical models and experiments. The new dynamic model establishes a basis for analysis of a more realistic fluid-structure interaction, which could, in the future, include ice block acceleration due to suction force and ice block impact loading on the propeller blade and shaft.

KEYWORDS

Marine Propulsion, Panel Methods, Unsteady Loading, Ice-Propeller Interaction

1 INTRODUCTION

Ice class propeller performance evaluation is a relatively new topic in marine hydrodynamics. A *3-D* panel method was presented by Shih & Zheng (*1993*) to examine the thrust and torque jump for a propeller interfacing with a rectangular ice block under a fixed proximity condition. Both physical and numerical models in two-dimensional space were established by Newbury et al. (*1993*). These models are all for non-contacting ice-propeller interaction. Veitch (*1995*) established a physical model for ice-contact load on a propeller blade to simulate an ice-milling process. A numerical structural load simulation was also implemented (Veitch et al. *1997*). Bose (*1996*) used a *3-D* panel method to evaluate the induced hydrodynamic load fluctuation between a milled ice block and the propeller blade leading edge profile. A propeller performance evaluation software, PROPELLA, was developed based on Liu (*1996*) and used to evaluate 6 different propellers for induced load under fixed proximity conditions (Liu et al. *2000*). The ice-contact load model by Veitch (*1995*) was modified by reformulating the pressure regimes and by incorporating the hydrodynamic and structural loads into an ice/propeller blockage impact model. These changes were implemented to PROPELLA to predict ice-propeller contact load (Doucet et al. *1998*). In all the above-mentioned studies, the gaps between the propeller leading edge and the ice-blockage are fixed. The next logical development is to determine the hydrodynamic loading fluctuation as a propeller approaches an ice-blockage, i.e., the gap is a function of time. The current study was initialized for this investigation.

2 ABOUT THE METHOD

A time-domain, low order panel method is developed in the computer code PROPELLA. Fundamentals of the panel method can be found in the work by Katz and Plotkin (*1991*). The current method, however, has the following features:

1. The formulation of the method is in the time domain. The propeller progresses with time steps and so does the shed wake. This enables both multi-body and multi-path treatment. The fluid is static so it allows objects in the flow domain to move in different paths (Liu *1996*).
2. A fast yet stable iterative matrix solver, the Bi-Conjugate Gradient Stability method is implemented in the code. An earlier version of the code has an option to store the matrix and solve the unknowns row by row (Liu *1996*).
3. A stable and suitable Kutta condition for multi-body and multi-path flow was established by implementing the Broyden iterative pressure procedure (Liu & Bose *1998*).
4. The code design and implementation take flexibility of the object surface into account so that the solid boundary varies with time when required (Liu *1996*).
5. Both hyperboloidal and flat quadrilateral panel formulations are implemented in the code to handle hyperboloidal (see Morino & Kuo *1974* for formulation and Liu & Bose *2000* for an implementation note), flat quadrilateral and triangular panels (Newman *1986*) by giving appropriate doublet and source influence coefficient matrices as required.
6. An optional prediction of the load and pressure distribution on blades of a propeller with sheet cavitation by a novel and simple model may also be considered. (to be published).

In the ice-propeller induced load model, the ice blockage is fixed in space. Exact simulation of a static object in the flow domain gives a singularity of the pressure coefficient on the ice-blockage surface since the velocity of the blockage is zero. To prevent the singularity, a close-to-zero velocity such as *0.0001 m/sec* is taken for the motion of the ice block. Water head is also taken into account so the pressure coefficient value at corresponding points on each blade are then different, even if the flow is steady in nature.

3 RESULTS AND DISCUSSION

There are *20* chordwise panel intervals and *14* equal radial intervals on each blade. Chordwise intervals are arranged by using a semi-cosine function. When a sinusoidal function to control the panel intervals is used, the leading edge panels often have a very large length to width ratio for a dense grid. This affects the accuracy of the results and gives a poor grid connection between the blade root and the hub at the leading edges of a propeller. A remedy was taken to limit the lowest length in the chordwise direction to no less than a predefined value such as *2.5%* of the local chord. The semi-cosine function for the chordwise panel arrangement is used for a better resolution when the prediction of the cavity detachment point is required in the cavitation simulation model.

The blade sectional shape, panel arrangement on an expanded blade plan, face side and right side views of the R-class propeller are shown in figure 1. The geometry details of the propeller may be found in Walker (*1996*).

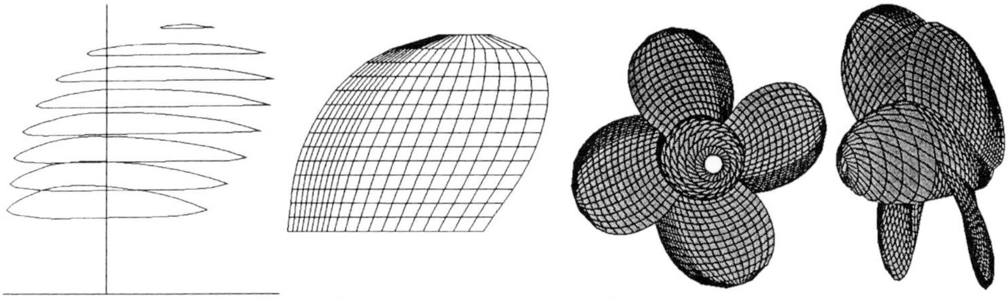

Figure 1: Sectional shape panel layout of the ice class propeller in expanded plan and the rear and side views of this propeller. The pitch ratio at $r=0.7R$ is *0.78* and the hub to diameter ratio is *0.30*. This propeller has a rake angle of *10*° and minor skew

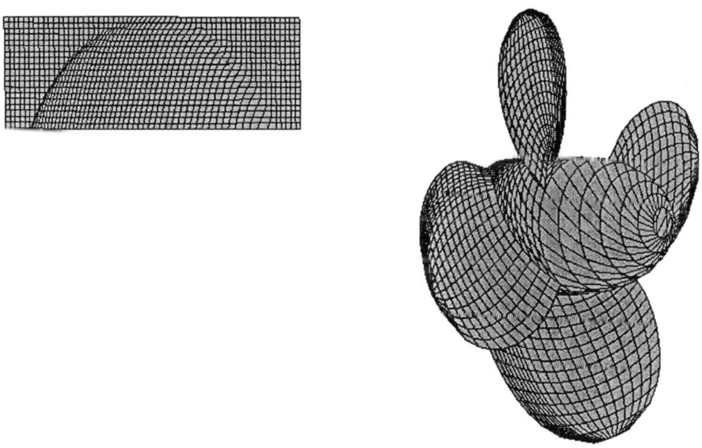

Figure 2: Initial position of the ice-class propeller relative to the blade-leading-edge contoured ice wall. The ice wall is fixed and the propeller is approaching the wall at a forward speed of *0.8 m/sec*

Figure 2 shows the ice-class propeller and the ice blockage contour-fitted to the blade leading edge. The ice blockage width is the same as the propeller diameter and its height is ¼ of the diameter. The top of the ice block is aligned with the maximum height of the blade tip. The rotational speed of the *200*-mm propeller model is *10* revolutions per second and the forward speed is *0.8 m/sec* so the advance coefficient J is *0.4*.

Figure 3 shows the model validation for open water calculations and measurements (Walker 1996) of the thrust and torque coefficients.

In the current method, a numerical Kutta condition is applied. For such a relatively small pitch ratio propeller, implementing the numerical Kutta condition to the current panel method gives a good overall load prediction in a wide range of J values. Results agree well with the experimental data.

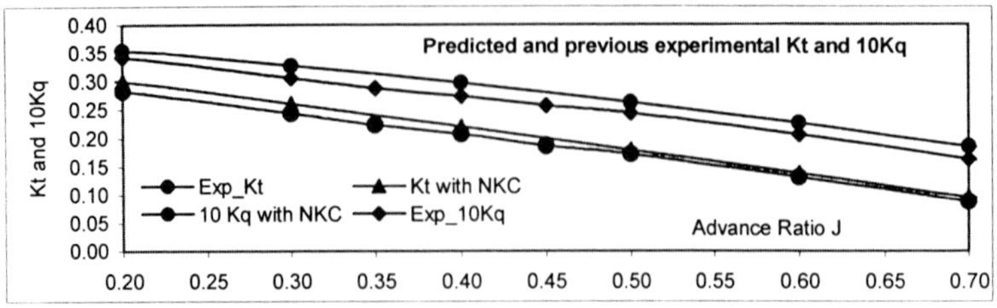

Figure 3: Validation of the prediction by the current method and the previous experimental results in a wide range of load conditions from *J=0.2* to *0.7* which correspond to the load factor of from *J/[p/D]=0.256* to *0.897*

Figure 4 shows a comparison of the predicted and measured thrust coefficient in an open water condition and the fixed proximity condition. Experimental data is taken from Doucet et al. (*1998*). All data are computed at *J=0.4*. The normalized gap values are based on the radius of the propeller model (*100mm*). Therefore the real distance from the ice to the propeller is *1* mm at a gap value of *1%R*. The predicted thrust under the proximity condition is slightly lower than the measurement. At a very small gap value, the predicted results are closer to the second set of experimental data than the first.

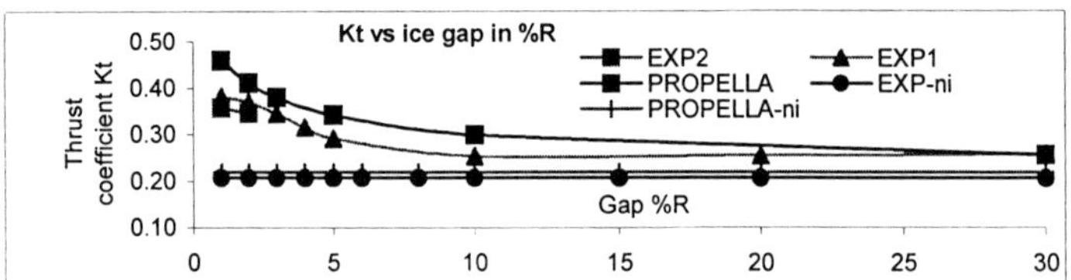

Figure 4: Two sets of comparison of thrust coefficient. Set one under open water condition contains two straight lines with + sign for prediction and • for measurement. Set two under proximity condition has three lines for prediction, experimental 1 and 2

Figure 5 shows a torque coefficient comparison with experimental values also taken from also Doucet et al. (*1988*). For open water flows, the current method gives a slight under prediction on torque. Under the proximity flow condition, the predicted torque agrees well with the first set of experiments until the gap becomes very small, where the prediction agrees well with the second set of experiments.

Comparison of the thrust coefficient under fixed proximity and variable proximity conditions is shown in figure 6. At large gap values, where the propeller is relatively far from the ice blockage, the two thrust coefficients are close. When the gap values decrease, predicted thrust under the variable proximity condition is substantially lower. This indicates that the fixed proximity gives a more conservative estimation of the load fluctuation.

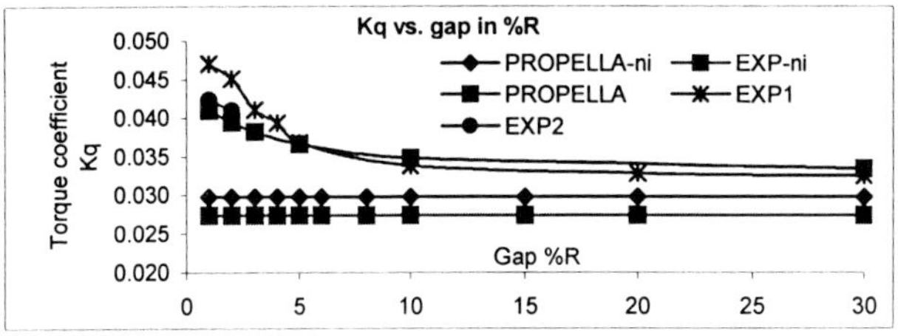

Figure 5: Two sets of comparisons on shaft torque coefficient: one for open water and another for the proximity condition

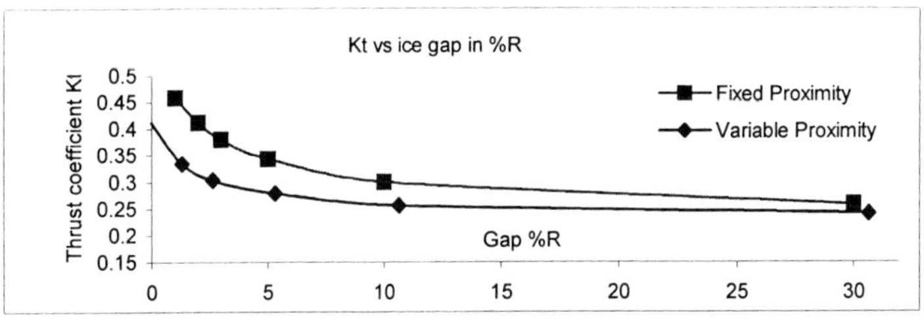

Figure 6: Comparison of thrust coefficient for fixed proximity and variant proximity computations

A comparison of the torque coefficients under two conditions is shown in figure 7. It can be seen that throughout the ranges of gap values, the fixed proximity condition yields much larger load fluctuations, which is conservative.

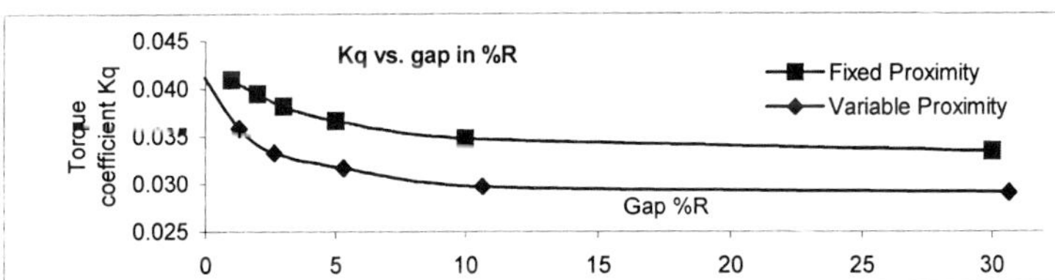

Figure 7: Comparison of torque coefficient for the fixed proximity and variant proximity computations

Load fluctuations on the key blade in terms of thrust and torque coefficient at the last two revolutions (revolution two and three) are presented in figure 8. The last time step of the first revolution corresponds to a gap value of *81.32%R* (with an absolute value of *0.074818 m*) at which value the key blade is pointing upward. At the last time step of the first revolution, i.e., the 119th time step, the key blade is fully blocked by the ice wall so both torque and thrust are higher than that of the adjacent steps even though the propeller was closer to the ice wall. At the 170th time step corresponding to *gap=13.31%R*, the key blade entered the ice blockage domain and at the 180th time step the propeller is at the center of the ice wall with a gap values of about 0%R. In this range of interaction, both the thrust and torque coefficients increase dramatically.

Figure 9 shows the pressure distribution at $r=0.7287R$ of the key blade at the 179th time step with a gap value of *1.31%R*. The Cp values at the trailing edge could be closer if the pressure difference allowance was set smaller (*0.05* in the current computation) for the iterative Kutta condition. However, too small a value of the target value would make the iteration divergence-prone.

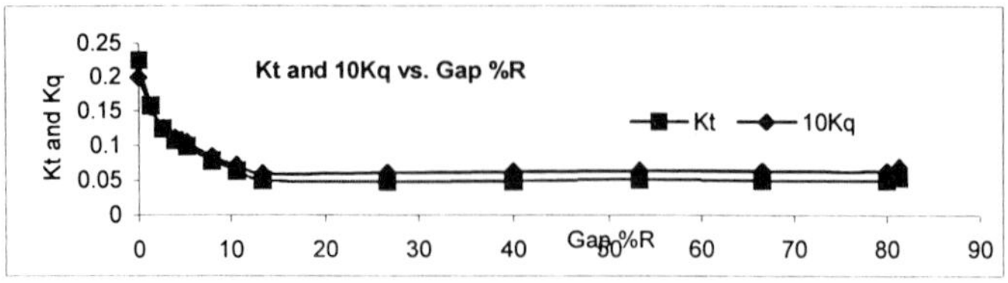

Figure 8: Kt and Kq fluctuations on the key blade of the ice-class propeller approaching an ice wall

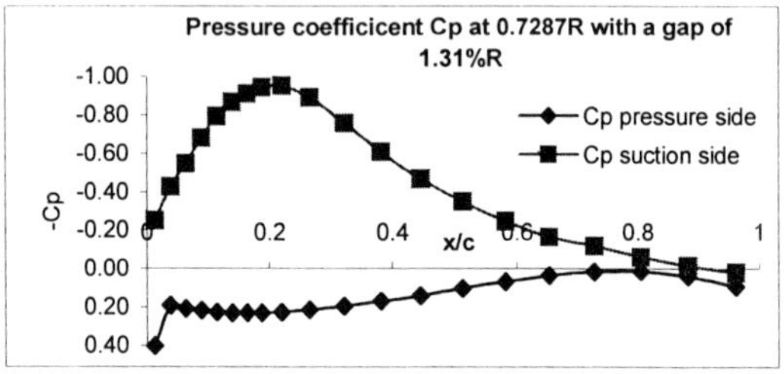

Figure 9: Pressure distribution of the key blade at the *179*th time step (*1.31%R*) at a radial location of *r=0.7287R*.

In computations for the variable proximity case, three revolutions with *60* time steps for each revolution were set. The numerical Kutta condition was used, and for a *500* MHz Xeon Intel processor with *2* GB of RAM on a DELL workstation, a computation took about *36* hours. This run with the above motion and geometry parameters took about 400 MB of RAM, with each additional revolution adding *50* MB. The real time for the model in this simulation is about *0.33* seconds.

4 CONCLUSIONS

A transient model for ice-propeller induced hydrodynamic load fluctuation is established. The model is validated with experimental data for open water and fixed proximity conditions and comparisons of results give good agreement. The model is extended to predict load fluctuations with varying proximity and the results indicate a lower load fluctuation compared to the fixed proximity model. At very small gap values, both models give a very high load jump at about the same magnitude.

Acknowledgement

The authors would like to thank the National Research Council Canada for its supports.

References

1. Bose N. (1996). Ice Blocked Propeller Performance Predictions Using a Panel Method. *Transactions of the Royal Institute of Naval Architects* **138**, 213-226.
2. Doucet J.M., Liu P., Bose N. and Veitch B. (1998). Numerical Prediction of Ice-Induced Loads on Ice-Class Screw Propellers Using a Synthesized Contact/Hydrodynamics Code. *OERC Report of Ocean Engineering Research Centre, Memorial University of Newfoundland* **OERC-1998-004**
3. Katz J. and Plotkin A. (1991). *Low-Speed Aerodynamics*, McGraw-Hill, New York

4. Liu P. (1996). A Time-Domain Panel Method for Oscillating Propulsors with both Chordwise and Spanwise Flexibility. *Doctoral Thesis, Memorial University of Newfoundland, Canada*
5. Liu P. and Bose N. (1998). An Unsteady Panel Method for Highly Skewed Propellers in Non-Uniform Inflow. *22nd ITTC Propulsion Committee Propeller RANS/Panel Method Workshop, 5-6 April, Grenoble, France*, 343-349.
6. Liu P, Doucet J.M., Veitch B., Robbins I. and Bose N. (2000). Numerical Prediction of Ice Induced Hydrodynamic Loads on Propellers Due to Blockage. *Oceanic Engineering International* **4:1,** 31-38.
7. Liu P. and Bose N. (2000). Hydrodynamic Characteristics of a Screw-Nozzle-Rudder Assembly. *Journal of Computational Fluid Dynamics of Japan* **9:1**
8. Morino, L. and Kuo, C.-C. (1974). Subsonic Potential Aerodynamics for Complex Configurations: A General Theory. *AIAA Journal* **12:2**, 191-197.
9. Newbury S., Shih, Y.L., Browne R.P., Revill C.R., Kenny S. and Zheng Y. (1993). Experimental and Theoretical Evaluation of Hydrodynamic Pressure during Non-Contact Propeller/Ice Interaction. *Institute Report of Institute for Marine Dynamics, National Research Council Canada* **IR-1993-15**
10. Newman, J.N. (1986). Distributions of sources and normal dipoles over a quadrilateral panel. *Journal of Engineering Mathematics* **20**,113-126.
11. Shih L.Y. and Zheng Y. (1993). Application of 3-D BEM to Time-Dependent Potential Flow Over a Propeller with Ice Blockage at Proximity Condition. *Proceedings of the Inaugural Conference of the CFD Society of Canada, PQ, Canada.*
12. Veitch B. (1995). Prediction of Ice Contact Forces on a Marine Screw Propeller during the Propeller-Ice Cutting Process. *Acta Polytechnica Scandinavica, Mechanical Engineering Series, Espoo, Finland* **118**
13. Veitch B., Bose N., Meade C. and Liu P. (1997). Prediction of Hydrodynamic and Ice Contact Loads on Ice-Class Screw Propellers, *Proceedings of the 16th International Conference on Offshore Mechanics and Arctic Engineering, American Society of Mechanical Engineers* **4,** 119-125.
14. Walker D. (1996). The Influence of Blockage and Cavitation on the Hydrodynamic Performance of Ice Class Propellers in Blocked Flow, *Doctoral Thesis, Memorial University of Newfoundland, Canada*

Practical Design of Ships and Other Floating Structures
You-Sheng Wu, Wei-Cheng Cui and Guo-Jun Zhou (Eds)

INVESTIGATIONS OF PODDED DRIVES IN A LARGE CAVITATION TUNNEL

J. Friesch

Hamburg Ship Model Basin (HSVA), Hamburg, Germany

ABSTRACT

Podded propulsion is increasingly being recognised by ship owners and ship yards as a perfect solution for a broader range of applications. More and more powerful podded propulsion systems – mainly for large cruise liners – have been developed in the last years. Outputs range from 5 to 30 MW per unit. Primary benefits of pod propulsion systems are space saving design, improved efficiency, optimised manoeuvrability and distinctly lower levels of vibration and noise compared with conventional marine propulsion systems. An efficiency higher than with a conventional system is reported to be achieved. The improved noise behaviour is mainly the result of more homogenous inflow conditions to the propeller and of the possibility of using a reduced propeller diameter. Especially for cruise ships the highest level of comfort on board must be guaranteed. Therefore normally extensive model tests are carried out not only for the propulsion behaviour but also for the cavitation and noise behaviour of the pod and its propeller(s).

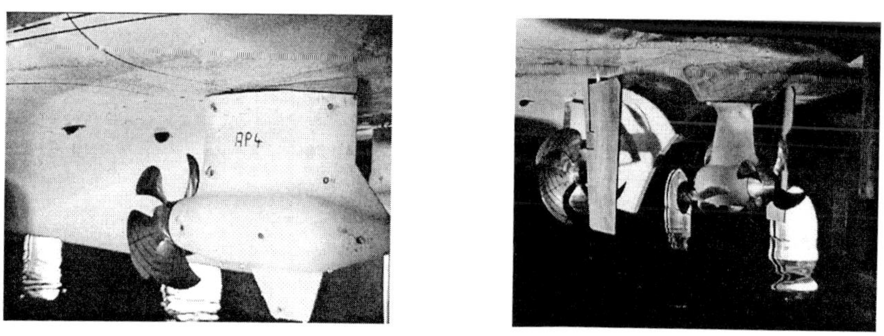

Figure1: Different Types of Podded Drives in the Test Section of the Cavitation Tunnel HYKAT

In the paper results of wake field measurements and of cavitation observations combined with pressure fluctuation and noise measurements in HSVA's large cavitation tunnel HYKAT for different pod arrangements are presented. The test arrangement, mainly the dynamometers used and the test procedures are described. The tests were performed with large propeller models (up to 250 mm in diameter) and with high rotational speeds (up to 35 revs per second) behind whole ship models of 8 to 11 meters in length. The results are compared with data of cruise ships with conventional propeller-shaft arrangements. The data show the excellent possibilities of pod propulsion systems mainly from the cavitation and noise point of view.

KEYWORDS

Model test techniques, Cavitation, Pressure pulses, Noise, Podded drives, HYKAT

1 INTRODUCTION

Propeller induced vibration can cause passenger discomfort or operational restrictions. The modern design trend for cruise liners is to place important public areas near the aft end of the ship, where they are particularly vulnerable to excitation by the propeller. High vibration can cause crew fatigue in warships like frigates or even aircraft carriers, where high power and lightweight structure are needed to deliver operational requirements. One of the largest problems in the design of commercial and naval vessels is to predict the vibration and noise behavior at low frequencies. Twin screw vessels have been build for many decades, years ago the shaft supports have been enclosed in bossings, which produced pronounced wake distributions and were therefore not desirable for high speed vessels. Nowadays many modern Ro/Ro ships, cruise liners, fast ferries and navy ships have an open shaft arrangement. Large brackets, the struts, support the shafting and are located in front of the propeller and very often V-shaped. But still there is a change in the inflow velocity distribution because of these struts which is influencing the cavitation behavior of the propeller. Also the danger of cavitation on the struts themselves and the related noise needs to be considered.

The aim of this paper is to show, how the use of podded propulsors will improve the cavitation, vibration and noise behavior of fast and highly sophisticated ships. Since the launching of the 14 MW pod units for the two *Elation Class* cruise liners, podded propulsion became accepted in the commercial shipbuilding world. From the hydrodynamic point of view, the primary advantages are:

- power savings
- low levels of propeller induced vibration and noise
- improvements of the manoeuvring behaviour.

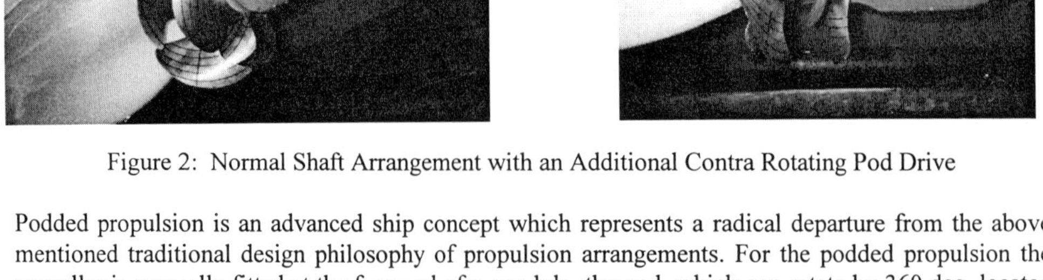

Figure 2: Normal Shaft Arrangement with an Additional Contra Rotating Pod Drive

Podded propulsion is an advanced ship concept which represents a radical departure from the above mentioned traditional design philosophy of propulsion arrangements. For the podded propulsion the propeller is normally fitted at the fore end of a gondola, the pod, which can rotate by 360 deg, located very close to the ship stern with the electrical engine located inside the shell of the gondola. Since the beginning in the late eighties, starting with the application on single screw supply vessels, artic tankers and icebreakers, the system proved to be efficient and reliable. The success followed in the nineties with the application on a lot of twin screw cruise ships with better hydrodynamic performance concerning propulsion efficiency, manoeuvering behavior, propeller cavitation, vibrations and noise compared to a conventional twin screw ship propeller arrangement. Fitting of podded drives can have significant impact on the aftward ship arrangements and influence on the hydrodynamic design of the stern of the vessel. In the former years, hull forms were usually those of an existing ship design, designed to be fitted with shaft lines, rudders, struts and stern thrusters. But hull form optimization work shows that even more improvements can be reached by a complete new design of the ships' aftbody.

Depending on the mission profile of the ship, the demand for passenger/crew comfort and for the specific requirements of the different pod manufacturers, a wide variety of arrangements are possible (Kurimo et al.,1997), ranging from single via multiple pure pods to combinations with conventional shafts which can create a contra-rotating set-up. Propellers are arranged in a pulling or/and pushing position. This presentation will consider the cavitation behavior of different pod arrangements (Fig. 1, 2 and 3).

2 MODEL TESTS

Figure 3: Pod Arrangement in HYKAT for a Modern Cruise Vessel

For model tests with podded propulsors use is made of an advanced measuring device developed specifically for those tests. The shape of the gondola and the streamlined strut are build around the mechanical, fully azimuthing z-drive unit with an integrated torque and thrust meter and a revolution pick-up. The pod total unit thrust is measured additionally with a force balance mounted on the vertical shaft. The reproducibility of the data lies within the range of 1% - 1.5%. The units will be fitted to an adjustable frame which allows to tilt and shift them within the models, such allowing to find the optimum "rudder" angle, the best lateral position and the optimum fore-aft tilt angle of the pod unit. The units can be used for both, tests in the towing tank and in the large cavitation tunnel HYKAT. Model tests in the towing tank include both, open water test and self propulsion tests. Open water tests need to be performed with the propellers alone and with the propeller/pod unit (Mewis, 1998). In the cavitation tunnel HYKAT the cavitation behavior and the vibration and noise behavior of the units are tested behind the whole ship model which is used in the towing tank self propulsion tests. The drive units allow high rotational speeds up to 35 revs per second which together with the large model propellers (diameters between 210 and 260 mm) allow to reach as high Reynoldsnumbers as for comparable conventional propeller shaft arrangements, thus fulfilling the recommendations of the ITTC for cavitation tests and pressure fluctuation measurements. A detailed description of the test facility HYKAT and the test procedures used, is given by Friesch, 1998 and 2000.

3 PROPELLER OUTLINE

One significant aspect of pod propulsion is directly connected to the working environment of the propeller. Propeller-hull clearance is often about 35% of the propeller diameter. For the pulling type pod, the velocity distribution in the propeller plane is very uniform, the propeller works nearly in open water condition which means that the change in angle of attack for the blade profiles varies less, compared to a conventional shaft arrangement (see Fig. 4a). It is therefore possible to reduce for example propeller skew and bring more loading to the propeller tip. Therefore redesign of a propeller my involve some economic benefits. For a propeller design task, always two opposite aspects have to be taken into account: efficiency on one side and cavitation and pressure pulses on the other. For a modern cruise vessel or a fast ferry with high comfort levels the second aspect is the most important one. In order to reduce the amount of cavitation and the related pressure levels it is necessary to act on both, the inflow conditions to the propeller (the wake) and the propeller geometry itself.

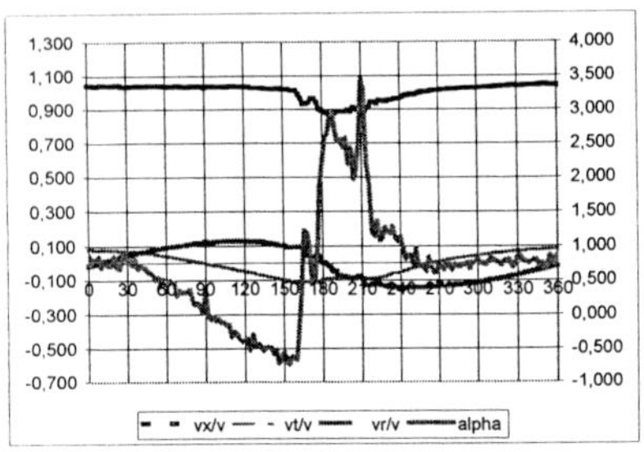

Figure 4a: Conventional Shaft Arrangement

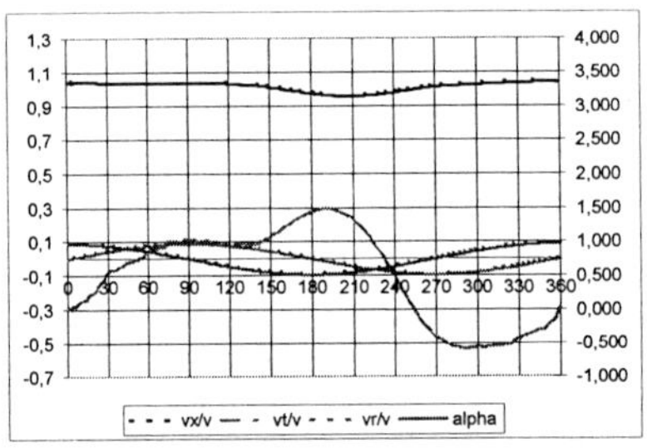

Figure 4b: Pod Arrangement

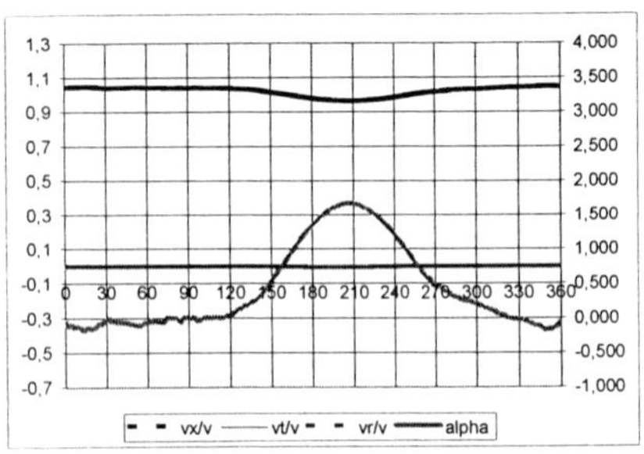

Figure 4c: Pod Adjusted to the Streamlines

3.1 The Wake

The wake on a conventional twin screw ship is mainly affected by two factors, the hull form and the appendages. In the design of a conventional twin screw vessel the stern lines design is strongly influenced by the need of fitting the bossings for the shafts and keeping a certain required hull- propeller clearance. Dealing with a ship with podded propulsion, bossings are no longer necessary, clearance can be increased by moving the pod propeller as aft as possible since there are no restrictions in combination with the shaft length.

The limitation concerning the pod position is strongly related to the possibility to rotate the pod by 360° and no part should protrude aft of the transom. In addition thick skegs to locate thrusters are no longer necessary and therefore an additional improvement of the wake distribution can be achieved. It is well known that for a twin screw vessel the strongest influence on the wake field comes from brackets, bossings and shafts, and efforts are always made to optimize their positions and orientations in order to reduce the wake peaks and irregularities. Fig. 4a shows the typical wake distribution on a specific radius for a conventional twin screw shaft arrangement. The peaks in the axial component of the velocity and the changes in radial and tangential components influence the angle of attack alpha of the propeller profiles dramatically. Getting rid of all parts in front of the propeller reduces the changes in angle of attack (Fig. 4b) and when additionally the propeller and gondola can be installed with a shaft inclination which is orientated to the direction of the flow lines the wake characteristics are clearly improved (Fig. 4c).

3.2 Propeller Design

The design of propellers for pushing pods is comparable to the design of propellers for conventional propulsion systems, with shafts, struts and bossings in front of the propeller. On the contrary, the pulling type pod requires to carry out the propeller design with particular emphasis to the interaction of pod body and strut, located behind the propeller plane. Therefore the design of pulling propellers has a strong need for a standard design procedure taking into account the interaction between pod housing, strut and propeller. This can be done by use of CFD – methods. Fig. 5 shows the pressure distribution on a pod housing and the flow around the housings due to the working propeller (right side of Fig.5) by use of particle tracing.

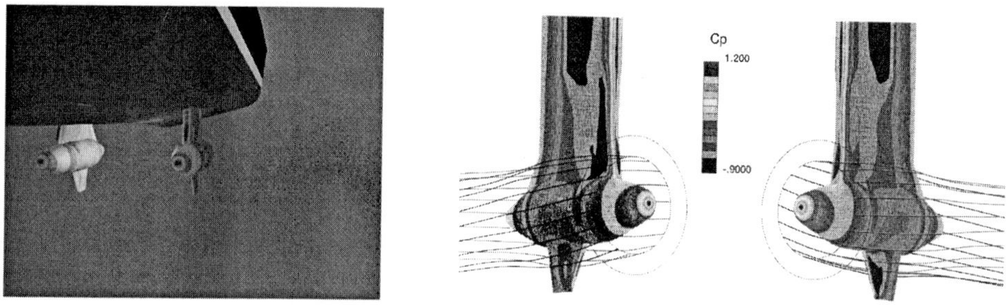

Figure 5: Pressure Distribution and Flow Around and Behind the Pod Housing

The most significant difference for the hydrodynamic design of a pulling podded propeller consist in the need to consider the influence on the pressure field of the pod rudder (strut) and the pod housing. Due to the reduced clearance between propeller and strut it can be seen with a CFD calculation that the pressure fields created in front of the leading edge of the strut and the pod housing are superimposed to the pressure field of the propeller blades. This leads to a change of the characteristic curves of k_t and k_q of a propeller in open water conditions. If we compare the k_t and k_q curves measured in open water for the propeller alone and those obtained with the propeller attached to the pod housing and strut like on the HSVA "propeller boat" (Mewis, 1998)) we see that at the same J value these data are higher. As the influence of the pod is not easily calculated the problem of a proper calculation of the right k_t and k_q curves can be solved by experimental results of suitable stock propeller tests (Mewis, 1998). Additionally this pressure field will influence the cavitation behavior of the propeller blades on the face side and should therefore be taken into account when designing the propeller. During the propeller design it is possible to act on a series of geometrical parameters in order to find the right loading distribution. When the propeller is working in a nearly homogeneous wake field with smooth gradients and low peaks the right choice of blade geometry and load distribution can achieve an optimized cavitation behavior with an improved behavior of the exciting forces. The selection of the radial pitch distribution, the skew, the chord length (blade area) together with the number of blades is the basis to reduce the cavity volume, and based on that, the hull pressures and also the bearing forces will be minimized. For pushing propellers the movement of the pod during manoeuvering conditions implies higher variations of the angle of attack on the propeller blade profiles and therefore thicker sections with wider cavitation buckets should be used.

4 CAVITATION BEHAVIOUR, INDUCED VIBRATIONS AND NOISE

For a pod driven vessel different factors contribute simultaneously to the improvement of the propeller- hull excitation: a new designed hull geometry, a more homogeneous wake field and the possibility to increase the tip clearance. All these factors allow to design a propeller with reduced skew, smaller blade area and less tip unloading and therefore better efficiency without disadvantages from the cavitation and vibration excitation point of view, therefore overcoming the larger resistance of the pod, compared to V-brackets.

Experimental facilities allow the observation of cavitation phenomena and the measurements of fluctuating pressures on the surface of full length ship models. The principal focus is on narrowband excitation at low multiples of propeller blade passing frequency (the blade rate), which lies normally between 8 and 15 Hz on a large twin-screw ship having a length and beam of about 250 and 30 meters respectively, and on the effect of broadband excitation in the range up to 50 Hz. The frequencies are low enough to affect a large part of the ship structure. Current numerical models for the prediction of the cavitation behavior are only capable of predicting excitation at multiples of propeller blade passing frequency, empirical data from model tests are needed to estimate the broadband excitation that results from different types of cavitation like clouds, bursting tip vortices and fluctuating sheet cavities. The achievable correlation between numerical predictions of fluctuating hull pressure near the propeller and experiments, and the correlation between experimental data at model and full scale, is encouraging at blade rate and higher harmonics (Friesch, 2000). It becomes progressively poorer at much higher multiples of blade passing frequencies and broad banded excitation where the excitation is very sensitive to small errors in changes of the cavity volume with time.

The good cavitation, vibration and noise behavior of mainly pulling (tractor) pods was proven in extensive test series for large cruise liners with different pod arrangements in HSVA's large cavitation tunnel HYKAT. In this tunnel, the whole ship models used in the towing tank tests, equipped with azimuthing drives and additional conventional single propellers (Fig. 1 and Fig. 2) were installed. For the described tests, the torque and the number of revolutions were measured directly at the propeller shaft, while net force and side force were measured by use of a force balance mounted watertight in the ship model. The reproducibility of the measured data was in the range of 1% to 1.5 %.

4.1 Results for Pulling Pod Arrangements

For all pulling propellers tested in the cavitation tunnel HYKAT the blade surface is nearly free from any kind of cavitation. As a consequence of the very good wake field the cavitation pattern consists in a very small area of sheet cavitation along the suction side leading edge merging smoothly and gradually to a very often rather thin cavitating tip vortex only visible few revolutions downstream (Fig. 6). From an erosion point of view the blades are free from any risk of erosion damage even for higher values of the pod rudder angle. In general, the cavitation phenomena as observed in all tested conditions are quite satisfactory. They show less fluctuations and very stable behavior. In some cases also a small stripe of stable sheet cavitation on the face of the propeller blades could be observed.

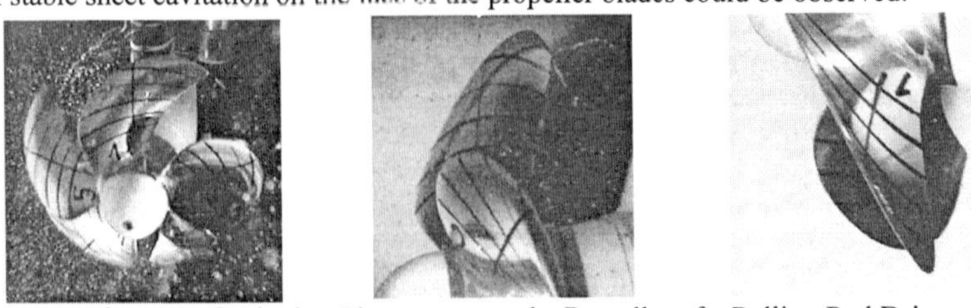

Figure 6: Cavitation Phenomena on the Propeller of a Pulling Pod Drive
$V_S = 25$ kts $k_T = 0.164$ $\sigma_n = 0.169$

In the case of podded ships the possibility is given to fit the pod in the most convenient position related to the hydrodynamic aspects. Fitting the propeller as far aft as possible allows to improve the total hull efficiency and to increase the tip clearance between propeller and hull. For a conventional twin screw ship the standard tip clearance varies between 0.25 and 0.30. For a podded ship these values can be increased till 0.35-0.38 with a corresponding reduction of the pressure pulses. In Figs. 2 and 3 the typical arrangement of up to 25 pressure transducers in the ship model can be seen. For every tested condition the harmonic analysis of the pressure pulse signals has shown that the main contribution is due to blade rate decreasing to very low values for the second and the further harmonic orders. These pressure pulses are mainly related to the dynamic displacement effect of propeller rotation, which is characteristic if very less energy comes from the growing and collapsing of the sheet cavities. This is in good accordance with the observed cavitation phenomena. For all tested ships with pulling pods very low levels of hull pressure fluctuations were predicted. As an average values between 0.6 and 1.3 kPa were measured in the worst condition in the most excited area which is in most cases represented by the pressure pick up just above the propeller in the pod center line. According to the calculated forces, the pod propeller develops a vertical total force which is rather small and which should guarantee an acceptable vibration behavior of the vessels. Fig. 7 shows the improvement in blade rate pressures of a pulling pod arrangement versus a comparable twin screw ship with conventional shaft arrangements at 24 knots. Fig. 8 shows the improvement in broadband excitation for the same configurations and ship speed.

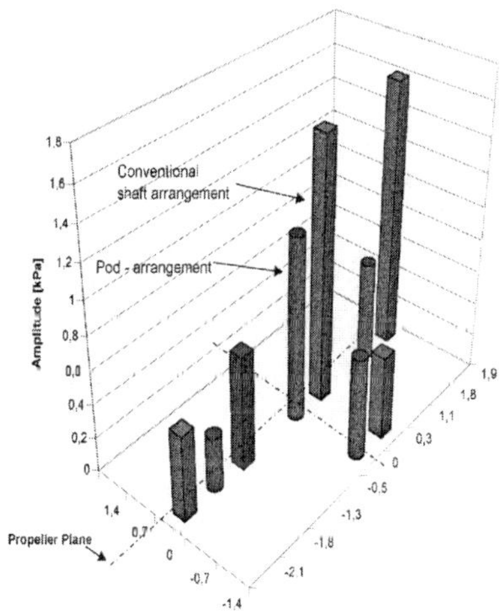

Figure 7: Comparison of Blade Rate Pressures

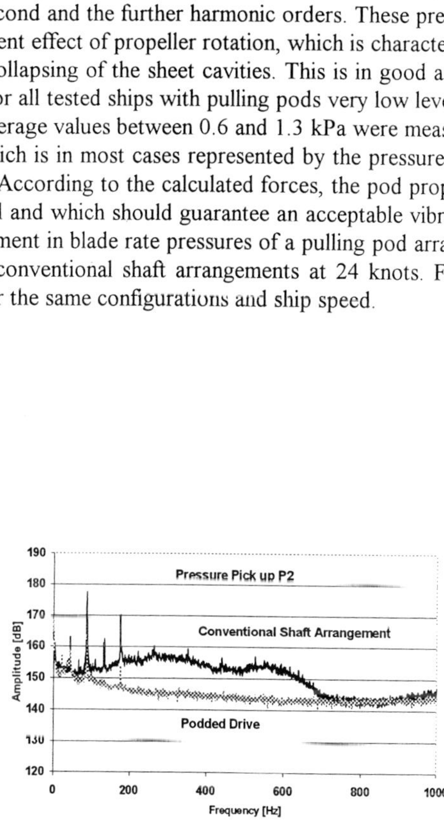

Figure 8: Comparison of Broadband Excitation

4.2 Results for SSP and Contra-Rotating Arrangement

For the SSP arrangement (Siemens-Schottel, 1997) shown in the right picture in Fig. 1 a quite comparable cavitation behavior as reported above, could be found for different loading conditions. Both propellers showed mainly tip vortex cavitation and only some small patches of sheet cavitation on the suction side. On the face of the blades no sign of cavitation could be detected (Fig. 9).

The CR-Arrangement in Fig. 2 represents the propulsor arrangement for a high speed research project with cruising speeds of up to 32 kts. For the high speed condition on both propellers sheet cavitation could be observed which covered the blade between 0.8R and the tip. Especially on the aft ward propeller this cavitation was more fluctuating and the tip vortex was quite strong with a tendency to bursting. No face side cavitation was observed on both propellers. Concerning the pressure fluctuations, mainly the transducers above the aft ward propeller showed a rather chaotic behavior. This is related to the fact, that they were hit by pulses created not only by the cavitation of the aft ward propeller but also at the same time by the cavitating tip vortex of the forward propeller. This resulted in an increase of mainly at the same time by the cavitating tip vortex of the forward propeller. This resulted in an increase of mainly the higher harmonics which were higher by a factor of two compared with the values known for conventional twin screw vessels. This behavior is not only related to the higher loading but also to the worse inflow conditions to the propellers, compared with a twin screw ship.

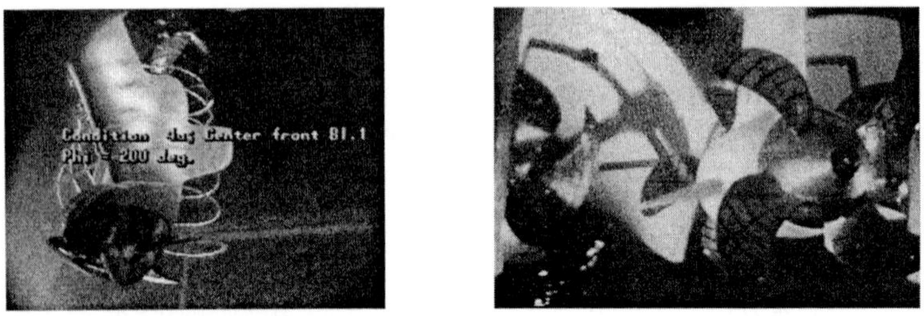

Figure 9: Cavitation Behavior of an SSP on Back and Face Side at $V_s = 22{,}5$ kts

5 CONCLUSIONS

From the large series of model tests (cavitation observations, pressure fluctuation and noise measurements) carried out on a series of cruise liners with different types of podded drives in HYKAT the following considerations can be made: As the wake field for at least the twin screw podded vessels proved to be very smooth and nearly uniform which results in low excitations induced by the propellers - both in terms of hull pressure fluctuations and shaft forces - an advantage for podded propulsion concerning the vibration and noise behavior can be stated. This may lead to a reduction of steel manufacturing problems and therefore even sometimes to an improvement in production cost. But is has to be mentioned, that podded propulsion by itself will not reduce automatically all vibration problems and the cases were a pod should be applied should carefully be considered before the final decision is made. This holds mainly true for propulsion arrangements which are combinations of podded drives with conventional propeller shaft arrangements.

References

Friesch J. (2000). Ten Years of Research in the Hydrodynamics- and Cavitation Tunnel HYKAT of HSVA. *NCT'50 International Conference on Propeller Cavitation.*

Friesch J. (1998). Correlation Investigations for Higher Order Pressure Fluctuations and Noise for Ship Propellers. *Third International Symposium on Cavitation, Grenoble.*

Kurimo R, Poustoshniy A. and Syrkin E. (1997). Azipod Propulsion for Passenger Cruisers: Details of the hydrodynamic development and experience on the propeller design for "Fantasy"-class cruise liners. *Kvaerner Masa-Yards.*

Mewis F. (1998). Podded drives im Vormarsch - Hydrodynamische Aspekte. *Schiff & Hafen* **11/98**, 50-53.

Siemens-Schottel (1997). New Podded Propulsive Drive. *HANSA* **No. 11,** 64-68.

Practical Design of Ships and Other Floating Structures
You-Sheng Wu, Wei-Cheng Cui and Guo-Jun Zhou (Eds)

TRIPLE POD PROPULSION IN THE WORLD'S LARGEST EVER CRUISE LINER

Raimo Hämäläinen[1] and Jaap van Heerd[2]

[1] Kvaerner Masa-Yards Inc., Turku New Shipyard
P.O. Box 666, FIN-20101 TURKU, Finland
[2] Maritime Research Institute Netherlands (MARIN)
P.O. Box 28, 6700 AA Wageningen, The Netherlands

ABSTRACT

Passenger comfort, operational safety and efficiency were the main targets in the hydrodynamic design of the world's largest ever Cruise Liner for Royal Caribbean International. The Shipyard made extensive design studies for several pod alternatives from twin pod to triple pod arrangements before contract. The paper describes the hydrodynamic advantages and disadvantages between several propulsion arrangements. Several CL skeg alternatives were studied in order to reach an excellent wake field for the pusher type Fixed Pod configuration.

The revolutionise Wave Damping Afterbody was designed for the first time on a large Cruise Liner by using the Shipyard's Shipflow and Model Basin's Rapid CFD-programs. The earlier experiences from fast ferries has shown big power level improvements for this new aftbody form. The paper describes this optimisation process and compares calculations for model and full scale.

Several thruster configurations with scallops and grids inside the bow thrusters were tested, and the selected solution resulted in a remarkable reduction in resistance.

Cavitation observations and hull pressure measurements were performed for two loading conditions and for various setting angles of the pods.

A lot of different types of turnable and fixed skegs were studied to find the optimal combination between manoeuvrability and heeling angles.

The extensive full-scale measurements were made to verify the calculations and tests made during the design and construction phase of the vessel.
The scaling of model test results are investigated. The interesting comparison between model test results and full-scale measurements are shown.

1 INTRODUCTION

The world largest Cruise Liner, from which we are discussing in this paper, has a maximum passenger capacity of 3,840 and a crew of 1,180. It has a length of 311 m, a maximum beam of 47.4 m, and a height from keel to funnel top of 72.3 m.
The whole vessel is a big entertainment centre with multiple options where passengers can spend their time. There is the Royal Promenade from deck 5 up to deck 9 with an artificial lit ceiling, a wide selection of shops, restaurants and entertainment areas fronting on street. Studio B shows lounge with its ice skating rink for public performance, as well as passenger use. Furthermore, a lot of sporting activities are present, such as a nine-hole golf putting course, a roller-blade track, a full size volley ball court, a rock climbing wall.
The vessel has two Azipod azimuthing drive units and a central fixed pod. Pod propulsion gives a lot of additional benefits to diesel-electric system and it makes a new comfort class possible to reach with good design.

2 POD CONCEPT

An AZIPOD (AZImuthing POdded Drive) is a 360-deg. azimuthing main propulsion unit for use in ships, which have an electric power plant (R. Kurimo, A. Poustoshniy, E. Syrkin). The unit incorporates an electric AC motor, which is located inside the pod. The motor directly drives a fixed pitch propeller, which can be at either end of the pod. Both pusher and tractor versions of the Azipod are possible. In the world largest Cruise Liner, the Voyager of the Seas, one Fixipod pushing unit amidships is flanked by two steerable Azipod pulling units. The simple principle of the Azipod is shown in figure 1.

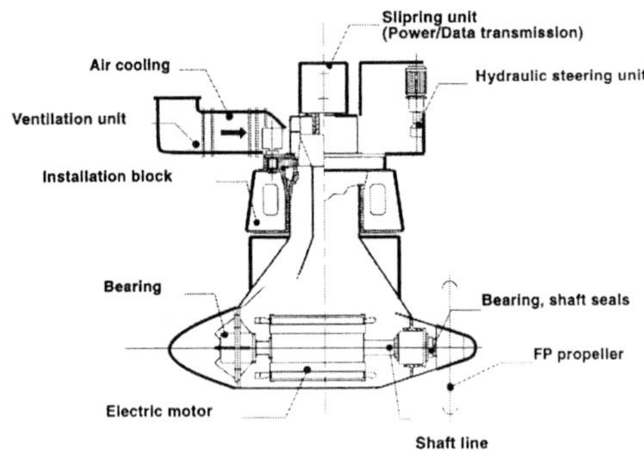

Figure 1: Azipod concept

The power and control data are transferred from the non-rotating part to the rotating part via slip rings. The air cooling system provides cooling air to the motor from the aft compartments. The steering system consists of two to four hydraulic motors, which activate the steel evolvement tiller ring. The Azipod control system includes an electric control subsystem on the bridge including all necessary main panels.
In addition to the basic advantages of diesel-electric drives the Pod propulsion gives additional benefits:

- the space for conventional propulsion motors can be saved
- the steering capacity in the aftbody is significantly greater than for conventional rudder stern thruster system
- reversing capacity and steering during astern navigation are excellent
- there is more freedom to the ship designer in the machinery arrangement
- reduced number of main components
- late installation possibility, only a few weeks before launching
- the units are environmentally clean and there is practically no risk for oil leakage
- excellent manoeuvrability, specially at low speed and berthing operations

- fuel saving through good hydrodynamic efficiency
- propeller hull excitations are very low due to the pulling propellers working in an excellent wake field, resulting in low noise and vibration levels

3 RULES OF THUMB IN HYDRO-DYNAMIC DESIGN OF A CRUISE LINER

The rules of thumb related to different hydrodynamic questions are a very useful starting point for a project, because they give a rough idea of the critical points and help the designer to concentrate carefully on the correct and most important critical subjects (R. Hämäläinen & J.v. Heerd).
In the world largest Cruise Liner project a lot of critical hydrodynamic items have to be taken into account so that the whole ship behaves well and can compete well with other smaller Cruise Liners. Some of these requirements are:

- low resistance behaviour (optimum lines),
- propeller hull excitations (vibration, noise, new harder Comfort-class, etc.),
- redundancy (2 or 3 propellers, etc.),
- good seakeeping ability (slamming behaviour, motions, etc.), and
- good manoeuvrability (course stability, harbour crabbing, etc.)

Special attention has to be paid to aft ship slamming. A pram-type aft ship with or without semi-tunnel is a generally applied aft hull form for all kinds of vessels today, owing a good resistance behaviour and hard stability requirements of this form. Vertical buttock angles are typically from 8 to 15 deg and close to the transom almost zero or negative (= wedge form) is quite normal. The height of the transom with regard to the construction waterline also depends on the design speed, but it is recommended to adopt a value between 0.5 to 1.0 m above the CWL for this type of vessels. Typical values for the radius between aftship buttocks and flat bottom vary from 0.4 of L_{PP} to 1.2 of L_{PP}. A larger radius is generally recommended to ensure that pressure field changes fluently into the transition location. A zero flare angle and tunnel shape ("pocket shape") in the aftship are sensitive to slamming and noise at zero and under 5 knots forward speed in following sea conditions.
A big overhang with large flat form close to the construction waterline is very sensitive for aftship slamming and whipping behaviour. One should pay a lot of attention to the construction and damping in that kind of situation. In figure 2 is presented a review of vertical forces for "Panamax-size" cruise liners with different aftbody form at zero speed in following sea condition.

Recommendation for aftbody frame section and buttock angles are equal to or more than 5 degrees with continuously varied (increased, convex form) angles from aft to fore direction. Anyway it is not possible to totally avoid aftship slamming in following seas at low speed or in stationary waves from all directions.

4 TARGET IN VOYAGER'S HYDRO-DYNAMIC DESIGN

The Shipyard's earlier experience of fitting Pod drives to nine vessels gave a good platform for the development work. Passenger comfort, operational safety and efficiency were the main targets in the hydrodynamic design of the vessel. The various items to consider were hull form design, resistance, cavitation, propeller excitation, manoeuvring and seakeeping.

4.1 Several Aftbody Alternatives Before Order

The Shipyard made extensive design studies for several pod alternatives from twin pod to triple pod arrangements before the contract. In the table 1 hydrodynamic advantages and disadvantages between several propulsion arrangement alternatives are described.

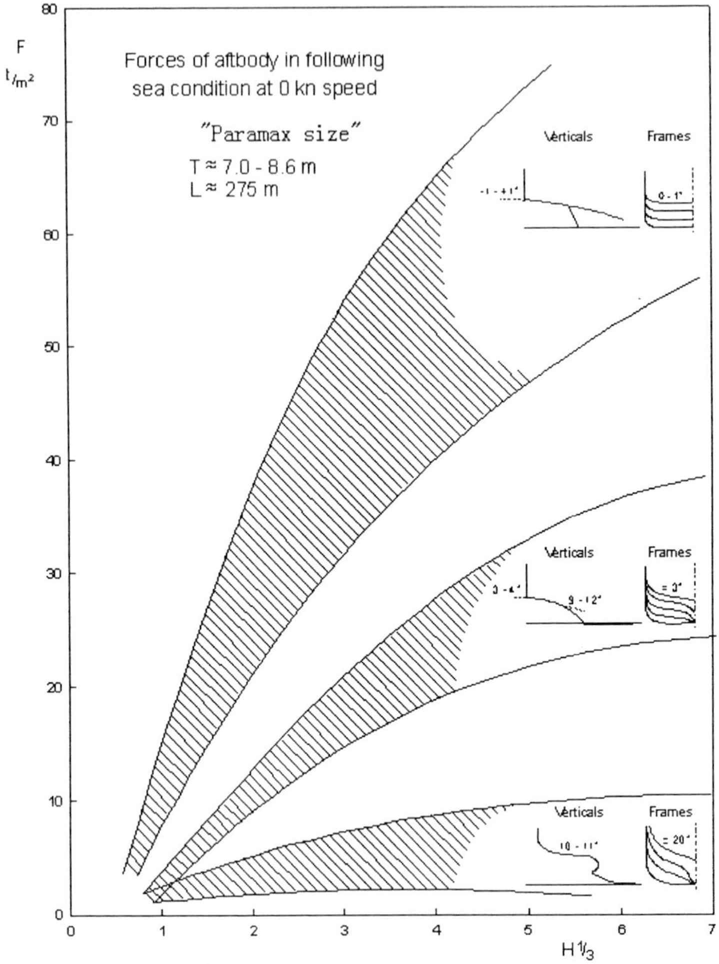

Figure 2: Review of slamming forces for "Panamax-size" cruise liners.

Alternative 1: Two turnable pulling Pods, one fixed pushing Pod in gondola and one small course keeping rudder at the CL. The Fixed pod is integrated into a gondola. See figure 3.

Alternative 2: Two turnable pulling Pods, one fixed pushing Pod and one small course-keeping rudder at the in CL.

The Fixed Pod is separately placed at the CL and a simple CL skeg is located ahead of it. See figure 4.

Alternative 3: Two turnable pulling Pods, one fixed pulling Pod and one small course-keeping rudder at the CL. The Fixed Pod is separately at the CL and a simple CL skeg ahead of it. See figure 5.

Alternative 4: Two turnable pulling Pods and a normal simple CL skeg. See figure 6.

Alternative 5: Two fixed pitch propellers in an open shaft arrangement, two flap type rudders behind the propellers and a normal simple CL skeg. See figure 7.

In table 1 a short comparison is given between the alternatives.

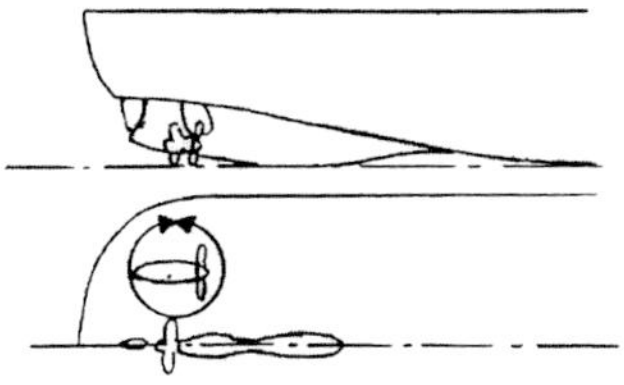

Figure 3: Alternative aftbody 1

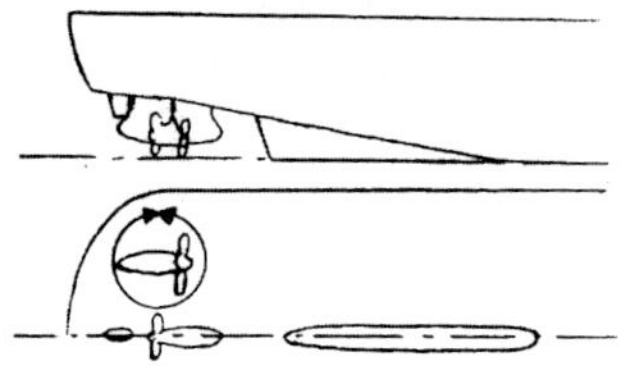

Figure 4: Alternative aftbody 2

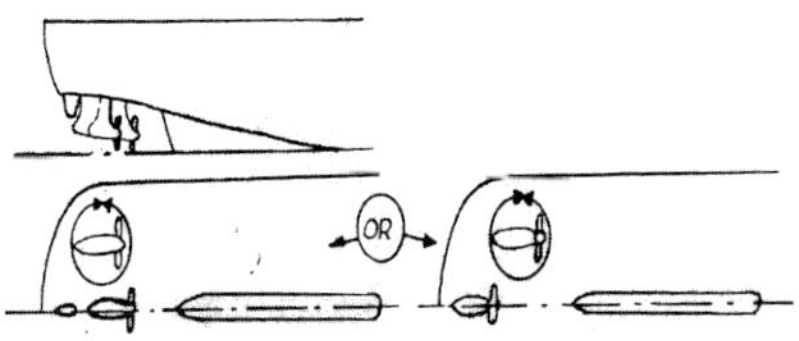

Figure 5: Alternative aftbody 3

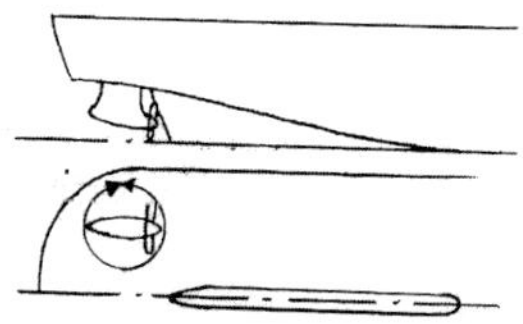

Figure 6: Alternative aftbody 4

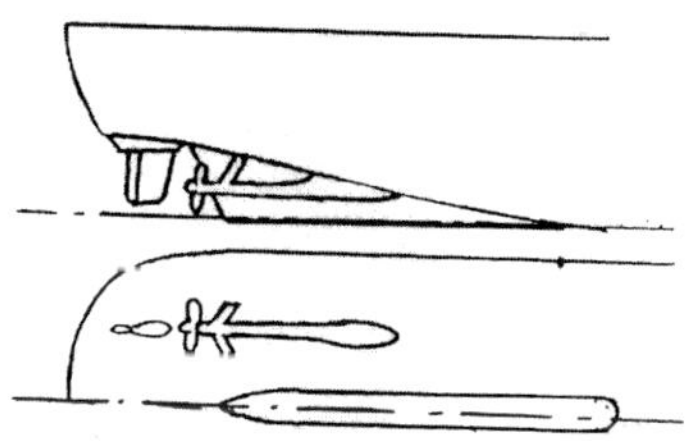

Figure 7: Alternative aftbody 5

TABLE 1

COMPARISON BETWEEN ALTERNATIVES

	Alternatives				
	1	2	3	4	5
Hull excitations	high wake peak 0.4 - 0.55	good wake peak 0.3 -0.45	good wake peak 0.25 - 0.3	good wake peak 0.25 - 0.3	moderate wake peak 0.35 - 0.40
Comfort class	not easy to reach	Easy	easy	no clear, propeller loading is high	not easy to reach, propeller loading high
Power level	3-10% better than conventional (alt 5) 3-5 % less power than in alt 4	1-4 % better than in alt 1	3-5 % less power than in alt 4	0-10 % better than conventional alt 5	well known alt worse than in alt 4
Manoeuvring	excellent course keeping ok!	Excellent course keeping	excellent course keeping?	excellent course keeping ok!	good course keeping ok!
Crabbing	excellent	Excellent	excellent	excellent	complicate
Redundancy	good	Good	good	moderate	moderate
Cost	more expensive than alt 4 and 5	more expensive than alt 4 and 5	more expensive than alt 4 and 5	more expensive than alt 5	cheapest

Alternatives 1, 2 and 5 were selected for further examinations, see figures 8 through 10.

Figure 8: Alternative 1 with conventional gondola

Figure 9: Alternative aftbody 2

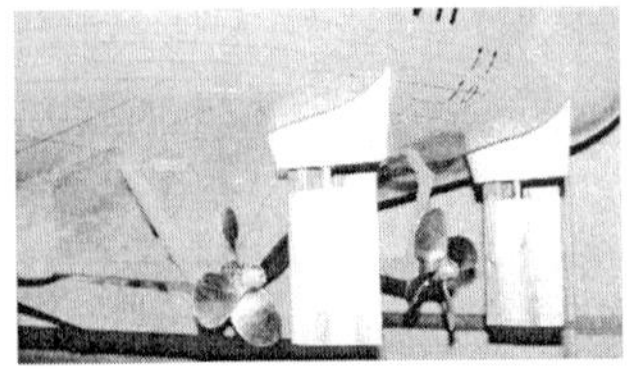

Figure 10: Alternative aftbody 5

Figure 11 shows axial wake fields to alternative 1 (thick gondola M-322 and slim gondola M-322C) and alternative 2 (= M322B).

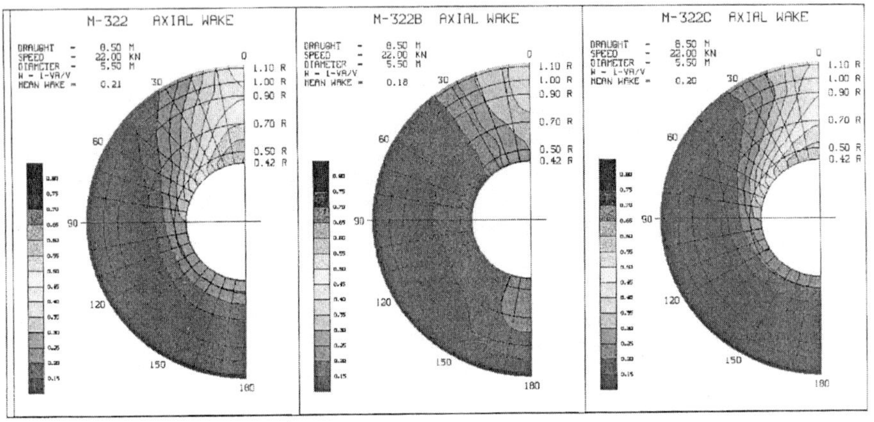

Figure 11: Axial wake fields for thick gondola in alternative 1, alternative 2 and slim gondola in alternative 1.

In table 2 the resistance comparison is presented between these alternatives: CL-skeg / Thick gondola, Slim / Thick gondola and CL-skeg / Slim gondola.

TABLE 2

Vs (kn)	CL-skeg / Thick g.	Slim g. / Thick g.	CL-skeg / Slim g.
18	0.95	0.97	0.98
20	0.97	0.97	1.00
22	0.98	0.97	1.01
24	0.97	0.98	0.99
26	0.96	0.97	0.98

Alternative 5, conventional propulsion showed from 3 to 6 per cent better results if compared with alternative 1. Main reasons were the better aftbody hull form design, better propeller design and a big interaction between cruiser stern transom and pod units. The power levels after the model tests were 3*14.7 MW. It was much more than the wanted "standard size" level of 14 MW.

4.2 Optimisation After Contract

The contract to build the world's largest Cruise Liner was signed on January 7 1997 between RCCL and Kvaerner Masa-Yards. The owner required the maximum redundancy, high comfort class and standard Pod size (14 MW). That means a big improvement excertion for the R&D project, because at the same time the vessel's size was increasing a quite a lot compared to the original R&D size. The Shipyard started an

intensive optimisation process with alternative 2 to reach wanted "standard power size" (= 14 MW) level by using Shipflow CFD (= Computational Fluid Dynamics) program and trying to reach the revolutionising Wave Damping Aftbody (= WDA) principle to the vessel's aftbody. The Shipyard developed the WDA originally for ferries sailing at relatively high speeds (R. Hämäläinen and J. v. Heerd). By using the Shipflow software, the shipyard calculated as many as 15 different versions to reach the best WDA effect. A lot of compromises had to be made taking into account standing in DP, passenger comfort, aftbody slamming etc.
The positioning of the Pod units to separate main fire zones and watertight compartments affected the locations of the Pod units (figure 12). A 0.2*Dp safety margin for the Azipods to avoid collision with bunker barges was used. The rule of thumb: as aft as possible and as far as possible from the centreline, was also here a good advice for turnable Pod units.

The shipyard's two best optimised versions were used in the model basin for Rapid calculations. MARIN calculated 7 versions before reaching the final ideal compromise version between high and low speed. In the following figure 13 a comparison between Shipflow calculation, Rapid calculation and model test is given.

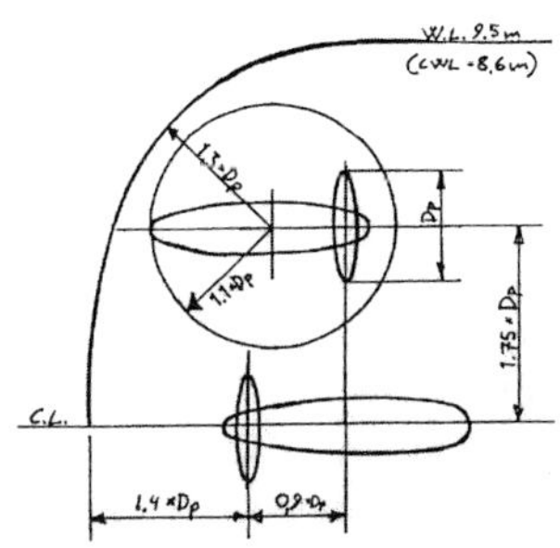

Figure 12: Pods location arrangement.

5 CALM WATER BEHAVIOUR

The model tests were carried out at the MARIN model basin in the Netherlands. The excellent wave behaviour was observed in the bare hull resistance tests. The WDA idea worked well for this large Cruise Ship. In figure 14 the tested aftbody forms are shown. A ducktail was added to the typical cruiser stern to achieve a good WDA effect in the aftbody. A ducktail with a big rounding in the corner significantly improved the aft body wave system by decreasing the wave and pressure resistance at the ship ending.
The region of the four big bow thruster tunnels was observed accurately. Several thruster configurations (Figure 15) with scallops and grids inside the bow thruster have been tested.
In the table 3 the reduction in resistance is presented.

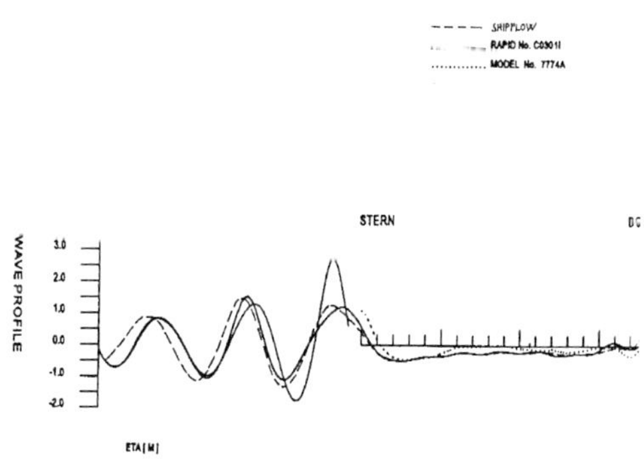

Figure 13: Wave profile comparison between Shipflov Rapid and model test.

TABLE 3

Ship speed (knots)	Resistance in percentages			
	Tunnels only	Most afterwards fairing	Afterwards gratings	Gratings at all tunnels
22	100	99.0	99.4	98.0

Cavitation observations and hull pressure measurements were performed on the side propellers (Azipod) and the centre propeller (Fixipod) for two loading conditions. Also the Azipods in turned condition were tested. It can be said that for the Cruise Ship equipped with three Azipod units, the cavitation behaviour is excellent. The excellent cavitation behaviour means low pressure pulses and low excitation forces on the hull. Azipod propulsion units are a good choice for vessels with very high demands regarding low noise and vibration levels.

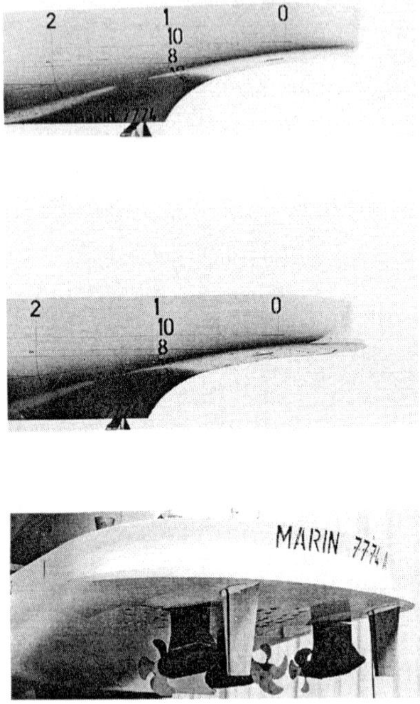

Figure 14: Tested aft bodies with WDA effect

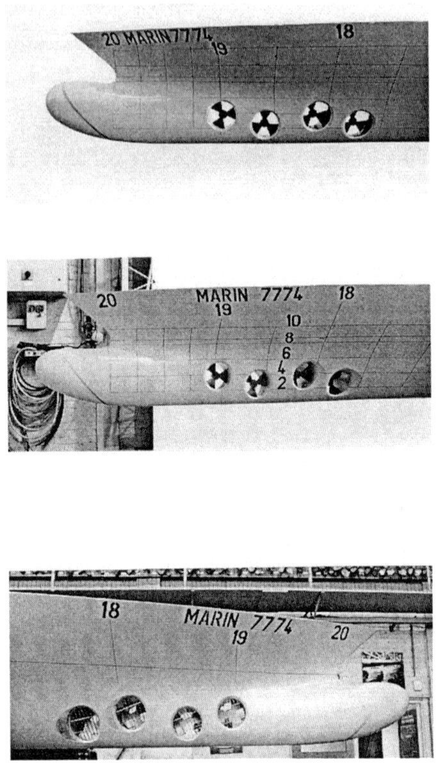

Figure 15: Tested bow thruster openings

6 MANOEUVRING BEHAVIOUR

Comfort and safety of passengers during different manoeuvring cases was the most important aspect in the tests. That is why the Shipyard concentrated much to heeling angles of the ship during several manoeuvring cases. The heeling angle should be less than 4 degrees or this large vessel. Several alternatives were tested to find the most optimum combination. In figure 16 a turning circle is shown in the model basin.

Figure 16: Turning circle test in model scale

The emergency steering was tested extensively to confirm the highest safety of vessel in different situations in archipelago manoeuvres.

7 SEAKEEPING BEHAVIOUR

Many calculations and tests were carried out to ensure a good comfort for the passengers. The main purpose of seakeeping performance tests and calculations were to document ship operability characteristics and limitations related to: passenger comfort and safety onboard related to motions and accelerations; speed loss in waves; ship safety related to extreme motions, i.e. large parametric resonance roll motions and handling features local structural loads related to slamming; bigger bilge keels than normally were installed according to test results.

8 FULL-SCALE TRIALS

During the ten-day sea trial extensive full-scale measurements were conducted to verify the calculations and tests made during the design and construction phase of the vessel. Typical manoeuvring tests were carried out including turning circles, zigzags, pull out, spiral tests, turning tests with side thrusters, coasting stop, astern trial, thruster efficiency tests, crabbing tests etc. The best way to stop the vessel at high and low speed was looked systematically. Cavitation of propellers and Pods were observed through ten windows and hull pressures by several pressure transducers.

- the manoeuvring behaviour of vessel was excellent:
- very good steerable, especially at low speeds
- the stopping distance is much smaller than in conventional cases
- 360 degrees directed thrust makes the vessel easy to handle
- harbour manoeuvring should be easy

Some differences between model and full scale were observed: The vessel was "stiffer" than expected (figures 17 and 18). That was against earlier observations and expectations. Normally you can expect similar or worse course stability.

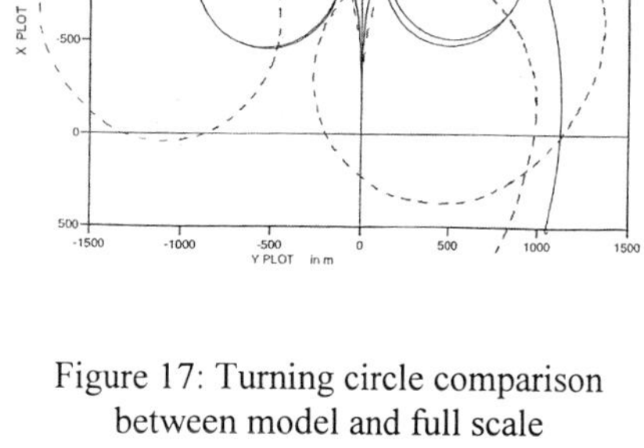

Figure 17: Turning circle comparison between model and full scale

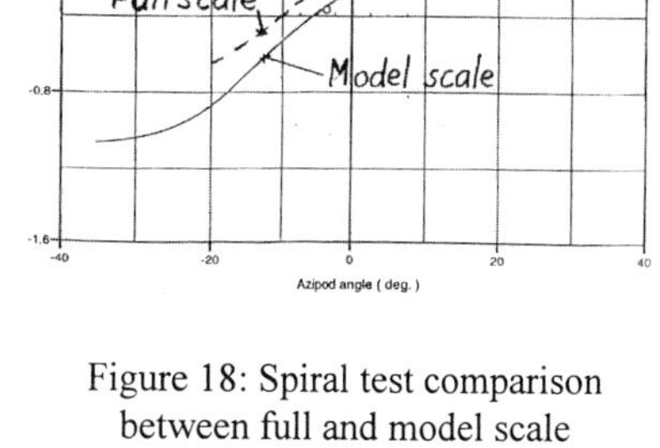

Figure 18: Spiral test comparison between full and model scale

Different kind of explanations were studied from big bow thruster openings to resistance differences. Main reason for the differences are more likely related to the small 5 m model (scale 1/55) for such a large cruise liner. The speed of the triple screw vessel was higher than expected after model tests (figure 19). The reasons for higher speed could be: excellent hull surface quality of the Shipyard, the scale effect corrections on Pods used during that time were rather old fashioned, and the scale effects due to the interaction between Azipods and Fixipod for this triple screw were not known. The correlation allowance from model tank power to trial power for such big vessel has to be carefully investigated.
However, the pressure pulse measurements showed the same low excitations from propellers than in model scale. The pressure pulse level stays low level at normal steering Azipod angles.

Figure 19: Wave damping aftbody effect in full scale.

9 CONCLUSIONS

The Pod propulsion concept gives many additional benefits to classical diesel-electric system.
Podded propulsion makes new comfort class of classification societies easier to reach.
The rules of thumb are very useful for prototype builder.
R&D works before final hydrodynamic design makes succession easier.
A combination of experience, numerical tool data and good co-operation give good possibilities for success story, as was the case in the Voyager-class vessels.
There is a lot of work in Pod propulsion project scaling to reach same accurate level with conventional cases.
The Wave Damping Aftbody revolutionises the aftbody design for all types of vessels.

References

R. Kurimo, A. Poustoshniy, E. Syrkin. (1997). Azipod propulsion for passenger Cruiser. *NAV & HSMV International Conference.*

R. Hämäläinen and J.v. Heerd. (1998). Hydrodynamic development for a large fast monohull passenger ferry. *The Society of Naval Architects and Marine Engineers.*

R. Hämäläinen, J. v. Heerd. (2000). Wave Damping Aftbody revolutionize aftbody design for all types of vessels, *IMAM 2000 conference.*

Practical Design of Ships and Other Floating Structures
You-Sheng Wu, Wei-Cheng Cui and Guo-Jun Zhou (Eds)

HYDRODYNAMIC TRENDS IN HULL LINES OF PODDED DRIVEN LARGE CRUISE VESSELS

Roger Lepeix

Chantiers de l'Atlantique, France

ABSTRACT

The propulsion of the large cruise vessels is under rapid change, in many aspects. One of the most important one is the present regular use of azimuthing pods, where the electric propulsion motors are installed outside the hull, in a kind of torpedo, suspended under the main hull. There are many advantages in using such pods, including better power/speed curve and manoeuvring qualities, improvements in building process, and room release in the general arrangement. This last aspect, combined with the absence of aft appendages, give to the ship designer new freedom in selecting the hull lines. There are no clear solution yet to optimise these new possibilities, and progressive modification of the afterbody lines will take place in the next years. The present experience with the newly delivered large cruise ships propelled by pods can already give ideas on the trend to follow for these ships. They also show the new limits, and where we cannot go. The first elements of this know how in development are presented here, on basis of the two first series of large cruise vessels with pods under building at Chantiers de l'Atlantique, in France.

KEYWORDS

Hull lines, Cruise vessel, Pods, Hydrodynamics, Propulsion, Slamming.

1 INTRODUCTION

The number of large cruise vessels propelled by azimuthing pods is rapidly increasing. Most of the cruise vessels presently delivered by all the european shipyards specialized in such ships are now equipped with pods, and it will continue as well in the next years. The ships already delivered in the last two years have now given some experience to the designers, and we can start discussing the new trends, in particular regarding the hull lines.

The market of cruise vessels is very sensitive, due to the special cargo these ships are transporting. The technical innovation is a permanent objective, and new lines and new ship types are presented each year, to attract new people, and to propose renewed entertainment. Also, new technics are needed to improve operability of these ships: entrance in new harbours, improved manoeuvrability, higher speeds, increase in size, in protection of environment, and finally in benefits for the owners and operators.

The pod propulsion is now under rapid development. Of course this propulsion have been used for decades on many ship types, but the application to large cruise ships is only possible since the last years. The first ones have been delivered with two azimuthing pods, which is now the classical solution, but the largest ones have now three pods, including one fixed pod in the centerline, and the largest project in order book will have two fixed and two azimuthing pods.

All problems concerning this new propulsion type are not yet solved, and many studies are on the way; it is then time to present and discuss the first results, and the comparisons between conventional and podded-driven cruise ships.

1.1 The Azimuthing Pods for Cruise Ships

Most of the aspects are not really new in the azimuthing pods presently installed on large ships; all the elements were already existing. What is now new is the possibility to install the electric engines outside the vessel, in a limited volume, for power levels up to about 20 MW, at least. The knowledge and experience coming from azimuthing propellers, electric tranmission, and the progress in electric engines have been combined and are now giving commercial products.
Several main pod suppliers can propose industrial solutions, using similar basic principles. The comparisons between them is still open, but elements are available.

The main interest for using azimuthing pods lies in the following aspects :

- the power/speed curve is improved, due to the absence of appendages: rudders, shafts, brackets, and aft lateral thrusters. The additional power coming from the pod volume is only a part of the saved one due to removing of the conventional appendages. The final improvement is not yet easy to predict, but is sufficient for considering pods as an attractive solution
- improvement of the manoeuvring qualities of the ships, both in open seas and in crabbing situations, both close to the quay or in dynamic positioning. This is coming from the possibility to use the propeller thrust in any direction
- freedom in the general arrangement of the ship, in utilization of the volume left by the electric engines pushed outside. The cruise ship designer is able to install additional cabins, by reorganizing the internal layout
- reduction of assembling works ans delay, by installation of the pod as a whole later in the building process of the ship. Also replacement of pods is made easy by their external position.

These gains by using pods have been considered large enough for arousing interest in shipbuilding industry and also in operating business. What is important is now to study the other consequences of this choice, and how to design ships equipped with pods.

1.2 New Trends in Hull Lines

The problem of selecting the best lines for a ship with pod propulsion is very important, both for industrial reasons, in a very competitive market, and for a good estimation of the gains which can be reached with pods. The main questions to be solved are :

- how to use the new freedom coming from the absence of shafts and the new location of electric engines, for the improvement of the hull lines ? These engines were already installed in conventional ships with electric transmission, but the removing of appendages is totally new, and nobody can at that stage give optimised solutions
- how to install the pods themselves, for a good combination with the hull, the best possible efficiency, and with a full 360° rotation ?
- are the manoeuvring qualities really improved ? How to operate the pods for using this improvement, and is it leading to installation requirements ? What will happen with course-keeping and crabbing capacities, and how to handle them ?

- are we decreasing the propeller-induced pressure fluctuations on the hull, and have we to change the propeller design ?
- do we expect modifications in seakeeping, and have we to take care of them in hull design ?
- how to control the optimisation process of the hull lines, including technical and economical aspects, for a special ship and for the ships to be designed in the future, and how to integrate industrial and economic considerations ?

Of course the answers have to be found both in research studies, and in real design work. We are only at the beginning of this process, and we can only give general tendancies.

1.3 Present Research Studies

There are many ongoing research studies on pods. We will mention only the projects where our shipyard Chantiers de l'Atlantique (Alstom Marine) is involved in, and where hull design is a large part of the work :

- OPTIPOD is a large project organised at european level, and supported by the European Commission. It brings together 14 partners from 7 countries; Chantiers de l'Atlantique is the provided coordinator, and the swedish model basin SSPA the technical manager. The main part of this project is devoted to the improvements of hull lines in introducing pod propulsion, and to all hydrodynamic performances. Other aspects are also included, as the optimisation of the general arrangement, the operation of the vessels, the environment aspects, and the economic ones. Four ships are used as reference vessels : a cruise liner, a ropax, a supply ship, and a general cargo. The last one has only one pod. First results will be available in the next months, and will be distilled in specialized conferences
- Pods in Service is another large project, organized as a Joint Industry Project. It is coordinated by MARIN, and also supported by the European Commission. The aim of this study is to understand and measure the forces and moments on the pods in service conditions, and the stresses in the seating and in the structure around. Another objective is to determine in which conditions the forces and moments on the moments are maximum, and to reproduce these conditions at model scale. This will help the designers of the pods and of the podded driven ships by giving them some guidelines, and reference values
- The Cooperative Research club, managed by MARIN, settled down also a specialized Working Group for pod study. The chairman of this group is Masa Yards, and the objectives are to assess the hydrodynamic qualities of the ships propelled by pods, compared to conventional ones, and to give indications on the design trends
- an open cooperation has also been organized under Euroyards, between the 5 shipyards members, all involved in building or designing ships with pods. The exchange of results is needed to check the predictions in hydrodynamic performances, which needs enough different vessels. These informations are confidential and only for mutual benefit.

All these studies, and others not mentionned, are often overlapping, which is good for redundancy. They cover the main new domains open by the pod propulsion, and are very useful for the developments of ship design. Of course many other aspects are covered by ongoing studies on pods, not indicated here: electrical engines, vibrations, mechanical aspects, etc...

1.4 Shipyard Experience

Chantiers de l'Atlantique are now regularly delivering large cruise ships with pod propulsion. Two main series must be identified: the Radisson/Festival series, derived from the conventional ship Mistral, propelled by 2 pods in the range of 7 to 10 MW each, and the Millennium series with 2 pods of about 20 MW each. These two series are coming from different histories, and have different aftbody lines.

The conventional lines of the Mistral have been only slightly modified by introducing a small flat part at the pod location, and a fairing on the lines. With this minimum change, we are not progressing in using the new freedom in hull lines, but we get in power/speed curve the improvements coming from the removal of the appendages, and the full use of pods for manoeuvring.
At the opposite, the Millennium series is not coming from corresponding conventional form, and then the afterbody lines are new. This form has been obtained by extrapolating the previous tendancies in pram direction, and we tried to open the aftbody as much as possible to take advantage in speed. This excursion outside usual aftbody lines has not been fully successfull; the extension of the flat part led to some aft slamming at low speed and small sea waves.
This is explained in more detail in the seakeeping part hereunder. Our shipyard is now working on intermediate trends, combining the hydrodynamic qualities of the conventional aftbodies and the gains in power and manoeuvrability.

2 HYDRODYNAMIC PERFORMANCES OF THE LARGE PODDED DRIVEN CRUISE SHIPS

2.1 Power/Speed Curve

As indicated above, the main gain on power/speed curve, coming from the use of pods, is due to the removing of all appendages, only partly compensated by the possible additional resistance of the pod itself.
The present possibilities of predictions are limited, due to the small number of ships in service. Many people are presently working on this question, including the model basins and the ITTC. Extrapolation procedures are under development for pods, and empirical coefficients will be prepared after enough speed tests will have been performed during sea trials of newbuildings. We will also have to identify the new types of disturbances directly connected to pod propulsion. For instance, we have to evaluate the importance on the measured speed of the fluctuations of the pod angle; this could give more additional resistance than the corresponding values for rudders.
The main new geometrical parameters influencing the power/speed curve are presented hereunder.
The location of the pod under the afterbody can influence the performance. It seems that the best location is the maximum aft, which is in general the more aft until the pod tails reach the transom position; in lateral direction, it is less clear.
One important point is the local form of the afterbody, allowing the full rotation of the pod. As discussed above, there are two main situations :

- installation of pods on a conventional hull form, using previously propellers with normal shafts. In this case, the solution is to flatten locally the afterbody lines, to allow the pod rotation; not flattening is also possible if the pod is designed with enough space at top edges for allowing the rotation. Our shipyard choose the local flattening for the Radisson/Festival series. An important parameter is the angle of the pod axis with the vertical. If inclined axis is possible, the integration of the flat part in the conventional afterbody is quite simple. If not, there must be a kind of headbox, which will give additional resistance
- design of a new afterbody, directly for a propulsion with pods. The temptation is here to extrapolate the flat part needed for the pods all around, to improve the power/speed curve. In case the afterbody becomes very horizontal, like in our original "Millennium", this could lead to stern slamming on low waves (see hereunder). It is possible to keep this solution if the slamming remain moderate, or if it can be solved by limited modifications.

Regarding the pod parameters, we can provisionally say that :

- the pod itself must be studied in details by the pod supplier, for minimum resistance, optimum interaction between the propeller(s) and the pod, and to decrease the risk of cavitation on the part of stay behind the pulling propeller. Taking into account the rotation of the propeller, this could lead to an asymmetric geometry on this part

- the alignment of the propeller axis with the flow is often necessary. On large cruise ships, the optimum angles are defined on model tests; current angles are about 3 to 5 degrees in vertical direction and 1 degree in lateral one.

2.2 Ship Wake and Pressure Fluctuations

The absence of aft appendages, and the use of pulling propellers, have a very positive action on the wake around the propeller. The only contribution giving small disturbance in the wake pattern is coming from the boundary layer of the vessel (see figures 1 & 1bis). With the influence of Reynolds number, the disturbed part of the wake pattern is further decreased in full scale compared to model scale. The propeller-induced pressure fluctuations on the hull are also improved, leading to reduced noise and vibrations on board.. This have been confirmed on Millennium series during sea trials, where maximum levels of about 1.0 to 1.5 kPa have been recorded at 23 knots, for the first harmonic. Observation of cavitation on the propeller blades on the first Millennium has also shown a large improvement compared to conventional vessels, for pod angles up to about 5 degrees, which is generally enough for course keeping. Of course the cavitation is increasing for larger angles. Other observations are scheduled on the third vessel of the series, in connection with the european project "Pods in Service".

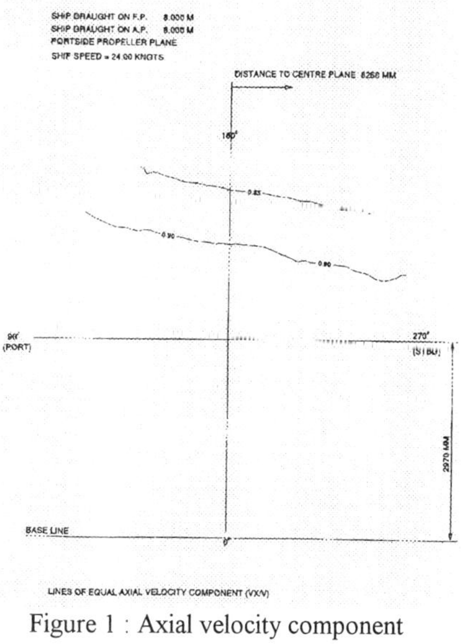

Figure 1 : Axial velocity component

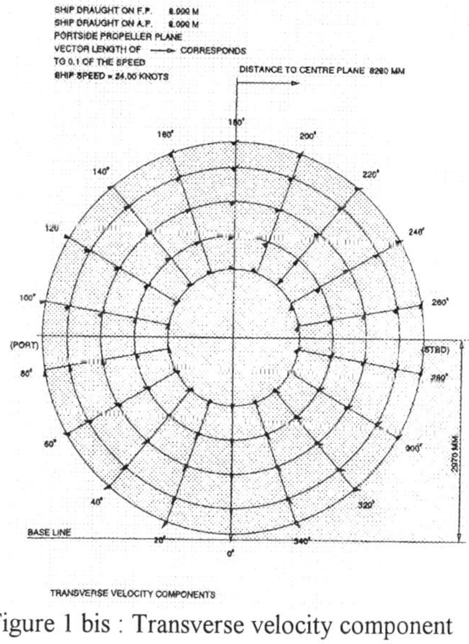

Figure 1 bis : Transverse velocity component

2.3 Manoeuvring Performances

The interest of using azimuthing pods is very clear in manoeuvring domain. The performances of the vessel in open sea manoeuvring or in crabbing in harbours are improved, even if the captains have to operate the ship in a different way.

2.3.1 Course-keeping

There are many possibilities in course keeping for podded driven ships. They have to be studied and decided at design stage: use of one pod, of two pods in parallel, of additional flaps or small rudders on the pods or behind the central skeg, etc... For safety reasons, the most common way is to use the two pods like rudders. The performance of the pod, in comparaison with rudders, is connected to the lift force generated on the connecting arm and on the torpedo of the pod. This is very dependant on the pod type and design. It is important to have enough lift at small angles, to avoid the cavitation on the propeller blades. The main difficulty is coming from this huge mass of more than 200 tons, in permanent rotation of few degrees. Technical solutions have been found by the main pod suppliers.

2.3.2 Manoeuvrability in open sea

The large cruise vessels have generally no manoeuvring problems in open sea, in particular those of Panamax size, due to the large L/B ratio. Moreover, they often have large central skegs, for building and drydocking reasons. Then, these vessels are stable, or marginally unstable. These good manoeuvring qualities are still increased by the use of azymuthing pods, thanks to the possibility to choose the propeller thrust direction. The performances in standard manœuvres are then much better than the IMO criteria.
These performances have been compared at model scale for the Festival/Radisson series, between conventional flapped rudders and azimuthing pods. The results for the tactical diameter are given in Figure 2. They show that the same diameter is reached with the rudder option at 35 degrees angle, and for the pods at only 26 degrees.
We must add that an eventual drawback of this low diameter could be a non acceptable heel angle in turning; in fact, this has not been observed in model or ship scale. Reasonable maximum angles have been recorded, probably due to the drop of speed during the first part of the turning manoeuvre, the added resistance of the turned pod being much higher than for the rudders. The first comparisons show a lower maximum heel angle for pods compared to rudders, but this must be confirmed on more vessels.

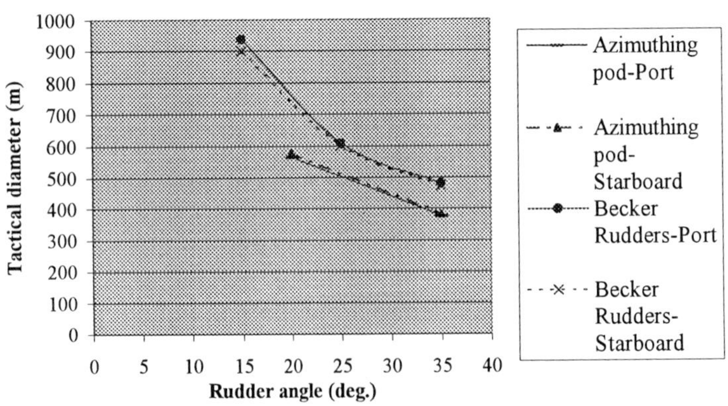

Figure 2 : Comparison tactical diameter Becker rudders/Azimuthing pod

3 AFT SLAMMING ON "MILLENNIUM"

3.1 Hull Lines

The afterbody hull lines of the "Millennium" have been selected according to :

- modern trends of lines, leading to improved power/speed performances, with a nice flow around the afterbody and transom, and enough stability margin
- operation of the azimuthing pods (360° turning), inside ship geometrical limits, and with manoeuvring qualities much improved, compared to conventional propulsion

- possibility to remove the pod afloat if needed.

To check hydrodynamic performances of the ship, model tests have been performed in 1998, during the design phase of the ship :

- in MARIN basin (Wageningen) for propulsion, cavitation, and crabbing
- in MARINTEK basin (Trondheim) for manoeuvring and seakeeping.

In the seakeeping program, pressure panels were installed on the model, both on forebody and afterbody, to check the slamming risks. These tests showed the good performances of the vessel at service speeds, including slamming aspects.

3.2 Aft Slamming on Ships

The slamming is defined as an interaction between the ship hull and the water waves, able to create local or global efforts in the ship, which must be taken into account in the design or scantling. The slamming phenomena must be studied if both the relative velocity between hull and water, and the total concerned area, can be large.
There are different levels of phenomena involved, with corresponding variables :

- local pressures, which can be recorded by transducers at model or ship scales ; they are mainly dependant on the relative local speed between hull and water
- total force on forebody or afterbody, not easily measured at ship scale. They can be measured at model scale by using a segmented model and recording of vertical strains between the different parts. Of course, the total force on model is influenced by the vibration of the model itself, which can be globally reproduced
  In a slamming event, the total force increases with the local pressure and with the total concerned area, at each moment
- ship vibration excitation (whipping). A brief and large shock on the hull can excite the vibration modes of the ship structure. This can be easily measured on board ship if it occurs, by using a number of accelerometers, or strain gauges. It is also possible at model scale
  The ship response depends on the total force and on the variation of this force with time (duration of signal, total time over a limit) ; this relation is generally not simple. The phase differences between the local impacts on the afterbody have a strong influence on the total duration of the signal, and then on the whipping amplitude
- local vibration in the ship, connected to comfort on board. The vibration levels are easily measured by accelerometers, and the results can be compared to admissible limits, if any.
  These levels depend on the ship whipping vibration, the location of the point in the ship, and the local environment (dampers...).

3.3 Aft Slamming on "Millennium"

The application to the Millennium of the above considerations gave the following :

- due to flat afterbody, the waves for which both the total slamming area (at the same time) is maximum, and the phase differences between local pressures are minimum were the very small waves, with a flat sea. This flat sea can then hit the flat ship afterbody on a large area at the same time. This is then the most defavorable situation. Fortunately, the relative speed between the hull and the water must be in such cases generally small
- only when very long waves, mainly from bow, can create a pitch motion, we could have some relative speed with a quasi flat water; this relative speed was small
- if the wave height increased, the local pressures also increased, due to increase of relative speed; the total slamming area had to be generally reduced, and the phasing scattered; the water cannot remain flat, for higher waves

- more generally, we can say that strong slamming can occur when a wave with the same form than the hull can hit the hull with a large speed. The corresponding waves can be different from one ship to another.

3.4 Observations on "Millennium"

The sea trials of the "Millennium" confirmed that aft slamming could occur at low wave heights, in special conditions. Tests with different kinds of measurements have been organized, to estimate the values of the variables (pressures, vibrations, noise, stresses), and to study the influence of operational and environmental parameters.

These measurements showed that :

- aft slamming can occur only at low ship speeds. The excitations and responses decrease with speed increase
- whipping responses were mainly limited to the 2-node global vibration, with a frequency around 1 Hz. The vibration levels in worst cases were under the limits in most common criteria
- no real noise was present in connection to slamming, and no questions were raised there
- the most spectacular aspect of the aft slamming was the lateral jets of water on both sides of the afterbody. These jets have not been considered as dangerous.

The measurements carried out at model scale (with original lines of Millennium) have been compared to corresponding ones coming from other vessels of similar size or commercial profile. Many alternative afterbodies have also been tested in segmented models, with local changes of the original afterbody, or with large modifications.
The objectives were directed towards two directions :

- to improve comfort onboard at zero or low ship speeds, for limited waves (up to about 3 meters significat height)
- to explore the maximum bending moments values in the vessel, in slamming conditions, with a combination of the wave part and the whipping part.

The general conclusion is that the slamming of "Millennium" was not really higher than the other vessels, if we consider all conditions, including forebody slamming at high ship speeds. Finally, the comfort aspects have been solved by welding additional volumes on the flat afterbody: two side skegs, one trim wedge, and one central wedge (from the second ship only) (see figure 3). The quality of the modified vessel in total bending moment has been certified by the classification society; after many full scale measurements in all kinds of waves, all possible limitations of operation were removed. These full scale measurements culminated in the Atlantic crossing of the "Infinity", second ship of the series, where stresses were recorded in the most sensible points on waves up to about 9 or 10 meters significant height, corresponding to individual waves up to about 18 meters.
On this special ship, where the original afterbody was very flat, the additional volumes led to large improvements :

- the central wedge is very effective when the afterbody is entering the water with high relative speed, the positive effect on the slamming impacts being different, depending on the heading of the waves
- the aft trim wedge can improve the power/speed curve, and also the slamming aspects for the stern waves. The angle of this trim wedge is very sensible; for economic reasons, the optimisation has been done mainly for powering aspects
- the lateral skegs can help to increase the duration of the impact; some air is trapped below the afterbody, smoothening the impact. This effect is amplified by horizontal holes drilled in the skegs, which can act as a pierced cushion, and then increase artificially the air compressibility.

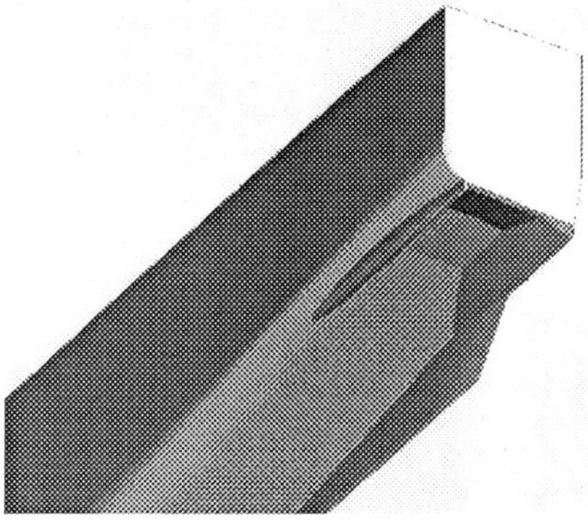

Figure 3 : Millenium –Additional volumes

4 CONCLUSIONS

The first studies of azimuthing pod propulsion applied on large cruise vessels, and the experience gained in building and delivering such vessels, show that the pod propulsion can be a very attractive solution. The improvement of performance is important, even if it is yet to be precised, with the future ships.

The hull lines selected for the two actual series of concerned ships built in our shipyard have followed different ways, and the new possibilities and limits have been explored. In particular, partly due to the evolution of the hull structures, special attention must be given to interaction of the ships with the sea, in moderate and high waves.

New developments are needed to clarify all the open questions. Many studies are presently running at all levels, and will give complementary design guidelines in the next years.

Practical Design of Ships and Other Floating Structures
You-Sheng Wu, Wei-Cheng Cui and Guo-Jun Zhou (Eds)

SIMULATING THE SELF-PROPULSION TEST BY A COUPLED VISCOUS/POTENTIAL FLOW COMPUTATION

S.K. Chou[1], C.Y. Hsin[2], S.W. Chau[1], W.C. Chen[1]

[1] United Ship Design and Development Center, Taiwan, China
[2] Department of System Engineering and Naval Architecture, National Taiwan Ocean University, Taiwan, China

ABSTRACT

A coupled viscous/potential flow computation for simulating the self-propulsion test will be presented in this paper. The flow around ship is first calculated by using a viscous flow approach, and the calculated viscous wake is used as the propeller inflow. The propeller forces are computed using a potential flow method. The rotating speed of propeller blade is determined by balancing the propeller thrust and ship resistance. The viscous flow around ship is then calculated by treating propeller forces as body force terms. The flow computation is performed iteratively until convergent results are obtained. The interaction between ship hull and propeller thus can be predicted for the simulation of self-propulsion test. The present results for a container hull form have been validated by a proper comparison with experimental data. It is demonstrated that the present method is quite promising in evaluating the self-propulsion characteristics of ships for practical use.

KEYWORDS

Self-propulsion test, Propeller-hull interaction, RANS calculation, Equivalent body force, Composite grid, Nominal wake, Effective wake.

1 INTRODUCTION

Currently, the determination of effective wake and thrust deduction coefficient is mainly derived from model experiment, namely, open water test and self-propulsion test. The propeller open water test has been successfully simulated by various computational methods such as vortex lattice method or boundary element method. However, the simulation of self-propulsion test has not been fully established yet. The reason is due to the difficulty of calculating the effective wake. The effective wake resulted from the interaction between propeller and ship hull has been studied by many naval hydrodynamic researchers. Huang and Groves(1976), Schetzr and Favin(1977) pioneered the work, and Stern and his colleagues (1988) first coupled the potential and viscous methods to calculate the effective wake. Kerwin and his colleagues (1994) decomposed the flow field into axisymmetric and non-axisymmetric parts, and developed the "equivalent body force" concept.

At the United Ship Design & Development Center (USDDC), great effort has been put on the development and application of CFD technology for long. The computational code presented in this paper, UVW (USDDC Viscous-Wave flow code), is based on an explicit finite volume method (Jaw & Chou, 1994) to calculate the viscous flow around a ship and the free surface effect is also considered, Chou (1998). The composite grid technique is incorporated to handle complex geometries, Chau & Chou (1998), and has been applied to the computation of flow around ship with appendage, Chau et al. (2000). The UVW code has become a feasible tool for improving the hydrodynamic characteristics of hull forms developed at USDDC. Furthermore, the coupled viscous/potential flow scheme, Hsin et al. (2000) and Chou et al.(2000), has been adopted to investigate the effective inflow in the propelled ship problem. In this research, the interaction between the hull, the rudder and the propeller for a container hull form is first investigated. Application of the proposed method for simulating self-propulsion test is then performed and validated by comparison with experimental data. It is shown that most features of the wake field can be captured by the present approach, which provides important information for the propeller design process.

2 COMPUTATIONAL METHOD

The 3-D RANS equations with pseudo-compressibility (Chorin, 1967) are employed to analyze the flow field of a ship moving at constant forward speed in calm water. The application of finite volume method to discretize the governing equations in a curvilinear coordinate space leads to the transformed equations in the computational domain. The convective terms are evaluated by central difference approximation and a 4-th order background dissipation term is added for the numerical stability of scheme and the elimination of non-physical oscillations, Jameson and Turkel (1981). The multi-stage Runge-Kutta explicit scheme is adopted to calculate the time integration. Local time step and residual smoothing technique are also employed for acceleration of convergence. The algebraic eddy-viscosity model proposed by Baldwin and Lomax (1978) is used to evaluate the turbulent viscosity. The modification for adverse pressure gradient is introduced for stern flow calculations, Johnson and King (1985).

2.1 Propeller Effect

The propeller effect is simulated by a coupled viscous/potential flow method. The propeller flow is calculated by a potential flow code, MIT-PSF-2 (Greeley and Kerwin, 1982), and then coupled with the viscous flow calculation described above to couple the propeller effect into the ship flow. MIT-PSF-2 is a vortex lattice method based on the lifting surface theory, and it has been widely used by the ship industry to calculate the steady propeller hydrodynamic characteristics.

The coupled viscous/potential method used here is mainly based on the "equivalent body force" concept developed by Kerwin (Kerwin et al, 1994), and this method has a very important assumption such that the flow due to the propeller-hull interaction is assumed to be axisymmetric. This assumption is made because that the coupling of viscous and potential calculations can be carried out consistently only under the condition of taking circumferential mean value, and this assumption is valid because that the induced velocity upstream of a propeller is usually dominated by the circumferential mean value. Therefore, when considering the propeller-hull interaction, the total velocity, V, can be decomposed into two parts, namely, the circumferential mean of the effective inflow, V_e, and the circumferential mean value of the propeller induced velocity, U_i :

$$V = \overline{V}_e + \overline{U}_i \tag{1}$$

In the coupling calculation, the propeller-induced velocity and loading are calculated by the potential flow code, and the "equivalent body force" is calculated by the total velocity and the propeller loading. With the "equivalent body force", the total velocity is obtained by the viscous flow code. The detailed description of the theory is discussed by Hsin (2000) and Chou et al. (2000). Since the equation (1) is a nonlinear equation, and an iterative calculation is necessary.

2.2 Free Surface Effect

Although the UVW solver is capable to calculate the viscous flow field with free surface effect, the complexity and computational resource involved in such calculation are not yet feasible for daily design work. In this research, the wave-making resistance is evaluated by an in-house potential code USDDC-WAVE, which is based on a Dawson-type high-order Rankine panel method (Lu & Chou, 1994). The resistance related to viscous effect is calculated by UVW in double-model flow. These two resistances are summed to represent the total resistance. Fig.1 shows the predicted total resistance coefficient C_T and corresponding towing tank result. At the design speed (18 knots), the prediction error is smaller than 0.5%, while large prediction error up to 10% occurs at low ship speed (12 knots).

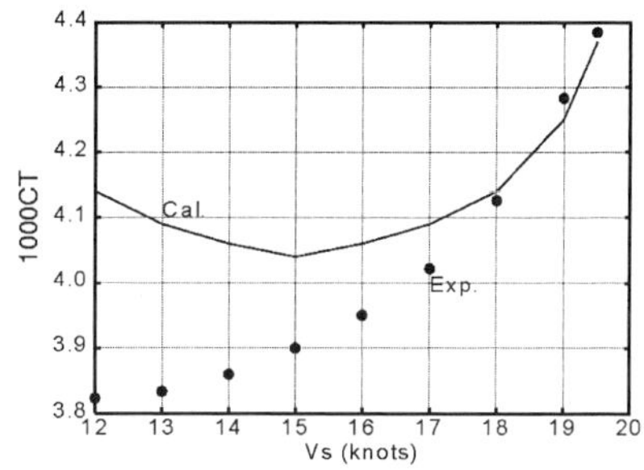

Figure1: Predicted and measures C_T val-

2.3 Composite Grid Approach

In order to incorporate body forces at propeller plane and rudder effect into viscous ship flow, the composite grid approach is adopted, Chau and Chou (1998). The propeller and rudder are treated in two different foreground grids, while the bare hull without any appendages is constructed in a single background grid. All these grids are separately generated, but similar grid point distribution is applied in overlapping regions. The body forces are defined at the computational cells, which are located at the mid-chord of propeller blade in the propeller block. In this way, body forces are more precisely allocated and good grid quality is easily achieved to reduce numerical errors and difficulties. For different propeller and rudder arrangement, only geometrical transformation of original grids is necessary.

2.4 Self-Propulsion Simulation

Following the "equivalent body force" theory and the numerical procedure to calculate the effective inflow, a numerical procedure to simulate the self-propulsion test is developed as follow:

1. The viscous flow around a ship hull without the propeller in operation is first calculated by the RANS code UVW, and the hull resistance is also obtained;
2. The circumferential mean value of the flow velocity at the propeller plane is calculated from the solution of UVW, and then used as the inflow of the propeller analysis program MIT-PSF-2.
3. With the calculated hull resistance, the propeller loading can be determined by a modified version of MIT-PSF-2, such that the rotational speed of the propeller is adjusted to obtain the self-propulsion point;
4. The circumferential mean values of the propeller induced velocity and the propeller force is calculated, and the "equivalent body force" is then computed and transferred into UVW;
5. The flow field and the hull resistance are calculated by UVW including the propeller effect again;
6. The flow velocity at the propeller plane is extracted and the circumferential mean is computed, i.e. the total velocity $\overline{V}$. The circumferential mean of the effective inflow, $\overline{V_e}$, can be calculated by subtracting the circumferential mean of the propeller induced velocity, $\overline{U_i}$, calculated in the last iteration (equation (1));

7. The calculated effective inflow is then used in the modified MIT-PSF-2 for the propeller flow computation as in 3.
8. Step 4 to 7 are then repeated until the convergence of the hull resistance is reached, and the self-propulsion point is thus obtained.

3 RESULTS AND DISCUSSIONS

3.1 Hull-Propeller-Rudder Interaction

The hull-propeller-rudder interaction of the model for a 145m long container ship is first investigated. The numerical computation is first performed for bare hull condition without propeller and rudder. The ship is forward advancing at design speed, and the propeller rotation speed is given at a fixed value. Therefore, the ship total resistance and propeller thrust are not in balance. This calculation only serves as a preliminary study of the effect of propeller and rudder on ship hull, especially from the resistance aspect. Table 1 summarizes the resistance and side force change for different rudder angles and propeller effect. All values in table are normalized with the bare hull resistance. This shows that the interaction among hull, propeller and rudder is strong, and should be considered as an integrated system for the prediction of the self-propulsion characteristics.

TABLE 1
RESISTANCE CHANGES DUE TO DIFFERENT CONFIGURATIONS OF HULL, PROPELLER AND RUDDER

Hull-Propeller-Rudder Configuration					Force on Ship Hull		Force on Rudder		Total Force	
Case	Ship Hull	With rudder	Rudder Angle	With Prop.	Resistance	Side Force	Resistance	Side Force	Resistance	Side Force
1	Yes	No	-	No	1.000	0.000	-	-	1.000	0.000
2	Yes	No	-	Yes	1.110	0.000	-	-	1.110	0.000
3	Yes	Yes	0°	No	0.957	0.000	0.047	0.000	1.004	0.000
4	Yes	Yes	5°	No	1.030	0.009	0.099	0.303	1.129	0.303
5	Yes	Yes	0°	Yes	1.042	0.017	0.111	0.041	1.154	0.058
6	Yes	Yes	5°	Yes	1.292	0.012	0.145	0.367	1.438	0.335
7	Yes	Yes	-5°	Yes	1.292	0.045	0.137	0.415	1.429	0.459

3.2 Numerical Self-Propulsion Test

The numerical simulation of self-propulsion test at design speed 18 knots (corresponding to 1.956 m/s in model scale) for the previously described ship is discussed in the section. The calculated wake at propeller plane of bare hull is given in Fig.2. Good visual agreement is observed, but some difference still occurs in wake fraction, Table 2. Fig.3 depicts the grid arrangement for flow field calculation of hull-propeller-rudder configuration.

TABLE 2
MEASURED AND CALCULATED WAKE FRACTION OF THE STUDIED SHIP

Bare Hull	V_m(m/s)	w	$C_T \bullet 10^3$
Measured	1.956	0.352	4.126
Calculated	1.956	0.290	4.140

The self-propulsion point of ship is reached through balancing the propeller thrust and the ship resistance. Therefore an iterative process is performed until the difference between the propeller thrust and the ship resistance is smaller than a given tolerance. In this case, 10 times of global iterations for RANS computation is necessary for a converged result. Fig.4 and Fig.5 show the predicted wake and

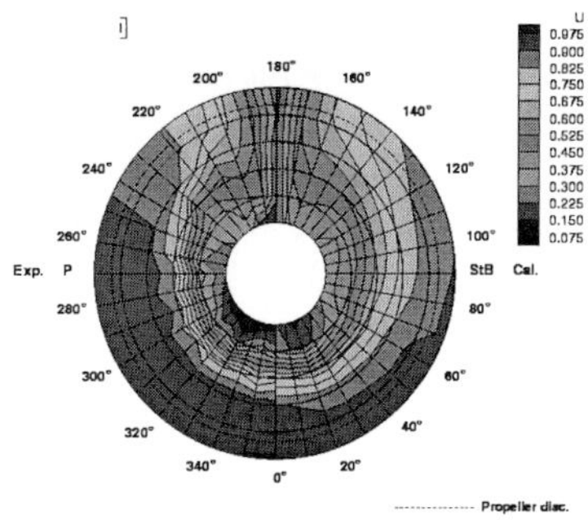

Figure2: Predicted and measured normal wake distribution

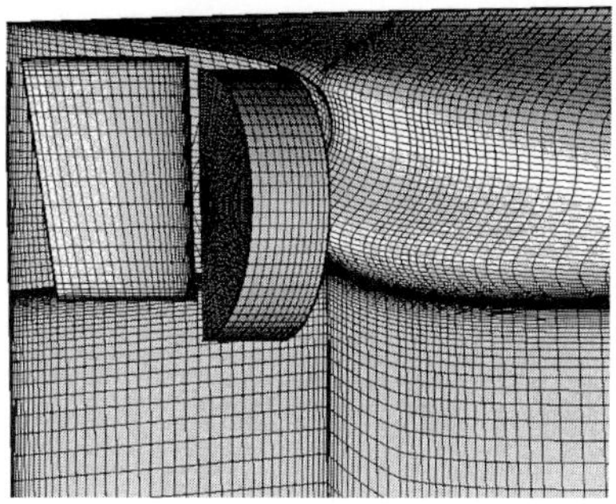

Figure3: Numerical grid for self-propulsion computation

cross flow vectors at propeller plane. Due to the propeller effect, axial velocity is obviously accelerated inside the propeller diameter. Besides, the cross flow rotates clockwise, which is corresponding to the propeller rotation direction. In reality, the flow at propeller with low axial velocity obtains more flow acceleration through rotating propeller. But it does not reflected in the numerical computation because of the axisymmetric assumption adopted in the present "equivalent body force" theory. Fig.6 shows the axial velocity distribution at propeller plane for different radius. Except the region close to the hub, axial velocity is larger than 80% ship speed for most propeller disk region. Table 3 summarizes the measured and calculated self-propulsion characteristics. Because the potential propeller code PSF2 can give quite good hydrodynamic characteristics prediction for propeller performance, the quality of self-propulsion prediction mainly depends on the prediction of ship total resistance with propeller and rudder installed and the effective wake. Although UVW gives quite accurate resistance prediction without propeller effect, it gives 4% resistance over-prediction than the measurement for propeller installed. After the force balance process, it leads to a smaller advance coefficient J , based on model ship speed, and hence larger thrust coefficient K_T and torque coefficient K_Q. Besides UVW seems too diffusive for wake prediction including propeller effect, and a larger effective wake fraction is obtained. The prediction of self-propulsion characteristics for propeller (i.e. t, J, K_T and K_Q) is relatively large, but the prediction accuracy of global system characteristics (i.e. η_H and η_O) is almost not influenced. The improvements of prediction accuracy for the effective wake (not only the wake fraction value but also the velocity distribution) and the total resistance are all necessary for obtaining more precise results for model scale simulation, and finally for the full scale prediction.

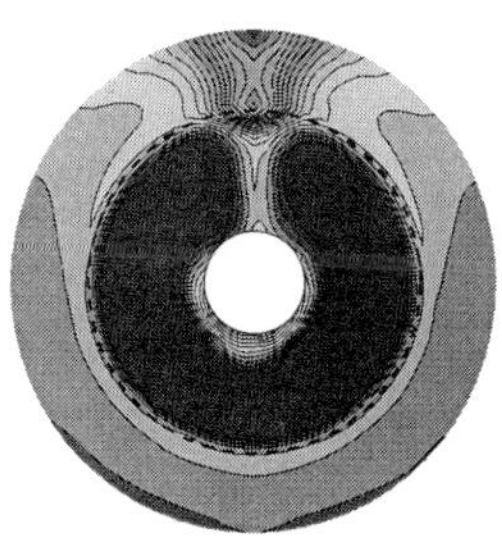

Figure 4: Predicted wake distribution at propeller plane

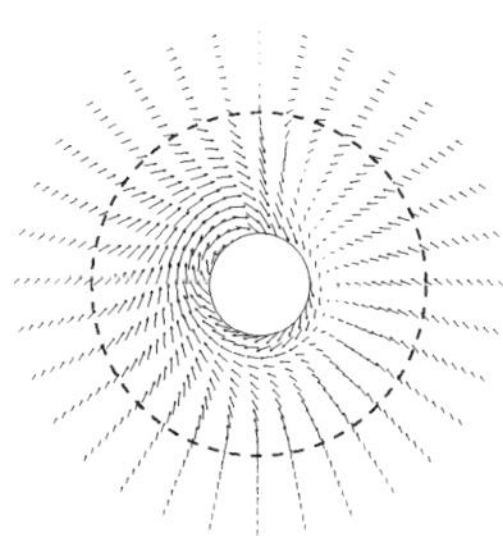

Figure 5: Predicted cross flow vectors at propeller plane

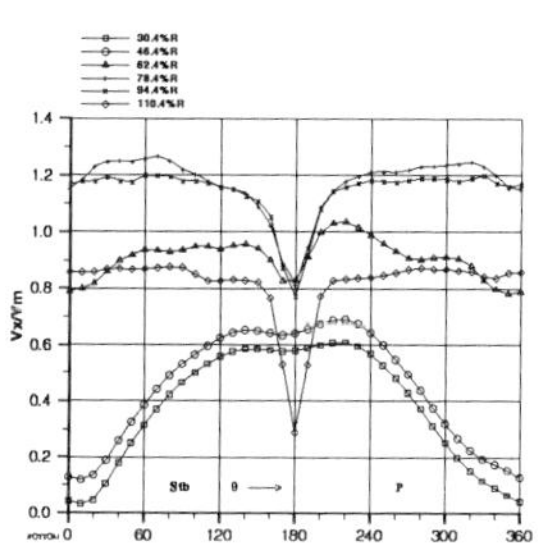

Figure 6: Axial velocity distribution at different propeller radius

TABLE 3

MEASURED AND CALCULATED SELF-PROPULSION CHARACTERISTICS OF THE STUDIED SHIP

With Prop.	V_m(m/s)	t	w	J	K_T	K_Q•10	C_T•10^3	η_H	η_O
Measured	1.956	0.182	0.284	0.8523	0.1753	0.261	4.91	1.142	0.652
Calculated	1.956	0.232	0.303	0.7256	0.1991	0.301	5.10	1.102	0.645
Error(%)	-	27.4	6.69	14.9	13.6	15.3	3.87	3.50	1.07

4 CONCLUDING REMARK

A computational method for simulating the self-propulsion test has been presented in this paper. The scheme is based on the balance of propeller thrust obtained by a potential flow method, and ship resistance calculated by a viscous flow approach. Through the validation of calculated results for a container hull form, the interaction between the hull, the rudder and the propeller is investigated and most features of wake field are captured by the present approach. Although there are rooms for further improvements, this initial work has demonstrated some encouraging results and shown to be a promising method in evaluating the self-propulsion characteristics of ships for providing important information for the propeller design process.

Acknowledgements

This research was funded by the Ministry of Economic Affairs under Grant Number 90EC2A- 170221.

References

Baldwin, B.S., Lomax, H. (1978). Thin-layer Approximation and Algebraic Model for Separated Turbulent Flows, *AIAA Paper*. **No.78-257**

Chau, S.W., Chou, S.K. (1998). Composite Grid Approach for Solving N-S Equations: Part 2, *Proceeding of the 12th National Conference on Naval Architecture and Marine Engineering*, 605-614.

Chau, S.W., Chou, S.K., Lu, C.Y.(2000). Application of Composite Grid Approach in Computing Flow around Ship with Appendage, *Trans. of the Aeronautical and Astronautical Society of the Republic of China*. **32:1**, 81-87

Chou, S.K. (1998). Computation of Ship Flow by Free Surface RANS Code, *Proceeding of 11th National Conference on Naval Architecture and Marine Engineering*, 274-284

Chou, S.W., Chau, S.K., Chen, W.C., Hsin, C.Y. (2000). Computations of Viscous Flow around Commercial Hull Forms with Free Surface or Propeller Effect, *Proceeding of Gothenburg 2000 - A Workshop on Numerical Ship Hydrodynamics*

Chorin, A.J. (1967). A Numerical Method for Solving Incompressible Viscous Flow. *Journal of Computational Physics*. **2**, 12-26

Greeley, D.S. and Kerwin, J.E. (1982). Numerical Methods for Propeller Design and Analysis in Steady Flow, *SNAME Transactions*. **90**

Jameson, W. S., Turkel, E. (1981). Numerical Solution of the Euler Equations by Finite Volume Methods Using Runge-Kutta Time Stepping Schemes. *AIAA paper*. **No.81-1259**

Hsin, C.Y., Tzeng, I.W., Chang, C.Y. (2000). Propeller Analysis and Design Using a Coupled Viscous/Potential Flow Method. *Proceeding of the 4th International Conference on Hydrodynamics*

Huang, T. T., Wang, H.T., Santelli, N., Groves, N.C.(1976). Propeller/Stern Boundary Layer Interaction on Axisymmetric Bodies:Theory and Experiment. *Technical Report DTNSRDC 76-0113*

Jaw, S.Y., Chou, S.K. (1994). Prediction of An Uniform Flow past HSVA Tanker with Two-layer k-ε turbulence Model, *Proceeding of CFD Workshop TOKYO 1994* **1**, 304-310

Johnson, D.A., King, L.S. (1985). A Mathematically Simple Turbulence Closure Model for Attached and Separated Turbulent Boundary Layer. *AIAA J.* **23**, 1684-1692.

Kerwin, J.E., Keenan, D.P, Black, S.D.B (1994). A coupled viscous/potential flow design method for wake-adapted, multi-stage, ducted propulsors using generalized geometry. *SNAME Transactions.* **102**, 23-56.
Lu, C.Y., Chou, S.K. (1994). A Computational Method for Calculating Ship Wave Resistance, *Proceeding of CFD Workshop TOKYO 1994.* **1**, 135-139
Schetz, J.A., Favin, S. (1977). Numerical solution for the near wake of a body with Propeller. *Journal of Hydronautics.* **11:4**, 136-141
Stern F.S., Kim H.T., Patel V.C. and Chen H.C. (1988). A viscous flow approach to the computation of propeller-hull interaction, *Journal of Ship Research.* **32:4**, 246 - 262
Stern, F.S., Kim, H.T., Patel, V.C. and Chen, H.C. (1988). Computation of Viscous Flow Around Propeller-Shaft Configuration, *Journal of Ship Research.* **32:4**, 263 - 284
Stern, F.S., Kim, H.T., Zhang, D.H., Toda, Y., Kerwin, J.E., Jessup, S. Jessup. (1994). Computation of Viscous Flow Around Propeller-Body Configurations : Series 60 Cb=0.6 Ship Model. *Journal of Ship Research.* **38:2**, 137-157

Practical Design of Ships and Other Floating Structures
You-Sheng Wu, Wei-Cheng Cui and Guo-Jun Zhou (Eds)

NUMERICAL COMPUTATION OF SHIP'S EFFECTIVE WAKE AND ITS VALIDATION IN LARGE CAVITATION TUNNEL

*J.W. Park, *J.J. Kim, *D.S. Kong, **J.M. Lew

*Samsung Heavy Industries, Shipbuilding & Plant research institute
**Chungnam National University, Department of Naval Architecture and Ocean Engineering, Korea

ABSTRACT

To presume more accurate stern flow field at propeller working condition for practical hull form, a series of velocity filed measurements were performed in large cavitation tunnel with LDV. Stern wake field around propeller plane was scanned in propeller working/not working condition about the two large container vessel. Additionally, tunnel wall effect and Reynolds number effect on velocity field were also investigated. To verify the possibility of application currently proposed theory to actual design of propeller & hull form design, computation results of potential solver with simple vortex ring theory and RANS solver with body force distribution method were compared with experimental results. The computational result of simplified vortex ring theory showed good agreement with experimental data and computation with RANS equation also showed reasonable agreement with measurement.

KEYWORD

Ship, Propeller, Wake, Interaction, Effective velocity, LDV, Vortex ring, RANS, Body force,

1 INTRODUCTION

The accurate prediction of effective velocity distribution around propeller is a crucial factor for the design of wake-adaptive propeller and stern hull form to obtain good propulsive performance. And total velocity data around propeller is the essential part in prediction of effective wake. Because it is the only practicable data in experiments, total velocity is used as verification of computational approach. But experimental data in propeller working condition are very few [2], especially for practical hull form. In this paper showed the experimental measurement results of practical container hull form. And with that, verified two computational approach which can be used in initial propeller/hull design stage.

2 EXPERIMENTS

Two models for large size container vessels were used. As the body plans shown in Figure 1, hull forms were similar to each other. Each hull had unique designed propeller. Except expanded area ratio, those propellers were also alike. Principle dimension of hull and propeller were shown in TABLE 1. The stern wake field was measured in large cavitation tunnel SCAT (Samsung CAvitation Tunnel) by LDV (Laser Doppler Velocimetor). On two forward plane (-0.40R, -0.28R) and two backward plane (+0.28R, +0.40R) from propeller plane, axial velocity was scanned. Each plane has 7 radial positions and 35 angular positions, hence a total of 255 measuring points. The measurement was performed in four setting condition (with and without propeller, rudder), But the rudder effect on wake filed was out of this paper scope, test results related with rudder were omitted. Since the laser beam radiated from bottom of the tunnel wall, there was a blank zone in the shadow of shaft. Simplified system diagram was showed in Figure 2. All measurement was performed in non-cavitating condition and the same tunnel velocity (6.5m/s) except the test for comparison with that of towing tank. Details were tabulated as TABLE2.

TABLE 1

PRINCIPLE DIMENSION OF HULL & PROPELLER

	Container A	Container B
L/B	6.916	6.671
B/T	3.170	3.170
Cb	0.663	0.656
(P/D)mean	0.948	0.935
Ae/Ao	0.943	1.005

TABLE 2

TEST CONSITION & MEASURING POSITION

Hull	Propeller	Position					V_T (m/s)
		- 0.40R*	- 0.28R	0.00R	+ 0.28R	+ 0.40R	
Container A	-	O	O	O	O	O	6.50
	O	O	O	-	O	O	6.50
Container B	-	-	-	O	-	-	2.10
	-	O	O	O	O	O	6.50
	O	O	O	-	O	O	6.50

* Stern direction positive

3 EXPERIMENTAL RESULTS AND DISCUSSION

To verify the tunnel wall effect, nominal velocity in cavitation tunnel was compared with towing tank result measured at same speed. The flow acceleration by the rigid plate as a substitution for free surface, appeared in the upper part of wake plane, but they showed good similarity, This comparison was shown in Figure 3. Reynolds number effect due to velocity difference, as compared in Figure 4 showed the same tendency with that performed in other large cavitation tunnel [1].

Figure 5 displayed wake field with and without propeller. It was looking upstream direction. Figures were non-dimensionalized value by reference tunnel velocity and dashed line stood for propeller diameter. Acquiring velocity information of one point needed 40~60 seconds, so all of data were fully time-averaged. Some of measured data were printed in Figure 5. The results of ship A and B were very similar, so this paper showed only results for ship B.

In propeller not working condition, (first row in Figure 5) the flow passing by ship showed gradually increased velocity in lower part. In propeller working condition (second row in Figure 5) due to propeller operating, velocity was accelerated at propeller upstream region. The difference between portside and starboard existed but that was not so much. However at propeller downstream velocity was accelerated about 0.3~0.4 and difference between portside and starboard is clearly showed within propeller disc. Further more, It showed more circumferencial smoothness in slipstream region. And, as expected, there was relatively small deceleration in outside of propeller disc.

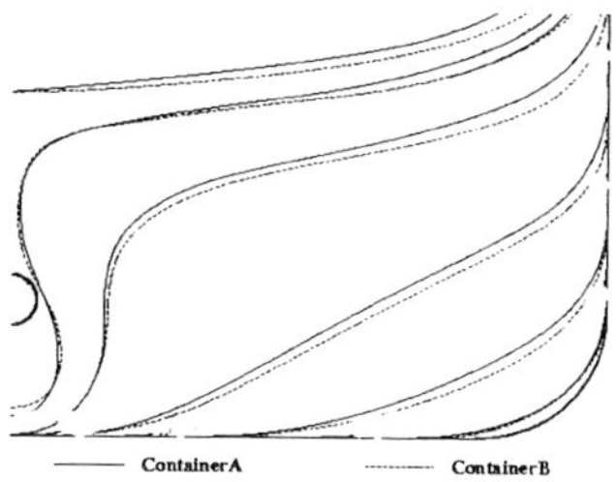

Figure 1 : Body plan for each hull

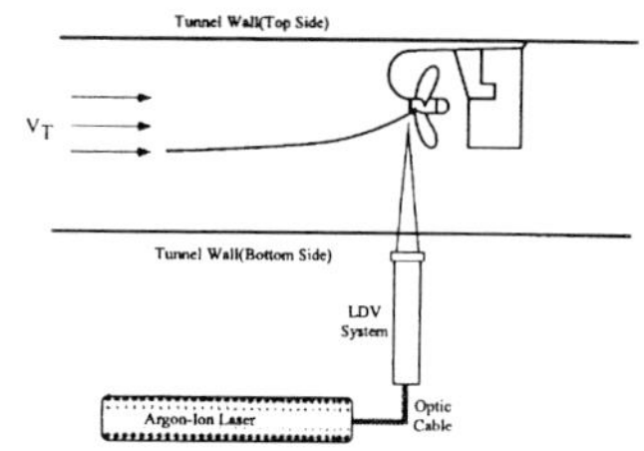

Figure 2 : Measurement system in SCAT

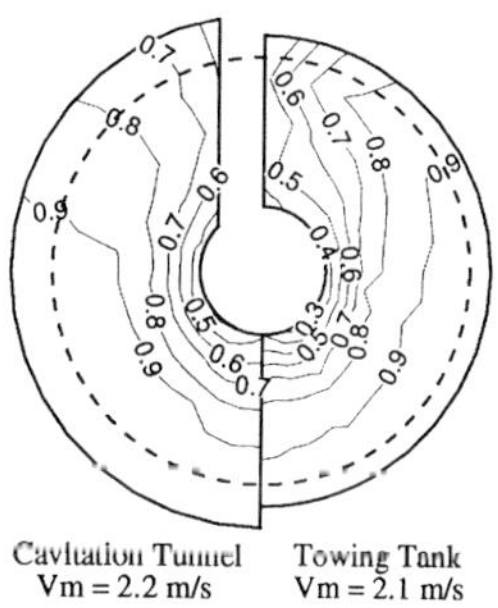

Figure 3 : Comparison CT with TT

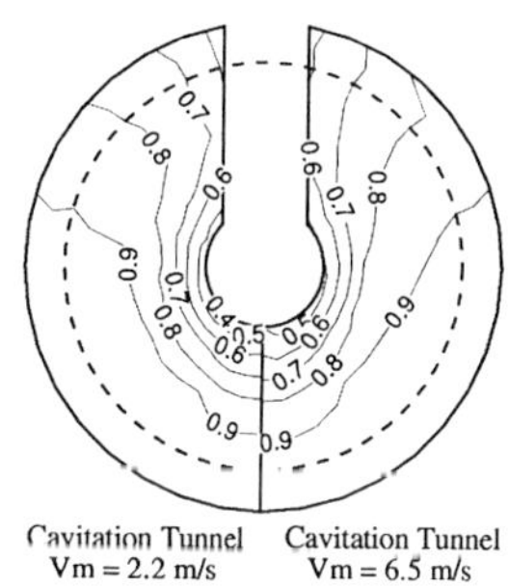

Figure 4 : Reynolds number effect due to velocity difference

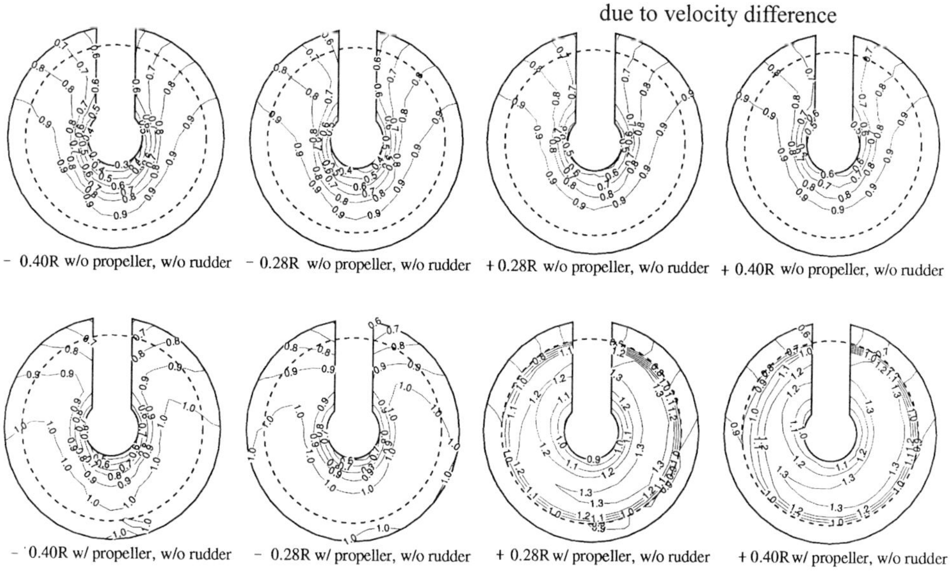

Figure 5 : Wake field in propeller working / not working condition

4 OUTLINE OF NUMERICAL METHODS

4.1 Effective Velocity Computation by Simplified Vortex Ring Theory

According to ITTC propulsor committee [2] Effective velocity u_e, which calculated from the interaction among hull, propeller and rudder, is defined as equation (1).

$$\underline{u}_e(x) \equiv \underline{u}_t(x) + \underline{u}_i(x) \qquad (1)$$

Here u_e, u_t, and u_i, are effective velocity, total velocity, and induced velocity by propeller respectively. With assumption of inviscid and incompressible flow, effective velocity can be expressed as following triple integral form.

$$\underline{u}_e(x) = \iiint_V \underline{w}_Q \times \nabla P\left(-1/_{4r\pi}\right) dV + \underline{U} \qquad (2)$$

To solve the equation (2) the vorticity w_Q has to be known beforehand. But it varies according to propeller action, so it can be initialized by nominal wake distribution and requires iterative computation procedure to obtain converged effective velocity distribution

In previous study, C.S.Lee and D.S.kong et al [3] have computed effective velocity distribution using slipstream contraction model, and iteration procedure with streamline tracing method. The former compared the computation results with measurement result for axi-symmetric body, the later compared with towing tank results based on thrust identity principle.

To investigate the possibility for the application of this methodology to propeller design, total velocity measurement results of large container ships were compared with computation results. Computation was made with assumption of steady and circumferencial mean velocity basis. Detailed numerical procedure and calculation example were described in [4]

4.2 RANS Solver with Body Force

In this approach, the propeller was replaced with an actuator disk and its effect reflected as body forces (axial and tangential direction) which were obtained by same method described in reference [5].

As an input of propeller performance program, effective velocity distribution was provided at the propeller plane. Since effective velocity term is directly related to body force distribution of an actuator disk, nominal velocity distribution without propeller should be simulated by RANS(Reynolds Averaged Navier-Stokes) equation to provide a input for propeller performance computation at first stage. This is the major difference of two methods. As an initial input data, vortex ring theory uses given nominal velocity distribution and RANS solve uses computed nominal velocity.

In order to obtain converged solution, iteration procedure was also required and the flow chart of computation was show Figure 9. During the computation, propeller induced velocity and body force distribution were simulated by propeller unsteady performance code (unsteady V.L.M method).

The viscous flow method solves the incompressible three-dimensional continuity and RANS equations in conjunction with the pressure Poissson equation and the modified Baldwin-Lomax turbulence model for mean velocity components.

The governing equations were expressed

$$\frac{\partial u_k}{\partial x_k} = 0 \qquad (4)$$

$$\frac{\partial u_i}{\partial t} + u_j \frac{\partial u_i}{\partial x_j} = -\frac{\partial}{\partial x_i}\widehat{P} + \frac{1}{R_{eff}}\nabla^2 u_i + \frac{\partial v_t}{\partial x_j}\left(\frac{\partial u_j}{\partial x_i} + \frac{\partial u_i}{\partial x_j}\right) + Fb_i \qquad (5)$$

Where, $$R_{eff} = \frac{1}{\frac{1}{\mathrm{Re}} + v_t}$$

And ***Fb*** ,***Re*** and ν_t means body force , Reynolds number and eddy viscosity, respectively.
The present method employed the O-H type non-staggered grid system and used second order finite differences for the spatial and four-stage Runge-Kutta time stepping scheme for the temporal discretizations. The numerical analysis of propeller-hull interaction was carried out for two practical container ships. The corresponding Reynolds number was 1.0e7, which was similar that of towing tank test. The computational domain extended one ship length L downstream and 0.6L in both depth and width. Then, the inlet of computation domain located mid ship.

5 NUMERICAL RESULTS AND DISCUSSION

Total velocity computed with two methods compared experiment at propeller upstream and down stream position. Results of simple vortex ring method were based on steady computation with circumferencial mean values. For the brief introduction of numerical modeling example, discrete vortex ring in transition area and propeller induced velocity with consideration of interaction effect were shown in Figure 6. Total velocity comparison at propeller upstream and down stream condition with/without propeller shown in Figure 7,8. Although simplified vortex ring theory was applied, the computation results showed good agreement with measured results even in propeller working condition. Figure 10 shows axial body force distributions which were obtained propeller performance code with computed nominal wake. In RANS solver, effective velocity was converged with four iterations. Velocity contours were shown Figure 11 and 12 for two locations. Size and location of axial total velocity contours were reasonable compared with measured data in Figure 5. In Figure 13, circumferencial mean axial velocity profiles were compared with experimental data at two difference locations.

6 CONCLUDING REMARKS

The LDV measurement around propeller in propeller working condition provided valuable information to study hull-propeller interaction and more accurate wake–adapted propeller design. Since all measured data were averaged value, that could not show phase or velocity fluctuation caused by rotation of propeller blades. Tangential and radial velocity field measurement and phase measurement would be further study topic. It would be also important to the study of unsteady behavior in interaction study. Although simplified vortex ring theory has critical limitation of proceeding measured or predicted nominal velocity, it's results showed good agreement with measured data. As alternative approach at initial design stage, viscous approach with body force showed reasonable agreement with experiment. If it calculated in more high Reynolds number such as cavitation tunnel condition, the accuracy would be improved but it need more computation time.

References

[1] Report of Propeller Committee. (1984). *Proceedings of 17th International Towing Tank Conference.*
[2] Cordier S., Briancon-marjollet L., Laurens J.M., Raulo J. (1995). Effect of wake scaling on the prediction of propeller cavitation. DCN Bassin d'Essais des Carenes VAL DE REUIL. *International Symposium on Cavitation.* CAV95
[3] Lee C. S., Kim Y. G. and Ahn. J. W., Interaction Between a Propeller and the Stern Shear Flow The *Proceeding of Korea-Japan Joint Workshop On Hydrodynamic In Ship Design.* Seoul
[4] Kong D. S., Kim Y. G. and Lew L.M. (1998). On the Practical Computation of Propulsion Factors of Ships. *Practical Design of Ships and Mobile Units.*
[5] Stern F., Kim H.T., Zhang D.H., Toda Y., Kerwin J. and S. Jessup. (1994). Computation of Viscous Flow Around Propeller Body Configurations: Series 60 $C_B = 0.6$ Ship model. *Journal of Ship Research.* **38:2**, 137~157

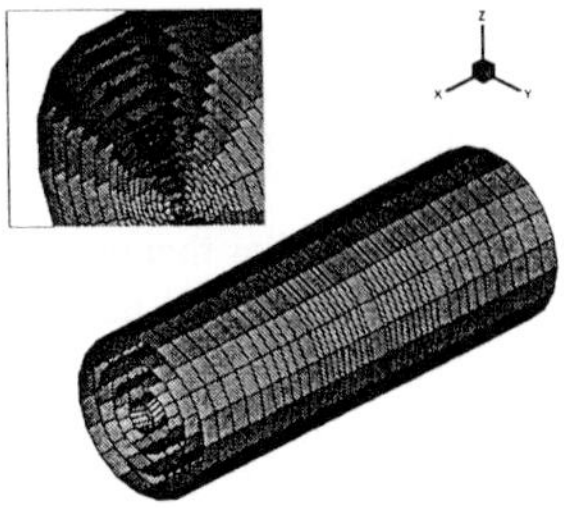

Figure 6 : Contracted vortex tube (vortex ring)

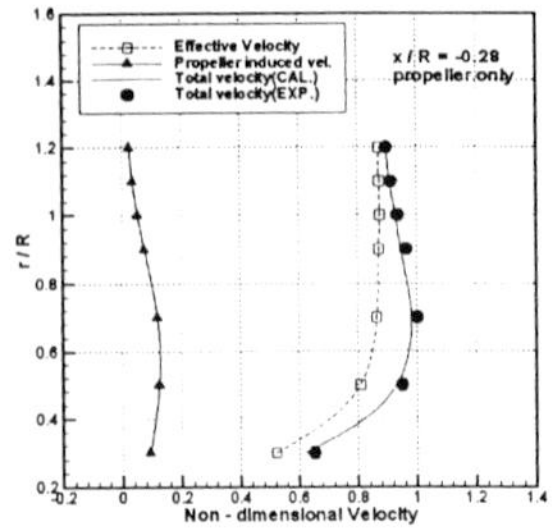

Figure 7 : Total velocity comparison (vertex ring)

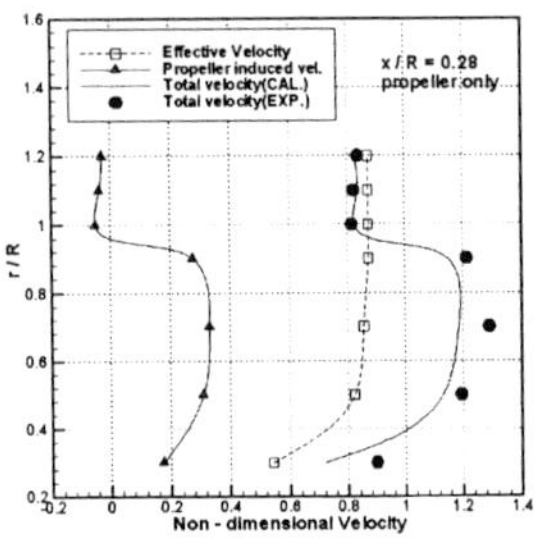

Figure 8 : Total velocity comparison (VLM method)

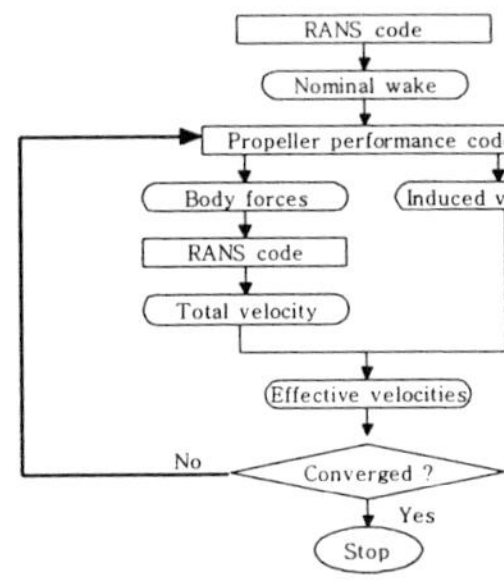

Figure 9 : Process flow for RANS solver

Figure 10 : body force distributions at + 0.0R

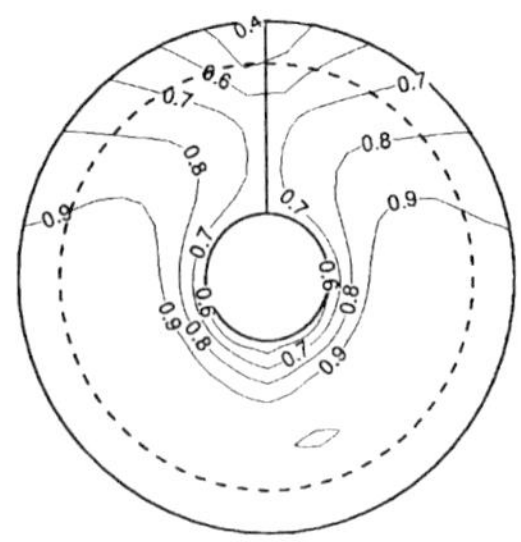

Figure 11 : total velocity contour W/propeller –0.28R (RANS solver)

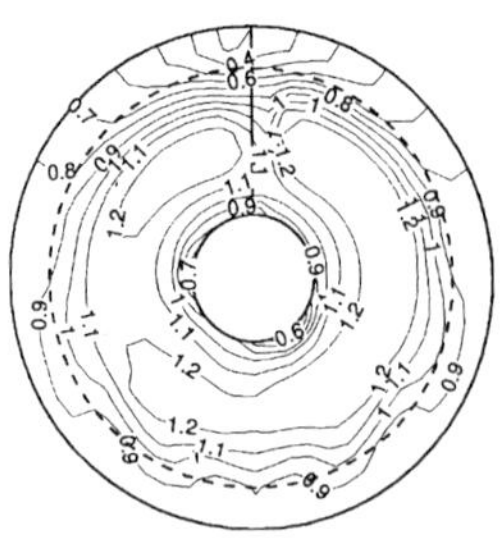

Figure 12 : total velocity contour W/propeller +0.28R (RANS solver)

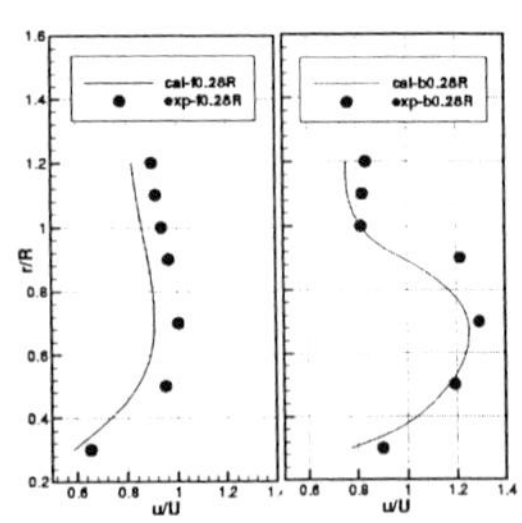

Figure 13 : total velocity comparison W/propeller (RANS solver)

Practical Design of Ships and Other Floating Structures
You-Sheng Wu, Wei-Cheng Cui and Guo-Jun Zhou (Eds)

WAKE FIELDS PREDICTION ON THE PROPELLER PLANE BY NEURAL NETWORK

H. J. Shin and S. M. Hwangbo

Hyundai Maritime Research Institute, Hyundai Heavy Industries Co., Ltd.,
1 JeonHa-dong, Dong-ku, Ulsan, Korea

ABSTRACT

Neural network technique that is emerging as a new knowledge over a wide range of fields nowadays is tried to estimate the wake fields on propeller plane. Neural network is well known as one of prospective and representative analysis method for prediction, classification, diagnosis of real complicated world problem, and it is widely applied even in the engineering fields. Further to its stable and effective system structure, generalization of input training patterns into different classification or categorization in training can offer more systematic treatments of input part and more reliable result.
There is no necessity to use logical programming and it can flexibly handle the incomplete information that is not easy to make a definition clear, because neural network has an ability to learn the knowledge through the external information.
For this study three-dimensional stern hull forms and nominal wake values from a model test are structured as processing elements of input and output layer respectively and Back-Propagation method is used for error correction.

KEYWORDS

Norminal wake, Propeller plane, Neural network, Back-propagation, Statistical analysis

1 INTRODUCTION

Afterbody shape of a merchant vessel has a big influence on the flow fields on the propeller plane and consequently on the self propulsion factors such as wake factor, thrust deduction factor, relative rotative efficiency as well as resistance characteristics. And this flow field on the propeller plane is directly connected with propeller design and governs its cavitation behavior and propulsive efficiency. So high wake peak or inhomogeneous wake patterns should be avoided by a good qualified hull form design. Every hull form designer try to image or infer flow patterns of his designing hull forms in the course of all design procedure because they want to keep a certain limitation of homogeneity of the flow fields within the given restriction.
Of course, CFD(Computational Fluid Dynamics) is one of good candidates to carry out this kind of estimation and simulation, another emerging approach to predict the flow fields or hydrodynamic behavior is based on the previous experimental experience and systematically accumulated data.

Neural network is today expanding it's application ranges and capabilities by adopting new technologies such as fuzzy theory, neural network technique and genetic algorithm in comparison with a previous rigid regression method. Even in ship design fields, new advanced techniques are already being introduced, but they are just in the limited ranges so far. They normally employ artificially formulated parameters (such as L/B ratio, B/d ratio, etc) as input variables and simply draw outputs for the estimation and optimization.
However, we carried out a study to use offset data themselves as input variables that are similar to the input hull form data of CFD computation to predict the wake fields on the propeller plane as output results. The offset datum at a certain point is simply converted to the angle value at the point converged on the vertical line of the propeller plane. By this simple treatment it is evaluated that all desired output points on the propeller plane and input offset points over hull surface are well connected to have physical meaning and it is one of the important factors to get better results.
Input and desired output data are constructed from 57 vessels built actually in HHI and measured wake data in Hyundai Maritime Research Institute including the various kinds and ranges from very large full VLCC to high speed slender container carrier.

2 BACK PROPAGATION

The neural network technique is structured to attempt artificial modeling of the human brain's problem solving capabilities such as calculation, recognition and training. The physical networking transfers a proper signal or stimulus by interaction between biological neurons, which will be referred as a processing element in the artificial system. The processing element(PE) has many input paths and combines, usually by a simple summation. The combined input is then modified by a transfer function. This transfer function can consist of special kinds of mathematical functions like sigmoid, hyperbolic tangent or sine. When input values are connected to each other, the proper weights that correspond to the synaptic strength of neural connection is multiplied and added to reach the desired output processing elements. The output value manipulated by the transfer function is generally passed to the output layer, but some hidden layers can be added between the input and output layer for the improvement of the network accuracy in a certain complex case.
The weights are modified continuously through a number of iteration to have appropriate correlation between input and output processing elements.
Back propagation network, which is also sometimes referred to as a multi-layer perceptron, is currently used as a common neural network paradigm. Back propagation network achieves its generality by the gradient-descent technique that is analogous to an error-minimization process. Error minimization is an attempt to fit a closed-form solution to a set of empirical data points.

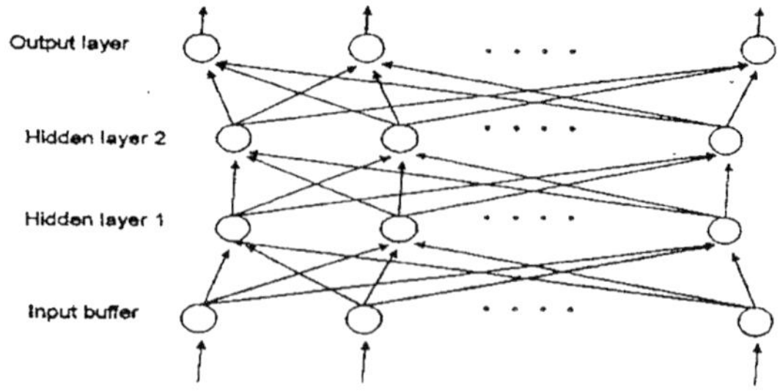

Figure1: Typical back-propagation processing element

A back-propagation element transfer its inputs as follows:

$$x_j^{[s]} = f\left(\sum_i (w_{ji}^{[s]} \cdot x_i^{[s-1]})\right) \qquad [1]$$

$$= f(I_j^{[s]})$$

$x_j^{[s]}$ Current output state of jth neuron in layer s
$w_{ji}^{[s]}$ Weight on connection joining i^{th} neuron in layer(s-1) to j^{th} neuron in layer s
$Ij^{[s]}$ Weighted summation of inputs to j^{th} neuron in layer s

Where f is traditionally, the sigmoid function but can be any differentiable function. The sigmoid function is defined

$$f(z) = \left(1.0 + e^{-z}\right)^{-1} \qquad [2]$$

If some global error function E associated with it which is a differential function of all the connection weights in the network, the critical parameter that is passed back through the layer s is defined by

$$e_j^{[s]} = -\partial E / \partial I_j^{[s]} \qquad [3]$$

Using the chain rule twice in succession gives a relationship between the local error at a particular processing element at level s and

$$e_j^{[s]} = f'(I_j^{[s]}) \cdot \sum_k (e_k^{[s+1]} \cdot w_{kj}^{[s+1]}) \qquad [4]$$

If f is the sigmoid function as defined in [2], then its derivative can be expressed as a simple function of itself as follows:

$$f'(z) = f(z) \cdot (1.0 - f(z)) \qquad [5]$$

Therefore, from[1], equation [4] can be rewritten as

$$e_j^{[s]} = x_j^{[s]} \cdot \left(1.0 - x_j^{[s]}\right) \cdot \sum_k \left(e_k^{[s+1]} \cdot w_{kj}^{[s+1]}\right) \qquad [6]$$

Provided the transfer function is a sigmoid. The summation term in [6] which is used to back-propagate errors is analogous to the summation term in [1] which is used to forward propagate the input through the network. Thus the main mechanism in a back-propagation network is to forward propagate the input to the output layer, determine the error at the output layer, and then propagate the error to the input layer using [6] or more generally [4]. The multiplication of the error by the derivative of the transfer function scales the error.

The aim of the learning process is to minimize the global error E of the system by modifying the weights. Given the current set of weights $w_{ij}^{[s]}$, increment or decrement of them is determined to decrease the global error. This can be done using a gradient descent rule as follows:

$$\Delta w_{ji}^{[s]} = -lcoef \cdot (\partial E / \partial w_{ji}^{[s]}) \qquad [7]$$

where lcoef is a learning coefficient. In other words, it changes each weight according to the size and direction of negative gradient on the error surface. The partial derivatives in [7] can be calculated directly from the local error values discussed in the above, because, by the chain rule and [1]:

$$\begin{aligned} \partial E / w_{ji}^{[s]} &= \left(\partial E / \partial I_j^{[s]}\right) \cdot \left(\partial I_j^{[s]} / \partial w_{ji}^{[s]}\right) \\ &= -e_j^{[s]} \cdot x_i^{[s-1]} \end{aligned} \qquad [8]$$

$$\Delta w_{ji}^{[s]} = -lcoef \cdot e^{[s]} \cdot x_i^{[s-1]} \qquad [9]$$

In the above the existence of a global error function has been assumed without actually specifying it. This function is needed to define the local errors at the output layer so that they can be propagated back through the network. Let o denote the actual output produced by the network with its current set of weights. If a vector $\underline{i}$ is presented at the input edge layer of the network and the desired output d is specified by a teacher, then the global error in achieving from the desired output is given by

$$E = 0.5 \cdot \sum_k \left((d_k - o_k)^2 \right) \qquad [10]$$

$$\begin{aligned} e_k^{(o)} &= -\partial E / \partial I_k^{(o)} \\ &= -\partial E / \partial o_k \cdot \partial o_k / \partial I_k \\ &= (d_k - o_k) \cdot f'(I_k) \end{aligned} \qquad [11]$$

E at the [10] defines the global error of the network for a particular ($\underline{i}$, $\underline{d}$). An overall global error function can be defined as the sum of all the pattern of specific error functions. Then each time a particular ($\underline{i}$, $\underline{d}$) is shown, the back-propagation algorithm modifies the weights to reduce that particular component of the overall error function.

3 HULL FORM DESCRIPTION FOR NEURAL NETWORK

In order to embody a real problem well in the network system, it is most important to represent the problem as the appropriate input data to have physical meaning. Offset data of a hull form themselves have a very native physical meaning because it indicates wideness at a certain position and this has a very close connection with flow speed on the propeller plane. Raw and simple representation may simulate the real problem effectively. It's the main reason why the offset data of hull form are simply used as input parameters in this neural network system.

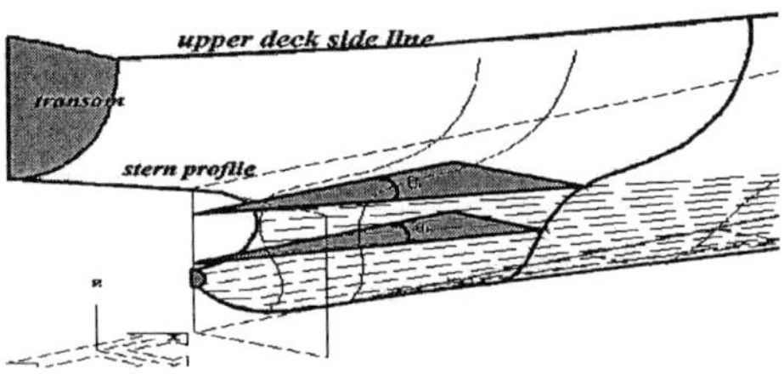

Figure 2 : Offset definition by angle value

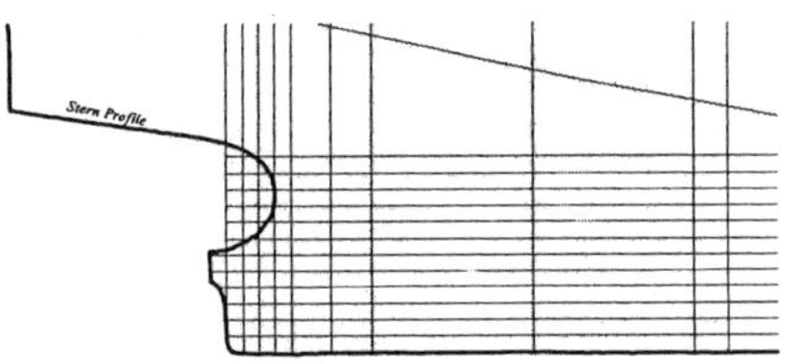

Figure 3 : Section division on the stern hull

But there is still one difficulty to make the network system recognize effectively the three-dimensional offset data even if they mean wideness at a certain point. To overcome this problem, lengthwise and breadthwise values of offset data are converted to an angular value converged horizontally into one vertical line at the propeller plane shown in Figure2. It gives physical meaning that two-dimensional value is expressed by one implicative value.

Furthermore, entire input data are obtained at the particular x and z position because each vessel has different dimensions respectively such as propeller diameter, ship length and breadth. When it is considered wake value is defined as the flow velocity induced on propeller plane, it is a highly meaningful approach to generate newly non-dimensional section and water line by propeller diameter size.

Stern hull form near the propeller plane is sliced up in order to reflect the affect of stern profile shape. Eleven new stations (0.1,0.2,0.3,0.4,0.5,0.75,1,2,3,4,5 x Diameter) and twelve new waterlines (from 0.1 times propeller dia. level to 1.2 times propeller dia. level by each 0.1 times propeller dia. increment) are generated and crossed to get the input data values, and it is good enough to represent afterbody of the hull which affect wake field on the propeller plane. Figure 3 shows a typical division of the stern part.

Measured wake values on the propeller plane are set as a desired output of the system, they are divided by every 10 degrees over a half side of propeller plane and 19 divisions in clockwise direction according to each angular position and radial direction. In addition, five radial divisions from 0.4 r/R to 1.0 r/R by 0.15 r/R steps are added and in consequence total 19 x 5 points of wake values are adopted as output parameters. The wake velocity is of cause non-dimensional value by ship's speed. The 60 hull forms are used for this prediction, and they are widely dispersed by block coefficient, L/B, B/d, and Froude number, which may cover the whole range of merchant vessel.

In general the reliability of the trained neural network system is quite dependent on the number of data group because the correlation between input and output data are more enhanced if more cases are shown to network.

The commercial program "NeuralWorks Professional II/PLUS" developed by "NeuralWare, Inc." is used for prediction of the wake field on the propeller plane in this paper.

4 TRAINING RESULTS

The neural network by back propagation results in 86.5% correlation and 0.056 RMS (Root Mean Square) errors during 50,000 the training number. After the system training, the additional 4 stern hull forms with various kind of dimensions are tested for the confirmation of the level of network learning. An axial wake velocity and it's harmonic wake distribution of a VLCC are shown as one example in the Figure 4, the harmonic wake distribution of a slender high speed large container carrier are shown in the Figure5. The wake field of the stern hull form with skeg under stern bulb is well predicted as shown in the figure 6. It is estimated that the predicted results are so analogous to measured original wake fields that it is good enough to be applied in the design stage before the model tests. Moreover, a very precise depiction of local vortices on the full afterbody hull forms can be stated very prospective application in the stern hull form design. To make a comparison with the wake predictions by CFD, very recent CFD results by KRISO(Korea Research Institute of Ships & Ocean Engineering), which have simulated relatively realistic hook-like vortex shape, is introduced here in Figure 7.

It is not easy to be directly stated which one is better or worse, but Figure 7 show that the present study is unlikely inferior to the CFD in the prediction of wake fields on the propeller plane for stern hull form design quantitatively as well as qualitatively.

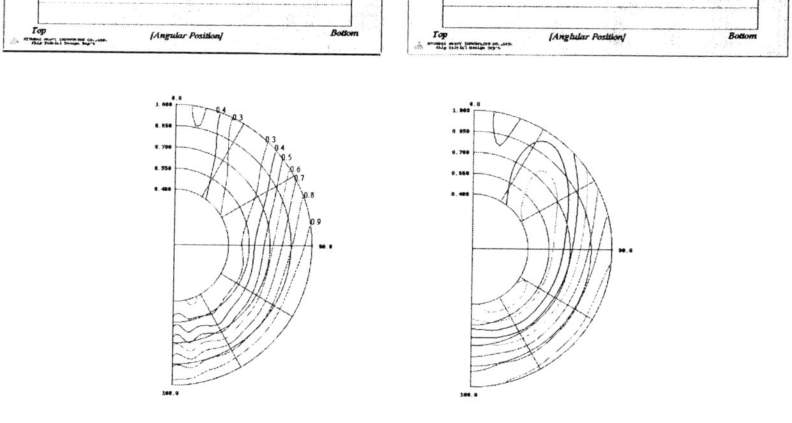

Figure 4: Axial & Harmonic wake distribution of Measured(left) and Predicted(right) wake fields for VLCC

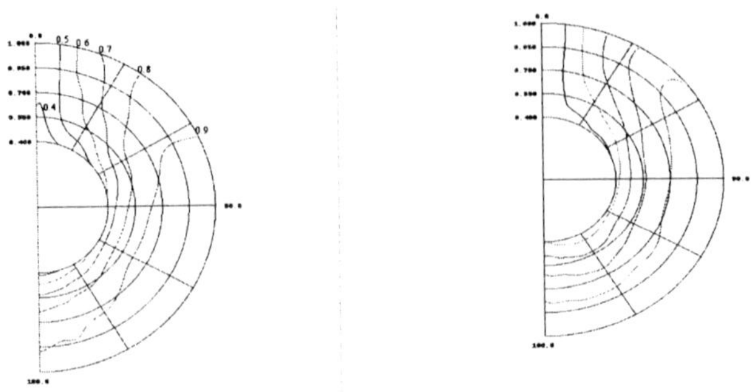

Figure 5: Measured(left) and Predicted(right) wake fields for high speed large container

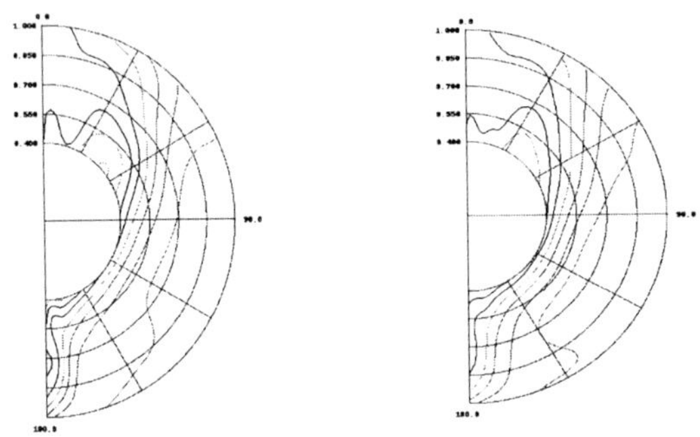

Figure 6: Measured(left) and Predicted(right) wake fields
for stern hull form with skeg under stern bulb

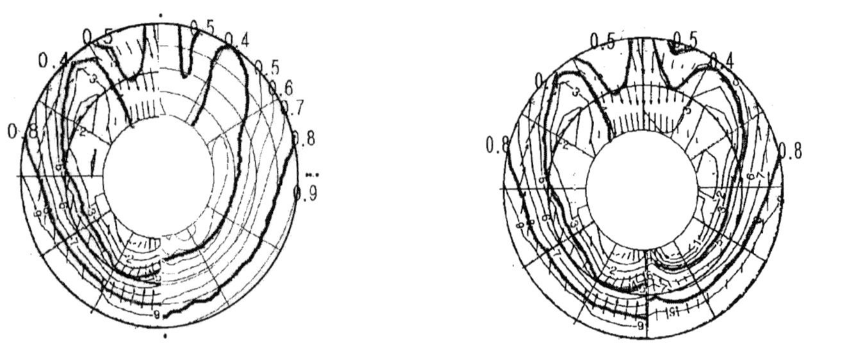

Figure 7: The measured and The predicted by Neural network(left) vs The measured and The calculated by CFD(right)

5 CONCLUSIONS

Simulation of the wake fields by neural network is carried out and prospective results are obtained. Many methods such as fuzzy or genetic algorithm have been applied in a very limited fields of the ship design, the flow fields prediction by back-propagation method of neural network in this paper may give advanced information to hull form designer for evaluation and optimization of stern hull form by prediction of a ship's wake distribution at the initial design stage.

Just offset data is enough to get the available information, and it is not needed to take a lot of time-consuming processing for obtaining target values because once the learning of system is finished additional time consumption is not necessary for the wake field predictions of another new vessels. Furthermore, if the learning data are conglomerated more, the reliability of the system will be upgraded.

Even though the training and basic theory of the system itself have no physical sense of the flow pattern but the simulation results after the training indicate that flow pattern is physically connected with the stern hull form as good as imagination. Particularly, stern hull form near the propeller plane is sliced up in order to reflect the affect of stern profile shape.

If further additional reinforcement through precise adjustment of hidden layer and processing elements can be done, the neural network may produce the much-enhanced results.

This study can be extended to predict and simulate all kinds of hydrodynamic matters. The form factor, wave resistance and all self-propulsion factors can be predicted through the neural network system with basic input parameters of offset data. It is regarded this system will be further developed to the entire optimization routine for the hull form design, which will be contributed to anyone who wants to approach optimization problem.

References

1. S.M.Hwangbo, H.J.Shin and H.J.Kim. (2000). The statistical prediction of wake fields on the propeller plane by neural network using back propagation. *IMCD 2000 Proceeding,* 593-603
2. Neural Ware, Inc. (1996). *Neural Computing*, Technical Publications Group of Neural Ware,Inc.
3. Skpura, D. M. (1996). *Building Neural Networks*, the ACM press
4. Kim. D. S. (1997). *Neural Networks Theory and Applications(I)*, Hi-tech Information
5. Kim. S. Y.,(1998). Neuro-Fuzzy Modeling for Initial Hull Form Design, *Asian Fuzzy Systems Symposium, proceeding*, 585-590
6. Kim. S. Y.,(1995). Hull Form Generation by Using TSK Fuzzy Model, *Fuzzy Logic and its Application, information,* 197
7. Kim. W. J., Kim. D. H. and Van. S. H. (1999). Comparison of turbulence models for the prediction of stern wake of VLCCs with different stern shape, *Proc. of the Annual Autumn Meeting, SNAK, Teajon*, 304-307

Practical Design of Ships and Other Floating Structures
You-Sheng Wu, Wei-Cheng Cui and Guo-Jun Zhou (Eds)

EFFECT OF VERTICAL PRE-SWIRL STATOR VANES ON THE PROPULSION PERFORMANCE OF A 300K CLASS VLCC

Jiman Yang[1], Kihyun Park[1], Kwang Kim[1], Jungchun Suh[1], Hyochul Kim[1],
Seunghee Lee[2], Jungjoong Kim[3] and Hyoungtae Kim[3]

[1]Dept. of Naval Architecture and Ocean Eng., RIMSE, Seoul National Univ., Republic of Korea
[2]Dept. of Naval Architecture and Ocean Eng., Inha Univ., Republic of Korea
[3]Dept. of Naval Architecture and Ocean Eng., Chungnam National Univ., Republic of Korea

ABSTRACT

In the present study, computational investigations are carried out to look at the effects of the two-bladed, pre-swirl, stator vane in simple vertical arrangement, on the propeller-hull interaction of the low speed full ship. Pre-swirl stator vanes were designed for a 300K class VLCC using the surface panel method based on the potential flow theory. In order to design an optimum stator vane, it is essential to know the flow around the vane, which is mainly influenced by the hull wake under the propeller action. A comprehensive viscous-flow method for the computation of propeller-hull interaction could be used to predict the total and effective flows around the stator vane. The stern flow of the model in self-propulsion condition was calculated by the viscous-flow method based on the Reynolds-averaged Navier-Stokes equations, in which the body-force distribution was used to represent the propeller action. The present study provided useful information about the effectiveness of the two-bladed, pre-swirl, stator vane for the improvement of the propulsive efficiency of the low speed full ship and may lead to a practical method for designing optimum pre-swirl vanes when it were combined with experimental methods for validating hydrodynamic performances of the vane-propeller system.

KEYWORDS

Two-bladed vertical pre-swirl stator vane, VLCC, Propeller-hull interaction, computational methods

1 INTRODUCTION

Most of pre-swirl propulsor systems normally comprise some 6~8 bladed stator vanes with a ring duct [Takekuma (1981); Van (1993); Lee (1994); Kim (1994)]. In order to achieve fully the advantage of the systems over other propulsion systems, it is necessary to know in detail local stern flows in design stage in the viewpoint that the inflow angles relative to the stator vanes are different for each radial station. As a consequence, their design is highly dependent on ships to be fitted with the system. Furthermore, the diameters of the stator vanes as well as the ring duct are usually of the order of

10~15% greater than the propeller diameter. The initial investment and operation costs for manufacturing and maintenance are compatible, even more than, with those of single propellers. In the Seoul National University Towing Tank(SNUTT), a research program is under progress for a pre-swirl stator vane propulsor system which can be more easily designed and manufactured than the pre-swirl stator vanes used until now[Yang (2000)]. In this work, we consider a relatively simple pre-swirl stator vane system comprising only two vertically located stator blades of rectangular wing type.

2 POTENTIAL FLOW ANALYSIS OF THE PROPULSOR SYSTEM

The flow analysis for the system was performed by the potential-based panel method [Pyo (1998)]. The interactions between the stator vane and the propeller are addressed by including the induced velocity on inflow velocity in an iterative manner. The potential flow analysis for the propeller can be directly applied for the stator analysis with the onset flow velocity plus the propeller induced velocity at the stator vane plane, since the stator vane can be regarded as a propeller without a rotational speed. The induced velocity is taken in a circumferentially averaged sense. This process is repeated until both the stator and the propeller loadings are converged within an allowance limit. In the loading calculations, the empirical formulas for the drag coefficient were used to take account into viscous effects [Hoerner (1965)].

Due to its initial development, the stator vane was designed for fitting to the propeller of an existing vessel of 'KTTC 300K class VLCC' (see Table 1 and 2 for their principal particulars), although it is necessary to perform the propeller redesign to prevent the propeller RPM-mismatching. The diameter (span) and the inflow angle of the stator vane were determined in such a way that the propulsion efficiency in open-water condition is increased as much as possible. This condition would be reasonable for the present stator vane because it is vertically located at the symmetrical center plane of ships. The propulsion efficiency is predicted with varying the inflow angle from 5° to 15°. Figure 1 shows that the efficiency gain is obtained by compensating the increase of the stator drag with the decrease of the rotational energy losses. Additional gain does not appear furthermore when the angle is too large (say over 12° in this case). Eventually, we take 12° as the inflow angle (pitch angle) of pre-swirl stator vane relative to onset flow. The computed results shown in Figure 2 imply that the effective diameter of the stator vanes will be of order of 0.9D(90% propeller diameter). A similar feature is also found in literature [Yang (2000)], in which the effective diameter of 0.85D is proposed. Neglecting the small difference in efficiency between such two choices, we take 0.85D as the stator diameter with our preference for a smaller one.

For the non-uniform flow (either measured by model tests or calculated by CFD method), we follow the corresponding iterative process used in the open-water condition. The predicted thrust and torque are used as the body force term in Navier-Stokes equations for the CFD calculation. Presumably predicted effective flow is then used as an upgraded onset flow in the potential flow analysis for the propulsor system.

TABLE 1
PRINCIPAL PARTICULARS OF KTTC 300K CLASS VLCC

	Scale Ratio	Design Speed	LBP (m)	LWL (m)	Breadth (m)	Depth (m)	Draft (m)	WSA (m^2)	Volume (m^3)	C_B
Ship	1	15.5 (knots)	320.0	325.5	58.0	30.0	20.8	27320.0	312737.5	0.8101
Model	100	0.7973 (m/s)	3.200	3.255	0.580	0.300	0.208	2.7320	0.3127	0.8101

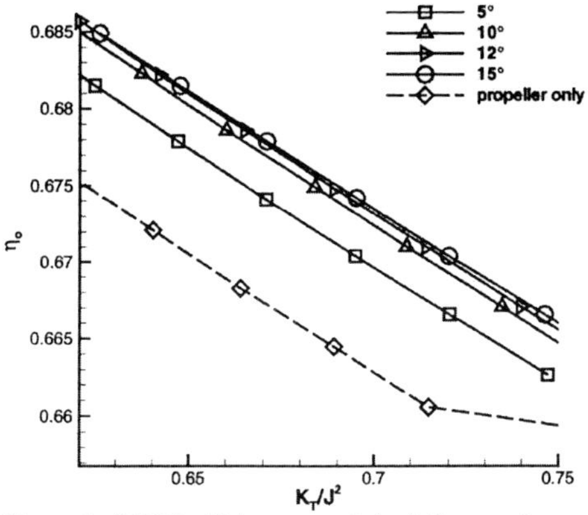

Figure 1 : P.O.W. efficiency .vs. stator inflow angle

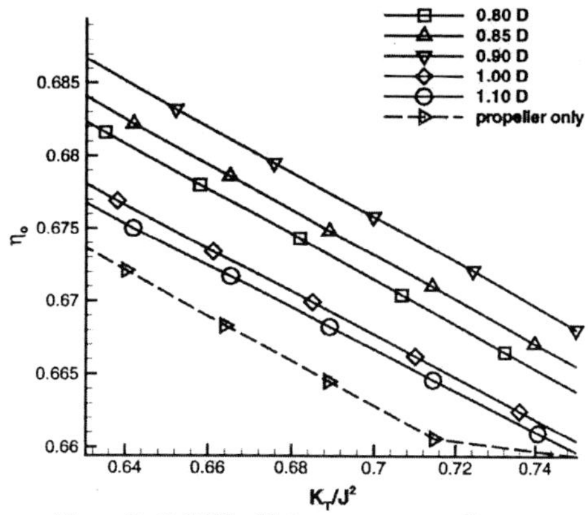

Figure 2 : P.O.W. efficiency .vs. stator diameter

TABLE 2
PRINCIPAL PARTICULARS OF AN EXISTING PROPELLER

D (m)	Blade No.		
9.8	4		
$(P/D)_{ROOT}$	$(P/D)_{0.7R}$	$(P/D)_{TIP}$	$(P/D)_{MEAN}$
0.5765	0.7212	0.6510	0.6896
A_E/A_O	R_H/R	$(C/D)_{0.7R}$	$(T_0/D)_{0.7R}$
0.4250	0.1550	0.2338	0.0156

3 VISCOUS-FLOW COMPUTATION OF PROPELLER-HULL SYSTEM

3.1 Overview of Computational Method

A comprehensive viscous-flow approach [Stern et al. (1988)], in which a numerical method for calculating the viscous flow around the ship's stern had been coupled with a potential flow method for calculating the propeller-performance in an interactive and iterative manner to predict the propeller-hull interaction, was used in the present study. However, there were number of changes in the detailed computational method. The finite analytic method [Patel et al. (1988)] for solutions of the Reynolds-averaged Navier-Stokes (RANS) Equations was replaced with the finite difference method [Sotiropoulos and Abdallah (1992)]. The governing equations are written in Cartesian coordinates and the modified Balwin-Lomax turbulence model [Kim and Kim (1997)] are used with the no-slip wall condition. The RANS equations are discretized using a non-staggered grid and the high-order finite difference method, and solved using a four-stage Runge-Kutta scheme. The constraint of a divergence-free velocity filed is enforced by solving a discretized pressure-Poisson equation to compute the pressure with eliminating the checkerboard problem associated with the use of a non-staggered grid and a MAC-type algorithm is used for velocity-pressure coupling. The equations are solved in unsteady form. However, for present steady-flow applications, time simply serves as an iteration parameter. The computational grid is generated by the use of the three-dimensional GRAPE method [Sorenson (1982)]. In order to locate the grid points over the propeller disk, the grid is generated in two blocks, so that the interface of which becomes the boundary with fixed grid points in the propeller plane. In towing condition, only the half of the whole domain is taken for the computation of the flow around a ship model using the center-plane as the symmetry boundary, while the whole domain is taken for the computation of the flow around the model in self-propulsion condition. For the computation of the viscous flow around a ship model in self-propulsion condition, a body-force distribution is used to represent the propeller action through the addition of body-force terms in the source functions of the momentum equations. The body force is obtained from a propeller-performance program for specified propeller geometry and operating conditions including the effective inflow. Since the body force depends on the effective inflow, which in turn depends on the body force, the complete solution must be obtained iteratively. However, iterative calculations were not performed in the present study, because the computed nominal wake was not accurate enough to justify such iterative calculation procedure to obtain the effective inflow.

3.2 Computational Results

Computations were carried out for predicting the stern and wake flows of a 300K VLCC model in towing as well as in self-propulsion conditions. Calculations were performed for the conditions corresponding to the experiments of the model tests in the towing tank, i.e., $Re = 4.6\times10^6$ and $J_A = 0.47$. Figure 3 shows a partial view of the computational grid used for the viscous-flow calculations. The inlet, exit, and outer boundaries are located at the mid-ship, the ship length downstream from the stern, and 0.6 times the ship length outward in radial direction, respectively. The first grid points off the hull surface are located at about $y^+ = 10$ and 66 axial, 35 radial, and 31 circumferential grid points for the towing condition, or 61 circumferential grid points for self-propulsion condition, were used. The grid for the interpolation of the body-force is also shown in the figure. The inlet boundary conditions were specified using the flat-plate turbulent boundary layer profiles. In Figure 4, the velocity distributions are shown over the propeller disk at the cross plane just upstream of the propeller for the towing and self-propulsion conditions, respectively. The influence of the propeller action was relatively weak at the upstream position. The axial velocity was generally under-predicted so that its volumetric mean value becomes smaller than the experimental one. The cross plane velocity was not predicted well either, especially for the bilge vortices. Therefore, the propeller-performance and consequently the propeller-hull interaction would not be analyzed correctly in the viscous-flow approach of Stern et al. [Stern et al. (1988)] if the computed nominal velocity were used. Figure 5 shows the velocity distributions at the cross plane just downstream of the propeller, respectively, for the towing and self-propulsion conditions. The influence of the propeller action became strong at this downstream location and the resultant propeller slipstream was developed more at further downstream.

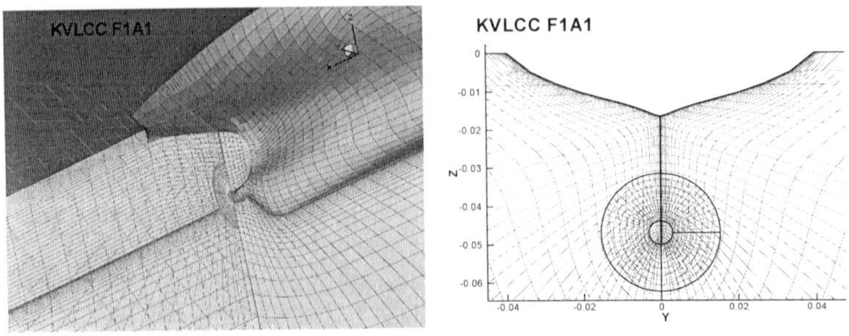

Figure 3 : Viscous-flow computational grid

Figure 6 shows the surface pressure distributions of the ship model for the towing and self-propulsion conditions. The influence of the propeller action appears locally only at the stern region close to the propeller. A little discrepancy is found in the pressure distribution on the starboard and port sides of the hull surface. But presently it is not clear whether the cause of the discrepancy is due to the different propeller effects on each side or just numerical errors.

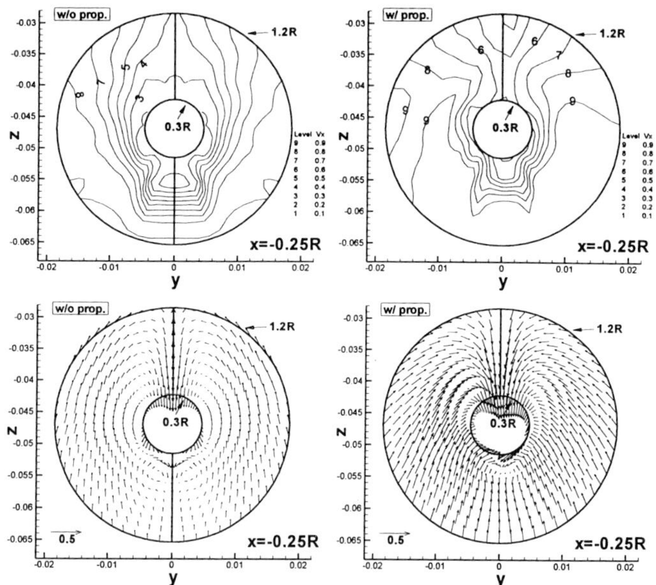

Towing condition Self-propulsion condition

Figure 4 : Velocity distributions over the propeller disk just upstream of the propeller

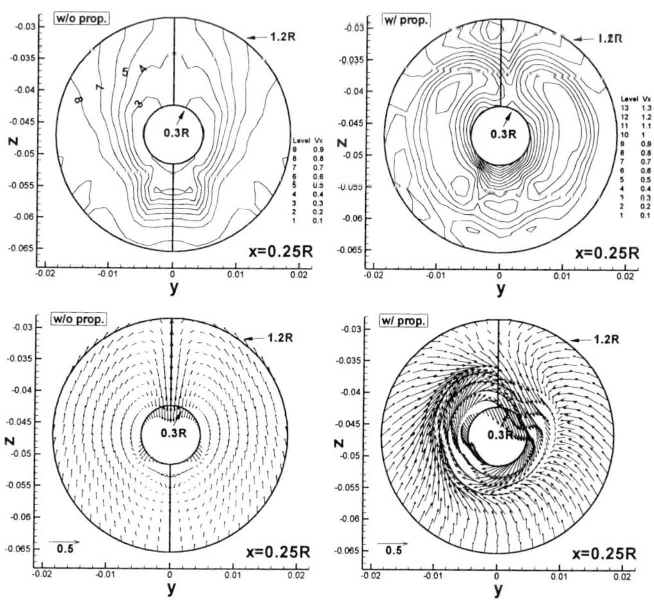

Towing condition Self-propulsion condition

Figure 5 : Velocity distributions over the propeller disk just downstream of the propeller

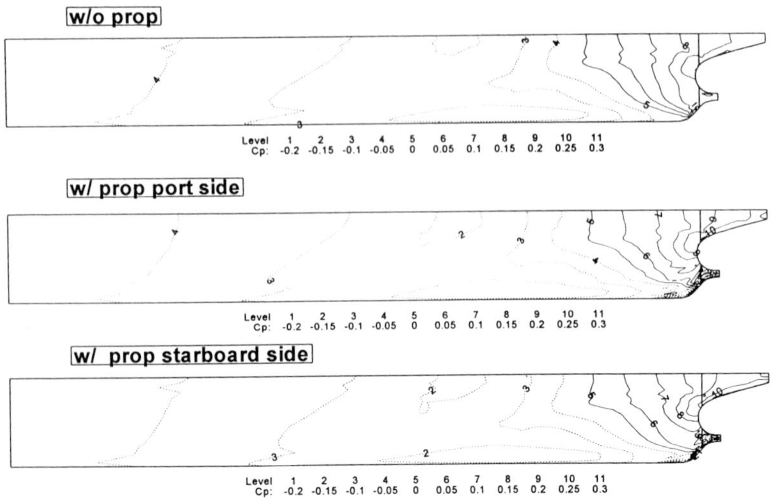

Figure 6 : Pressure distributions on the hull surface

4 PRELIMINARY TEST RESULTS

The preliminary resistance and self-propulsion tests were performed for another 300K class VLCC with the same arrangement of the pre-swirl stator vane [Yang (2000)], (see Table 3). The model propeller of diameter 170mm was used, whose geometry is given in Table 2. The hull form is slightly different from that of the VLCC mentioned in section 2 and 3, but the beneficial effect of the stator vane on the performance was expected to occur. Resistance and self-propulsion tests were carried out at Seoul National University towing tank. Propulsion performance is estimated by the 1978 ITTC performance prediction method and the modified method (B-1 method) proposed by Van [Van (1996)]. Comparison of self-propulsion factors with and without pre-swirl stator vane is given in Table 4. The predicted factors remarkably depend on the prediction method taken. A rigorous method for precise prediction is needed to capture this feature. Meanwhile, due to the pre-swirl stator, the hull resistance leads to about 0.02% increase in EHP and the preswirl flow created by its reaction can be expected to reduce DHP by about 1%.

5 CONCLUDING REMARKS

Computational investigations were carried out to look at the effects of the two-bladed, vertical pre-swirl stator vane on the propeller-hull interaction of the low speed full ship. Pre-swirl stator vanes were designed for a 300K class VLCC using the surface panel method based on the potential flow theory. In addition, the stern flow of the model in self-propulsion condition was calculated by the viscous-flow method based on the Reynolds-averaged Navier-Stokes equations. The present study provided useful information about the effectiveness of the two-bladed, vertical pre-swirl stator vane for the improvement of the propulsive efficiency of the low speed full ship and may lead to a practical method for designing optimum pre-swirl vanes when it is combined with experimental methods for validating hydrodynamic performances of the vane-propeller system. This work was supported by the BK21 Project.

TABLE 3
PRINCIPAL PARTICULARS OF KOMAC 300K CLASS VLCC

	Scale Ratio	Design Speed	LBP (m)	LWL (m)	Breadth (m)	Depth (m)	Draft (m)	WSA (m^2)	Volume (m^3)	C_B
Ship	1	15 (knots)	320.0	325.5	58.0	31.0	20.8	27784	314200	0.814
Model	60	0.9961 (m/s)	5.3333	5.425	0.9667	0.5167	0.3467	7.7177	1.4546	0.814

TABLE 4
COMPARISON OF SELF-PROPULSION FACTORS AT DESIGN SPEED

	t	w_M	w_S	η_R	η_O	η_H	η_D	DHP(HP)
w/o vane ITTC '78	0.206	0.383	0.303	0.975	0.662	1.138	0.735	17514
w/ vane ITTC '78	0.214	0.419	0.323	0.984	0.648	1.160	0.740	17409
w/ vane ITTC 21st B-1	0.214	0.419	0.359	0.984	0.623	1.226	0.751	17144

References

Hoerner S. (1965). *Fluid Dynamics Drag*, Brictown, N.J.

Kim Ki-Sup et al. (1994). A Pre-swirl Stator-Propeller System as a Reliable Energy-Saving Device, *Propeller/Shafting '94 Symposium*, 9.1-9.16.

Kim J.J. and Kim H.T. (1997). Application of Modified Baldwin-Lomax Model to Turbulent Flow of a Double Body Tanker Model, *Proceedings of Annual Spring Meeting SNAK*, 316-319.

Lee Jin-Tae et al. (1994). Development of a Pre-swirl Stator Propulsion System for a 300K VLCC, *Journal of SNAK* **31:1,** 1-13.

Patel V.C. et al. (1988). Ship Stern and Wake Flows: Solutions of Fully-Elliptic Reynolds-averaged Navier-Stokes Equations and Comparisons with Experiments, *Iowa Institute of Hydraulic Research, The University of Iowa, IIHR Report No. 323.*

Pyo S.-W., Kim K. and Suh J.-C. (1998). Steady/Unsteady Analysis of Ducted Propellers by Using a Surface Panel Method, *3rd International Conference on Hydrodynamics.*

Sorenson R.L. (1982). Grid Generation by Elliptic Partial Differential Equations for a Tri-Element Augmentor-Wing Airfoil, *Numerical Grid Generation (ed. Thomson. J.F.)*

Sotiropoulos F. and Abdallah S. (1992). A Primitive Variable Method for the Solution of Three-Dimensional Incompressible Viscous Flows, *Journal of Computational Physics* **103**.

Stern F. et al. (1988). A Viscous Flow Approach to the Computation of Propeller-Hull Interaction, *Journal of Ship Research*, **32:4,** 246-262.

Takekuma et al. (1981). Development of Reaction Fin as a Device for Improvement of Propulsive Performance of High Block Coefficient Ships, *Journal of SNAJ* **150.**

Van Suak-Ho et al. (1993). Some Remarks on the Powering Performance Prediction Method for a Ship Equipped with a Preswirl Stator-Propeller System, *20th International Towing Tank Conference* **2,** 81-84.

Van Suak-Ho et al. (1996). A Powering Performance Extrapolation Method for a Preswirl Stator-Propeller System, *21st International Towing Tank Conference* **2,** 192-198.

Yang Jiman et al. (2000). Effect of Pre-swirl Stator Vane on the Propeller Hull Interaction of a Full Ship, *Proceedings of Annual Autumn Meeting SNAK*, 188-191.

Practical Design of Ships and Other Floating Structures
You-Sheng Wu, Wei-Cheng Cui and Guo-Jun Zhou (Eds)

DEVELOPMENT AND EXPERIMENTAL STUDY OF A NOVEL SUBMARINE GUIDE VANE PROPELLER SYSTEM

Hui-Zhi Yao Hong-Cui Shen

China Ship Scientific Research Center, Wuxi, Jiangsu 214082 China

ABSTRACT

Submarine Guide Vane Propeller is a new device developed by CSSRC to improve powering performance and reduce propeller-radiated noise of submarine. It is composed of propeller and guide vane, the latter is installed at an appropriate position between stern appendages and propeller. By interactions among stern appendages, main hull and the guide vane, the inflow of propeller is adjusted, so that the energy losses in propeller and the unsteady forces and radiated noise can be reduced. In the research it is shown that the new device can increase submarine maximum speed by 0.3~0.5 knot and reduce propeller radiated noise by 2~4 dB.

KEYWORDS

Submarine, Guide vane, Powering performance, Noise, Propeller, Propulsion

1 INTRODUCTION

In comparison with the total cost of a submarine, the hydrodynamic part is very small, however it plays an important role for the total performance of the submarine. For example, the resistance must be as small as possible, which is important for the endurance of the submarine. It must be manoeuvrable, the propeller must be as silent as possible and have a high efficiency, and the radiated noise must be as small as possible etc.

The development of submarine indicates increased maximum speed in submerged condition on the one hand and strong demands for low noise levels at operating speed on the other hand. From hydrodynamic point of view, these two demands call for a more careful design of the submarine hull form and propeller.

With the research of submarine design and performance, improvements have been made in propulsion devices and in hull form refinements. At the same time, several kinds of stern arrangements and unconventional propulsors have been investigated all over the world, such as the "X" type stern wings and pumpjet propulsion system and so on.

As simple to construct, easy to manufacture, convenient to install, Guide vane (pre-swirl stator) has been considered as a promising energy saving device for ships during the last decade, this energy saveing device has also been used to prevent cavitation erosion.

In 1991, CSSRC investigated the potential gain in propulsive efficiency by adopting such a device in submarine design, energy saving rates over 6.5 percent were achieved. However, it required a new propeller designed for the proper horsepower/rpm relationship. In addition, noise radiation reduction has become more and more important for up-to-date submarines. In 1995, CSSRC initiated research to develop a new efficiency improving as well as noise reducing device that is suitable for submarine application. This resulted in the new concept of integrated Submarine Guide Vane Propeller(SGVP) system. The SGVP is a novel submarine stern arrangement in which the guide vane is fitted into the submarine hull at a proper position between the stern wings and propeller. It is expected to improve the propulsive performance of full form submarine and to reduce the radiated noise of the propeller.

Then CSSRC carried out long term systematic research in the following years, which consisted of the following contents:
(1) hydrodynamic mechanism of SGVP
(2) flow structure investigation around the guide vane
(3) interaction between the guide vane and the propeller
(4) hydrodynamic performance and radiated noise investigation
(5) hydrodynamic design optimization

Two typical imaginary submarines were selected as the original hull forms on which the research of the SGVP was to be conducted, since they had been intensively investigated and provide very good hydrodynamic performance. After extensive investigations, including a series of model tank tests, a series of cavitation and noise experiments, a series of flow field measurements, as well as a series of physical analyses, an important hydrodynamic breakthrough came in the form of a larger diameter of guide vane and a smaller diameter of propeller system which showed very encouraging performance characteristics. Figure 1 shows the stern arrangement of SGVP.

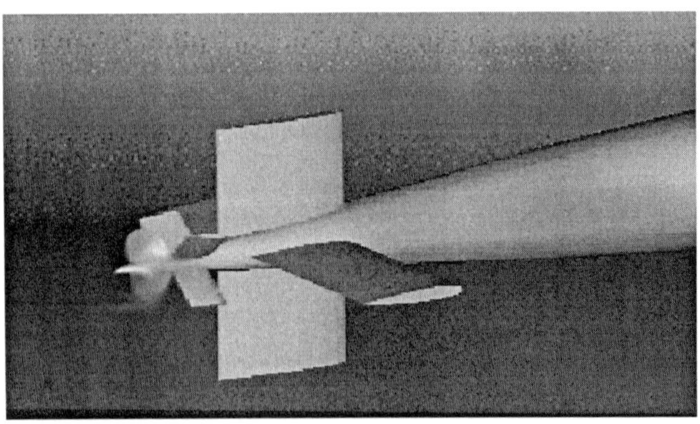

Figure 1: Layout of an imaginary submarine with guide vane

In this paper, the design concept and hydrodynamic features of SGVP are described, followed by the mechanism analyses of this novel system, some conclusions are presented in the end of the paper as well.

2 BASIC DESIGN CONCEPT

As is well-known, guide vane (pre-swirl stator) used as an energy saving device for surface ships has been widely investigated in many ship hydrodynamics institutes. From these researches we can find that it is very difficult to make the stator adapt to the complicated wake of surface ships. And even in this case, energy saving rates of about 4~12 percent could be gained. Then, for an axisymmetric body with appendages as a submarine, the wake is relatively simple, one may think it is easier to achieve improved performance. This is not true of course, in particular when we take low radiated noise of propeller into consideration at the same time.

In our research, in order to improve the powering performance of a submarine, we choose proper section form, suitable size, appropriate pitch angle and blade number of the guide vane. The basic purpose of such arrangement is to produce a swirling flow opposed to the direction of rotation of the propeller, thereby annulling the swirl induced by the propeller and at the same time increasing the relative tangential velocity of the propeller blades. In such a way the rotational kinetic energy losses in the propeller slipstream can be cancelled. Of course, the drag increment due to guide vane must be remained rationally small.

For low radiated noise of propeller, we should consider the interactions among the submarine hull, stern appendages, guide vane and propeller as a whole. Following measures have been taken into account: First, the diameter of guide vane should be larger than that of the propeller so that the tip vortices generated by the guide vane can be moved off the propeller disc. Then the number of guide vane blades should be chosen so that the least common multiples of the number of propeller blades and guide vane blades as well as that of the number of propeller blades and guide vane blades plus stern appendages are large. In our research, 4-blade guide vane was used commonly. Also we must consider the circumferentially relative position between guide vane and stern appendages and guide vane pitch angle. Even more important consideration is to select the proper guide vane longitudinal position between stern appendages and propeller. By suitable arrangement of these parameters the inflow uniformities of the propeller can be improved thus the unsteady blade force and radiated noise can be reduced.

3 INVESTIGATIONS OF THE WAKE CHARACTERISTICS

In order to understand the hydrodynamic mechanism of the SGVP and to optimize its design, a series of wake measurements were conducted in the wind tunnel of CSSRC.

The wakes of a typical submarine with and without guide vane are given here. Figure 2 shows the wake contour at the propeller plane of a full form submarine without guide vane, from it we can find that the flow field at the propeller plane is circumferentially non-uniform, which results in spatial and temporal fluctuations of propeller blade angle of attack. These angle-of-attack fluctuations result in unsteady blade loadings and the generation of noise.

The flow non-uniformities are caused by upstream appendages and the interaction between them and the hull boundary layer. Also this phenomena has been intensively investigated by many researchers in recent years, the flow behind an appendage or appendages is characterized by a complex velocity field having velocity excesses at the inner radii where the propeller operates and velocity deficits at the outer radii.

Figure 3 shows the wake contour at the propeller plane for a full form submarine with guide vane. The flow nonuniformities are significantly decreased in the upper region of the propeller disc compared with the case without the guide vane.
We also investigated the evolution of the flow structure along the aft portion of the model. Figure 4 shows axial velocity contour in a plane before propeller disc. The differences between it and that of the propeller disc can be easily found, these might be caused by the strong viscous effect and the pressure distribution there. So we can adjust the relative position among appendages, guide vane and propeller and other parameters to improve the uniformities of propeller inflow.

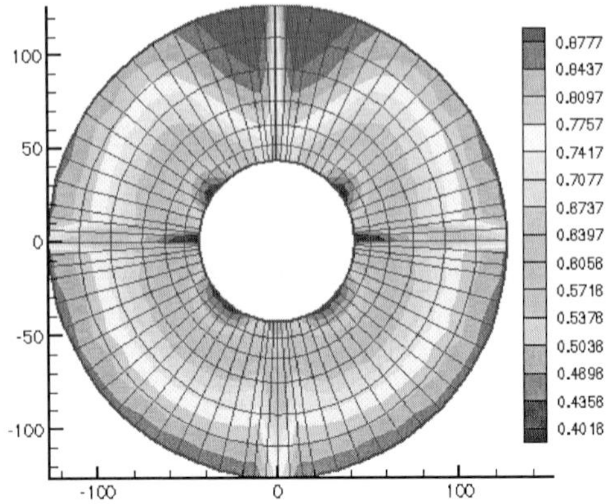

Figure 4 Axial velocity contour in a plane before propeller disc

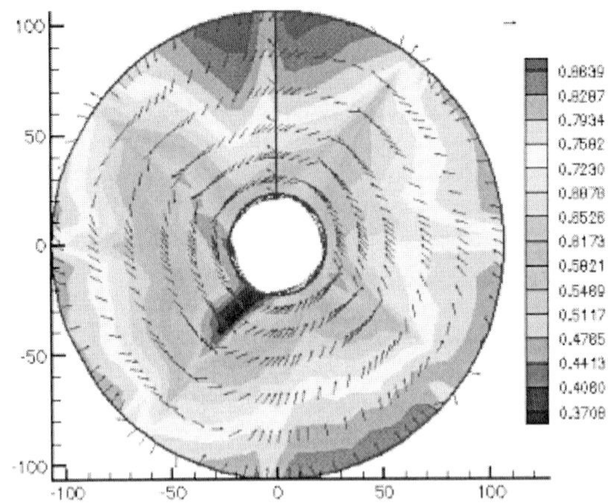

Figure 5 Transverse vectors and axial velocity contour in the propeller plane

3.1 Powering Performance Investigations

As mentioned above, CSSRC carried out SGVP studies on two basic hull forms. For form A with high hull diameter to length radio, we developed 3 guide vanes, then experiments were conducted in the towing tank of CSSRC, which included resistance tests with and without guide vane, open water tests for the propeller and self-propulsion tests with and without guide vane. The geometrical parameters of guide vane and experimental results are shown in Table 1 and Table 2 respectively.

Another series of model experiment of SGVP were conducted on the basis of form B with a low diameter to length ratio of the hull. Also 3 guide vanes were design for the given hull form. The geometrical characteristics of the Guide Vanes and experimental results are also shown in Table 1 and Table 2 .

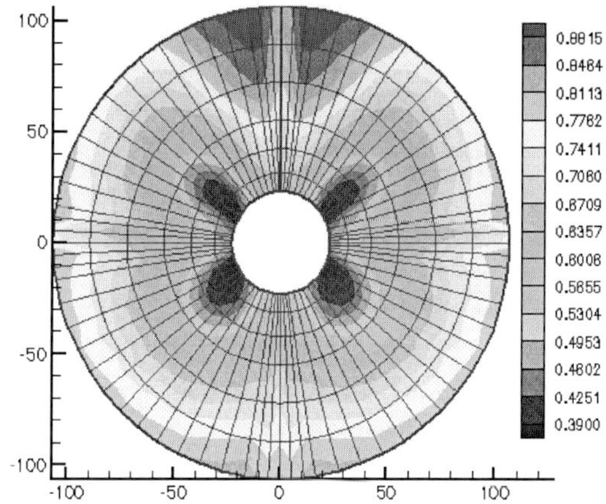

Figure 2 Wake contour at the propeller plane of a full form submarine without guide vane

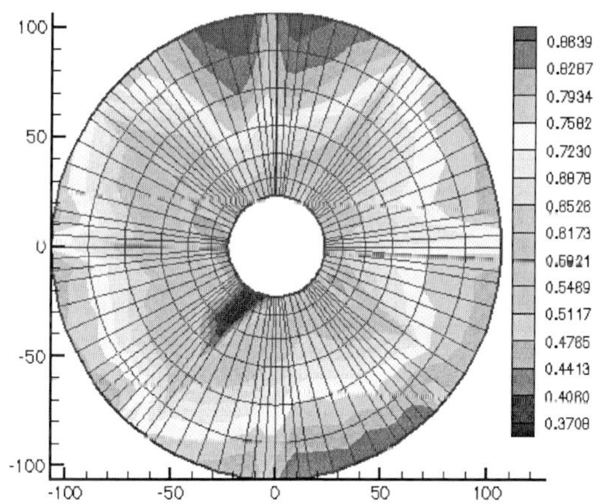

Figure 3: Wake contour at the propeller plane for a full form submarine with guide vane

Figure 5 presents the flow vectors as well as the axial velocity contour in the propeller plane. Obviously guide vane creates circumferential flow in the propeller disc, also we can see the tip vortices in the outer radii region. So in the later design the diameter of the stator was about 10 to15 percent greater than the propeller.

4 DESIGN AND EXPERIMENTAL INVESTIGATIONS OF SGVP

The investigations consisted of the following stages:
(1) hydrodynamic design and powering performance experiments
(2) low noise design and acoustic performance investigations
(3) final design and experimental verification

We are going to describe each of them briefly here.

TABLE 1

GEOMETRICAL PARAMETERS OF GUIDE VANES

No. of Guide Vane	Number of blades (Zg)	Angle (αg)	Wetted area (1+Sg/So)
A+No.1	4	9.2°	1.003637
A+No.2	4	6.7°	1.003473
A+No.3	6	6.7°	1.005348
B+No.4	4	5.0°	1.002700
B+No.5	4	11.0°	1.002706
B+No.6	4	15.0°	1.002708

Table 3 shows the increment rates of resistance coefficients and propulsion factors. From these data it can be found that the increment of form drag coefficient with guide vane is small, for N0.4 guide vane there was even a negative increment, the causes of this resistance decrease are the viscous interactions between the guide vane and main body.

It is also clear that guide vane makes the wake fraction increase, whereas the thrust deduction fraction and the relative rotation efficiency varies a little.

TABLE 2

EXPERIMENTAL RESULTS OF GUIDE VANES

No. of Guide Vane	$\Delta Cr \times 10^3$	ΔWg	Δt_g	$\Delta \eta_r$
A+No.1	0.198	0.136	0.020	0.012
A+No.2	0.153	0.099	-0.005	0.013
A+No.3	0.193	0.136	-0.004	0.007
B+No.4	-0.02	0.062	0.001	-0.005
B+No.5	0.02	0.110	0.010	-0.005
B+No.6	0.03	0.165	0.012	-0.005

TABLE 3

RELATIVE VALUES OF RESISTANCE COEFFICIENTS AND PROPULSION FACTORS

No. of Guide Vane	$1+\Delta Cr/Cr$	$1+\Delta Wg/Wo$	$1+\Delta tg/to$	$1+\Delta \eta rg/\eta ro$	K%
A+No.1	1.2750	1.447	1.164	1.012	4.2
A+No.2	1.2125	1.326	0.960	1.013	6.7
A+No.3	1.2681	1.447	0.967	1.007	8.1
B+No.4	0.9961	1.181	0.993	0.995	2.3
B+No.5	1.0339	1.322	1.073	0.995	5.6
B+No.6	1.0508	1.483	1.087	0.995	11.7

From the experimental results we can find that above 4 to 11 percent of efficiency increases can be gained by the SGVP, the increases of efficiency are produced by the cancellation of the swirl flow of the propeller and change of direction and distribution of the force on the propeller.

3.2 Acoustic Performance Investigations

Since guide vane was found to be very effective for the improvement of propulsive performance of the body of revolution with appendages, investigations were made to cover various hydrodynamic aspects, especially the hydroacoustic performance.

The acoustic tests were carried out in the cavitation tunnel of CSSRC . In the first stage, only guide vane No.5 was selected, and the same propeller with and without guide gane No.5 was operated in the non-uniform flow simulated to that of the form B. As shown in Figure 6, the difference between the two conditions is small, and the noise with the guide vane is even higher than that without it. Certainly this is not desirable.

Then some refinements were made. First according to velocity measurement research, guide vane tip vortices impinged on the propeller blades, it certainly causes noise. In order to avoid its occurrence we decided to increase the diameter of guide vane to 10 percent greater than that of the propeller, so that we can move the tip vortices off the propeller disc to reduce the velocity fluctuations in the propeller disc.

On the other hand, from experimental results of the first phase, the propeller designed according to the inflow condition without guide vane, would produce the required thrust at a different advance coefficient when it operates behind Guide Vane. Therefore a new propeller was designed according to the propeller/guide vane combination effects. The new system obtained remarkable noise reduction. The experimental results are shown in Figure 7.

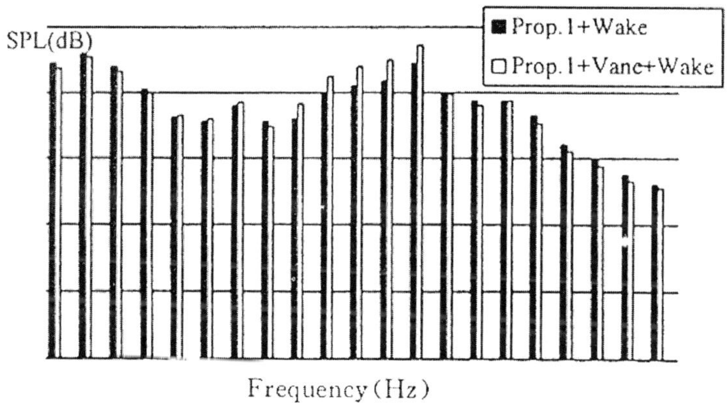

Figure 6: Effect of guide vane No. 5 on acoustic performance

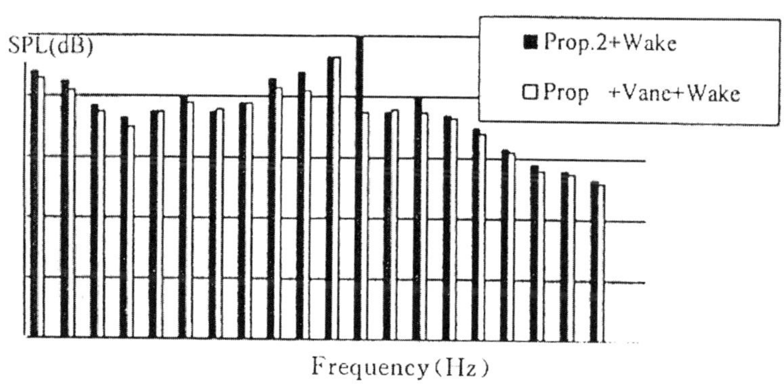

Figure 7: Effect of guide vane No.7 on acoustic performance

3.3 Final Design and Experimental Verification

Maintaining the original horsepower/rpm relationship, a new guide vane No.7 and a new propeller for form B were designed on the above knowledge. In these case the diameter of guide vane was 12 percent greater than that of the propeller and the angle between guide vane and flow was 8 degree. Experimental results show that the increments of speed are about 0.4~0.6 knots and the radiated noise reductions are about 3dB.Table 4 and Table 5 show the experimental results.

TABLE 5

POWERING PERFORMANCE EXPERIMENTAL RESULTS OF OPTIMIZATION DESIGN

Performance	ΔJ	$\Delta \eta_o$	$\Delta \eta_D$	ΔV_s(kn)
Increment	-0.077	-0.020	0.068	0.5

TABLE 5

RADIATED NOISE REDUCTIONS OF OPTIMIZATION DESIGN

Speed	7 kn	9 kn	13kn
Noise reduction	0.9dB	3.7dB	3.1dB

4 CONCLUSIONS

A novel submarine stern arrangement called submarine guide vane propeller(SGVP) system was developed for improving powering performance and reducing radiated noise of the propeller. Based on the model experiment results, about 5-10 percent saving of DHP and 2~4dB noise reduction are expected. Moreover, guide vane can be regarded as an appendage of the submarine,which is simple to construc, convenient to install, and has remarkable energy saving rate and low noise level, it should be regarded as a promising practical device for submarines.

References

Shen Hongcui et al.(1997). Submarine Guide Vane Propeller for Increasing Efficiency and Reducing noise. *J. Ship Mechanics.* **1:1**,1-7.

Neely et al.(1991). Asymmetric pre-swirl stator for US Coast Guard island class patrol boats. *DTRC report /SHD-1335-02.*

Smith et al.(1992). Propeller Erosion Reductions with an Asymmetric Preswirl Stator. *23rd ATTC.*

Ukon et al.(1992). Prevention of Root Erosion by Pre-swirl Propeller Fin. *Proc. ISPC.*

Cui Jigang et al.(1995). Experimental Research on Powering Performance of Submarine with Guide Vane. *CSSRC Technical Report No.* 95318.

Ye Yongxing et al.(1996). Experimental Research on the Acoustic Performance of a Submarine model with Guide Vane. *CSSRC Technical Report No.* 96010.

Zhu Xiqing et al.(1995). Experimental Study of Acoustic Performance of a Submarine Model with Guide Vane. *CSSRC Technical Report No.* 95404.

Shen Hongcui et al.(1997). Development of Guide Vane as a Device for Inproving Powering Performance and Reducing Noise for Body of Revolution with Appendages, *Proc. China-Korea Marine Hydrodynamics Meeting,* 1997

Shen Hongcui et al. (2000) Hydrodynamic Study of SGVP, *CSSRC Technical Report No.*00413

C. M. Lee.(1999) Final Report of the 22nd ITTC Specialist Committee on Unconventional Propellers. *Final Report of the 22nd ITTC.*

Practical Design of Ships and Other Floating Structures
You-Sheng Wu, Wei-Cheng Cui and Guo-Jun Zhou (Eds)

DEVELOPMENT AND APPLICATION OF A HIGH SPEED VIDEO SYSTEM IN HSVA'S LARGE CAVITATION TUNNEL HYKAT

C. Johannsen

Hamburg Ship Model Basin (HSVA), Hamburg, Germany

ABSTRACT

Cavitation tests in model scale are still the most reliable tool to avoid subsequent problems concerning cavitation erosion or propeller induced hull vibrations in full scale. The latest generation of large cavitation tunnels allows the performance of those tests in the realistic three-dimensional wake field of complete ship models at high Reynolds Numbers. In opposite to this very advanced testing technique the procedure for cavitation observations did basically not change over the last hundred years: The movement of the propeller blades is frozen by flash light synchronized with the propeller shaft speed, resulting in one picture of the cavitation pattern per revolution. The temporal development of the cavitation, which is responsible for vibration excitation and propeller erosion damages, can not accurately be studied by this conventional technique. To overcome this shortcoming, a High Speed Video (HSV) system has been established at HSVA for the large cavitation tunnel HYKAT. After years of development this system became easy to handle and is now suitable as a valuable addition to standard cavitation tests. It provides 4,500 high quality pictures per second, i.e. about 180 pictures of the cavitation development per propeller revolution. By simultaneous monitoring of the actual hull pressure values deep insight into the hydrodynamic phenomena can be achieved, which are responsible for vibration excitation and erosion. The paper presents both a detailed description of the recording system as well as first results and video records obtained by the new technique.

KEYWORDS

Testing Technique, Cavitation Tests, Pressure Pulse Measurements, High Speed Video Recording

1 INTRODUCTION

For more than hundred years the performance of model tests in a cavitation tunnel has been an indispensable tool for the prediction of cavitation erosion and propeller induced vibration excitation. This holds more than ever since fast and highly powered passenger vessels, ferries and container ships are demanded worldwide.

The performance of cavitation tests with complete ship models in large facilities like HSVA's Hydrodynamics and Cavitation Tunnel HYKAT (Fig. 1) was a considerable step forward to increase the reliability of those tests. Friesch, Johannsen, Payer (1992). The closed test section of HYKAT allows cavitation observations at high tunnel water speeds, i.e. high Reynolds Numbers, in the realistic three-dimensional wake field generated by the ship model itself. Propeller induced hull pressure pulses can be measured directly in the spatially curved after body contour of the model. In this regard the cavitation testing technique has reached a level, where further improvements are hardly imaginable within the next years. However, this does not hold true for the commonly used evaluation techniques. Worldwide it is still normal procedure to perform cavitation observations under stroboscopic light and to judge propeller induced hull pressure variations just by consideration of the amplitudes resulting from the Harmonic Analysis of the hull pressure time function. This is a pity since a large amount of additional information is hidden in the relation between particularities of the hull pressure time function and the corresponding cavitation development. Especially for the judgement of cavitation aggressiveness with respect to erosion, for the elaboration of remedies against too high pressure pulses or for deeper insight into phenomena like tip vortex bursting this information would be valuable.

To make this additional information available in future, a High Speed Video observation technique has recently been established at HSVA within a research project sponsored by the German Ministry of Research. The certain challenges in this development resulted from the high rotational model propeller speed normally used in HYKAT (around 25 revs per second) and especially from the fact that - to achieve closest possible observation distances - the whole system had to operate from inside the ship model, i.e. inside the water filled model trunk above the test section.

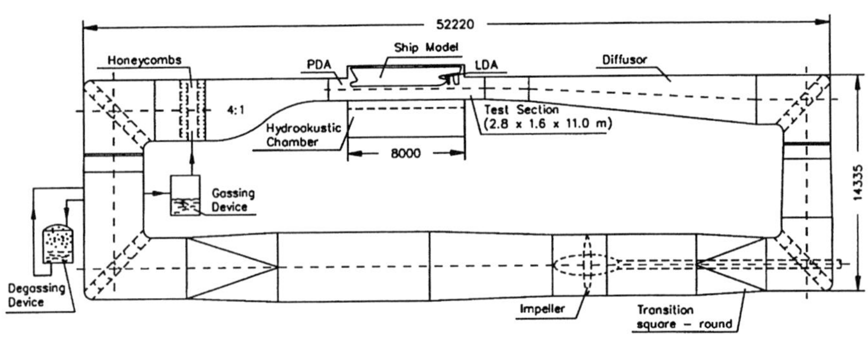

Figure 1: Hydrodynamics and Cavitation Tunnel (HYKAT)

2 INSUFFICIENCIES OF THE CONVENTIONAL CAVITATION TESTING TECHNIQUE

Basically the technique for the observation of cavitating model propellers didn't change during the last hundred years: The general idea was to freeze the movement of the propeller blades by a flash light synchronized with the propeller shaft speed. While at the beginning this had been achieved by constant light reflected from a revolving mirror, Burrill (1951), nowadays stroboscopic lights are applied for this purpose.

For documentation tasks hand made sketches, conventional video recording with 25 frames per second and/or photographs are in use worldwide. The strobe light control system additionally allows to release a flash not exactly once per revolution but every 361° for example, resulting in the illusion of a slowly turning propeller. This illusion can be further improved by subsequent edition of the video record with a digital video editor, eliminating all the dark pictures between the strobe light flashes and this way removing the flicker from the record. Nevertheless, the result is just an illusion of slow motion. In reality these records still provide only one picture of the cavitation behavior per revolution in a certain angular position of the propeller blade. Often the sheet cavitation behavior looks rather unstable under stroboscopic light. The blade area covered by cavitation grows, shrinks and even jumps rapidly in a constant propeller blade position, leading to a very restless cavitation appearance. Indeed, the amount of cavitation grows and shrinks when the propeller blade enters and leaves the ship's 12 o'clock wake peak. But this is a continuous and rather periodic process reappearing every propeller revolution again. What can be seen in the strobe light observation is just one distinct moment of always new periods of this process. Consequently the so called intermittent character of a sheet cavitation results from a non-periodicity of the cavitation process and not from a restless behavior within one period as it may look like. It is known that propeller blades can be damaged by cavitation collapsing heavily or with cloudy or bubbly character in a certain phase of the propeller revolution. This collapse, which has to be judged by the experimentalist, is illuminated by one strobe light flash if at all. For a serious judgement this is poor indeed.

It is normal procedure to perform measurements of the propeller induced hull pressure variations simultaneous to the cavitation observation. Therefore the actual hull pressure value is monitored by pressure pick-ups in the tunnel wall or in the model hull with a sampling rate high enough to provide proper hull pressure time functions as shown in Fig. 2 for one revolution of a five bladed propeller. These time functions can be analyzed by Fast Fourier Transformations (FFT) or Harmonic Analysis to provide time averaged information about the hull pressure amplitudes generated by the propeller under consideration. These amplitudes are sufficient for comparison of different propellers or as an input for a vibration analysis of the steel structure. Nevertheless, the simultaneous strobe light cavitation observation provides just one picture of the cavitation behavior, belonging to one single cross in Fig. 2. This does not allow correlation between certain peculiarities of the cavitation behavior (e.g. bursting of the cavitating tip vortex) and distinct peaks of the hull pressure time function. This correlation could be the basis to elaborate remedies against too strong pressure variations.

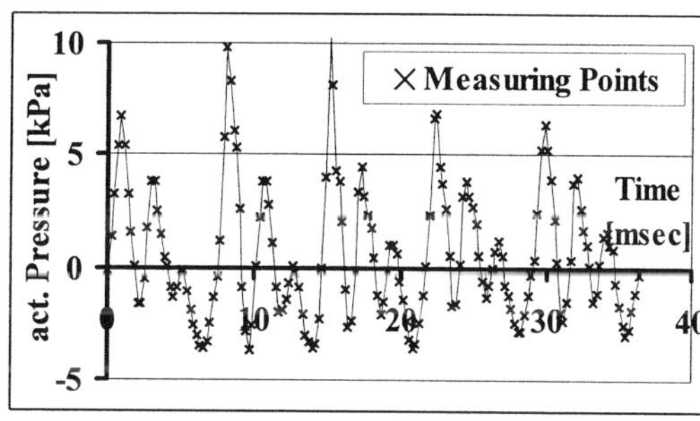

Figure 2: Hull Pressure Time Function for One

Beside these disadvantages the conventional strobe light observation also bears the risk of a frequently occurring miss-interpretation among inexperienced observers, which the author would like to clarify in this context: Under stroboscopic light the helical trace of developed tip vortex cavitation appears stationary in the ship's wake field as shown in Fig. 3. To inexperienced observers these helical traces pretend to be flow lines, e.g. hitting or even "piercing" the rudder blade in distinct and stationary locations. Indeed, the propeller generates a downstream swirl of load depending strength. But this swirl is much weaker

Figure 3: Frozen Movement of a Cavitating Propeller

and rotates in the opposite direction than may be suggested by the vortex traces. From one strobe light picture to the next a water particle does not cover one loop of the helical vortex trace but moves downstream a distance *d* almost axially as indicated by the white arrow in Fig. 3. Here *Z* means the number of propeller blades.

To remove the above problems and insufficiencies a High Speed Video system has been established at HSVA, which will be explained in detail in the next section.

3 DESCRIPTION OF HSVA'S HIGH SPEED VIDEO SYSTEM

The basis of HSVA's High Speed Video system is a *KODAK* EKTAPRO HS Motion Analyzer, Model 4540, consisting of a camera, the so called imager, and a video processor. These and all the other system components are described below. A structural diagram and a photograph are given in Figs. 4 and 5 respectively.

3.1 The Imager

The imager is a camera of 8.5 x 8.5 x 19 cm size, equipped with a video chip of 256 x 256 pixels. The latter provides 256 gray levels per pixel, i.e. one image covers 64 kbyte of information. To increase its sensitivity the camera provides an optional +6 db gain setting. For high speed cavitation observations in HYKAT the imager is combined with a 18 - 90 mm C-mount zoom objective. All functions of this objective, i.e. focus, focal length and aperture, are remotely controlled. To realize a close observation distance, the imager is installed inside the ship model, which therefore has to be equipped with a perspex window in its stern region. HYKAT is operated with a closed test section without a free surface to allow high tunnel water speeds. The so called trunk above the test section, where the ship model is installed inside, is

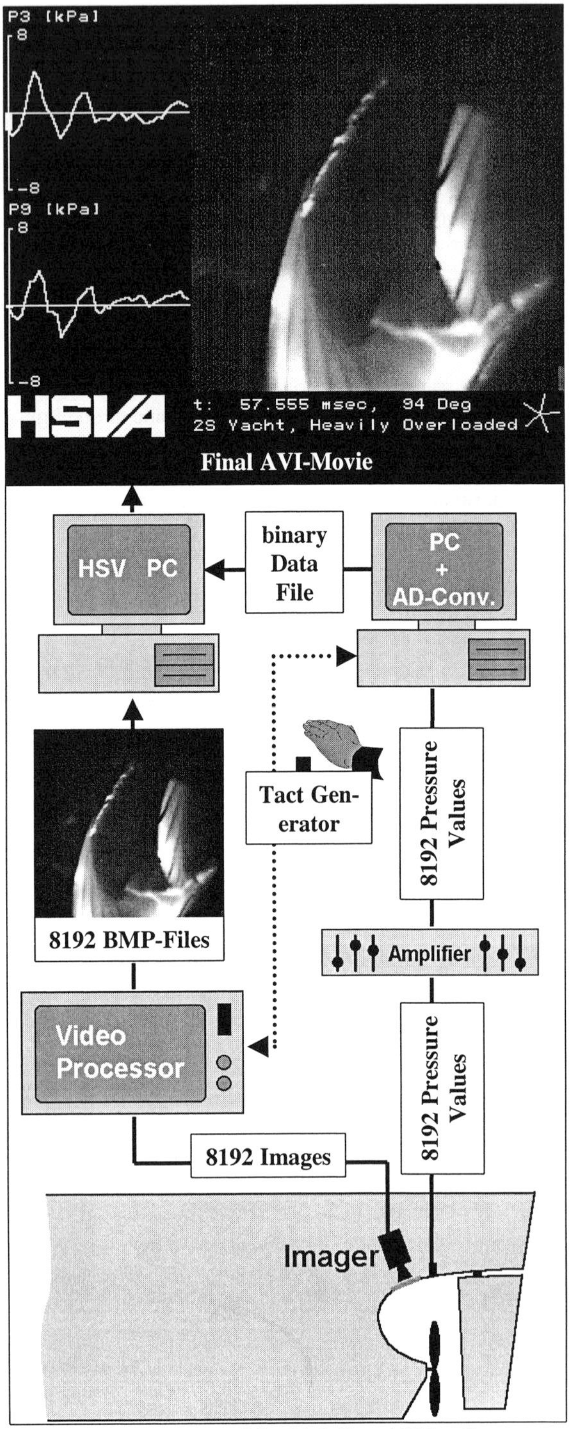

Figure 4: Structure of the High Speed Video System

completely filled with water to avoid air surging into the test section and to allow quick pressure variations. Weitendorf, Friesch, Song (1988). This concept requires operation of the high speed video imager in a water tight housing. This housing basically consists of a 40 cm long brass tube of 13 cm diameter. It is linked to the trunk side wall by a plastic hose containing all the cabling. The housing is mounted on a watertight remotely controlled spindle drive allowing axial adjustment of the observation distance.

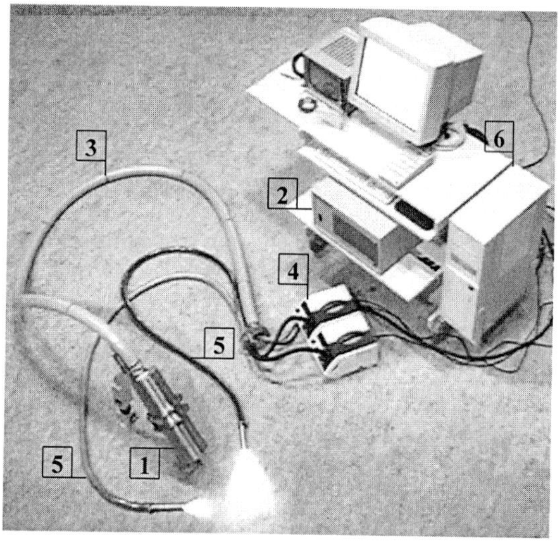

Figure 5: Components of the High Speed Video System

Water Tight Imager Housing (1), HSV Processor (2), Connecting Hose (3), Cold Light Sources (4), Light Guides (5), HSV PC (6)

3.2 Illumination

The high speed video recording of a cavitating propeller requires an intensive illumination. Since the test section of HYKAT is wide (2.8 m) and the model propeller normally operates in its center plane, an illumination from outside through the observation windows is difficult. Initial attempts with two huge theater spotlights heated the perspex tunnel windows to a critical degree but provided just a poor illumination of the model propeller. Furthermore large shady areas remained on the propeller blades. To overcome this problem two 2.5 kW cold light sources were purchased, providing spot light via fiber glass light guides of 3 m length. While the light sources are located outside the tunnel the light guides are separately coated by water tight plastic hoses allowing variable arrangements inside the ship model.

3.3 The Video Processor

The video processor is linked to the imager by a 55 core cable. It has a capacity of 512 Mbyte, which is enough to store 8192 images in its video memory. Its performance allows data acquisition with a recording rate of 4,500 video frames per second, i.e. the complete data acquisition takes only 1.8 seconds. At the first glance this seems to be a very short observation time. But with a typical model propeller speed of 25 revs per second in HYKAT the capacity of the video processor allows the continuous recording of about 46 propeller revolutions with an angular solution of pictures every 2°. A further increase of the recording rate would be possible but requires an excessive illumination and reduces the number of pixels per frame. The recording process can either be controlled by an external trigger signal or internally by the processor itself. After the data acquisition the memory of the processor can be read out to a conventional video recorder or to a PC hard disc drive. The latter generates one 66 kbyte data file per frame in Windows bitmap format. To handle this large amount of incoming data a 600 MHz Pentium III processor (256 Mbyte RAM) in combination with a 38 Gbyte hard disc is utilized, running under Windows NT.

3.4 Hull Pressure Data Acquisition

For the acquisition of the hull pressure data basically the same measuring equipment is used as for conventional HYKAT cavitation tests. Up to 32 *Kulite* pressure pick-ups with a diaphragm diameter of

3.7 mm can be installed in the model hull above the propeller. The pressure signals are amplified and than fed into a *Data Translation* analog-digital (AD) converter board, which is operated in a Windows 95 PC. Using 32 channels, this AD board allows simultaneous data acquisition with a sampling rate up to 13 kHz. As the video processor also the AD converter can be triggered externally. The large number of output values is stored binary in a raw data file in the PC. Although no trigger signal from the propeller shaft is required for High Speed Video purposes, it is monitored as long as a measuring channel is free.

3.5 Synchronization

One major issue is the exact synchronization between the High Speed Video recording and the simultaneous acquisition of the hull pressure values. This synchronization ensures that for each cross in Fig. 2 a corresponding video frame can be associated. Not only the sampling rates but also begin and end of video and hull pressure data acquisition need to be synchronized. For this purpose a little electronic circuit has been developed by HSVA. Its main parts are a tact generator adjusted to 4.5 kHz tact frequency and a counter IC opening a gate for exactly 8192 pulses after pressing a starter button. This circuit supplies both the video processor and the AD converter. To allow previous adjustment of camera and light settings the circuit can alternatively be switched to pass the propeller shaft trigger signal (also used for strobe light control) to the video processor. In this mode the High Speed Video camera continuously provides one live picture per propeller revolution.

3.6 The Procedure

The final result of a High Speed Video Observation is a movie clip in AVI format. Fig. 6 shows every second picture of a very short piece of such a clip. Beside the original 256 x 256 pixel High Speed Video frame (Fig. 6, top, item 1) this movie additionally includes graphical elements, visualizing the actual hull pressure value optionally measured in two different locations (2,3), the history of the hull pressure time signal (4) and the angular orientation of the propeller (5). Furthermore the elapsed time (6) and a caption are included. The procedure to obtain this final movie consists of two phases:

After adjustment of all light and imager settings phase one (the recording phase) starts when the operator releases the tact generator to provide 8192 pulses. Than the imager generates 8192 video frames temporarily stored in the video processor. Simultaneously the AD converter generates a

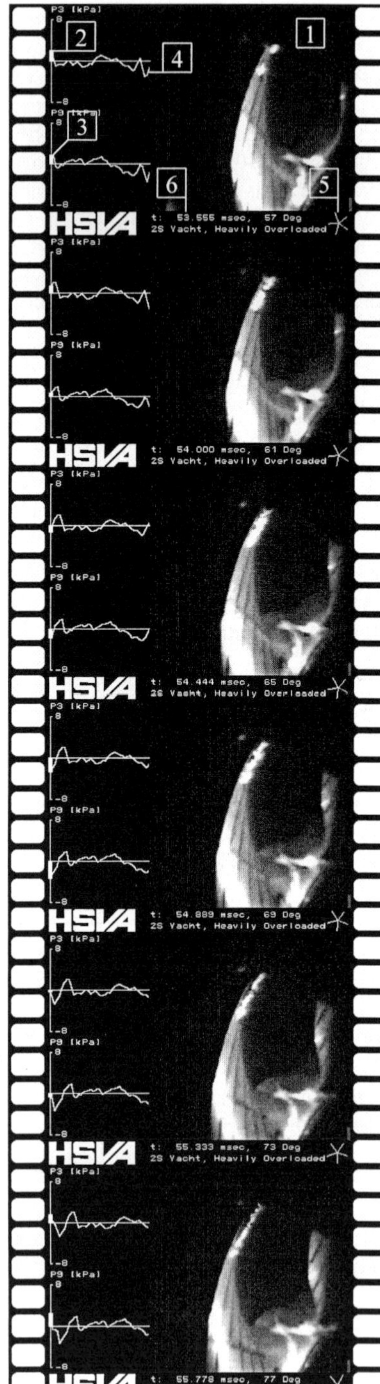

Fig. 6: Overloaded Yacht Propeller

binary data file containing the corresponding hull pressure values.

In phase two (the data processing phase) the memory of the video processor is read out, i.e. 8192 BMP files are generated on the High Speed Video PC, each containing one video frame. Than two self made FORTRAN programs manage the following tasks:

- Reading the binary data file, transferring AD converter units into pressure values by means of a calibration array, calculating actual angular propeller orientations for each frame by means of the strobe signal (if recorded), output to an ASCII file for other purposes.

- Manipulation of the BMP files piece by piece, i.e. generation of left side and bottom margin with the graphical elements mentioned above.

After that a commercial video editor composes the AVI movie file from the 8192 manipulated single pictures. The whole process takes about half an our until the final video clip can be handed out on a CD ROM.

3.7 Synchronization Check

The control of video processor and AD converter by the same set of 8192 pulses already ensures synchronized operation of these units. Nevertheless, to make sure that no delay is caused by the rest of the measuring chain, especially by the measuring amplifier, the complete HSV system was used in a pre-test to record the excitation of a pressure pick-up installed in a water glass, which was exposed to the single beat of a hammer. The result is given as a short video sequence in Fig. 7. As can be seen, the pressure signal shows the excitation exactly in the moment when the hammer touches the water glass. This was a good verification for the whole system.

4 FIRST RESULTS

Fig. 6 gives a short piece of the High Speed Video movie clip obtained for a 110 m long twin screw yacht. Of course this printed picture is only a poor substitute for the very impressive motion picture handed to the customer and shown during the oral presentation of the paper on the PRADS 2001 symposium. Nevertheless, one can see the large excitation of the pressure pick-ups right in the moment when the cavitation grows and develops into tip vortex cavitation. From former investigations carried out with the model of a container ship in HYKAT it is known that even

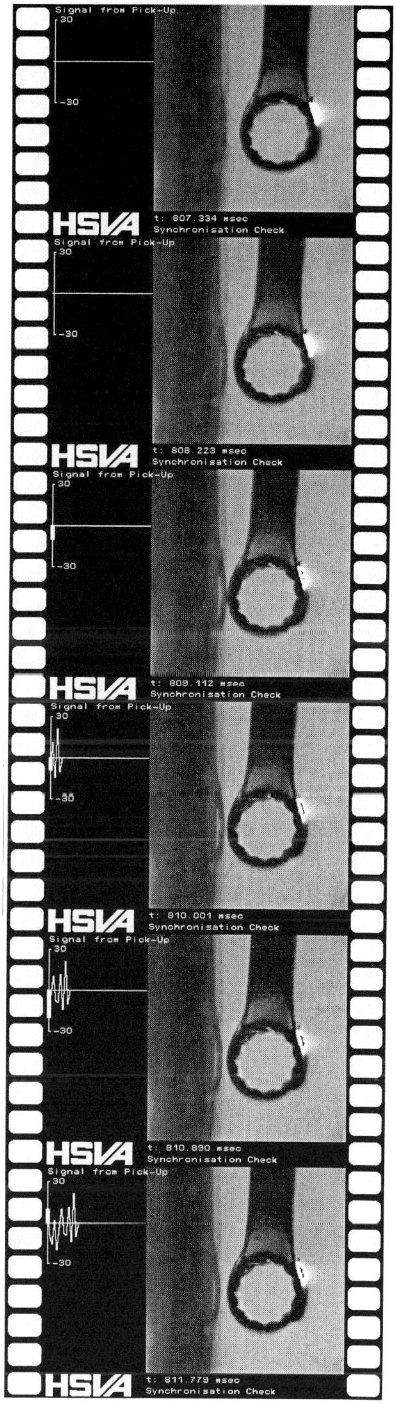

Fig. 7: Synchronization Check

container ship in HYKAT it is known that even much stronger excitations are produced in the moment when this vortex bursts. Johannsen (1998). In the present case such a bursting could not be observed, although the propeller was heavily overloaded already.

An other phenomenon could be investigated instead: The periodicity of the cavitation behavior in Fig. 6 was very poor, leading to an unstable appearance in the strobe light observation. While on a first glance small differences in the geometry of the propeller blades were made responsible, a more careful High Speed Video analysis proved the responsibility of air bubbles in the tunnel water. Supposition for the development of cavitation as in Fig. 6 was the presence of an air bubble in the vicinity of the blade tip just in the angular blade position where the vapor pressure was under-passed. Whenever no bubble was present at that time in that location, no cavitation occurred at all. This observation is in line with the theory. Water can sustain pressure levels far below vapor pressure (and even negative pressure, i.e. tensile stress) as long as there is no partition surface to serve as starting point for cavitation bubble growth. Isay (1989). At least in sea water those partition surfaces are available at any time, since sea water contains air bubbles and other nucleis to a sufficient extent. In opposite to this full scale situation a sufficient content of partition surfaces is not self-evident for the water of a cavitation tunnel. If the water is very clean and the gas content is low, the cavitation behavior in model scale may consequently suffer from an unrealistic non-periodicity as reported in the present case. To avoid this, a sufficient air bubble content should be ensured at any time. Weitendorf, Tanger (1995).

The above findings are not completely new and should be understood more as an example for the potential that is inherent in the High Speed Video technique for future research. This technique has been established for HYKAT. It is now easy to handle and efficiently to use. It is ready for both scientific and customer-orientated use in the field of hydrodynamics.

5 CONCLUSIONS

Basically two major things are necessary to achieve reliable model test results for cavitation occurrence and vibration excitation. On one hand an advanced cavitation test facility is required, which ensures full scale similarity of the phenomena to be studied. HYKAT, with its large test section and high tunnel water speed, has proven its membership to this group of facilities more than once. On the other hand advanced observation and evaluation techniques are required to make optimum use of such a facility. The establishment of a High Speed Video system with simultaneous hull pressure measurement was a considerable step forward in this regard. So the tools are available, waiting for frequent use.

References

Burrill L.C. (1951). Sir Charles Parsons and Cavitation. *Transactions of the Institute of Marine Engineers* **LXIII:8,** 149 ff.

Friesch J., Johannsen C., Payer H.G. (1992). Correlation Studies on Propeller Cavitation Making Use of a Large Cavitation Tunnel. *Transactions of the Society of Navel Architects and Marine Engineers (SNAME),* 2-1 ff

Isay W.H. ((1989), *Kavitation*, Schiffahrts-Verlag "Hansa", Hamburg, Germany (in German)

Johannsen C. (1998). Investigation of Propeller-Induced Pressure Pulses by Means of High-Speed Video Recording in the Three-Dimensional Wake of a Complete Ship Model. *Twenty-Second Symp. on Naval Hydrodynamics (ONR),* 189 ff

Weitendorf E.A., Friesch J., Song C.C.S. (1988). Considerations for the New Hydrodynamics and Cavitation Tunnel (HYKAT) of the Hamburg Ship Model Basin (HSVA). *Schiffstechnik* **Bd. 35,** 59 ff

Weitendorf E.A., Tanger H. (1995). Pressure Fluctuation Investigations in Two Conventional Tunnels and the Hydrodynamics and Cavitation Tunnel HYKAT. *Proceedings of the PROPCAV 95 Symp., Newcastle.*

Practical Design of Ships and Other Floating Structures
You-Sheng Wu, Wei-Cheng Cui and Guo-Jun Zhou (Eds)

UNCERTAINTY ANALYSIS OF TOWING TEST

Mo-Qin He, Hong-Cui Shen and Shu-Long He

China Ship Scientific Research Center, Wuxi, Jiangsu 214082, China

ABSTRACT

A towing tank measurement system is described, based on this system a model resistance repeated tests and a propeller open water repeated tests are conducted. Uncertainty analysis results of those two towing tests have been drawn and discussed.

KEYWORDS

Uncertainty Analysis, Towing Test, Measurement

1 INTRODUCTION

The efforts for improving the accuracy of ship model tests and hence to ensure the reliability of the test results have been continuously enhancing for years in many towing tanks in the world. In recent years, uncertainty analyses of measurements and model tests attracted more attention, and were addressed by the Propulsion Committee and Resistance Committee of 22nd ITTC (1996~1999)[1,2]. To accurately determine full-scale ship power performance, the first and the most important action among all the efforts is evidently to improve the measurement accuracy of the model test, though data representation and model-ship correlation methods are also very important. This paper presents uncertainty analyses of a resistance test and an open water test by a newly built measurement system in the towing tank of China Ship Scientific Research Center (CSSRC).

2 DESCRIPTION OF MEASUREMENT SYSTEM

To assess the uncertainties of model tests at the towing tank, the facilities and the corresponding measurement systems are briefly described in this section.

The measurement system newly built at CSSRC towing tank is composed of a personal computer, 4 A/D converters, and various kinds of transducers. Each A/D converter has 16 channels for analogue signal input and two counters for plus signal input. The resolution is 12 bits, and the maximum sampling rate is 100kHz. The diagram of the measurement system is shown in Figure 1.

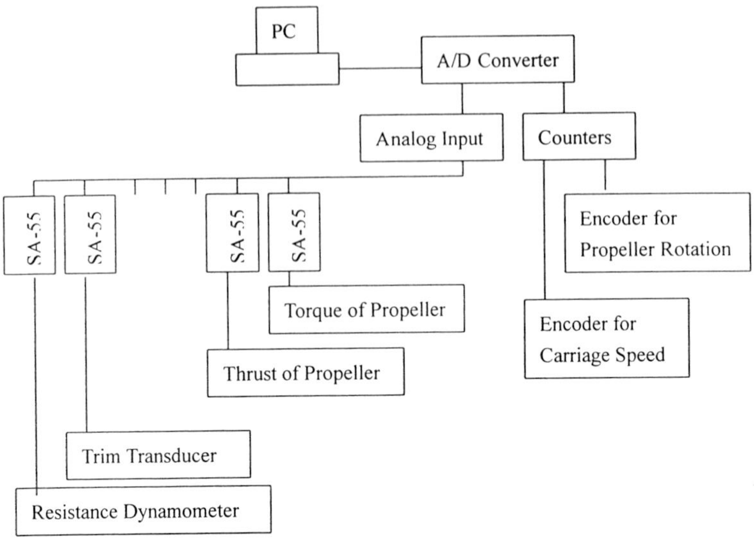

Figure1: Diagram of Measurement System

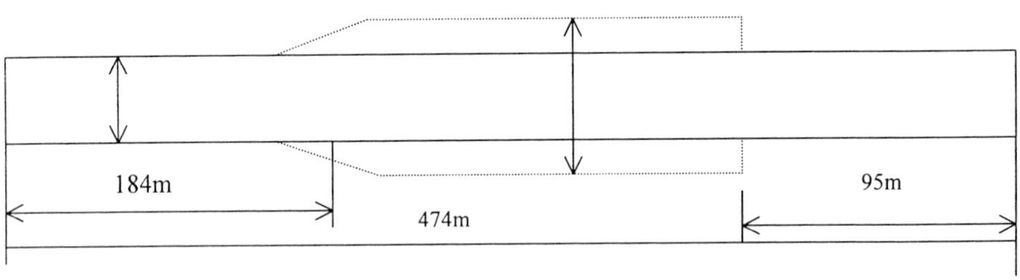

Figure 2: Sketch of Towing Tank

2.1 Towing Tank

Towing tests are conducted in a towing tank in CSSRC. The tank is 474 meters long and is divided into three sections, acceleration section, test section, and retardation section. Those three sections are 95, 195, and 184m respectively. Both acceleration section and retardation section are 7 meters wide, while the width of the test section is 14m. The water depth in the towing tank is 7 m. There are two rows of wave damper along with tank walls in the test section. A diagram of the towing tank is given in Figure 2.

2.2 Carriage and Carriage Speed

The test model is towed by a carriage, which runs at a controlled speed along the tank. The measurement system of the carriage speed (Ux) consists of an optical encoder fixed to a driven wheel, a pulse counter, and a computer with an AD card for data sampling. The driven wheel used for measuring the running speed is elastically mounted under the carriage for keeping touching the top surface of the rail, and is driven by the relative movement between the carriage and the rail. Its circumference is 500 mm and the axis is connected to the optical encoder. The optical encoder is perforated around its circumference with 5000 equally spaced windows. As the encoder follows the driven wheel to rotate, these windows provide optical pulse signals. Each pulse signal received by the pulse counter indicates a 0.1 mm movement of the carriage. The pulse numbers are counted 10 times a second providing the speed change every 0.1 second. After converting the pulse numbers to the carriage speed, the digitals below 1 mm/s are omitted, resulting in the truncated error of 0.5 mm/s [3]. The measurement system of carriage speed is calibrated ever one year. Procedures of the calibration are conducted under the guide of Quality System Document (Q702).

2.3 Rotation Rate of Propeller Measurement

In open water test and self-propulsion test, a DC motor through a gearbox drives the propeller, and its rotation rate defines the propeller rotation rate. The measurement device of the rotation rate is also an optical encoder, which is similar to the carriage speed measurement system, except that it is perforated around its circumference with only 1000 equally spaced windows. When the encoder rotates with the motor, these windows are counted with a pulse counter. Each pulse signal in this device means the motor rotate of 1/1000 revolution/s.

2.4 Force and Moment Measurement

The resistance measurement system is composed of a 20kg strain-gage type load cell and a signal conditioner (SA-55) for amplifying and filtering resistance data. The load cell is mounted on a R63 type Resistance Dynamometer manufactured by K&R Company. The same load cell measures both the model resistance during resistance tests and the captive forces of the model during self-propulsion tests.

Thrust and torque are measured by a R25 self-propulsion dynamometer made by K&R Company. Transducers of the dynamometer are of strain-gage type, signals are picked up and amplified by a conditioners (SA-55) before transmitting to the computer controlled A/D card for data sampling.

2.5 Transducers Calibration

The measurement system is always checked on the carriage before and after every test. The check procedure is more or less similar to the calibration procedure. The calibration is carried out once every year for the whole system from end to end, including the linearity quality analysis of the resistance measurement system.

The end-to-end calibration is to establish the functional relationship between the load on the transducer and the output of the whole measurement system. The repeatability of the calibration results for different period of time is also examined. This calibration is conducted by simulating the drag force using the standardized weights at different increments, and by recording the output of the load cell. The procedure is to carefully orient the central line of the transducer and the loading line of the weights in the vertical direction. The weights are placed on the tray in both ascending procedure from zero to the maximum load, and descending procedure from the maximum load to zero. For each load the output of the transducer is recorded for 4 seconds at the sampling rate of 25Hz. A linear regression curve fit is made based on the average values of the forces measured within the 4 seconds.

2.6 Measurement of the other parameters

The value for water density is interpolated from ITTC (1963) based on the water temperature measured 300 mm under water surface in the test section of the tank. The gravity acceleration (g) is computed from the latitude of the CSSRC lab site. The wetted surface area of the model is taken by integration method.

3 UNCERTAINTY ANALYSIS

3.1 Resistance Test

According to the Manual of Quality Control [2] of CSSRC, the error allowances of model length and model surface area are 0.1%. The qualification records of the model manufacture indicate that the geometry of the test model is qualified, and the maximum bias of model length L and wetted surface area S are as follows:

B_L=0.006m (per unit length, 1 m)

B_S=0.006m×m (per unit wetted surface area, 1 m^2)

From the carriage speed system described above, the model towing speed is measured by the distance over the time interval of carriage movement (Ux = Distance/Time Interval). The distance transferred from pulse numbers by the encoder is of 0.1 mm in accuracy. Compared to the distance measurement, the timer accuracy is much higher, therefore the bias of Ux measurement is mainly due to the data truncation:

B_{Ux} = 0.0005 m/s (per unit speed, 1m/s)

The bias of total resistance R mainly consists of bias of transducer, bias of amplifier and bias of A/D converter. Suppose all tests are conducted just after an end-to-end calibration, so the bias of each measurement system is mainly from their non-linear index and the A/D converter truncation error. In this system, non-linear index for transducer and amplifier are 0.1%. The A/D converter used in the system is the one with 12 bits, corresponding to the relative resolution of 1/4096. Finally the bias of the total resistance is:

B_R = 0.29 N (for 200 kN transducer)

The water density ρ and viscosity ν is not measured during each test. Only measured is the water temperature, and the density is read from a chart. If the chart is accurate enough compared to the water temperature measurement, the bias of water density and viscosity can be considered as mainly caused by the temperature measurement. The thermometer employed in the temperature measurement has a resolution of 0.1℃, which may cause differences up to 0.3 % in water density and 0.2 % in viscosity around 25℃, namely

B_ρ = 0.003

B_ν = 0.002e-6

A summary of the error sources is listed in Table 1.

For a given ship model with the length L, wetted surface area S, the total resistance coefficients C_T and the residuary resistance coefficients C_r are obtained by measurements, and are represented by the measured towing speed Ux and model resistance (towing force R). Prediction of bias and precision limits of the final experimental results of C_T and C_r involve the evaluation of elemental error sources

TABLE 1

SUMMARY OF ERROR SOURCE

Parameter	R	Ux	L	S	ρ	ν
Bias	0.029	0.0005	0.006	0.006	0.0030	0.2e-8

for individual measurement devices, the estimation of bias and precision limits for individual variables (for example, L, S, R, U_x etc.), and the errors induced in data reduction methods.

As far as the measurement systems and the test conditions are concerned, the error sources can be divided to four sectors:

model geometry, L, S;

carriage speed U_x;

resistance dynamometer, providing R;
and water temperature, giving ρ and ν.

Bias and precision for each measured variable can be estimated based on analysis or statistics. In order to determine the bias and precision limits of the final test results, the errors of all the measured variables and their propagation into the final results have to be analyzed. The data-reduction equations are used to analyze the propagation of errors, and hence to derive the bias and precision limits of the final results. The bias and precision limits from different sources are combined in the root-sum-square (RSS) form to derive overall uncertainties with 95% confidence level.

In reference [4], uncertainty analysis of a 6 meters Series 60 model resistance test has been made. The analysis was based on five repeated resistance tests of the same ship model. The total resistance coefficient C_T is calculated by following equation:

$$C_T = R/(0.5\,\rho\,Ux^2 S) \tag{1}$$

For the force R measurement system, bias is associated with the calibration standard, and the linear regression curve fit for the load cell calibration. The bias of the carriage speed measurement U_x is associated with the pulse counter. The bias of the water density is associated with the water temperature measurement. The bias for the model geometry mainly due to the difference between the integrated model wetted surface area based on the offset data and the actual wetted surface area in the test.
From equation (1), the bias of the total resistance coefficient is the combination of the above-mentioned four parts, and can be described as follows

$$\left[\frac{B_{C_T}}{C_T}\right]^2 = \left[\frac{B_R}{R}\right]^2 + \left[\frac{B_\rho}{\rho}\right]^2 + 4\left[\frac{B_{U_X}}{U_x}\right]^2 + \left[\frac{B_s}{S}\right]^2 \tag{2}$$

In order to easily transfer the experimental results to those of the full scale ship performance, they are represented with regard to Froude numbers Fr. For different Fr, the bias of wetted surface areas are supposed to keep the same, but biases of resistance and speed measurements vary with carriage velocity. Table 2 is the summary of C_T and Fr uncertainty analysis results, and Table 3 is the summary of biases given by reference [4].
Report of Resistance Committee of 22nd ITTC [1] gave a comparison of uncertainty analyses of resistance tests carried out at seven facilities. In the seven tests, three were conducted with Series 60 type models. The results showed that the bias limits varied from 0.3 % to 1.8 %. The precision limits

TABLE 2

SUMMARY OF C_T AND FR UNCERTAINTY ANALYSIS

Fr	0.10	0.20	0.32
C_T	3.894e-3	3.548e-3	4.997e-3
Preci. Index	1.3%	0.4%	0.3%
Bias	4.1%	1.1%	0.3%
Uncertainty C_T	4.3%	1.2%	0.5%
Uncertainty Fr	0.12%	0.11%	0.10%

TABLE 3

BIAS OF CR

Fr	0.10	0.20	0.32
Bias of C_T	0.16e-3	0.04e-3	0.02e-3
Bias of Cf	0.0010e-3	0.0009e-3	0.0008e-3
Bias of (1+k)	0.03	0.03	0.03
Bias of Cr	0.28e-3	0.13e-3	0.10e-3

varied from 0.3 % to 0.9 % for multiple tests and 1.1 % to 2.9 % for single test. The total uncertainties varied from 0.8 % to 1.8 % for multiple tests and from 1.4 % to 2.9 % for single test. By average of the results provided by the seven facilities, the total uncertainty was 1.1 % for multiple tests and 2.1 % for single test. The uncertainty of C_T in the present repeated tests are 1.2 % at Fr = 0.2 and 0.5 % at Fr = 0.32. They are similar compared to the ITTC results. However for Fr = 0.1, the uncertainty of C_T goes up to 4.3 %, greater than the average value of the ITTC results. This may be caused by the fact that at low Fr the measured quantities in the present test were much smaller than the total measuring range of the dynamometer. This introduced larger measurement errors.

TABLE 4

PARAMETERS OF PROPELLER MODEL

$B_{0.75R}$	Blade	D_p	H/D_p	A_e/A_0	t/C	Rake
0.06425m	4	0.224m	1.075	0.55	0.045	6^0

TABLE 5

BIAS SOURCES OF THE OPEN WATER TEST

Bias Sources	H29, SA-55	A/D	Calibration Coefficient	Zero Setting	Bias Limited
Propeller Thrust	0.566	0.006	0.080	0.000023	0.572
Hub Thrust	0.566	0.006	0.080	0.000023	0.572
Propeller Torque	0.021	0.00023	0.00588	0.0000009	0.022
Hub Torque	0.021	0.00023	0.00588	0.0000009	0.022

3.2 Open Water Test

Open water test [5] was conducted in the towing tank with the same measurement system. A propeller model of B4-55 series and a propeller dynamometer H29 was used in the open water test for uncertainty analysis. The propeller model parameters are listed in Table 4.

In addition to the parameters described in the uncertainty analysis of resistance test, some more error sources need to be considered. Propeller thrust and torque are measured by force and moment cells in the H29 dynamometer. The signals generated in the dynamometer were picked up and amplified by SA-55, then converted to digital by the A/D card. The results of the bias sources of the open water test are listed in Table 5.

The bias limit of the propeller diameter measurement comes from the errors of model setting and the data reading, they are respectively 0.02 mm and 0.0025 mm by estimation. The bias limit is 0.02 mm. Four error sources of propeller rotation rate have been taken into consideration. They are the truncation error 0.01, reading error 0.001, fluctuation error 0.002, and setting error 0.02. The bias limit is 0.01.

Table 6 gives the results of uncertainty analysis of repeated open water test for the B4-55 propeller model.

TABLE 6

SUMMARY OF UNCERTAINTY ANALYSISOF OPEN WATER TESTS

	J	0.2034	0.4071	0.6100	0.8148
J	U_J	0.0018	0.0015	0.0025	0.0038
	U'_J	0.009	0.004	0.004	0.005
	K_T	0.4214	0.3444	0.2571	0.1666
K_T	U_{KT}	0.0075	0.0027	0.0032	0.0033
	U'_{KT}	0.018	0.008	0.012	0.020
	K_Q	0.0658	0.0553	0.0436	0.0308
K_Q	U_{KQ}	0.0012	0.0004	0.0005	0.0006
	U'_{KQ}	0.018	0.008	0.012	0.019

4 CONCLUSIONS

Based on the newly built towing tank measurement system, a model repeated resistance test and a propeller repeated open water test were conducted. The uncertainty analyses for both measurements are described in this paper. The comparison of the present uncertainty analyses of resistance tests with those provided by the Resistant Committee of ITTC'1999 show that the results in CSSRC towing tank is in good agreement with other towing tanks except at very low Froude numbers. By changing the load cell or resetting the amplify coefficients may certainly improve the accuracy of the present measuring system used for the towing tests at the low Froude numbers.

References

Report of Resistance Committee, *22nd ITTC Proceedings.*
Report of Propulsion Committee, *22nd ITTC Proceedings.*
Specification of Velocimeter for The Towing Tank Carriage, 1983, CSSRC.
Shen Hongcui, He Moqin and Zhou Yi. (1999). Uncertainty Analysis of Resistance Test, *Journal of Ship Mechanics*, **3:6**.
He Shulong. (2000). Uncertainty Analysis of A B4-55 Propeller Open Water Test, *CSSRC Report 00077.*

Practical Design of Ships and Other Floating Structures
You-Sheng Wu, Wei-Cheng Cui and Guo-Jun Zhou (Eds)

TRANSIENT FLOODING IN A DAMAGED FERRY

José M. Riola & Jesús Valle

Canal de Experiencias Hidrodinámicas de El Pardo, Madrid, Spain

ABSTRACT

The Ro-Ro ferries damage stability is one of the main topics of great interest in the maritime community in the last years. The first stage in the initial flooding after collision is one of the most interesting effects to study.
The main aim of these tests carried out at El Pardo Model Basin is to contribute to a better understanding and to acquire data and experimental knowledge about the effect on stability and survivability of the water inflow in the lateral tanks, double bottom ducts, machinery spaces and garage deck in the first transient period after a side damage.

KEYWORDS

Ro-Ro, Transient, Flooding, Collision, Survivability, Damage, Stability, Roll, Damping, Analysis, Capsize, Basin, Test

1 INTRODUCTION

Ro-Ro ferries are very popular around the world and due to fast pier work produce great economic benefits. But their navy design origins which is a risk assumed in such a context, keep an inherent danger flooding condition of stability. Several accidents in the latter years of the century demonstrate that a strong effort in this Maritime Security field is necessary.

After a collision or damage in the hull, the ingress of water will create a heeling moment that will produce a roll response. If the heel is enough to permit the pass of water onto the Ro-Ro deck, the heeling moment can overcome the restoring one and the balance damping can cause the ship to capsize.

The flooding effect in a ship due to the side damage is influenced by the inertia and heeling moment. Both moments are produced by the weight of the water and the buoyancy changes suffered by the different ship wetted surfaces.

The first step to avoid the problem of the heel after a damage is the design of symmetric lateral tanks to equalize the flooding effects as fast as possible through the down flooding ducts.

$$\text{Linear} = B_1 \frac{d\Phi}{dt}$$

$$\text{Non-linear} = B_1 \frac{d\Phi}{dt} + B_2 |\Phi| \frac{d\Phi}{dt} + B_3 \left|\frac{d\Phi}{dt}\right| \frac{d\Phi}{dt}$$

$$\text{Non-linear} = B_1 \frac{d\Phi}{dt} + B_2 \Phi^2 \frac{d\Phi}{dt} + B_3 \frac{d\Phi}{dt}^3$$

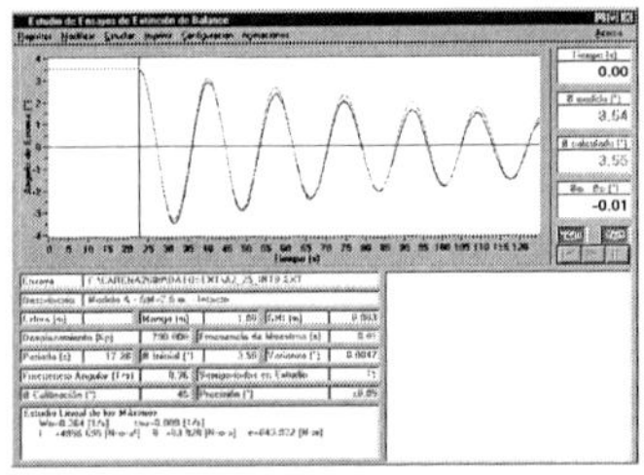

Figure 6: Computer output.

The inertial coefficient is a function of roll and time variables however this inertia increment can be obtained from the flooding tanks inertia values when filling up.

The damping coefficient can be taken from the damaged ship decay test in addition to calculating the restoring moment with the non-constant displacement and metacentric height because both are proportional to the ingress of water. Therefore all the equation inner moment coefficients can be obtained.

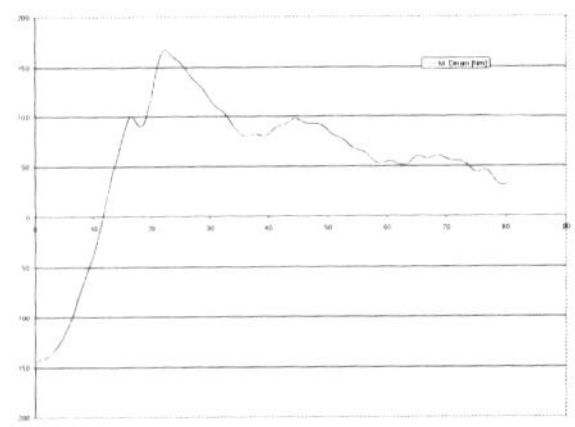

Figure 7: Dynamic response

Using a naval architecture program static righting moments can be calculated with different tank water levels.

The dynamic effect on the roll response can be obtained as a difference between the total moment value and the static external ones.

This dynamic effect exists in reality but it is not taken into account in the naval architecture programs because they are static calculations. The determination of these effects from a large number of tests helps to obtain a dynamic mathematical model.

6 TEST MATRIX

The model was tested with different metacentric heights of values 2.0, 2.5 and 3,0 meters for her intact characteristics in order to find the limiting survival conditions. The testing ship characteristics and the equalizing symmetrical tank data were the following:

TABLE 1

SHIP CHARACTERISTICS

	Ship
Displacement (ton)	21500
Block coefficient	0.65
Residual freeboard (m)	2.0

TABLE 2

FILLING LEVEL

Water Level (m)	*Filling (%)*
0	0
1,0	5,7
2,0	14,9
3,0	25,8
4,0	37,4
5,0	49,3
6,0	61,2
7,25	75,0

A series of test were carried out in order to compare the importance of the residual freeboard related to GM in the balance response. Additional tests with different widths in the down ducts (2, 4 and 6 m^2) and ventilation tube diameters (from 0 to 20 mm model scale) were done.

The two flooding cases: total flooding in the machinery space and partial flooding only in lateral tanks, were done in order to compare the maximum roll value and the elapse time to reach the equilibrium.

Several conclusions, presented at the final of this paper, were obtained from the comparison of the temporal series of these testing cases.

7 SCALE EFFECTS

One of the most important problems testing models is to take into account the scale effects. Data from models is commonly extrapolated to full scale considering that there are no scale effects however it is well known that viscous effects are very important in roll motions.

The assumption of equal damping coefficients in the model and the real vessel has to be studied and revised because the existence of these scale effects can be evidenced even in models.

Future tests using the same model with different scales will be quite valuable in order to determine the correct extrapolation procedure.

8 CONCLUSIONS

- Evidence of the importance of this transient phenomenon is that in a GM condition the model capsizes in this transient flooding test when she passed a survivability standard test of four meters significant height waves.

- The additional heel produced by wind and waves should worsen the roll results.

- Residual freeboard and metacentric height are the two main parameters with influence in the flooding roll response. A lower GM and a lower freeboard facilitate higher roll angles and the ingress of water onto the Ro-Ro deck.

- The response time from zero to the maximum heeling value caused by flooding is similar to the natural roll period of the ship with consequent dangerous dynamic amplification.

- The importance of the scale effects in the down flooding ducts and ventilation tubes are demonstrated. The stabilization time depends mainly on down flooding duct and ventilation tube areas.

- In the comparison between the roll response to the total machinery space flooding and the lateral tanks only, the time series shows a similar maximum first peak roll value but the time needed to get the upright position again is quite longer in the lateral case flooding.

- There is a great influence of the roll in the equalizing tank filling and hence these areas need a careful design, which takes into account the dynamic process.

- The roll damping analysis shows a large dynamic amplification effect over the static one and must be taken into account in a preliminary Ro-Ro ferry design process.

References

[1] Valle J., Pérz-Rojas L. & Carrillo E.J. (2000). Influences of Test Parameters on Roll Damping Coefficients. *STAB'2000*. Australia.

[2] Riola J.M. (1999). Stability After Collision. Revista General de Marina.

[3] Valle J. (1998). Experimental Theoretical Study of Damping Non-Linearities in the Ship Roll Movement. *Doctoral Thesis*. ETSIN-UPM.

[4] Journée J., Vermeer H. & Vredevelt A. (1997). Systematic model experiments on flooding of two Ro-Ro vessels. *STAB'97*. Varna, Bulgaria. September.

[5] Riola J.M. & Marón A. (1997). Survival tests of a damaged ferry vessel. *STAB'97*. Varna, Bulgaria.

4. STRUCTURES AND MATERIALS

Practical Design of Ships and Other Floating Structures
You-Sheng Wu, Wei-Cheng Cui and Guo-Jun Zhou (Eds)

PREDICTION OF WAVE-INDUCED ROLLING RESPONSES BY A TIME-DOMAIN STRIP THEORY

Zhao-Hui Wang[1], J. Juncher Jensen[2] and Jin-Zhu Xia[3]

[1]Offshore Division, Clough Engineering Limited,
22 Mount Street, Perth, WA6000, Australia
[2]Department of Mechanical Engineering, Technical University of Denmark,
Studentertorvet, Building 101E, 2800 Lynby, Denmark
[3]Centre for Oil & Gas Engineering, The University of Western Australia,
Crawley, WA6009, Australia

ABSTRACT

A time domain strip theory formulation is presented for predicting wave loads and 5-DOF motions of a ship at constant forward speed and arbitrary wave heading. The hydrodynamic memory effect due to the free surface is approximated by a set of higher-order ordinary differential equations, which can automatically include the non-linear memory effect and the 'momentum slamming' force. The formulation is generalised from the time-domain strip theory for vertical wave loads and ship responses (Xia *et al*, 1998, *Marine Structures*). Based on the time-domain strip theory formulation, a linear method for predicting wave-induced anti-symmetric motions and loads of rigid-body ships is developed. Numerical calculations are presented for a panamax container ship for the roll motion and torsional moment and are compared with experimental results.

KEYWORDS

Wave Loads, Time-Domain, Strip Theory, Ship Responses, Roll.

1 INTRODUCTION

In a previous study published by Xia *et al* (1998), a time-domain strip theory was developed for predicting vertical wave loads and ship responses. It takes into account memory effects through the use of a higher-order differential equation. The linear case of the theory is coincident with the classical linear strip theory formulation (Salvesen *et al*, 1970). Its semi-empirical generalisation, however, produces the 'momentum slamming' force and models a non-linearity in memory effects. Comparisons with model tests of different kinds of ships (Xia *et al*, 1998; Wang *et al*, 2000) show very favourable agreement for vertical motions and loads.

The purpose of the on-going research is to complete the task of formulating a time-domain strip theory, which covers both symmetric and anti-symmetric wave-induced ship responses. In the present paper the time-domain strip theory for symmetric problems is generalised to model wave loads and ship motions in five degrees of freedom, assuming the ship travels at a constant speed and arbitrary wave heading. Following the general formulation, a linear method for predicting anti-symmetric responses of rigid ships is given. The ship motions are represented in a principal coordinate system; the same methodology may be easily applied to flexible ships (Bishop and Price, 1979).

One might expect that a time-domain strip theory would treat the non-linear anti-symmetric ship motions as efficiently as for the symmetric ones. However, in matters of detail the anti-symmetric problem is by far the more difficult. It introduces a number of complications that were not encountered previously. These fresh obstacles have to be investigated separately. The present paper reports the progress achieved. A single degree of freedom problem is solved numerically for roll motion and torsional moment of monohull ships. Numerical calculations are presented for a panamax container ship and the results are compared with the model experiments (Tan, 1972).

2 TIME-DOMAIN STRIP THEORY FORMULATION

The Cartesian coordinate system is defined as an 'equilibrium' set of axes translating with the mean forward speed U of the ship in the $+x$-direction. The origin is at the stern. The z=0 plane corresponds to the calm water level, and z is positive upwards. The x-z plane is coincident with the centre plane of the ship hull.

The relative displacement between a ship section and the wave surface can be expressed as

$$\bar{z}_l(x,t) = w_l(x,t) - \bar{\xi}_l(x,t), \qquad l = 2, 3, 4 \tag{1}$$

where l = 2 and 3 denote the lateral and vertical movements; l = 4 denotes the rotation about the x-axis. $w_l(x,t)$ is the displacement component of the ship section and $\bar{\xi}_l(x,t)$ is the average displacement component of the wave at the ship section and is related to wave elevation with the Smith correction (Bishop and Price 1979).

The time-domain sectional hydrodynamic force vector $\mathbf{F}(x,t)$ consisting of the lateral, vertical and rotational components may be expressed by (Wang, 2000)

$$\begin{cases} \mathbf{F}(x,t) = \dfrac{D\mathbf{I}}{Dt} \\ \sum_{j=0}^{J}\left(\mathbf{B}_j\mathbf{I} - \mathbf{A}_j\dfrac{D\bar{\mathbf{z}}}{Dt}\right)^{(j+1)} = 0 \end{cases} \tag{2}$$

where $\mathbf{I}(x,t)$ represents both the impulsive and memory vectors in the hydrodynamic momentum; D/Dt is the total derivative with respect to time t, $\dfrac{D}{Dt} = \dfrac{\partial}{\partial t} - U\dfrac{\partial}{\partial x}$, with U being the forward speed of the ship; $()^{(j)} = \dfrac{\partial^j}{\partial t^j}$; and $\mathbf{A}_j(x)$ and $\mathbf{B}_j(x)$ are the so-called frequency-independent hydrodynamic coefficient matrices derived by a rational approximation from the frequency dependent added-mass $\mathbf{m}(x,\omega)$ and damping coefficients $\mathbf{N}(x,\omega)$

$$- m_{kl} + \frac{iN_{kl}}{\omega} = \frac{\sum_{j=0}^{J} A_{j,kl}(i\omega)^j}{\sum_{j=0}^{J} B_{j,kl}(i\omega)^j}, \qquad k, l = 2, 3, 4 \tag{3}$$

where i is the imaginary unit and ω is the wave frequency.

Equation 2 represents a time-domain counterpart of the frequency-domain strip theories, for example, Salvesen *et al* (1970). It can be extended empirically to include the non-linear loading effects due to the time variation of the wetted ship surface by assuming that the coefficients $\mathbf{A}_j(x)$ and $\mathbf{B}_j(x)$ vary with not only the longitudinal position but also the instantaneous submergence, i.e. $A_{j,kl} = A_{j,kl}(x, \bar{z}_3, \bar{z}_4)$ and $B_{j,kl} = B_{j,kl}(x, \bar{z}_3, \bar{z}_4)$. For a ship with lateral symmetry, $m_{kl} = m_{lk}$ and $N_{kl} = N_{lk}$ under the still water-line, thus $A_{j,kl} = A_{j,lk}$ and $B_{j,kl} = B_{j,lk}$.

By integrating the higher-order differential equation in Equation 2, the sectional hydrodynamic force vector $\mathbf{F}(x,t)$ can be rewritten as

$$\mathbf{F} = -\overline{\mathbf{m}}\frac{D^2\bar{\mathbf{z}}}{Dt^2} + U\frac{\partial\overline{\mathbf{m}}}{\partial x}\frac{D\bar{\mathbf{z}}}{Dt} - \frac{\partial\overline{\mathbf{m}}}{\partial t}\frac{D\bar{\mathbf{z}}}{Dt} - \frac{D\mathbf{q}_J}{Dt} \tag{4}$$

where $\overline{\mathbf{m}}(x, \bar{z}_3, \bar{z}_4)$ is the added mass matrix of the ship section when the oscillating frequency tends to infinity; $\frac{D\mathbf{q}_J}{Dt}$ accounts for the hydrodynamic memory effect with $\mathbf{q}_J(x,t) = \{q_{J,2}, q_{J,3}, q_{J,4}\}^T$ governed by the following set of differential equations

$$\begin{cases} \dfrac{\partial \mathbf{q}_j}{\partial t} = \mathbf{q}_{j-1} - \mathbf{B}_{j-1}\mathbf{q}_J - (\mathbf{A}_{j-1} + \mathbf{B}_{j-1}\overline{\mathbf{m}})\dfrac{D\bar{\mathbf{z}}}{Dt} \\ \mathbf{q}_0 = 0 \end{cases} \qquad j = 1, 2, ..., J \tag{5}$$

The third term of $\mathbf{F}(x,t)$ in Equation 4 provides the momentum slamming force. It vanishes in the linear problem. It is noted that $\mathbf{I}$ in Equation 2 and $\mathbf{q}$ in Equations 4 and 5 are vectors instead of matrices as used in Wang (2000), which is not rational in expressing the coupling between the displacement components.

3 LINEAR TIME-DOMAIN ANTI-SYMMETRIC MOTIONS AND WAVE LOADS

For a ship with lateral symmetry, anti-symmetric motions (sway, roll and yaw) are not coupled with symmetric motions (surge, heave and pitch) in linear theories (see e.g. Salvesen *et al*, 1970). The symmetric and the anti-symmetric motions and wave loads may be treated separately. In this section a brief introduction to the formulation of the anti-symmetric responses of a rigid ship is given, whereas a detailed derivation can be found in Wang (2000).

3.1 Anti-symmetric Relative Motion and Wave Forces

The relative displacement for anti-symmetric motions may be expressed as

$$\begin{cases} \bar{z}_2(x,t) = \eta_2(t) + v_6(x)\eta_6(t) - \xi_2(x,t) \\ \bar{z}_4(x,t) = \eta_4(t) - \xi_4(x,t) \end{cases} \tag{6}$$

where $\eta_2(t)$, $\eta_4(t)$ and $\eta_6(t)$ are the generalised co-ordinates corresponding to sway, yaw and roll; $v_6=1-x/x_g$ accounts for the influence of yaw with x_g being the x-coordinate of the centre of gravity of the ship.

The total anti-symmetric force acting on a section can be expressed as

$$Z_k = F_k + R_k + E_k - J_k, \qquad k = 2, 4 \tag{7}$$

where $Z_2(x,t)$ is the sectional horizontal shear force and $Z_4(x,t)$ is the sectional torsional moment about the x–axis; F_k is the sectional hydrodynamic force given by Equation 4; E_k, R_k and J_k are respectively the sectional hydrostatic force, restoring force and inertial force, which may be expressed as

$$\begin{cases} E_2(x,t) = -\rho g S(x)\bar{\xi}_4(x,t) \\ E_4(x,t) = -\rho g \bar{z}_4[\dfrac{B(x)^3}{12} + z_b(x)S(x)] \end{cases} \tag{8}$$

$$\begin{cases} R_2(x,t) = 0 \\ R_4(x,t) = \mu g z_c(x)\eta_4(t) \end{cases} \tag{9}$$

$$\begin{cases} J_2(x,t) = \mu(\ddot{w}_2 - z_c\ddot{w}_4) \\ J_4(x,t) = i_x\ddot{w}_4 - \mu z_c\ddot{w}_2 \end{cases} \tag{10}$$

where ρ is the density of the water and g is the gravitational acceleration; $S(x)$ denotes the immersed area of the ship section; $z_b(x)$ and $z_c(x)$ are the vertical positions of the buoyancy and gravity center of the section; $\mu(x)$ is the longitudinal mass distribution of the ship structure and $i_x(x)$ is the sectional mass moment of inertia about the x-axis.

3.2 Equations of Linear Anti-symmetric Motions

Integration of Equation 7 over the ship length yields the equations of anti-symmetric rigid body motions for the ship hull as

$$\begin{cases} (\mathbf{a} + \mathbf{A})\ddot{\eta} + (\mathbf{B} + \mathbf{B}^*)\dot{\eta} + \mathbf{C}\eta = \mathbf{W} + \mathbf{Q}(\mathbf{q}) \\ \dot{\mathbf{q}} = \Phi\mathbf{q} - \Psi \end{cases} \tag{11}$$

with the initial conditions

$$\eta(0) = \dot{\eta}(0) = \mathbf{q}(x,0) = 0 \tag{12}$$

In Equation 11, $\eta(t)$ is the generalized coordinate vector, $\eta=\{\eta_2, \eta_6, \eta_4\}^T$; $\mathbf{q}(x, t)$ is a vector used in the integration of the higher-order hydrodynamic equation, $\mathbf{q} =\{q_{1,2}, q_{1,4}, q_{2,2}, q_{2,4}, \ldots, q_{J,2}, q_{J,4}\}^T$; $\mathbf{a}$ is the generalized structural mass matrix, $\mathbf{A}$, $\mathbf{B}$ and $\mathbf{C}$ are the generalized fluid added mass, damping and restoring force coefficient matrices; , $\mathbf{B}^*$ is the viscous damping matrix; $\mathbf{W}(t)$ is the generalized wave force vector and $\mathbf{Q}(\mathbf{q})$ is the generalized hydrodynamic 'memory' force vector. Expressions of the matrices and vectors in Equation 11 can be found in Wang (2000). It is noted that $\mathbf{q}$ in Equations 11 and 12 is a vector instead of a matrix as used in Wang (2000).

3.3 Discussion about Hydrostatic Forces

The sum of the sectional forces E_4 and R_4 (in Equations 8 and 9 respectively) is slightly different from the expression given by Bishop and Price (1979), which is

$$E_4 + R_4 = -\rho g \bar{z}_4 [\frac{B^3}{12} + (z_b - z_c)S] - z_c E_2 \tag{13}$$

Because of Equation 13, the roll restoring coefficient in Bishop and Price (1979) is written as

$$C_{44} = \rho g \int_0^L [\frac{B^3}{12} + (z_b - z_c)S)]dx \tag{14}$$

which is different from the roll restoring coefficient based on Equations 8 and 9,

$$C_{44} = g \int_0^L [\rho(\frac{B^3}{12} + z_b S) - \mu z_c]dx = \overline{GM} \cdot \Delta \tag{15}$$

where L is the ship length, $\overline{GM}$ is the metacentric height and Δ is the weight of the water displaced by the ship.

It is seen that Equations 14 and 15 are identical only if

$$\int_0^L \rho z_c S dx = \int_0^L \mu z_c dx \tag{16}$$

which is the case if the buoyancy force equals the gravitational force in every section, i.e., $\rho S(x) = \mu(x)$, or if $z_c(x)$ has the same value in every section. However, that is not the case for general ships.

4 NUMERICAL METHOD

The singularity distribution method (Yeung, 1973) is used to determine the two-dimensional frequency-domain hydrodynamic coefficients (added mass **m**, added damping **N**) for sway and roll and the coupling between them.

Equation 3 can be exactly satisfied at 2J-1 frequencies, but in order to achieve the best approximation within a given frequency range, the frequency-independent hydrodynamic coefficients $A_{j,kl}$ and $B_{j,kl}$ are obtained by minimizing $\sum_{r=1}^{K} \{\mathrm{Re}(\varepsilon_r)^2 + \mathrm{Im}(\varepsilon_r)^2\}$ with

$$\varepsilon_r = \sum_{j=0}^{J} (i\omega_r)^j [A_{j,kl} + (m_{kl} - \frac{iN_{kl}}{\omega_r})B_{j,kl}] \tag{17}$$

where ω_r, r = 1, 2, ..., K are discrete circular frequencies within the given ship-dependent frequency range. Generally, J = 3 suffices for most sectional shapes except for roll and roll-to-sway of the extreme U-shaped sections. In the case of J = 4, the approximation of the added mass and damping is better for the extreme U-shaped sections, but less satisfactory for heave of general sections. And problems appear with numerically induced oscillations which makes the time-domain simulation unstable. Therefore, J = 3 is adopted in the time-domain simulation in this paper.

The differential equations for the coupled fluid-structure system are integrated in the time domain by using the Adams predictor-corrector scheme. The fourth order Runge-Kutta method is used to predict the function values at the first four time steps.

5 CALCULATIONS FOR A PANAMAX CONTAINER SHIP

A computer program has been developed according to the above numerical method. In this presentation only the roll motion and the corresponding torsional moment are simulated, due to a numerical stability problem in the time-domain simulation with sway and yaw. Here numerical results are demonstrated for a Panamax container ship and are compared with the experiments carried out at MARIN in The Netherlands. A detailed description of the experiments including the body plan and the main particulars of the ship can be found in Tan (1972). The computation and comparison of the vertical motions and structural responses for the same ship were given in Wang *et al*, (2000) and Wang (2000).

The model tests were conducted in seven wave headings: β = 25, 45, 65, 180, 205, 225 and 245 degrees (180 degrees denote head waves). Seven wavelength to ship length ratios are used: λ/L = 0.35, 0.5, 0.6, 0.7, 0.9, 1.1 and 1.4. The wave amplitude a was kept constant at $L/120$. A range of ship speeds between Fn=0.22 and Fn=0.27 was investigated. The model was kept on course by an autopilot controlling the rudder during the test. Here comparisons are only made for Fn=0.245, since the responses only vary slightly with the speed in the considered speed range. It is well known that the roll motion in the near-resonance region is strongly affected by viscous damping. The roll damping and the related cross-coupling coefficients may be calculated in different ways. In the present calculation, a proportion of the critical damping is used $B^{*}_{44} = 2d\sqrt{(a_{44} + A_{44})C_{44}}$, where $0.08 \le d(\beta) \le 0.4$. The viscous damping affects mainly in the low frequency range.

Comparisons of the frequency response functions of roll $\phi_a/(ka)$ and torsional moment $M_T/(\rho gaBL^2)$ at six headings are shown in Figures 1 and 2, where k is the wave number and B is the ship breath. It is seen that the agreement is acceptable for the roll motion, and the roll motion is much larger in stern quartering seas than that in head quartering seas. However, the agreement for the torsional moment is less satisfactory.The disagreement may be mainly attributed to the neglecting of coupling effects with sway and yaw. It is noted that the torsional moment is about the x-axis.

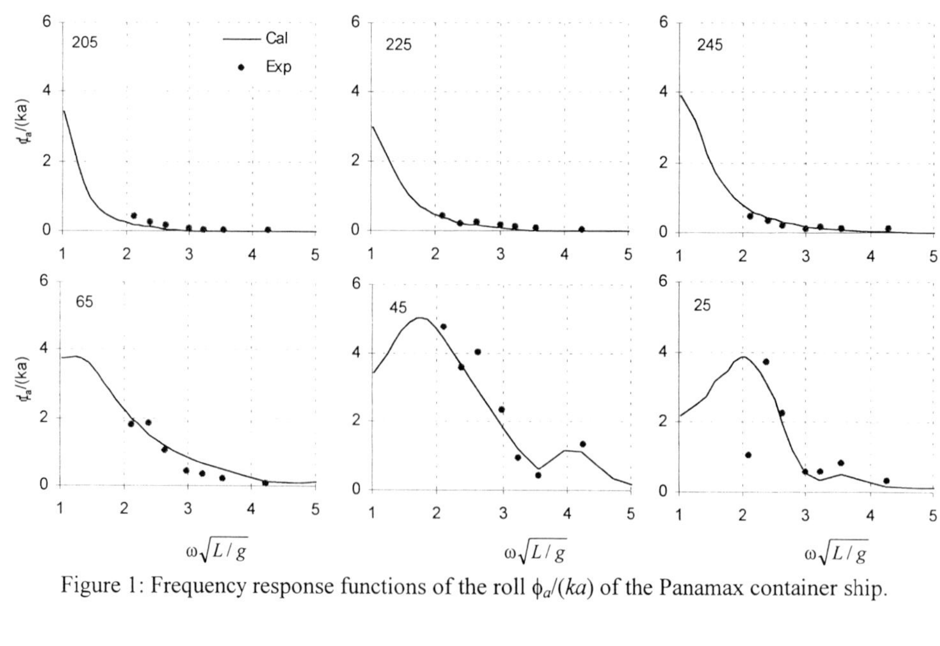

Figure 1: Frequency response functions of the roll $\phi_a/(ka)$ of the Panamax container ship.

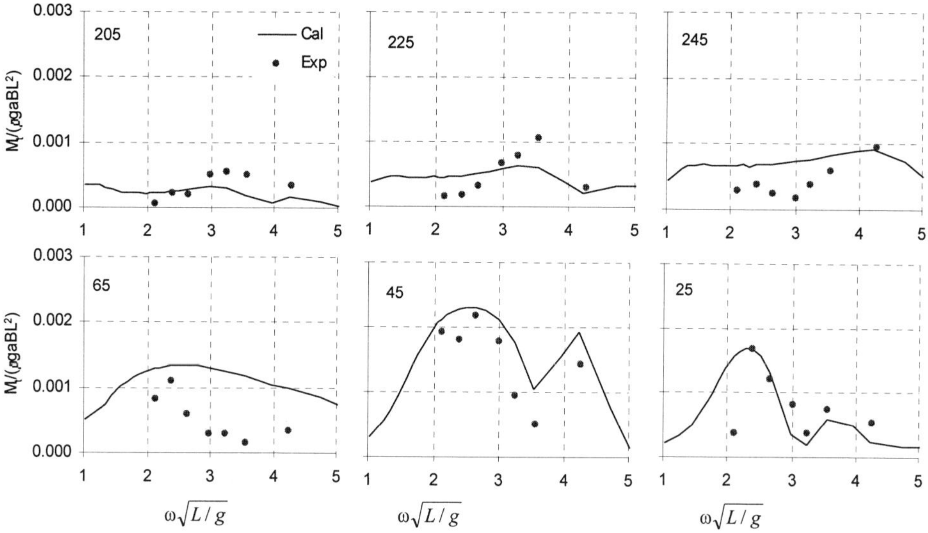

Figure 2: Frequency response functions of the midship torsional moment $M_T/(\rho gaBL^2)$.

Acknowledgements

The research work done in this paper was financially supported by a Ph.D. Scholarship of the Technical University of Denmark awarded to the first author and a UWA Small Research Grant of the University of Western Australia granted to the third author. The financial supports are gratefully acknowledged.

References

Bishop, R.E.D. and Price, W.G. (1979). *Hydroelasticity of ships*, Cambridge University Press.

Salvesen, N., Tuck, E.O. and Faltinsen, O.M. (1970). Ship Motions and Sea Loads. *Trans. SNAME.* **78,** 250-287.

Tan, S.G. (1972). Wave Load Measurements on a Model of a Large Container Ship. Netherlands Ship Research Centre, Report 173S.

Wang, Z. (2000). Hydroelastic analysis of high speed ships, *Ph.D. Thesis.* Department of Naval Architecture and Offshore Engineering, Technical University of Denmark.

Wang Z., Xia J., Jensen J.J. and Braathen A. (2000). Prediction of Vertical-Plane Wave Loading and Ship Responses in High Seas. *Proc. 23rd Symposium on Naval Hydrodynamics, ONR.*

Xia J., Wang Z. and Jensen J.J. (1998). Nonlinear Wave Loads and Ship Responses by a Time-Domain Strip Theory, *Marine Structures.* **11:3,** 101-123.

Yeung R.W. (1973). A Singularity Distribution Method for Free-Surface Flow Problems with an Oscillating Body. University of California Berkeley, College of Engineering, REP.NO. NA 73-6.

Practical Design of Ships and Other Floating Structures
You-Sheng Wu, Wei-Cheng Cui and Guo-Jun Zhou (Eds)

METHODS TO REDUCE THE EFFECTS OF IRREGULAR FREQUENCIES IN HYDRODYNAMIC ANALYSIS OF VESSELS WITH FORWARD SPEED

S.X. Du [1,2], D.A. Hudson [2], W.G. Price [2] and P. Temarel [2]

[1] China Ship Scientific Research Centre, P.O. Box 116, Wuxi, Jiangsu 214082, China.
[2] School of Engineering Sciences, Ship Science, University of Southampton, Highfield, Southampton, SO17 1BJ, United Kingdom.

ABSTRACT

This paper investigates the occurrence of irregular frequencies in the hydrodynamic analysis of moving ships by using translating, pulsating source methods. This is achieved through a theoretical analysis of a rectangular box travelling and oscillating in waves, together with numerical calculations. These analyses indicate the importance of accurate evaluation of the waterline integral term and a technique to improve this evaluation for a rectangular box is presented. Investigations into the hydrodynamic data for a Series 60 hull form highlight the importance of elements close to the waterline in the appearance of an oscillation in value in these data. Practical numerical methods to reduce the effects of this oscillation are introduced. Their effectiveness is demonstrated through calculations performed for the Series 60 hull form.

KEYWORDS

Forward speed, seakeeping, irregular frequency, Series 60, Green's function.

1 INTRODUCTION

In the zero speed formulation of the boundary value problem employing a distribution of singularities over the wetted body surface of a surface piercing vessel to determine velocity potential functions, the solution of the developed integral equation at certain frequencies is not unique or does not exist. These irregular frequencies were first discussed by John (1950) and subsequently studied by Inglis and Price (1981). John showed that these irregular frequencies relate to the eigenvalue problem of the imaginary interior fluid motion inside the body with the same free surface condition as the external fluid surrounding the body. In a numerical scheme of study, singularity distribution methods use a discrete form of the integral equation to derive singularity strengths subject to appropriate body-boundary conditions. Whilst John (1950) analytically demonstrated that the integral equation becomes singular at discrete, the numerical implementation of the integral equation is never, or very rarely, actually singular at any discrete frequency. However, its determinant does approach zero over a narrow range of frequencies either side of, and including, the precise irregular frequency. Numerically the matrix is ill conditioned. The presence of an irregular frequency is noted by a very rapid change in the calculated values of the fluid actions, hydrodynamic coefficients, etc. over a narrow range of wave frequencies. For the majority of ship-like structures these irregular frequencies lie outside the range of practical interest for rigid body motions. They can, however, create difficulties for the analysis of flexible bodies, multi-hulled vessels and in using frequency domain data to create impulse response functions for subsequent simulation of vessel response in the time domain.

Although the zero speed irregular frequency problem has received much attention in analysing two- and three-dimensional fluid-structure interactions, it is slightly surprising that there is little discussion of irregular frequencies when the forward speed of the surface piercing vessel is non zero. In recent validation studies of two independent translating, pulsating source methods, the hydrodynamic data for several hull forms display an oscillation in value at a similar frequency to the irregular frequency for a zero speed (pulsating source) method (Du et al. 1999, 2000). This oscillation is different in appearance, occupying a wider frequency range and being much smoother in nature. Comparisons with experimental results indicate that it is not a physical phenomenon (Bailey et al. 2001). Bruzzone et al. (2000) indicate a similar oscillation in hydrodynamic data in calculations performed for a frigate form using a translating, pulsating source method.

In this paper, the occurrence of irregular frequencies for a travelling surface piercing vessel is discussed by examining the behaviour of eigenvalues, ill-conditioning of matrices, etc. associated with a rectangular box of ship-like proportions. These numerical calculations highlight the importance of elements closing to the waterline in the appearance of the oscillation in the hydrodynamic data. The waterline integral term in the translating pulsating source Green's function is often approximated and a numerical technique to improve evaluation of this term for a rectangular box is presented. Finally, numerical techniques to reduce the effects of the oscillation for practical calculations are introduced. Calculations performed for a Series 60 hull form indicate its effectiveness in reducing the amplitude of the oscillations in the hydrodynamic data.

2 INVESTIGATIONS FOR A RECTANGULAR BOX

2.1 Mathematical Formulation

The formulation of the boundary value problem for a ship travelling and oscillating in waves has been described many times in the literature (see for example, Newman 1978, Faltinsen 1990) and will not be repeated here. Suffice to say, the velocity potential for the radiation and diffraction problems may be expressed as an integral equation, i.e.,

$$
\begin{aligned}
c\phi(\mathbf{x}) = \iint_{S_b} \left[\phi(\xi) \frac{\partial G(\mathbf{x},\xi;-U)}{\partial n(\xi)} - G(\mathbf{x},\xi;-U) \frac{\partial \phi(\xi)}{\partial n(\xi)} \right] dS + \frac{2U\omega_e i}{g} \oint_c \phi(\xi) G(\mathbf{x},\xi;-U)\, d\eta \\
+ \frac{U^2}{g} \oint_c \left[\phi(\xi) \frac{\partial G(\mathbf{x},\xi;-U)}{\partial \xi} - G(\mathbf{x},\xi;-U) \frac{\partial \phi(\xi)}{\partial \xi} \right] d\eta
\end{aligned}
\quad (1)
$$

$$
c = \begin{cases} 0, & \mathbf{x} \notin \Omega \cup S_b \\ 2\pi, & \mathbf{x} \in S_b \\ 4\pi, & \mathbf{x} \in \Omega \end{cases} \quad for\ \phi = \phi^e; \quad c = \begin{cases} 0, & \mathbf{x} \notin \Omega^i \cup S_b \\ -2\pi, & \mathbf{x} \in S_b \\ -4\pi, & \mathbf{x} \in \Omega^i \end{cases} \quad for\ \phi = \phi^i;\ \mathbf{n} = (n_1, n_2, n_3)
$$

The line integral on the waterline is taken positive in the clockwise direction and the unit normal vector's position direction points into the exterior fluid field domain Ω, and a solution sought in order to satisfy a boundary condition imposed on the body surface. When a source distribution method is employed this integral equation may be expressed as,

$$\iint_{S_b} G(\mathbf{x},\xi;-U)\sigma(\xi)dS + \frac{U^2}{g}\oint_c G(\mathbf{x},\xi;-U)\sigma(\xi)n_1^2 / \sqrt{n_1^2+n_2^2}\,dl$$

$$= 4\pi\phi^e(\mathbf{x}) \qquad x \in \Omega \cup S_b$$

$$-2\pi\sigma(\mathbf{x}) + \iint_{S_b} \frac{\partial G(\mathbf{x},\xi;-U)}{\partial n(\mathbf{x})}\sigma(\xi)dS + \frac{U^2}{g}\oint_c \frac{\partial G(\mathbf{x},\xi;-U)}{\partial n(\mathbf{x})}\sigma(\xi)n_1^2 / \sqrt{n_1^2+n_2^2}\,dl \tag{2}$$

$$= 4\pi\frac{\partial \phi^e(\mathbf{x})}{\partial n(\mathbf{x})} \qquad x \in S_b$$

and for the corresponding interior fluid Dirichlet problem as,

$$\iint_{S_b} G(\mathbf{x},\xi;-U)\frac{\partial\phi^i(\xi)}{\partial n(\xi)}dS - \frac{U^2}{g}\oint_c G(\mathbf{x},\xi;-U)\frac{\partial\phi^i(\xi)}{\partial n(\xi)}n_1^2 / \sqrt{n_1^2+n_2^2}\,dl = 0;$$

$$\iint_{S_b} \frac{\partial G(\mathbf{x},\xi;-U)}{\partial n(\mathbf{x})}\frac{\partial\phi^i(\xi)}{\partial n(\xi)}dS - \frac{U^2}{g}\oint_c \frac{\partial G(\mathbf{x},\xi;-U)}{\partial n(\mathbf{x})}\frac{\partial\phi^i(\xi)}{\partial n(\xi)}n_1^2 / \sqrt{n_1^2+n_2^2}\,dl = 0;\ \mathbf{x} \in S_b \tag{3}$$

Let us know examine Eqn. 2. If at a certain frequency, there occurs a non-unique or no solution of source strength, the determinant of the resultant matrix should be zero or at least ill-conditioned when a discrete numerical approach is employed. It may be observed that the source distribution integral equations in Eqn. 2 have the same form as Eqn. 3 of the interior Dirichlet eigenvalue problem. Thus, when a discrete numerical approach is employed for both problems, the coefficient matrices associated with Eqns. 2 and 3 must be the same. It therefore follows that the irregular frequency of the source distribution method corresponds to the Dirichlet eigenvalue problem of the interior fluid field.

For simple geometric structures such as a rectangular box with length L, Breath B and Draft T, the solution to the Dirichlet interior eigenvalue problem may be determined theoretically. In the case of a translating box a similar solution procedure may be adopted. The principal difference being the free-surface boundary condition now imposed on the solution. The frequencies corresponding to this problem are given as (Du et al. 2001),

$$\omega_e = \left|Um \pm \sqrt{gk\,coth\,kT}\right|,\ k^2 = n^2 + m^2, k > 0 \tag{4}$$

where $\phi^i(x,y,z) = \sin(mx)\sin(ny)\sinh k(z+T),\ n = 2\pi i / B, m = 2\pi j / L$. For a surface-piercing box having zero forward speed the frequencies corresponding to this problem are given as (Wu and Price 1986),

$$\omega_e = \sqrt{gk\,coth\,kT},\ k^2 = n^2 + m^2, k > 0 \tag{5}$$

2.2 Numerical Results

From equations 4 and 5 it is possible to predict the occurrence of irregular frequencies in the source distribution method applied to a rectangular box with or without forward speed. As discussed previously the determinant of the matrix $[\partial G / \partial n]$ should approach zero in the numerical method at an

When formulating the translating pulsating source Green's function method the upstream far-field radiation condition may be satisfied automatically by introducing a Rayleigh viscous coefficient to the free-surface boundary condition and seeking the solution as this coefficient tends to a small positive value. If this approach is examined then it is apparent that by introducing the Rayleigh viscous coefficient to satisfy the upstream radiation condition, the magnitude of the Green's function is decreased when the field point is upstream and beneath the source point, but when the field point is downstream of the source point the magnitude of the Green's function is increased artificially. Although the Rayleigh viscous coefficient tends to a very small positive number, its influence in determining the value of the Green's function numerically appears to be important.

In order to artificially damp the value of the Green's function and its derivatives when the field point is downstream and above the source, a Rayleigh viscous damping coefficient may be introduced in a similar manner to that used in the formulation of the problem. In Figures 5 and 6, the hydrodynamic coefficients for the Series 60 form are shown, including the original results together with those predicted where the influence of certain source-field point combinations was simply ignored, those when the source points are moved down slightly and those calculated when the artificial damping technique is employed. It can be seen that all sets of calculations produce smoother hydrodynamic coefficients than the original results. Although in general both the results by damping method and shift method are not as smooth as those by part panel ignored method, however, as shown in Figure 6, the anti-symmetric coefficients are predicted more accurately. The motivation for pursuing the damping technique is that it has the potential to be more generally applicable than the other techniques, since decisions dependent on hull form, frequency of oscillation, forward speed, panel location, etc. should not need to be taken for each application of the method. However, much work remains to be done to determine a suitable value of the artificial damping coefficient and to ensure that such an approach does not adversely affect predictions at other frequencies.

3 CONCLUSIONS

A theoretical analysis of the problem of a rectangular box moving with forward speed indicates that irregular frequencies exist in this case and that their occurrence may be predicted using Eqn. 4. Numerical calculations performed using a translating pulsating source method indicate the existence of irregular frequencies for very low forward speeds. As forward speed is increased the numerical method indicates that irregular frequencies do not occur. Re-formulating the line integral term in the integral equation as a surface integral allows the approximations inherent in the method to be reduced. Results calculated using the re-formulated line integral, for a rectangular box travelling at Fn=0.2, indicate the presence of an irregular frequency.

Calculations of hydrodynamic coefficients for a Series 60 hull form exhibit oscillations in the value of the hydrodynamic coefficients in the region of a predicted irregular frequency. The origin of this oscillation in value caused by the influence of source points located upstream and beneath the waterline panels in the stern region of the vessel being unduly large.

Three different practical techniques are employed to reduce the size of this influence. Namely, ignoring the contributions of certain panels in the calculation of source strengths, moving the field points close to the waterline down slightly and introducing the concept of artificial damping when the source point is upstream and beneath the field point. Each of these techniques reduces the size of the oscillation in value of the hydrodynamic coefficients and produces smoother trends in these data with frequency. This improves agreement with experimental measurements for the symmetric coefficients. For the anti-symmetric coefficients, agreement with experimental data at some frequencies is reduced, since the contributions of waterline panels to the overall hydrodynamic quantities is greater in these case.

Before proposing a general technique to reduce the oscillations in the value of the hydrodynamic coefficients further work needs to be done applying the concept of artificial damping to other hull discretisations, hull forms and forward speeds. Criteria for the inclusion of the influence of panels on waterline panels will also be investigated as will their incorporation into the numerical schemes of study.

References

Bailey, P.A., Hudson, D.A., Price, W.G. and Temarel, P. (2001). Comparisons between theory and experiment in a seakeeping validation study. *Transactions RINA* **143.**

Bruzzone, D., Gualeni, P. and Sebastiani, L. (2000). Applications of Green's function methods to the seakeeping computations for high speed vehicles. *NAV 2000: Thirteenth International Conference on Ships and Shipping Research,* 9.1.1.-9.1.11, Venice.

Du, S.X., Hudson, D.A., Price, W.G. and Temarel, P. (1999). Comparison of numerical evaluation techniques for the hydrodynamic analysis of a ship travelling in waves. *Transactions RINA* **141**, 236-258.

Du, S.X., Hudson, D.A., Price, W.G. and Temarel, P. (2000). A validation study on mathematical models of speed and frequency dependence in seakeeping. *Proceedings of the Institution of Mechanical Engineers, Part C* **214**, 181-202.

Du, S.X., Hudson, D.A., Price, W.G. and Temarel, P. (2001). An investigation into irregular frequencies in speed dependent Green's functions, *Submitted for Publication.*

Faltinsen, O.M. (1990). *Sea Loads on Ships and Offshore Structures,* Cambridge Ocean Technology Series. Cambridge University Press. UK

Gerritsma, J. and Beukelman, W. (1964). The distribution of the hydrodynamic forces on a heaving and pitching ship model in still water. *Fifth Symposium on Naval Hydrodynamics,* 219-257.

Inglis, R.B. and Price, W.G. (1981). Irregular frequencies in three dimensional source distribution techniques. *International Shipbuilding Progress* **28**, 57-62.

John, F. (1950). On the motion of floating bodies. *Communications on Pure and Applied Maths* **3**, 45-101.

Newman, J.N. (1978). The theory of ship motions. *Advances in Applied Mechanics* **18**, 221-283.

Todd, F.H. (1953). Some further experiments on single-screw merchant ship forms- Series 60. *Transactions SNAME* **61**, 516-589.

van Leeuwen. G. (1964). The lateral damping and added mass of an oscillating ship model. *Report of Shipbuilding Laboratory, Delft University of Technology* **23.**

Wu, X.J. and Price, W.G. (1986). An equivalent box approximation to predict irregular frequencies in arbitrarily-shaped three-dimensional marine structures. *Journal of Applied Ocean Research* **8:4**, 223-231.

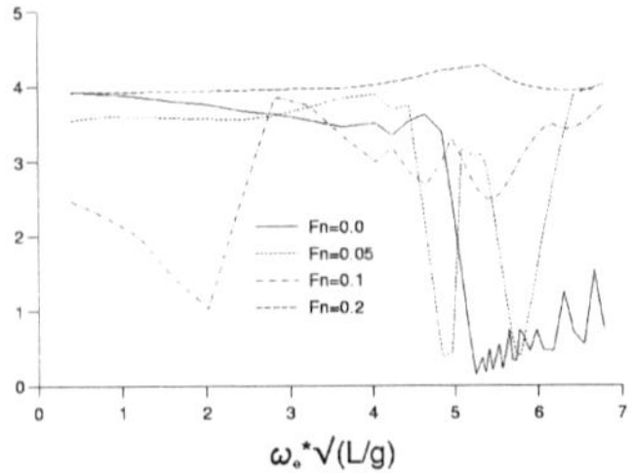

Figure 1: Minimum of the amplitudes of the eigenvalues of matrix $\left[\partial G / \partial n\right]$

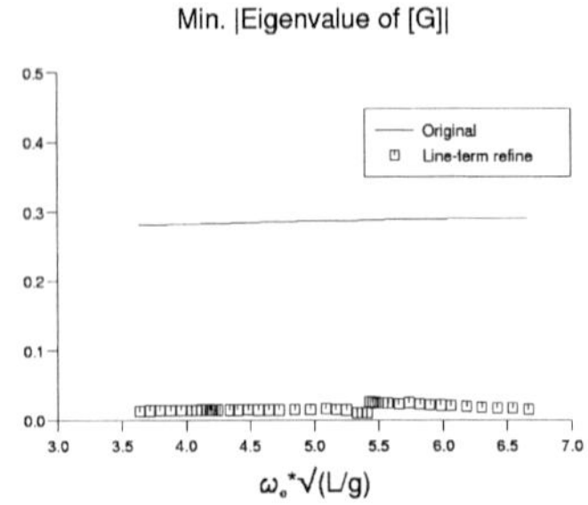

Figure 2: Minimum of the amplitudes of of the eigenvalues of matrix $[G]$

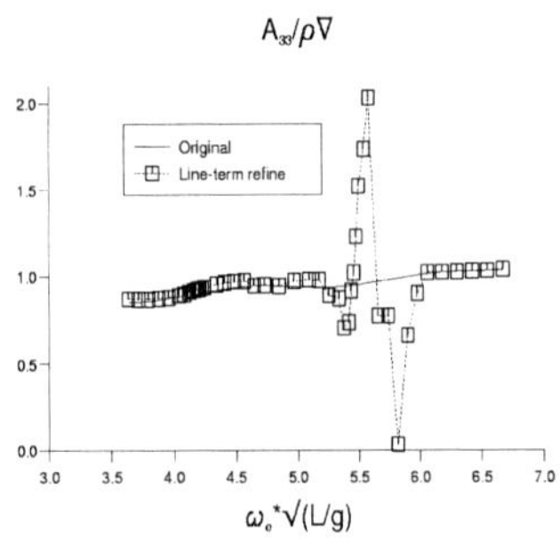

Figure 3: Comparison of the heave added mass Coefficients

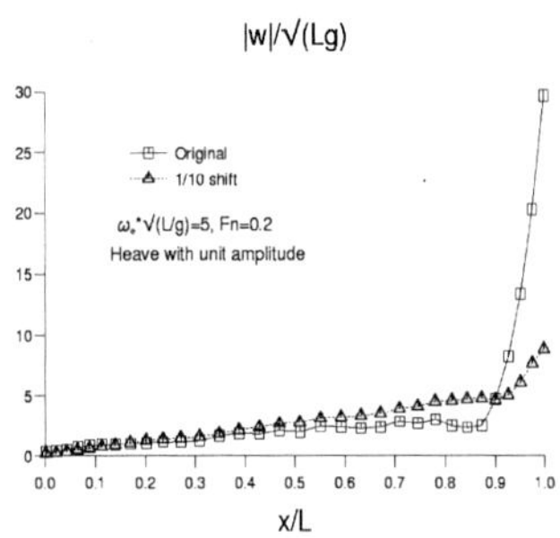

Figure 4: Vertical disturbed velocity along waterline panels

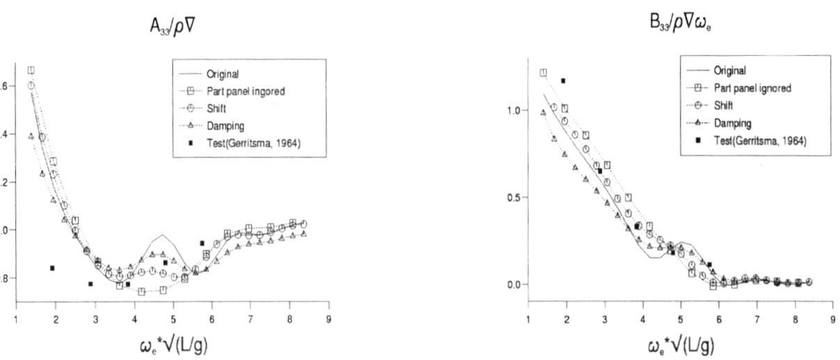

5 (a) 5 (b)

Figure 5: Comparisons of the heave hydrodynamic coefficients obtained by deferent methods

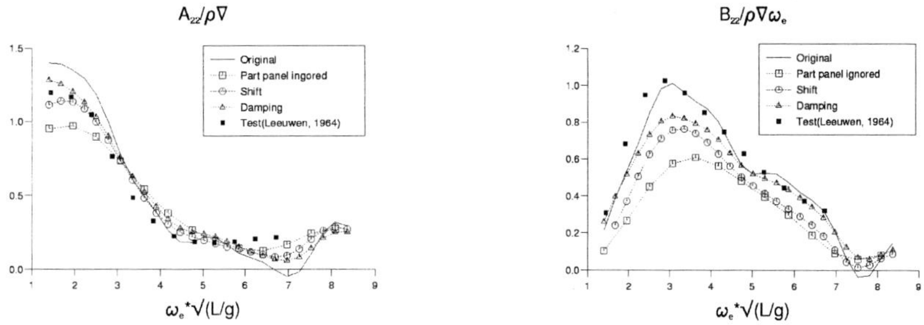

6 (a) 6 (b)

Figure 6: Comparisons of the sway hydrodynamic coefficients obtained by deferent methods

Practical Design of Ships and Other Floating Structures
You-Sheng Wu, Wei-Cheng Cui and Guo-Jun Zhou (Eds)

THE EFFECTS OF FORWARD SPEED ON HYDRODYNAMIC PRESSURE AND STRUCTURAL RESPONSE OF SHIPS IN WAVES

Chih-Chung Fang[1] Hua-Tung Wu[1] Hoi-Sang Chan[2] Chung-Yung Lu[1]

[1] United Ship Design and Development Center, Taiwan, China
[2] Department of Marine Technology, University of Newcastle, UK.

ABSTRACT

In this study, two different three-dimensional source distribution techniques, three-dimensional translating-pulsating source (3DT) and three-dimensional pulsating source (3DP) are used for the hydrodynamic load calculation of a high-speed monohull ship (RD-200) advancing in oblique waves. The numerical results are compared with experimental values obtained from wave load tests carried out at the SSPA. It is observed that the difference between the theoretical results obtained by these two modellings is insignificant for motion responses. However, the 3DP method demonstrates an erroneous fluctuation of wave load amplitudes in bow seas. The comparison of hydrodynamic pressure distribution obtained from two mathematical models has been discussed with different forward speeds. There are some discrepancies between two numerical models on the wave pattern and pressure distribution of ship. Some comparison results of structural analysis based on the hydrodynamic pressures calculated by 3DP and 3DT techniques have been shown and discussed.

KEYWORDS

Oblique waves, Three-dimensional source technique, Unsteady wave pattern, Hydrodynamic pressure, Structural response.

1 INTRODUCTION

To establish a confidence in the reliability of the hydrodynamic calculations for structural response analysis, it is important to investigate the influence of modelling for our current analytical method. The dynamic wave loads are the most significant ones, which are the ultimate criterion for the hull design of a ship. In order to ensure structural safety of ships in extreme hazard environments, it is necessary to estimate structural responses to waves in the design procedure.

With the advent of powerful computers, it is possible nowadays to produce the direct design method when ship travelling in waves. This design method can apply the hydrodynamic load directly on ship structural members for stress response analysis. In this paper, emphasis is given to the determination of dynamic loads rather than the structural FEM (finite element method) analysis procedure because structural analysis practices are well established.

The three-dimensional Green function method is the most popular technique to overcome insufficiencies of two-dimensional strip method in seakeeping analysis. Application of three-dimensional source distribution techniques to solving the ship-wave interaction problems was presented by Chan(1990). It was found that the correlation between experimental results and numerical values obtained by the three-dimensional pulsating source distribution method was satisfactory for the motions and sea loads except some discrepancies which were believed to be caused by the neglect of forward speed effects on the free surface. To investigating the effect of forward speed on ship structure, Fang et al.(1998) used two different 3D source distribution techniques for wave load calculation on a high speed monohull (RD-200) advancing in regular waves. It is found that the difference between those results of ship motions obtained by these two modelling is insignificant. However, the simplified three-dimensional oscillating source method demonstrates erroneous fluctuation of wave load amplitudes in bow seas.

In this study, using two different three-dimensional source distribution techniques, three-dimensional translating-pulsating source (3DT) and three-dimensional pulsating source (3DP), numerical investigations have been carried out for a monohull ship (RD-200) advancing in oblique waves with different forward speeds. The numerical results are compared with experimental values obtained from tests carried out at the SSPA. The comparison for the unsteady wave pattern and hydrodynamic pressure distribution obtained from two mathematical models has been discussed. Some comparison results of structural analysis based on the hydrodynamic pressures calculated by 3DP and 3DT techniques will be shown.

2 THEORETICAL BACKGROUND

In order to simplify the problem which a ship moving in oblique regular waves of frequency ω_0 at an angle of wave attack β (180^0 at head seas) undergoes oscillatory and lateral drift motions, we assume that the body maintains its nominal course at constant forward speed U and oscillates harmonically with encounter frequency ω. Figure 1 shows the right-hand coordinate system o-xyz which moves in the same direction and speed as the moving body. The x-axis is pointing upstream parallel to the longitudinal plane of the body and the z-axis is pointing vertically upward through the center of gravity of the body with the origin in the plane of the mean free surface. The body is assumed rigid and oscillates in six degrees of freedom about its mean position with complex amplitudes ξ_j (j=1,2,...6). Here j=1,2,3,4,5,6 refer to surge, sway, heave, roll, pitch, and yaw modes of motion respectively.

For dynamic equilibrium, the total wave-induced forces must be equal to the mass inertia forces and the coupled linear equations of motion of the rigid body can be written as

$$\sum_{k=1}^{6}\left\{(M_{jk}+A_{jk})\ddot{\xi}_k+B_{jk}\dot{\xi}_k+C_{jk}\xi_k\right\}=F_j^W+F_j^V \qquad j=1,2\ldots6 \qquad (1)$$

where $\ddot{\xi}_k$ and $\dot{\xi}_k$ are motion acceleration and velocity respectively; M_{jk} is the mass matrix; A_{jk} is the added mass; B_{jk} is the damping; C_{jk} is the restoring coefficient. F_j^W is the wave exciting force; F_j^V is the excitation force due to viscous effects. The indices j and k indicate the direction of the fluid force and the mode of motion respectively.

Within the framework of potential theory, the fluid is assumed to be ideal and incompressible. The unsteady motions of the ship and the fluid are assumed to be small so that the unsteady body boundary and free surface conditions can be linearised. The solution of the linearised unsteady forward motion is constructed by means of the three-dimensional Green's function for solving the boundary integral equation. The Green's function satisfies the three-dimensional Laplace's equation, the linearised free surface condition, the sea bottom condition and the far-field radiation condition. Chan(1990) presented a three-dimensional translating-pulsating source distribution technique (HSC-3D) which is satisfied the classical linearised free surface condition. Both divergent and transverse wave systems are account for.

A high-frequency assumption may be made that the frequency of oscillation ω is much higher than the differential operator $U\partial/\partial x$ in the free surface boundary condition which leads to the use of a simplified three-dimensional pulsating source distribution technique (3DP). For oscillating source model the Green function $G(p;q)$ is a zero speed limit of the translating-pulsating source function (3DT) and the water-line contours in the integral disappear.

The global structural loads of a vessel to waves, such as shear forces and bending moments, arise from imbalance between the inertia and the external forces. After the linear coupled equations of motion are solved in the frequency domain, the three-dimensional motion-induced radiation pressure distribution can be obtained together with quasi-hydrostatic pressure, the Froude-Krylov and diffraction pressure distributions. The dynamic global wave loads can be obtained by the integration of these hydrodynamic pressures and inertia loads.

3 HYDRODYNAMIC RESPONSE IN OBLIQUE WAVES

The ship motion and global wave load are the integrated effects with which a lot of factors not only hydrodynamical but also mechanical have been involved. The hydrodynamic pressure distribution and wave elevation around a ship will be more hydrodynamic phenomena than the resulting body motion and global force.

Hydrodynamic pressure $p_T(\vec{x};t)$ on the instantaneous wetted body surface is calculated by the Bernoulli equation for the unsteady flow:

$$p_T(\vec{x};t) = -\rho(\Phi_t + \frac{1}{2}\nabla\Phi \bullet \nabla\Phi - \frac{1}{2}U^2 + gz) \tag{2}$$

Expanding this pressure in a Taylor series about the undisturbed position of the hull and linearising by neglecting higher than first-order terms and ignoring steady pressure terms. This hydrodynamic pressure $p_T(\vec{x};t)$ can be splitted into four parts: The Froude-Krylov pressure $P_{FK}(\vec{x})$, diffraction pressure $P_D(\vec{x})$, radiation pressure $P_R(\vec{x})$ and dynamic restoring pressure $P_{qs}(\vec{x})$ as follows:

$$p_T(\vec{x};t) = (P_{FK}(\vec{x}) + P_D(\vec{x}) + P_R(\vec{x}) + P_{qs}(\vec{x}))e^{-i\omega t} \tag{3}$$

Than, the linearised hydrodynamic pressure over the body surface can be in the form

$$p_T(\vec{x};t) = \rho(i\omega + U\frac{\partial}{\partial x})\phi_j e^{-i\omega t} - \rho g(\xi_3 + \xi_4 y - \xi_5 x)e^{-i\omega t} \quad j=0,1,\ldots 7 \tag{4}$$

The model tests were carried out with a free-running 1:25 scale hull model in the SSPA Maritime Dynamics Laboratory, which has a basin with the dimensions of 88 * 39.5 * 3.5 meters. The model was self-propelled, controlled by an auto-pilot and totally free to move in any of the six degrees of freedom. Some of the main particulars are given in Table 1.

Figures 2 and 3 show some selected motion responses and global wave loads of the RD-200 at a nominal speed of 24 knots (Fn=0.42) in the regular waves. It is observed that the difference between the theoretical results obtained by these two modellings is insignificant for motion responses as shown in Figure 2. However, the 3DP method demonstrates an erroneous fluctuation of wave load amplitudes in bow seas (Figure 3). Results for the vertical bending moment along the ship are presented in Figure 4. There are some discrepancies between two numerical models observed around the middle part of ship. It shows that the 3DP technique gives lower vertical bending moment prediction for the RD-200 at V=24knots; $\omega_0 = 1.0 rad/\sec; \beta = 170^0$ but higher prediction at $\omega_0 = 1.2 rad/\sec; \beta = 170^0$ than the 3DT technique.

In order to illustrate the forward speed effects on the oscillating body, the free surface disturbances of the RD-200 due to the pulsating source (3DP) and the translating-pulsating source (3DT) are shown in Figures 5 (V=10knots; $\omega_0 = 1.0 rad/\sec; \beta = 170^0$). The wave contours are contained within the incident waves, radiation waves and diffraction waves. Figure 5 (right) indicates the free surface flow patterns behave like a pure oscillating with the 3DP model. The surface waves are simply the ring waves radiating outward in all horizontal directions. The effects of forward speed are clearly absent in this model. Since the 3DT technique can account both divergent and transverse wave systems, we may find the Kelvin wave patterns grow behind the body as shown in Figure 5 (left). Moreover, the transverse waves are more visible than the divergent waves in the illustrated contours.

An investigation of wave pattern and hydrodynamic pressure distribution is conducted by using 3D panel codes with different forward speeds. Firstly, a numerical validation has been carried out for the hydrodynamic pressure and wave elevation round the ship in Figures 6 (V=0.0knots; $\omega_0 = 1.0 rad/\sec; \beta = 170^0$). It is shown that the translating-pulsating Green function (3DT) will represent as the oscillating source when the Froude number F_n is equal to zero. The pressure distributions between two numerical techniques are very consist as shown in Figures 6(upper).

Numerical results of the wave pattern and pressure distribution of RD-200 on the ship hull are presented in Figure 6 (V=0,10,24 knots; $\omega_0 = 1.0 rad/\sec; \beta = 170^0$). There are some discrepancies observed on the comparison of wave pattern and pressure distribution of ship in oblique waves. The 3DP gives higher-pressure gradient distribution than the 3DT for the RD-200 at V=24knots; $\omega_0 = 1.0 rad/\sec; \beta = 170^0$. Furthermore, the hydrodynamic pressure distribution obtained from 3DP shows a significant fluctuation along the intersection of the free surface and the hull surface. From the physical point of view, the sophisticated 3DT modelling is much closer to reality than the simplified 3DP modelling. The 3DT technique can account both divergent and transverse wave system of ship advancing in waves.

3.1 Structural Analysis of the Global Coarse Mesh

The stress analyses have been made for a ship travelling in regular waves by using the finite element method (FEM) with different ship speed, wave direction (β) and wave frequency (ω_o). To investigate the effect of forward speed on the responses of ship hull structure in waves, three ship speeds (0, 10 and 24 knots) with one wave condition (β=170^0, ω_o =1.0 rad/sec) are selected to calculate the dynamic wave loads and perform the structural analyses. While both 3DP and 3DT source distribution techniques were adapted to calculate the dynamic wave loads for the structural analyses.

Figures 7 to 9 show the instantaneous ($\omega_e t=0$) dynamic stress distributions of the RD-200 with dynamic wave loads calculated by 3DT under various ship speeds (0, 10, 24 knots) respectively. In this study, two nodes within the high stress zone of main deck and bottom shell are selected to investigate the numerical results of structural responses with various dynamic loads. The amplitudes of dynamic longitudinal stresses of selected nodes are summarized in Table 2. Figure 10 shows the trend of main deck and bottom stresses with respect to ship speeds. The significant differences of calculated structural responses between 3DP and 3DT have been obtained as the ship speed increases.

4 CONCLUSIONS

Two different three-dimensional linearised potential techniques have been used for the calculations of ship motions and sea loads in oblique waves with different forward speeds. It is found that the difference between those results of ship motion responses obtained by two numerical modelling is insignificant. However, the simplified three-dimensional pulsating source method demonstrates erroneous fluctuation of wave load amplitudes in bow seas. For the fully three-dimensional problem of a moving ship in waves, the translating-pulsating source exactly satisfies the linearised free surface condition without further assumption in terms of the forward speed.

In this numerical investigation, there are some discrepancies observed from the wave pattern and pressure distribution for RD-200 travelling in oblique waves when the forward speed increases. The 3DP technique gives higher pressure gradient prediction than the 3DT method at V=24 knots. Furthermore, the hydrodynamic pressure distribution obtained from 3DP shows a significant fluctuation along the intersection of the free surface and the hull surface. The stress analyses of structure have been performed for the RD-200. It has been found that the significant differences of calculated structural responses between 3DP and 3DT will be obtained as the ship speed increases.

In order to establish a confidence in the reliability of the hydrodynamic calculations for structural response analysis, it is important to precise validate our current analytical method by measuring the wave flow or hydrodynamic pressure in the near future.

References

Chan, H. S. (1990). A Three-dimensional Technique for Predicting First and Second Order Hydrodynamic Forces on a Marine Vehicle Advancing in Waves, *Ph.D. Thesis*, Department of Naval Architecture and Ocean Engineering, University of Glasgow, UK.

Fang, C.C., Chan H.S. and Huang P.C. (1998). Prediction of Motions and Wave Loads of a Mono-hull Vessel in Oblique Waves, *Proc. Twelfth Asian Technical Exchange and Advisory Meeting on Marine Structures*, TEAM'98 KANAZAWA, Japan, 6-9 July, 177-185.

Wang, W.H., Chen, T.L. and Fang, C.C. (1999). Fatigue Life Analysis on Structural Hot Spot of a High-Speed Craft, *Proc. Thirteenth Asian Technical Exchange and Advisory Meeting on Marine Structures*, TEAM'99 KEELUNG, Taiwan, China 18-21 October,397-406.

Hua-Tung Wu, Fang, C.C. and Chan H.S. (2000). Dynamic Loadings for Structural Analysis of a Large Container Ship Based on 3D Green Function Method, *Proc. Fourteenth Asian Technical Exchange and Advisory Meeting on Marine Structures*, TEAM'2000 VLADIVOSTOK, Russia, 18-21 September.

Fang, C.C., Chan H.S. and Lu, C.Y. (2000). Numerical Study of Hydrodynamic Pressure on a High Speed Displacement Ship in Oblique Waves, *Proc. The Fourteenth Asian Technical Exchange and Advisory Meeting on Marine Structures*, TEAM'2000 VLADIVOSTOK, Russia, 18-21 September.

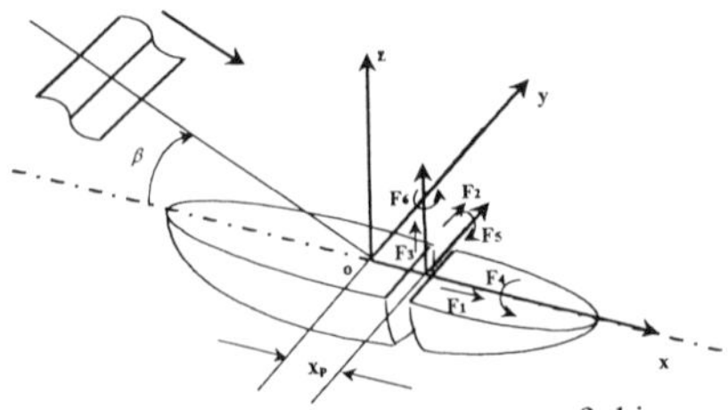

Figure 1: Coordinate system of ship

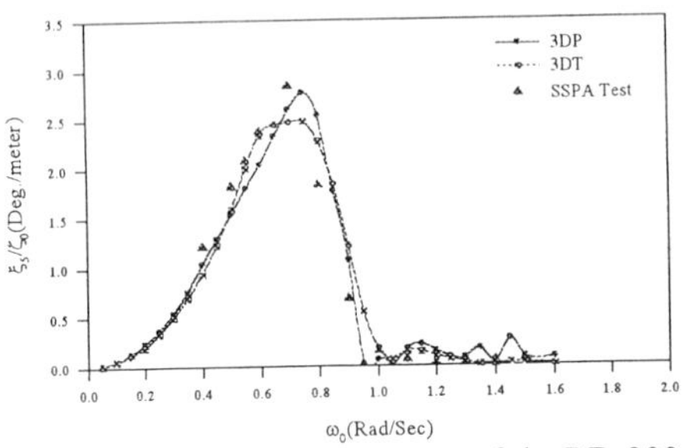

Figure 2: Pitch motion response of the RD-200 at V=24knots $(\beta = 170^0)$

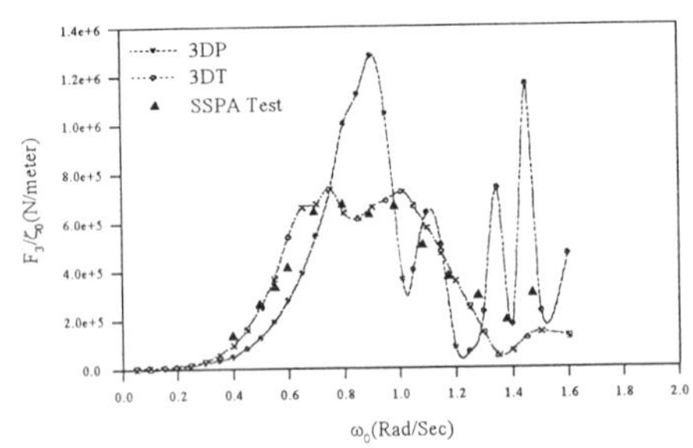

Figure 3: Vertical shear force of the RD-200 at V=24knots(x=50m from A.P.; $\beta = 170^0$)

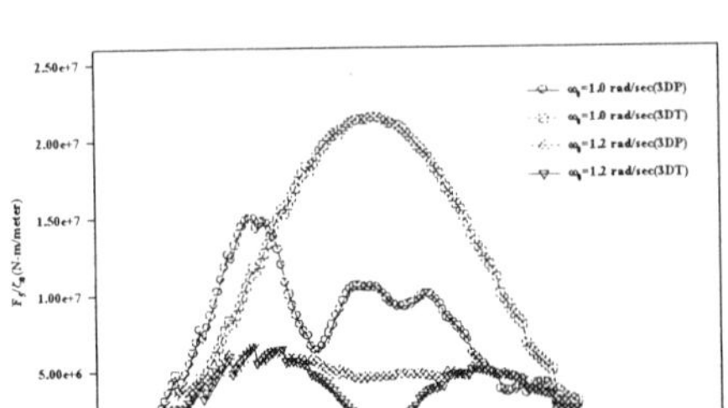

Figure 4: Bending moment along the RD-200 length at V=24knots($\beta = 170^0$)

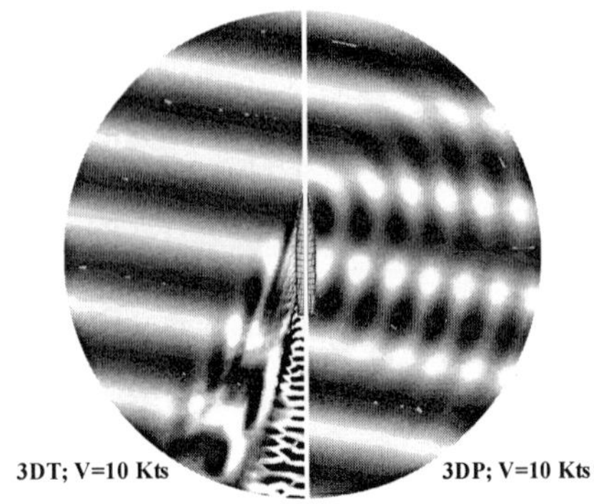

Figure 5: Wave pattern of the RD-200 (3DP &3DT;V=10kts,ω_0=1.0rad/sec,β=170^0;t=0.0 sec)

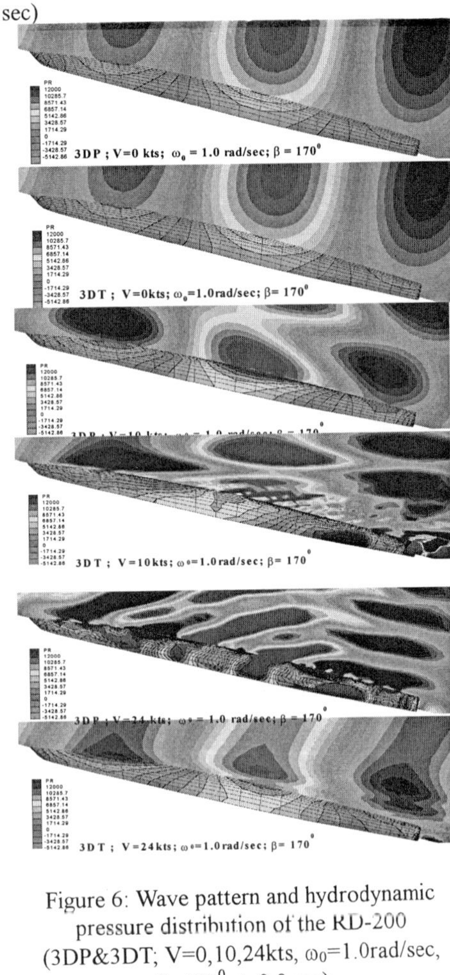

Figure 6: Wave pattern and hydrodynamic pressure distribution of the RD-200 (3DP&3DT; V=0,10,24kts, ω_0=1.0rad/sec, β=170^0; t=0.0 sec)

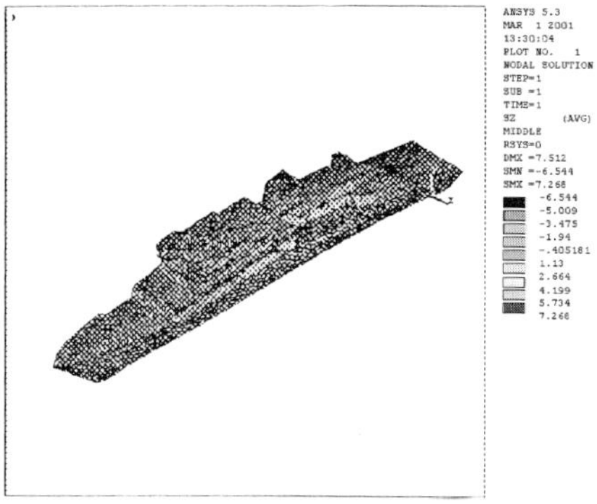

Figure 7: Stress distribution of the RD-200 (3DT, Vs=0kts, ω_0=1.0rad/sec, $\beta=170^0$, $\omega_e t=0$)

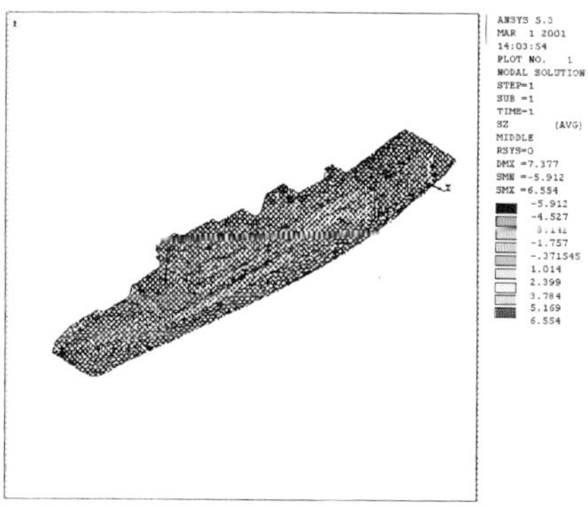

Figure 8: Stress distribution of the RD-200 (3DT, Vs=10kts, ω_0=1.0rad/sec, $\beta=170^0$, $\omega_e t=0$)

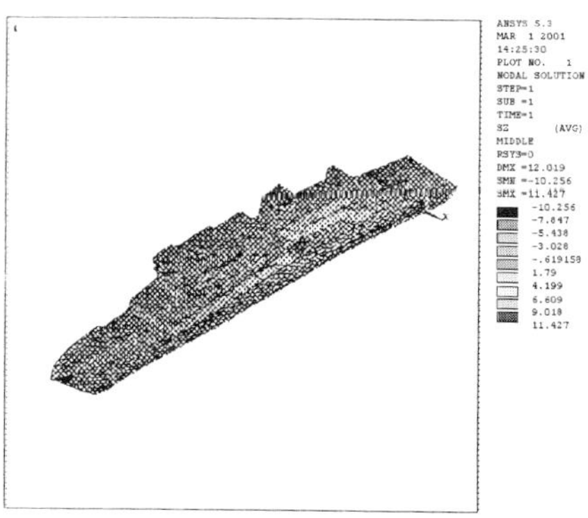

Figure 9: Stress distribution of the RD-200 (3DT, Vs=24kts, ω_0=1.0rad/sec, $\beta=170^0$, $\omega_e t=0$)

TABLE 1

PRINCIPAL PARTICULARS OF THE RD-200

	RD-200
LBP, L (m)	**90.00**
Beam overall, B (m)	**13.20**
Draught, D (m)	**3.55**
Displacement (tonnes)	**1840.50**
VCG (m)	**1.95**
LCG aft of F.P.(m)	**46.92**

TABLE 2

LONGITUDINAL STRESS AMPLITUDE (N/MM^2) OF THE RD-200 (ω_0=1.0RAD/SEC, $\beta=170^0$)

Vs	3DP		3DT	
(knots)	Deck	Bottom	Deck	Bottom
0	7.560	6.616	7.560	6.616
10	10.917	9.560	10.613	8.746
24	15.184	13.605	11.584	10.293

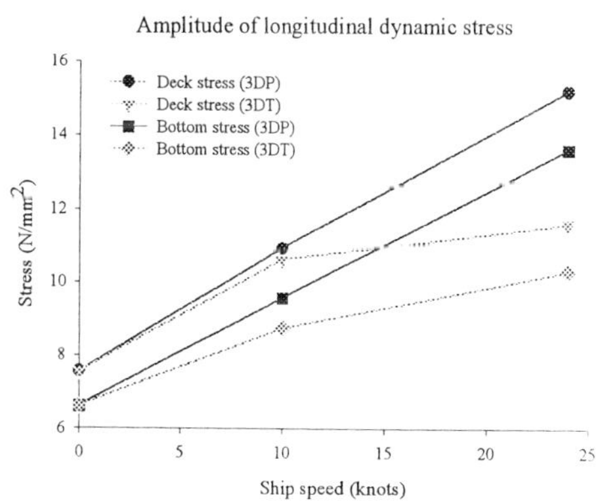

Figure 10: Comparison of longitudinal stresses of the RD-200 with different forward speed (ω_0=1.0rad/sec, $\beta=170^0$)

Practical Design of Ships and Other Floating Structures
You-Sheng Wu, Wei-Cheng Cui and Guo-Jun Zhou (Eds)

SHIP MOTIONS AND SEA LOADS BY A 3D RANKIN PANEL METHOD

Li Xu[1], Wei-Xing Zhang[1], Chen-Bi Zhao[2], Fa-Yan Xu[1] and You-Fang Chen[1]

[1]China Classification Society(CCS), Beijing 100006, China
[2]Wuhan Science and Technology University, Wuhan 430063, China

ABSTRACT

The paper is concerned with ship motions and sea loads by a 3D Rankine Panel Method to assess the ship structural strength. The method was used to solve the linearized ship motions and sea loads for a ship with a steady forward speed in quasi-random incident waves. The solution algorithm involves the discretization of the ship hull and part of the free surface by panels, over which a B-spline approximation is employed. The method is for the solution of the wave-body interactions in frequency domain and will be developed in time domain to solve the non-linear problem in the near future. In the paper, results are shown for a Wigley hull and a large container ship with and without a forward speed. The hydrodynamic pressures exerted on the ship are found in remarkable agreement with related experimental measurements carried out by China Classification Society or the other referenced results, demonstrating that the solution scheme is available for the 3D flow and sea loads.

KEYWORDS

Sea loads, 3D Rankine Panel method, B-spline, Free surface

1 INTRODUCTION

3D frequency and time domain methods based upon the potential theory are widely applied to solve the wave loads. Over the past 20 years, Rankine source panel methods have been successfully employed to obtain the results of free-surface boundary value problem ranging from the linearized wave resistance and ship motion problems to nonlinear breaking wave simulations. But the geometry of ship is complex and it is important to describe the hull exactly to avoid the divergence of numerical simulation for the 3D flow of the ship and sea loads. In the paper, the free-surface and body boundary conditions are linearized about the double-body flow at the presence of a forward speed and the solution algorithm involves the discretization of the ship hull and part of the free surface by panels, over which a rational non-uniform B-spline approximation is employed. The method is for the solution of the wave-body interactions in frequency domain and will be developed in time domain to solve the non-linear problem in the near future.

2 GOVERNING EQUATIONS

2.1 Velocity Potentials

The total velocity potentials can be decomposed into three parts as follows:

$$\Phi_T = \Phi + \phi + \varphi \tag{1}$$

Where φ is unsteady potential as:

$$\varphi = \mathrm{Re}\left\{ e^{i\omega t} \left[A(\overline{\varphi}_0 + \varphi_7) + \sum_{j=1}^{6} \xi_j \varphi_j \right] \right\} \tag{2}$$

Φ is the double-body potential as:

$$\Phi = -Ux + \phi^s \tag{3}$$

ϕ^s is the disturbed potential and satisfied as follows:

$$\begin{aligned} &[F] \qquad \frac{\partial \phi^s}{\partial z} = 0 \qquad z = 0 \\ &[B] \qquad \frac{\partial \phi^s}{\partial n} = Un_1 \end{aligned} \tag{4}$$

The linearized conditions of free surface for the steady and unsteady disturbed potentials are respective as follows: (Nakos1990)

$$\begin{aligned} \nabla\Phi \cdot \nabla(\nabla\Phi \cdot \nabla\phi) + \frac{1}{2}\nabla(\nabla\Phi \cdot \nabla\Phi) \cdot \nabla\phi + g\phi_z - \Phi_{zz}(\nabla\Phi \cdot \nabla\phi) = \frac{1}{2}\nabla(\nabla\Phi \cdot \nabla\Phi)\nabla\Phi \\ -\frac{1}{2}(U^2 - \nabla\Phi \cdot \nabla\Phi)\Phi_{zz} \qquad On \quad z = 0 \end{aligned} \tag{5}$$

$$\begin{aligned} \varphi_{tt} + 2\nabla\Phi \cdot \nabla\varphi_t + \nabla\Phi \cdot \nabla(\nabla\Phi \cdot \nabla\varphi) + \frac{1}{2}\nabla(\nabla\Phi \cdot \nabla\Phi) \cdot \nabla\varphi + g\varphi_z - \Phi_{zz}(\varphi_t + \nabla\Phi \cdot \nabla\varphi) = 0 \\ On \quad z = 0 \end{aligned} \tag{6}$$

2.2 Boundary Value Problem

Based on the above equations and the linearized conditions, the boundary value problems of steady, radiation and diffraction potentials are obtained respectively as follows.
For steady potential, the boundary value problem formulation can be written below:

$$\begin{aligned} &[L] \qquad \nabla^2 \phi = 0 \\ &[F] \qquad \left.\frac{\partial \phi}{\partial z}\right|_{z=0} = S_x\phi_x + S_y\phi_y + S_{xx}\phi_{xx} + S_{yy}\phi_{yy} + S_{xy}\phi_{xy} + SS \\ &[B] \qquad \frac{\partial \phi}{\partial n} = 0 \end{aligned} \tag{7}$$

For radiation potential and diffraction potential, the boundary value problem formulations can also be written same as one of the steady potential. The formulations and the other nomenclature except the paper can be seen in the reference[5] because of the maximum length for the paper. The integral equation to determine the different velocity components is derived as follows:

$$\varphi(\vec{x}) - \iint_{FS} \frac{\partial\varphi(\vec{x}')}{\partial z} G(\vec{x},\vec{x}')d\vec{x}' + \iint_{FS+B} \varphi(\vec{x}') \frac{\partial G(\vec{x},\vec{x}')}{\partial \vec{n}} d\vec{x}' = \iint_{B} \frac{\partial\varphi(\vec{x}')}{\partial \vec{n}} G(\vec{x},\vec{x}')d\vec{x}' \tag{8}$$

2.3 Motion Equations and Forces

The excited forces can be obtained from the above formula:

$$X_i = -\rho \operatorname{Re}\left\{\iint_B [i\omega(\overline{\varphi}_0 + \varphi_7) + \nabla\Phi \cdot \nabla(\overline{\varphi}_0 + \varphi_7)] n_i ds\right\} \tag{9}$$

Then the motion equation can be derived as follows:

$$\sum_{j=1}^{6} [-\omega^2(m_{ij} + a_{ij}) + i\omega b_{ij} + c_{ij}]\xi_j = X_i \qquad i = 1,\cdots,6 \tag{10}$$

The pressures and forces are obtained below:

$$p - p_a = -\rho\left[\frac{1}{2}\nabla(\Phi+\phi)\cdot\nabla(\Phi+\phi) - \frac{1}{2}U^2 + gz\right]_{\vec{x}\in\overline{B}} - \rho(\varphi_t + \nabla\Phi\cdot\nabla\varphi)_{\vec{x}\in\overline{B}} - \rho\left[(\vec{a}\cdot\nabla)(\frac{\nabla\Phi\cdot\nabla\Phi}{2} + gz)\right]_{\vec{x}\in B} \tag{11}$$

$$F_i(t) = \operatorname{Re}\left\{e^{i\omega t}\left[X_i + \sum_{j=1}^{6}\xi_j(\omega^2 a_{ij} - i\omega b_{ij} - c_{ij})\right]\right\} \quad i = 1,\cdots,6 \tag{12}$$

3 NUMERICAL METHODS

3.1 Hull Expressions

The paper applies rational non-uniform B-spline functions defined by parameters to express the hull surface, i.e.

$$\vec{P}(u,v) = \sum_{i=0}^{M_B}\sum_{j=0}^{N_B} \vec{B}_{ij} N_{i,k}(u) N_{j,l}(v) \tag{13}$$

Where $\vec{P}(u,v)$ is defined as the point of hull surface described by (u,v).

$\vec{B}_{ij}$ is the control grid of the B-spline functions.

$N_{i,k}(u)$ and $N_{j,l}(v)$ are the elementary function of k and l order of B-spline as follows:

$$N_{i,k}(t) = \begin{cases} 1 & t \in [t_i, t_{i+1}] \\ 0 & Other \end{cases} \tag{14}$$

$$N_{i,k}(t) = \frac{t - t_i}{t_{i+k} - t_i} N_{i,k-1}(t) + \frac{t_{i+k+1} - t}{t_{i+k+1} - t_{i+1}} N_{i+1,k-1}(t) \tag{15}$$

So the $\vec{P}(u,v)$ can be derived by De Boor algorithm and the hull surface can be expressed by skinning method shown as the following figures 1-2.

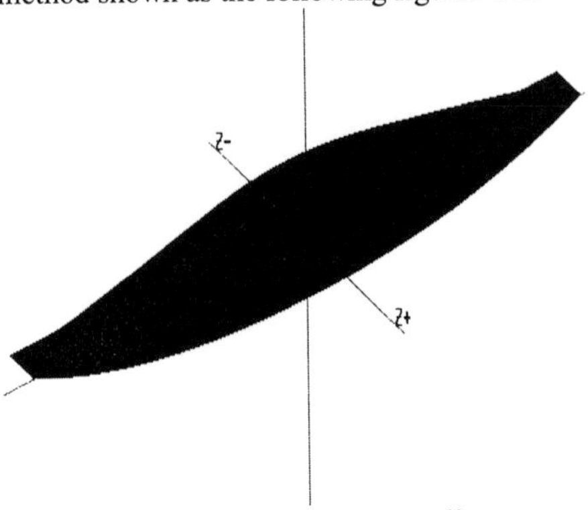

Figure 1: 3D Wigley Hull

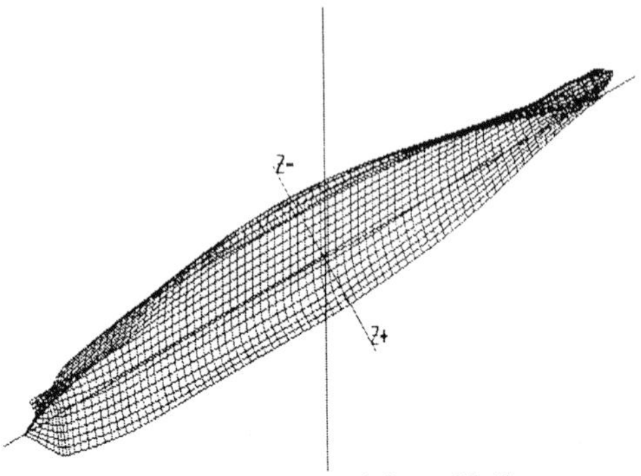

Figure 2: 3D Flokstra Hull

3.2 Solution for Velocity Potentials

Based upon the rational non-uniform B-spline function for hull surface, the disturbed potential ϕ^s can be discretized as follows:

$$\phi^s(u,v) = \sum_{i=0}^{M_\phi}\sum_{j=0}^{N_\phi}\Psi_{ij}N_{i,g}(u)N_{j,h}(v) \tag{16}$$

So the integral equations can be derived below:

$$\sum_{i=0}^{M_\phi}\sum_{j=0}^{N_\phi}\Psi_{ij}f_{ij}(u_s,v_t) = b(u_s,v_t) \qquad \begin{array}{l} s = 0,1,\cdots,M_\phi \\ t = 0,1,\cdots,N_\phi \end{array} \tag{17}$$

Where $f_{ij}(u_s,v_t)$ and $b(u_s,v_t)$ can be seen in the reference[5].
The disturbed potentials are also derived as well as the integral equations:

$$\sum_{i=0}^{M^F}\sum_{j=0}^{N^F}\Psi_{ij}^F f_{ij}^F(x,y) + \sum_{i=0}^{M^B}\sum_{j=0}^{N^B}\Psi_{ij}^B f_{ij}^B(x,y) = b(x,y) \tag{18}$$

$$\sum_{i=0}^{M^F}\sum_{j=0}^{N^F}\Psi_{ij}^F f_{ij}^F(u,v) + \sum_{i=0}^{M^B}\sum_{j=0}^{N^B}\Psi_{ij}^B f_{ij}^B(u,v) = b(u,v) \tag{19}$$

The radiation conditions corresponding to the above equations are following:

$$\frac{\partial\phi^F(x,y)}{\partial x} = \sum_{i=0}^{M^F}\sum_{j=0}^{N^F}\Psi_{ij}^F\frac{\partial N_{i,q}(x)}{\partial x}N_{j,r}(y) = 0 \qquad x = x_0 \tag{20}$$

$$\frac{\partial^2 \phi^F(x,y)}{\partial^2 x} = \sum_{i=0}^{M^F}\sum_{j=0}^{N^F} \Psi_{ij}^{\prime F} \frac{\partial^2 N_{i,q}(x)}{\partial^2 x} N_{j,r}(y) = 0 \qquad x = x_0 \tag{21}$$

The other potentials based on the B-spline functions and the radiation conditions can be also derived.

If the collocated points are available on the hull and free surfaces to set up a set of closed linear equations, the velocity potentials can be determined effectively. The results of the collocated points for Wigley and Flokstra container ship are shown in the table 1.

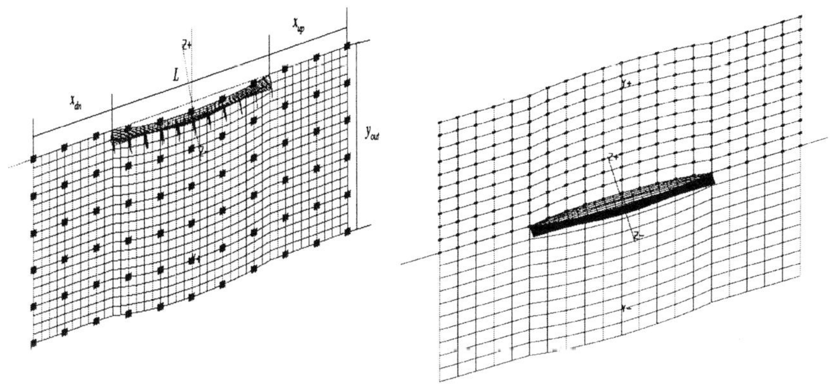

Figure 3: The collocated points for Flokstra container ship

4 CALCULATION RESULTS AND CONCLUSIONS

The parameters of Wigley and Flokstra Hull are shown in table below.

TABLE 1
PARAMETERS FOR NUMERICAL CALCULATION

Parameters	Wigley Hull	Flokstra Hull
Truncation distance x_{up} /L	0.5	0.5
Truncation distance x_{dn} /L	0.5	0.5
Truncation distance y_{out} /L	1.0	1.0
Collocated point number on free surface	21x11	21x11
Collocated point number on hull surface	21x6	21x6
Froude number F_r	0.3	0.27
Froude number of longitudinal collocated points F_h	1.342	1.208
B-spline order number of velocity potential of hull surface $g \times l$	3x3	3x3
B-spline order number of velocity potential of free surface $q \times r$	3x3	3x3

Some of the calculation results are shown for a Wigley hull and Flokstra container ship. The coefficients of added mass, damping and hydrodynamic pressures exerted on the ship are found in remarkable agreement with related experimental measurements.

Based on the numerical results, the conclusion can be drawn that the 3D Rankine Panel method for large container ship is available and effective, and the solution scheme can be developed to long-term prediction of the sea loads and the non-linear calculations which have been carried out in CCS.

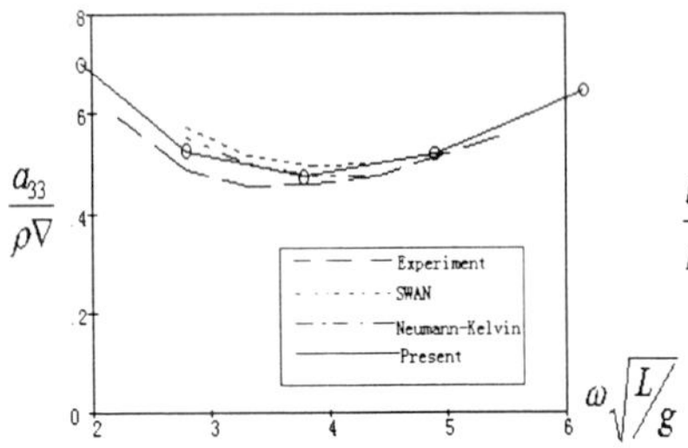

Figure 4: The coefficient a_{33} of added mass for Wigley

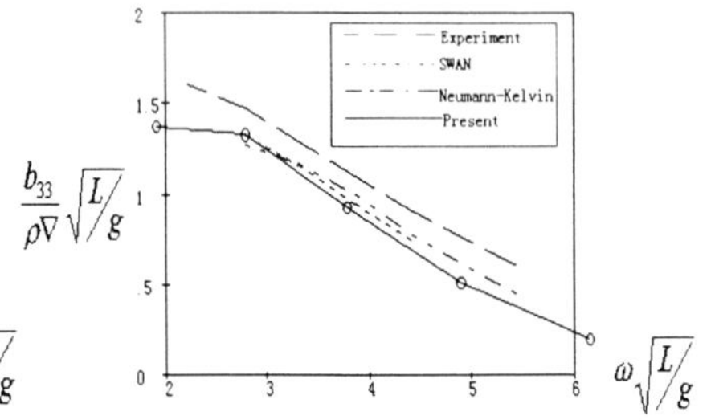

Figure 5: The coefficient b_{33} of damping for Wigley

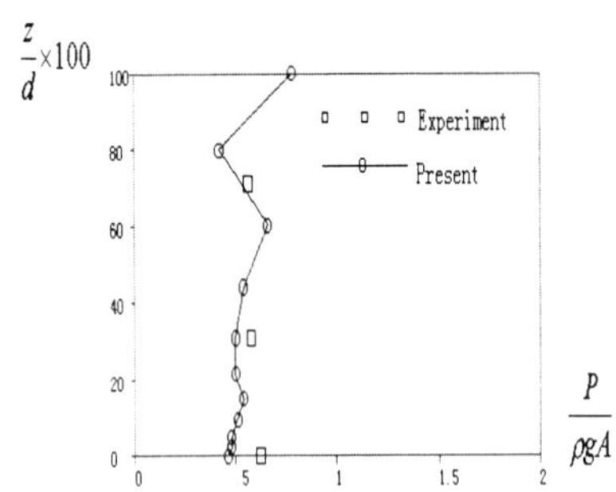

Figure 6: The pressure distribution at station 5 for Flokstra (λ /L=0.7 2A/L=0.028)

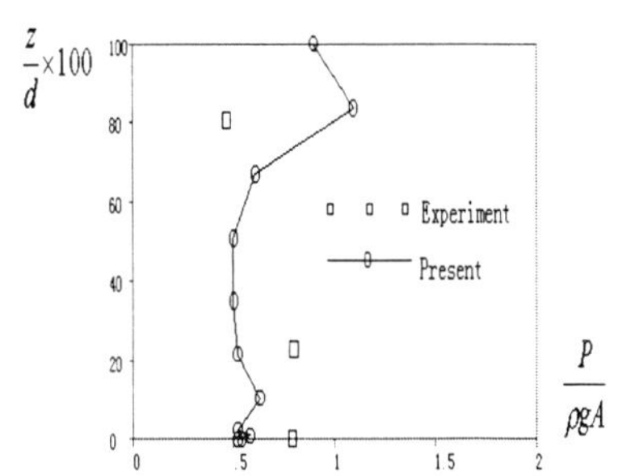

Figure 7: The pressure distribution at station 16 for Flokstra (λ /L=0.7 2A/L=0.028)

References

Beck, R.F. (1996). Application of Modern Numerical Method in Marine Hydrodynamics, *ANAME* **Vol. 104.**

Kouh, J-S, Ho C-H. (1996). A High Order Panel Method based on Source Distribution and Gaussian Quadrature. *Ship Tech.Res.* **Vol.43.**

Maniar, H.D.(1995). A 3-D Heigher Order Panel Method based on B-Spline, *PH.D. Thesis MIT*

Nakos D.E., Sclavounos P.D. (1990). Steady and Unsteady Ship Wave Patterns, *J.F.M* , **Vol.215.**

Zhao C. B. And Zou Z. J. (1999). A 3D Patential Flow Computing Method based on NURBS, *14th International Workship on W.W & F.B.*

You-Sheng Wu, Wei-Cheng Cui and Guo-Jun Zhou (Eds)

EXPERIMENT ON EXTREME WAVE LOADS OF A FLEXIBLE SHIP MODEL

Rui-Zhang Chen Shuang-Xing Du You-Sheng Wu
Ji-Ru Lin Jia-Jun Hu Ya-Lin Yue

China Ship Scientific Research Center, Wuxi, Jiangsu 214082, China

ABSTRACT

A self-propelled, flexible model of the S175 container ship made of elastic materials was manufactured and tested. The purpose of the test is to study the wave-induced loads and responses of the ship in severe waves, mainly focusing on the non-linearity of the loads with respect to the wave heights, the biasing of sagging relative to hogging bending moments. The test data has been analyzed and compared with the numerical results. It is concluded that to thoroughly assess the non-linear behavior of the wave loads, the flexible effect of the ship hull girder should be carefully considered, and the wave tests of a flexible ship model would be a useful approach to gain more insight towards this purpose.

KEYWORDS

Extreme wave loads, Hydroelasticity, Model test, Flexible model

1 INTRODUCTION

Non-linear effects on wave-induced loads and motions are often significant for a ship traveling in severe waves. These in turn are very important in design and strength assessment of ships in service. Design codes and rules for wave loads are initially determined based on both theoretical and experimental investigations.

For several decades comparative studies on theoretical and experimental results of the S175 ship models have been widely carried out in the world for verification of theoretical predictions, and for development of empirical design methods. The results have also been provided for validation of experimental procedures adopted at different wave tanks. Four organizations, namely IHI, CSSRC, NKK and SHI carried out model tests of the S175 container ship, and contributed the results to the 16th ITTC Seakeeping Committee (16th ITTC report, 1981). At the 18th Seakeeping Committee (18th ITTC), 23 organizations participated the comparative model tests of the S175 ship and investigated the test procedures and data analysis methods, among which 19 organizations submitted the results to the committee report. Comprehensive database of the experimental results covering different wave conditions and the corresponding responses of the S175 models is available for

validation of linear procedures. Rather large scatter in the results of heave and pitch transfer functions in head seas was observed. It appeared that the scatter was not simply random, but correlated to a function of wave steepness (O'Dea, et al., 1992).
In order to determine the non-linearity in vertical ship motions, systematic tests of a S175 model with the length of 3.5 meter were carried out at the Hydromechanics Laboratory of the U.S. Naval Academy (O'Dea, et al., 1992). Heave, pitch motions and vertical acceleration at bow were measured in various wave conditions. The variations of the heave and pitch transfer functions with regard to the wave amplitude were found to be non-linear in the wave frequency range. In order to predict this nonlinear behavior, it was mentioned that a Volterra series expansion to the cubic order, or a full nonlinear approach without any serious approximation should be used. To assess the influence of the wave amplitude on the transfer functions of heave, pitch and vertical bending moment, Tao and Incecik (1998) carried out a segmented model test of S175 running with two Froude numbers in regular head waves. The transfer functions were given as functions of non-dimensional wave frequencies and steepness. With the same purpose, Fonseca and Soares (1998) also carried out a model test of a segment S175 model both in regular and irregular head waves.
The first experiment on flexible ship model to investigate the longitudinal strength was carried out in Japan by Sato (1954). The model was constructed of brass plate with its longitudinal members being scaled in more or less simplified form. Heave and pitch motions, stresses and wave height were measured. Akita and Ochi (1955) used the same material for the model test of ship strengths in waves. It is noticed that in order to obtain a measurable stress, rather large model scale was necessary in the case of a metal model.
De Does (1960) used fiberglass polyester laminates to construct three 2.479m long ship models for measuring bending moments in waves. Fukasawa (1981) chose foamed vinyl chloride to make the elastic model of a large bulk carrier. Later on, synthesized resin and foamed urethane, and aluminum plates attached at both sides of the ship model were employed for the elastic models of a bulk carrier and a patrol ship (Watanabe, et al., 1984, 1985). In order to obtain information about the effects of bow flare to the deck wetness and asymmetry phenomenon of the vertical wave bending moments, Watanabe (1989) produced two 4.5m long flexible S175 models with different bow flare forms using synthetic resin and foam urethane, and tested in regular and irregular waves.
The application of plastic materials enables an entirely elastic ship model, with acceptable not too large scale, being manufactured to allow for the satisfaction of the similitude principle. This provides the possibility of measuring and obtaining much more information than the segmented model about the structural responses over the whole hull of the ship, including the bending moments, shearing forces, torques at any cross sections, and stresses at local hot spots. Generally speaking, the elastic ship model satisfies the geometric similarity of the hull form, hydrodynamic similarity, together with the structural similarity with regard to the global vertical bending and shearing, and hence may be used to predict the hull girder wave loads, motions and the global structural responses. The material chosen for this kind of model should basically have the following properties (Lin, et al., 1991):
(a). Young's modulus of the material should be less than that of steel by an order of 10^{-2} and its Poisson ratio should be close to that of steel.
(b). The material is isotropic and has a comparatively large region of linear strain-stress relationship.
(c). Within the linear region, the material has stable mechanical properties and no distinguishable creepage in the usual atmospheric temperature.
(d). Formalizing and adhering may be easily performed.
A kind of ABS702 material was fund to be of acceptable mechanical properties, as shown in Table 1 (Lin, et al., 1991). Lin, et. al. (1991, 1996) tested two flexible models made of ABS702 in the wave basin of CSSRC. In the present work a self-propelled flexible S175 model of the same material was tested in the towing tank of CSSRC. The main purpose of the test is to study the wave loads, especially focusing on the non-linearity of the loads with respect to the wave heights, the biasing of sagging relative to hogging bending moments. The test data is compared with the numerical results.

2 EXPERIMENT

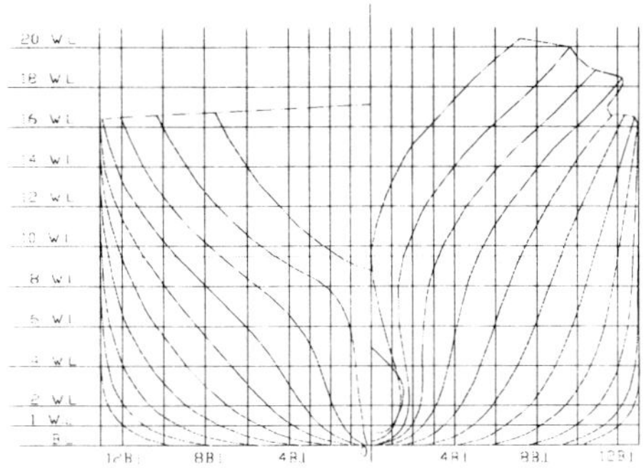
Figure1: Body plan of the model

The principal particulars and the body profile of the model are shown respectively in Table 2 and Fig.1. The length between perpendiculars of the model is 3.6m, the thickness of plastic plate for the hull is 2mm, while the thickness of the plastic plate for bulkhead, keel and girder located on deck is 4mm. All together 21 transverse bulkheads setting at each station (from $0^{\#}$~$20^{\#}$) were used to reinforce the transverse stiffness of the model and to supply the places to locate the ballast blocks. The weight and locations of the ballast blocks were suitably arranged to satisfy the similitude of the weight distribution along the length and the location of the gravity center of the ship.

TABLE 1
MECHANICAL PROPERTIES OF ABS702

Item	Value	Temperature condition
Young's Modulus	2.84×10^9 N/m^2	8°–29° C
Poisson ratio	0.343	8°–29° C
Density	1.09 g/cm^3	
Rate of water absorption	1.69%	25° C
$\lvert E_{mx} - E_{my}\rvert / E_{mx}$	< 2.61%	8° -29° C
Linear region	ε<2000$\mu\varepsilon$	

TABLE 2
PRINCIPAL PARTICULARS OF THE S175 MODEL

Items	Prototype	Model
L_{pp} (Length between perpendiculars, m)	175.00	3.6
B (Beam, m)	25.40	0.523
D (Depth, m)	15.40	0.317
T (Draught, m)	9.50	0.195
Δ (Displacement mass)	24742t	215.25kg
K_{yy} (Radius of longitudinal gyration)	$0.236L_{pp}$	$0.236L_{pp}$
FG (Longitudinal center of gravity from forward perpendicular, m)	90.0	1.851
GM (Metacenter height, m)	1.0	0.021
KM (Transverse metacenter above keel, m)	10.52	0.216
EI (Bending rigidity at midship, kg·mm^2)	2.28×10^{18}	10.66×10^{9}
f (2-node frequency, Hz)	1.60	12.57

According to the requirement of Extreme Hull Girder Wave Load Committee of ISSC'2000, the measuring items were vertical bending moments, vertical acceleration at the FP, and heaving, pitching motions. The 2-node frequency of the model floating in still water, and the relative wave making vertical bending moment at a certain traveling speed were also measured. The model test was carried out in the towing tank of CSSRC in order to have longer record during each run. The experiment

conditions are listed in Table 3, including both regular and irregular waves.
The model was self-propelled, free to surge, heave, pitch and vertical structural vibrations, however restrained for sway, roll and yaw. Heave and pitch were measured by means of 4-component motion measuring device. A Servo needle type wave height meter was attached to the towing carriage in a fixed position at about 4.5 meters in front of the model. The bending moments were measured using

TABLE 3
EXPERIMENT CONDITIONS

Fn = 0.275			Head sea
Case	Wave	Wave height(mm)	Wave length (λ /L_{pp})
1	regular	$2a_w$=L/50=72	0.6, 0.7, 0.8, 0.9, 1.0, 1.1, 1.2, 1.3, 1.5, 2.0
2	regular	$2a_w$=L/40=90	0.5, 0.6, 0.7, 0.8, 0.9, 1.0, 1.1, 1.2, 1.3, 1.4, 1.5, 2.0
3	regular	$2a_w$=L/30=120	0.6, 0.7, 0.8, 0.9, 1.0, 1.1, 1.2, 1.3, 1.4, 1.5, 1.8, 2.0
4	regular	$2a_w$=L/20=180	1.0, 1.1, 1.2, 1.3, 1.4, 1.5, 1.6, 1.7, 1.8, 1.9, 2.0, 2.2
5	irregular	a ITTC wave spectrum of $H_{1/3}$=0.11 m and T_1=1.20 sec	

strain gauges longitudinally distributed on the starboard side-deck plate. The relationships between the measured strains and the vertical bending moments were obtained by the static calibration method. The locations of these measurement points were:

M1 near Station 5 (95 cm from AP); M2 at Station 7.5;
M3 near Station 10 (185 cm from AP); M4 at Station 12.5;
M5 near Station 15 (265 cm from AP).

Before each wave test when the model was at rest in still water, the strain signals were set to zero. Therefore, the measured bending moments of the model traveling in waves were regarded as the differences from the static bending moments in still water, however with the wave-making components being included. Thus the sagging and hogging phenomena were easily distinguished. The time history of the irregular waves was generated by a white noise wave generating system with the truncated frequency 2.0Hz. The vertical bending moments at Mj (j =1~5) along the model in irregular waves were obtained from the strain measurements.

3 RESULTS AND DISCUSSIONS

The non-dimensional amplitudes of pitch ϑ, heave Z, vertical acceleration $\ddot{Z}$ and bending moments are respectively represented as:

$$C_\theta = \frac{\vartheta}{ka},\quad C_Z = \frac{Z}{a},\quad C_A = \frac{L\ddot{Z}}{ga},\quad C_{Mj} = \frac{M}{D}$$

where L denotes ship length (L_{pp}); g, gravity acceleration; a, wave amplitude; k, wave number. The subscript Mj (j=1~5) indicates the measurement points. M and the denominator D denote respectively any component of the vertical bending moment at position Mj, and the corresponding parameter used for non-dimensionalization defined as

$D=gL^2Bh_W$, for $M=M^{(1)}$ (the double amplitude of the first harmonic bending moment);

$D=\rho gL^2Ba^2/d$, for $M=M^{(2)}$ (the amplitude of the 2-nd harmonic bending moment);

$D=\rho gL^2Ba^3/d^2$, for $M=M^{(3)}$ (the amplitude of the 3-rd harmonic bending moment);

$D=\rho gL^2Ba$, for $M=M^{2\text{-nod}}$ (the amplitudes of bending moment at 2-node flexible mode frequency), M^{sag} and M^{hog} (sagging and hogging bending moments) respectively;

where B, d, ρ and h_w represent respectively ship breadth, draft, fluid density, and wave height.

The 2-node flexible mode frequency and the structural damping ratio of the model floating in still water were determined as 9.4Hz and 0.067 respectively from the time history of amidships

impulse-exciting responses as showed in Figure 2.
The distribution of the wave-making bending moments along the length of the model traveling with the forward speed Fn=0.275 in still water was measured and shown in Figure 3.

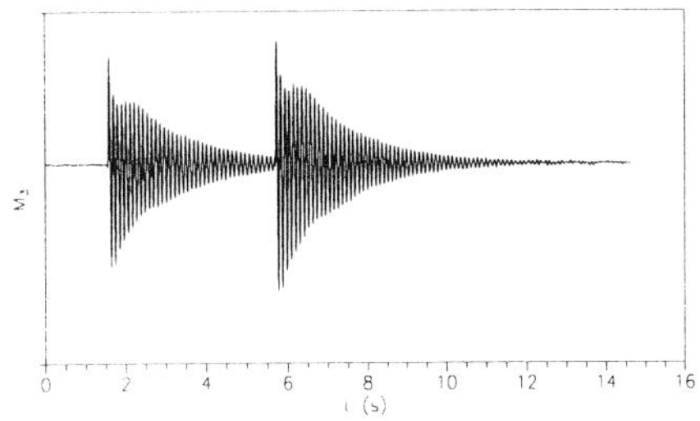

Figure2: Time history of the impulse response moments along the ship (a/L=1/60)

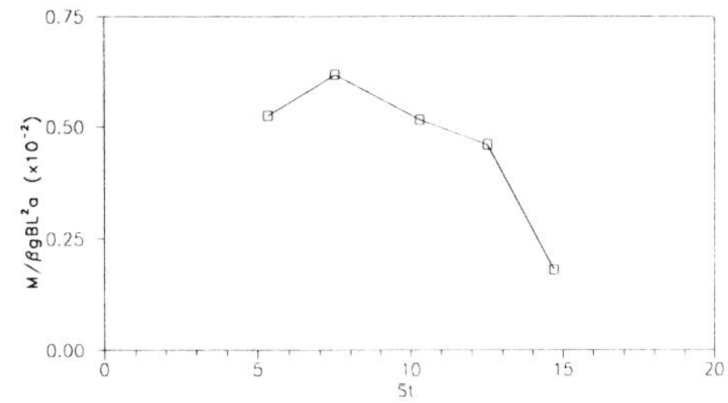

Figure3: Distribution of wave-making bending

In the regular wave tests, nearly ten complete cycles were analyzed for each wave frequency and amplitude combination. The frequency analysis of the measured time records in regular head seas was performed using FFT. The corresponding frequency domain expressions were separated into several groups at the first, second and third harmonics of the wave encounter frequency, as well as the higher frequency near the 2-nodes flexible mode frequency of the hull respectively. These were again transformed to time domain expressions adopting IFFT. The amplitudes of the first, second and third harmonic frequency response functions with respect to wave steepness, as well as the sagging and hogging bending moments are presented and compared in the present paper in Figures 4~8.

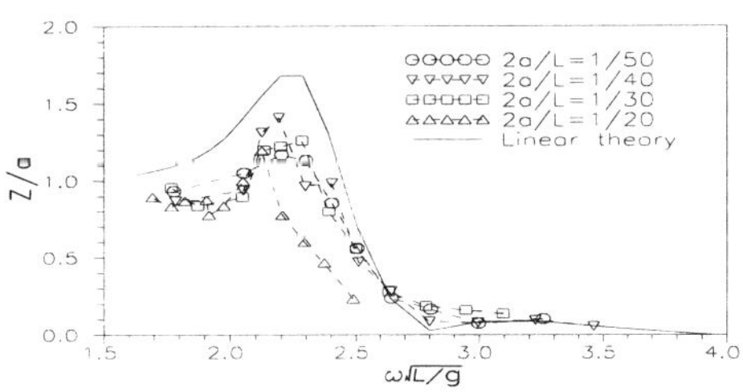

4(a) Heave

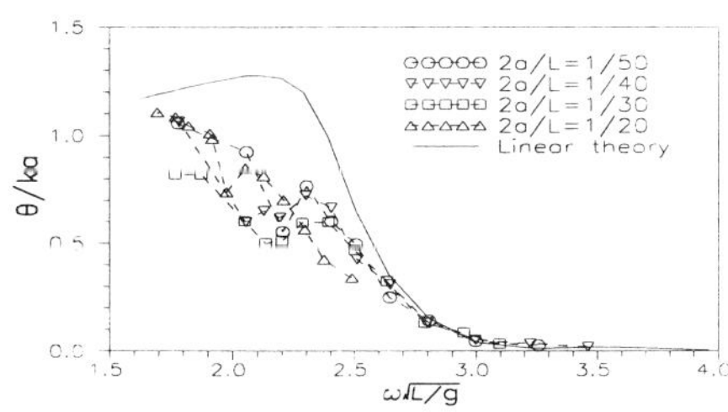

4 (b) Pitch

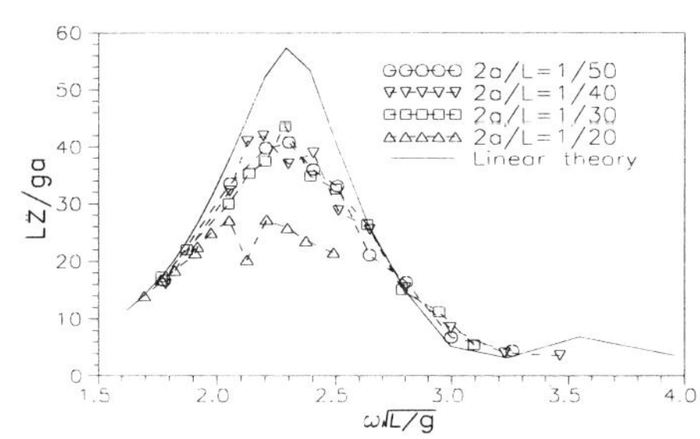

4(c) Vertical Acceleration
Figure 4: Comparisons of motion responses

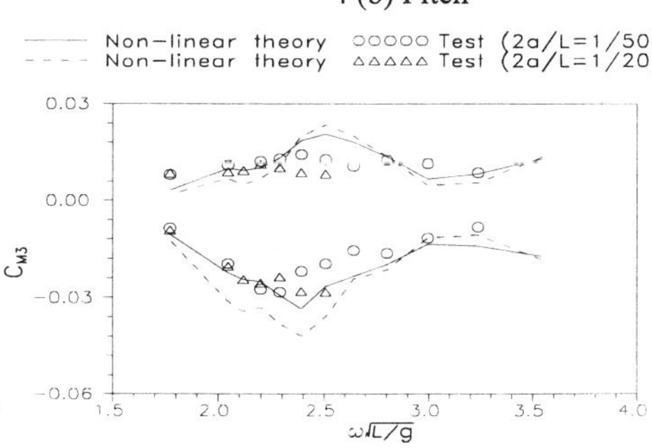

Figure 5: Comparisons of the Sagging and hogging bending moment (amidships)

Figure 4 shows the comparisons of the two dimensional linear numerical results and the experimental results for the first harmonics of heave, pitch and acceleration at bow. It can be seen that in the non-dimensional frequency range of 1.5 to 2.5, the numerical results are greater than the experimental results up to nearly 30%. It may mainly due to omission of the viscous effect on the ship motion in the numerical results. The same tendency can also be found in references (Watanabe, et al., 1989). The measured heave, pitch and vertical acceleration at bow for the low and medium wave height cases (2a/L=1/50~1/30) are close to each other, even there exists a slight divergence. In high wave height cases (2a/L=1/20), the measured results are quite different from the results of other cases in the non-dimensional wave frequency region 2.2~2.6 (λ/L=1.3~0.9), showing the strong non-linear behavior induced mainly by the flare impact and green water effects. Indeed, huge amount of green water on deck and severe water spray at the bow were observed during the model test in this test condition.

Figure 5 shows the comparisons of the 2-D non-linear numerical results and the measured data for the sagging and hogging bending moments. The component of wave-making vertical bending moment due to the forward speed effect was excluded in the predictions. The numerical results coincide well with the experimental data with maximum 20% discrepancy in the sagging moment near λ/L=1.0, and show the evident sagging to hogging asymmetry. Figure 6 exhibits the comparison of amidships measured non-dimensional sagging and hogging bending moments with respect to wave frequencies and wave heights. Similar to the motion responses, the apparent differences occur among the results for different wave heights in the non-dimensional wave frequency region 2.2~2.6, corresponding to λ/L=1.3~0.9, where the ship experienced severe flare impact and bow immersing. The maximum divergence of the non-dimensional sagging moments for the four wave heights is over 0.01. The peak values of sagging moments for different wave heights occur at different wave frequencies, while their maximum values appear at $\omega\sqrt{L/g}=2.3$ and λ/L = 1.2.

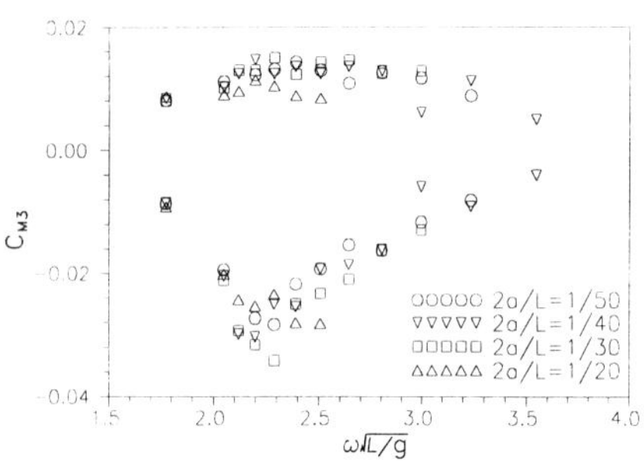

Figure 6: Sagging and hogging bending moments measured amidships

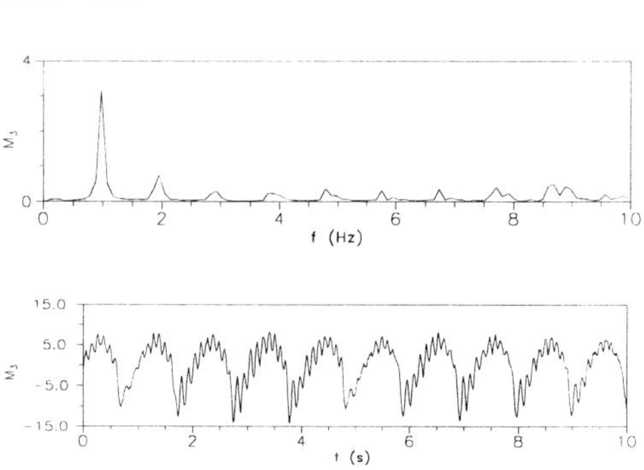

Figure 7: Time history and its FFT of bending moment measured amidships (λ/L=1.2, 2a/L=1/30)

The non-linear nature of the bending responses may also be investigated from their components of the first and higher order harmonics. Figure 7 shows the time history of the amidships measured bending moment and its FFT transformation. It can be seen that even if the model is responding to regular waves and the first harmonic component of its bending moment is dominant, the higher order components, such as at $2\omega_e$, $3\omega_e$ harmonic frequencies and the 2-nodes flexible mode frequency of the hull, are also of certain amount. Figure 8 gives the amplitudes of these components with respect to the non-dimensional wave length. The measured first harmonic components of the bending moments for different wave heights are close to the 2-D linear numerical results, which are over-estimated in the wave region λ/L=0.7~1.2 and slightly under-estimated in the region of λ/L=1,3~2.5. This indicates that the linear 2-D numerical method can give the reasonable predictions for the first harmonic components of the wave-induced vertical bending moments. In general, the amplitudes of the higher harmonic components of the bending moments are one order smaller than the first harmonic components, and increase when the wave height becomes larger, as clearly shown in Figures 8(b, c).

It is interesting to note that the bending moment components at the 2-nodes flexible mode frequency of the wetted hull increase rapidly with the increasing wave height. Their peak values usually occur in the region λ/L=1.0~1.5. By comparison of Fig 8(a) and Fig8 (d), it can be found that for the low wave height (2a/L=1/50), the contributions of the 2-nodes flexible mode components to the total bending moments are relatively low, of about 15% of the total. For the case of high wave height (2a/L=1/20), the amplitude of the bending moment due to the response at the 2-nodes flexible mode frequency is nearly of the same order as the first harmonic component. It may be concluded that to thoroughly assess the non-linear behavior of the wave loads, the flexible effect of the ship hull girder should be carefully considered, and the wave tests of a flexible ship model would be a useful approach to gain more insight towards this purpose.

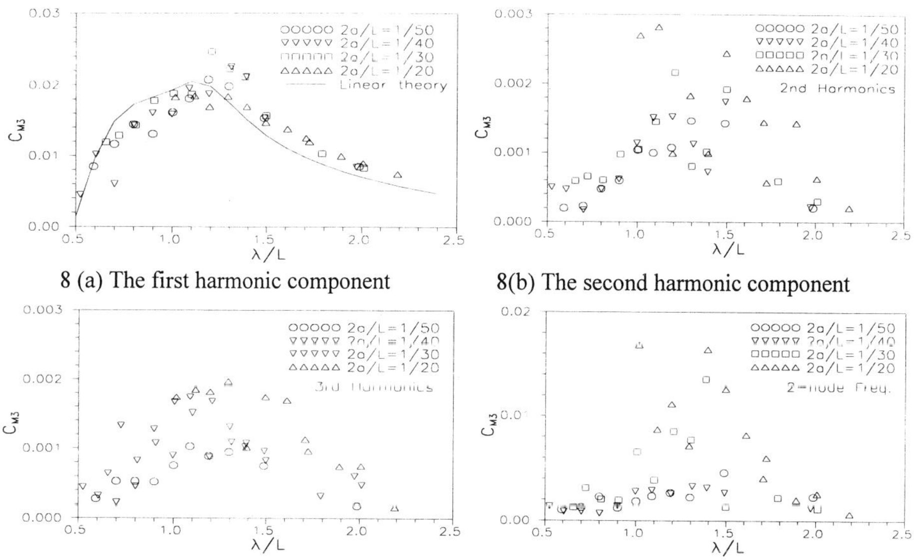

8 (a) The first harmonic component

8(b) The second harmonic component

8(c) The third harmonic component

8(d) The 2-nodes flexible mode component

Figure 8: The first and higher order components of the bending moment at the measurement point M3 (amidships)

In the model test, 30 runs in irregular waves were carried out. The wave generated in the towing tank corresponded to a wave spectrum of $h_{1/3}$=0.11m and T_1=1.2s, which was of good agreement with the target ITTC wave spectrum, as shown in Figure 9. The measured distribution of the significant double amplitude values of the vertical bending moments at Mj (j=1~5) is shows in Figure 10.

4 CONCLUSIONS

A flexible model of the S175 container ship was manufactured and tested to investigate the non-linear wave loads in severe seas. The experimental data clearly indicate the presence of the non-linear behaviors of both motion and structural responses. The non-linearity is not only reflected in the biasing of sagging relative to hogging bending moments, and is also revealed in the appearance and behaviors of the response components at the high order harmonics of the wave encounter frequency and the 2-nodes flexible mode frequency. The amplitudes of these components evidently increase when the wave gets higher. Their contributions to the total bending moments seem to be an important factor of introducing the non-linear asymmetric responses of ship structures in extreme wave conditions. The application of a flexible ship model made of plastic material may de useful to further investigate this kind of phenomenon.

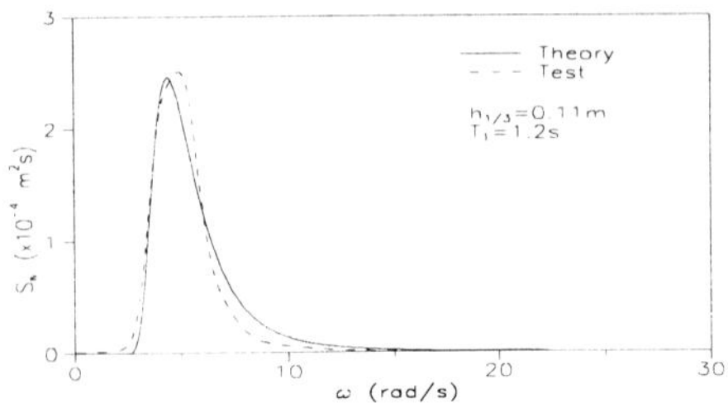

Figure 9: Comparison of measured and target wave spectrum

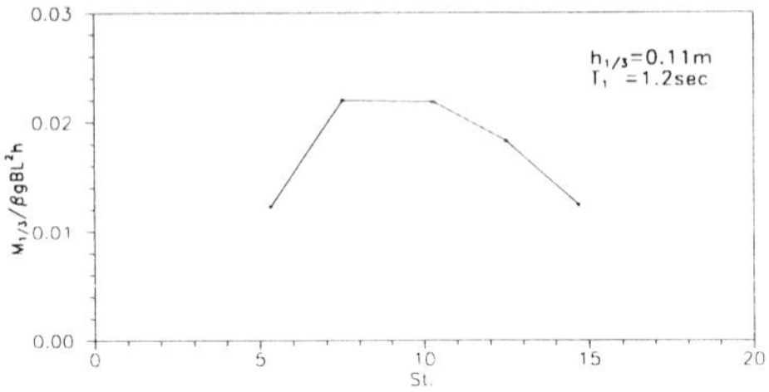

Figure 10: Distribution of the significant double amplitude values of the bending moments

References

1 16th ITTC Seakeeping Committee Report, *Proceeding of the 16th ITTC*, Leningrad, Sep.1981.

2 18th ITTC Seakeeping Committee Report, *Proceeding of the 18th ITTC*, Kobe, Japan, Oct.1987.

3 O'Dea J., Powers E. and Zselecsky J. (1992). Theoretical and Experimental Study of the Nonlinearities in Vertical Plane Ship Motions. *Proceedings of the 19th Symposium on Naval Hydrodynamics*, Seoul, Korea

4 Tao Z. and Incecik A. (1998). Time Domain Simulation of Vertical Ship Motions and Loads in Regular Head Waves. *17th OMAE*, Lisbon

5 Fonseca N. and Soares C. G. (1998). Time Domain Analysis of Large-amplitude Responses of Ship in Waves. *Proceedings of PRADS'98*, The Hague, The Netherlands

6 Soto M. (1954). Model Experiments on the Longitudinal Strength of Ships Running Among Waves, *E.T.T. Report* No. 614.

7 Akita Y. and Ochi K. (1955). Model Experiment on the Strength of Ships Moving in Waves, *Trans. S.N.A.M.E.* 203-236.

8 Does J. Cb. De. (1960). Experimental Determination of Bending Moments for Three Models of Different Fullness in Regular Waves, *International Shipbuilding Progress*, **7:68.**

9 Fukasawa T., Yamamoto Y., Fujino M. and Motora S. (1981). Motion and Longitudinal Strength of a Ship in Head Sea and the Effects of Non-linearity (4th report) – Experiment. *Journal of the Society of Naval Architects of Japan*, **150.**

10 Watanabe I., et al. (1984). Measurement of Wave Impact Pressure on Patrol Ship 'Shikine' Using Elastic Similar Model. *Proceedings of Annual Presentation Meeting of SRI,* **44**

11 Watanabe I. and Sawada H. (1985). Effects of Elastic Responses to the Longitudinal Bending Moment in Bi-direction Irregular Waves, *Journal of the Society of Naval Architects of Japan*, **158.**

12 Watanabe I., Ueno M. and Sawada H. (1989). Effect of Bow Flare Shape to the Wave Loads of a Container Ship, *Journal of the Society of Naval Architects of Japan*, **166.**

13 Lin J., et al. (1991). Experiment of an Elastic Ship Model and the Theoretical Predictions of its Hydroelastic Behavior, *Proc. of 1st Intl. VLFS*, Honolulu, Hawaii, USA

14 Lin J., et al. (1996). The Comparative Study on the Elastic Ship Model Test and the Full Scale Ship Trial. *Ship Behavior Research*, **3.**

Practical Design of Ships and Other Floating Structures
You-Sheng Wu, Wei-Cheng Cui and Guo-Jun Zhou (Eds)

ESTIMATION OF NONLINEAR LONG-TERM EXTREMES OF THE VERTICAL BENDING MOMENTS IN SHIPS

Gro Sagli Baarholm[1] and Torgeir Moan[1]

[1]Department of Marine Structures, NTNU, N-7491 Trondheim, Norway

ABSTRACT

This paper deals with estimation of the long-term extreme of hull girder loads in ships for a given return period. In principle, this response is obtained by combining the response in all the sea states. The long-term response for a linear system can be effectively obtained by determining the response for each sea state, specified by the significant wave height, H_s, and the peak period, T_p, in the frequency domain. However, if the response is nonlinear, time domain simulation would be required and long time series is required to limit the statistical uncertainty. Long-term analysis becomes complicated and time consuming, and it is crucial to improve the efficiency in the calculation. In this work it is shown that the long-term extremes can be estimated by considering only a few short-term sea states. A long-term analysis based on identifying the most important sea state, defined by the coefficient of contribution, using linear analysis is applied. An iteration procedure is thereafter used to find the nonlinear long term extreme values. It is concluded that only a limited number of sea states is necessary to get an acceptable estimate of the nonlinear *D*-year response as long as the most important sea states are included, *i.e.* the sea state with maximum coefficient of contribution.

KEYWORDS

Ship, Vertical bending moment, nonlinear, Long term, simplified

1 INTRODUCTION

Safety requirements to ship hulls are specified in Ship Rules issued by Classification Societies who harmonise their rules through the International Association of Classification Societies, IACS, *e.g.* Nitta et al (1992). Traditionally extreme wave loads are obtained from Ship Rule formulae. The rule book approach meet well the busy designers' need and preference for rules which are simple to use, omitting the possibility to quantify the effect of a proposed design change on the subsequent safety on that design. As new designs are entering the arena, the ship rules may become outdated since the formulae are based on data for various ship types and would not generally be optimal for a given vessel. The variation and complexity of novel hull concepts, possible influence of structural dynamics as well as nonlinear load effect, suggest that *direct calculation* of the hull girder load effects are preferable in such cases.

The most common way to establish design loads by direct calculations is by means of long term description of the response itself. This requires that a wide range of sea states must be analysed and that the final distribution is obtained as a weighted sum of these short term distributions, considering various wave heading, velocities and operational requirements, *i.e.* velocity and heading profiles. If linear response analysis are conducted, the response in a given sea state is defined by the transfer function, while time domain simulations are in general required to obtain the nonlinear response. This leads to time consuming analysis. A more efficient and practical feasible method to estimate the long term design values is therefore desired. In Soares (1984), Farnes (1990), Farnes & Moan (1994), Videiro (1998), Videiro & Moan (1999) application of the direct long-term approach Gaussian and non-Gaussian response is outlined.

2 LONG TERM DISTRIBUTION

The long term peak distribution of maxima is obtained by summation of short term probabilities of exceedance in all possible combinations of mean wave periods, significant wave heights, heading angles and speeds, *e.g.* Bishop & Price (1979). Only one heading and forward speed are considered in this work. These two variables are independent in addition to independence between forward speed and significant wave height. The probability of exceedance is given as

$$Q_{LT}(r) = \int\limits_{H_s\,T_p}\int Q_R(r\,|h,t,\beta_0,u_0) f_{H_s,T_p}(h,t)\bar{w}\,dh\,dt \qquad (1)$$

β_0 and U_0 is the heading angle and forward speed, respectively. $\bar{w} = \bar{w}_{h,t,u,\beta}$ is a weighting function and $f_{H_s,T_p}(h,t)$ is the long term joint probability distribution of significant wave height and peak periods as suggested *e.g.* by Haver & Nyhus (1986). $Q_R(r\,|h,t,\beta_0,u_0)$ is the short term cumulative probability distribution. The contribution to the probability of exceedance $Q_{LT}(r_D)$ from a sea state is given as, *e.g.* Farnes & Passano (1989), Larsen & Passano (1990), Videiro (1998),

$$C_R(h,t) = \frac{Q_R(R > r_D\,|h,t,\beta_0,u_0) f_{H_s,T_p}(h,t)\bar{w}\,dh\,dt}{Q_{LT}(r_D)} \qquad (2)$$

There is a limited area of the scatter diagram that contributes to the long-term response. Most parts of the scatter diagram have $C_R(h,t)$-values close to zero. The scatter diagram can be divided into two regions. That is a contributing region $\tilde{s}$ and a non-contributing region outside $\tilde{s}$. The contributing areas could be selected by choosing the sea states $si=(h,t)$ where $C_R(si)>a$. Applying this information, Eqn. 1 can be discretized as

$$Q_{LT}(r_D) = \sum_{i=1}^{N_{si}|C_R(si)>a} Q_R(R > r_D\,|\,si,\beta_0,u_0) f_{H_s,T_p}(h_{si},t_{si})\bar{w}\,\Delta h_i\,\Delta t_i \qquad (3)$$

$N_{si}|C_R(si)>a$ is the number of discrete, contributing sea states within $\tilde{s}$, Δh and Δt are the grid sizes. The weight is defined by the zero-up-crossing wave period, T_z, for the sea state si and the average zero-up-crossing wave period wave period, $\bar{T}_z$, as $\bar{w} = \bar{T}_z / T_{z,si}$. The weight function $\bar{w}_{si}$ has to be recalculated for the sea states within $\tilde{s}$. The long term value r_D for a given return period of D years is given by $Q_{LT}(r_D)=1/N_D$, where N_D is the number of extremes during D-years.

2.1 Iteration Approach

The procedure applied to estimate the nonlinear, long-term response could be split into two steps. The first step is based on linear frequency domain analysis and used to locate an initial sea state applied in the time domain simulation. Step two is an iteration approach based on several time domain simulations. An illustration of the procedure is also shown in Figure 1 and more details can be found in Sagli (2000), Baarholm & Moan (2000).

2.2 Effect of Operational Restriction

A change in the route of the ship can occasionally be made in order to avoid severe weather situations. Boundaries limiting the operability are obtained by combining short-term statistics and sea keeping criteria. These are for instance factors like slamming, vertical motion and acceleration that are commonly estimated using linear response analysis. The limiting curves are presented as envelope curves as a function of the significant wave height and peak period, $H_s^{\lim}$ & T_p. According to Sagli (2000) reasonable estimate of the long term response is obtained by approximating the operability limiting boundary by $\tilde{H}_s = \min H_s^{\lim}$.

Limited operability implies a modification in the probability of encountering very severe sea conditions. Consequently, there is a change in the effective scatter diagram. This is studied in Sagli (2000), Baarholm & Moan (2000) where the weather restriction is accounted for by modifying the simultaneous distribution, given as $F_{H_s,T_p}(h,t) = F_{H_s}(h)F_{T_p|H_s}(t|h)$, describing the probability of storms. The distribution of the significant wave height is approximated by the Weibull distribution for the largest wave heights, Haver & Nyhus (1986). By modifying the Weibull parameters such that the *100*-years sea state is equal to the upper operability limit, the weather restriction is taken into account. Three different modification of the original, simultaneous distribution $F_{H_s,T_p}(h,t) = F_{H_s}(h)F_{T_p|H_s}(t|h)$ into $\hat{F}_{H_S,T_P}(h,t) = \hat{F}(h)F_{T_P|H_S}(t|h)$ will be briefly discussed in the following. These modifications are truncation, rescaling and reshaping of the Weibull distribution.

Truncation may be compared to a situation where the ship is never present in sea states with significant wave height larger than $\tilde{H}_s$. However, the ship will in general also be present in larger sea states. This encourages establishment of distribution with tail. Rescaled and reshaped Weibull distribution describes the situation where the shipmaster maneuvers into calmer areas and some variability around the limiting value, $\tilde{H}_s$, is allowed for. The rescaling and reshaping of the Weibull distribution is done by finding a new scale and shape parameter in the Weibull distribution, respectively. The new parameters might be determined by letting $\tilde{H}_s$ represent the D-year sea state. Meaning that in the modified scatter diagram, the significant wave height will only exceed $\tilde{H}_s$ once every *D*-year. Comparisons of the three different modifications are shown in Figure 2 together the original Weibull distribution. The tail-part is zoomed in. $\tilde{H}_s = 10.0m$ is chosen for the purpose of illustration. The figure shows that a rescaled distribution has smaller density in the tail then the reshaped distribution. This implies that the reshaped distribution allows for more severe sea states. Accordingly, larger long-term responses are expected. For theoretical details see Sagli (2000), Baarholm & Moan (2000).

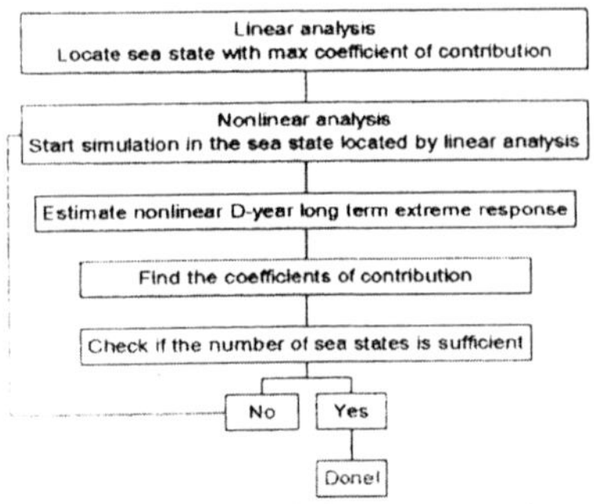

Figure 1: Procedure for the approximate long term analysis

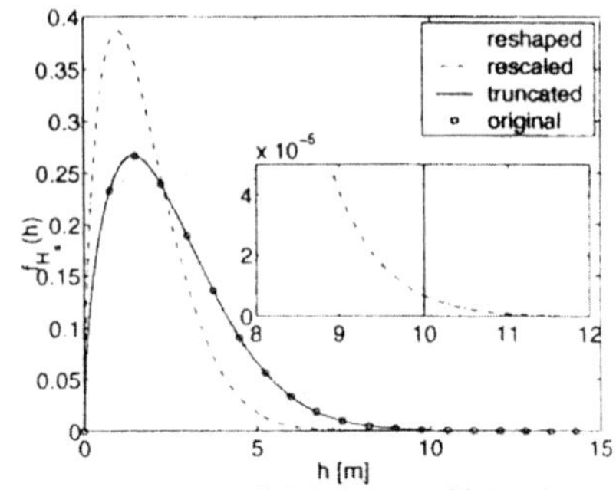

Figure 2: Comparison of the original Weibull density and the rescaled, reshaped and truncated Weibull. The operability limit is equal to $\tilde{H}_s = 10.0m$.

3 CASE STUDIES

The procedure in Figure 1 is applied to find the nonlinear vertical bending moment on three different vessels. The nonlinear simulations are obtained by using a quasi-nonlinear theory described in Wu & Moan (1996). Validation based on a model test is presented in Wu et al (1996). The chosen vessels are the S-175 container ship, a tanker and destroyer. Description of the hull form and the weight distribution for the S-175 container ship is given in Salvesen (1983) and in Dalzell (1964) for the tanker and destroyer. The long term extreme value of the vertical bending moment midship for the S-175 container ship is calculated at two forward speeds. Those are Froude numbers *Fn=0.0* & *Fn=0.275*. In addition, similar analysis were performed with the tanker and the destroyer at Froude numbers *Fn=0.0* & *Fn=0.2*. Only head sea cases are studied herein. The three parameter-Weibull distribution was used to fit the maxima samples, Sagli (2000). The short-term wave process is described by the Pierson-Moskowitz wave spectrum in all cases. The long-term wave process is characterised by the joint probability density distribution for the significant wave height, H_s, and peak period, T_p, for the Northern North Sea as proposed by Haver & Nyhus (1986). The return period is *D=100* years.

4 RESULTS AND DISCUSSION

The results for the S-175 container ship, tanker and destroyer are presented in TABLE 1. The estimated nonlinear, long-term vertical bending moments midship are made dimensionless using the main dimensions of the vessels. The nonlinear effects tend to increase the sagging and hogging moments for the S-175 container ship and the tanker as the forward speed increases. The destroyer experiences reduced sagging and hogging moments as the forward speed increases. As expected the tanker is influenced by small nonlinear effects, while the container ship experiences a rather large change in sagging and hogging as a result of the speed change.

Figure 3 show the distribution of the $C_R(si)$ -coefficients for the nonlinear, sagging moment midship for the S-175 container ship at forward speed *Fn=0.275*. The $C_R(si)$-contours For the S-175 container ship sagging has two distinct peaks where $C_R(si)$ reaches a maximum. The maxima are almost equivalent in value. The maxima are close, and they are easily detected by studying the $C_R(si)$ -

gradient. It is reasonable to believe that the number of maxima detected in the response cases, are related to the degree of nonlinearity in the response. Extensive results are shown in Baarholm & Moan (2000) for the S-175, destroyer and the tanker in both sagging and hogging condition.

TABLE 1

100-YEAR NONLINEAR VERTICAL SAGGING MOMENT MIDSHIP, r_{100}^{nl}, ESTIMATED USING *APPROXIMATE* LONG TERM ANALYSIS. $r_{100,1(lin)}^{nl}$ AND $r_{100,1(nl)}^{nl}$ ARE CALCULATED USING THE SINGLE SEA STATE WITH MAXIMUM $C_R(si)$ IN LINEAR AND NONLINEAR LONG-TERM RESPONSE ANALYSIS, RESPECTIVELY. THE VALUES ARE MADE DIMENSIONLESS WITH THE VESSELS' MAIN DIMENSIONS AND THE NONLINEAR LONG-TERM EXTREME VALUE, r_{100}^{nl}.

		Sagging			Hogging		
Ship	Fn	$\left(r_{100}^{nl}/\rho gBL^3\right)\cdot 10^{-3}$	$r_{100,1(lin)}^{nl}/r_{100}^{nl}$	$r_{100,1(nl)}^{nl}/r_{100}^{nl}$	$\left(r_{100}^{nl}/\rho gBL^3\right)\cdot 10^{-3}$	$r_{100,1(lin)}^{nl}/r_{100}^{nl}$	$r_{100,1(nl)}^{nl}/r_{100}^{nl}$
S-175	0.0	1.49	0.84	0.87	0.64	0.88	0.92
	0.275	2.35	0.80	0.87	0.66	0.79	0.94
Tanker	0.0	0.86	0.85	0.89	0.73	0.85	0.88
	0.2	0.87	0.85	0.86	0.80	0.84	0.89
Destroyer	0.0	2.00	0.87	0.95	0.83	0.90	0.95
	0.2	1.76	0.89	0.94	0.78	0.92	0.95

As long as the sea state with maximum $C_R(si)$ -value is captured, there is no need to perform a large number of simulations. This can be illustrated by using Figure 3. From this figure one may select a number of sea states and thereafter find the total sum of the contributing sea states, *i.e. total contribution*= $\Sigma_{Nsi}\, C_R(si)$. The chosen sea states can then be applied to estimate the desired nonlinear long-term bending moment, $r_{D,Nsi}^{nl}$. The deviation from the *true* answer with $Nsi = \infty$, can be expressed as the ratio $r_{D,Nsi}^{nl}/r_D^{nl}$. The ratio $r_{D,Nsi}^{nl}/r_D^{nl}$ versus $\Sigma_{Nsi}\, C_R(si)$ is shown in for the S-175 container ship. Both zero forward speed and forward speed are represented.

Since the distribution of the $C_R(si)$ -values for the S-175 at Froude number $Fn=0.275$ has two maxima, one could wonder if both maxima must be included to get a reasonable long-term estimate. For the sagging, both maxima-regions have $\Sigma_{Nsi}\, C_R(si)\approx 50\%$. Including only the sea state with maximum $C_R(si)$ -value, enclosed by a circle in Figure 4, causes a reduction of approximately *8%* and *14%* for hogging and sagging respectively. The small maximum in hogging has very little effect on the long- term response. Excluding this maxima region, causes a reduction of less than *0.5%*. According to Sagli (2000) inclusion of *50%* of the $C_R(si)$ -values cause reduction of approximately *5%* for both the tanker and the destroyer.

The long-term estimates of the sagging and hogging mid ship using a single sea state are presented in TABLE 1. These sea states correspond to the sea state with maximum coefficient of contribution in linear and nonlinear long-term analysis. Using the sea state with maximum $C_R(si)$ in nonlinear analysis, the underestimation is smaller than using the sea state located by using linear analysis. In the latter case, the largest underestimation is approximately *20%*. This occurs for sagging and hogging responses at Froude number $Fn=0.275$ for the S-175 container ship. The sea state with maximum $C_R(si)$ for linear response is located in between the two maxima located by nonlinear analysis. For the other response cases, the maxima located by linear and nonlinear analysis are quite close in in both significant wave height and peak period, implying that the underestimation is smaller using the sea state with $C_R(si)$ -maxima from linear analysis. One may question the reason for locating the maximum $C_R(si)$ using nonlinear analysis, since a rather good estimate of the long term extremes can be obtained by using the sea state located by linear analysis. The answer is that, - if the nonlinear $C_R(si)$ -maxima is not found, you can not be sure of the accuracy of your estimated long-term extremes. Simply because you do not know the location of this sea state relative to the sea state with $C_R(si)$ -maxima for nonlinear response.

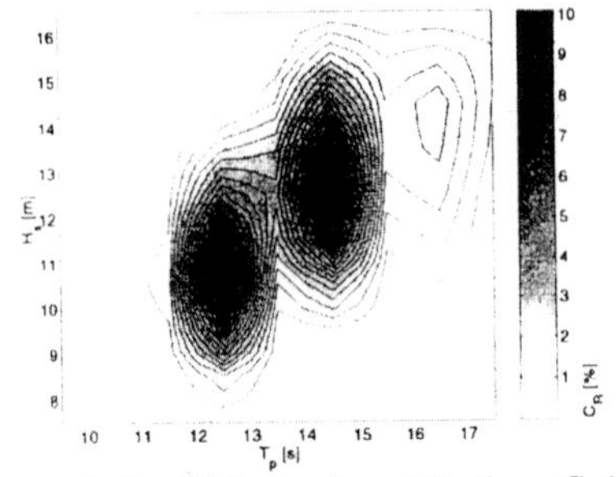

Figure 3: Coefficient of contribution, $C_R(si)$, for the nonlinear vertical bending moment midship. S-175 container ship at Froude number $Fn=0.275$.

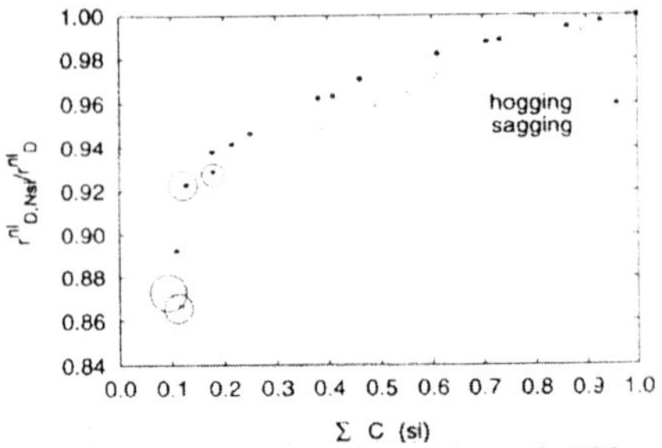

Figure 4: Influence on the estimated *100*-year value of the vertical bending moment midship due to a reduced number of included sea states. All speeds are included. A circle encloses results using the single sea state with maximum coefficient of contribution.

4.1 Restricted Service

Comparisons between the nonlinear long-term value in restricted versus unrestricted service are given in TABLE 2. Effect due to restricted service is taken into account by applying a rescaled scatter diagram for an operability limit at $\tilde{H}_s = 10m$. A *30%* reduction of the sagging moment midship is experienced for the S-175 container ship when severe sea states above *10m* are avoided. The restricted service has an effect of *15-20%* on the hogging moment midship.

TABLE 2

100-YEAR NONLINEAR VERTICAL SAGGING AND HOGGING MOMENT MIDSHIP FOR THE S-175 CONTAINER SHIP, $r_{100}^{nl,rest}$. EFFECT DUE TO RESTRICTED SERVICE IS TAKEN INTO ACCOUNT BY APPLYING A *RESCALED* SCATTER DIAGRAM FOR AN OPERABILITY LIMIT AT $\tilde{H}_s = 10m$. THE RESULT USING THE SINGLE SEA STATE WITH MAXIMIM $C_R(si)$, $r_{100,1}^{nl,rest.}$, ARE INCLUDED. THE VALUES ARE MADE DIMENSIONALESS *W.R.T.* THE NONLINEAR VALUE, r_{100}^{nl}, IN UNRESTRICTED SERVICE.

	Fn	$r_{100}^{nl,rest.} / r_{100}^{nl}$	$r_{100,1}^{nl,rest.} / r_{100}^{nl}$
Sagging	0.0	0.71	0.62
	0.275	0.71	0.64
Hogging	0.0	0.82	0.76
	0.275	0.85	0.78

4.2 Final Comments

The coefficient of contribution might be used to define a design sea state. Estimation of the most probable maximum response value in this sea state, with duration as given by the scatter diagram, will give a reasonable estimate of the long term extreme value within an error of *10-15%,* as shown in Figure 4. Including more sea states, improves accuracy. The method is still rather time consuming since the most critical sea state must be identified to ensure the grade of accuracy. However, the method can be used for verification of more simplified methods used to estimate the long-term extreme values, see Sagli (2000).

5 CONCLUSIONS

The long-term extremes can be estimated by considering only a few short-term sea states as long as the most important sea states, defined by the coefficient of contribution, are identified.

References

15th & 16th ITTC Seakeeping Committee. (1983). Summary of Results Obtained with Computer Programs to Predict Ship Motions in Six Degree of Freedom and Related Responses (1976-1981).

Baarholm G. S. and Moan T. (2000). Estimation of Nonlinear Long Term Extremes of Hull Girder Loads in Ships. *Marine Structures*. **13**, 495-516

Bishop & Price. (1979). *Hydroelasticity of Ships*. Cambridge University Press, Cambridge, U.S.A.

Dalzell J. F. (1964). An Investigation of Midship Bending Moments Experienced in Extreme Regular Waves by Models of a Tanker and a Destroyer. *Technical report. SSC-156. Ship Structure Committee*. Washington, D.C.

Farnes K. A. and Passano E. (1989). *LOSSTA User's Manual*. SINTEF, Trondheim, Norway.

Haver S. (1987). On the Joint Distribution of Heights and Periods of Sea Waves. *Ocean Engineering*. **14:5**, 359-376.

Farnes K.A. (1990). Long-term Statistics of Response in non-Linear Marine Structures. *Doctoral dissertation*. Norwegian Institute of Technology, Trondheim, Norway.

Farnes K. A. and Moan T. (1994). Extreme Dynamic, non-Linear Response of Fixed Platforms Using a Complete Long-Term Approach. *Applied Ocean Research*. **15**, 317-326

Haver S. and Nyhus K. A. (1986). A Wave Climate Description for Long Term Response Calculation. *Proc. of OMAE'86*.

Haver S., Sagli G. and Gran T. M. (1998a). Long Term Response Analysis of Fixed and Floating Structures. *Proc. of Wave'98-Ocean Wave Kinematics, Dynamics and Loads on Structures. Int. OTRC Symp*. Houston, USA, May-April.

Haver S., Sagli G. and Gran T. M. (1998b). Long Term Response Analysis of Fixed and Floating Structures. *Technical report. 98s97*4715*. STATOIL, Stavanger, Norway.

Larsen C. M. and Passano E. (1990). Extreme Response Estimation for Marine Risers. *Proc. of OMAE'90. ASME*.

Nitta A., Arai H. and Magaino A. (1992). Basis of IACS Unified Longitudinal Strength Standard. *Marine Structures*. **5**, 1-21.

Sagli G. (2000). Model Uncertainty and Simplified Estimates of Long Term Extremes of Hull Girder Loads in Ships. *Doctoral dissertation*. Norwegian University of Science and Technology, Trondheim, Norway.

Soares C. G. (1984). Probabilistic Models for Load Effects in Ship Structures. *Doctoral dissertation*. Norwegian Institue of Technology, NTH, Trondheim, Norway.

Videiro P. and Moan T. (1999). Efficient Evaluation of Long-Term Distributions. *Proc. of OMAE'99. ASME*. paper no. 6014.

Videiro P. M. (1998). Reliability Based Design of Marine Structures. *Doctoral dissertation*. Norwegian University of Science and Technology, Trondheim, Norway.

Winterstein S. R., Ude T. C., Cornell C. A., Bjerager P. and Haver S. (1993). Environmental Parameters for Extreme Response: Inverse form with Omission Factors. *Proc. of IOSSAR'93*. G. I. Schuëller, M. Shinozuka and J. T. P. Yao, Eds. A. A. Balkema, Rotterdam, 77-84.

Wu M.-K. and Moan T. (1996). Linear and Nonlinear Hydroelastic Analysis of High-Speed Vessels. *J. of Ship Research*. **40:2**, 149-163.

Wu M.-K., Aarsnes J. V., Hermundstad O. A. and Moan T. (1996). A Practical Prediction of Wave-Induced Structural Responses in Ships with Large Amplitude Motion. *Proc. of the 21st Symposium on Naval Hydrodynamics*.

A DIRECT CALCULATION APPROACH OF DETERMINING EXTREME COMBINED BENDING MOMENTS FOR FAST FINE FORM SHIPS

Xue-Kang Gu and Jin-Wei Shen

Department of Marine Structures, China Ship Scientific Research Center, Wuxi, Jiangsu 214082, China

ABSTRACT

For fast fine form ships, direct calculation approaches play more and more important roles towards rational structural design. In this paper, the coupling effects between the higher order harmonic components and whipping components of nonlinear vertical bending moments are described. To further investigate this phenomenon and other nonlinear dynamic effects of wave loadings, a nonlinear time domain strip theory is developed and validated. The nonlinear wave-induced bending moments and nonlinear combined bending moments including whipping components are obtained. Considering the coupling effects and the speed losses of ships in waves, a direct calculation approach of determining extreme combined bending moments for fast fine form ships are presented. Taking the combined bending moments as the dominant load parameter in head sea condition, the matched ship motions, nonlinear wave-induced wave loads and design operational conditions (design wave and ship speed) are obtained at the instant when the combined bending moments reach their maximum to ensure that the most unfavorable responses are achieved. Finally, an application example for a fast containership is presented.

KEYWORDS

Nonlinear wave-induced bending moments, Higher order harmonic component, Whipping response, Coupling effects, Speed loss, Design operational condition, Dynamic loading approach.

1 INTRODUCTION

For structural designs of conventional ships, classification rules and regulations are still the main criteria because of their successful practice and feedback experience, simplicity in applications and familiarity to designers, although the resultant structures are usually not the most optimum ones for specific ships. For modern ships, especially for newly designed large size or fast ships with fine hull forms, the structural designs will be beyond the limits of traditional rules. For example, the design formula of vertical bending moments in IACS UR-S11 do not explicitly include whipping bending moments. However, effects of the whipping components on structural design can not be ignored for ships with flat bottoms and large bow flares. For merchant ships, these effects are mostly implicitly reflected in the longitudinal distributions of section modulus except for rules of Germanischer Lloyd and ABS-SafeHull with whipping bending moments expressions. Furthermore, although the sagging and hogging bending moments are distinguished in UR-S11, they are just influenced by block coefficient. The nonlinear loading characteristics of modern ships with fine hull forms are not fully reflected in the formula. All of these can only be properly considered in direct calculations.

The designs based on direct calculations of ship motions, wave loads and structural finite element analyses play more and more important roles toward rational structures and meanwhile widely accepted by different classifications. Among all the aspects of direct calculation methods, the determination of wave loads and ship motions is of critical importance to achieve successful ship structures. Up to now, linear frequency domain wave load theories have been the main tools in calculating ship responses in waves in many well-established direct calculation approaches and associated commercial codes.

In 1992, American Bureau of Shipping developed a dynamic loading approach (DLA) (Liu, et al,1992) for VLCC. The loads acting on ship structures are explicitly considered in dynamic sense and all key combinations of primary and secondary loadings for structural design including cargo holdings, headings, environmental conditions are provided. Since then, it has also been applied to bulk carriers, containerships and LNG ships. It is a remarkable step toward rational designs of ship structures, although the responses (motions, loads and extreme values) of ships are still based on linear strip theory and the truly dynamic responses, i.e. slamming induced responses, are not taken into account. With the Large Amplitude Motion Program (LAMP), the DLA method is modified to incorporate nonlinear responses, including whipping component that is crucial to fine form ships (Shin, et al, 1997). It is not clear, however, from the paper how to apply the nonlinear wave loads and slamming induced loads simultaneously for structural designs.

Hachmann (1991) presented a revised design wave concept to distinguish sagging and hogging design bending moments based on a revised strip theory. The wavelength at peak of response amplitude operator (RAO) of linear vertical bending moments is set as design wavelength, whereas the design wave amplitude results from linear long-term design value (probability level of 10^{-6} and a 10% reduction) divided by the peak value of the RAO. No whipping was considered, although the author also realized that it is significant for large-sized vessels keeping on sailing at high speed in extreme waves.

In this paper, the newly found coupling effects between the higher order harmonic components and whipping components of nonlinear vertical bending moments are described. To further investigate this phenomenon and other nonlinear dynamic effects of wave loadings, a nonlinear time domain strip theory is developed and validated. The nonlinear wave-induced bending moments and nonlinear combined bending moments (CBM) including whipping components are obtained. Considering the coupling effects and the practical speeds of ships in waves, a direct calculation approach of determining extreme combined bending moments for fast fine form ships are presented. Take the combined bending moments as the dominant load parameter in head sea condition, the matched ship motions, nonlinear wave-induced wave loads and design operational conditions (design wave and ship speed) are obtained at the instant when the combined bending moments reach their maximum to ensure that the most unfavorable loadings are achieved. Finally, an example for a fast containership is presented.

2 COUPLING EFFECTS BETWEEN HIGHER ORDER HARMONIC AND WHIPPING COMPONENTS

In addition to usually concerned wave-induced components with wave encounter frequencies and whipping components with hull girder natural vibration frequencies in vertical bending moments in moderate or high amplitude regular waves, there are higher order harmonic components with frequencies which are multiples of wave encounter frequency (Watanabe, 1989, Adegeest, 1994, Shin, et al, 1997). These harmonic components are induced by non-sinusoidally changed hydrodynamic forces on non-wall-sided ship hulls and cannot be revealed by linear theories.

Gu, Shen & Moan (2000) analyzed the vertical bending moments using FFT transformation based on the model test results of a fast containership with large bow flare and low rigidity. Some sample results are displayed in Figure 1. Where, λ, H, L and Fn are wavelength, wave height, ship length and Froude number, respectively. The wave encounter frequency is 0.9Hz in this case and the first natural frequency is 5.4Hz. Although the motions of the ship are almost monochromatic, the higher order harmonic bending moments have non-negligible contributions to the nonlinearity of combined bending moments (CBM) and mainly have frequencies between the wave encounter and the first hull natural frequency. It was found that there are coupling effects between higher order harmonic and whipping components. That means that the higher order harmonic components, especially those have frequencies close to hull natural frequencies, are influenced by whipping component due to hydroelastic effects; and the whipping components are significantly amplified when the first hull natural frequency equals a multiple of wave encounter frequency.

Because of these coupling effects, some advanced direct load calculation procedures, such as dynamic loading approaches (Liu, et al, 1992, Shin, et al, 1997) and design wave method (Hachmann, 1991), should be replenished if the whipping components are to be rationally included.

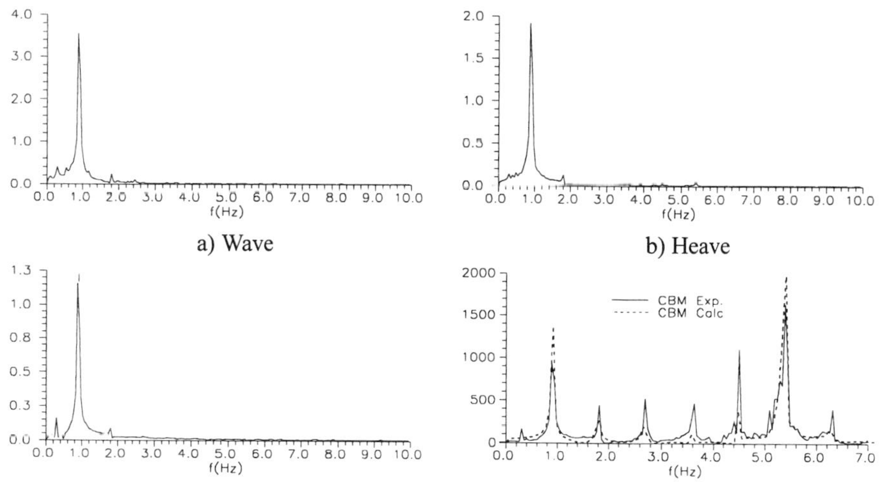

c) Pitch d) Bending moment at midship

Figure 1: Frequency characteristics of wave, motions and bending moments from FFT transformations on measured time histories (λ/L=1.0, H/L=0.04, Fn=0.244)

3 TIME DOMAIN SIMULATION OF NONLINEAR RESPONSES OF SHIPS IN WAVES

The importance of nonlinear responses of ships in waves has been generally recognized (ISSC, 2000). However, the applications to various nonlinear theories and associated codes in direct load calculations for ship structural designs to incorporate these nonlinear effects and whipping components are limited.

To predict nonlinear vertical ship motions and structural responses in severe waves and include these effects in direct design of ship structures, a nonlinear time-domain strip theory is developed based on linear strip theory (Salvesen, et al, 1970). The restoring, Froude-Krylov forces and all hydrodynamic coefficients are varied with the instantaneous draughts of hull sections. Also the bottom impact, bow flare slamming and green water forces are included in ship motion predictions. The flexible modes of ship hull girders are accounted for by a Timoshenko beam theory (Gu, 1999).

The predicted results are validated for various ship types such as S-175 containership (Watanabe & Guedes Soares, 1999), destroyers and bulk carriers. Some sample comparisons of the combined bending moments (CBM) predicted with that of the containership model tests in regular waves with different wave heights are shown in Figure 2. In small or moderate amplitude waves, the predicted bending moments agree very well with test results; In large amplitude waves, the combined bending moments increase nonlinearly with the wave heights and the predicted values are larger than that of the tests.

The relations between bending moments and wavelengths in large amplitude waves compared with the test results are shown in Figure 3. As expected, there are more than one peaks in calculated bending moments, the first one is at λ/L=1.0 and the second one is at λ/L=1.25. The corresponding ratios of the first natural frequency to wave encounter frequency are about 6.0 and 7.0 for these two peaks, respectively. This phenomenon means that when the first natural frequency equals a multiple of the wave encounter frequency, the combined bending moment will reach their extreme values.

The frequency characteristics of combined bending moments predicted are also shown in Figure 1d (dashed line). The energy distribution on frequencies is similar to the experimental results but the influences of whipping component on higher order components are generally reflected.

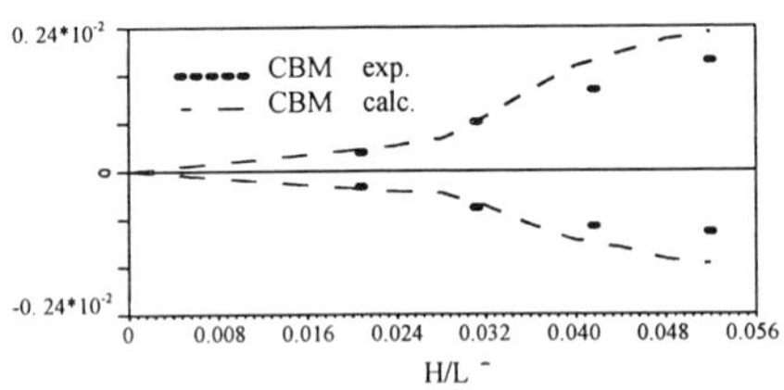

Figure 2: The relation of bending moments and wave height (λ/L=1.0,Fn=0.244)

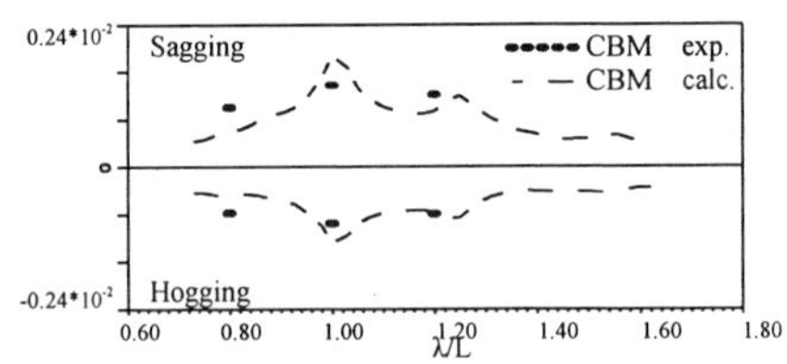

Figure 3: Comparison of predicted amplitudes of bending moments and that of model test in regular waves (H/L=0.04, Fn=0.244)

4 A MODIFICATION TO DYNAMIC LOADING APPROACH

Dynamic loading approach (DLA) (Liu,et al, 1992, Shin, et al, 1997) is a practical direct load calculation method with many innovative concepts such as dominant load parameters, structural load cases, equivalent wave systems and so on. Although some improvements, e.g., reasonably including nonlinear wave-induced and whipping bending moments in equivalent waves, have to be made, the overall framework is flexible enough to admit further modifications.

Because of the existence of the higher order harmonic bending moments and their coupling effects with whipping components, the predicted combined bending moments in equivalent regular waves are very sensitive to wavelength and ship speed. The operational condition, here means the combination of wavelength, wave height and ship speed, should be carefully considered to ensure that the most unfavorable responses are achieved. Moreover, when the speed loss of a ship must be taken into

account in long-term linear load predictions, it is difficult to determine the ship speed in equivalent design waves. Hence, in order to rationally include whipping components in direct calculations of wave loads, a modification to Dynamic Loading Approach is proposed as following:
(1) Choose the long-term wave data statistics and associated spectra;
(2) Choose ship speed loss expression;
(3) Long-term predictions of linear wave-induced bending moments to obtain linear design loads corresponding to appropriate probability of exceedence.
(4) Calculate a set of response amplitude operators (RAOs) of the ship in head sea with different ship speeds, the wave lengths at the peaks of the RAOs are consider as a set of candidate wave lengths.
(5) A set of candidate wave amplitudes are equal to the quotients of the linear design loads divided by peak values of RAOs.
(6) Every ship speed and the associated candidate wave length and amplitude forms a operational condition. The nonlinear wave loads are calculated in every condition, then the curves of sagging and hogging wave-induced and combined loads are obtained. Usually there are several peaks on the combined curve, the loads, wave parameters and ship speed at maximum peaks (sangging and hogging) of combined bending moment are design values.
(7) Other responses such as ship motions and hydrodynamic pressures are obtained at same instants when the combined bending moment reaches its maximum (sagging) or minimum (hogging).

5 CASE STUDY

For the containership mentioned in above sections, the direct load calculation is applied to illustrate the determination of the design parameters.

Choose the North Atlantic wave data scatter diagram and the ITTC two-parameter wave spectrum as long-term environment.

The voluntary speed loss of the ship in waves are estimated as following (Gu, 1999):
(1) Take the bow bottom slamming (at 0.15L aft of FP, the threshold water entry velocity $V/\sqrt{gL}=0.096$) with frequency of 70 slams per hour as the reason and criterion for speed reduction.
(2) Take the most probable wave periods and associated wave heights in the wave data scatter diagram as the most commonly seen sea states.
(3) With the nonlinear response theory in previous section and its code, one-hour ship motions in head seas with different forward speeds are simulated, and then the number of slams is counted.
(4) It was observed in the model tests that for the full-scale containership there were about 58~70 obvious whippings per hour happened in operational condition $T_2\sqrt{g/L}=2.14$, $H_s/L=0.03$, $F_n=0.244$ and 75 whippings in condition $T_2\sqrt{g/L}=2.23$, $H_s/L=0.044$, $F_n=0.122$ judged from vertical bending moments at midship. The predicted number of slams should be amended for the difference between theoretical and experimental results.
(5) It is assumed that the ship speed will linearly decreased with heading angle increasing. In following sea ($H_s/L=0.0385$), it is not less than 90% of maximum design speed and linearly decreased with the wave height increasing.
Then, the estimated maximum permissible speeds, Fn, of the ship in waves are:

$$\frac{F_n}{F_{nm}}=\frac{180-\beta}{180}\left(1-2.6\frac{H_s}{L}\right)+\frac{\beta}{180}\left(1-0.549\frac{H_s}{L}-259(\frac{H_s}{L})^2\right) \tag{1}$$

Where, F_{nm} is the maximum design speed; β is heading angle, 0° (following sea)$\leqslant\beta\leqslant$180° (head sea); H_s and T_2, g are significant wave height, zero-crossing wave period and gravitational acceleration, respectively. The ship speed is not less than 5 knots; otherwise the heading angle will be changed.

The linear long-term wave-induced bending moments of the containership are calculated in short-crested waves. The ratio between them and the design bending moments according to IACS UR-S11 is shown in Figure 4. The long-term predicted values with constant ship forward speed are also shown in the figure. It can be seen that for this ship the predicted values considering speed loss are close to that with constant speed of two third maximum design speed.

Take the bending moment at probability of 10^{-5} as linear design value which is 97% of the mean value of IACS sagging and hogging design bending moments, divide it by peaks of RAOs in head sea with different forward speed, the relations of candidate wave lengths and heights with different speeds are shown in Figure 5.

The nonlinear bending moments in different operational conditions are shown in Figure 6. There are two obvious peaks existed in combined bending moments.

To further explain the reason for the peaks of combined bending moments, the coincidence, D, of the hull natural frequency with higher order harmonic frequencies is shown Figure 7. Here D is defined as the absolute value of the difference between the remainder after the natural frequency is divided by wave encounter frequency and the nearest integer (0 or 1), and is a degree of coupling between whipping component and a harmonic component. $D \approx 0$ means that the natural frequency is about a multiple of a higher order harmonic component, and in this case the combined bending moments will achieve their extremes as shown in Figure 6.

Choose the maximum combined bending moments as design moments (the second peak in Figure 6), then the associated wave and ship speed are considered as design wave and speed in the calculation, see Table 1. The bending moments are non-dimensionalized by linear design bending moment. Other responses are obtained at the same instant of time during the simulation of the combined bending moments in design waves.

TABLE 1

DESIGN WAVE, SHIP SPEED AND DESIGN BENDING MOMENTS

Wavelength/ship length λ /L	0.94
Wave height/ship length H/L	0.04
Ship speed Fn	0.215
Linear design bending moment	1.0
Nonlinear wave-induced bending moments (sagging/hogging)	1.210/0.761
Nonlinear combined bending moments (sagging/hogging)	2.773/1.916

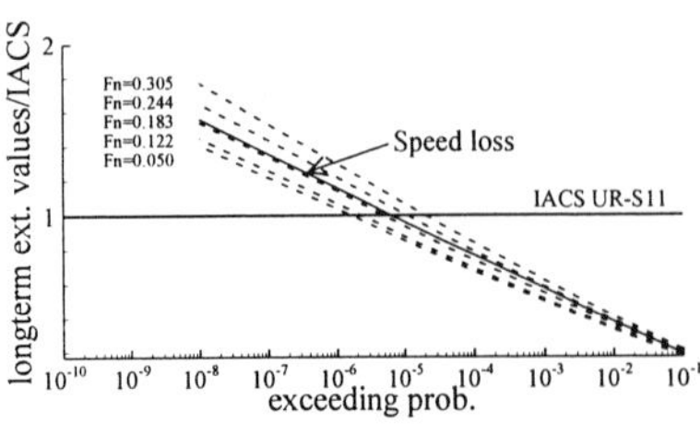

Figure 4: Long-term linear wave-induced bending moments

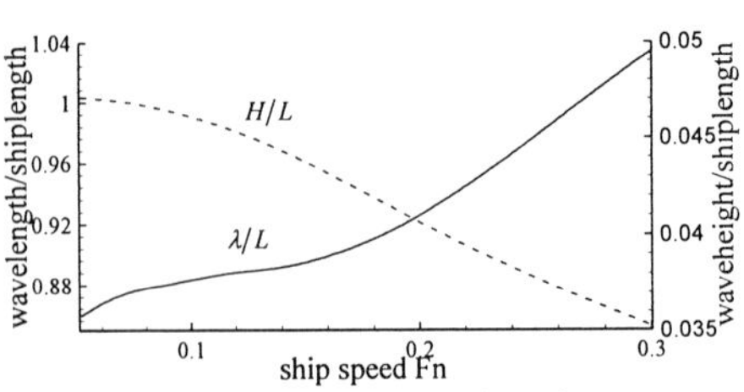

Figure 5: Candidate wavelengths, wave heights and ship speeds for nonlinear response calculations

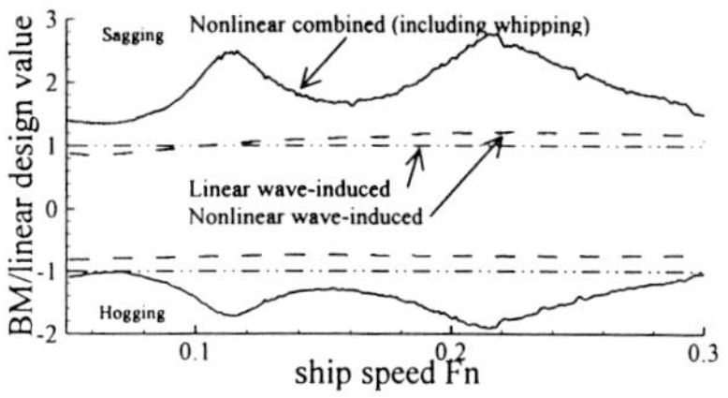

Figure 6: The nonlinear vertical bending moments for candidate operational conditions

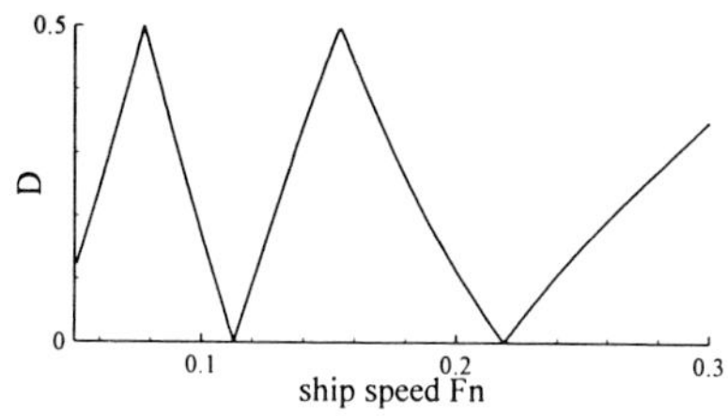

Figure 7: Coincidence of the first natural frequency with higher order harmonic frequencies

6 CONCLUDING REMARKS

To rationally consider nonlinear wave-induced and extreme combined bending moments including whipping components in direct load calculations for fast ship structural designs, a modification to dynamic loading approach is presented. The nonlinear responses are predicted by a nonlinear time-domain strip theory. The coupling effects between higher order harmonic and whipping bending moments, the speed loss of ships in severe waves are well considered in determining the design operational conditions to ensure the most unfavorable responses are achieved.

The modification to dynamic loading approach is suitable for alternative nonlinear ship response theories to provide more comprehensive information and high accuracy for direct design loads calculations.

References

Adegeest L.J.M. (1994). Nonlinear Hull Girder Loads in Ships. *Ph.D. Thesis, Delft University of Technology.*

Gu X. K., Shen J. W. and Moan T. (2000). Experimental and theoretical investigation of higher order harmonic components of nonlinear bending moments of ships. *Ship Technology Research, Schiffstechnik*, **47:3**, 143-154.

Gu X. K. (1999). Direct Design of Ship Structures. *Ph.D. Thesis, China Ship Scientific Research Center.*

Hachmann D. (1991). Calculation of Pressures on a Ship's Hull in Waves. *Ship Technology Research, Schiffstechnik*, **38:**, 111-133.

ISSC (2000). Report of Special Task Committee VI.1-Extreme Hull Girder Loading. *Proceedings of the 14th International Ship and Offshore Structures Congress*, Nagasaki, Japan.

Liu D., Spencer J., Itoh T., Kawachi S. and Shigematsu K (1992). Dynamic Load Approach in Tanker Design. *Transactions of Society of Naval Architect and Marine Engineers*, **100**, 143-172.

Salvesen N.; Tuck E.O.and Faltinsen O.M. (1970), Ship motions and sea loads, *Transaction of Society of Naval Architect and Marine Engineers*, **78**, 250-287.

Shin Y.S., Chung J.S., Lin W.M., Zhang S. and Engle A. (1997). Dynamic Loadings for Structural Analysis of Fine Form Container Ship Based on a Non-linear Large Amplitude Motions and Loads Method. *Transactions of Society of Naval Architect and Marine Engineers*, **105**, 127-154.

Watanabe I. and Guedes Soares, C. (1999). Comparative Study on the Time Domain Analysis of Nonlinear ship Motions and Loads. Marine Structures, **12:3**, 153-170.

Watanabe I., Keno M. and Sawada H. (1989). Effect of Bow Flare on Shape to Wave Loads of a Container Ship. *Journal of the Society of Naval Architects of Japan*, **166**, 259-266.

Practical Design of Ships and Other Floating Structures
You-Sheng Wu, Wei-Cheng Cui and Guo-Jun Zhou (Eds)

FLUTTER OF HYDROFOIL IN VISCOUS FIELD

Can Sima, Xiao-Ci Zhang and You-Sheng Wu

China Ship Scientific Research Center
Wuxi, Jiangsu 214082, China

ABSTRACT

A Boundary Element Method is introduced to solve the coupled responses of an elastic two-dimensional hydrofoil in viscous flow. A modified Oseen's equation is employed in the present paper to describe the viscous field. The application of the Oseen's equation is extended to a much larger range of Reynolds number by introducing the turbulent viscosity in the numerical analysis. A fluid-structure interface boundary condition is presented. The coupled problem is therefore represented as a boundary integral method, with Oseenlets being the fundamental functions. The two-dimensional hydrofoil is modeled as an elastically supported rigid body with identical NACA2421 cross-section and two degrees of freedom. Newmark integral method is used to solve the boundary integral equation. The theoretical results of the viscous flow induced flutter of the hydrofoil are described in time and in frequency domain. The corresponding phenomena and the factors, which affected the flutter are discussed.

KEYWORDS

Hydroelasticity, Flow-induced flutter, Boundary element method, Oseenlet, Hydrofoil, Turbulent viscosity.

1 INTRODUCTION

Improper design of an appendage of a ship, especially a thin hydrofoil, stabilizing fin, ruder, or a propeller, for example, may introduce unexpected vibrations or flutter, resulting in structural damage and serious noise. This did happen in some fast surface ships, for example hydrofoil crafts, and some submersibles. This phenomenon requires further investigations about the coupled responses of a thin wall structure and the surrounding viscous flow.

In the present paper a Boundary Element Method is introduced to the coupled responses of an elastic two-dimensional hydrofoil responding to viscous flow excitation. By introducing the viscous fundamental solutions (Oseenlets) of Oseen's equation as the basic functions, the viscous flow field can also be described by a boundary integral. Analogous to the usual boundary element methods widely employed in potential theories, by distributing the Oseenlets over the body's wetted surface, the fluid disturbances induced by the presence and motions of the body, solely depends on the values of the strengths of the Oseenlets. The strengths are solved from the boundary integral equation to ensure the body boundary conditions.

In the present paper a two-dimensional hydrofoil is analyzed. It is approximately represented as a rigid NACA2421 foil elastically supported at a rotating center to allow for two degrees of motion freedom, namely the vertical motion simulating the bending mode, and the rotation simulating the torsion mode. Newmark integral method is employed to solve the boundary integral equations for the strengths. The coupling with the foil is reflected in the boundary condition of the hydrofoil-wetted surface. FFT method is used to translate the results from time domain to frequency domain. The responses of the hydrofoil are then presented by the response frequency spectrum and limit cycles.

2 OSEENLET BOUNDARY ELEMENT METHOD

The Oseen's equation is the simplification of Navier-Stokes equation.

$$\begin{cases} \nabla\cdot\vec{V}=0 \\ -U\dfrac{\partial\vec{V}}{\partial x}+\upsilon\nabla\times\vec{\omega}-\rho^{-1}\nabla p=0 \end{cases} \tag{1}$$

Let $\dfrac{q}{\rho}=\dfrac{p}{\rho}+\dfrac{|\vec{V}'|^2}{2}$, where $\vec{V}'$ is the disturbing velocity, $\vec{V}=U\vec{i}+\vec{V}'$, U is the flow velocity, and $\vec{i}$ represents the direction of coming flow. Then the equation (1) can be rewritten

$$\begin{cases} \nabla\cdot\vec{V}'=0 \\ -\nabla\dfrac{q}{\rho}-U\dfrac{\partial\vec{V}'}{\partial x}+\upsilon\nabla\vec{V}'=0 \end{cases} \tag{2}$$

The solution of equation (2) is

$$\begin{cases} V_{11}^{*}=\sigma\left\{\dfrac{x}{r^2}-ke^{kx}\left[K_0(kr)+\dfrac{x}{r}K_1(kr)\right]\right\} \\ V_{12}^{*}=\sigma\left[\dfrac{y}{r^2}-ke^{kx}\dfrac{y}{r}K_1(kr)\right] \\ V_{21}^{*}=V_{12}^{*} \\ V_{22}^{*}=\sigma\left\{-\dfrac{x}{r^2}-ke^{ke}\left[K_0(kr)-\dfrac{x}{r}K_1(kr)\right]\right\} \end{cases}, \quad \begin{cases} q_1=-\sigma\rho U\dfrac{x}{r^2} \\ q_2=-\sigma\rho U\dfrac{y}{r^2} \end{cases} \tag{3}$$

where, $r=\sqrt{x^2+y^2}$, σ is the strength of Oseenlet, K_0 and K_1 are the Bessel functions respectively, p are the pressure.

The body surface is divided into n boundary elements $\Gamma_j(j=1,2,\cdots,n)$, taking the middle points of the boundary elements as the nodes. The strength of an Oseenlet is assumed to be constantly distributed over each element. An integral equations for the unknown strengths of the lift and drag Oseenlets, i.e. σ_{1j},σ_{2j} are transformed into a set of linear simultaneous equations. These equations are solved in time domain, starting with a stationary position of the foil. At the first time step, the velocities of the wetted surface are zero. Using this wetted surface boundary condition the strength of the Oseenlets at t = 0 can be determined by the equation (4).

$$\begin{cases} \sum\limits_{j=1}^{n}\sigma_{1j}\int_{\Gamma}V_{11}^{*}\,d\Gamma+\sum\limits_{j=1}^{n}\sigma_{2j}\int_{\Gamma}V_{12}^{*}\,d\Gamma=-U \\ \sum\limits_{j-1}^{n}\sigma_{1j}\int_{\Gamma}V_{21}^{*}\,d\Gamma+\sum\limits_{j-1}^{n}\sigma_{2j}\int_{\Gamma}V_{22}^{*}\,d\Gamma=0 \end{cases} \tag{4}$$

The turbulent velocities of flow and q, representing the pressure acting on the surface, may be written as

$$\begin{cases} u' = \sum_{1}^{n} \sigma_{1m} \int V_{11}^{*} ds + \sum_{1}^{n} \sigma_{2m} \int V_{12}^{*} ds \\ v' = \sum_{1}^{n} \sigma_{1m} \int V_{21}^{*} ds + \sum_{1}^{n} \sigma_{2m} \int V_{22}^{*} ds \end{cases} \tag{5}$$

$$q = -\rho U \int (\sigma_1 \frac{\overline{x}}{\overline{r}^2} + \sigma_2 \frac{\overline{y}}{\overline{r}^2}) ds - \rho\pi U (l\sigma_1 + \sigma_2) \tag{6}$$

It is well known that the Oseen's equation is limited to a very low Reynolds number. In order to overcome this difficulty, and to make the applicable range of Reynolds numbers as large as possible, the turbulent viscosity is introduced into the equation. The kinematic viscosity coefficient υ is replaced by the turbulent viscosity coefficient υ_t. In fact, many appendages of ship hull (such as fin, hydrofoil, rudder and propeller) are submerged in turbulence of ship wake. It seems reasonable to assume the turbulent viscosity coefficient υ_t as a constant scalar. In this paper $\upsilon_t = 0.025Ud$, where d is the characteristic length of the hydrofoil.

3 TWO DIMENSIONAL EQUATIONS OF BENDING AND TORSION COUPLED MOTIONS

The equations of motion of the hydrofoil's bending and torsion are

$$\begin{cases} m\ddot{h} + S_\alpha \ddot{\alpha} + k_h h = Q_h \\ S_\alpha \ddot{h} + I_\alpha \ddot{\alpha} + k_\alpha \alpha = M_\alpha \end{cases} \tag{7}$$

where, Q_h, M_α denote hydrodynamic force and moment respectively, determined by q solved from Oseenlet boundary element method. m is the mass of the hydrofoil section, S_α, I_α, k_h and k_α are respectively the first and the second static inertial moments, the bending and torsion stiffness.
The coupled dynamic responses between the fluid and the structure are analyzed taking the both parts as an entire system. The correction of viscous flow boundary conditions caused by the surface displacement and velocity should be accounted for. These calculations are carried out by the constant step-by-step method and Newmark integral method. The wetted surface boundary condition is changed with time. In every time step, the position of Oseenlet keeps changing. Thus the term at the right hand side of equation (4) (-U and 0) are replaced by $-U\cos\alpha - u_x$、$-v_y$, where u_x, u_y are the components of the hydrofoil surface velocities, and α is the torsion angle of hydrofoil.

4 RESULTS OF CALCULATION AND DISCUSSIONS

The original parameters of the NACA2421 two-dimensional hydrofoil are showed in Table 1, 2 and 3, where x_d and x_b are respectively the distance between the rotating center and the leading edge, the distance between the gravity center and the rotating center; f_h and f_α are respectively the bending and torsion natural frequencies of the hydrofoil in vacuum.
Table 4 gives the summarized results. Figure 1 is the typical time history of vibration; Figure 2 shows the frequency spectrum curves; Figure 3 is the limit cycles of bending and torsion.
It is found that under the viscous flow excitation, flutter may happen on the NACA2421 two-dimensional hydrofoil in three different ways. This can obviously be seen from the typical time history curves of the vibrations (Fig.1) and the limit cycles of its bending and torsion modes (Fig.3). The three flutter phenomena are respectively as follows: the first is of converged amplitude (Fig.3 (a)); the second shows diverged amplitude (Fig.3(c)), and the third appear to have equal-amplitude (Fig.3 (b)). The critical velocity domains of the three kinds of flutter (divergent vibration,

equal-amplitude vibration and convergent vibration) of the hydrofoil can be clearly distinguished from the calculation results. Using FFT method, the calculated time domain results are translated to frequency domain. It is exhibited by Fig.2 that when the hydrofoil performs a convergent vibration, no peak values appear on the frequency spectrum curves of the response amplitude. In the case of a divergent vibration or an equal-amplitude vibration, peak values clearly appear in the frequency spectrum curves of the amplitudes, especially at the frequencies 7.8Hz and 12.7Hz. The peak value of divergent vibration is bigger than that of equal-amplitude vibration. This flutter is particularly harmful to produce serious under-water noise, and in some cases is the cause of the structural damage.

TABEL 1

MATERIAL AND GEOMETRIC PARAMETERS OF NACA2421

Type of hydrofoil	NACA2421
The length of chord	c=0.12 m
The maximum thickness	t=0.025 m
Density	$\rho=2.7\times10^3$ kg/m^3
Mass	$m = 0.56$ kg
Distance between the center of gravity and leading edge	x_z=0.0576 m

TABLE 2

TORSION CENTER PARAMETERS

	center I	center II	centerIII
x_d (m)	0.02	0.03	0.04
$x_b = x_z - x_d$	0.0376	0.0276	0.0176
S_α (kg·m)	0.02105	0.0154	0.00985
I_α (kg·m^2)	0.1347	0.0982	0.0729
$x_\alpha = S_\alpha/(c/2)$	0.418	0.257	0.164

TABLE 3

THE NATURAL FREQUENCY OF HYDROFOIL

	torsion center I		torsion center II		torsion center III	
k_h (N/m)	240.0		240.0		240.0	
f_h (Hz)	3.295		3.295		3.295	
$\lambda = f_h/f_\alpha$	k_α (N/m)	f_α (Hz)	k_α (N/m)	f_α (Hz)	k_α (N/m)	f_α (Hz)
0.4	360.8	8.237	263.0	8.236	195.27	8.237
0.6	160.36	5.491	116.9	5.491	86.78	5.491
0.8	90.2	4.118	65.76	4.118	48.82	4.118
1.0	57.73	3.295	42.1	3.295	31.24	3.295
1.2	40.09	2.746	29.2	2.744	21.7	2.746

The results explain that viscous flow-induced flutter is not the same as the flutter induced by air. The lowest critical velocity of the airflow for the flutter to happen is just corresponding to the case when the value of λ, the ratio between the bending and torsion natural frequencies, is close to one. However a NACA 2421 hydrofoil may flutter at a lower fluid velocity when λ is about 0.6, as is shown in Table 4. The main factors which affect viscous flow induced flutter are: (1) the position of torsion axis (the rotating center), (2) the ratio of bending to torsion frequencies; and (3) the ratio of hydrofoil density to water density.

TABLE 4

THE RESULT SUMMERY

<table>
<tr><th>x_d (m)</th><th>vibration λ
V(m/s)</th><th>0.4</th><th>0.6</th><th>0.8</th><th>1.0</th><th>1.2</th></tr>
<tr><td rowspan="4">0.02</td><td>< 0.05</td><td>C</td><td>C</td><td>C</td><td colspan="2" rowspan="4">C</td></tr>
<tr><td>0.05~0.1</td><td>E</td><td rowspan="3">D</td><td rowspan="2">E</td></tr>
<tr><td>0.1~0.2</td><td rowspan="2">D</td></tr>
<tr><td>> 0.2</td><td>C</td></tr>
<tr><td rowspan="4">0.03</td><td>< 0.05</td><td>C</td><td>C</td><td colspan="3" rowspan="4">C</td></tr>
<tr><td>0.05~0.1</td><td rowspan="2">E</td><td>E</td></tr>
<tr><td>0.1~0.4</td><td rowspan="2">C</td></tr>
<tr><td>> 0.4</td><td>C</td></tr>
<tr><td>0.04</td><td colspan="6">C</td></tr>
</table>

C—converged vibration, D—diverged vibration, E—equal-amplitude vibration

5 CONCLUSIONS

(1) It is shown that the boundary integral method can be extended to solve the viscous flow induced fluid-structure interaction problems with certain success. Although the previous application of Oseen's equation showed limitations in the Reynolds number range, it is still possible to extend its range of application to allow for much larger Reynolds number to be calculated for some engineering problems, like the flutter of an elastic hydrofoil. The turbulent viscosity introduced in this paper has shown its advantage in overcoming the limitations of the Oseen's equations.

(2) The flutter phenomena of the NACA2421 hydrofoil can be exhibited from the time histories, the limit cycles, and the frequency spectra.

(3) The analysis method developed in the present paper may be employed to predict the possible occurrence of the flutter of the appendages of a ship.

References

1 Himeo Y. (1970). On Viscous Flow about Bluff Body at High Reynolds Number. *J. of the Naval Architects of Japan*, **127.**

2 Zhang Xiaoci, et al. (2000). The Analysis of the Vortex Induced Vibration of Hydrofoil in Turbulent Field. *Ship Mechanics of China*, **65.**

3 Zhang Xiaoci, et al. (2000). Numerical Simulation and PIV Experimental Display of Vortex Induced Vibration for Hydrofoil. *ICHD 2000*, Yokohana, **423.**

4 Sima Can, et al. (1999). Applications of Oseenlet Boundary Element Method in Viscous Fluid-structure Coupling Motion Problems. *Ship Mechanics of China*, **38.**

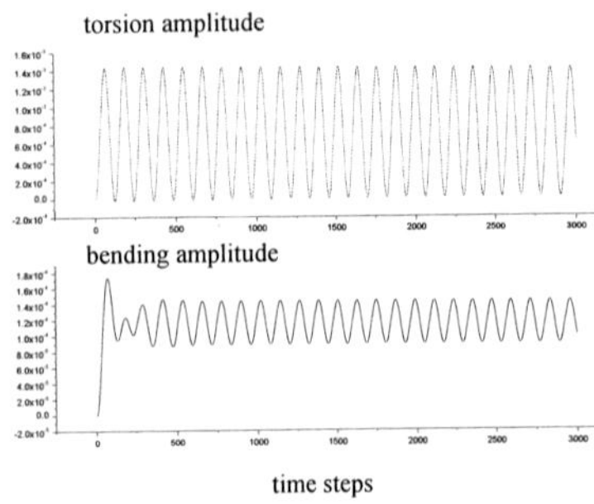

(a) Torsion center I , $\lambda = 0.8$, $u = 0.1m/s$

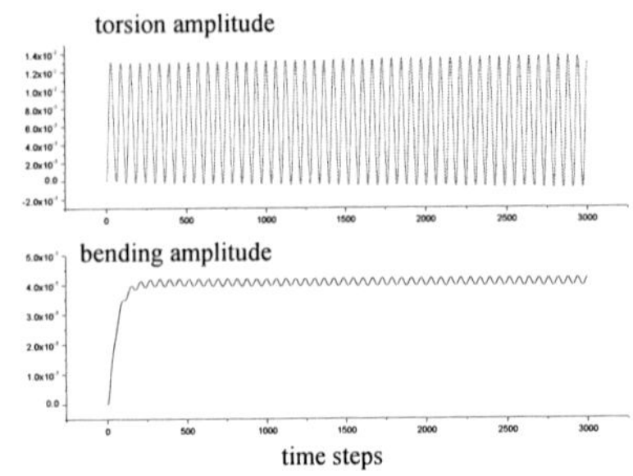

(b) Torsion center I , $\lambda = 0.4$, $u = 0.6m/s$

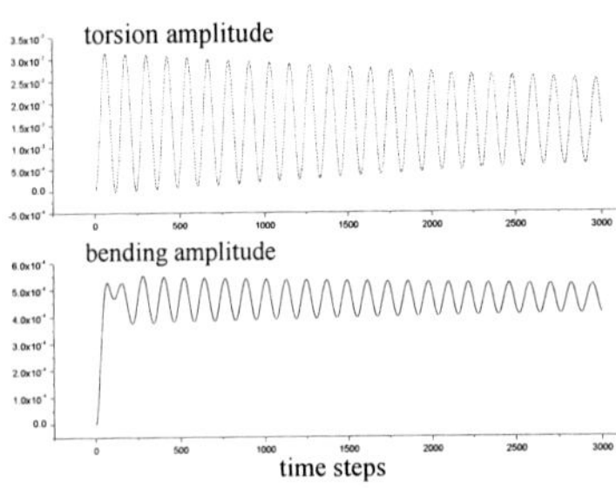

(c) Torsion centerIII, $\lambda = 0.8$, $u = 0.2m/s$

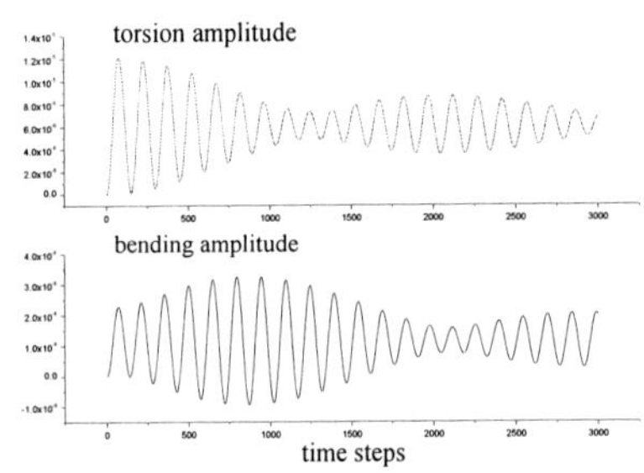

(d) Torsion centerIII, $\lambda = 1.0$, $u = 0.01m/s$

Figure1: The typical time history curves of vibration

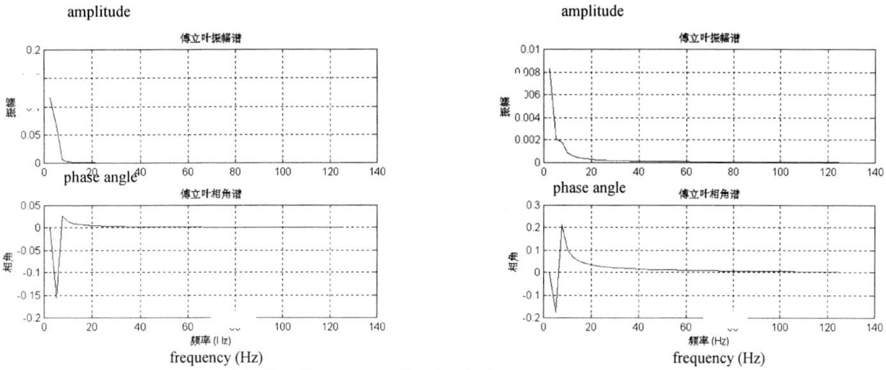

(a) Torsion center Ⅰ, λ=1. 2， $u = 0.1m/s$

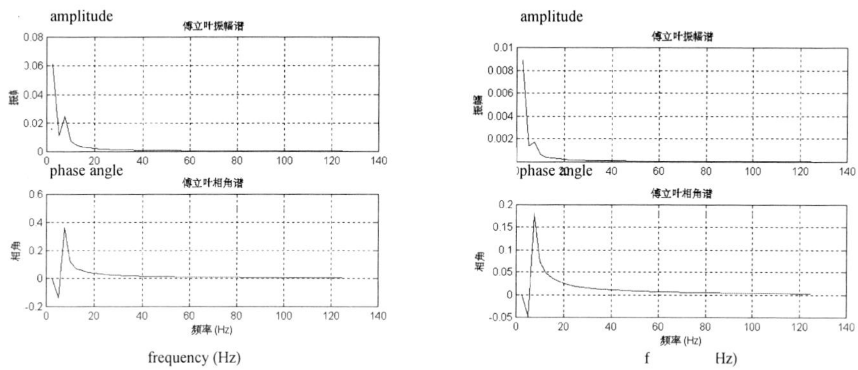

(b) Torsion center Ⅰ, λ=1.8， $u = 0.1m/s$

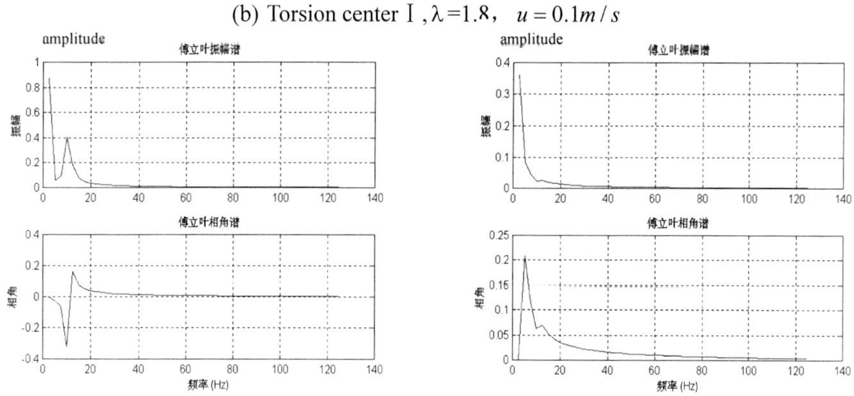

(c) Torsion center Ⅰ, λ=0.4， $u = 0.8m/s$

Figure 2: The frequency spectrum

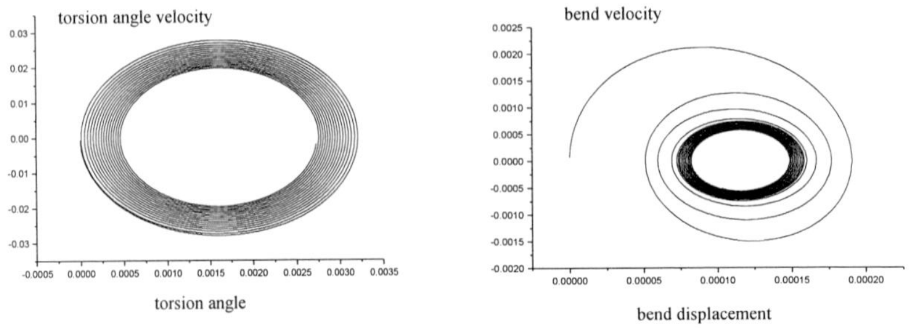

(a) Torsion center I , λ=1. 2, $u = 0.1m/s$

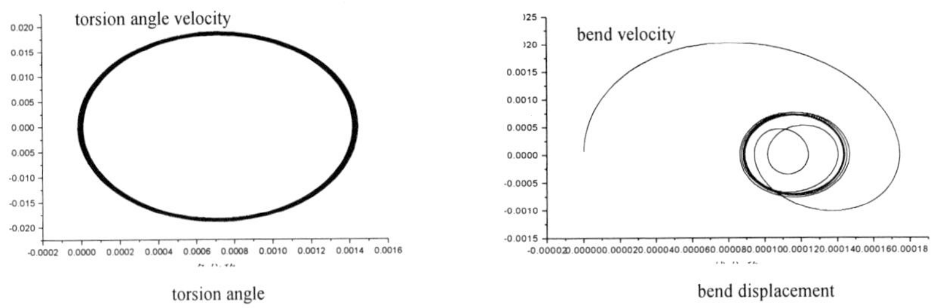

(b) Torsion center I , λ=1. 8, $u = 0.1m/s$

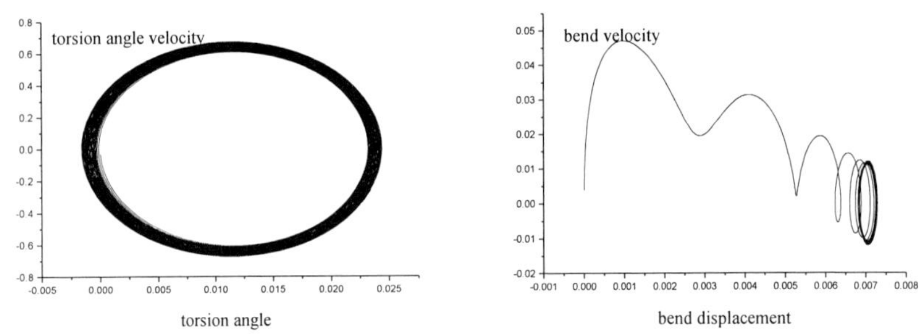

(c) Torsion center I , λ=0. 4, $u = 0.8m/s$

Figure 3: The limit cycles of bend and torsion

Practical Design of Ships and Other Floating Structures
You-Sheng Wu, Wei-Cheng Cui and Guo-Jun Zhou (Eds)

SYMMETRIC AND ANTISYMMETRIC HYDROELASTIC ANALYSIS OF A BULKER IN WAVES

S.E. Hirdaris, W.G. Price and P.Temarel

School of Engineering Sciences, Ship Science, University of Southampton, Southampton, SO17 1BJ, UK

ABSTRACT

The influence of structural configurations (such as large deck openings and double skins encountered in containerships, bulk carriers and LNG carriers) on predicting the global dynamic behaviour of these flexible vessels in waves is of particular interest. This is due to the options available for structural modelling, namely two-dimensional beam and three-dimensional FE idelisations. Two-dimensional idealisations offer a fast and efficient means of simulating the dynamic behaviour in waves whilst three-dimensional models are time and effort consuming, even when a relatively simple idelisation is used. The high rate of bulk carrier casualties in recent years makes this type of vessel a suitable example for investigating the influence of modelling on the fluid structure interactions and subsequent loads and responses in waves. Therefore, in this paper two- and three-dimensional hydroelasticity theories are applied to predict and compare the dynamic behaviour of a bulk carrier hull in regular oblique waves. Both symmetric and antisymmetric distortions are incorporated in these investigations.

KEYWORDS

Bulk carriers, Dynamic loads, FE idealisation, Hydroelasticity, Wave-induced dynamic response

1 INTRODUCTION

The evaluation of wave-induced motions and loads are two of the most important aspects in ship design. Traditionally, prediction of wave-induced loads, in the main, relies on quasi-static and/or empirical methods. Overall, such methods proved to be quite adequate over the years with problems arising dealt by relying on accumulated experience. Nevertheless, the international amendments for bulkers, tankers and LNG carriers brought about by IACS and IMO are a measure of the concern related to the behaviour of such vessels and associated risks involving loss of human life, environmental pollution etc. (IACS 1997; Tustin 1998).

During the past ten years, more than 700 crewmen have died as a result of bulk carrier casualties. Following a rush of 25 bulk carrier losses in 1990 and 1991 five more bulk carriers and 74 lives were lost in 1993, followed, in 1994, by the tragic loss of six bulkers and 100 lives (IACS 1997; Tustin 1998). Research undertaken by IMO and IACS suggests that some of the possible causes for 90% of

the losses could be hull girder failure, loss of stability and loss of reserve buoyancy (IACS 1997). However there still remains 10% unexplained disasters, which could be attributed to structural failure, although concrete evidence is hard to come by. The accurate prediction of seaway induced dynamic loads and their consequences may be able to shed light in explaining these disasters.
Hydroelasticity theory, unlike empirical and quasi-static methods, provides a rigorous analysis whereby the fluid-flexible structure interaction is formulated and the wave induced dynamic loads and responses (e.g. stresses, bending moments, etc) are evaluated. The two-stage approach of *dry* or *in vacuo* and *wet* analyses leads to a direct evaluation of dynamic loads and responses for a vessel travelling at arbitrary heading in regular waves and irregular seaways. This theory is unified in the sense that it incorporates rigid body motions (i.e. conventional seakeeping) as well as distortions. The concept of hydroelasticity theory gained significant momentum in the mid-seventies, when Bishop and Price (1979) established the basic principles of a theory for flexible beam-like hulls subject to steady state or transient (e.g. slamming) wave-induced loads, combining *Timoshenko beam* and *strip* theories. The method was applied successfully to a variety of beamlike merchant and naval ships and its capabilities in simulating symmetric, antisymmetric and unsymmetric dynamic behaviour in waves was demonstrated. Comparisons with available experimental and full-scale measurements, such as those by Bishop et al (1984) and Aksu et al (1993), confirmed the validity of the hydroelastic approach.
The three-dimensional hydroelasticity theory was developed to simulate the wave-induced dynamic behaviour of non-beamlike vessels in waves (Bishop et al 1986). This analysis makes use of two- or three-dimensional finite element models of the structure and a singularity distribution (such as pulsating source) over its wetted surface. The application to SWATHs, using crude and refined structural models, has demonstrated the versatility and potential of this method (Bishop et al 1986, Price et al 1994). Recent applications to trimarans showed good agreement with flexible model experiments for wave induced loads, such as prying moment, for a range of speeds and headings (Miao et al 1997). Applications to non-slender, mono-hulled vessels, such as sailing yachts, showed that the method is capable of simulating accurately the dynamic behaviour of such vessels and identifying correctly areas where structural problems may arise (Louarn et al 1999). The analysis has also been applied to beamlike mono-hulls (neglecting any small deck openings), showing good agreement between wave-induced loads and responses predicted by two- and three-dimensional models (Price et al 2001).
The objectives of the current investigations are to examine the influence of three-dimensional structural modelling on the dynamic behaviour of slender beamlike vessels in regular waves and, in particular, vessels with large deck openings. Along these lines a vessel with principal dimensions and characteristics similar to those of MV Derbyshire is used in this paper (Bishop et al 1991; Hirdaris et al 2000). Two fluid-structure interaction models, incorporating both symmetric and antisymmetric motions and distortions, are used, namely (i) **beamfd** where Timoshenko beam idealisation is combined with a suitable strip theory (Bishop and Price 1979) and (ii) **shell3d** where a shell FE idealisation is used in conjunction with a three-dimensional pulsating source distribution over the wetted surface of the hull. Comparisons between the *dry* and *wet* hull characteristics, such as principal mode shapes, natural frequencies and wet resonance frequencies, and wave induced loads in oblique regular waves predicted by these models are presented.

2 FLUID-STRUCTURE INTERACTION MODELLING

Two- and three-dimensional structural models were generated to idealize the structural and inertia characteristics of a vessel based on the OBO MV Derbyshire. The length of this vessel is L_{OA}=294m ($L=L_{BP}$=282m) and the remaining dimensions and properties can be found in Bishop et al (1991).
For the three-dimensional model (shell3d) shell finite elements were used to incorporate major structural elements such as deck, side, inner/outer bottom, hopper spaces, bulkheads, major longitudinal girders etc. The material contributions of other stiffeners and transverse frames were subsumed into the properties of adjacent shell elements. For simplicity of modelling the superstructure

as well as the stiffness of the hatch covers were not included, thus resulting in a deck with a number of large openings. Element SHELL63 with four nodes and six degrees of freedom per node was selected, considering only membrane deformations (Swanson Ltd. 2000). The three-dimensional FE model, shown in Fig.1(a), consists of 6439 nodes and 3673 shell elements and is in the form of 46 sections along the vessel. This facilitates the determination of the mass distribution along the vessel, by manipulating the mass density of elements in a section, as well as the acquisition of properties required for the beam model. This method was preferred to using lump mass elements so that relevant numerical problems during modal analysis, as noted by Price et al (1994), can be avoided. The omission of transverse frames and longitudinal stiffeners from structural idealisations of mono- and multi-hulled vessels has been dealt successfully in the past, as illustrated for example by Miao et al (1997), Louarn et al (1999) and Price et al (2001), by incorporating *fictitious* bulkheads of negligible mass and thickness (e.g. 1mm for the current application), where appropriate. Omission of such elements results in mode shapes containing localised distortions, which are not physically representative of the global dynamic behaviour of the vessel. A *fan* shape form, clearly seen in the detail of the FE model shown in Fig.1(b), was adopted for all actual and fictitious transverse bulkheads to avoid using excessive number of elements.

A finite difference formulation for a non-uniform Timoshenko beam (beamfd) was used to provide a comparator for the three-dimensional model (shell3d). The lengthwise variations of structural and inertia properties for the beam model were calculated from model shell3d, to allow for meaningful comparisons between the two idealisations. Properties such as mass, moments of inertia, second moments of area, effective shear areas, torsional stiffness, warping stiffness, shear centre were evaluated at every section of model shell3d, with the saw-tooth character being a result of the loading condition (for mass properties) and the large deck openings (for structural properties). These properties were illustrated by Hirdaris et al (2000) and, hence, not repeated here. However, the section moduli at deck and keel, relating to vertical bending, and at side, relating to horizontal bending, are shown in Fig.1(c), as they are used when comparing wave-induced loads. Previous two-dimensional structural models (e.g. Bishop et al 1991) adopted smooth variations of structural characteristics. It should be noted that although the 46 sections along the shell3d model are of unequal length, as seen in Fig.1(a), beamfd model was discretised using sections of equal length along the hull.

The mean wetted surface of the vessel, shown in Fig.1(d), is idealised using 952 four-cornered panels. For convenience a one-to-one correspondence between shell finite elements and panels on the mean wetted surface has been adopted. This implies that triangular elements, used to overcome warpage constraints (i.e. limitations imposed on four nodes of an element being coplanar), are also idealised as four-cornered panels with, for example, one node assigned to two corners. A pulsating source is allocated to each panel.

3 DRY AND WET ANALYSES

The *dry* hull natural frequencies obtained from models beamfd and shell3d are shown in Tables 1(a) and 1(b), for the symmetric and antisymmetric modes respectively. The dominant distortions of antisymmetric principal modes are highlighted. Views of the first few principal mode shapes obtained from the shell3d model are shown in Fig.2, highlighting the dominant distortion. The vertical deflections for the first two symmetric mode shapes are shown in Figs.3(a,b). The first antisymmetric (torsion dominant) and second antisymmetric (bending dominant) mode shapes are shown in Figs.3(c,d,e,f), represented by the horizontal deflection and the twist, respectively. For the shell3d model the horizontal deflections at both deck (junction with side plating) and keel (centreline) are used. From Figs.3(a-f) and Tables 1(a) and 1(b) it can be seen that there is good overall agreement between the two structural models used, in terms of natural frequencies and mode shapes, for the first few mode shapes. Nevertheless as the complexity of the mode shapes increases so do the differences. It should be noted that inclusion of warping effects in the beamfd model is essential for agreement in the antisymmetric modes.

The calculations in regular waves were carried out for the bulker proceeding at 7.41 m/s in regular waves encountered at a heading of $\chi = 135^{O}$, thus allowing inclusion of both symmetric and antisymmetric mode shapes. For the preliminary investigation a set of 8 flexible mode shapes, shown in Tables 1(a) and 1(b), was included in the wet analysis. The resonance frequencies for the *wet* hull can be identified by examining the variation of corresponding principal coordinate amplitudes with encounter frequency ω_e. The differences between the (wet) resonance frequencies predicted by two- and three-dimensional fluid-structure interaction models, also shown in Tables 1(a) and 1(b), follow, in general, the same trends as the dry hull natural frequencies.
The modal internal actions and wave-induced loads are obtained in the form of bending moments, shear forces and torsional moment for the beamfd and in the form of direct and shear stress components for the shell3d model. Vertical and horizontal bending moments and longitudinal direct stress can be related, for a beamlike vessel, using the following relationships:

$$\sigma_{deck} = \frac{VBM}{S_{deck}} \quad or \quad \sigma_{keel} = \frac{VBM}{S_{keel}} \quad and \quad \sigma_{side} = \frac{HBM}{S_{side}} \tag{1}$$

where S_{deck}, S_{keel} and S_{side} denote the corresponding section moduli, shown in Figure 1(c).The vertical and horizontal bending moments obtained directly from model beamfd and from the relevant stresses of model shell3d using the above relationships are shown in Figs.4(a) and 4(b), respectively, for $L/\lambda=1$, λ denoting the length of the ship. Contributions from symmetric modal direct stresses only were used for the vertical bending moment and antisymmetric modes only for the horizontal bending moment to allow for direct comparisons with the relevant bending moments obtained from model beamfd. There are some differences between the vertical bending moment obtained from the direct stresses on the deck (junction with side plating) and keel (centreline). Nevertheless, there is good overall agreement between the two models for the vertical bending moment magnitudes along the vessel. The agreement is not as good between the horizontal bending moment variation along the hull obtained from model shell3d using direct stresses on the side plating, either at nodes near the centroid or the junction with deck, and predictions from beamfd, the latter being considerably larger in the vicinity of amidships. It has been observed that the horizontal bending moment magnitudes predicted by model beamfd are rather sensitive to the values of torsional moment of inertia (per unit length). This discrepancy is under further investigation.
As a further comparison, the variation of the wave-induced direct stresses on the deck (junction with side plating) and keel (centre line) are shown in Figs.5(a,b) for $L/\lambda=1$. The stresses from the three-dimensional model shell3d were calculated including and excluding the antisymmetric mode shapes, to investigate their contributions. Comparisons are also made with corresponding stresses obtained from model beamfd using the relationships of equation (1). As expected the antisymmetric modes contribute negligibly to the stresses on the keel centre line. Nevertheless, they have a significant contribution to the direct stresses obtained at the junction of deck and side plating. The differences observed between two- and three-dimensional direct stresses at keel and deck are similar to those observed for the vertical bending moment in Fig.4(a).

4 CONCLUSIONS

- This paper focuses on comparing the influences of different levels of refinement in fluid-structure interaction modelling on the dry hull dynamic characteristics and wave-induced loads and stresses for ships with large deck openings, such as bulk carriers, containerships, LNG carriers etc.. The results of this investigation are important from the point of view of selecting the detail to be incorporated into a model when working at various stages of the design process, such as concept and preliminary design.

- The dry hull analysis showed good overall agreement between beamfd and shell3d models for the first few symmetric and antisymmetric mode shapes; however, differences increase with modal complexity.
- The wave-induced vertical bending moment and direct stress show good overall agreement between the two models, taking into account the saw-toothlike nature of relevant second moment of area and section moduli. However, the agreement is not good between the horizontal bending moment values predicted by both models. The contributions from antisymmetric modes to stresses on the deck, at the junction with side plating, are noted.
- Further investigations are required to assess the influence of structural properties with saw-toothlike variation, a consequence of the large deck openings, on the dry hull characteristics and, consequently, wave-induced loads. The effects of wetted surface idealisation on wave-induce loads also merit further investigation.

References

Aksu, S, Price, W.G., Suhrbier, K.R. and P. Temarel, P. (1993). A comparative study of the dynamic behaviour of a fast patrol boat travelling in rough seas, *Journal of Marine Structures* **6,** 421 - 441.

Bishop, R.E.D. Clarke, J.D., and Price, W.G. (1984). Comparison of full scale and predicted responses of two frigates in severe weather trial, *Trans. RINA* **126,** 153 - 166.

Bishop, R.E.D. and Price, W.G. (1979). Hydroelasticity of Ships, *Cambridge University Press*, UK.

Bishop, R.E.D., Price, W.G., and Temarel, P. (1991). A theory on the loss of MV Derbyshire, *Trans. RINA* **127,** 169-186.

Bishop, R.E.D, Price, W.G. and Wu, Y. (1986). A general linear hydroelasticity theory of floating structures moving in a seaway, *Phil. Trans. Royal Soc. London* **316:A,** 375 - 426.

Hirdaris, S.E., Price, W.G. and Temarel, P. (2000). Hydroelastic analysis of a bulk carrier, (keynote lecture), *International Conference on Ship and Shipping Research NAV2000,* **1,** 5.6.1 - 5.6.12, Venice, Italy.

IACS (1997). Bulk carriers - guidance and information on bulk cargo loading and discharging to reduce the likelihood of over-stressing the hull structure, *IACS publications*, London, UK.

Louarn, F.H. and Temarel, P. (1999). An investigation of the structural dynamics of a racing yacht, *In Proceedings of the 14th Chesapeake Sailing Yacht Symposium*, 123-142, USA.

Miao, S.H., Price, W.G. and Temarel, P. (1997). The hydroelastic behaviour of multi hulls travelling in a seaway, *Proceedings of the 3rd International Conference for Advances in Marine Structures*, Dunfermline.

Price, W.G., Temarel, P. and Keane, A.J. (1994). Hydroelastic analysis of a SWATH in waves, *In International Conference of Hydroelasticity in Marine Technology*, 231 - 243, Trondheim, Norway.

Price, W.G., Salas Inzunza, M.A. and Temarel, P. (2001). The dynamic behaviour of a mono-hull in oblique waves using two- and three-dimensional fluid-structure interaction models, *Accepted for publication RINA*.

Swanson Ltd. (2000). The ANSYS 5.6 users manual.

Tustin, R.D. (1998). Existing bulk carrier safety – on the evaluation of the foremost hold structure under conditions of hold flooding, *RINA International Conference on the Design and Operation of Bulk Carriers*, 48 - 68, London, UK.

TABLE 1

DRY HULL NATURAL (ω_r) AND WET RESONANCE (ω_{er}) FREQUENCIES (RAD/S) FOR (a) SYMMETRIC AND (b) ANTISYMMETRIC DISTORTIONS

(VB = VERTICAL BENDING; HB = HORIZONTAL BENDING; T = TORSION; **HB**,**T** = DOMINANT HB, T)

(a) Mode type	beamfd		shell3d	
	ω_r	ω_{er}	ω_r	ω_{er}
2 node VB	4.42	3.03	4.53	3.25
3 node VB	9.25	6.62	9.01	6.62
4 node VB	14.24	9.03	13.24	10.09
5 node VB	17.62	12.76	15.92	12.61

(b) Mode type	beamfd		shell3d	
	ω_r	ω_{er}	ω_r	ω_{er}
1 node HB – 1 node **T**	5.22	4.41	5.00	4.97
2 node **HB** - 2 node T	5.86	4.96	5.55	5.01
2 node HB – 2 node **T**	10.51	8.71	12.63	9.30
3 node **HB** – 3 node T	11.14	9.58	10.71	9.11

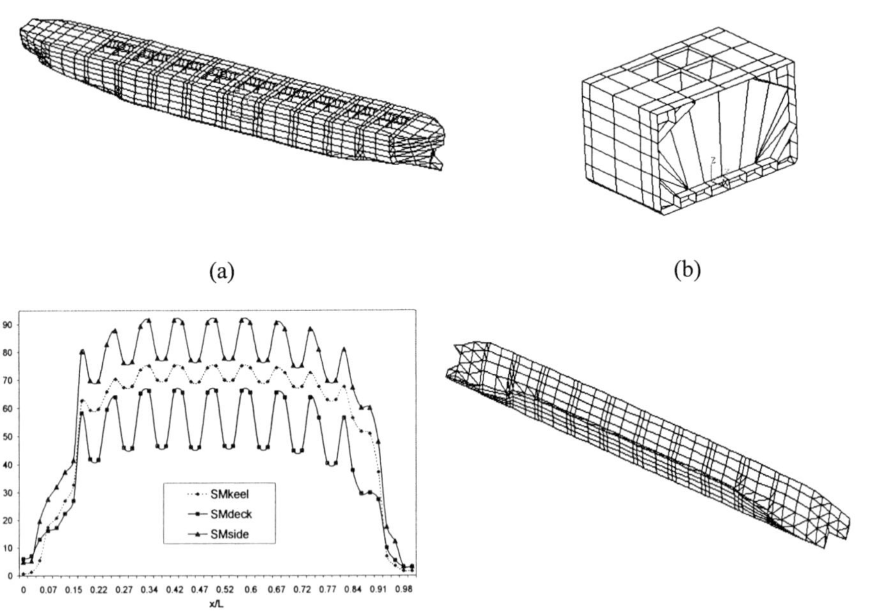

Figure 1: (a) Three dimensional FE model ; (b) Transverse cross-section in the parallel body ; (c) Section modulus (m^3) variation along deck, keel and side of the vessel ; (d) Idealisation of the mean wetted surface.

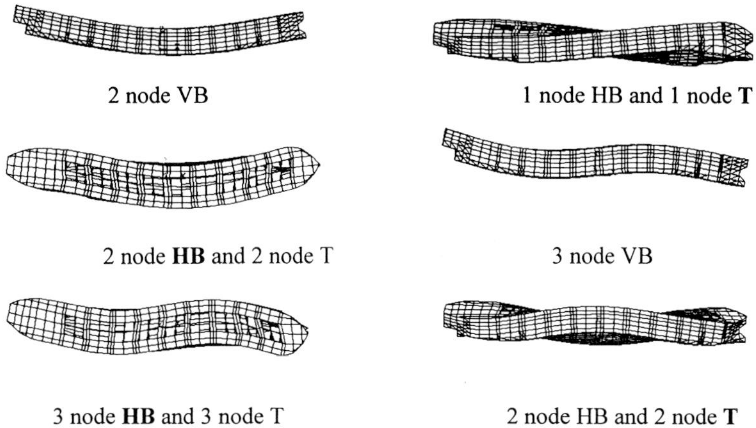

Figure 2: Principal mode shapes of model shell3d (VB: vertical bending; HB: horizontal bending; T: torsion, **HB,T**: dominant HB, T).

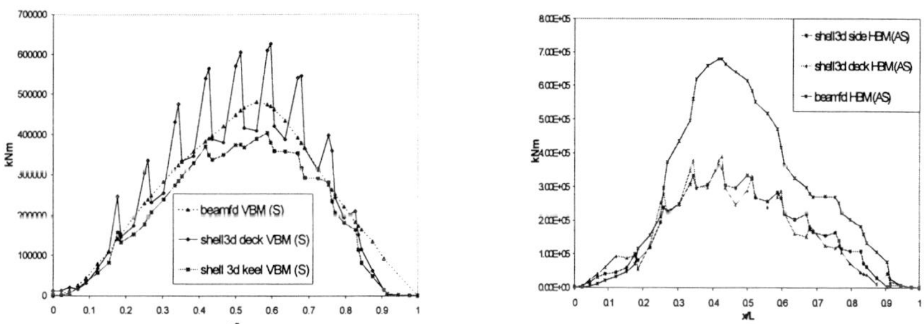

(a) (b)

Figure 4: Variation of (a) vertical and (b) horizontal bending moments (kNm) along the ship obtained from models beamfd and shell3d (keel centre and deck side junction) when symmetric modal stresses (S) and antisymmetric modal stresses (AS) are taken into account, for $L/\lambda = 1.0$.

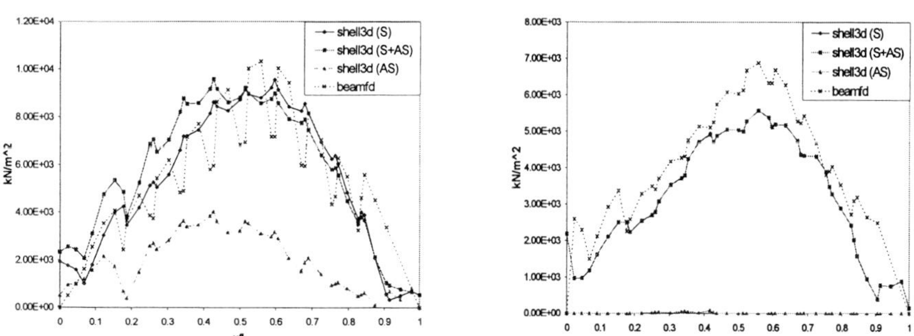

(a) (b)

Figure 5: Longitudinal direct stress variations (kN/m^2) along the ship obtained from models beamfd and shell3d when symmetric only (S), symmetric and antisymmetric (S+AS), antisymmetric only (AS) modal stresses are considered for $L/\lambda = 1.0$ (a) Direct stress along deck-side junction; (b) Direct stress along keel centerline.

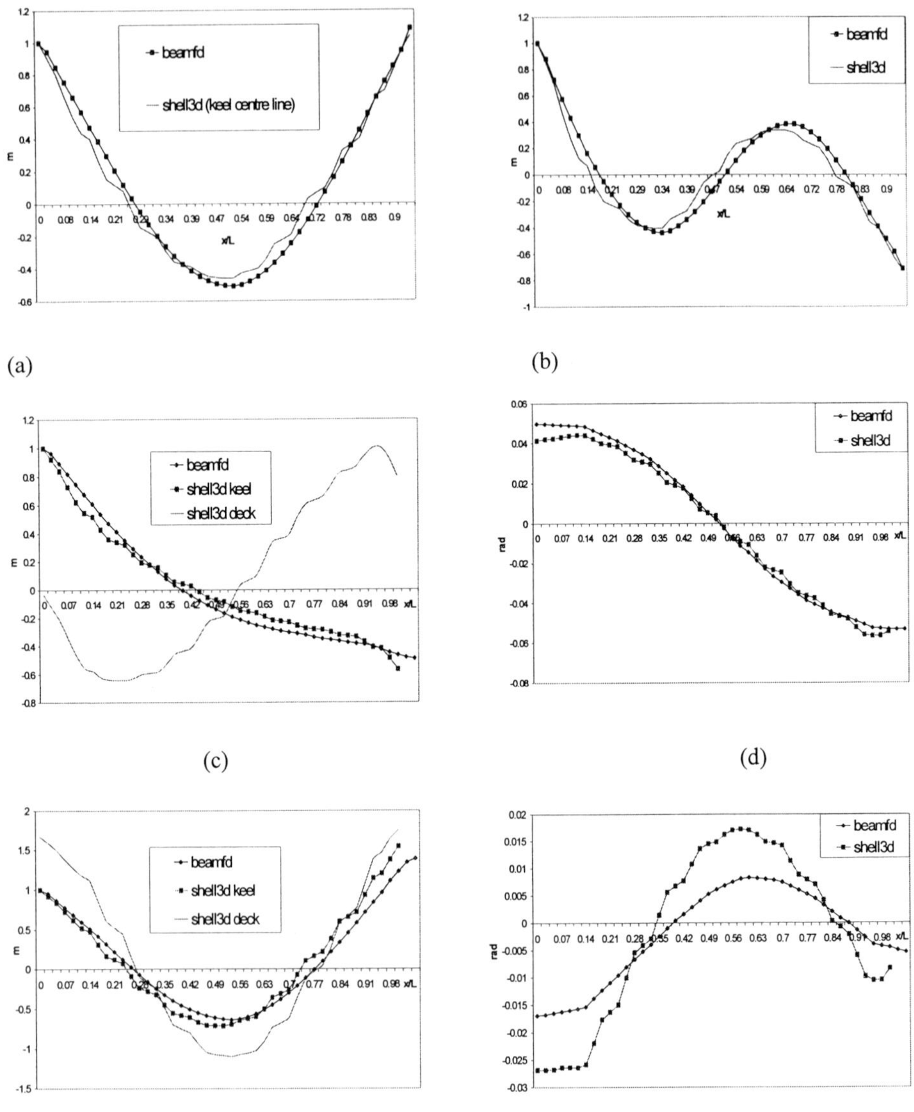

Figure 3: Principal mode shapes for: (a,b) 2 and 3-node symmetric vertical deflections ; (c,e) 1 and 2-node horizontal deflections (keel centerline and deck side junction for shell3d) and (d,f) twist angles for 1-node torsion dominant and 2-node HB dominant antisymmetric modes, respectively.

HYDROELASTIC MODEL FOR BOTTOM SLAMMING

A. Bereznitski [1] and V. Postnov [2]

[1] Ship Structures Laboratory, Delft University of Technology,
Mekelweg 2, 2628 CD, Delft, The Netherlands
[2] Ship Structures Laboratory, St-Petersburg State Marine Technical
University, Lotsmanskaia 3, 190008, St-Petersburg, Russia

ABSTRACT

The effect of hydroelasticity is studied extensively for a 2D problem of bottom slamming. A number of wedge-shaped structures are considered for water penetration at small deadrise angles. The effect of hydroelasticity is studied as a function of different parameters influencing the impact interaction. These parameters include various stiffness of the structure, presence of entrapped air between the structure and the water surface, and water penetration at different deadrise angles. The primary objective is establishment of range, within which hydroelasticity plays an important role. This range will show when the hydroelasticity should be taken into account and in which cases it can be skipped. The problem is studied with application of MSC.Dytran code. At the same time for verification purpose a specific case is compared with results obtained with application of a self-developed code.

KEYWORDS

Slamming, Hydroelasticity, Hydrodynamic loads, Water impact, Fluid-structure interaction

1 INTRODUCTION

The importance of the hydroelasticity is still not studied well with respect to the dynamic fluid-structure interaction problems. The question "Should be hydroelasticity considered or it can be neglected?" is still not answered. The correct choice of suitable models and approaches for solving of water impact problems remains a great challenge. A large number of scientists work on this problem (SSC-385 (1995), Arai and Miyauchi (1998), Faltinsen (1997), Hermundstad, Wu, and Moan (1994), Korobkin (1998), Kvalsvold and Faltinsen (1993), Price and Temarel (1994)). It was shown in our recent publications (2000a, 2000b) that effect of hydroelasticity plays a very important role for a flat impact, where the bottom is parallel to the free water surface. The entrapped air between the bottom and water was not considered. The present paper introduces a new model based on application of MSC.Dytran code. A detailed study of different parameters influencing the fluid-structure interaction is performed for a 2D problem of bottom slamming. Water penetration of a wedge-shaped structure is studied for various structural stiffness and different small deadrise angles (Fig. 1). The influence of entrapped air is also investigated.

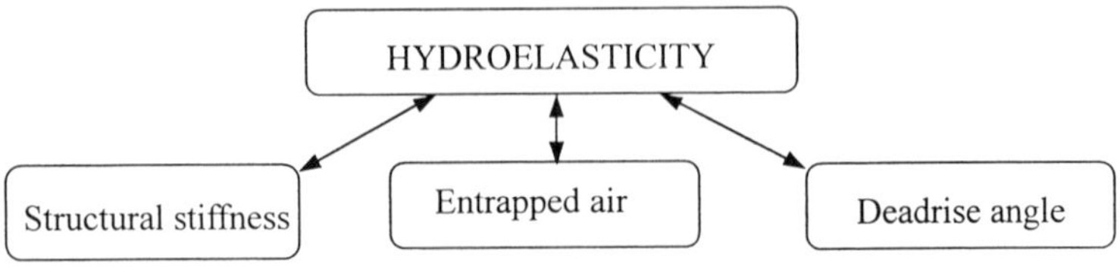

Figure 1: Hydroelasticity

2 THE MODEL

The generalized model for the bottom slamming problem is shown in Fig. 2. A wedge-shaped body with flexible bottom penetrates into water at a constant speed V_{rb}=2.5 m/s. α is a deadrise angle. The water domain has width of 5.6 m and depth of 2.3 m, which corresponds to the experimental set-up of a series of drop tests performed in our laboratory. The width of the body is 2.5 m. The problem is solved with application of MSC.Dytran code. This software belongs to a 3D-analysis code based on an explicit integration procedure, which can be used for dynamic fluid-structure interaction problems. For the structural analysis a finite element formulation is used with application of a Lagrange mesh. The fluid analysis is performed with use of finite volume formulation and application of an Eulerian mesh. The Eulerian mesh is fixed in space but material can flow through the mesh. Several different materials can be defined within the Eulerian mesh. A special coupling procedure introduced within the code provides the fluid-structure interaction at each time increment. The loads obtained from the fluid analysis are applied to the structure and the structural response is the boundary condition for the fluid analysis. This allows modeling of real hydroelastic interaction. The mesh of MSC.Dytran model is shown in Fig. 3 (because of symmetry only half is considered). The top part is initially filled with air and the rest with water. MSC.Dytran can handle only solid elements for Eulerian mesh. In order to build a 2D model a thin slice of water and air is considered having only one solid element in thickness. A rigid wall boundary condition is applied at the sides of the air-water region. The analysis starts when the air gap between the body and the water surface is about 10 cm. It was found from an additional series of calculations that this is the optimal distance. A bigger distance is more time consuming for the simulation, while there is no visible change of air pressure until the distance is reduced to about 10 cm.

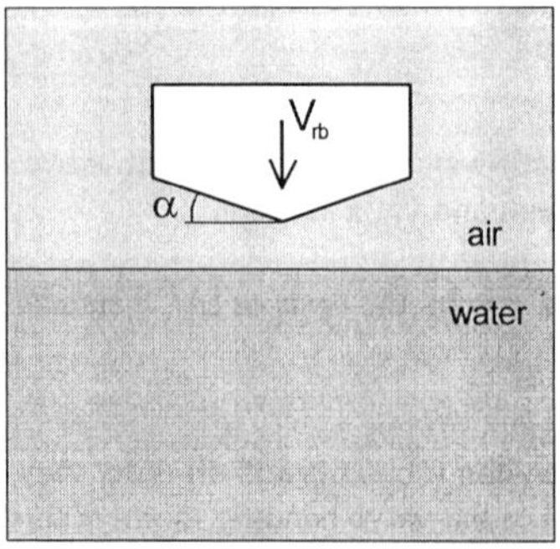

Figure 2: The model

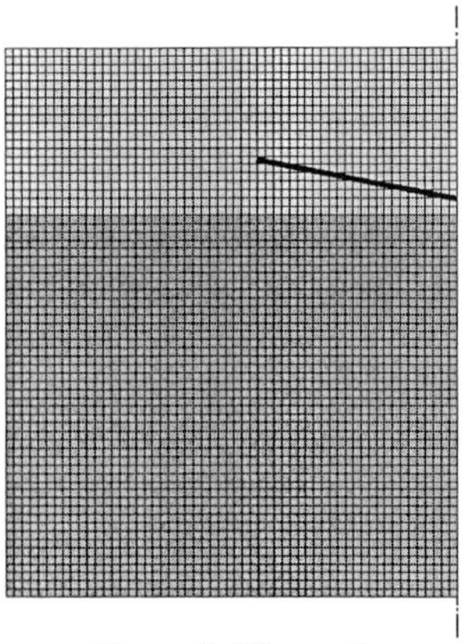

Figure 3: The mesh

Water is described by a polynomial equation of state, which for the linear case can be rewritten as:

$$p = a_1(\rho/\rho_0 - 1) \tag{1}$$

Here p is pressure; ρ is overall material density; $\rho_0 = 1000$ kg/m^3 is reference density; $a_1 = 2.25$ GPa is bulk modulus. Air is described by a gamma law equation of state:

$$p = (\gamma - 1)\rho e$$

Here $\gamma = 1.4$ is ratio of specific heats; e is specific internal energy unit mass. Initial density of air was assumed $\rho_0 = 1.2887$ kg/m^3 and $e = 196000$ J/kg, which corresponds to the atmospheric pressure $p = 0.101$ MPa.

The water and air regions are subdivided into Eulerian solid elements. But the strict boundary between the water and air elements exists in the beginning of the analysis only. Later on, some elements become empty of water because they are covered by the structure. Also a water-air mixture is formed near the free surface and the impact area. This means that some elements can be partly filled with air and water at the same time. If the model includes both water and air, atmospheric pressure is initially applied to both domains. If no air is modelled (the top part of the mesh consists of so called "void" elements) the cavitation pressure in water is assumed $p_{cav} = -0.101$ MPa. The flexible bottom is modeled using one row of shell elements and represents a beam. A schematic drawing of the beam is shown in Fig. 4. The beam penetrates into the water with constant speed V_{rb}=2.5 m/s driven at the sides. In order to study the influence of the structural stiffness on the hydroelasticity, four different beams are considered (Table 1). The moment of inertia and the mass per unit length of each beam are chosen so that the first frequency of free vibration of the beam would vary linearly with respect to the mass per unit length and the beam number (Fig. 5). The moment of inertia of the beam as a function of the first frequency is also shown in Fig. 5. Parameters of the beam N1 and the beam N3 correspond to real stiffness and mass parameters of the bottom involved in the test set-up of the experiment conducted in our laboratory. The displacement of the center of the beam W(t) (Fig. 4) is monitored during the analysis.

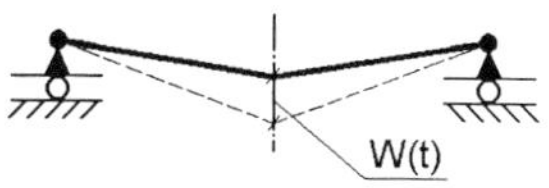

Figure 4: The beam

TABLE 1

PROPERTIES OF BEAMS

Beam	First frequency of free vibration, Hz	Moment of inertia, m^4	Mass per unit length, kg/m	Modulus of elasticity, GPa
N1	19.0	1.61e-6	59.3	210
N2	27.2	5.42e-6	97.1	210
N3	35.4	1.27e-5	134.8	210
N4	43.6	2.47e-5	172.5	210

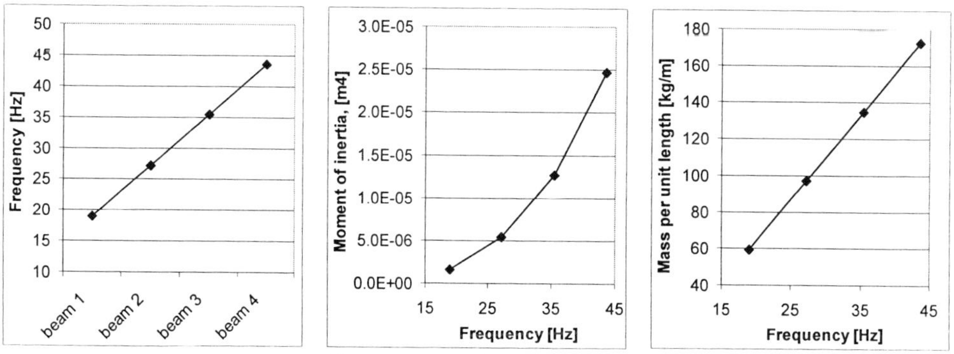

Figure 5: Properties of the beam

3 VERIFICATION OF THE MODEL

As the first step in development of the model a simple case was considered for the beam N3. The MSC.Dytran model is compared with the hydroelastic model described in our recent publications (2000a, 2000b), which is based on application of our self-developed code. Water penetration is studied for zero-deadrise angle without consideration of the entrapped air. No cavitation in the water is modeled (pressure can be negative).

A history plot of displacement in the center of the beam is shown in Fig. 6. In spite of a big distinction between the two codes, the results are in quite a good agreement. The difference in the maximum displacement and the period of vibration comes up to 10-15% only. This difference can be explained by the following factors:

- First of all, in MSC.Dytran the structure is modeled with application of the Lagrange mesh introducing the nonlinear geometrical behavior, while the self-developed code operates with linear behavior.
- Secondly, there is a big difference in the models. The self-developed code model consists of 2D rectangular water elements and 2D beam elements for the bottom structure. While the MSC.Dytran model operates with solid water elements and shell elements for the structure. Moreover the fluid in MSC.Dytran is modeled with application of the finite volume formulation, while the self-developed code is based on the finite element method.

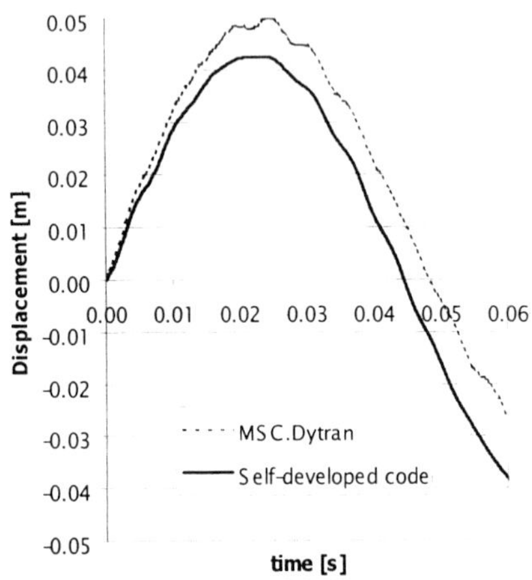

Figure 6. Displacement as a function of time

4 RESULTS OF CALCULATIONS

An advanced series of calculations has been carried out in order to study different parameters influencing the fluid-structure interaction. For each of the four beams (Table 1) the following deadrise angles: $\alpha = 0°, 2°, 5°, \text{and} \, 10°$ are considered. For each of these cases two variants are investigated with account of entrapped air and without air. In order to evaluate the effect of hydroelasticity for each model two simulations are performed and the maximum displacement in the center of the beam is monitored. In the first case the hydroelastic problem is solved. This means that the hydrodynamic loads obtained from the fluid analysis are applied to the structure and the structural response is the boundary condition for the fluid analysis (Fig. 7). This interaction takes place for each time increment. In the second case the fluid analysis is performed assuming that the structure is rigid. The pressure loads are stored. As soon as the fluid analysis is completed the structural analysis starts and the stored pressures are applied to the structure (Fig. 8). So there is no accounting of the structural response in the fluid analysis.

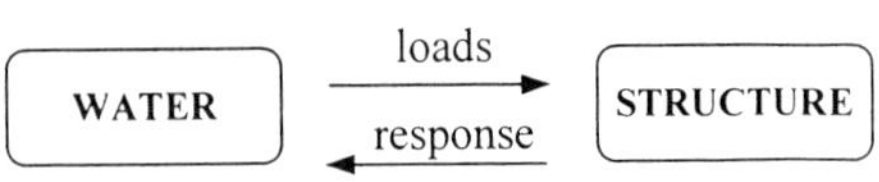

Figure 7: Hydroelastic interaction ("elastic")

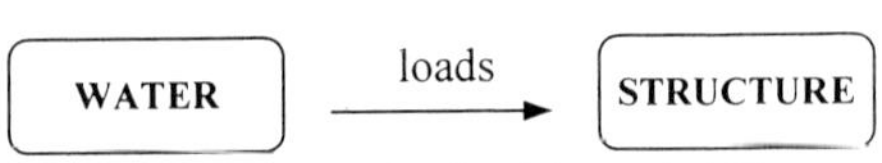

Figure 8: No hydroelasticity ("rigid")

A typical history plot for displacement in the center of the beam is shown in Fig. 9 for the beam N3 impacting the water surface with zero-deadrise angle (with consideration of entrapped air and

5 CONCLUSIONS

An extensive study of different parameters influencing the bottom-slamming phenomenon is discussed in this paper based on results of numerical analyses for more than 64 models. The following conclusions can be made:

- It was found that effect of hydroelasticity is extremely important for the bottom-slamming problem when the deadrise angle is close to zero. With increases of the angle, the structural response predicted through application of the rigid body water penetration approach differs less from the results of the hydroelastic solution.
- The effect of hydroelasticity is more significant for flexible structures and becomes less important as the structural stiffness increases.
- Entrapped air between the structure and the water surface leads to reduction of impact loads for very small deadrise angles ($0° - 5°$) being a quite effective cushion. For bigger angles the effect of the entrapped air is not important and can be skipped. The most dangerous case is observed for a two-degree dead-rise angle when some air can escape the interface area, thus making the impact loads higher. At the same time, for zero-angle the loads are a little bit lower because more air is entrapped.

The conclusions given above are based on results of quite simple models. Only a 2D case is studied and the constant speed of penetration is considered. For more complicated cases probably different results might be expected.

References

Arai M. and Miyauchi T. (1998). Numerical Study of the Impact of Water on Cylindrical Shells. Considering Fluid-Structure Interactions. *The 7th International Symposium on Practical Design of Ships and Mobile Units, (PRADS'98)*, The Netherlands.

Bereznitski A., Boon B. and Postnov V. (2000a). Influence of the Ship Structure Flexibility on the Hydrodynamic Impact Loads Due to Bottom Slamming. *Proceedings of the 7th International Marine Design Conference*, Korea.

Bereznitski A., Boon B. and Postnov V. (2000b). The Effect of Hydroelasticity on the Impact Pressure Due to Bottom Slamming on Ship Structure. *Proceedings of 10th International Offshore and Polar Engineering Conference*, USA.

Faltinsen O.M. (1997). The Effect of Hydroelasticity on Ship Slamming. *Phil. Trans. R. Soc. Lond.* A(1997) **355**, 575-591.

Hermundstad O.A., Wu M.K. and Moan T. (1994). Hydroelastic Response Analysis of a High Speed Monohull. *Hydroelasticity in Marine Technology*, Proceedings of the International Conference, 245-259.

Hydrodynamic Impact on Displacement Ship Hulls. (1995). Ship Structure Committee, (SSC-385), USA.

Korobkin A. (1998). Unsteady Hydroelasticity of Floating Plates. *Hydroelasticity in Marine Technology*, Proceedings of the Second International Conference, 109-117.

Kvalsvold J. and Faltinsen O.M. (1993). Hydroelastic Modelling of Slamming Against the Wetdeck of a Catamaran. *FAST'93, Proceedings of the Second International Conference on Fast Sea Transportation*. Yokohama Japan, 13-16.

Price W.G. and Temarel P. (1994). Prediction of Wave Induced Responses, loads and Stresses Using Hydroelasticity Theory. *International Conference on Ship and Marine Research*. **1**.

without). The maximum displacement corresponds to the "rigid" case without consideration of entrapped air. The smallest displacement is found for the "elastic" case with consideration of air. Figures 10-13 represent results of all calculations. Here the maximum displacement is printed as a function of deadrise angle for all four beams. The surface diagram (Fig. 14) shows the maximum displacement as a function of two parameters: the deadrise angle and the first frequency of free vibration of the beams (or in other words the beam stiffnesses). This figure represents results for the "elastic" case with consideration of entrapped air. A summary of all results is also plotted in the surface diagram of Fig. 15. The most accurate model ("elastic" case with consideration of entrapped air) predicts the smallest displacements.

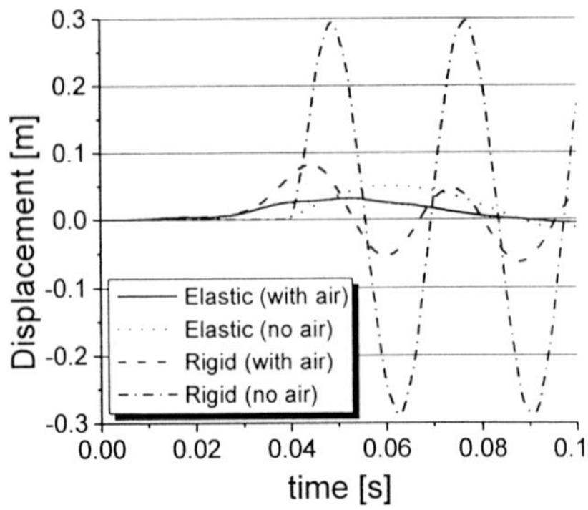

Figure 9: Displacement as a function of time (beam N3, $\alpha = 0°$)

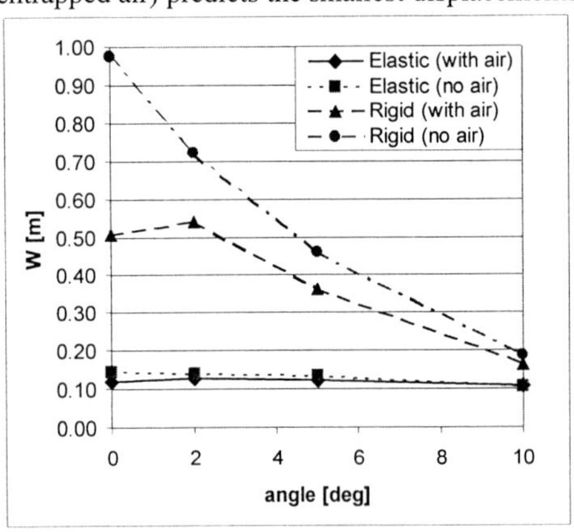

Figure 10: Maximum displacement as a function of angle (beam N1)

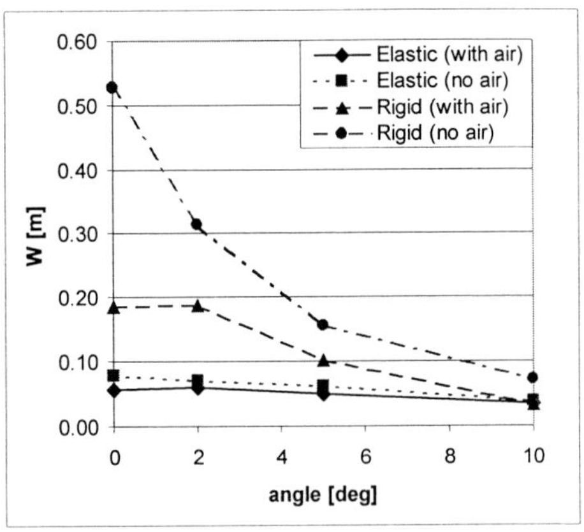

Figure 11: Maximum displacement as a function of angle (beam N2)

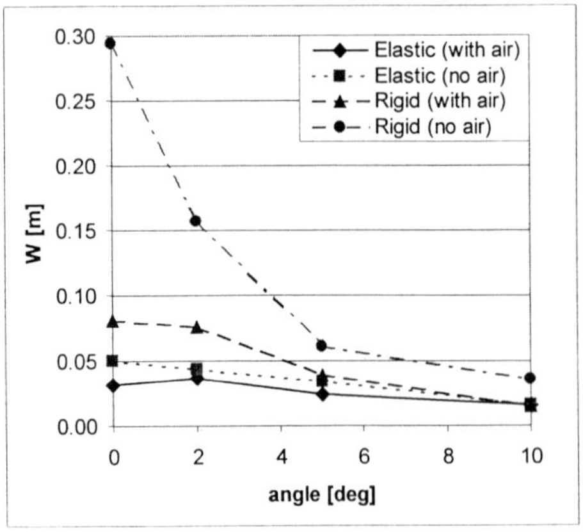

Figure 12: Maximum displacement as a function of angle (beam N3)

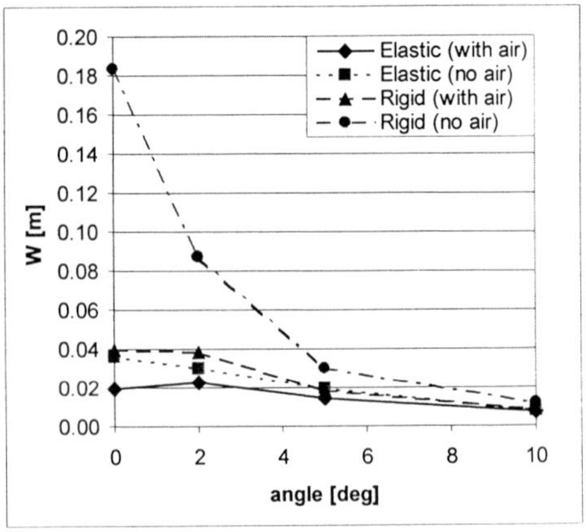

Figure 13: Maximum displacement as a function of angle (beam N4)

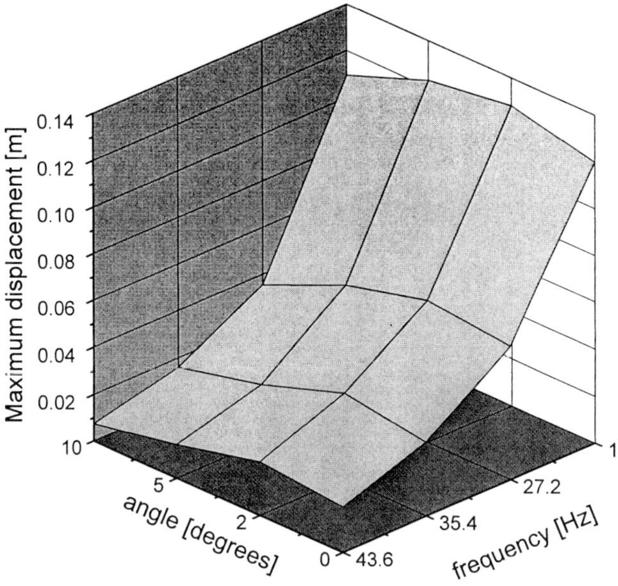

Figure 14: Maximum displacement of the beam ("elastic" with air)

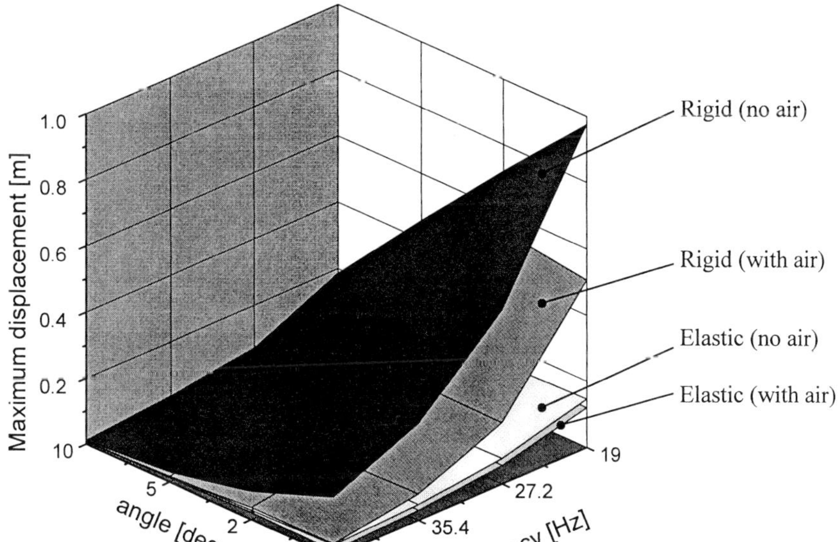

Figure 15: Maximum displacement of the beam (all results)

Practical Design of Ships and Other Floating Structures
You-Sheng Wu, Wei-Cheng Cui and Guo-Jun Zhou (Eds)

HYDRODYNAMIC IMPULSIVE LOADS ACTING ON SHIP-HULL PLATES

Gang Wang

China Classification Society, Shanghai 200135, China

ABSTRACT

A numerical method is given to calculate the hydrodynamic impulsive pressure acting on side plates and bottom plates at the bow of ship by using Mixed Eulerian-Lagrangian Method with time marching procedure. The ship structure is assumed to be a 2-D elastic body. Based on Green's theorem and the motion Equations of the plate with the large deflection, The motion Equations of fluid-solid coupling based on the displacement of plate structures and fluid potential function are first derived. Considering the dynamic boundary conditions, the numerical solutions are obtained by using the methods of the nonlinear boundary element and the finite difference. Then water entries of the elastic wedge bodies with various deadrise angles and plate thickness are calculated. The calculation results indicate that impulsive loads vary with the changes of plate thickness and hydro-elastic effect compared with rigid body's.

KEYWORDS

Impulsive loads, Fluid-Solid coupling , Hydro-elastic effect, Boundary element method, Finite difference method,Ship structure

1 INTRODUCTION

With the increasing of the size and speed of ship, the structure at the bow of ship suffers from a great impulsive loads and damage. Therefore, it is the most important to determine the hydrodynamic impulsive pressure acting on side plates and bottom plates of the bow of ship at the design stage.
Several analytical methods, such as Wagner's water entry theory, similarity flow theory, asymptotic expansion method or Laplace transformation method. With the development of modern computational technique, the numerical methods, such as FEM, FDM(including MAC, PIC or VOF method), BEM, are widely applied to calculate the hydrodynamic pressure. BEM is more effective method for calculating flow field, its computational domain is limited to the boundary of the flow field, therefore its advantages are attributed to decrease the number of the computational elements and improve the calculation accuracy. Geers(1977)firstly calculated water-entry bodies' fluid added masses and hydrodynamic impulsive pressures by the boundary element method . Based on the free surface 's nonlinear flow, Greenhow(1988) analyzed the problem about vertical water entry and exit of two

dimensional wedge bodies and cylindrical shells. Zhao&Faltinson(1992) presented a new method to calculate the "jet phenomena " of wedge's water-entry at high speed under assumption of one dimensional flow in the "jet flow". Falch(1986) calculated accurately nonlinear deformation of free surface and hydrodynamic impulsive pressures at deferent time steps. In my nation, many important and significant works (Qian(1994); Hou&Lu(1998)) as same as above have also being done.

Actually, the ship hull is elastic structure. As ship hull entering water, the deformation of ship structure will occur and also influence the motion of the fluid, i.e. coupling effect between structure and fulid will occur. Early stage results of test and theory computation have proved such a truth that the elastic character will change the duration time and distribution of the hydrodynamic impulsive pressure. Therefore, the elastic effect of the structure has to be considered. In order to study the hydro-elastic effect of ship hull and catch the truly fluid-solid coupling procedure during water entry. Two shortcomings in the past works have been improved in this paper, which are (1) rigid wedge bodies are replaced by elastic wedge bodies , (2) the effect of fluid-solid coupling between elastic bodies and the water is deeply considered. The computational model is set up, which the ship structure is assumed to be a elastic body composed of two 2-D elastic plate beam structures, and it enter water with a high speed. According to Green's theorem and the motion Equations of the plate beam structures with the large deflection, a new unifying motion Equation of fluid-solid coupling based on the displacement of ship structures and fluid potential function are first applied to calculate impulsive loads. Results of several conditions are compared with the rigid body's, and analyzing conclusions are given.

2 BASIC THEORY

Consider a two-dimensional body penetrating a water surface with a initial velocity $\vec{V}_0$, based on the axial symmetry, only the half fluid domain is dealt with computation while other is taken as image Ω^*. As shown in Fig.1, xoy is a fixed coordinate system, $x'o'y'$ is a moving coordinate system; β is

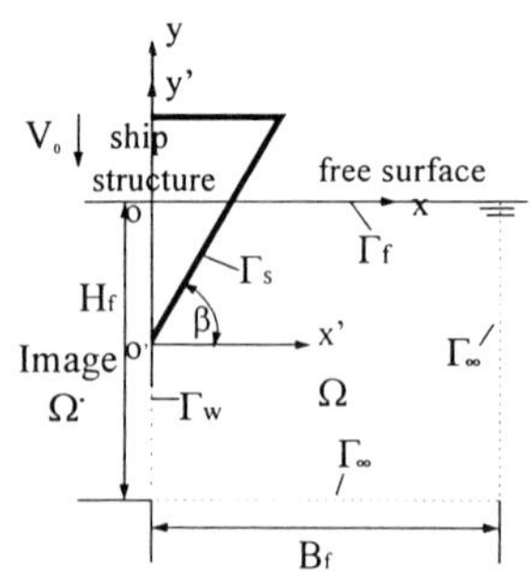

Figure1: computational domain

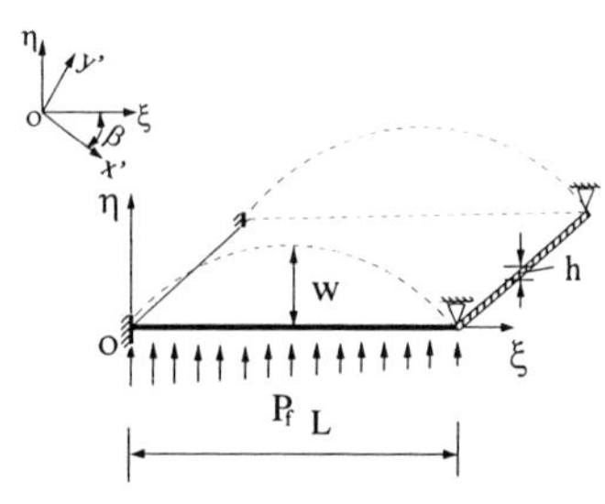

Figure 2: simplified mechanic model of ship hull

deadrise angle; Ω is computational domain of fluid field, Γ is boundary which is made of the body surface Γ_s, the free surface Γ_f, the axial symmetric boundary Γ_w or the infinite boundary Γ_∞; H_f and B_f are depth and breadth of fluid field respectively.

Assume that $\vec{V}'$ is the relative velocity of fluid to body surface, $\vec{V}_e$ is the bounded velocity between $x'o'y'$ and xoy; $\vec{V}_c$ is the velocity of body surface in $x'o'y'$, so absolute velocity of fluid is

$$\vec{V} = \vec{V}' + \vec{V}_e + \vec{V}_c \tag{1}$$

It is assumed that the fluid is incompressible and inviscid and effect of air cushion is neglected when $\beta > 3^\circ$; density of fluid is constant ρ_f It is also assumed that the flow of fluid is irrotational, and the velocity $\vec{V}$ may be expressed by a velocity potential function $\varphi(x,y,t)$ as

$$\vec{V} = \nabla\varphi \tag{2}$$

and $\varphi(x,y,t)$ satisfies a Laplace Equation

$$\nabla^2\varphi = 0 \qquad (x,y)\in\Omega \tag{3}$$

the boundary conditions are as follows:

$$\frac{\partial\varphi}{\partial n} = \vec{V}_s\cdot\vec{n}_s \qquad (x,y)\in\Gamma_s \tag{4}$$

$$\frac{\partial\varphi}{\partial n} = 0 \qquad (x,y)\in\Gamma_w,\Gamma_\infty \tag{5}$$

where $\vec{V}_s$ and $\vec{n}_s$ are fluid velocity and normal vector on body surface, respectively.
the hydrodynamic pressure of fluid is

$$P = P_0 - \rho_f\left[\frac{d\varphi}{dt} - \vec{V}\cdot(\vec{V}_e+\vec{V}_c) + \frac{1}{2}\vec{V}\cdot\vec{V} + gy\right] \qquad (x,y)或(\xi,\eta)\in\Gamma_s \tag{6}$$

According to Green's theorem and the boundary conditions, boundary integral Equation is given as follows

$$\theta\varphi = \int_{\Gamma_s}\left(\vec{V}_s\cdot\vec{n}_s G - \varphi\frac{\partial G}{\partial n}\right)ds + \int_{\Gamma_f}\left(\frac{\partial\varphi}{\partial n}G - \varphi\frac{\partial G}{\partial n}\right)ds - \int_{\Gamma_w+\Gamma_\infty}\varphi\frac{\partial G}{\partial n}ds \tag{7}$$

where θ is angle at source point; function G is Green fuction。
It is further assumed that ship hull plates at bow are simplified to be a plate-strip beam with one ends fitted and another one rested on elastic hinge as shown in Fig.2, where $\xi o\eta$ is local coordinate system fixed on the body surface and L denotes its breadth, h thickness, ρ_s density, E modulus of elasticity and μ Poisson's ratio(taken 0.5 for normal), respectively. w is the local displacement in the local coordinate system, respectively. P_f is hydrodynamic impulsive pressure.
Assume that the entry into water of ship hull is a constant-velocity procedure, the equation of finite deformation of the plate-strip beam is given as follows:

$$\rho_s h\ddot{w} = P_f - \left(\frac{EI_h}{1-\mu^2}\right)\frac{\partial^4 w}{\partial\xi^4} + \frac{hE}{2(1-\mu^2)L}\frac{\partial^2 w}{\partial\xi^2}\int_0^L\left(\frac{\partial w}{\partial\xi}\right)^2 dx \tag{8}$$

where $I_h = h^3/12$ is moment of inertia.
Based on the formula (6) and (8), and $P_f = -(P-P_0)$, so fluid-soild coupling motion Equation is obtained

$$\begin{cases}\rho_s h\ddot{w} = \rho_f\left[\dfrac{d\varphi}{dt} - \vec{V}\cdot(\vec{V}_e+\vec{V}_c) + \dfrac{1}{2}\vec{V}\cdot\vec{V} + gy\right] + f_p & 0\le\xi/L<C_w \\ \rho_s h\ddot{w} = f_p & C_w\le\xi/L\le 1\end{cases} \tag{9}$$

where C_w is wetted length ratio of the plate beam; f_p is stress function.

3 NUMERICAL COMPUTATION

The body's motion has a great effect on fluid field within the limits of 2.5 twice of body's effective length, so boundary network is to be subdivided within the limits of 2.5L and more larger distance of network is to be taken at the outside. Based on the theorem, network for computational domain boundary of fluid field is determined $n = n_s + n_f + n_b$. The finite difference method about forms of center difference is applied to solve the motion Equation of plate beam. the plate beam is divided uniformly into $N_s - 1$ segments along axis ξ, $\Delta\xi = L/(N_s-1)$.

Based on the method of time step proceeding in time domain, the basic form for all Equations are expressed as type of one order forward difference, and the value of step $k+1$ is obtained by step k, Δt is time interval. Then n order linear Equations for fluid-solid coupling may be expressed as

$$\{x\} = [A]^{(k)^{-1}} \cdot \{[B]^{(k)} \cdot \{y\}\} \tag{10}$$

where vectors $\{x\} = \left[\varphi_1^{(k+1)} \quad \cdots \quad \varphi_{n_s}^{(k+1)} \quad \frac{\partial \varphi_{n_s+1}^{(k+1)}}{\partial n} \quad \cdots \quad \frac{\partial \varphi_{n_s+n_f+1}^{(k+1)}}{\partial n} \quad \varphi_{n_s+n_f+2}^{(k+1)} \quad \cdots \quad \varphi_n^{(k+1)} \right]^T$ and

$$\{y\} = \left[U_1^{(k)} \quad \cdots \quad U_{n_s}^{(k)} \quad \frac{\rho_f}{h\rho_s} \varphi_{n_s+1}^{(k+1)} + U_{n_s+1}^{(k)} \quad \varphi_{n_s+1}^{(k+1)} \quad \cdots \quad \varphi_{n_s+n_f+1}^{(k+1)} \quad 0 \quad \cdots \quad 0 \right]^T$$

(U is pressure function), respectively; $[A]$ and $[B]$ are computational matrix.

4 EXAMPLES

According to the numerical solution, the values of initial variables are given as follows:

$$n_{ms} = 100, \quad n_f = 300, \quad n_b = 100, \quad N_s = 50$$
$$\rho_f = 1 \times 10^3\ kg/m^3, \quad g = 9.8 m/s^2,$$
$$E = 2.06 \times 10^{11}\ Pa, \rho_s = 7.8 \times 10^3\ kg/m^3,$$
$$\Delta t = 1 \times 10^{-6} \sim 2 \times 10^{-5} (s)$$

According to ships' particulars,all computational parameters as shown in Fig.3, are given in Table 1.

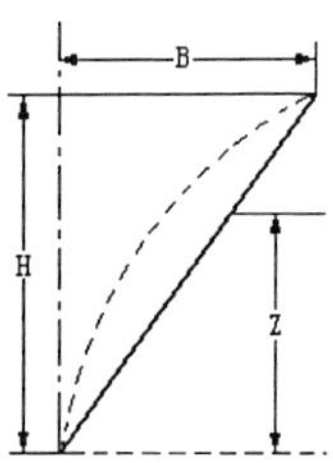

Figure 3: sketch of elastic

TABLE 1

MAIN PARTICULARS OF ELASTIC BODY

No	β	H(m)	Z(m)	h(m)	V_0 (m/s)
1	30°	0.285	0.158	0.005, 0.01, 0.02	10
2	45°	0.57	0.32	0.01	5,10,15
3	60°	0.636	0.21,0.37	0.005, 0.01, 0.02	10

Fig4. Shows water entry of elastic body with velocity $V_0 = 10 m/s$, deadrise angle $\beta = 30^\circ$ and different thickness $h = 0.005、0.01、0.02(m)$. Fig5. Shows water entry of elastic body with different velocity, $V_0 = 5、10、15(m/s)$ deadrise angle $\beta = 45^\circ$ and thickness $h = 0.01(m)$. Fig6. Shows water entry of elastic body with velocity $V_0 = 10 m/s$, deadrise angle $\beta = 60^\circ$ and different thickness $h = 0.005、0.01、0.02(m)$.

From Fig.4 to Fig.6 , compared with water entry of rigid body, we find that

(1) maximum impulsive pressure of elastic body is less than rigid body, this is mainly caused by the elastic effect which decreases the entry velocity. But elasticity character have hardly effect on distribution of impulsive pressure on the body;

(2) slamming of rigid body happen earlier than elastic body. From the results of the elastic body with the different deadrise angle ,the time of slamming is more earlier with the decrease of the deadrise angle .especially as deadrise angle reaches 30°, this difference is more obviously.

(3) impulsive pressure on elastic body is depended on the deadrise angle, entry velocity and rigidity of elastic body. It is as same as impulsive pressure on rigid body , which increase with the increasing of the entry velocity and decrease of deadrise angle. But we have noted that because of the elastic vibration of body's self ,the change of impulsive pressure on elastic body have fluctuant character and

maximum impulsive pressure is more increasing with decrease of the elastic body's thickness .

5 CONCLUSIONS

Elastic character of ship structure is effect on ship slamming. Compared with the rigid body, Elastic effect decrease the magnitude of impulsive pressure and change the variety regularity of impulsive pressure which have fluctuation ,although it can't change the distribution of impulsive pressure. The results have shown that Elastic effect of ship structure is depended on their physical parameters. In practice, many design of ship structures on slamming is based on designer's experience. With development of high-speed vessels , It is necessary for ship designer to select optimization structure's parameters in order to decrease the impulsive pressure, so numerical computational method presented in this paper is applicable to calculate the hydrodynamic loads on ship and other works such as floats of seaplanes , missiles water entry and so on.

References

Falch S. (1986). A Numerical Study of Slamming of two Dimensional Bodies. Dr.ing. thsis, Division of Marine Hydrodynamics. The Norwegian Institute of technology.

Geers T.L.and LodenW.A. (1977). Finite Element. Boundary Integral Analysis of Fluid-Solid Impact. LMSC-D563591. Lockheed Palo Alto Research Lab., Palo Alto, California

Greenhow M. (1998). Water Entry and Exit of a Horizontal Circular Cylinder. *Applied Ocean Research.* **10:4**, 191-198

Hou Y. and Lu C.H (1998). The Water Entry of Wadge by the Boundary Integral Method. *Scientific Report of China*. ShangHai.

Qian Q. (1994). Numerical Simulation of Water Entry of Elastic Body. Dr.ing. thsis. Division of Mechanic. The Huazhong Institute of technology

Zhao R. and Faltisen O. (1992). Water Entry of two Dimensional Bodies. *Journal of Fluid Mechanics.* **246**, 593-612

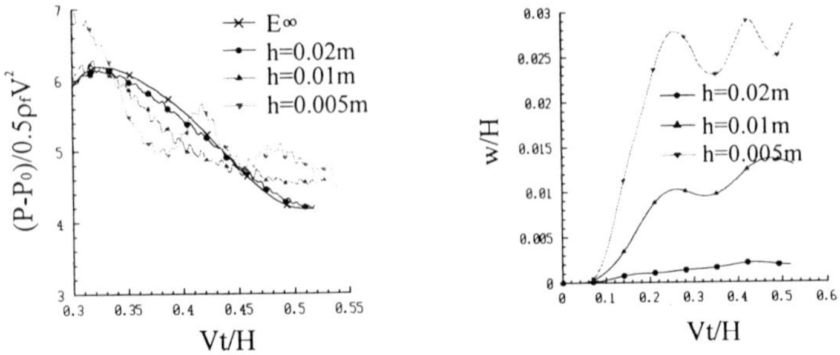

(a)history of impulsive pressure at Z (b)history of plate displacement at Z

Figure 4: water entry of body with $V_0 = 10m/s$, $\beta = 30^\circ$ and $h = 0.005、0.01、0.02(m)$.

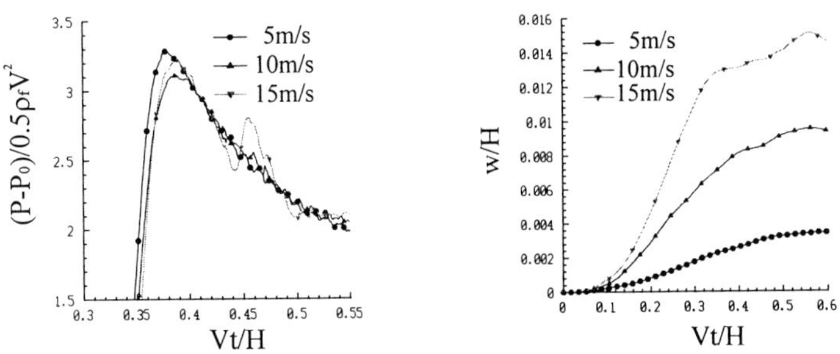

(a)history of impulsive pressure at Z (b)history of plate displacement at Z

Figure 5: water entry of body with $V_0 = 5、10、15(m/s)$, $\beta = 45^\circ$ and $h = 0.01(m)$.

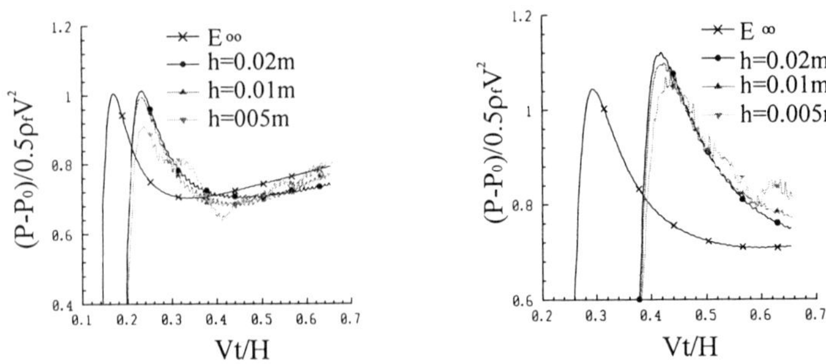

(a)history of impulsive pressure at Z1 (b)history of impulsive pressure at Z2

Figure 6: water entry of body with $V_0 = 10m/s$, $\beta = 60^\circ$ and $h = 0.005、0.01、0.02(m)$.

Practical Design of Ships and Other Floating Structures
You-Sheng Wu, Wei-Cheng Cui and Guo-Jun Zhou (Eds)

RISK ANALYSIS APPLIED TO OCCURRENCE OF MAXIMUM WAVE BENDING MOMENT

Emil Aall Dahle[1], Dag Myrhaug[1] and Hanne Therese Wist[1]

[1]Faculty of Marine Technology, Norwegian University of Science and Technology, N-7491 Trondheim, Norway

ABSTRACT

In the rules of classification societies, the operating life of a ship is often taken as 20 years. Regarding the longitudinal bending moment, a probability of exceedance of 10^{-8} is indicated, i.e. one extreme wave encounter amongst the total of 10^8 waves (with a mean period of 6sec). The maximum hogging moment will occur with the maximum wave crest amidships in a wave with the same length as the ship. Correspondingly, the maximum sagging moment occurs when the same wave has the trough amidships. Extensive analysis of wave records to evaluate the relation between wave heights and wave periods has been carried out in Norway. On this basis, a method for assessment of the probability of occurrence of high waves with a distance between them corresponding to the ship length is outlined, and sample numerical results are presented for one sea area (North Atlantic) in a sinusoidal wave for a barge form. A simple method for obtaining a specified level of probability of exceedance of the longitudinal strength is outlined on this basis, and an example is presented to illustrate the approach.

KEYWORDS

Risk analysis, longitudinal strength, reliability design, wave load

1 INTRODUCTION

The longitudinal strength of ships constitutes the capability of the hull girder to withstand the demand, defined as the loads imposed by the sea and by the still water bending moment which is imposed by the containment of cargo, ballast and bunker. The latter component is well under control by standard calculations. The former is depending upon the dynamic loads imposed by the sea encountered by the ship in its operating life.

Risk analysis of the design should ensure that the demand would only exceed the capability with a pre-determined, low probability. Formerly, a static approach was employed to calculate the wave bending moment. The ship was positioned in equilibrium in a sinusoidal wave with the same length as the ship, normally with a height/length relationship of 0.09 to 0.10, corresponding to a commonly adopted near-breaking criterion for ocean waves. The wave crest was positioned at amidships for calculation of the hogging moment, and correspondingly, the trough was at amidships for calculation of the sagging moment. The still water bending moment was added or subtracted as the case might be.

As specified in the Load Line Convention, elaboration of rules for strength of ships is a task for acknowledged ship classification societies. So, using maximum allowable stress in the deck and bottom plating as stipulated in the rules, the amidship section could be designed. Then, checking for buckling resistance was carried out, and the job was so far done.

Now the simple procedure outlined above has been abolished, and two options are open. First, deterministic, detailed equations are given in the rules for calculation of the wave bending moment. Second, a probabilistic method can be employed, if all relevant statistical distributions are at hand (sea loads, material strength etc.). Both options should ideally provide a probability of exceeding 10^{-8} during the lifetime of a ship (about 20 years), operating in the North Atlantic. This figure corresponds to one critical wave encounter in a ship lifetime (20 years), using a wave period of 6 seconds.

In the following, the former basic approach is re-iterated in a modified form as follows:

- The maximum wave moment occurs when there is close correspondence between the wave length λ and the ship length L, i.e. $\lambda \approx L$.
- For this wave length, the probability of exceedance can be calculated for any wave height H up to the maximum height of 0.09 to 0.10 times λ (the near-breaking wave criterion).
- Then, the scantling can be designed with a capability to withstand the maximum wave moment that corresponds to the chosen probability of exceedance.

The probability of exceeding wave heights for given lengths can be obtained from wave statistics for the sea area under consideration. Long-term wave statistics must be employed. In this paper, both hogging and sagging moments are considered in the sense that the same critical wave is assessed to produce both moments upon passing the ship. The moments can be determined for a specific route, while for unrestricted service, data from e.g. the North Atlantic can be employed. The approach may be suited for a relatively simple way of evaluating alternative designs. As such, the method could find its place in risk analysis, where simple methods are often applied to evaluate alternative solutions. The approach outlined above is illustrated by a sample calculation for a barge form, for unrestricted service. The classification rule risk level of 10^{-8} probability of exceedance in 20 years (or about $5 \cdot 10^{-10}$ per year) is commented upon.

2 LONG TERM WAVE STATISTICS FOR SPECIFIED SHIP LENGTHS

The probability of exceeding wave heights within a narrow band of wave lengths corresponding to some typical ship lengths has been calculated from available wave statistics. The purpose is to obtain data for risk analysis for ships that are restricted to operate in specific trades. Such restrictions are presented in the rules of classification societies, and they must be noted in the ship certificates. It is believed that use of relevant wave statistics may yield a more transparent view of the probability of exceedance. For the analysis, typical wave/ship lengths of 100, 150, 200, 250m with a ± 5% band have been selected, and the probabilities of wave height exceedance within each wave/ship length band have been found.

The probability of exceedance P in Figure 1 has been calculated as follows:

$$P = P\{(H > h) \cap (T_z - \Delta t \le T \le T_z + \Delta t) | T_z\}$$
$$= \int_h^{\infty} \int_{T_z-\Delta t}^{T_z+\Delta t} \int_0^{\infty} f_{HT}(h,t|H_s,T_z) \cdot f_{H_sT_z}(H_s,T_z) dH_s \, dt \, dh \quad (1)$$

So, P is the probability of the wave height H to exceed the value h with the wave period T in the interval $T_z \pm \Delta t$ for given values of the mean zero-crossing wave period T_z. The wave period interval Δt corresponds to $\Delta\lambda = 0.05\lambda$ using the deep water dispersion relationship for linear waves, $T = 0.8\sqrt{\lambda}$ (sec); $f_{HT}(h,t|H_s,T_z)$ is the joint probability density distribution of H and T for given values of T_z and significant wave height H_s; and $f_{H_sT_z}(H_s,T_z)$ is the joint probability density distribution of H_s and T_z. Here the Myrhaug and Kjeldsen (1984) joint distribution of H and T is used together with a joint frequency table of H_s and T_z from the North Atlantic (Sea area 9 in Hogben et al., 1986). The results are presented in Figure 1, showing P versus h for given values of L and T_z. It should be noted that although the integration of H is from h to infinity, there is a physical limit of H due to breaking.

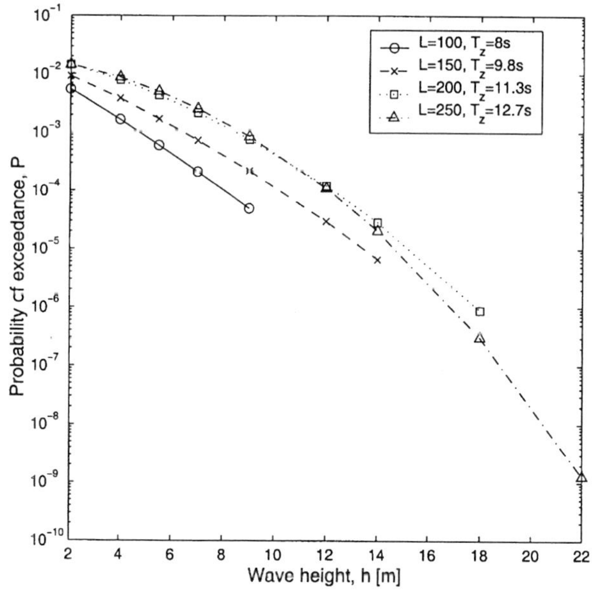

Figure 1: Probability per year of exceeding wave heights related to wave lengths of $\lambda = 100$, 150, 200 and 250m within a $\pm$ 5% band. Sea area selected for this example is the North Atlantic (Sea area 9, Hogben et al., 1986).

Waves with a H/λ relation above 0.09 to 0.10 are non-existing because waves break at about this level. The "ancient" way of calculating the wave bending moment with this maximum H/λ relation was probably based upon this fact. However, Figure 1 allows for selection of probability level of design. In severe weather, the shipmaster normally choose a heading almost against the waves to avoid heavy rolling. As only relatively severe weather contributes to the level of probability, the question of ship heading is of less importance.

3 PHYSICAL WAVE BENDING MOMENT

To illustrate the use of the wave statistics from Figure 1, a barge with a rectangular amidships section, $C_B = 1$ is used, with a sufficiently high freeboard to allow full effect of waves. For such a shape, the hogging and sagging wave moments are identical in magnitude, physically as well as in the classification rules. A sinusoidal wave with the same length as the ship length is used in the following for calculation of the physical moment, which is found by simple, analytical integration. The result is:

$$M_{WO(Physical)} = 0.253 \cdot L^2 \cdot B \cdot H_0 \qquad \text{(kNm)} \qquad (2)$$
$$H_0, L = \text{ wave height and ship length} \qquad \text{(m)}$$

In Table 1, the actual probabilities of maximum wave heights, based upon the near-breaking criterion of $H_{0(\max)} / L \approx 0.09$ are shown. The probability of exceedance per year has been taken from Figure 1.

TABLE 1
$H_{0(\max)}$ AND CORRESPONDING PROBABILITY PER YEAR P OF ENCOUNTERING $H_{0(\max)}$
BASED UPON $H_{0(\max)} / \lambda \approx 0.09$

L (m)	100	150	200	250
$H_{0(\max)}$ (m)	9	14	18	22
P (from Figure 1)	$4.9 \cdot 10^{-5}$	$6.6 \cdot 10^{-6}$	$8.9 \quad 0^{-7}$	$1.3 \quad 0^{-9}$
No. of waves encountered per year (T_z = 6sec)	257	34	4	0.006

The probabilities of meeting maximum waves are very different. While the 100m ship can expect to encounter 257 extreme waves per year, the 250m ship will encounter such a wave only once in 166 years.

4 THE RULE WAVE BENDING MOMENT

Advanced, three-dimensional computer programs can simulate and integrate the pressures on the ship hull to provide the physical wave bending moment. However, a simplified approach is presented by classification societies, e.g. by Det Norske Veritas (2000). There, the capability is given as maximum allowable stress in the deck and bottom plating as:

$$\sigma_a \leq 175 \text{ N/mm}^2$$

For normal ship steel, NV-NS, the yield point is $\sigma = 235$ N/mm^2. Thus, a material safety factor against yield of about 1.35 is built into the rules.
The still water rule bending moment is also presented in the rules, but is neglected in the following. The demand, or the rule hogging and sagging wave moment M_{WO} amidships, for a barge form with a unity block coefficient and for unrestricted service is given as:

$$M_{WO\,(Rule)} = 0.19 \cdot C_W \cdot L^2 \cdot B \qquad \text{(kNm)} \qquad (3)$$

C_W = wave coefficient (empirical, given in the rules, depending on L)
B = breadth of ship (m)

This wave moment can be compared with the simplified physical wave moment for the barge form, i.e. by taking $M_{WO(Physical)} = M_{WO(Rule)}$ and using Eqs. (2) and (3):

$$H_0 = 0.751 \cdot C_W \tag{4}$$

C_W is given in the rules. The wave heights calculated from Eq.(4) together with the probabilities of exceedance found from Figure 1 are given in Table 2.

TABLE 2
PROBABILITY P OF EXCEEDANCE OF ALLOWED STRESS LEVEL PER YEAR DEDUCED FROM THE RULES, DET NORSKE VERITAS (2000)

L (m)	100	150	200	250
C_W (from rules)	8	8.9	9.8	10.4
H_0 (from Eq.(4)) (m)	6.0	6.7	7.4	7.8
P per year (from Figure 1)	$4.2 \cdot 10^{-4}$	$8.6 \cdot 10^{-4}$	$1.8 \cdot 10^{-3}$	$1.7 \cdot 10^{-3}$
H_0 / L	0.060	0.045	0.037	0.031

As can be seen, the probabilities of exceedance per year are quite different, and are much higher than the goal of $5 \cdot 10^{-10}$ per year as indicated in the rules.

5 RISK ANALYSIS APPROACH WITH A NUMERICAL EXAMPLE

5.1 Risk Analysis Approach

To illustrate the procedure of application of risk analysis principles in the establishment of amidships scantlings, the box-shaped vessel is used .The calculation procedure is outlined in Table 3.

5.2 Numerical Example

The commonly used figure of P=10^{-4} per year is adopted as acceptance criterion because the wave height for the classification society probability of exceedance (5 0^{-10} per year) cannot be achieved due to the breaking criterion for waves. The dimensions of the ship (barge) are L = 150m, B = 25m. The trade is the North Atlantic. In relevant weather, the ship will head against the waves.
From Figure 1, H_0= 10m (the rule figure is 6.7m, the near-breaking criterion gives 13 to 14m).
From Eq. (1), $M_{WO(Physical)} = 0.253 \cdot L^2 \cdot B \cdot H_0$ =$1.42 \cdot 10^6$ kNm.
Finally, $Z = M_{WO(Physical)} / \sigma_a = 8.1\text{m}^2$.
As Z is proportional to the wave height, the rule wave height gives lower Z, while the breaking criterion gives a higher Z. As already mentioned, the material safety factor of 1.34 is included in the rule wave moment. If this factor is excluded, the critical rule wave height is H_0= 6.7 · 1.34 = 9.0m, which is close to H_0 based upon the 10^{-4} criterion used in this numerical example.

6 CONCLUSIONS

Ship classification rules as examplified by the rules of Det Norske Veritas (2000), present deterministic formulas for amidships wave bending moments. The moment is indicated to give a probability of exceedance of the rule bending moments of once in the lifetime of the ship (20 years).

Because waves are restricted in height by the near-breaking criterion $H/\lambda \approx 0.09$, the criterion of 10^{-8} is not realistic for ship lengths below about 200m. For the purpose of risk analysis, the scantlings can be based upon a more realistic and commonly used acceptance criterion of exceedance of 10^{-4} per year. An example is presented for the North Atlantic, for which this criterion is used. Similar statistics for other sea areas can be produced by the method described in the paper.

TABLE 3
CALCULATION PROCEDURE FOR ASSESSMENT OF AMIDSHIPS SCANTLINGS BASED ON RISK ANALYSIS PRINCIPLES

Choose the probability of exceedance per year, *P*. (This figure is the risk acceptance criterion for the selected trade)

⇓

From Figure 1, find H_0 that corresponds to P

⇓

Calculate physical wave bending moment $M_{WO(Physical)}$ from Eq. (2).
(Simple numerical calculations can be used for ship forms. Minimum still water bending moment can be calculated as given in the rules)

⇓

Choose material (safety) factor, e.g. 1.35, and calculate allowable stress:
$\sigma_a = \sigma_{yield}/1.35 = 175\,\text{N/mm}^2$

⇓

Calculate amidships section modulus $Z = M_{WO(Physical)}/\sigma_a$

⇓

Design amidships scantlings based on Z

⇓

For comparison (if relevant), repeat procedure for other sea areas or trades, with same, or different ship length

Acknowledgement

This work was partly funded by Anders Jahres Foundation for the Advancement of Science.

References

Det Norske Veritas (2000). *Rules For Ships.*
Hogben, N., Duncan, N.M.C. and Olliver, G.F. (1986). *Global Wave Statistics.* Unwin Brothers, London, UK.

Myrhaug, D. and Kjeldsen, S.P. (1984). *Parametric Modelling of Joint Probability Density Distributions for Steepness and Asymmetry of Deep Water Waves.* Applied Ocean Research, Vol. 6, pp. 207-220.

Practical Design of Ships and Other Floating Structures
You-Sheng Wu, Wei-Cheng Cui and Guo-Jun Zhou (Eds)

FUZZY RELIABILITY ANALYSIS OF A SHIP LONGITUDINAL STRENGTH

J. M. Yang[1] and J. Y. Huang[1]

[1]Department of Naval Architecture and Marine Engineering
National Cheng Kung University
Tainan, Taiwan, China

ABSTRACT

Basically, the nature of structural strength itself is fuzzy or vague and it should not be defined as a single value. In order to obtain a rational strength region of a ship structure and to attain the integrity of a ship, the fuzzy arithmetic approach is needed in analyzing a ship's structural strength. In this paper, the fuzzy reliability in which the fuzziness in both loads and resistance is taken into consideration is used in assessing the strength of single hull tankers. The performance function in the reliability method, based on the ultimate strength of a ship's hull girder, is evaluated by means of the advanced second moment (ASM) method. A triangular fuzzy membership function is applied to get approximate solutions that have generally wider interval containing the exact ranges. Furthermore, two tankers, one is made of mild steels and the other is made of high tensile steels, are selected as examples to examine the effect of high tensile steels on fuzzy reliability. The results of this study may be used in reliability-based design or the development of reliability-based design criteria.

KEYWORDS

Fuzzy Reliability, Membership Function, α-cut, Advanced Second Moment Method, Fuzzy Arithmetic

1 INTRODUCTION

During the past two decades, the reliability analysis of structures grew rapidly and was used to improve and modify the design specifications [1,2]. The traditional design approach still works quite well and remains simple for routine design practice. However, in order to achieve an optimal ship design, the ultimate object, which is within an acceptable degree of reliability, becomes necessary to apply certain probabilistic approach in design to account for the randomness of sea loads and various uncertainties in strength analysis. There are two types of uncertainties involved in the structural strength analysis, subjective (cognitive) uncertainty and objective (non-cognitive) uncertainty. The objective uncertainty is usually treated by means of stochastic analysis. However, the subjective uncertainty may be treated by traditional probability theory or modeled by fuzzy set theory [3,4,5].

In this investigation, the reliability assessment of ship structures based on a fuzzy definition of failure is studied.

The wide use of high tensile steels on ship structures has resulted in a great amount of weight-reduction on ships. However, the buckling strength and fatigue strength of these ships may also be reduced since the thickness of the plates on ships becomes relatively thinner. In this paper, two tankers, one is made of mild steels and the other is made of high tensile steels, are selected as examples to examine the effect of high tensile steels on the fuzzy reliability. The results of this study may be used in reliability-based design or in the development of reliability-based design criteria.

2 FUZZY SET THEORY AND FUZZY FAILURE FUNCTION

A fuzzy set is a set whose boundary is not well defined and can be characterized by membership grade function. The membership function or " characteristic function" of a fuzzy set $\tilde{A}$ can be denoted by $\mu_{\tilde{A}}(x)$, i.e.

$$x \in X \quad \mu_{\tilde{A}}(x): X \rightarrow [0,1] \tag{1}$$

where X is a universal set. The membership function $\mu_{\tilde{A}}(x)$ of the fuzzy set $\tilde{A}$ maps x from X to the real interval [0,1], which is also the domain of the membership function $\mu_{\tilde{A}}(x)$. An α-cut of a fuzzy set $\tilde{A}$ can be defined as

$$\tilde{A}_\alpha = \{ x \in X \mid \mu_{\tilde{A}}(x) \geq \alpha, \quad \text{for } \alpha \in [0,1] \} \tag{2}$$

α is denoted as a confidence level or a "threshold". $\tilde{A}_\alpha$ is defined as a set has grades that are more than or equal to some minimum value α. Furthermore, an α-cut is able to defuzzify a fuzzy set to a single or more crisp values [6].

A probability function can be characterized by certain ensemble averages, such as mean and variance. The cognitive uncertainty in a design variable can be described by membership function for its fuzziness and its α-cuts [7]. In this study, the mean value of a random variable is considered to be fuzzy and its variance is non-fuzzy. There are several different kinds of membership functions that could be used. The common triangle-shape membership function is adopted in present investigation. The fuzzy failure function can be written as

$$\tilde{M} = R(x) - S(x) \pm (1-\mu(x)) \cdot d_R \mp (1-\mu(x)) \cdot d_s = 0 \tag{3}$$

where $\tilde{M}$: fuzzy failure function; $\mu(x)$: membership function;
$R(x)$: strength function; d_R : allowable tolerance of strength;
$S(x)$: load function; d_s : allowable tolerance of load;

In general, the definition domain of distributions may not be closed interval. For examples, $(-\infty, +\infty)$ or $[o, +\infty]$, a significant interval instead of the whole domain of definition needs to be carefully selected [7]. The significant interval in this investigation is chosen as $[\mu - 2\sigma, \mu + 2\sigma]$ since the random variables used in this study are normal distributions $N(\mu, \sigma)$ or equivalent normal distributions.

3 ULTIMATE STRENGTH OF A SHIP HULL

There are two major aspects in analyzing the ultimate strength of ship structures: (1) failure due to yielding and plastic flow and (2) failure due to instability and buckling. The former one is recognized as global failure of a ship hull and the latter case may be considered as failure of a local principle member or a grilliage buckling case.

The following limit states and failure modes are considered in the fuzzy reliability analysis:
(1) Initial yield failure mode

$$M_I = SM_e \times \sigma_y \qquad (4)$$

(2) Torsional/flexural buckling of stiffeners (tripping)

$$M_{ut} = SM_{eff} \times \sigma_{cr} \qquad (5)$$

(3) Hull girder collapse mode

$$M_u = (-0.172 + 1.548\,\phi_{cp} - 0.368\,\phi_{cp}^{\;2}) \times SM_e\,\sigma_y \qquad (6)$$

Detailed definitions of each parameter in the limit states equations for above modes of failure are given in Mansour [2,8].

4 STILL-WATER AND WAVE BENDING MOMENT

4.1 Still Water Bending Moment M_s

In designing a ship, only the maximum load during its lifetime, which is about 20 years of service, is considered. The maximum allowable still-water bending moment of a ship can be estimated by Dnv rules (1993). In the present investigation, the still-water bending moment is taken as a random variable [2], which is normally distributed with a mean value that equals to sixty percent of the value, obtained by Dnv rules. The C.O.V. of it is equal to 0.4.

4.2 Wave-Induced Bending Moment M_w

As described in reference [9], when the basic information, such as the length, the beam the draft of a ship, the Froude number and the significant wave height are given, the root-mean-square (RMS) value of the wave-induced bending moment of a ship can be estimated by the interpolation procedure through seakeeping tables. The extreme value distribution function of wave-induced bending moment can be expressed as follows [1,2]:

$$F_w(w) = \exp\left(-N \exp\left(\frac{-w^2}{2\lambda_0}\right)\right) \qquad (7)$$

$$\mu_w = \sqrt{2\lambda_0 \ln N} + \frac{0.5772}{\alpha_N} \qquad \sigma_w = \frac{\pi}{\sqrt{6}\alpha_N} \qquad \text{where} \qquad \alpha_N = \sqrt{\frac{2\ln N}{\lambda_0}}$$

N is the expected number of wave bending peaks. A peak normally occurs once in every 8~10 seconds, and hence in a 3 hours storm, N can be estimated as: $N = 3\text{x}60\text{x}60/10 \approx 1000$ (peaks/3hr). The notation λ_0 is the mean square value obtained from sea-keeping tables, μ_w is the mean of the extreme wave-induced bending moment and σ_w is the standard deviation.

5 ADVANCE SECOND MOMENT MOTHOD

The aim of reliability analysis is to measure the ability of certain engineering system to fulfill the design purpose for a specific time period. It starts with quantification and treatment of uncertainties, which are considered as random variables that can be described by mean values, variances and probability distribution functions.

The safety/reliability index β is defined as the shortest distance from the origin to the failure surface, $g(\overline{X})=0$, in the reduced Y-coordinate system [10]. The safety index can be obtained by the following equation.

$$\beta = \min(\sum_{i=1}^{n} y_i^2)_{y_i^*}^{1/2} \qquad (8)$$

More specifically the safety index can be determined by solving the following set of non-linear equations iteratively:

$$\beta = \frac{-\sum_i y_i^* (\frac{\partial G}{\partial y_i})_{y_i^*}}{\left[\sum_i (\frac{\partial G}{\partial y_i})_{y_i^*}^2\right]^{1/2}} = -\sum_i \alpha_i y_i^* \qquad \alpha_i = \frac{(\frac{\partial G}{\partial y_i})_{y_i^*}}{\left[\sum_i (\frac{\partial G}{\partial y_i})_{y_i^*}^2\right]^{1/2}} \qquad (9)$$

where $G(\bar{Y}^*) = G(y_1^*, y_2^*, \ldots, y_n^*)$ is the failure surface in the reduced space, and y_i^* is the coordinate of the point closest to the origin in the reduced space. The probability of failure can be expressed as follows:

$$P_f = \Phi(-\beta) \qquad (10)$$

where Φ is the standard normal distribution function.

6 CASE STUDY

In this study, a tanker of 260000 tones built with high tensile steel is selected. The probabilities of failure for various failure modes of the above tanker are compared to those of Mansour and Hovem's study [2]. The main characteristics of tankers in the present study and those of Mansour's study are listed in Table 1.

TABLE 1

MAIN CHARACTERISTIC OF TANKERS

Main characteristics	Present study	Mansour (1994)
Displacement (tons)	260000	149000
Length overall (m)	329	273
Length between Perpendiculars (m)	315	260
Beam (m)	58	42
Depth (m)	28.7	23.5
Draft (m)	19.5	16
C_b	0.8194	0.71
Modulus of elasticity at desk (m-cm^2)	7.174×10^5	4.658×10^5
Mean value of yield strength (Mpa)	345	259

Through the main characteristics of those two tankers are quite different, the sizes of principal members on decks of both tankers are almost the same as shown in Figure 1.

The performance function (limit state function) g ($\bar{X}$) in the present study is expressed as:

$$g(\bar{X}) = \tilde{x}_u \, S\tilde{M} \, \tilde{\sigma} - \tilde{x}_{sw} \, \tilde{M}_{sw} - \tilde{x}_w \, \tilde{x}_s \, \tilde{M}_w \qquad (11)$$

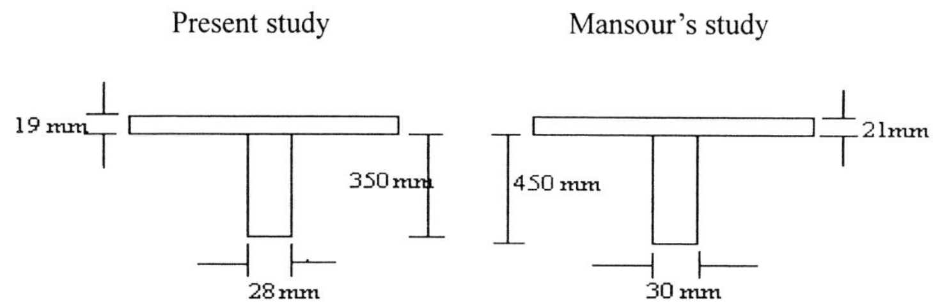

Figure1: Dimensions of principle member

where $\tilde{SM}$: the section modulus. Proper section modulus should be selected based on failure mode.

$\tilde{\sigma}$: yield stress or critical stress

$\tilde{x}_u$, $\tilde{x}_{sw}$, $\tilde{x}_w$ and $\tilde{x}_s$ are random variables representing uncertainties are listed in Table 2. The first term right after the equal sign of the equation (11) may be considered as strength function, whereas the second and the third term of the equation (11) represent the load functions in the fuzzy failure function.

The r.m.s values of wave-induced bending moment (assumed sea state H_s=12.2m) of the tanker are calculated according to the reference [2] as shown in Table 2. The distributions, the mean values and the C.O.V. of modeled uncertainties are treated as random variable in fuzzy reliability analysis and are also listed in Table 2.

TABLE 2

DISTRIBUTION, MEAN AND C.O.V. OF RANDOM VARIABLES

Random Variable	Distribution	Mean	C.O.V.
$\tilde{x}_u$	Normal	1.0	0.15
$\tilde{x}_{sw}$	Normal	1.0	0.05
$\tilde{x}_w$	Normal	0.9	0.15
$\tilde{x}_s$	Normal	1.15	0.03
Plastic S.M.	Log Normal	7.67×10^5 m-cm^2	0.04
Elastic S.M.	Log Normal	7.174×10^5 m-cm^2	0.04
Effective S.M.	Log Normal	6.981×10^5 m-cm^2	0.04
Hull girder S.M.	Log Normal	7.174×10^5 m-cm^2	0.04
$\tilde{M}_{sw}$	Normal	4.00×10^6 kNm	0.40
$\tilde{M}_w$	Extreme	11.45×10^6 kNm	0.09
$\tilde{\sigma}_y$	Log Normal	34.5 kN/cm^2	0.07
Column buckling $\tilde{\sigma}_{cr}$	Log Normal	31.35 kN/cm^2	0.07
Tripping $\tilde{\sigma}_{cr}$	Log Normal	22.7 kN/cm^2	0.07
Hull girder collapse $\tilde{\sigma}_{cr}$	Log Normal	25.435 kN/cm^2	0.07

6.1 Results and Discussions

Figures 2-4 show the results of the fuzzy reliability analysis for initial yield failure mode, tripping failure mode and hull girder collapse failure mode. The initial yield failure mode and the hull girder collapse failure mode are considered as globe failure modes. The tripping failure mode is considered as local failure modes. The tripping failure mode has the highest probability of failure.

A comparison between the results of a tanker made by mild steel [2] and the results of a tanker made by high tensile steel on the fuzzy reliability analysis for various failure modes is shown in Figures 5-7. It is found that the probability of failure of the tanker made by high tensile steel is about twice or three times higher than that of the tanker made by mild steel.

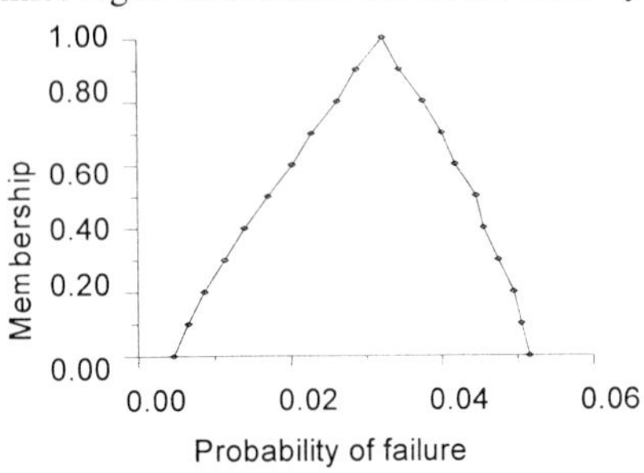

Figure 2: Deck initial yield failure

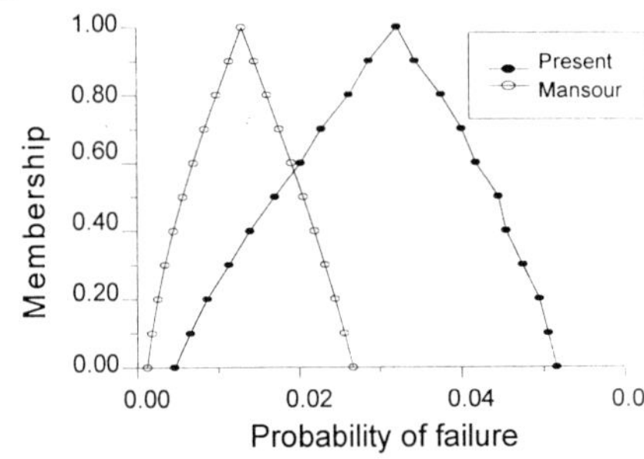

Figure 5: Deck initial yield failure

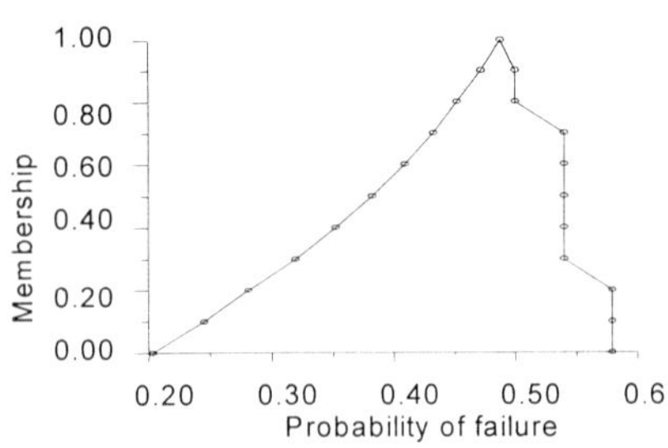

Figure 3: Tripping failure

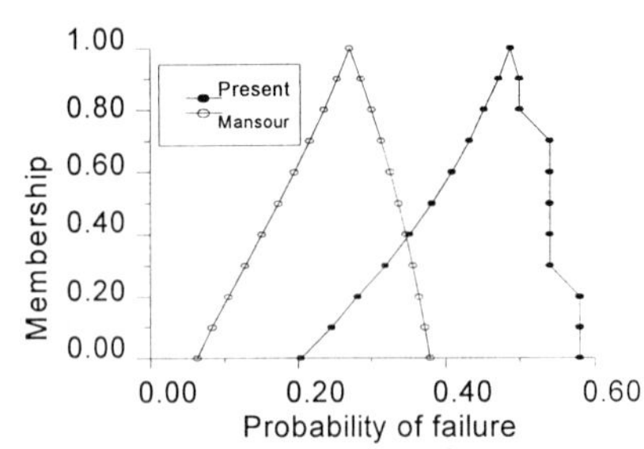

Figure 6: Tripping failure

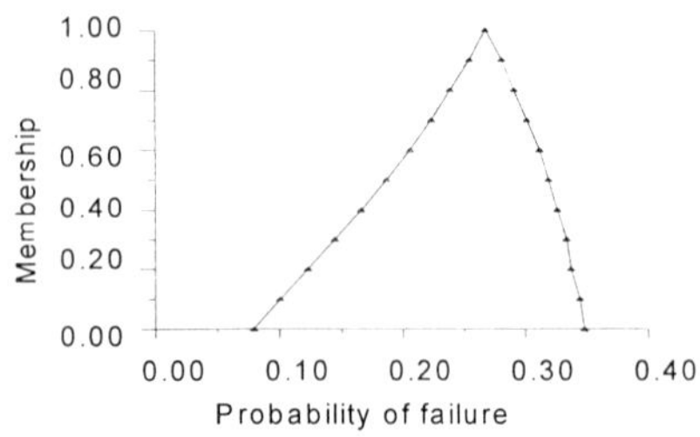

Figure 4: Hull girder buckling collapse

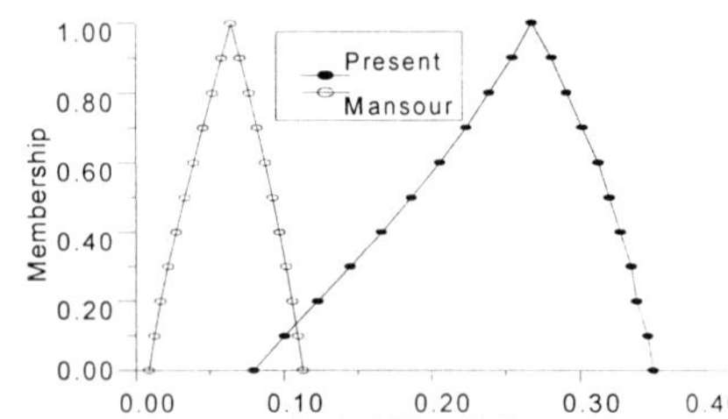

Figure 7: Hull girder buckling collapse

7 CONCLUSIONS

In this study, approximate analysis models for predicting the fuzzy reliability of ship hull girder are presented. These models take into account for the randomness and fuzziness. Probabilities of failure associated with global and local failure modes are calculated. The resulting probabilities of failure imply that tripping failure of the principal member of the tanker is the governing mode in failure. In general, the larger a vessel is, the higher safety level the vessel can get [1]. However, the results obtained in this study reveal that larger vessels, which are made of high tensile steel, are not necessary to have higher safety levels. Because the use of high tensile steel in ship design generally reduces the thickness of ship hulls, this may reduce the buckling strength and fatigue strength of a ship. Therefore, more attention should be given to the detail design and the working procedure to avoid the premature failure of the ship hulls.

Acknowledgement

The National Science Council, Taiwan, China, under Grant NSC 88-2611-E006-015, supported this work. The authors would like to extend their gratitude to NSC for the support.

References

[1] Mansour A.E. (1990). An Introduction to Structural Reliability Theory, *Ship Structure Committee Report*, SSC-351,

[2] Mansour, A.E., and Hovem L. (1994). Probability Based Ship Structural Safety Analysis. *Journal of Ship Research* **38:4**, 329-339.

[3] Zadeh, L. A. (1965). Fuzzy Sets. *Information and Control*, **8**, 338-353.

[4] Ayyub B.M. (1992). Structural Reliability Assessment with Ambiguity and Vagueness in Failure. *Naval Engineering Journal*, 21-35.

[5] Shao W.J. (1995). Structural Fuzzy Reliability Analysis and Its Applications in Strength Problems for Ships. *OMAE*, Volume II, 343-347.

[6] Klir G.J. and Yuan B. (1995). *Fuzzy Sets and Fuzzy Logic- Theory and Applications.* Prentice Hall PTR, New Jersey.

[7] Chao R.J. (1995). *Structural and Reliability Analysis of Ships with Fuzzy and Random Variables*. Ph.D. Dissertation, Department of Civil Engineering, University of Maryland.

[8] Mansour A.E. (1985). Approximate Formulae for Preliminary Design of Stiffened Plates. *ICOSSAR '85*, 4th International Conference on Structural Safety and Reliability, Kobe, Japan

[9] Loukakis T.A. and Chryssostomidis C. (1975). Seakeeping Standard Series for Cruiser Stern Ships. *Trans. SNAME*, **83**, 67-90.

[10] Hasofer, A.M. and Lind N.C. (1974). Exact and Invariant Second Moment Format. *Journal of the Engineering Mechanics Division, ASCE* **100:EMI**, 111-121.

Practical Design of Ships and Other Floating Structures
You-Sheng Wu, Wei-Cheng Cui and Guo-Jun Zhou (Eds)

RELIABILITY-BASED REQUALIFICATION OF EXISTING OFFSHORE PLATFORMS

Torgeir Moan [1] and Ole Tom Vårdal [2]
[1] Norwegian University of Science and Technology, Trondheim, N-7491, Norway
[2] Aker Offshore Partner ASA, P.O. Box 1, Sandsli, N-5861, Bergen, Norway

ABSTRACT

This paper deals with the philosophy and principles that should be applied in developing procedures for re-qualifying existing platforms, for an extended service period, with particular view of crack growth, based on the results of in-service inspections, weld improvement and possible increased inspection program. These principles have been applied to assess existing North Sea platforms. A probabilistic fracture mechanics model is applied towards this aim. Application of this methodology requires determination of target safety levels, which are consistent with the safety regulations to comply with as well as the reliability method to be used. In this paper particular attention is paid to establishing the target level.

KEYWORDS

Requalification, Offshore platforms, Fatigue, Reliability, Target level

1 INTRODUCTION

Requalification means the process of demonstrating fitness for further use of an installation that may be necessary to undertake due to e.g. desired change of use or extended service life or occurrence of damages. Basically, requalification involves the same assessments as carried out during initial design, except that it is carried out with more specific information obtained through monitoring and inspection during fabrication and operation. Moreover, improvement of the safety level of an existing structure is much more expensive than for new structures. This fact justifies the use of more advanced analyses of loads, responses, resistances as well as use of reliability analysis, and more elaboration on target safety level than implied by design codes.

The focus of this paper is on requalification of offshore platforms for extension of service life in view of crack growth, based on results from in-service inspections, weld improvement and possibly an increased inspection program. The principles for requalification that are outlined in ISO standards (1994, 1998) are general and do not represent an industry standard for this kind of assessment of offshore structures. This paper deals with the development of methodology towards this aim.

While current fatigue design procedures are based on a semi-probabilistic SN- approach, a probabilistic fracture mechanics approach needs to be adopted to properly account for the effect of inspection results and the inherent uncertainty in prediction of crack growth. Such a method also allows for treatment of the interaction between design, inspection, maintenance and repair requirements.

In the present assessment a reliability measure of the crack growth is obtained by using and extending the probabilistic fracture mechanics tool, PIA (Probabilistic Inspection Analysis), Kirkemo (1988, 1989), Moan et al.(1999) and Johannesen et al. (2000). When a probabilistic approach is used, the target level also needs to be expressed in probabilistic terms.

Published test results for fatigue life improvement methods such as grinding, hammer-peening etc. document a high potential for increased capacity against fatigue crack growth and may be utilised to achieve an extended fatigue life.

The probabilistic method applied is outlined in an accompanying paper, Johannesen et al.(2000). The focus here is on establishing the target level for the reliability analysis as well as describing how the desired safety level is achieved by the inspection result: "No Crack Growth Detection", post weld treatment, and possibly an extended inspection program.

2 IN-SERVICE VERSUS DESIGN FATIGUE ASSESSMENT

Table 1 summarizes characteristic features of the in-service and design fatigue assessment process. As long as it is extremely complicated/expensive to change topology and member dimensions for an existing structure, it is beneficial to search for other measures than traditional reconstruction to improve the safety level. In-service observations by Magnetic Particle/Eddy Current (MPI/EC) inspections and/or strain measurements may be used to reduce the uncertainty of the estimated fatigue life (fatigue potential). A reduced uncertainty in the estimated fatigue life may be used to reduce the required safety margin. The safety margin for the design fatigue assessment procedure is defined by the selected SN-curve and the regulatory requirements in terms of Design Fatigue Factors (DFF= $T_{fatigue}/ T_{service}$, where T denotes life). Table 2 shows the values specified by NPD.

The design SN-curve represents the 97.7 % quantile, which corresponds to a fatigue life approximately one third of the mean fatigue life. Implementation of the DFFs in Table 2 gives an additional multiplication factor on the fatigue life. Depending upon the access for inspection and the severity of failure consequences, the design fatigue life, therefore, will be between 3 and 30 times the expected fatigue life. The estimated initial fatigue lives are normally also conservative since each simplification and assumption made in the fatigue analysis, is intended to be slightly conservative. This is due to the approaches used for calculating hydrodynamic loads (which are the primary source of crack growth), the corresponding load effects and crack growth resistance.

The actual safety margin of the design fatigue assessment methods may be obtained by use of performance data i.e. by systematic comparison of crack growth observed and estimated fatigue life. For instance, the studies of jackets by Vårdal and Moan et al (1997-1999) resulted in the following main conclusions:

- The number of predicted fatigue cracks are 3 to 15 times higher than observed
- 2-5 % of the observed fatigue cracks occurs in joints without any significant fatigue potential according to the fatigue assessment methods.
- The in-service observations indicate that the crack growth starts from large initial defects.

The studies indicate that fatigue life estimates based on "standard" fatigue assessment from 1985-1990 have conservative mean values, but at the same time the level of uncertainty is even higher than assumed.

The availability of in-service observations makes it possible to achieve significant reductions in the level of uncertainty, (Vårdal and Moan, 1997). A sufficient amount of EC/MPI inspection with the result "No detection of a propagating crack" may be used to document an upper limit for the potential for crack growth.

Fatigue analyses are normally based on a long-term distribution of stress-ranges and SN data. The fatigue resistance is modelled by Paris-Erdogan's crack propagation law. The stress intensity factor is described as a function of crack depth a, with assumed crack aspect ratios, c/a. Further details about the probabilistic crack growth model used for steel-plated and tubular joints may be found in Johannesen et al (2000) and Moan et al (2000), respectively. Particular efforts were made in the former paper to establish K-factors for the weld toe of as-welded and ground welds, recognising membrane and bending stress conditions. Various models for POD curves are discussed in the quoted papers as well as in Moan et al., (1997).
On the other hand, the fatigue resistance obtained by the fracture mechanics (FM) model is sensitive to the model of initial crack size and stress intensity factor. It is, therefore, essential that the FM-model be validated when adopted in a regulatory context. This especially includes comparisons between probabilistic SN and FM prediction (Moan et al., 1993, Johannesen et a.l, 2000).

3 EQUIVALENT PROBABILISTIC REGULATORY FATIGUE EQUIREMENTS

3.1 General

According to NPD (1998) design fatigue factors (DFF) are specified according to the accessibility and failure consequences of the detail, see Table 2. It is noted that consequences are defined in relation to whether the structure survives a 100-year wave condition after fatigue failure of the relevant structural component. Access for inspection implies that maintenance by inspection and potential repairs are carried out. However, due to the lack of specification of the quality and extent of in-service inspections required in NPD (1998), the target reliability level is deduced from the requirement of DFF values for components that are not accessible for inspection.
This derived target level can then be met by designing the structure with a certain resistance against crack growth, as well as by using inspection. It is then important that design criteria are chosen so that the target level can be reached by reasonable efforts during operation i.e. a practical amount of maintenance by inspection and repair. The DFF factors in Table 2, for components accessible for inspection, define design criteria that combined with inspection intend to achieve the target reliability level for the service lifetime.

3.2 Target Reliability Level

Consider relevant welded joints in a platform that has been operating for a time T_{past} and is intended to be operating for another period T_{future}. The target level for the fatigue reliability in the remaining period T_{future} can then be obtained by considering a fictitious new weld at $T_{future}= 0$ that satisfies the relevant regulatory requirement in Table 2. For a weld which is not inspected, this means a DFF of 10 or 3, depending upon the failure consequences. The "deterministic" safety measure (DFF) may then be transformed into a reliability measure by using the relevant probabilistic fracture mechanics approach.
It is crucial that the implied target levels are determined by the same reliability methodology that will be used later to demonstrate compliance with the target values. This is because the results are sensitive to the assumptions made in the reliability analysis. With a required remaining in-service time of 20 years and a required design fatigue factor of 10, the target reliability level as at the end of T_{future} is found to be $\beta_{t,DFF=10} = 3.5$, based on the methodology in Johannesen et al.(2000). The target reliability level is independent of the initially specified service life.
The target level for the case when the consequences of failure are minor (i.e. the structure fulfills the ALS requirements according to NPD (1998), should correspond to DFF=3. The corresponding target level is found in a similar manner to be $\beta_{t,DFF=3}=2.5$.
The reliability approach outlined above, by considering a single component can be made more refined by introducing a systems reliability approach that incorporates the following failure modes: fatigue failure of a component, followed by ultimate systems failure. The target level is then inferred from the

NPD regulation, e.g. requiring DFF=10 for statically determinate systems, i.e. with a conditional system failure probability of 1.0 given component failure. The target of fatigue probability $P[FF_i]_T$ level for a single component is then

$$P[FF_i]_T \cdot P[F_{SYS} \mid FF_i] = P[F_{SYS}]_T \tag{1}$$

where $P[F_{SYS}|FF_i]$ is the system failure probability given fatigue failure for the actual platform. Any presence of residual resistance after a component failure would then imply a relaxed fatigue reliability target level compared to the target value given above. This approach is described more in detail by Moan et al. (1999).

The reliability target values obtained by current analysis methodology are notional values. A systematic comparison of the theoretical prediction of and observed fatigue crack growth are to be used for an adjustment of the theoretical model, as outlined e.g. by Vårdal, Moan et al. (1997-2000). Based upon the authors' knowledge such data are not available for floating offshore structures. As long as the same model is applied for estimating the reliability target values and the components reliability level, the conclusions based on use of "notional" reliability values will give reasonable results.

4 CASE STUDY: THE VESLEFRIKK B PRODUCTION PLATFORM

The Veslefrikk B platform is a semi-submersible platform (Fig 1) and was initially constructed in 1985 and modified to a production unit and installed on the Veslefrikk field in 1989. In 1999 the platform was modified to increase the payload and to improve the fatigue life of the corner areas of pontoon/column connections. Additional payload capacity was achieved by sponsons on the pontoon. Two additional columns and sponsons on the existing columns were required for the purpose of stability, see Fig. 1.

A detailed FEM analysis also revealed a high fatigue potential for the two horizontal braces.

Traditional deterministic SN-curve estimates identified that the fatigue life of the most critical area was 5-10 years. Upon 10 years in-service and a requirement of an additional 20 years in operation, motivated the use of a refined analysis method for estimating the potential of fatigue crack growth.

4.1 Inspection after 10 Years In-service.

An inspection program with nearly 100% EC- and 30% UT-inspection of the welds in the horizontal braces was carried out before the platform was taken to the yard. The cracks detected by EC were examined by microscope and classified as hydrogen cracking. These non-propagating cracks were repaired. The UT-inspection revealed 11 crack sites that need to be monitored by inspection during in-service.

The inspection program in the repair dock mainly focused on external MPI, one propagating crack was detected, at a site with initial fatigue life of 15 years. This crack is assumed to have started from major internal fabrication defects. The crack was weld repaired and the detection resulted in an extended UT-inspection program.

The corrosion protection system included 71 anodes attached to the skin of the horizontal bracing by 284 doubling plates which represented a high crack growth potential. All doubling plates were removed, replaced by bracelets for support of anodes and inspected. Afterwards the welded areas were ground and hammer-peened.

4.2 Inspection and Weld Improvements Activities Required

Sufficient capacity against fatigue crack growth, can be documented by using inspection results "No Crack Growth Detection", weld improvements methods as well as more frequent inspections.

Figure 2 shows an example of a hot spot where weld improvements methods are used in addition to the result "No Crack Growth Detection" achieved by EC/MPI inspection. A hot spot with a fatigue life of 7 years and inspected after T_{past} = 10 years with the result "No Finding" implies a reliability level corresponding to β=1.0 in a T_{future} = 20 years period. By including weld improvement by grinding, the reliability is increased to $\beta_{DFF=2}$ = 2.5, equal to $\beta_{DFF=3}$. However, a target reliability level $\beta_{t,DFF=10}$ = 3.5 can only be reached by an inspection program defined according to the procedure described below.
The assessment of the actual hot spot comprises a comparison of the reliability level at the end of a total service lifetime of say 30 years, with the target level. Figure 3 presents a more suitable format for identifi-cation of recommended activities. It is noted that for initial fatigue lives (based on SN-approach) of 0-15 years, a favourable inspection result, grinding as well as inspection in the last 20 years in-service would be required. (Obviously this situation should in principle not occur if normal fatigue design practice would require an initial fatigue life of at least 20 years.) For welds with initial fatigue lives which exceed 15 years, the target reliability can be reached by a "No Find" during inspection and grinding. It is seen that grinding improves the reliability significantly. When no grinding is applied, it would be necessary to have an initial fatigue life of 35 years or more to reach the target level when a favourable inspection is made after 10 years as well as by inspections that are scheduled in the last 20 years.
The improved fatigue capacity by grinding and hammer-peening is restricted to the surface of the weld. Any internal fabrication defects may upon weld improvement be the critical area for crack growth. An extensive program for UT-inspection (Ultrasonic Testing) was carried out to verify that the internal defects were sufficiently small, as mentioned above.

4.3 In service Inspection Program for the Remaining 20 Years

To reach the target level it may be necessary to carry out inspections in the period after the time of requalification. Recommended time for inspection is defined as the year when the component reliability curve and the target reliability level, intersects. Table 3 indicates the amount of in-service inspections required for fulfillment of the reliability target level. To get a more cost optimal maintenance procedure, additional inspections are specified for early detection of crack growth. Early detection of crack growth gives the possibility for grinding repair and increased inspection frequency for avoiding costly weld repairs.

5 CONCLUSIONS

This paper deals with extension of the service life of offshore platforms in view of crack control based on a probabilistic fracture mechanics approach.

It is demonstrated that "No Crack Growth Detection" during extensive inspections after 10 years in-service, use of weld improvement methods by burr grinding and frequent future in-service inspection have made it possible to extend the service life of of the platform without the significant structural modification required for reduction of the load effect in the horizontal braces. The alternative approach would have been to modify the overall design at a much larger costs.

Currently, there exists no industry standard for reliability-based requalification of service-life. The work presented in this and the accompanying paper, by Johannesen et al. (2000), contributes towards establishing such a standard. In particular, an approach to define a target safety level that is consistent with the current fatigue design criterion of NPD has been developed. This procedure may be applied to establish target levels consistent with other regulations or standards.

Acknowledgments

The authors would like to thank Statoil and DNV Classification for their constructive co-operation.

Practical Design of Ships and Other Floating Structures
You-Sheng Wu, Wei-Cheng Cui and Guo-Jun Zhou (Eds)

DETERMINISTIC AND PROBABILISTIC ASSESSMENT OF FPSO HULL GIRDER STRENGTH

A. Incecik and Y. Pu

Department of Marine Technology, University of Newcastle
Newcastle Upon Tyne, NE1 7RU, UK

ABSTRACT

FPSOs are good candidates for marginal and deep water oil fields. Due to its unique operational scenarios, such as operating generally in a location for quite long time, constant change of its loading conditions, possible harsh environments etc, the design criteria for conventional ships are not necessarily applicable to FPSOs. The objective of this paper is to develop a new assessment procedure based on the prediction of global hull girder loads and stresses from first principle. In the paper stillwater and wave-induced loads on a typical FPSO were predicted by both first principle and rule-based calculations. The long-term prediction of hull-girder loads was also carried out for two sites. The hull girder strength was then assessed by both deterministic and probabilistic approaches. In deterministic analysis the primary and local strength of the FPSO is assessed by rule-based formulations. The progressive collapse analysis was used to predict the ultimate strength of the hull girder. In probabilistic analysis, reliability of three failure modes of the hull girder was estimated by the first order second moment method. It is found that the rule-based wave-induced load prediction method yields much smaller values than the first principle calculation. It is believed that wave-induced loads should be obtained using a first principle approach.

KEYWORDS

Wave-induced loading, Design of FPSOs, Long-term prediction, Reliability analysis, Reliability-based design, Probabilistic analysis.

1 INTRODUCTION

The number of FPSOs has been increasing rapidly over the world recently. It has certain advantages over the other types offshore structures for marginal and deepwater oil fields. Due to its unique operational scenarios, such as operating generally in a location for quite long time, constant change of its loading conditions, possible harsh environments etc, the design criteria for conventional ships are not necessarily applicable to FPSOs.

The most significant structural failure would be the loss of hull girder integrity which would lead to loss of containment of crude oil, flooding and potentially the loss of lives and the facility. This would also have serious consequences for the marine environment. For these reasons a study sponsored by Shell UK Exploration and Production Ltd was carried out with participation of the University of Newcastle Upon Tyne and Armstrong Technology Associates Limited in early 1998. The objectives of the study are:

- To develop a new assessment procedure based on the prediction of global hull girder loads and resulting stresses from first principle calculations.
- To correlate different hull girder assessment methods recommended by classification societies with a reliability based procedure developed in item (1)
- To recommend a process of prescribing environmental loadings appropriate to the field area, design life and mode of operation.
- To develop a design acceptance criterion for the hull girder.

The major findings in this project are reported in this paper. In addition, reliability analysis was carried out for three failure modes of the hull girder of the FPSOs.

2 PROBABILISTIC PREDICTION OF HULL GIRDER LOADINGS

Hull girder loads were predicted by both rule-based formulae and first principle calculation. The extreme wave-induced hull girder loads can be predicted by various theoretical methods. The ship motions and hull girder loads under regular waves were evaluated by a two dimensional strip theory approach where the hydrodynamic force coefficients were obtained from the results of experiments carried out with various two dimensional cylinders by Vugts (1968). The method was benchmarked against results previously computed for a tanker form using 3D methods. It proved that the 2D method has good accuracy. Details of the theoretical formulations are given in Pu, et al (1998).

Sea environment is continuously changing. There are large number of possible wave directions and magnitudes. Therefore the assessment of hull girder strength of the FPSO needs to consider the responses not only in regular waves but also in irregular waves. The long-term prediction was used to estimate the extreme design load on the hull girder since the long term approach considers all wave conditions encountered by the floating structure regardless of their magnitudes and integrates all short term statistical responses on the basis of the long term joint probabilities of wave parameters and the environment directional probabilities.

In the present study the weight factor in the long term prediction is the joint probability of significant wave height and zero crossing period given in a 1000 year scatter diagram for the central North Sea (CNS) and west of Shetland (WOS) provided by Shell UK Exploration and Production.

The above methods were applied to two FPSOs, whose particulars were shown in Table 1. The long term extreme wave-induced vertical bending moment for a probability of exceedence (P.O.E.) of 10^{-8} is shown in Table 2, in which n_4 is the ratio of bending moment predicted by the first principles calculation to that by rules.

The wave-induced loads were also evaluated by formulae in classification society rules. The results are presented in Table 3, in which M_w is the maximum wave-induced bending moment, M_s is stillwater bending moment, M_t is total bending moment ($=M_w+M_s$).

TABLE 1

PRINCIPAL DIMENSIONS OF THE TWO VESSELS

	Aframax Tanker (Form A)	FPSO Barge (Form B)
Length overall	238.0 m	212.0 m
Length BP	228.0 m	200.2 m
Breadth	41.0 m	40.5 m
Depth	21.0 m	22.4 m
Draft (scantling)	15.0 m	15.4 m
Block Coefficient	0.825	0.9
Displacement	118,574 t	118,525 t

TABLE 2

VERTICAL BENDING MOMENT WITH A P.O.E OF 10^{-8}

			Return period (years)	BM (kN-m)	n_4
Barge	CNS	Full Load	19.5	3.886e6	1.43
		Ballast	18.2	3.586e6	1.32
	WOS	Full Load	36.4	4.953e6	1.83
		Ballast	34.1	4.514e6	1.67
Aframax	CNS	Full Load	19.3	3.625e6	1.1
		Ballast	19.1	3.826e6	1.16
	WOS	Full Load	34.8	4.705e6	1.43
		Ballast	34.8	4.838e6	1.47

TABLE 3

RULE BASED PREDICTION OF BENDING MOMENT

		M_w (kN-m)	M_s (kN-m)	M_t (kN-m)
Barge	Hogging	2.710×10^6	0.707×10^6	3.417×10^6
	Sagging	-2.783×10^6	-3.276×10^6	-6.059×10^6
Aframax	Hogging	3.301×10^6	2.327×10^6	5.628×10^6
	Sagging	-3.532×10^6	-1.091×10^6	-4.623×10^6

It can be seen that rules underestimate the wave-induced bending moment by as much as 83%. The difference between the first principle and rule-based calculations is due to the underwater form, the loading condition and location where the vessels are to be operated.

3 DETERMINISTIC ASSESSMENTS

In deterministic strength assessment, primary strength, local strength and ultimate strength of the hull girder were checked. For primary strength the section modulus of amidships section was calculated.

For local strength, three failure modes, namely plate failure, failure of the stiffener - column and torsional buckling of stiffeners, were considered in rule-based local strength calculation. Three classification society rules, those of the American Bureau of Shipping, Det Norske Veritas and Lloyd's Register, were used to compare the differences in ship safety margins. Because the difference of the results obtained from these rules is very small, only the results based on Lloyd's Register rules are presented in this paper.

Although the prediction of the ultimate strength of hull girder is not a compulsory task in a rule-based design process, it is recognised that the ultimate strength predictions can offer designers very useful information about the strength reserve of ships. Therefore progressive collapse analysis of the two ships was carried out by employing a procedure described by Rutherford and Caldwell (1990).

The results are shown in Table 4, in which $n_1 = \text{actual SM} / \text{required SM}$, where SM is section modulus of the cross-section. n_2 is safety margin for plate failure mode, which is defined as:

$$n_2 = \sigma_y / \sigma_i \qquad \text{in tension}$$

$$n_2 = S_{plt} / \sigma_i \qquad \text{in compression}$$

where S_{plt} is strength of plates, σ_i is stress in ith element. n_3 is safety margin for stiffened plates, is defined as:

$$n_3 = \sigma_y / \sigma_i \qquad \text{In tension}$$

$$n_3 = S_{st} / \sigma_i \qquad \text{in compression}$$

where $S_{st} = \min(S_{ca}, S_{ct})$, S_{ca} is critical column buckling stress of stiffeners, S_{ct} is critical torsional buckling stress of stiffeners. Also shown in Table 4 is n_5, which is defined as the ratio of ultimate strength predicted by progressive collapse analysis to total bending moment. From this table it can be seen that Barge FPSO is safe, but Aframax Tanker has local failure.

TABLE 4

STRENGTH RESERVES FOR DIFFERENT FAILURE MODES

	n_1	n_2	n_3	n_5	
Barge FPSO	1.43	1.40	1.67	1.28	rule-based
Aframax Tanker	1.24	1.40	0.93	1.42	load
Barge FPSO	1.05	1.03	1.22	0.94	P.O.E.= 10^{-8}
Aframax Tanker	0.97	1.03	0.73	1.13	load (WOS)

It is interesting to note that Aframax Tanker is safe under rule-based loading, but fails under loads from first principle calculation. For the barge it is marginally safe if one only considers the values of n_1, n_2 and n_3 under load from the first principle. But the total bending moment exceeds the ultimate strength of the ship. So the ship fails too. This shows the importance of carrying out progressive collapse analysis. In the local buckling strength check, the shift of neutral axis cannot be considered. So the safety factors obtained from local buckling analysis are overestimated. It means that whenever safety factors of local buckling are very close to the allowable values, a more rigorous analysis needs to be carried out to determine the strength reserves.

4 PROBABILISTIC ASSESSMENTS

Reliability analysis is a powerful tool to quantify the uncertainties involved in the design process. Due to the unique operational scenario of FPSOs, uncertainties in the design of this type of structures would be different from other conventional ships. So it is important to apply reliability techniques to analyse FPSOs.

For the reliability analysis of hull girders, failure probability of three failure modes, namely ultimate strength of the hull, initial yielding, failure of stiffened panels, need to be assessed. In this study First Order Second Moment Method (FORM) was used. The random variables and their uncertainties are shown in Table 5, where M_w is the wave-induced bending moment, M_s is the stillwater bending moment, σ_y^h is yield stress of high tensile steel, σ_y^l is yield stress of mild steel, E is Young's modulus of the material, SM_e is the effective section modulus, b_p and t_p are the width and thickness of the plate, h_w and t_w are the height and thickness of the web plate, b_f and t_f are width and thickness of the flange of the stiffener, a is the length of the stiffened plate, η is an as-welded parameter which defines the width = ηt of the yield residual tension zone in the plate elements each side of the weld. δ' is the ratio of the initial central deflection of the plate to the thickness of the plate, X_m is the model uncertainty of the strength formula of stiffened plates.

The mean and standard deviation of yield stresses and Young's modulus of steel material were chosen based on the published data (Atua, et al, 1996). Generally speaking there is not much arguments about the uncertainties involved in these random variables. For wave-induced bending moment, a value corresponding to 50% of probability of exceedence from long term analysis is used as the mean value as suggested by Mansour (1997). A 60% of the maximum stillwater bending moment at full load condition is used as the mean value of stillwater bending moment (Guedes Soares and Moan, 1988). A coefficient of variation of 25% is adopted. It should be pointed out that this is strictly speaking not applicable to FPSOs due to its continuous change of loading conditions. More work needs to be done in this aspect.

The limit state equations for ultimate strength, initial yielding and failure of stiffened plates are expressed respectively as:

$$G(x) = M_U - (M_W + M_S) \tag{1}$$

$$G(x) = \sigma_y^h SM_e - (M_W + M_S) \tag{2}$$

$$G(\mathbf{X}) = \sigma_y X_m \phi_m SM_e - (M_s + M_w) \tag{3}$$

where M_U is the ultimate strength of the hull. A 10% of coefficient of variation of M_U is assumed.

The reliability analysis was carried out only for the barge FPSO because the local failure of Aframax. The calculated respective reliability indices for three failure modes were 6.14, 11.8 and 5.78. Obviously the reliability level of the barge is quite high.

5 CONCLUSIONS

It may be concluded that the rule-based wave-induced load predictions yield much smaller results than the first principle calculations. There is not a simple relationship between these two. A first principle prediction technique should be employed in calculating wave-induced loads for ships used in fixed locations. It is recognised that extreme responses are expected to be highly non-linear due to physical limits such as the ship deck height. In other words wave loading should be assessed with non-linear methods and this is an area worthy of further in depth investigation using, for example, time domain simulations.

TABLE 5
RANDOM VARIABLES AND THEIR UNCERTAINTIES

Random variables	Mean	C.O.V. (%)	Distribution type
M_W (N.mm)	2.78282×10^{12}	15	Gumbel – I
M_S (N.mm)	1.31034×10^{12}	25	Normal
σ_y^h (N/mm^2)	433.1	9	Log-normal
σ_y^l (N/mm^2)	286.7	9	Log-normal
E (N/mm^2)	212992	2	Normal
SM_e (mm^3)	3.41612×10^{10}	5	log-Normal
b_p (mm)	907.5	1	Normal
t_p (mm)	19.5	2	Normal
h_w (mm)	292	2	Normal
t_w (mm)	11.5	2	Normal
b_f (mm)	57.5	2	Normal
t_f (mm)	28.0	2	Normal
a(mm)	4800	1	Normal
η	4.5	7	Normal
δ'	0.1	49	Log-Normal
X_m	0.986	12	Normal

Progressive collapse analysis is a very important complementary strength check, especially when safety factors of the section modulus requirement and local buckling strength are very close to the allowable values. The necessity is clearly shown in Table 4.

The reliability of the barge is quite high. Due to the special operational procedures of FPSOs, it is believed that the uncertainties of stillwater bending moments and wave-induced bending moments of FPSOs must be different from those of other types of ships. So in order to evaluate a more realistic reliability level of FPSOs, further research is required to estimate these uncertainties.

Acknowledgement

The authors want to thank Shell UK Exploration and Production Ltd and Dr N.W. Snedden for the financial support to the project on which this paper is further developed. We are grateful to Armstrong Technology Ltd, specifically Mr G Mackie and Mr D. Dunce, for providing data and technical support during the execution of the project. Thanks are also due to Dr. H.S. Chan for his assistance in long term prediction.

References

Atua K., Assakkaf I. and Ayyub M. (1996). Statistical Characteristics of Strength and Load Random Variables of Ship Structures. *Proceeding of the 7th Specialty Conference on Probabilistic Mechanics and Structural Reliability*, ed. by D.M. Frangopol and M.D. Grigoriu, Worcester, Massachusetts, Aug. 7-9, 106-109.

Guedes Soares C. and Moan, T. (1988). Statistical Analysis of Still-water Load Effects in Ship Structures. *SNAME Transactions*, **96**, 129-156.

Lloyd's Register of Shipping. (1997). User's Manual LR. PASS Personal Computer Programs. **2**, 'Direct Calculation'.

Mansour A., Wirsching P., Luckett M. and Plumpton A. (1997). Assessment of Reliability of Existing Ship Structures. Ship Structures Committee, Report Number SSC-398.

Pu Y., Aryawan I., Chan H.S., Dunce D., Mackie G. and Incecik A. (1998). Development of Generalised Design Methodology for Hull Girder Strength Assessment of Mono-hull FPSOs/FSUs. Final Report for SHELL UK Exploration and Production.

Rutherford S. and Caldwell J.B. (1990). Ultimate Longitudinal Strength of Ships: a Case Study. *SNAME, Annual Meeting*, **14**, 1-26.

Smith C. S. (1977). Influence of Local Compressive Failure on Ultimate Longitudinal Strength of a Ship's Hull. *PRADS – International Symposium on Practical Design in Shipbuilding*, Tokyo, 73 – 79.

Vugts J.H. (1968). The Hydrodynamic Coefficients for Swaying, Heaving and Rolling Cylinders on a Free Surface. Delft University of Technology, Report No. 194.

Practical Design of Ships and Other Floating Structures
You-Sheng Wu, Wei-Cheng Cui and Guo-Jun Zhou (Eds)

CONSISTENT CODE FORMULATION FOR SHIP STRUCTURAL DESIGN

A.E. Mansour[1], J.S. Spencer[2], P. H. Wirsching[3], J.E. McGovney[1], D. D. Tarman[1]

[1] University of California at Berkeley, USA
[2] Vice President, American Bureau of Shipping, Houston, TX, USA
[3] Professor, University of Arizona, Tucson, AZ, USA

ABSTRACT

An approach has been developed for formulating a consistent code for design of ship structures. The developed approach is based on partial safety factor format and a prescribed reliability level determined from calibration with existing ABS SafeHull Rules. The main advantage of the new formulation is that it is consistent and provides uniform reliability independent of ship size or magnitude of the design variables. Such consistency and uniformity cannot be attained from the usual safety factor format currently used in IACS and Classification Societies' ship design Rules. As a result of the inconsistencies in the current design Rules, and the deviation of the new formulation from them, it is feasible to reduce scantlings and weight in a certain range of design parameters governing the design of a structural member. Another avenue for weight reduction or adjustment is via comparing the magnitudes of the safety indices calculated for similar structural members subjected to the same loading environment.

As an example, the paper presents a consistent formulation for determining the minimum required section modulus of tankers. Two formulations are considered, one based on yield strength of the material, which is the criterion used in current design Rules; and the other is based on buckling and ultimate strength considerations. The results of either formulation show that, using this consistent procedure, it is feasible to reduce the required section modulus of some tankers whose design wave to stillwater bending moment ratio is low, without sacrificing their relative safety level.

KEYWORDS

Reliability based code formulation, Ship structural design

1 INTRODUCTION

Probability based design has the promise of producing well designed systems, i.e., structures having reduced weight and/or improved reliability relative to design based on conventional practices. The basic advantage of probability-based design is the inclusion of more information into the design process that is not generally included in conventional design procedures. Not only are basic geometric characteristics of the structure considered, but also information about the uncertainties associated with the strength of various elements and the loading are also considered. Probability based design allows the engineer, through good judgment, to design on the basis of a target reliability for the structure thus designing a more efficient structure.

Partial safety factors (PSF) for both strength (resistance) and load account for the variability associated with these quantities. Generally, the higher the uncertainty, the greater the partial safety factor will be for loads. For strength variables, greater uncertainty yields smaller partial safety factors. A base probability of failure, or reliability index, must be established as a prerequisite to PSF selection.

While probability based design has the potential to benefit engineers in every field, this paper specifically focuses on the application of probability based design to establish a consistent code for tankers. The goal is to provide a more complete design guideline for shipbuilding not only to save weight in design, but also to provide a baseline reliability that every ship will achieve regardless of size or operation.

2 APPROACH

The approach developed consists of a sequence of steps that begin with adopting a general partial safety factor format. The general format for the code is:

$$\phi R \geq \sum_{i=1}^{n} \gamma_i k_i f_i$$

where f_i = nominal value of load component i
k_i = load combination factor for load component i
R = nominal resistance or strength
γ_i = partial safety factor for load uncertainty i
φ = partial safety factor for resistance uncertainty

This general format is applicable to most code provisions, although in some cases the adopted format deviated slightly from above. Next, a limit state function, g(x), is developed form each code provision. The limit state function, g(x), is established such that: $g(\underline{x}) < 0$, represents failure state; $g(\underline{x}) > 0$, represents safe state; $g(\underline{x}) = 0$, represents the limit state equation (boundary between failure and safe states). And $\underline{x}$ = main design variables represented as random variables

The nominal values (or characteristic values) are then defined in terms of average (or mean) values and the standard deviations of the design variables.

A target reliability is then assigned. The target reliability is determined using a calibration procedure with current Rules at at least one set of the design variables or a non-dimensional value of their combination. For example, in the case of the hull girder section modulus formulation, the calibration is made at a specific value of the ratio $r = M_w/M_s$ where M_w is the design wave bending moment and M_s is the design stillwater moment (i.e. the nominal values). At this specific ratio, the section modulus determined from the new code formulation is equated to the section modulus determined from the existing code formulation (IACS). For the selected ratio r, the implied safety index and the corresponding partial safety factors are determined using first order reliability method and a simulation technique developed by Wu (1998) for automating the calculation of partial safety factors. The computed safety index taken as the target reliability is kept constant in the remaining computation of the partial safety factors in order to develop a consistent formulation.

By varying r and maintaining constant target reliability, a set of partial safety factors is determined for each new r. These partial safety factors can be plotted versus the ratio of the nominal wave to stillwater bending moment, r. The ratio R of the minimum section modulus according to the new formulation and the minimum section modulus per the current IACS formulation can be calculated for the range of practical interest of r (1.2 to 2.1 for tankers). In the range of R<1, weight savings can be accomplished without falling below the minimum target reliability adopted in the calibration procedure.

3 HULL SECTION MODULUS FOR TANKERS

3.1 yield Strength Formulation

Formulation of a consistent code based upon reliability requires a limit state function to be established. In accordance with the approach outlined above, the limit state equation for yield strength Rule formulation is:

$$g(\underline{x}) = f_y SM - M_T$$

where f_y = the yield strength of the material considered as a random variable
$M_T = M_s + M_w$ = the total bending moment (equals the stillwater bending moment plus the wave bending moment, both considered as random variables)
SM = section modulus of the ship considered deterministic

IACS gives the following yield strength formula for global total strength assessment (see ABS SafeHull Rules ref. 1):

$$SM = \frac{M_T}{f_b} = \frac{M_s + M_W}{f_b} \tag{1}$$

with variables defined by their nominal values
f_b = permissible bending stress

The reliability based formulation for yield strength based on the above g-function gives:

$$SM \geq \frac{\gamma_1 M_S + \gamma_2 M_w}{\varphi f_y} \tag{2}$$

where γ_1, γ_2, φ = partial safety factors applied to stillwater bending moment, wave bending moment, and yield strength respectively

A target reliability is then assigned. The target reliability is based on equating equations 1 and 2 and rearranging the terms to include α which is the ratio of the mean wave to stillwater bending moments. It is necessary to use the means of these values because the partial safety analysis program gives PSF based upon means. Using means also allows the random variables (stillwater bending moment, wave bending moment, and yield strength) to be represented by normalized probability distributions with means equal to unity. The section moduli resulting from equations (1) and (2) can be made equal at either a midpoint value or at an extreme value of the ratio α. The range of α is determined from several tankers currently in operation (see ABS data chart, Appendix A). Equation 3 below shows that for a certain value of α, specific partial safety factors can be found.

$$\frac{f_y}{f_b} = \frac{\gamma_1 + 0.6\alpha\gamma_2}{1 + 0.6\alpha} \tag{3}$$

with variables as defined above and,

$$\alpha = \frac{\mu_{M_w}}{\mu_{M_s}}$$

In order to relate the nominal values to the mean values, the biases applied to M_w, M_s, f_y are 1.0, 0.6, and 1.2 respectively.

Equating the section moduli given by Equ.1 and Equ.2 at the value chosen of α, a target reliability index is established. This target safety index will be applied for the remaining values of α in order to insure a consistent standard of reliability. Based on the established reliability index and the nominal bending moment ratio, r (r is based on the biases applied to α), a value representing the left hand side of Eqn.3 is found and compared with the left hand side number by IACS Rules. It must be noted that the biases given above should be applied to the partial safety factors given by the analysis program to allow the comparison to be made based on nominal values. R, which is the ratio of section modulus resulting from the reliability formulation (Eqn.2) to that of IACS (Eqn.3), denotes this comparison in the analysis.

3.2 Ultimate Strength Formulation

The reliability based code formulation for a ship's ultimate strength is similar to the formulation for yield strength. Ultimate strength formulation is based upon a g-function given by:

$$g(\underline{x}) = cf_y SM - [M_s + k_w(M_w + k_d M_d)]$$

where c = buckling knockdown factor; default values are 0.85 for mild steel, and 0.65 for high strength steel

f_y = yield strength of the material

k_w= load combination factor between stillwater bending moment and the combined wave and dynamic moments; default value is 1.0.

k_d= load combination factor between the wave and dynamic bending moments; default value is 0.7

M_s, M_w, M_d = stillwater bending moment, wave bending moment, and dynamic bending moment, respectively

SM = section modulus

IACS provides no formulation for the section modulus based on ultimate strength. However, for comparison purposes, Eqn.1 may be used.

Probability based partial safety factor design characterizes the ultimate strength requirement, based on the limit state equation above, as:

$$SM \geq \frac{\gamma_1 M_s + \gamma_2 M_w + k_d \gamma_3 M_d}{c\varphi f_y} \tag{4}$$

with variables described above and

γ_1, γ_2, γ_3, φ = partial safety factors applied to stillwater bending moment, wave bending moment, dynamic bending moment, and yield strength

The default value of k_w, which is 1.0, has been used.

Equating Eqn.1 and Eqn.4 so as to calibrate the reliability analysis on the basis of IACS standards:

$$\frac{cf_y}{f_b} = \frac{\gamma_1 + (0.6\gamma_2 + 0.063\gamma_3)\alpha}{\varphi(1 + 0.6\alpha)} \tag{5}$$

where α equals the ratio of mean wave to sillwater bending moments and all other variables are as described above. The dynamic bending moment is related to the wave bending moment through the correlation factor given above.

A similar procedure is used to develop the relation between r and R as described earlier for the case of yield strength formulation.

3.3 Yield Strength Results and Formulation

Because the formulation is based on random variables which include stillwater bending moment, wave bending moment, and material yield strength, probability distributions must be established for each quantity. This input, including mean, standard deviation, and distribution type are used in the PSF analysis program. All variables are normalized to give means equal to unity. The yield strength variable, f_y, is characterized by a lognormal distribution with a coefficient of variation (COV) of 0.08. The stillwater bending moment, M_s, is represented by a normal distribution with a COV of 0.25 while the wave bending moment, M_w, is characterized by an extreme value distribution with COV of 0.1.

The yield strength formulation is calibrated at two points. The first point corresponds to an "α" value of 2.78 equating to an "r" value of 1.67. This value is chosen because it represents the median nominal wave to stillwater bending ration of the 25 case study ships used. The second calibration point is at "r" of 2.00 or "α" at 3.33. From this, the fixed safety indices associated with each calibration point are 4.5 and 4.35 for calibration at 1.67 and 2.00 respectively.

TABLE 1

RESULTS FOR CALIBRATION AT R = 1.67

Calibration 1.67			**β=4.5**	**Pf=2.21E-5**		**Yield Strength**	
α	1.67	2.00	2.33	2.67	3.00	3.33	3.67
$r=M_w/M_s\vert_{nom}$	**1.00**	**1.20**	**1.40**	**1.60**	**1.80**	**2.00**	**2.20**
Φ	0.993	1.001	1.006	1.011	1.015	1.017	1.019
γ_1	0.870	0.826	0.794	0.770	0.754	0.736	0.725
γ_2	1.580	1.638	1.675	1.702	1.720	1.734	1.744
A	0.202	0.179	0.160	0.135	0.135	0.135	0.135
β	4.5	4.5	4.5	4.5	4.5	4.5	4.5
Ratio, R	**0.920**	**0.946**	**0.970**	**0.992**	**1.011**	**1.028**	**1.044**
Φ mean	0.8279	0.8339	0.8386	0.8426	0.8456	0.8475	0.8495
$\gamma_{1\ mean}$	1.45	1.376	1.324	1.284	1.256	1.227	1.209
$\gamma_{2\ mean}$	1.58	1.638	1.675	1.702	1.72	1.734	1.744

In both Table 1 and Table 2, the variables are defined as follows:

β = safety index

γ_1 = still water partial safety factor

γ_2 = wave bending moment partial safety factor

Φ = yield strength partial safety factor

R, Ratio = LRFD SM/ Safe-Hull SM, i.e. the ratio of the section modulus based on Load Resistance Factor Design (which is the same as partial safety factor design) to the section modulus based on ABS SafeHull design which is the same as IACS.

The partial factors of safety are plotted in Figures 1 and 2. They are plotted vs. "r" (defined as the ratio of nominal wave bending moment to still bending moment). The partial safety factor applied to the yield strength is slightly less than one as expected. The partial safety factor applied to wave bending moment is above one, which is also expected. However, the partial safety factor applied to the maximum nominal still water bending moment is less than one. The safety index associated with calibration 1.67 is 4.5. When the moment ratio is raised to 2.0, the safety index is reduced to 4.35.

TABLE 2

RESULTS FOR CALIBRATION AT R = 2.0

2.0 Calibration			β=4.35	Pf=3.35E-5		Yield Strength	
α	1.67	2.00	2.33	2.67	3.00	3.33	3.67
$r = M_w/M_{s|nom}$	**1.00**	**1.20**	**1.40**	**1.60**	**1.80**	**2.00**	**2.20**
Φ	0.997	1.005	1.011	1.015	1.018	1.021	1.023
γ_1	0.866	0.823	0.793	0.769	0.749	0.736	0.723
γ_2	1.537	1.595	1.630	1.656	1.675	1.687	1.689
A	0.207	0.184	0.164	0.148	0.135	0.124	0.115
β	4.35	4.35	4.35	4.35	4.35	4.35	4.35
Ratio, R	**0.900**	**0.924**	**0.946**	**0.967**	**0.985**	**1.00**	**1.01**
Φmean	0.8307	0.8375	0.8422	0.8458	0.8487	0.8509	0.8526
$\gamma_1$mean	1.443	1.371	1.322	1.281	1.249	1.226	1.205
$\gamma_2$mean	1.537	1.595	1.63	1.656	1.675	1.687	1.689

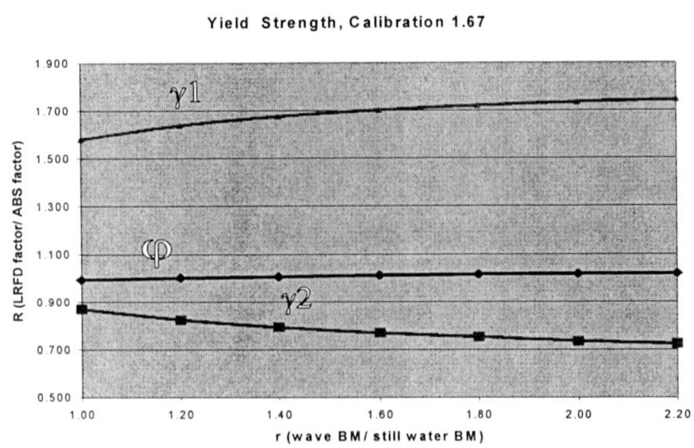

Figure 1: Partial Safety Factors for Calibration at 1.67

Figure 3 plots "R" vs. "r" (the section modules ratio vs. the moment ratio). Also included on this plot are the design moment ratios for 13 ships in both hogging and sagging conditions. The Table from which the nominal wave to stillwater bending moment ratios were calculated is included in Appendix A. This plot describes the differences between probability based design and conventional design codes. Based on the calibration point, where R equals 1.0, the probability based design is equivalent to the IACS Rules. All other points on the calibration curves represent ratios of reliability based section modulus to IACS defined section modulus as calculated using constant values of the safety index β. The "r" represents different design conditions of wave to still water bending moments. It follows from the formulation of the g-function that the curve will yield an acceptable design based on the calibrated probability of failure. The section modulus of the reliability-based design can be inferred from R. In some cases, this section modulus is less than that suggested by IACS (R less than 1.0). In other cases, the section modulus is greater than the IACS section modulus (R greater than 1.0). For any given ship moment ratio value, a comparison can be made between the two design methods.

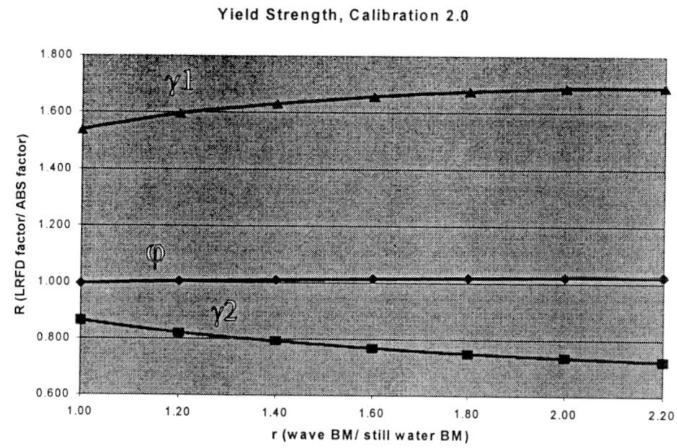

Figure 2: Partial Safety Factors for Calibration at 2.0

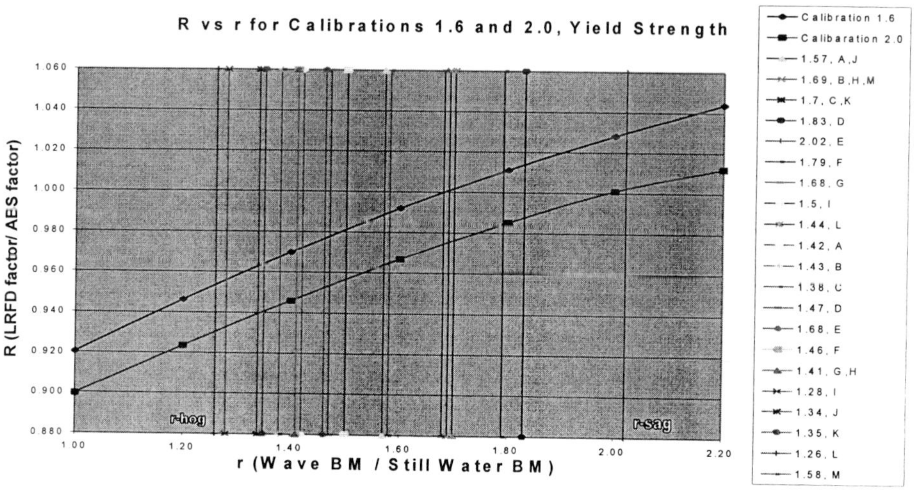

Figure 3: R vs r for Yield Strength Calibrations 1.67 and 2.0

When the probability based design code gives a section modulus less than required by IACS rules, weight reduction can be achieved while maintaining the design reliability level. For calibration at r equals 1.67, this means about half of the ship case studies can save weight while the other half would technically have to add weight to achieve the desired safety level. However, if the assertion is made that the ship with the highest nominal wave to stillwater bending moment ratio, r, is operating safely, then all case studies except that ship have the potential to save weight. If the ship at r equals 2.0 as a new calibration point is operating safely, and the probability of failure of that ship (which is higher than any other ship in the case study) is acceptable, then all other ships could have a reduced section modulus and operate at the same acceptable probability of failure (see lower line of Figure 3).

4 ULTIMATE STRENGTH RESULTS AND DISCUSSION

The distributions used to characterize yield strength, stillwater bending moment, and wave bending moment are identical to those give in yield strength results. Additionally, the dynamic bending moment is represented by an extreme value distribution with a COV equal to 0.3.

The ultimate strength formulation is first calibrated at a α value equal 2.67 or an r-value of 1.60. This value is chosen because it represents the median nominal wave to stillwater bending ration of the 13 case study ships used. The second calibration point is at an r value of 2.00 or α at 3.33. The fixed safety indices associated with each calibration point are 3.2 and 3.02 for calibration at 1.60 and 2.00 respectively.

TABLE 3

RESULTS FOR CALIBRATION AT R = 1.6

Calibration 1.6			**β=3.2**	**Pf=6.87E-4**		**Ultimate Strength**	
A	1.67	2.00	2.33	2.67	3.00	3.33	3.67
$r=M_w/M_{s|nom}$	**1.00**	**1.20**	**1.40**	**1.60**	**1.80**	**2.00**	**2.20**
Φ	1.029	1.033	1.038	1.041	1.044	1.048	1.049
γ_1	0.838	0.806	0.777	0.755	0.740	0.728	0.715
γ_2	1.251	1.290	1.321	1.340	1.354	1.368	1.375
γ_3	1.048	1.051	1.055	1.055	1.054	1.057	1.057
A	5.036	5.660	6.297	6.962	7.613	8.269	8.947
β	3.2	3.2	3.2	3.2	3.2	3.2	3.2
Ratio, R	**0.94**	**0.96**	**0.98**	**1.00**	**1.02**	**1.03**	**1.04**
Φmean	0.8575	0.861	0.865	0.8673	0.8699	0.8733	0.8738
$\gamma_1$mean	1.397	1.343	1.295	1.259	1.234	1.213	1.192
$\gamma_2$mean	1.251	1.29	1.321	1.34	1.354	1.368	1.375
$\gamma_3$mean	1.048	1.051	1.055	1.055	1.054	1.057	1.057

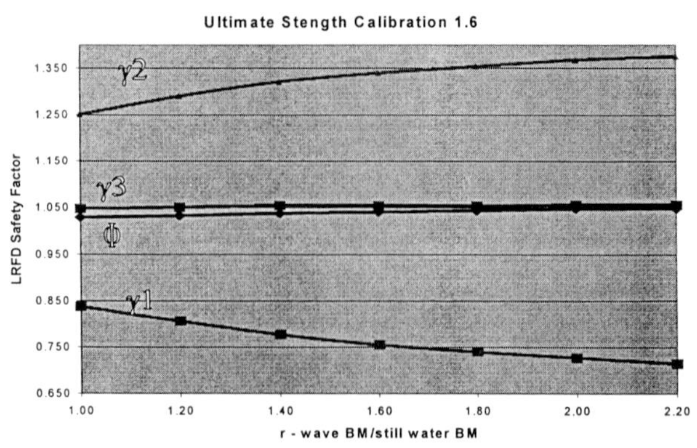

Figure 4: Partial Safety Factors for Calibration at 1.6

TABLE 4
RESULTS FOR CALIBRATION AT R = 2.0

Calibration 2.0			β=3.02	Pf=1.26E-3		Ultimate Strength	
α	1.67	2.00	2.33	2.67	3.00	3.33	3.67
$r=M_w/M_{s\|nom}$	**1.00**	**1.20**	**1.40**	**1.60**	**1.80**	**2.00**	**2.20**
Φ	1.037	1.041	1.043	1.047	1.049	1.053	1.053
γ_1	0.832	0.800	0.776	0.757	0.741	0.726	0.715
γ_2	1.221	1.258	1.281	1.301	1.314	1.328	1.335
γ_3	1.049	1.047	1.055	1.050	1.049	1.055	1.051
A	4.915	5.520	6.137	6.780	7.412	8.047	8.947
β	3.02	3.02	3.02	3.02	3.02	3.02	3.02
Ratio, R	**0.92**	**0.94**	**0.96**	**0.97**	**0.99**	**1.00**	**1.01**
Φmean	0.864	0.867	0.869	0.873	0.874	0.877	0.878
$\gamma_1$mean	1.387	1.333	1.293	1.262	1.235	1.210	1.191
$\gamma_2$mean	1.221	1.258	1.281	1.301	1.314	1.328	1.335
$\gamma_3$mean	1.049	1.047	1.055	1.050	1.049	1.055	1.051

As can be seen in Tables 3 and 4, the results for ultimate strength conditions are similar to those for the yield strength in Tables 1 and 2. Figures 4 and 5 are plots of the partial safety factors. The trends of these factors are identical to those in the yield case, with the addition of γ_3 reflecting the presence of dynamic bending moments. The dynamic bending moment safety factor is above 1.0 as expected.

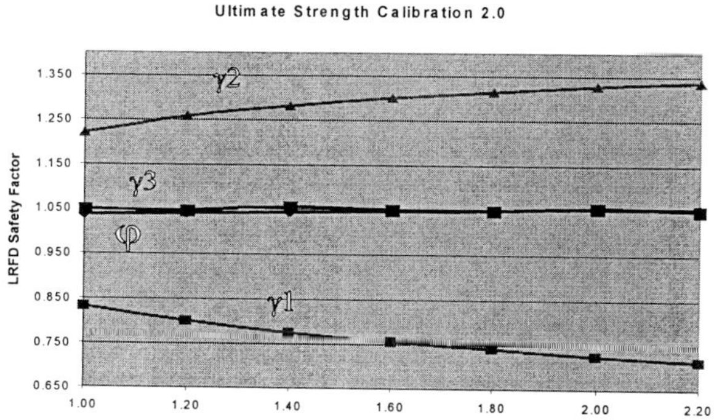

Figure 5: Partial Safety Factors for Calibration at 2.0

For the ultimate limit state, the safety indices are 3.2 and 3.02 for calibrations 1.6 and 2.0 respectively, i.e. smaller than for the yield case. Again, this is as expected because of the inclusion of the dynamic moment; and ships, which have hulls that are comprised of thin plates and long members, are more likely to fail in buckling than yield.

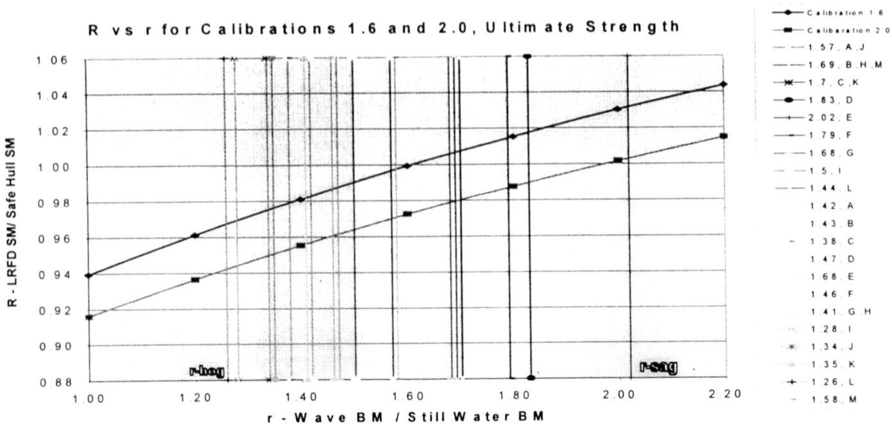

Figure 6: R vs r for Ultimate Strength Calibrations at 1.6 and 2.0

Figure 6, like Figure 3, is based on the intersection of the calibration curves with R=1.0. This again gives a baseline probability of failure value associated with a specific moment ratio, r. If a ship's r-value leads to a R-value less than one, the ship can again save weight while maintaining the prescribed safety index established during calibration. These observations are identical to those made in yield strength formulation.

5 CONCLUSIONS

An approach has been outlined for formulating a consistent code for design of ship structures based on partial safety factors and a target reliability level which is based on a calibration with existing ABS SafeHull Rules, and IACS formulation.

Two major advantages of probability-based design over conventional methods have been identified. The first advantage is that partial safety factor design provides a consistent reliability standard independent of ship size or magnitude of the design variables. The second advantage is the feasibility of weight reduction. As has been shown in this paper, weight savings can be achieved without sacrificing the relative safety by reducing the required section modulus of ships with low ratios of design wave to stillwater bending moments.

References

1.American Bureau of Shipping. (2000). *Rules for Building and Classing Steel Vessels, Part 5, Specific Vessel Types*. Houston, TX.

2.Mansour A., Wirshing P., White G. and Ayyub B, (1996). *Probability Based Ship Design: Implementation of Design Guidelines for Ship: A Demonstration.* SSC-392.

3.McGovney, Jodi. (2000). *Development of Reliability Based Classification Rules for Tankers.* University of California at Berkeley Thesis.

4.Wu, Justin Y-T. (Southwest Research Institute) and Xiaoping Du (The University of Iowa). (1998). Partial Safety Factor Analysis Program, version abs3b.

Practical Design of Ships and Other Floating Structures
You-Sheng Wu, Wei-Cheng Cui and Guo-Jun Zhou (Eds)

RELIABILITY OF STIFFENED SHIP DECKS

K.Rajagopalan [1]

[1] Ocean Engineering Centre, Indian Institute of Technology, Madras
CHENNAI 600 036, India

ABSTRACT

Stiffened plates are important components in ship structures. The reliability of stiffened ship decks must be known for characterizing the reliability of the entire ship hull. The collapse of the stiffened ship deck can be in several modes,the predominant being the inelastic buckling of the longitudinal stiffener alongwith an effective width of the attached plating. Classical solutions are available for the ultimate strength. More accurate solutions can be effected using the finite element nonlinear analysis methods. In this paper, the reliability of the stiffened ship deck is computed using the classical methods. Parametric studies have been carried out to bring out the sensitivity of the various parameters involved in the reliability picture.Randomness in yield stress seems very dominant. In addition it has been shown that the classical method used has an important bearing on the reliability.

KEYWORDS

Ship structures, Risk analysis, Stiffened plates, Buckling, Effective width, Stiffener tripping, Reliability, Safety index

1 INTRODUCTION

The stiffened deck is one of the most fundamental components of virtually all ship structures. The ultimate strength and reliability of the hull girder depends on the ultimate strength and reliability of the stiffened ship deck. Failure of cross-stiffened ship decks under in-plane compressive loads is primarily by inelastic buckling of the stiffened panel between transverse frames. However, when uncertainties are considered other buckling modes are also possible with different probability values. The ultimate strength evaluation for the stiffened deck itself is a complex task. Several analytical solutions have been proposed. Accurate non-linear finite element methods require enormous modeling effort and computing time.

The reliability of stiffened ship decks is evaluated in this paper using the point estimation technique (Rajagopalan 1993). The ultimate strength has been evaluated by three different methods namely, due to Shapir and White(1983), Hughes(1983) and Mikami and Niwa(1996). The loading considered is the in-plane compression generated by the overall sagging of the ship girder as a beam.

1.1 Methodology for Ultimate Strength

During the last decade many analytical or semi-empirical methodologies or formulae for the computation of ultimate strength of stiffened plates subjected to in-plane compressive loads have been proposed. Local buckling of the deck plate or the plate elements forming the stiffener does not necessarily represent the ultimate strength , since post-buckling strength is possible. Thus either the overall instability or panel instability with locally buckled plate elements is likely to dictate the ultimate strength. The stability analysis of a stiffened plate can be done by examining the following four failure modes shown in Figure 1:

(a) buckling of the longitudinal stiffener with its adjacent plate as a column

(b) tripping(lateral-torsional buckling) of the longitudinal stiffener

(c) local buckling of the plate between the stiffeners

(d) failure of the whole stiffened deck as an equivalent orthotropic plate

The last one is normally avoided by ensuring that transverse frames are of adequate size. The case has however to be analysed to insure that the overall buckling is indeed avoided.

The first mode is the most usual one and is due to excessive slenderness of the "column" (the longitudinal stiffener and associated effective plating after local plate buckling if any). Failure may be towards the plate or towards the stiffener depending on the column's initial shape and the type of loading considered. In continuous panels it is usual that failure is towards the plate in one span and towards the stiffener in adjacent span.

The second and third modes are local modes due to lack of torsional rigidity of the stiffener and due to stocky stiffeners respectively.

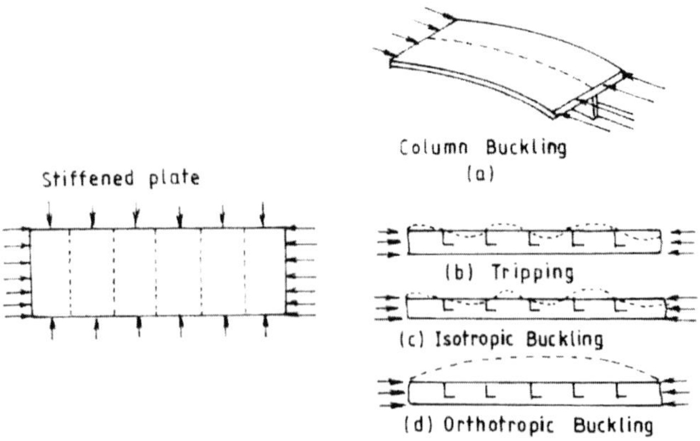

Figure 1: Failure modes of stiffened plates

2 METHODS OF RELIABILITY EVALUATION

Two well established methods for the evaluation of reliability are the FOSM and the AFOSM which however require the evaluation of the derivatives of the failure function.
In the FOSM method, the mean μ_z and the standard deviation σ_z of the failure function Z where

$$Z = R - S \tag{1}$$

are evaluated at the point represented by the mean value of the variables.The reliability index is then

$$\beta = \frac{\mu_Z}{\sigma_Z} \tag{2}$$

and the probability of failure is

$$p_f = \Phi(-\beta) \tag{3}$$

The most probable failure point is thus assumed to coincide with the mean values of the variables in the FOSM method. In the AFOSM method, the failure point is located by an iterative process.

3 THE POINT ESTIMATION TECHNIQUE

The PEM works on the analogy between the probability density function of a random variable and the distributed loading function on a beam with two supports as shown in Figure 2.

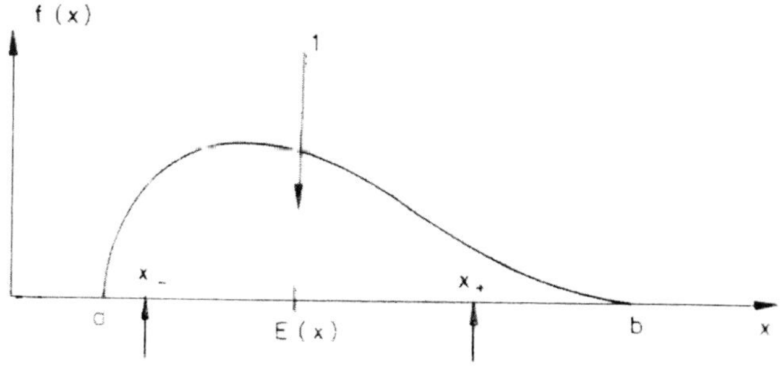

Figure 2: Basis of the Point Estimation Method

The mean of a random variable is given by the expression

$$E(X) = \int_a^b x f(x) dx \tag{4}$$

The total load is equal to unity since the area under any probability density function is unity. The reactions at the two support points x_- and x_+ are P_- and P_+ respectively where

$$x_- = E(X) - \sigma_x \tag{5}$$
$$x_+ = E(X) + \sigma_x \tag{6}$$

These reactions are called point estimates. It is clear from the beam analogy that

$$P_- + P_+ = 1 \tag{7}$$

$$P_- x_- + P_+ x_+ = E(X) \tag{8}$$

$$P_- x_-^2 + P_+ x_+^2 = E(X^2) \tag{9}$$

and so on.The concept of the PEM can be extended to a problem with many random variables.

4 RELIABILITY OF STIFFENED SHIP DECKS

The failure equation of a stiffened ship deck subjected to in-plane compressive load can be written as

$$Z = \sigma_u - \sigma_x \tag{10}$$

where the ultimate strength σ_u is evaluated in this paper using three methods.The stiffened deck shown in Figure 3 typical of a tanker is considered. The stiffener properties are shown in Figure 4.

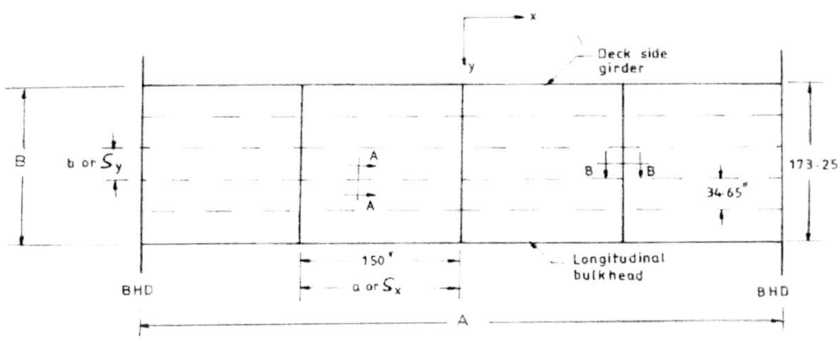

Figure 3: Details of the stiffened deck of a tanker used in the reliability study

The mean of the yield stress is 45000 psi (315 MPa) and the mean of the applied in-plane stress is 20000 psi (140 MPa). The reliability indices obtained from a computer program written for carrying out the reliability analysis are shown in Table 1.The computer program determines the mean ultimate strength $\mu_{\sigma u}$ and the standard deviation of the ultimate strength $\sigma_{\sigma u}$ using the point estimation technique[Rajagopalan(1993)].

The relevant ultimate strength equations described by Shapir and White(1983), Hughes(1983) and Mikami and Niwa(1996) are programmed and integrated with the point estimation technique. The input involves the stiffened ship deck geometry, the mean and the coefficient of variation(COV) in the yield stress and the mean and the coefficient of variation(COV) in the applied in-plane stress.The mean and the standard deviation of Z are then computed by the program using the relevant ultimate strength equations of the stiffened plate in the following manner:

$$\mu_z = \mu_{\sigma u} - \mu_{\sigma x} \tag{11}$$

$$\sigma_Z = \sqrt{\sigma_{\sigma u}^2 + \sigma_{\mu x}^2} \tag{12}$$

The reliability index is then calculated using equation(2).

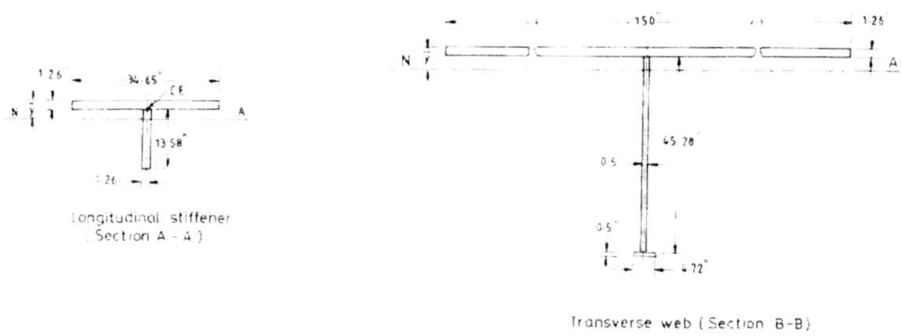

Figure 4 Details of Longitudinal and Transverse Stiffeners

TABLE 1
RELIABILITY INDICES WITH DIFFERENT STRENGTH METHODS

COV IN Yield Stress σ_y	COV IN Applied in-plane Stress σ_x	β (Shapir&White)	β (Hughes)	β (Mikami&Niwa)
5%	10%	8.269	6.838	8.271
10%	10%	5.096	4.447	5.550
15%	10%	3.573	3.175	4.252

5 CONCLUSIONS

Based on the research reported in this paper, the following conclusions are advanced:

The reliability analysis of stiffened ship decks can be carried out by the point estimation technique outlined in this paper.

As the ultimate strength equation of Hughes is conservative, the reliabilities obtained using this method are comparatively less.

Acknowledgements

Thanks are due to the Ministry of Surface Transport,Government Of India, for sponsoring an R&D Scheme on Reliability of Stiffened Ship decks to the Ocean Engineering Centre,Indian Institute of Technology,Madras for which the author is the co-ordinator.However the opinion expressed in this paper is of the author alone and not necessarily of these organisations.

References

Hughes O.F. (1983). *Ship Structural Design*, John Wiley, New York, USA

Mikami I and Niwa K. (1996). Ultimate Compressive Strength of Orthogonally Stiffened Steel Plates, *Journal of Structural Engineering.* **122:6**, 674-682

Rajagopalan K. (1993). Point Estimation Technique for the Reliability Analysis of Ship Structures, *Computers and Structures.* **48:5**, 835-841

Shapir Y and White G.J. (1983). An Analysis of the Ultimate Strength of Deck Structures under In-plane Loads, *Marine Technology.* **20:3**, 230-251

Practical Design of Ships and Other Floating Structures
You-Sheng Wu, Wei-Cheng Cui and Guo-Jun Zhou (Eds)

TOTAL ANALYSIS SYSTEM FOR SHIP STRUCTURAL STRENGTH

T. Yoneya[1], H. Kobayashi[1], M. Abdul Rahim[2], Y. Sasaki[1] and M. Irisawa[1]

[1] Technical Investigation and Information Department, Nippon Kaiji Kyokai
Chiba, Japan
[2] Singapore Office, Nippon Kaiji Kyokai, Singapore

ABSTRACT

This paper outlines a total analysis system for ship hull structures, which integrates a wide variety of analysis functions to realise practical applications of rational methods to the assessment of ship structural strength. It is based on direct calculations of wave-induced load as well as three-dimensional structural analysis of an entire-ship or hold structure. Three major analysis functions of the total system are ship motion and wave load analysis, ship structural analysis and statistical analysis. In addition to these major functions it is essential for realisation of a wave-induced response analysis of an entire-ship structure to develop a powerful load translator, which can efficiently link a hydrodynamic analysis model to a structural analysis model. The load translator is multifunction to be capable of dynamic load generation of liquid, granular and container cargoes and ballast water as well as wave pressure conversion in both frequency and time domains. There are pointed out some practical and essential features to be resolved when performing such total analysis using three-dimensional ship structural analysis. Typical examples of its practical application to double hull tankers and large container carriers are shown.

KEYWORDS

Total system, Wave load, Ship motion, Ship structure, Statistical prediction, Load transfer, Entire-ship structural analysis, Partial structural analysis, Hotspot stress analysis, Buckling strength, Fatigue strength, FEA

1 INTRODUCTION

Nippon Kaiji Kyokai, or ClassNK, has been establishing a comprehensive assessment system for ship structures aiming at realisation of rational and consistent assessment of ship structures over its lifetime. This includes the development of new ship structural rules and guidances based on advanced research and implementation of sophisticated and leading edge technologies in hull structural analysis as well as the development of computer program systems using the latest information technologies. These various advanced technologies have been brought together as an integrated group of systems and services under the umbrella of PrimeShip, which was first announced in 1995 (Hidaka, M, 1996). PrimeShip-ASSAS (Advanced Ship Structural Analysis and Validation System), which has been yielded in the process of establishing the comprehensive assessment system, is the most rational and practical analysis system for ship hull structures among PrimeShip systems.

Considering the sea, loading and navigation conditions, ASSAS, which is simply called "the system" hereinafter, performs ship motion analyses, structural analyses of the entire-ship structure, hold or part structure, and zooming analyses of stress concentration parts. This is followed by precise and complete evaluation of yielding, buckling and fatigue strengths of structural members. The strength evaluation is primarily based on statistical analysis using wave spectrum and long term wave statistics, though design wave approach can also be applied. Development of the system started in 1995 and completed in 1997, but it has been continuously upgraded with enhancement and improvement of its functions based on the latest technologies.

The system is widely used within the society. It is fully applicable to the design assessment of a ship with the class notation "DATA"(Designed by Application of Total Analysis concept) under the PrimeShip umbrella, as well as to the damage investigation, development of new structural rules and so on. In the process of establishing the comprehensive assessment system for ship structures, ClassNK has also developed and published a technical guide regarding the strength evaluation of hull structures (Nippon Kaiji Kyokai, 1999), which compiles technical backgrounds on new structural rules or guidelines for a consistent and rational design and assessment of hull structures as a fruit of the long time research project. The system has also been used as an analysis and evaluation tool for the development of new structural guidelines, which require considerable amount of calibration analyses on existing ships. Newly developed assessment methods and criteria will be implemented into the system in turn.

2 ASSESSMENT SYSTEM

2.1 Analysis Functions

The overall analysis flow of the system is shown in Figure 1. The system has three major functions namely load calculation, structural analysis and statistical analysis. The calculations of ship motion and dynamic load are carried out based on the latest strip theory. Structural analysis is performed for an entire ship, hold or part structures using the finite element method and the static and dynamic loads obtained from the load calculations. Statistical analysis includes the short and long term predictions of ship motion, dynamic load and structural response to waves.

In addition to the above-mentioned major functions, the followings are essential in practical application of the total analysis concept to ship structures:

1. Load translator to efficiently transfer the dynamic loads such as wave pressure, internal pressure acting on tanks, inertia force of the cargo, hull girder sectional forces, etc., into the entire-ship or hold structural models.
2. Displacement translator or zooming function to connect two different sizes of structural models – e.g. from entire-ship to hold structure or partial structure; or from hold or partial structure to finer models for hot-spot stress analysis, etc. – which facilitates translation of displacements obtained from the former into geometrical boundary conditions for the latter.
3. Strength evaluation function to efficiently process large amount of analytical results of stresses obtained from frequency response analysis, time-history analysis and statistical analyses etc.
4. FEM pre- and postprocessors to effectively model ship structures to suit the analysis requirements and postprocessor to clearly represent the structural analysis results.

As for the FEM pre- and post-processors, PrimeShip-NASTASS (NK Advanced Structural Analysis

Support System) can be used as a powerful exclusive modeller for ship hull structures in lieu of general-purpose commercial software packages for CAD/CAE. It has been developed by ClassNK for achieving efficient direct strength calculations of hold structures within a week (Tsutsui, Y, et al., 1996).

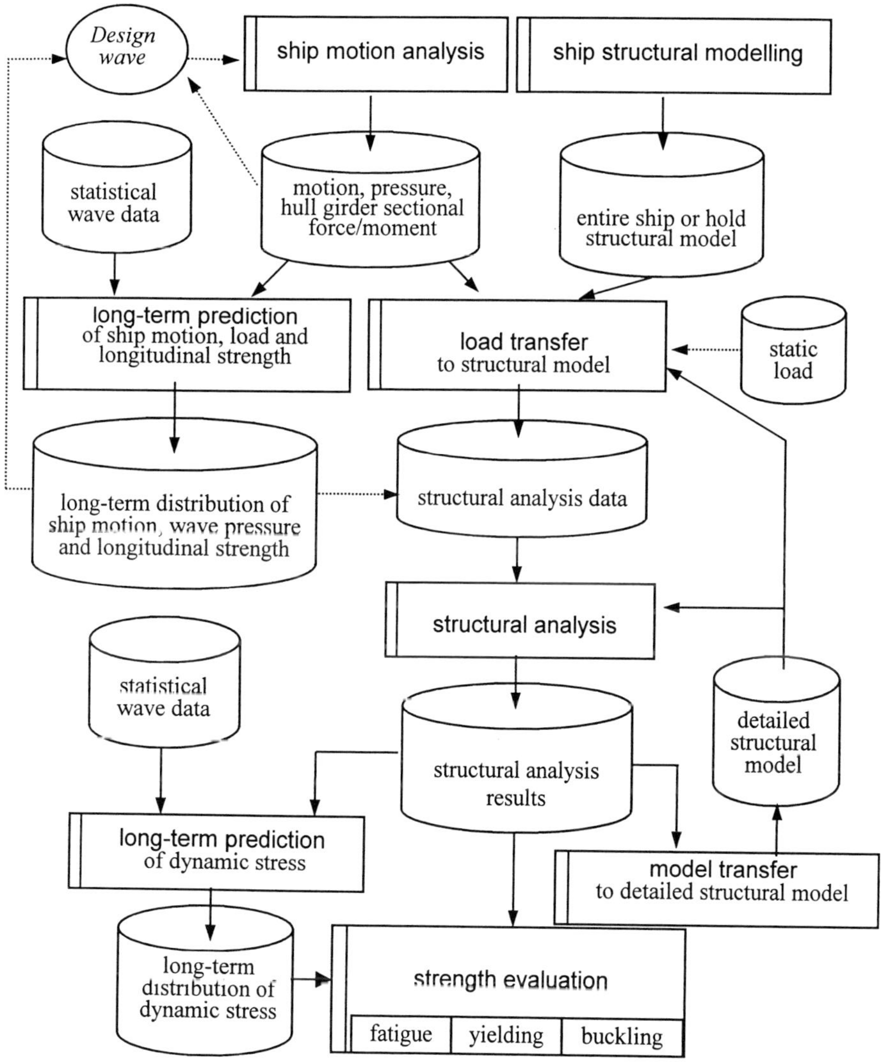

Figure 1: Overall analysis flow of the total analysis system

The principal method of analysis employed in the system is the frequency domain analysis of a linear dynamic response. This method is basically intended to be used for fatigue strength assessment based on the long term probability distribution of dynamic stress, where the lower range of linear dynamic stress has a significant role. When assessing yielding and buckling strength, the maximum stress during the design lifetime can be evaluated adding the static stress obtained from separate static analysis to the dynamic stress amplitude at a reference probability of exceedance.

Another method of analysis employed in the system is a kind of quasi-static method of non-linear time domain analysis to simulate the quasi-static response during one cycle of a regular wave, e.g. an extreme design wave. In this method, the non-linearity of ship motion and of internal and external pressure induced by modulating wave surface can be taken into account, which cannot be considered in the linear frequency domain analysis.

Based on the two methods of analysis mentioned above, the system can perform:

1. Whole-ship structural analysis under still water
2. Frequency response analyses of ship motions and wave loads in regular waves
3. Quasi-static analysis of ship structures (entire-ship , hold structure and part structure) to a regular wave
4. Quasi-static frequency response analysis of ship structures in regular waves
5. Short and long term predictions of ship motion, wave load and ship structural responses in irregular waves
6. Quasi-static non-linear time-history response analysis of ship structures during one wave cycle of a regular wave
7. Yielding and buckling strength evaluation based on maximum expected stress of primary structural members
8. Fatigue strength assessment based on the long term distribution of local stress amplitude at the stress concentration areas

2.2 Ship Motion and Wave Load Calculation

Figures 2 and 3 show some sample screens of ship motion and wave load analysis using the system. Figure 2 shows the operation of visually confirming the defined hull form of a container ship. Figure 3 shows the response amplitude operator (RAO) of the longitudinal bending moment in waves at the midship section. The menu window on the right side helps to display the moments at different sections at different ship speeds easily.

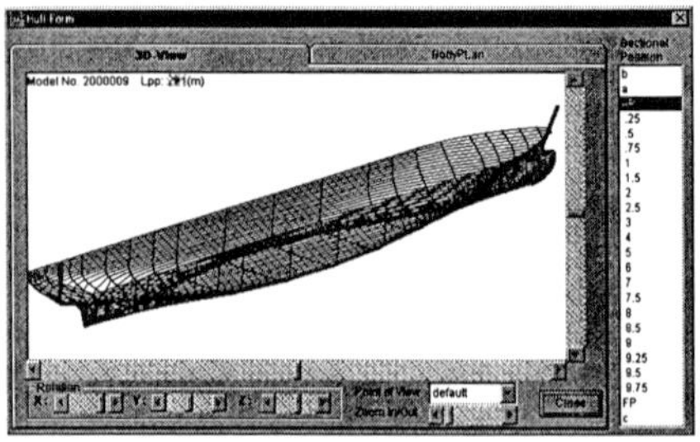

Figure 2: Display of a defined hull form

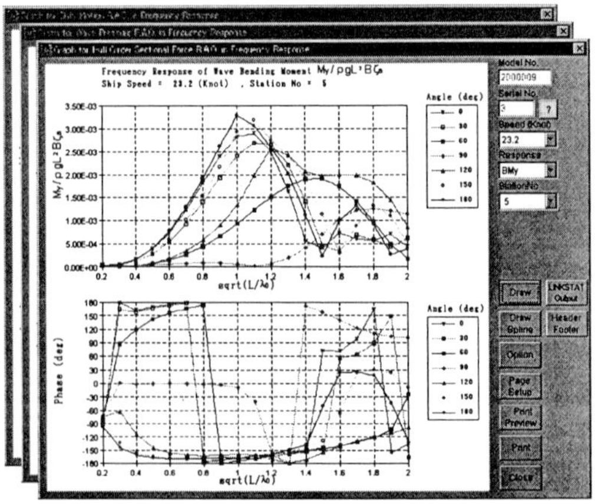

Figure 3: RAOs of wave bending moment

2.3 *Ship Structural Analysis*

In general, the ship structure is modelled for the following levels of analysis:

1. Entire-ship structural analysis
2. Hold structural analysis
3. Local structural analysis

The system can perform all of the above. The mesh size for the above analyses is normally one primary frame space for entire-ship structural analysis, one secondary frame space for hold structural analysis and one plate thickness for local stress analysis of stress concentration parts for fatigue assessment.

The entire-ship structural analysis is normally applied to the evaluation of global structural strength of the hull girder or the global deformation affecting the partial structural behaviour. However, with the tremendous increase in computing power, these days the same entire-ship structural model can be used for all the analyses by having appropriate fine mesh for the hold part, or very fine mesh for the stress concentration area within the same entire-ship model. Figure 4 shows an example of an entire-ship structural model with fine meshes for the three midship holds for stress evaluation. This huge finite element model has 184,542 elements and 102,039 nodes in total, where the three holds with fine meshes have 173,431 elements and 96,914 nodes.

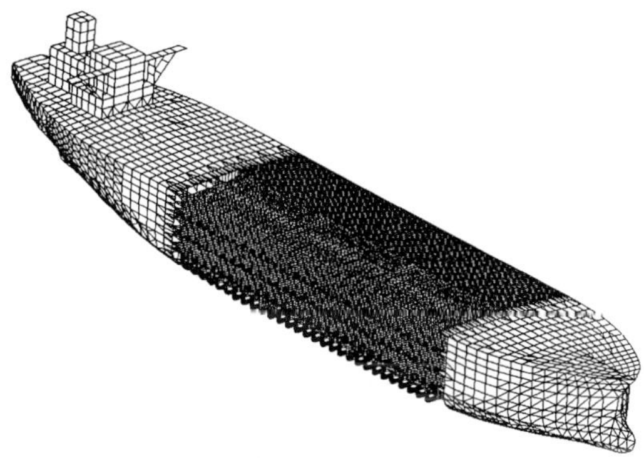

Figure 4: An entire-ship model of a double hull VLCC with fine meshes for the midship holds

2.3.1 *Applied load*

The wave pressure at each strip calculated based on the strip method is translated into nodal loads for the structural model. In non-linear time-history response analysis, the variation in waterline due to wave and ship motion is considered. The inertia force exerted by ballast water and liquid or granular cargo due to ship motion is considered as acting on the tank boundaries. The container load is assumed to be acting on the primary bottom and side members of the hold taking into account contact phenomenon.

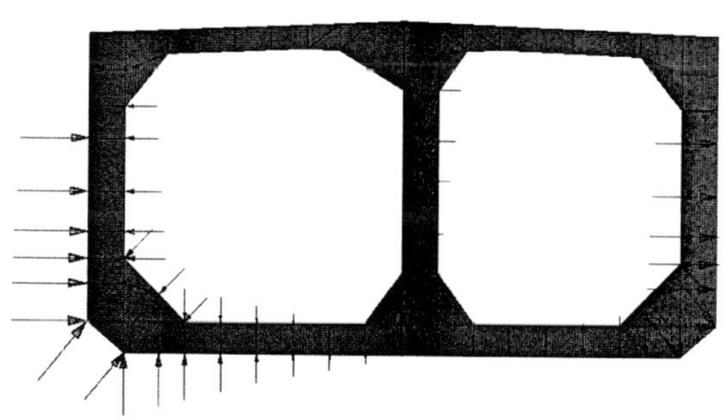

Figure 5: External wave pressure and tank inertia pressure simultaneously acting on a double hull tanker in waves

The hydrodynamic pressure obtained from ship motion analysis based on the strip theory can be transferred to the finite element model using the load translator with graphical user interface. Once both hydrodynamic and structural models are linked to each other, the hydrodynamic pressure is automatically converted to nodal forces for the finite element model. Figure 5 shows an example of simultaneous action of the wave load on the outer hull and inertia load of the liquid cargo on the tank boundaries in the FE model of an Aframax type of double hull tanker.

2.3.2 Boundary conditions

When applying the calculated results of dynamic loads by ship motion analysis to an entire-ship structural analysis, special attention should be paid on unbalanced forces over the entire-ship. The unbalanced forces may be caused by inaccuracy of the strip theory, difference of the load distributions between hydrodynamic and structural models and unbalance of tank inertia pressure resulting from simplified tank modelling. These undesirable unbalanced forces should be reduced to a minimum.

2.3.3 Strength assessment

A new buckling strength formulation (Nippon Kaiji Kyokai, 1999) has been adopted in the system. The interaction relationship takes into account five load components: compression and bending in two directions and in-plane shear. The buckling strength interaction relationship is given in its simplified form as:

$$\Gamma_B = C_1^{\alpha_1} C_2^{\alpha_2} \left[\left(\frac{\sigma_x}{C_1 \sigma_{xcr}} \right)^{\alpha_1} + \left(\frac{\sigma_y}{C_2 \sigma_{ycr}} \right)^{\alpha_2} - 1 \right] = 0$$

where C_1 and C_2 are functions of σ_{bx}, σ_{by}, σ_{bxcr}, σ_{bycr}, τ and τ_{cr}. σ_x and σ_y are the in-plane compressive loads acting on the plate in x and y directions, σ_{bx} and σ_{by} are the bending loads acting on the plate in x and y directions, and τ is the in-plane shear force acting on the plate. The subscript '*cr*' stands for critical values. C_1 and C_2 are functional variables of the aspect ratio of the plate.

The fatigue analysis based on the hotspot stress approach using one design S-N curve is adopted in the system in addition to the conventional nominal stress approach using different S-N curves according to the type of weld joint. In addition, the effect of structural mean stress and residual welding stress can be taken into account in a form of the following equivalent stress range:

$$\Delta\sigma_{eq} = \Delta\sigma^{\alpha}(\sigma_{mean} + \sigma_{res} + \Delta\sigma/2)^{1-\alpha}$$

where, $\alpha = 0.6485$, σ_{mean} is structural mean stress, σ_{res} is residual stress, and $\Delta\sigma$ is hotspot stress range.

3 ANALYSIS EXAMPLES

Application examples of the system with respect to a post-Panamax container ship and a double hull Aframax tanker are given here. Figures 6 and 7 show the results of global structural behaviour and local stress distribution by the same entire-ship analysis with finer meshes around the hatch corners of the post-Panamax container ship.

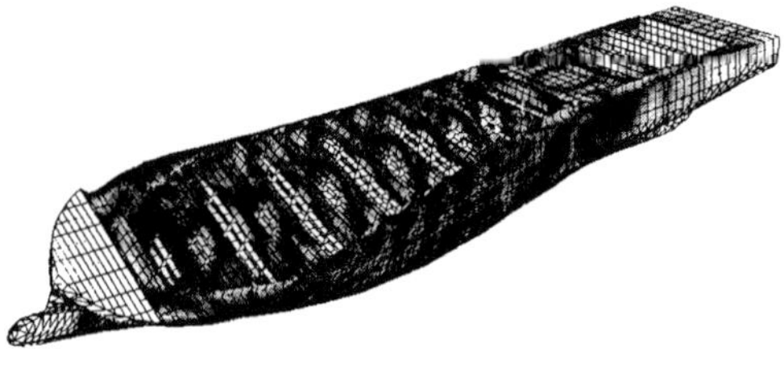

Figure 6: Torsional deformation and stress distribution of a post-Panamax container ship

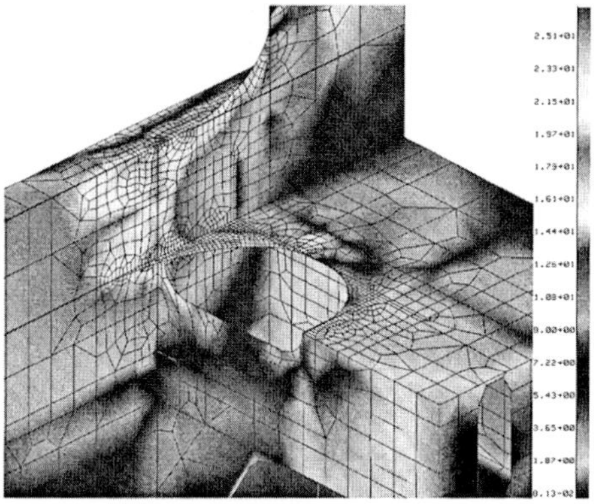

Figure 7: Local stress distribution of an entire-ship model around finer meshes embedded

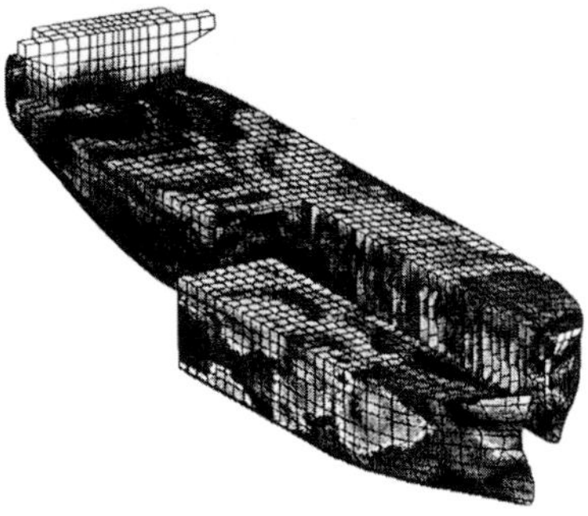

Figure 8: Dynamic stress distribution of a double hull Aframax tanker in waves

An example of a series of calculations to obtain stress response functions of a double hull Aframax tanker is given in Figure 8. In the calculations unbalanced forced were carefully checked and reduced as much as possible.

4 CONCLUDING REMARKS

The Advanced Structural Analysis and Validation System (ASSAS) is aimed at providing a practical tool to the rational assessment of ship structural strength considering actual environmental conditions as precisely as possible. With the advent of computing technology that can process massive amounts of data in a short time, it has become possible to accurately assess very complicated structures like ship hull structure. This paper outlines with examples the advanced analysis and validation technologies involved in the system which supports many important aspects of the technical calculations for damage investigation, new rules development, and plan approval and technical consultation services offered by ClassNK.

References

Hidaka, M (1996), "Role of Classification Societies – Introduction of Advanced Technology of ClassNK "PrimeShip"," *ClassNK Technical Bulletin, Vol. 14, pp 79-80.*

Nippon Kaiji Kyokai (1999), *Technical Guide Regarding the Strength Evaluation of Hull Structures.*

Tsutsui, Y, et al. (1996), "PrimeShip-NASTASS: NK Advanced Ship Structural Analysis Support System," *ClassNK Technical Bulletin, Vol. 14, pp 91-100.*

Practical Design of Ships and Other Floating Structures
You-Sheng Wu, Wei-Cheng Cui and Guo-Jun Zhou (Eds)

UNCERTAINTY AND SENSITIVITY ANALYSES IN THE PREDICTED CRITICAL BUCKLING STRENGTH OF A LONGITUDINALLY STIFFENED SUB-PANEL

Wei-Cheng Cui, Li-Juan Shi and Jin-Fei Zhang

School of Naval Architecture and Ocean Engineering
Shanghai Jiao Tong University, Shanghai 200030, China

ABSTRACT

Reliability methods have gained more and more popularity in ship structure community. In carrying out the reliability analysis of existing ship structures or implementing reliability based design rules for new ships, it is necessary to determine the statistical characteristics for both capacity and demand. For ship structures deck buckling under wave sagging bending moment is one of the most important failure modes to be considered and for this failure mode the structural resistance is often modeled by the buckling of a longitudinally stiffened sub-panel. In this paper the uncertainty in the predicted critical buckling strength and its sensitivities to the basic random variables are studied. A new method to calculate the mean and standard deviation of a function of random variables is introduced and a new sensitivity vector which can describe both parametric sensitivity and importance measure is proposed. The results would be useful for reliability analysis and design.

KEYWORDS

Uncertainty, Longitudinal stiffened sub panel, Flexural buckling, Stiffener tripping, Critical buckling strength, Sensitivity analysis.

1 INTRODUCTION

Structural design is always based on a comparison between demand and capacity, no matter whether it is a deterministic safety factor approach or a reliability based approach. Although the deterministic approach is still dominant in the current design rules of ship structures, reliability based design rules are expected to be implemented soon [Guedes Soares et al 1996].
In carrying out the reliability analysis of existing ship structures [Mansour et al 1997] or implementing reliability based design rules for new ships [Guedes Soares et al 1996], it is necessary to determine the statistical characteristics for both capacity and demand. Much work has been done and reported in the literature, e.g Hughes et al (1994). However, it is pointed out in [Cui 2001] that the concepts and methods in determining the uncertainty in both load and capacity are not very consistent. Some clarifications have been made. This paper is a further continuation of Ref. [Cui 2001].

According to Hughes et al (1994), uncertainties in structural engineering systems can be mainly attributed to ambiguity and vagueness in defining parameters of the systems and their relations. For the structural capacity considered in this paper, the vagueness related uncertainties also exist. Let us define the true capacity of a structure in the following way:

$$C_{true} = \frac{C_{true}}{C_{\exp}} \frac{C_{\exp}}{C_{pred}} C_{pred} \tag{1}$$

where C_{true} is the true capacity of the structure, C_{exp} is the structural capacity determined from experiments, C_{pred} is the structural capacity calculated from a theoretical method.

Then the determination of the capacity through experiments, C_{exp}, may contain a large amount of vagueness-related uncertainties. For example, the definition of the structural failure and the measurement of the actual structural state need human judgement and they are not unique in nature. However, these uncertainties are extremely difficult to quantify. As a matter of fact, structural engineers and researchers have dealt only with the ambiguity types of uncertainty.

The second ratio of the experimentally determined value over the predicted value is called the modeling uncertainty, while the uncertainty in the theoretically predicted value is called natural uncertainty. Determination of the modeling uncertainty for the present method has been addressed in Ref. [Cui 2001] with a mean bias of 0.95 and a standard deviation of 0.10. In this paper, we concentrate on the natural uncertainty. A new method to calculate the mean and standard deviation of a function of random variables is introduced and a new sensitivity vector which can describe both parametric sensitivity and importance measure is proposed. Based on these, the sensitivities of the critical buckling strength of a longitudinal stiffened sub-panel to some basic geometric and material parameters are studied.

2 PREDICTION METHOD FOR BUCKLING STRENGTH OF A LONGITUDINALLY STIFFENED SUB-PANEL

Various methods are available to calculate the structural capacity of ship structures corresponding to different types of failure modes. Generally these methods can be categorized as simplified analytical solutions, e.g. [Zhou et al 1997], semi-finite element methods such as ISUM [Paik 1993] and Smith's method [Smith 1977] and fully nonlinear finite element method [Chen et al 1983]. In this paper, we take a very typical example of deck panel buckling for discussion.

For many ship structures such as large oil tankers, deck buckling under wave sagging bending moment is the dominant ultimate strength failure mode. For this failure mode, the structural capacity can be estimated by the critical buckling stress of a longitudinal stiffener with an attached plating. The prediction formulas are given as follows [Zhou et al 1997]:

$$\sigma_{cr} = \sigma_y (1 - \frac{\sigma_y}{4\sigma_E}) \tag{2}$$

where σ_y is the yield stress of the material, MPa; σ_E is the minimum of elastic flexural buckling stress σ_{EB} and the elastic buckling stress due to stiffener tripping σ_{ET}, MPa.

The elastic flexural buckling stress σ_{EB} and the elastic buckling stress due to stiffener tripping σ_{ET} can be calculated by the following equations [Zhou et al 1997]:

$$\sigma_{EB} = \pi^2 EI(b_e^{'})/l^2 F(b_e) \qquad \sigma_{ET} = \frac{E}{I_P}[\frac{\pi^2 I_\varpi}{l^2} + \frac{Cl^2}{\pi^2 E} + \frac{GK}{E}]$$

$$\frac{b_e^{'}}{t} = \sqrt{\frac{E}{\sigma_{cr}}} \qquad I_\varpi = \frac{1}{12}b_f^3 t_f (d_w + \frac{t_f}{2})^2 + \frac{1}{144}b_f^3 t_f^3 + \frac{1}{36}d_w^3 t_w^3$$

$$\frac{b_e}{t} = 2\sqrt{\frac{E}{\sigma_{cr}}} - \frac{t}{b}\frac{E}{\sigma_{cr}} \qquad I_P = \frac{1}{12}b_f^3 t_f + \frac{1}{3}d_w^3 t_w + b_f t_f (d_w + \frac{t_f}{2})^2 \tag{3}$$

$$G = \frac{E}{2(1+\mu)} = \frac{E}{2.6}, \quad K = \frac{1}{3}b_f t_f^3 + \frac{1}{3}d_w t_w^3, \quad C = E/(\frac{3b}{t^3} + \frac{4d_w}{t_w^3})$$

where E is Young's modulus, MPa; b is the width of the attached plating, mm; t is the thickness of the attached plating, mm; b_e is the effective width of the attached plating, mm; $b_e^{'}$ is the tangent width of the attached plating, mm; l is the length of the longitudinal stiffener, mm; F is the sectional area calculated using effective with, mm^2; I is the sectional moment of inertia calculated with tangent width, mm^4. I_ω is the sectorial moment of inertia, mm^6; I_P is the polar moment of inertia of stiffener, mm^4; G is the shear modulus, MPa; K is St. Venant's torsional constant, mm^4; C is the spring stiffness of the attached plating, N; b_f is the flange width, mm; t_f is the flange thickness, mm; d_w is the web height, mm; t_w is the web thickness, mm.

3 A NEW METHOD TO DETERMINE THE NATURAL UNCERTAINTY OF A FUNCTION OF RANDOM VARIABLES

Mathematically, calculation of the mean and standard deviation of the critical buckling strength is to calculate the mean and standard deviation of a function of random variables. For this purpose, Cui et al (1998) have compared various methods and a new method has been proposed. The method is described as follows:

Let $X_1, X_2, \ldots, X_n$ be n uncorrelated basic random variables with means and standard deviations of μ_{xi} and σ_{xi} (i=1,2,…,n). G is a differentiable function of these basic random variables denoted by Y=G(X). We are asked to calculate the mean and standard deviation of Y.

Let us define

$$\begin{aligned}
y^* &= G(\mu_{X1}, \mu_{X2}, \cdots, \mu_{Xn}) \\
y_i^+ &= G(\mu_{x1}, \cdots \mu_{xi-1}, \mu_{xi} + \sigma_{xi}, \mu_{xi+1}, \cdots \mu_{xn}) \\
y_i^- &= G(\mu_{x1}, \cdots \mu_{xi-1}, \mu_{xi} - \sigma_{xi}, \mu_{xi+1}, \cdots \mu_{xn}) \\
\mu_i &= [y_i^+ + y_i^-]/2 \\
\sigma_i &= |y_i^+ - y_i^-|/2 \\
y_{ij}^+ &= G(\mu_{x1}, \cdots \mu_{xi-1}, \mu_{xi} + \sigma_{xi}, \mu_{xi+1}, \cdots, \mu_{xj} + \sigma_{xj}, \cdots, \mu_{xn}) \\
y_{ij}^- &= G(\mu_{x1}, \cdots \mu_{xi-1}, \mu_{xi} - \sigma_{xi}, \mu_{xi+1}, \cdots, \mu_{xj} - \sigma_{xj}, \cdots, \mu_{xn}) \\
\mu_{ij} &= [y_{ij}^+ + y_{ij}^-]/2 \; (i < j) \\
\Delta\mu_i &= \mu_i - y^* \\
\Delta\mu_{ij} &= \mu_{ij} - y^* - \Delta\mu_i - \Delta\mu_j
\end{aligned} \tag{4}$$

then the mean and standard deviation of Y is re-defined as [Cui et al 1998]

$$\mu_y = y^* + \sum_{i=1}^{n} \Delta\mu_i$$
$$\sigma_y^2 = \sum_{i=1}^{n} \sigma_i^2 + 2\sum_{i=1}^{n} (\Delta\mu_i)^2 + \sum_{i<j} (\Delta\mu_{ij})^2 \tag{5}$$

It has been proved [Cui et al 1998] that this new method has an accuracy of at least the second order Taylor series solution. In some cases it is even higher. This method will be used in the following section to calculate the sensitivities of each of the random variable and its parameters.

4 UNCERTAINTY AND SENSITIVITY ANALYSES IN THE PREDICTED CRITICAL BUCKLING STRENGTH OF THE LONGITUDINALLY STIFFENED SUB-PANEL

4.1 Uncertainty Analysis in the Predicted Critical Buckling Strength of the Longitudinally Stiffened Sub-panel

If we assume that each of the geometrical and material parameters are random variables with statistical characteristics given in Table 1, the mean and the standard deviation of the critical buckling stress using this new method together with the formulas in section 2 are 438.2 MPa and 32.28 MPa respectively. The coefficient of variation is 0.074.

4.2 Sensitivity Analysis Approaches

The critical buckling strength is a function of several random variables. It is certainly of great significance to know the sensitivities of the function to each of the random variables and rank the importance of the basic random variables through some measures. In general the sensitivity analysis needs to address two problems. One is the significance of a random variable to the function value while the other is how sensitive the function value to the change of the parameters of the random variables such as mean and standard deviation. The former is also called the importance measure while the later is called the parametric sensitivity factor (DNV 1992).

TABLE 1
STATISTICS OF BASIC RANDOM VARIABLES

Variable order	Variable name	Physical meaning	mean	s.d.	COV
1	σ_y (MPa)	Yield strength of the material	476.77	38.14	0.08
2	t (mm)	Thickness of the plating	8.0	0.24	0.03
3	E (MPa)	Young's modulus	200000.0	6000.0	0.03
4	l (mm)	Length of the stiffener	1500.0	15.0	0.01
5	B (mm)	Width of the plating	400.0	4.0	0.01
6	D_w (mm)	Web height	105.745	1.05745	0.01
7	t_w (mm)	Web thickness	6.5	0.065	0.01
8	b_f (mm)	Flange width	30.0	0.30	0.01
9	t_f (mm)	Flange thickness	14.255	0.14255	0.01

Different definitions for sensitivity analysis exist in the literature and there is no universally accepted one [e.g. Wirsching et al 1994; Mansour and Wirsching 1995]. In order to provide answers to both problems, Zhang et al (1998) suggested to define a sensitivity vector for each random variable to describe both the sensitivity of parameters and the significance of the random variable.

In this paper, we adopt a similar definition as Zhang et al (1998) to carry out the sensitivity analysis for the function of random variables.

Let $X_1, X_2, \ldots, X_n$ be n uncorrected basic random variables with means and standard deviations of μ_i and σ_i (i=1, 2, ..., n). G is a differentiable function of these basic random variables denoted as $Y=G(X_1,X_2,\ldots,X_n)$ with mean μ_y and standard deviation σ_y. The sensitivity vector of Y to X_i is defined as follows:

$$\bar{S}_i = (\delta_{\mu i}, \eta_{\mu i}, \delta_{\sigma i}, \eta_{\sigma i}, \lambda_i), \quad \delta_{\mu i} = (\frac{\partial \mu_y}{\partial \mu_i})\sigma_i, \quad \eta_{\mu i} = (\frac{\partial \mu_y}{\partial \sigma_i})\sigma_i$$

$$\delta_{\sigma i} = (\frac{\partial \sigma_y}{\partial \mu_i})\sigma_i, \quad \eta_{\sigma i} = (\frac{\partial \sigma_y}{\partial \sigma_i})\sigma_i, \quad \lambda_i = \frac{\sqrt{\delta_{\mu i}^2 + \eta_{\mu i}^2 + \delta_{\sigma i}^2 + \eta_{\sigma i}^2}}{\sum_{i=1}^{n} \sqrt{\delta_{\mu i}^2 + \eta_{\mu i}^2 + \delta_{\sigma i}^2 + \eta_{\sigma i}^2}} \quad (6)$$

where $\delta_{\mu i}$ is a measure of the sensitivity of the mean predicted strength to the change in the mean value of the random variable X_i; $\eta_{\mu i}$ is a measure of the sensitivity of the mean predicted strength to the change in the standard deviation of the random variable X_i; $\delta_{\sigma i}$ is a measure of the sensitivity of the standard deviation of the predicted strength to the change in the mean value of the random variable X_i; $\eta_{\sigma i}$ is a measure of the sensitivity of the standard deviation of the predicted strength to the change in the standard deviation of the random variable X_i; λ_i is the measure of the relative importance of the random variable X_i to the predicted strength Y.

4.3 Sensitivity Analysis Results of the Predicted Critical Buckling Strength

The method presented in this paper can be easily implemented into a computer program. Sensitivity analyses to the predicted critical buckling strength are carried out using the program and the results are given in Table 2.

TABLE 2
SENSITIVITY OF THE CRITICAL BUCKLING STRENGTH TO BASIC RANDOM VARIABLES

Sensitivity Variable	$\delta_{\mu I}$	$\eta_{\mu i}$	$\delta_{\sigma i}$	$\eta_{\sigma I}$	λ_I
σ_y	32.1859	-0.4847	-0.4277	32.2125	0.9290
T	-0.4941	-0.0239	-0.0615	0.0083	0.0102
E	1.0610	-0.0678	0.1610	0.0399	0.0219
L	-0.7823	-0.0079	-0.1182	0.0208	0.0161
B	-0.0728	0.0015	-0.0065	0.0002	0.0015
d_w	0.6935	-0.0216	0.1025	0.0167	0.0143
t_w	-0.0164	-0.0001	-0.0044	0.00001	0.0003
b_f	0.1420	-0.0016	0.0201	0.0007	0.0029
t_f	0.1792	-0.0016	0.0255	0.0011	0.0037

Using these sensitivity results ranking of the variables can be made. A summary of the ranking of the variables with respect to each of the sensitivity measure is given in Table 3. From table 3 it can be seen that the yield strength of the material is the dominant parameter which affects the uncertainty of the critical buckling strength. The next two important parameters are the Young's modulus and the length. After that comes with the web height and the thickness of plating. The rest of the variables are very insensitive to the uncertainty of the critical buckling strength.

TABLE 3
RANKING OF VARIABLES BASED ON THE SENSITIVITY STUDIES

Sensitivity Ranking	$\delta_{\mu l}$	$\eta_{\mu i}$	$\delta_{\sigma i}$	$\eta_{\sigma i}$	λ_i
1	Yield strength	Yield strength	Yield strength	Yield strength	Yield strength
2	Young's modulus	Young's modulus	Young's modulus	Young's modulus	Young's modulus
3	Length	Plating thickness	Length	Length	Length
4	Web height	Web height	Web height	Web height	Web height
5	Plating thickness	Length	Plating thickness	Plating thickness	Plating thickness
6	Flange thickness	Flange width	Flange thickness	Flange thickness	Flange thickness
7	Flange width	Flange thickness	Flange width	Flange width	Flange width
8	Plating width	Plating width	Plating width	Plating width	Plating width
9	Web thickness	Web thickness	Web thickness	Web thickness	Web thickness

5 SUMMARY AND CONCLUSIONS

In reliability analysis and design of ship structures it is necessary to determine the probabilistic characteristics for both the demand and the capacity. For ship structures deck buckling under wave sagging bending moment is one of the most important failure modes to be considered and for this failure mode the structural resistance is often modeled by the buckling of a longitudinally stiffened sub-panel. In this paper a new method to calculate the mean and standard deviation of a function of random variables is introduced and a new sensitivity vector which can describe both parametric sensitivity and importance measure is proposed. Based on these, the sensitivities of the critical buckling strength of a longitudinal stiffened sub-panel to some basic geometric and material parameters are studied. Using these sensitivity results ranking of the variables have been made. It can be seen that the yield strength of the material is the dominant parameter which affects the uncertainty of the critical buckling strength. The next two important parameters are the Young's modulus and the length. After that comes with the web height and the thickness of plating. The rest of the variables are very insensitive to the uncertainty of the critical buckling strength.

References

Chen YK, Kutt LM, Piaszczyk CM, Bieniek MP. (1983).Ultimate strength of ship structures. *Trans. SNAME*, **91**, 149-168.

Cui WC, Xu XD, Qiu Q. (1998). A fast method to calculate the mean and the standard deviation of the function of random variables. *J. of Ship Mechanics,* **2:6**, 50-60 (in Chinese).

Cui WC. (2001). On consistent determination of structural capacity statistics for reliability analysis, submitted to *Journal of Ship Mechanics.*

DNV Classification Notes No.30.6. (1992). *Structural reliability analysis of marine structures*. DET NORSKE VERITAS Classification AS.

Guedes Soares C, Dogliani M, Ostergaard C, Parmentier G, Pedersen PT. (1996). Reliability based ship structural design. *SNAME Transactions* **104**, 357-389.

Hughes O, Nikolaidis E, Ayyub B, White G, Hess P. (1994). Uncertainty in strength models for marine structures. *SSC-375*, Ship Structure Committee.

Mansour A, Wirsching P, Luckett M, Plumpton A et al. (1997). Assessment of reliability of existing ship structures. *SSC-398*, Ship Structure Committee.

Mansour AE, Wirsching PH. (1995). Sensitivity factors and their application to marine structures, *Marine Structures*, **8**, 229-255.

Paik, J.K. (1993). *ALPS/ISUM: A computer program for nonlinear analysis of large plated structures using the Idealized Structural Unit Method.* Pusan National University, Pusan, Korea.

Smith, C.S. (1977). Influence of local compressive failure on ultimate longitudinal strength of ship's hull, *Proc. Int. Symp. Practical Design in Shipbuilding (PRADS'77)*, Tokyo, pp.73-79.

Zhang W, Cui WC, Xu BH, Zhang SK. (1998). New method on calculating sensitivity factors, *J. of Shanghai Jiaotong University*, **32:11**, 26-29 (in Chinese).

Zhou GH, Shu CG, Lu ZG, Chen RZ, Cui WC. (1997). Introduction to reliability research on oil tanker hull structure and reliability analysis of longitudinal strength. *Science and Technology of Ship Survey*, **2**, 108-123 (in Chinese).

Practical Design of Ships and Other Floating Structures
You-Sheng Wu, Wei-Cheng Cui and Guo-Jun Zhou (Eds)

SENSITIVITY ANALYSIS ON ULTIMATE HULL BENDING MOMENT

Ph. RIGO [1,2], C. TODERAN [1] and T. YAO [3]

[1] Department of Naval Architecture (ANAST), University of Liege
Chemin des Chevreuils n°1 (B52/3), B-4000 Liege, Belgium
[2] National Fund for Scientific Research of Belgium – NFSR
[3] Dept. of Naval Architecture and Ocean Engineering (NAOE), Osaka University
2-1 Yamada-Oka, Suita, Osaka 565-0871, Japan

ABSTRACT

The authors performed for the Committee VI.2 of ISSC'2000 an assessment of the sensitivity of the "average strain-average stress curves (σ–ε)" on the ultimate hull bending moment (Mu) and the moment-curvature relationship (Mu-ϕ). This assessment concerns the ultimate bending moments obtained through standard progressive collapse analysis (Smith Algorithm). Such progressive approach requires for each stiffened panel a "σ–ε" curve. The objective is to quantify the variation of the ultimate bending moment (Mu) induced by the uncertainties associated to the "σ–ε" curves. Extended and updated results are available in this paper.

KEYWORDS

Ultimate strength, Stiffened panels, Strain and stress curve, Ultimate bending moment, Hull girder, Progressive collapse analysis, Sensitivity analysis.

1 INTRODUCTION

Buckling, tripping, yielding,... of plate, panels and longitudinal structural members affect the overall collapse behaviour of a ship hull cross-section under longitudinal bending moment. Figure 1 shows several typical "*average stress-average strain (σ–ε)*" relationships of a stiffened panel composed of a stiffener and attached plating under axial compression (beam-column). Even if all these curves consider the buckling effect they differ by the value of the "σ_u" compressive ultimate strength as well as by the shape of the post-collapse residual strength. The models used to get these curves are the following:

(M1) HULLST (*Yao 1991, 1999*): Rational numeric analysis of the "σ_u" ult. strength and "σ–ε" curve,

(M2) RAHMAN (*Rahman and Chowdhury 1996*) + HUGHES (*Hughes 1988*): Rational formulation for the ultimate strength "σ_u" and simplified rational "σ–ε" curve.

(M3) IMPERIAL COLLEGE (*Dowling 1991*) + PROCOL (*Rigo 1998*): Rational formulation for the ultimate strength "σ_u" and empirical "$\sigma-\varepsilon$" curve shape.

(M4) PAIK (*Paik 1995*) + PROCOL: Empirical formulation for the ultimate strength "σ_u" and empirical "$\sigma-\varepsilon$" curve shape.

(M5) ABAQUS©: Non linear FEM numeric analysis for "σ_u" ult. strength and "$\sigma-\varepsilon$" curve.

(M6) ADAMCHACK (*Adamchack 1984*): Rational formulation for the ultimate strength "σ_u" and rational formulation "$\sigma-\varepsilon$" curve [IRCN's model, France; *Quesnel et al. 2001]*.

(M7) ULSAP (*ULSAP 2000*) + PROCOL: Rational formulation for the ultimate strength "σ_u" and empirical "$\sigma-\varepsilon$" curve shape.

The models that did not provide a "$\sigma-\varepsilon$" curve (but only the "σ_u" compressive ultimate strength) have been linked with the PROCOL interface model which provides an empirical ($\sigma-\varepsilon$) curve shape for a given "σ_u" ultimate strength (*Rigo 1998*).

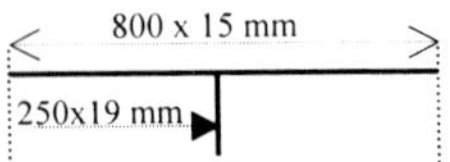

- Frame space = 4000 mm
- Yield stress = 235 N/mm²
- E = 205800 N/mm²
- Plate Imperfection : 5 mm
- Stiffener Imperfection : 2 mm
- Residual Stress : 17.6 N/mm²
- Lateral Pressure : 0.005 Mpa

16 configurations (of the box girder) using typical panels of commercial and passenger ships have been assessed (with different initial imperfection levels and lateral pression).

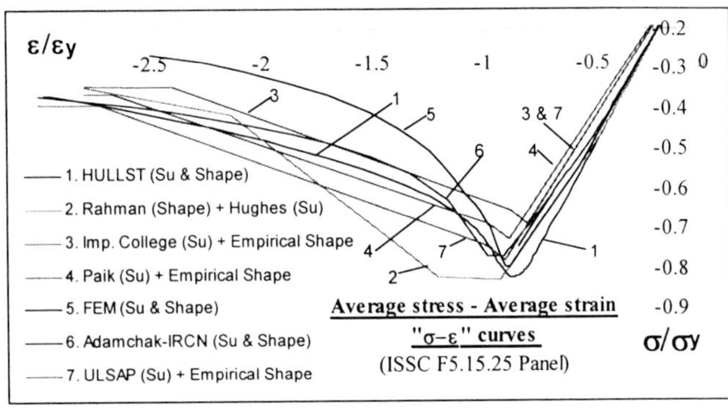

Figure 1: "$\sigma-\varepsilon$" curves of the reference element of a box girder with 7 different "STR" models.

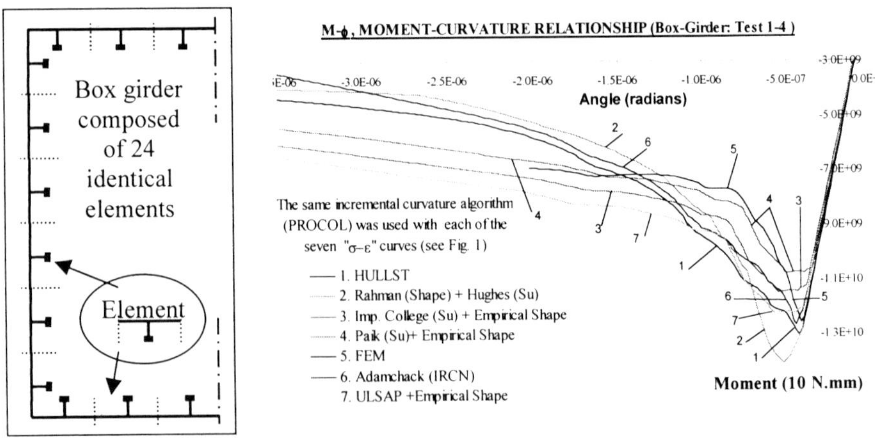

Figure 2: Moment-curvature relationships of a box girder subjected to longitudinal bending moment. Sensitivity of the "$\sigma-\varepsilon$" element curve (STEP 2) on the ultimate bending moment.

Figure 2 shows seven (Mu-ϕ) moment-curvature relationships of a simple rectangular box girder associated to the seven previous "$\sigma-\varepsilon$" curves (Fig. 1). The studied box girder is composed of 24 identical elements. No hardcorner element is considered. This means that each "$\sigma-\varepsilon$" curve shown at

Figure 1 is sufficient to perform the progressive collapse analysis and to provide an assessment of the sensitivity of the "σ–ε" curve on ultimate bending moment (Mu-ϕ). Using PROCOL the same *"PCA algorithm"* was used with each "σ–ε " curve (PCA = Progressive Collapse Analysis). This PCA algorithm is the incremental curvature procedure implemented in HULLST (*Yao and Nikolov, 1991*). Based on these tests, Table 1 provides a quantitative assessment of the sensitivity of the "σ–ε" curves (σ_u and shape) on the "Mu" ultimate bending moment. It is observed that the level of uncertainty on the "Mu" ultimate bending moment is at least equal to the uncertainty on the "σ_u" ultimate strength of the constitutive elements (e.g. 5.2 to 4.3% for M1; 3.9 to 3.9% for M3; 11.4 to 14.9% for M4 and 2.7 to 2.6% for M5). Moreover, using an "*unsuitable*" shape (even with an accurate "σ_u") can double or triple the COV(Mu). For instance, with the M2 model, we have COV(σ_u)=5.6% and COV(Mu)=13.7%. This is due to an excessive "ultimate strength's plateau" of the "σ–ε" curve (Figure 1). Based on this assessment we can assume that COV(σ_u) ≤ COV(Mu) ≤ 2.5 COV(σ_u).

TABLE 1

SENSITIVITY OF THE **"σ–ε" CURVES** ON THE ULTIMATE BENDING MOMENT OF THE BOX GIRDER (Fig.2) (Same mesh model [step 1] and same PCA model [step 3])

	Ultimate Strength of the Panel [σu/σu(average)]							**Ultimate Bending Moment of Girder [Mu/Mu(average)]**						
σ–ε Model	M1	M2	M3	M4	M5	COV	Standard Deviation	M1	M2	M3	M4	M5	COV	Standard Deviation
Test 1-1	1.1201	1.0368	0.9988	0.8444		7.8%	10.0%	1.0860	1.1013	0.9866	0.8260		9.4%	11.0%
1-2	1.0643	1.0707	0.9499	0.8856	1.0296	6.6%	7.2%	1.0480	1.1360	0.9360	0.8640	1.0160	8.0%	9.3%
1-4	1.0650	1.0727	0.9394	0.8916	1.0313	6.8%	7.2%	1.0501	1.1389	0.9289	0.8724	1.0097	7.9%	9.3%
1-5	1.0309	1.1234	0.9371	0.9509	0.9578	6.2%	7.0%	1.0221	1.1755	0.9114	0.9199	0.9710	7.9%	9.6%
1-6	1.1114	1.0345	1.0279	0.8263		8.7%	10.6%	1.0708	1.0954	1.0215	0.8123		9.4%	11.2%
1-7	1.0598	1.0709	0.9732	0.8699	1.0262	6.3%	7.3%	1.0387	1.1355	0.9613	0.8516	1.0129	7.5%	9.3%
1-9	1.0581	1.0707	0.9613	0.8743	1.0356	6.6%	7.3%	1.0414	1.1320	0.9444	0.8538	1.0285	8.1%	9.4%
1-10	1.0089	1.1116	0.9509	0.9271	1.0015	4.9%	6.1%	1.0095	1.1664	0.9345	0.9004	0.9891	7.0%	9.2%
2-7	1.0184	1.0399	0.9970	0.9447		2.9%	3.5%	0.9735	1.1457	0.9669	0.9139		7.3%	8.7%
3-7	1.0007	1.0823	1.0007	0.9163		4.2%	5.9%	0.9949	1.1516	0.9744	0.8790		7.6%	9.8%
4-7	1.0579	1.0521	0.9655	0.9245		5.5%	5.7%	1.0343	1.1127	0.9510	0.9020		7.4%	8.0%
5-7	1.0448	1.0470	0.9958	0.9278	0.9846	3.7%	4.4%	1.0777	1.0617	0.9615	0.8974	1.0016	5.6%	6.6%
6-7	1.0476	1.0130	1.0491	0.8903		5.5%	6.5%	1.0249	1.1388	0.9964	0.8399		8.2%	10.7%
7-7	1.0450	1.0465	1.0157	0.8928		5.4%	6.3%	1.0313	1.1284	0.9848	0.8554		8.0%	9.8%
8-2	1.0415	0.9990	1.0922	0.8388	1.0284	6.5%	8.6%	1.0469	1.1747	0.9590	0.7405	1.0788	12.0%	14.7%
8-7	1.0625	1.0289	1.1119	0.7660	1.0307	9.4%	12.1%	1.0676	1.2037	0.9996	0.6803	1.0488	12.8%	17.4%
	5.2%	5.6%	3.9%	11.4%	2.7%	6.0%	7.2%	4.3%	13.7%	3.9%	14.9%	2.6%	8.4%	10.3%
	COV = Mean Absolute Variation to Average					Average value		COV = Mean Absolute Variation to Average					Average value	

COV = Σ [*ABS*(m-Xi)]/N with **Xi** (i=1,N) and **m** the average of the Xi and *ABS*(x) the absolute value of x

2 A GENERIC FRAMEWORK FOR PROGRESSIVE COLLAPSE ANALYSIS

Usually progressive collapse analysis to evaluate the ultimate bending moment of hull girder is achieved in 3 steps and each step is characterised by a numerical model associated with theoretical assumptions. For each model, it is necessary to specify the assumptions and to assess their influence (sensitivity) on the result (ultimate bending moment of hull girder). This is particularly relevant as the paper's aim is to assess the sensitivity of the "σ–ε" curves on the ultimate bending moment (Mu).

At the beginning of a "Progressive Collapse Analysis" (PCA) the "*raw data*" are the same for each model/user and they concern:

- the scantling (plates, stiffeners, ..) of the midship section and the frame spacing,
- the initial imperfections (plate deflection, stiffener deflection and residual stress).,
- the lateral loadings (if considered by the "σ–ε" element model)

The 3 main steps of a complete "Progressive Collapse Analysis" (PCA) are the following:

STEP 1: Modelling (to decompose the structure in elements and to build the mesh model),

STEP 2: Evaluate the "σ–ε" element curve. This requires a stress-strain model called "STR" model.

- STEP 2.1: Assessment of the "σ_u" compressive ultimate strength of each element. This requires to survey all the failure modes (plate and stiffener induced buckling, tripping, yielding and local instabilities) and to model of the initial imperfections and boundary conditions.
- STEP 2.2: Definition of the "shape" of the "σ–ε" curves.

NB: Steps 2.1 and 2.2 can be performed together in the same routine (e.g.: HULLST, Yao and Nikolov, 1991) or independently (e.g.: Rahman and Chowdhury, 1996).

STEP 3: Perform the Progressive Collapse Analysis (using a "PCA" model that includes an incremental procedure of the curvature).

In order to choose a method to evaluate the ultimate strength of a hull girder, a designer have in fact to select 2 models, one for STEP 2 (the "STR" element model) and one for STEP 3 (the "PCA" model). Sensitivity of both has to be evaluated.

2.1 Sensitivity of the PCA Model (STEP 1 & STEP 3) on the Ultimate Bending Moment

Most of the available PCA models are based on the model of Smith (1977). This means that the background of all the current PCA models is the same. So, only the quality of the numerical procedure can generate some differences between the different PCA models. On the other hand, the accuracy and reliability of a progressive collapse analysis are strongly influenced by the mesh model "*quality*" (STEP 1). In fact, the mesh model of the considered structure (STEP 1) depends on the considered PCA model and the available elements: plates, stiffeners, stiffened plates, stiffened panels, curved elements, hard corners, etc. In short, we can say that a PCA model is "better" than another if the mesh model of the structure requires less simplifications/assumptions. For instance, the simplified Rahman progressive collapse model (*Rahman et al. 1996*) has the same background as the sophisticated HULLST model (*Yao, 1999*) but is less accurate as it requires a simplified modelling.

2.2 Sensitivity of the STR Model (STEP 2) on the Ultimate Bending Moment

To provide reliable information to select a relevant "STR" model (STEP 2), it is necessary to assess separately the sensitivity of STEP 2 on the ultimate bending moment. This job was initiated by Rigo and the ISSC'2000 Committee VI.2 through a series of analysis that quantify the sensitivity of the "σ–ε" curves on the ultimate bending moment (*Yao et al. 2000*). To achieve this goal, the ultimate bending moment of 3 ships studied by the ISSC Committee were re-evaluated with different STR models (STEP 2) but with the same PCA model (STEP 1 and STEP 3) using the PROCOL software. PROCOL links the HULLST's PCA model with several STR models developed by different authors (*Paik et al. 1997; Rahman et al. 1996; Dowling 1991; Yao and Nikolov 1991*). For each ship an identical mesh model is used, including 105 elements for the VLCC, 90 elements for Energy Concentration and 99 for the Container. Only the stress-strain curves differ. All the relevant data are available in Yao et al. (2000).

TABLE 2

SENSITIVITY OF THE «σ_u» ULTIMATE STRENGTH OF ELEMENTS [STEP 2.1] ON THE ULTIMATE BENDING MOMENT OF THE HULL GIRDERS (VLCC ; ENERGY CONCENTRATION AND CONTAINER SHIP)

(Same mesh model [step 1]; same "σ–ε" curves [step 2.2] and same PCA model [step 3])

SAGGING	**VLCC (105 elements)**				**ENERGY CONCENTRATION (90 elements)**				**CONTAINER SHIP (99 elements)**			
	S(ult) of elements		Ult. Bending Moment		S(ult) of elements		Ult. Bending Moment		S(ult) of elements		Ult. Bending Moment	
"σ–ε" Model (STR) (same shape)	Sm = Su(av)/Sy	Sm/Sm(av)	Mu 1000 MN.m	Mu/M(av)	Sm = Su(av)/Sy	Sm/Sm(av)	Mu 1000 MN.m	Mu/M(av)	Sm = Su(av)/Sy	Sm/Sm(av)	Mu 1000 MN.m	Mu/M(av)
model 1	0.823	1.035	21.02	1.059	0.874	1.022	16.32	1.043	0.756	1.000	6.01	1.025
model 2	0.820	1.031	20.94	1.055	0.871	1.019	16.43	1.050	0.776	1.026	6.08	1.036
model 3	0.757	0.952	18.01	0.916	0.837	0.979	14.73	0.941	0.753	0.996	5.63	0.960
model 4	0.782	0.983	18.71	0.951	0.838	0.980	15.11	0.966	0.740	0.979	5.74	0.979
	0.7955 = Sm(av)	**3.27%** Mean Variation	19.67 = M(av)	**6.66%** Mean Variation	0.855 = Sm(av)	**2.05%** Mean Variation	15.6475 = M(av)	**4.65%** Mean Variation	0.75625 = Sm(av)	**1.31%** Mean Variation	5.866 = M(av)	**3.09%** Mean Variation

S(ult) = "σ_U" compressive ultimate strength of an element
Sy = Yield Stress
Mu = Ultimate bending moment

Su(av)= Su average value of all the elements of the ship (for instance :105 for the VLCC)
Sm = Su(av)/Sy

HOGGING	**VLCC (105 elements)**				**ENERGY CONCENTRATION (90 elements)**				**CONTAINER SHIP (99 elements)**			
	S(ult) of elements		Ult. Bending Moment		S(ult) of elements		Ult. Bending Moment		S(ult) of elements		Ult. Bending Moment	
"σ–ε" Model (STR) (same shape)	Sm = Su(av)/Sy	Sm/Sm(av)	Mu 1000 MN.m	Mu/M(av)	Sm = Su(av)/Sy	Sm/Sm(av)	Mu 1000 MN.m	Mu/M(av)	Sm = Su(av)/Sy	Sm/Sm(av)	Mu 1000 MN.m	Mu/M(av)
model 1	0.823	1.035	28.60	1.013	0.874	1.022	18.71	1.023	0.756	1.000	6.472	1.016
model 2	0.820	1.031	29.29	1.037	0.871	1.019	18.64	1.019	0.776	1.026	6.5	1.020
model 3	0.757	0.952	27.26	0.965	0.837	0.979	17.76	0.971	0.753	0.996	6.26	0.982
model 4	0.782	0.983	27.82	0.985	0.838	0.980	18.05	0.987	0.740	0.979	6.26	0.982
	0.7955 = Sm(av)	**3.27%** Mean Variation	28.2425 = M(av)	**2.49%** Mean Variation	0.855 = Sm(av)	**2.05%** Mean Variation	18.29 = M(av)	**2.10%** Mean Variation	0.75625 = Sm(av)	**1.31%** Mean Variation	6.373 = M(av)	**1.77%** Mean Variation

In addition to these rationally-based "σ–ε" curves (HULLST, Rahman-Hughes, etc.), five simplified "σ–ε" curves are considered (see Figure 3, Shape 1 to Shape 5). These simplified curves are built by PROCOL and are composed of a perfect elastic deflection, a plastic deflection (*ultimate strength's plateau)* and a linear post-collapse deflection. All consider a buckling effect and have the same "σ_u" compressive ultimate strength. They only differ by the shape of their post-collapse residual strength:

Shape 1:	no residual strength after ultimate strength,
Shape 2:	a sharp reduction of the residual strength and no ultimate strength's plateau,
Shape 3:	a smooth reduction of the residual strength and no ultimate strength's plateau,
Shape 4:	a smooth reduction of the residual strength and a long ultimate strength's plateau,
Shape 5:	no reduction of the strength after the ultimate strength.

HULLST provides the reference "σu" compressive ultimate strength and reference "σ–ε" curve. To compare the different "σ–ε" curves of the same element, we classify the "σ–ε" curves based on:
- the element ultimate compressive average-stress and associated average-strain (noted "σ_u" and "ε_u"),
- the "shape" of the "σ–ε" curve.

2.3 Assessment of the Sensitivity of the "σ_u" Ultimate Strength (same shape)

The "σ_u" compressive ultimate strength of each element is evaluated with 4 different models. Then, 4 different "σ–ε" curves are obtained for each element. They have the same shape, i.e. same elastic regime, same "*ultimate strength's plateau*" (*plateau*) and post-collapse strength (e.g. curves 3 and 7 of Figure 1). They only differ one to the other by the level of "σ_u". Then, using the same PCA algorithm (model), these 4 sets of "σ–ε" curves are used to compute the (Mu) ultimate bending moment. Table 2 gives the sensitivity of the "σ_u" element ultimate strength on the ultimate bending moment (Mu). It is observed that a variation of 1% on the ultimate strength of the elements induces a variation of the ultimate bending moment of 1% and 2% for, respectively, hogging and sagging. This means that the uncertainty on the (Mu) sagging ultimate bending moment is the double of the uncertainty on the "σ_u" element ultimate strength.

2.4 Assessment of the Sensitivity of the Shape of the "σ–ε" Curve (Same "σ_u")

To assess the impact of the "shape" of the stress-strain curves on the ultimate bending moment (Mu), a second group of "σ–ε" curves is defined. They are characterised by the same "σ_u" element ultimate strength but different shapes. As 5 different standard shapes are considered (see figure 3), 5 sets of "σ–ε" curves are defined. Then, the sensitivity of the "shape" of the "σ–ε" curves on the (Mu) ultimate bending moment (see Table 3) are evaluated by using the 5 sets of "σ–ε" curves in the PROCOL model.

Table 3 shows that the shape sensitivity on (Mu) is higher for sagging than hogging (double). Between *shape 1* (no post collapse strength) and *shape 5* (perfect elastoplastic strength) the average variation is about 25% in sagging and 12% in hogging. Comparison between *shape 3* and *shape 4* shows that the length of the "*ultimate strength's plateau*" is the more sensitive parameter.

An overestimation of the length of the *plateau* (shape 4) increases the ultimate bending moment of 10 to 20 %. On the other hand, shape 2, which has no plateau, provides more conventional results (i.e. close to the reference value). Mu is similar for the VLCC, is underestimated by 3% for Energy Concentration and about 7% for the container. However, it should be noticed that the actual length of the *plateau* depends on the scantlings of the panel and the stiffeners.

TABLE 3

SENSITIVITY OF THE SHAPE OF THE "σ–ε" CURVES [STEP 2.2] ON THE ULTIMATE BENDING MOMENT OF HULL GIRDERS (VLCC; ENERGY CONCENTRATION AND CONTAINER SHIP)

(Same mesh model [step 1]; same element ultimate strength [σ_u ;step 2.1] and same PCA model [step 3])

SAGGING	VLCC (see Figure 4) 105 elements		ENERGY CONCENTRATION 90 elements		CONTAINER 99 elements	
"σ–ε" Curve Shape (same σ_U, ε_U)	Mu 1000 MN.m	Mu/M(ref)	Mu 1000 MN.m	Mu/M(ref)	Mu 1000 MN.m	Mu/M(ref)
M(ref) (HULLST)	20.42		16.84		6.72	
Shape 1	19.64	0.962	16.27	0.966	5.77	0.859
Shape 2	20.38	0.998	16.28	0.967	5.89	0.876
Shape 3	20.94	1.025	16.43	0.976	6.08	0.905
Shape 4	25.65	1.256	18.91	1.123	7.15	1.064
Shape 5	27.91	1.367	19.39	1.151	7.35	1.094
Mean Variation =		13.77%		7.32%		10.35%

M(ref) = Ultimate bending moment obtained with the HULLST software (Yao 1999)

HOGGING	VLCC (see Figure 4) 105 elements		ENERGY CONCENTRATION 90 elements		CONTAINER 99 elements	
"σ–ε" Curve Shape (same σ_U, ε_U)	Mu 1000 MN.m	Mu/M(ref)	Mu 1000 MN.m	Mu/M(av)	Mu 1000 MN.m	Mu/M(av)
M(ref) (HULLST)	28.88		19.03		6.72	
Shape 1	28.25	0.978	18.49	0.972	5.95	0885
Shape 2	28.89	1.000	18.49	0.972	6.27	0.932
Shape 3	29.29	1.014	18.69	0.982	6.50	0.967
Shape 4	31.03	1.074	19.87	1.044	7.06	1.050
Shape 5	31.26	1.082	20.41	1.072	7.11	1.058
Mean Variation =		3.86%		3.82%		6.46%

Figure 4 shows the (M-ϕ) moment-curvature relationships obtained for the VLCC with 5 different sets of "σ–ε" curves (simplified shapes) and the HULLST's set of curves (same mesh model, same "σ_u" element ultimate strength and same PCA algorithm). An example of these "σ–ε" curves for a typical element are presented at Figure 3. This element corresponds to the '*element type #2*' of the mesh model presented by *Yao et al*, 2000.

Sensitivity analysis on the ultimate bending moment with respect to various factors (plate and stiffener thickness', yield stress, residual stress, initial deflections and frame spacing) is included in *Yao et al.* (2000).

3 CONCLUSIONS

Different "σ–ε" curves have been used with the same (PCA) progressive collapse analysis model (STEP 3) and using the same structure mesh model (STEP 1). The comparison of the results assesses the sensitivity of the "σ–ε" curves on the (Mu) ultimate bending moment. Sensitivity of the "σ_u" element ultimate strength and the sensitivity of the shape of the "σ–ε" curves have also been performed. The results show that the main significant factor in the complete procedure of evaluation of the ultimate bending moment is the STEP 2.1: Evaluation of the "σ_u" ultimate compressive strength of the elements/components (Su). Moreover the results show that the "shape" (Step 2.2) is not negligible and particularly the length of the *ultimate strength's plateau*. Such a conclusion is not surprising as the main uncertainties and numerical assumptions are linked with STEP 2.1 (for instance: modelling of the initial imperfections and the boundary conditions, local buckling, tripping, bi-axial compression, lateral loading,...).

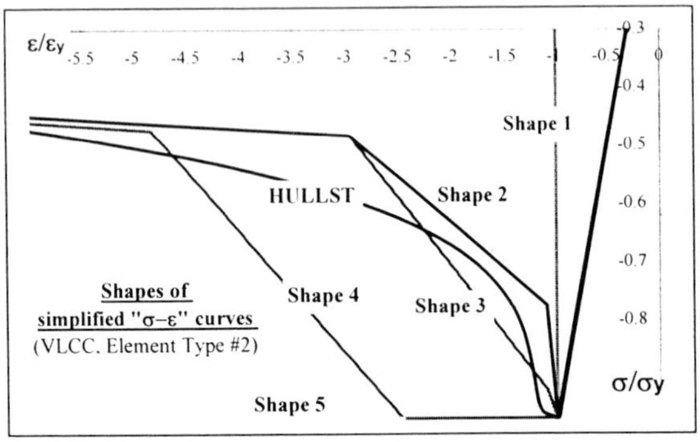

Figure 3: Shapes of the simplified "σ–ε" element curves (with same σ_u) for the sensitivity analysis of the shape of "σ–ε" curve (STEP 2.2).

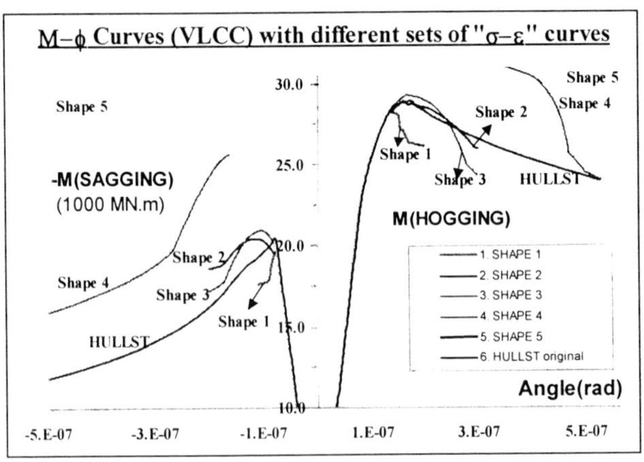

Figure 4: Moment-curvature relationships (VLCC) with different "σ–ε" curves (same mesh model, same "σ_u" element ultimate strength and same PCA model). Sensitivity analysis of the shape of "σ–ε" curve (STEP 2.2)

References

- Adamchack J.C. (1984). Approximate Method for Estimating the Collapse of a Ship's Hull in Preliminary Design, *Proc. Ship Structure Symposium '84*, SNAME (Ed.). 37-61.
- Dowling et al. (1991). Design of Flat Stiffened Plating: Phase 1 Report. CESLIC Report SP 9. Dept. of Civil Engineering. Imperial College, London.
- Hughes, O.F. (1988). *Ship Structural Design: A Rationally-Based, Computer-Aided Optimization Approach*. Edited by the SNAME, New Jersey, USA, 566p.

- Paik, JK, and Mansour AE. (1995). A Simple Formulation for Predicting the Ultimate Strength of Ships. *J. of Marine Science and Technology*. **1**, 52-62.
- Quesnel, Pradillon, Toderan, Rigo. (2000). Ultimate Strength of Longitudinally Stiffened Panels: Multi-criteria Comparative Analysis. *Proc. 8th Int. Symp. PRADS*, Elsevier Science Ltd., Shanghai, China, 7p.
- Rahman R.K., Chowdhury M. (1996). Estimation of Ultimate Longitudinal Bending Moment of Ships and Box Girders. *Journal of Ship Research*. **40:3**,.244-257.
- Rigo Ph. (1998). Constraints – Ultimate Strength – Hull Girder. Chap IV of LBR-5 an Optimization Model for Naval and Hydraulic Structures. Thesis: Agrégation Enseignement Supérieur, University of Liege, Belgium, 378p. (in French)
- Smith C.S. (1977). Influence of Local Compressive failure on Ultimate Longitudinal Strength of ship's hull, *Proc. Int. Symp. PRADS'77*, Tokyo, Japan. 73-79.
- ULSAP (2000). *ALPS/ULSAP: User's Manual*. Proteus Engineering (proteusengineering.com). USA, 46p.
- Yao, T, and Nikolov, PI. (1991). Progressive Collapse Analysis of a Ship's Hull under Longitudinal Bending (Part 1). *J. Soc. Naval Arch. of Japan*. **170**, 449-461. (Part 2 : **172**, 437-446 in 1992).
- Yao, T. (1999). Ultimate Longitudinal Strength of Ship hull Girder; Historical Review and state of Art. *Int. J. Offshore and Polar Engineering*. **9:1**, 1-9.
- Yao et al. (2000). Report of Committee VI.2: Ultimate Hull Girder strength, *Proc. 14th International Ship and Offshore Structures Congress*, Ohtsubo and Sumi (Eds), Elsevier. **2**, 321-391.

Practical Design of Ships and Other Floating Structures
You-Sheng Wu, Wei-Cheng Cui and Guo-Jun Zhou (Eds)

ASSESSMENT OF ULTIMATE LONGITUDINAL STRENGTH OF AGED TANKERS

A. Ikeda,[1] T. Yao,[2] O. Kitamura,[3] N. Yamamoto,[4] M. Yoneda,[2] and H. Ohtsubo[5]

[1]Maritime Bureau, Ministry of Land, Infrastructure and Transport,
2-3-1, Kasumigaseki, Chiyoda-ku, Tokyo 100-8918, Japan
[2]Graduate School of Engineering, Osaka University, 2-1, Yamadaoka, Suita, Osaka 565-0871, Japan
[3]Nagasaki Research & Development Center, Mitsubishi Heavy Industries Ltd.,
5-717-1, Fukahori-machi, Nagasaki, Nagasaki 851-0392, Japan
[4]Research Center, Nippon Kaiji Kyokai, 1-8-3, Ohnodai, Midori-ku, Chiba, Chiba 267-0056, Japan
[5] Graduate School of Engineering, University of Tokyo, 7-3-1, Hongo, Bunkyo-ku,
Tokyo 113-8656, Japan

ABSTRACT

The influence of thickness diminution due to corrosion on the ultimate longitudinal strength is investigated. At the beginning, the characteristics of corrosion damage in existing single hull tankers are investigated. It was found that the diminution ratio of the thickness at deck/bottom part is almost the same as that of the section modulus of the hull girder cross-section. Then, two series of progressive collapse analyses are performed on hull girders of eleven single hull tankers subjected to longitudinal bending, and the ultimate longitudinal strength was evaluated. Both the FEM and the Smith's method are applied for analyses. Almost linear relationship was found between the reduced section modulus and the ultimate longitudinal strength of aged single hull tankers with corrosion damage. It was concluded that the reduction of the section modulus should be within 15% of its original value to prevent break-up of the hull girder in case of single hull tankers.

KEYWORDS

Aged ship, Thickness diminution, Ultimate longitudinal strength, FEM, Smith's method, Progressive collapse analysis

1 INTRODUCTION

Casualty of M.V. Nakhodka in 1997 caused a serious environmental pollution by spilt oil along the coast of Japan Sea. The investigation committee was established by the Japanese government to find out causes of the casualty. It was found that one of the causes of this casualty was the thickness reduction in panels and stiffeners due to heavy corrosion, which led to the reduction in ultimate

longitudinal strength.
It was considered that an international regulation is necessary to prevent the occurrence of similar casualties. To provide technical data for the proposal of such regulation, a research panel was newly established in Japan, and the ultimate longitudinal strength of aged single hull tankers was investigated. This paper summarises the technical aspect of the results of investigations carried out in the panel.
First, the characteristics of thickness diminution due to corrosion was thoroughly investigated on the basis of measured data on existing tankers. The influence of thickness diminution on section modulus of the hull girder was analysed.
At the same time, progressive collapse analyses were carried out on three existing tankers applying the FEM and on eleven tankers applying the Smith's method both considering the influences of buckling and yielding of structural members. On the basis of the calculated results, the influence of thickness diminution on ultimate longitudinal strength was investigated. The calculated ultimate longitudinal strength was compared with the design bending moment specified by the IACS rule, and the safety of the aged tankers is discussed in relation to the thickness diminution due to corrosion.

2 CHARACTERISTICS OF CORROSION DAMAGE IN EXISTING SINGLE HULL TANKERS

2.1 Survey of Plate Thickness Measurement Record

Data of diminution of plate thickness were collected from the plate thickness measurement record. The data at 184,899 points in total were obtained from 307 measurement records on 144 tankers (JG 2000). Large scatter was observed in the measured data depending on the locations of measurements as well as object ships. Stochastic characteristics of measured data were investigated including the relationships among diminutions in different members and that between diminutions of structural members and section modulus of a hull girder.

2.2 Correlation between Diminution of Sectional Area and Diminution of Section Modulus

Figure 1 shows the correlation between diminution ratios of the flange sectional area and the section modulus. Here, flange is defined as the deck and/or bottom plating with longitudinal stiffeners and girders. Figure 1 includes data from 25 tankers which are classified into Group 1 (over 150 kDWT), Groupe 2 (80 to 150 kDWT), Groupe 3 (60 to 80 kDWT) and Groupe 4 (below 60 kDWT).
The correlation coefficient between two diminution ratios is 0.99 or more in both upper deck and bottom parts of the vessel, and the correlations are very good. This indicates the possibility to estimate the diminution ratio of the section modulus on the basis of the diminution ratio of the flange sectional area.

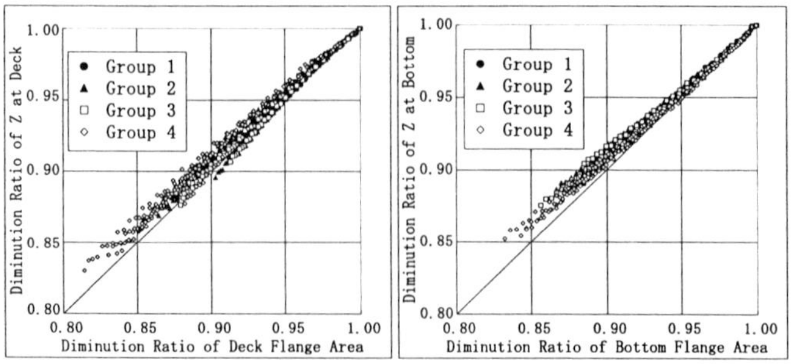

Figure 1: Correlation between diminution ratios of flange sectional area and of section modulus

3 SINGLE HULL TANKERS FOR STRENGTH ANALYSIS AND ASSUMED DIMINUTION

For strength analysis, 11 single hull tankers were chosen as object vessels, which are classified into 3 groups according to DWT. Principal dimensions and built years of individual vessels are indicated in Table 1 (JSRA 1999, 2000).

TABLE 1
PRINCIPAL PARTICULARS OF OBJECT SHIPS

Ship ID	Built Year	Lpp(m)	D(m)	DWT(ton)
A	1982	195	16.3	40,000
B	1989	265	23.8	146,000
C	1993	319	31.5	290,000
D	1992	315	30.53	281,018
E	1989	222	20.3	95,987
G	1991	215	20.4	70,887
H	1975	260	22.4	139,414
J	1975	265	21.5	134,970
L	1981	219	18.3	58,508
M	1979	200	18.7	59,535
N	1976	197	18.3	39,999

As for the degree of thickness diminution, an extreme case was considered, in which individual structural members suffer the diminution of their limit values specified by classification society rule. This diminution is denoted as Δt_{max}, and is called the case with full diminution. In this case, the diminution is roughly 25% of the original thickness at the deck and bottom plating, and is about 30% of the original for longitudinal stiffeners. The cases with diminution of a half of the extreme case and without diminution are also analysed.
The minimal guaranteed value of the yield strength was not used; instead, lower bound of the actual value was chosen for more realistic estimation.

4 METHODS OF ANALYSES

4.1 Finite Element Method

A series of detailed 3-dimensional finite element analysis was performed on Ships A, B and C applying an explicit finite element code, LS-DYNA3D, which was specially customised for the collapse simulation of the ship structure (JSRA 1999). In the present analysis, the panels were assumed to be accompanied by initial deflection of a hungry-horse mode. For the longitudinal stiffeners, initial deflections of flexural and torsional buckling modes were imposed. The welding residual stress was not considered. Combined bending moment and direct pressure load were applied.

4.2 Smith's Method

Another series of progressive collapse analysis was also performed on 11 vessels in Table 1 applying the Smith's method (Smith 1977). The computer code, HULLST (Yao *et al.* 1992) was used for this analysis (JSRA 2000).
In the Smith's method, the cross-section of a ship's hull is divided into stiffened plate elements composed of a stiffener and attached plating. At the beginning, the average stress–average strain relationships for individual elements are derived considering the influences of buckling and yielding. Then, progressive collapse analysis is performed assuming that the plane cross-section remains plane

during progressive collapse. A curvature is imposed incrementally. In this process, all the elements are assumed to behave following their average stress-average strain relationships, and the shift of a neutral axis of the cross-section caused by buckling and yielding of individual elements is considered.
In the present analysis, similar initial deflections as those in the FEM analysis were imposed on plates and stiffeners. In addition to this, welding residual stress was given in the panel, although the influence of direct pressure load on bottom and side shell plating was not considered.

5 RESULTS OF PROGRESSIVE COLLAPSE ANALYSIS AND DISCUSSIONS

5.1 Results of Analysis Applying the FEM

Figure 2 shows the collapse mode obtained by the FEM analysis on Ship B with full diminution under the sagging condition. It is seen that the buckling/plastic collapse takes place at a certain cross-section with the concentration of plastic deformation. At this cross-section, tripping of the deck longitudinals is observed as well as local collapse of the deck plating and overall collapse as a stiffened panel by buckling. On the other hand, no collapse is observed in the remaining part, and this results in the collapse of a jackknife mode. Analysis was performed for three cases assuming no, a half and full diminutions, and the evaluated ultimate strength is plotted by solid marks against diminution ratio of the section modulus of a hull girder in Fig. 4 together with those obtained by the Smith's method.

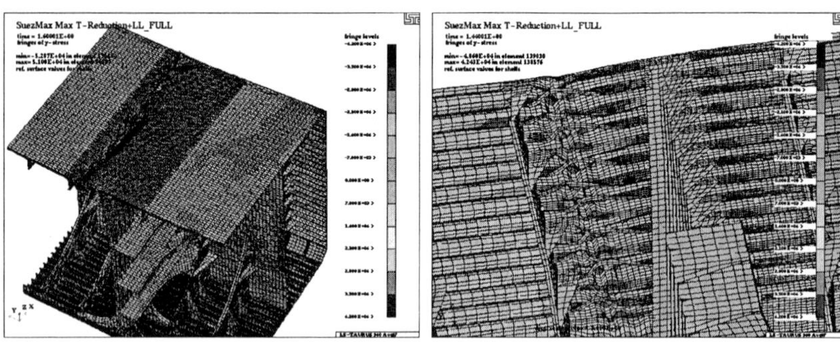

Figure 2: Collapse mode of Suezmax tanker in sagging

5.2 Results of Analysis Applying the Smith's Method

Figure 3 shows the relationship between the bending moment and the curvature for Ship D. It is seen that the flexural rigidity of the hull girder and the ultimate longitudinal strength are reduced in accordance with the magnitude of thickness diminution. With full diminution, the reduction ratio of the ultimate longitudinal strength is about 33%.
In sagging, the ultimate strength is attained soon after the deck has been collapsed by buckling. Contrary to this, in hogging, the buckling collapse of the bottom dominates the collapse of a hull girder under longitudinal bending.

5.3 Comparison of Ultimate Longitudinal Strength Obtained by Different Methods

The ultimate strength obtained by the FEM analysis is compared with that by the Smith's method in Fig. 4. A comparison of the ultimate strength proves that both results show good agreement each other by a mean difference of about 5%. The FEM analysis gives higher ultimate strength than the Smith's method in general. This can be attributed to the following causes: (1) The simplified analysis was performed under the assumption that there was no interaction between adjacent elements and that the cross-section of the hull girder remains plane, i.e., compressive strain level should increase linearly with respect to the distance from the neutral axis; (2) The FEM analysis has a tendency to give upper bound strength because of the relatively coarse mesh of practical degree. Consequently, it can be concluded that the actual ultimate longitudinal strength is intermediate between the results applying the FEM and the Smith's method.

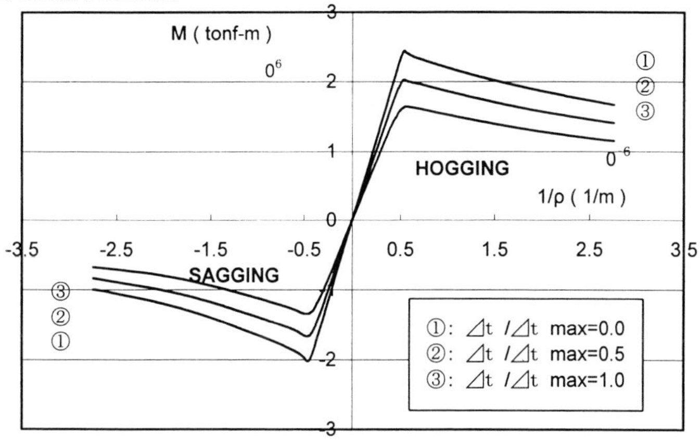

Figure 3: Bending moment-curvature relationships by Smith's method (Ship-D)

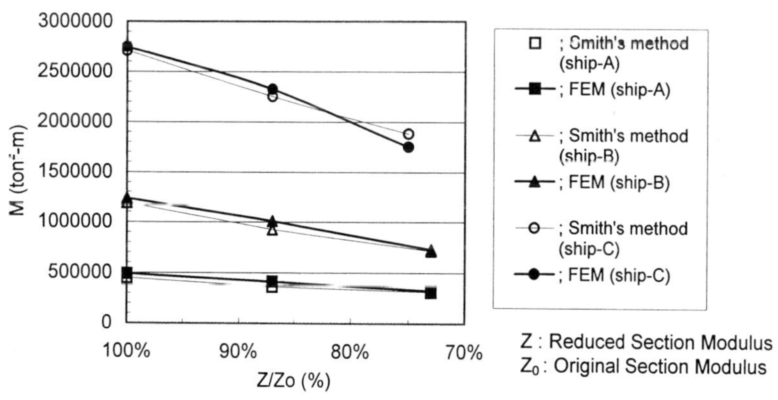

Figure 4: Comparison of the ultimate longitudinal strength applying the FEM and the Smith's method

5.4 Strength Assessment of Aged Single Hull Tankers

On the basis of the ultimate longitudinal strength evaluated by the Smith's method, the safety factor is calculated, which is defined as the ratio of the evaluated ultimate longitudinal strength to the maximal bending moment expected to encounter during the service period.

The safety factors are plotted against the diminution ratio of the section modulus in Fig. 5. It is shown that the upper limit of diminution ratio in terms of section modulus at which the ultimate longitudinal strength coincides with the maximal bending moment corresponds to approximately 15 % of the original value. Accordingly, the section modulus should not be less than 85 % of the original value from the viewpoint of ultimate longitudinal strength. This implies that the diminution of the flange area should be less than 15 % of the original value according to the results in Chapter 2.

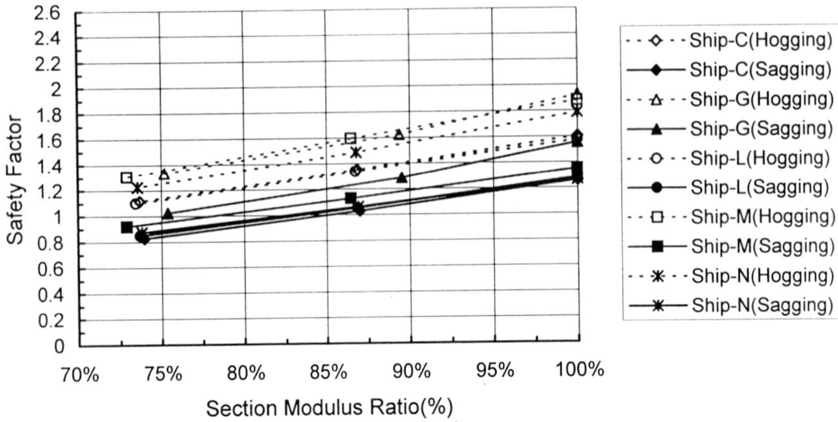

(a) 40-71 KDWT Tankers

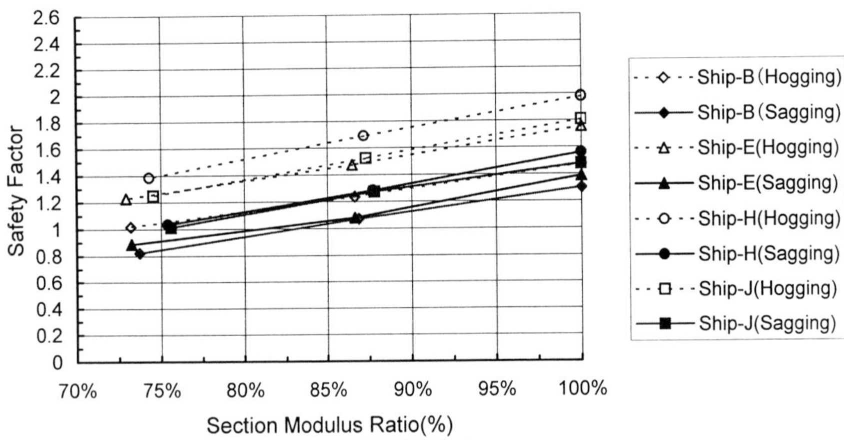

(b) 96-146 KDWT Tankers

Figure 5: Relationship between safety factor and diminution ratio of section modulus

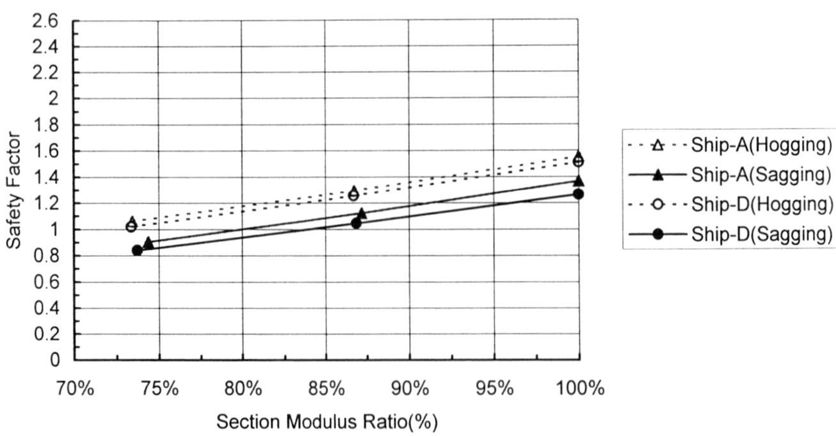

(c) 280-290 KDWT Tankers

Figure 5: Relationship between safety factor and diminution ratio of section modulus (continued)

6 CONCLUSIONS

In the present paper, results of the research to assess the ultimate longitudinal strength of aged single hull tankers are summarised. It has been found that:

(1) Diminution ratio of the section modulus of hull girder is approximately the same as that of the flange area of the deck and/or bottom plating with longitudinal stiffeners and girders
(2) The ultimate longitudinal strength evaluated by the Smith's method shows good agreement with that obtained by the FEM.
(3) Linear relationship is observed between the diminution ratios of the ultimate longitudinal strength and the section modulus.
(4) Diminution of 15% is allowable at the deck and the bottom plating from the viewpoint of ultimate longitudinal strength.

References

JG (Japanese Government) (2000). Annex 2: Diminution of Section Modulus of Tankers due to Corrosion in "Amendments to the Assembly Resolutions A.744 (18)." *IMO DE43/INF.3.*

JSRA (Japan Shipbuilding Research Association) (1999) Investigation into Structural Safety of Aged Ships. *Report of No.74 Rule and Regulation Committee* (in Japanese).

JSRA (Japan Shipbuilding Research Association) (2000) Investigation into Structural Safety of Aged Ships. *Report of No.74 Rule and Regulation Committee* (in Japanese).

Smith C.S. (1977). Influence of Local Compressive Failure on Longitudinal Strength of a Ship's Hull Girder. *Proc. Int. Symp. on PRADS*. Tokyo, Japan. 73-79.

Yao T. and Nikolov P.I. (1992). Progressive Collapse Analysis of a Ship's Hull Girder under Longitudinal Bending (2nd Report). *J.Soc. Naval Architects of Japan* **172**, 437-446

progressive collapse of a hull girder: simplified method based on *Smith* method (Yao et al, 1991,1992), ISUM (Paik, 1996), plastic node method (Bai, 1993) and nonlinear finite element method (Chen et al, 1983). In the first part of this paper, a simplified method using the formulas of effective width of plating by Guedes Soares(1988) and three distinct zones of stiffened panel by Rahman and Chowdhury(1996) is applied to analyze the hull girders. Based on literatures and the authors' experience, an ultimate equation is suggested. The results by the present simplified method and ultimate strength equation agree well with those by ISSC committee VI.2 members.

The fabrication imperfections and the degradation effects due to corrosion defects and fatigue cracks the hull experiences are of inherently random nature. It is therefore necessary to evaluate the ultimate hull girder strength based on reliability methods. Moreover, because the degradation effects are time-dependent, a time-variant reliability method should be applied. In carrying out the reliability assessment based on progressive collapse analysis, the limit state function is very complex and may only be expressed implicitly. Among the methods available in literatures for solving such a problem, the response surface method is an effective and powerful tool, in which the limit state function is approximated by a simple and explicit function at the sampling points (Bucher & Bourgound, 1990). In the second part of this paper, the time-variant reliability assessment relative to the ultimate strength of the ship's hull girders is conducted.

2 SIMPLIFIED METHOD AND ULTIMATE STRENGTH EQUATION

In this Section, a simplified method to calculate the maximum value of the bending moment of ship's hull girders is proposed. The basic assumptions of the method are summarized as follows:

- The stiffened panel and corner elements, into which the cross section is subdivided, are considered to act and behavior independently;
- The cross section is assumed to be plane and to remain plane when its curvature is increasing;
- The overall grillage collapse of the deck and bottom structures is avoided by using sufficiently strong transverse frames.

A stiffened panel is composed of a longitudinal stiffener with its attached plating. Following the approach of Rahman and Chowdhury (1996), three distinct zones in the whole range of the stiffened panel's loading-shortening behavior are considered. The ultimate strength of a stiffened panel is given by

$$\sigma_{pu} = \min(\sigma_{uf}, \sigma_{up}) \tag{1}$$

where σ_{uf} and σ_{up} are the ultimate beam-column failure values of the panel when lateral pressure causes compression of stiffener flange and plating respectively. According to Hughes (1983), a solution may be obtained by solving the following equations for stiffener failure,

$$\sigma_y = \sigma_{uf} + \frac{M_0 y_f}{I} + \frac{\sigma_{uf} A(\delta_0 + \Delta) y_f}{I} \Phi \tag{2}$$

and for plate failure,

$$\sigma_y = \sigma_{up} + \frac{M_0 y_p}{I_e} + \frac{\sigma_{up} A_e(\delta_0 + \Delta + \Delta_p / \Phi) y_p}{I_e} \Phi \tag{3}$$

where Δ is the initial eccentricity, δ_0 and M_0 are the maximum deflection and bending moment due to lateral load alone; Δ_p is the eccentricity caused by reduced stiffness of compressed plating; I and A are respectively the second moment and the sectional area of the panel, considering b_p (panel width) is fully effective where as I_e and A_e are the similar properties but for transformed section replacing b_p by b_{pe}(effective width); y_f is the distance from the panel neural axis to the stiffener flange and y_p to the plating of the transformed section; Φ is the magnification factor for combined loading. The effective width of the plating is calculated using Soares formulas (1988).

To predict the crack propagation and fatigue life, *Paris-Erdogan* equation is used

$$\frac{da}{dN} = C\Delta K^m \tag{4}$$

where a is the crack size, N is the number of cycles, ΔK is the stress range intensity factor and C and m are material parameters. The complete fatigue life T_f is equal to the sum of the time to crack propagation T_p with the time to crack initiation T_i,

$$T_i = kT_p \tag{5}$$

where k can vary from 0.1 to 0.15. The crack size is assumed to have a normal distribution with the mean and variance, see Guedes Soares (1999).

The corrosion effect is time dependent. It has been generally assumed that corrosion defect developments may be divided into three phases. In the first one, corrosion rate is zero because coatings provide corrosion protection. The second phase is initiated when the protection coating is damaged and the defect depth then increases. In the third phase, corrosion rate may be assumed to be constant. A model for predicting corrosion depth was suggested as (Sun and Bai, 2001):

$$d(t) = r_s[t - (\tau_i + \tau_t) + \tau_t e^{-\frac{t-\tau_i}{\tau_t}}] \tag{6}$$

where τ_i is the coating lifetime, τ_t is the transition time and r_s is the steady corrosion rate which should be fitted by inspection results. The coating lifetime τ_i may be assumed to follow Weibull distribution and r_s to be normal distribution.

From the viewpoint of engineering, a stiffened panel is considered ineffective when the crack size exceeds the critical value determined by CTOD-method (Ghose et al, 1995) or corrosion induced thickness reduction exceeds 25% of the nominal plate thickness.

The ultimate moment capacity obtained from the simplified method is the maximum value on the bending moment-curvature curve and so it is time consuming when reliability analysis relative to the ultimate strength failure mode is carried out by means of the simplified method. Therefore, some ultimate strength equations have been proposed based on various assumptions of stress distribution over the cross-section. For instance, Xu and Cui (2000) assumed a stress distribution in which the middle 1/3 of the hull depth is elastic while the rest of the hull depth is plastic/ultimate strength

The present authors suggest that the ultimate moment capacity M_u can be predicted by the following equation:

$$M_u = \sum_i \sigma^c_{u_i} A_{ps_i} z_i + \sum_j \sigma^t_{u_j} A_{ps_j} z_j + \sum_k \sigma^e_k A_{ps_k} z_k \tag{7}$$

where A_{ps} is the area of stiffened panels/hard corners and z is the distance to the neutral axis. Fig.1 shows a schematic diagram of stress distribution under sagging condition. The stress distribution used in Eqn.(7) does account for the ultimate strength of individual stiffened panels and hard corners, e.g. it is not uniformly distributed as Fig. 1.

In Eqn.(7), the compressive ultimate strength region, tensile ultimate strength region and elastic region are denoted by i, j, k respectively. $\sigma^c{}_u$ is the ultimate compressive strength which is obtained from Eqn(1) for stiffened panels or yield stress for hard corners. $\sigma^t{}_u$ is the ultimate tensile strength (yield stress). Elastic stress σ^e has a linear distribution around the neutral axis. Based on observations of stress distribution from more comprehensive numerical analysis, it is suggested by the present authors that the height of the elastic region is $H/6$ for the compressive region, and $H/3$ for the tensile region.

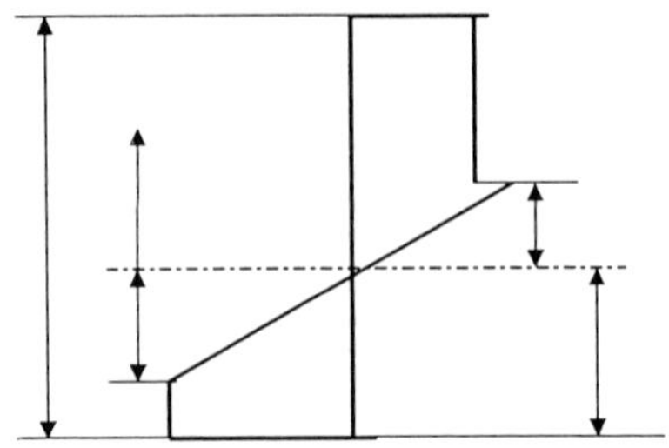

Based on Eqn.(7), the ultimate moment capacity of a hull girder may be estimated by following steps:

(1) Subdivide the cross section into stiffened panels and hard corners;
(2) Estimate the ultimate strength of each stiffened panel according to Eqn. (1) or other recognized formulas;
(3) Calculate the distance "g" from the ship bottom to the neutral axis by assuming the total force from the stress integration over the cross-section is zero;
(4) Calculate the ultimate moment capacity of the hull girder using Eqn.(7).

3 RELIBILITY ASSESSMENT

The limit state function at time t is given by

$$g(\mathbf{x}|t) = \chi_u M_u(t) - [\chi_s M_s(t) + \phi_w \chi_w M_w(t)] \tag{9}$$

where $M_u(t)$ is the ultimate moment capacity, $M_s(t)$ and $M_w(t)$ are the stillwater and vertical wave-induced bending moments respectively; χ_u , χ_s and χ_w represent the model errors in predicting the hull's ultimate moment capacity, SWBM and VBWM the ship experiences.

The failure probability at the time T, which is assumed as a uniform distribution, is expressed by

$$P_f(T_s) = \frac{1}{T}\int_0^T P_f(t)dt \tag{10}$$

where $P_f(t)$ is a conditional failure probability at time t,

$$P_f(t) = \int_{g(\mathbf{x}|t)<0} \cdots \int f_X(\mathbf{x}|t)d\mathbf{x} \tag{11}$$

SWBM can be modeled by a Poisson rectangular pulse process in a ship service life. Its cumulative distribution at any time is fitted by a normal distribution and given by

$$\Phi_{M_s}(M_s) = \frac{1}{\sqrt{2\pi}\sigma_{M_s}} \int_0^{M_s} \exp\left[-\frac{1}{2}\left(\frac{M_s - \mu_{M_s}}{\sigma_{M_s}}\right)^2\right] dM_s \tag{12}$$

where μ_{M_s} and σ_{M_s} are the mean and standard variance. When the COV of SWBM, $\xi_{M_s} = \sigma_{M_s}/\mu_{M_s}$, is determined, μ_{M_s} can be given at $1/v_s T_0$ exceedance level for the present calculation,

$$\mu_{M_s} = \frac{M_{s,0}}{1+\xi_{M_s}\Phi^{-1}(1-\frac{1}{\nu_s T_0})} \tag{13}$$

where ν_s is the mean arrival rate of one load condition and $M_{s,0}$ is the specified maximum SWBM in a design lifetime T_0 of 20 years, e.g. as given by IACS (1995).

VWBM is naturally stochastic process and may be described by either short-term or long-term statistics. It is generally accepted that the long-term VWBM can be modeled as a Poisson process and the peak of each individual VWBM, M_w, can be well approximated by a Weibull distribution.

$$F_{M_w}(M_w) = 1 - \exp\left[-\ln(\nu_w T_0)\left(\frac{M_w}{M_{w,0}}\right)^{h_w}\right] \tag{14}$$

where ν_w is the mean arrival rate of one wave cycle; h_w is the shape parameter varying from 0.9 to 1.1 and reasonably taking 1.0 as a representative value; $M_{w,0}$ is the maximum VWBM in the reference design period T_0=20 years. The Cumulative Distribution Function (CDF) for the largest of the individual VWBM can be found for a total of the $\nu_w T$ repetitions,

$$F_{M_{w,T}}(M_w) = [F_{M_W}(M_w)]^{\nu_w T} \tag{15}$$

The CDF of the combined bending moment $M_{t,T}$ can be obtained based on *Ferry-Borges* method and the load combination factor of VWBM, φ_w, is given by

$$M_{t,T} = M_s + \varphi_w M_{w,T}$$

In order to overcome the computational complexity in evaluating the failure probability using Eqn(11), the response surface method is applied. A response surface with a polynomial type function including squared terms but not cross terms is adopted:

$$G(\mathbf{x}) = a + \sum_{i=1}^{r} b_i x_i + \sum_{i=1}^{r} c_i x_i^2 \tag{16}$$

where r is the number of basic random variables; $\mathbf{x} = (x_1, x_2, \cdots, x_r)$ is the basic random vector; a, b_i and c_i are the unknown coefficients that can be determined using $2r$+1 sampling points. Once the response surface is defined, the failure probability is computed using modified Monte Carlo simulation technique (Sun & Chen, 1997).

4 ANALYSIS OF ISSC-2000 HULL GIRDERS

The ultimate moment capacity and time-variation reliability estimation of five hull girders used to ISSC-2000 benchmark study are evaluated considering the degradations of corrosion and fatigue. The relevant principal particulars are listed in Table 1. Table 2 summarizes variable reference measurements in this evaluation. The determination of the various variables is based on literatures and engineering judgment.

TABLE 1
PRINCIPAL PARTICULARS OF ISSC-2000 SHIP HULL GIRDERS

Description	Bulk Carrier	Container Ship	Double Hull VLCC	Single Hull VLCC	Frigate Model
Length, L (m)	285	230	315	313	18
Breadth, B (m)	50.2	32.2	58.0	48.2	4.2
Depth, H (m)	26.7	21.5	30.3	25.2	2.8
Block coefficient, C_B	0.82	0.60	0.86	0.83	—
Frame spacing, l (m)	5.22/2.61*	3.27	3.50	5.10	0.457

*Deck/bottom frame spacing

The calculation results of ultimate moment capacity for the hull girders are listed in Table 3. It may be observed that the results from the simplified method and ultimate moment capacity equation presented by the authors agree well with the ISSC committee report. Fig.2 shows the corresponding moment-curvature relationships for the hull girders.

TABLE 2
VARIABLE REFERENCE MEASUREMENTS

Variable	Distribution	Bulk Carrier	Container Ship	Double Hull VLCC	Single Hull VLCC	Frigate Model
Elastic modulus E	lognormal	2.06×10^5MPa (0.06)*	2.06×10^5MPa (0.06)	2.06×10^5MPa (0.06)	2.06×10^5MPa (0.06)	2.06×10^5MPa
Yielding stress σ_y	lognormal	392.0MPa[1] 352.8MPa[2] 313.6Mpa[3] (0.08)	352.8MPa[1] 313.6MPa[2,3] (0.08)	313.6MPa (0.08)	313.6MPa[1,3] 235.2MPa[2] (0.08)	245.0MPa
Stiffened panel eccentricity Δ/l	normal	0.001(0.1)				
Initial center deflection W/h_p	normal	0.01(0.1)			0.18(0.1)	0.01
Nominal steady corrosion rate r_s	normal	1.0×10^{-4}m/year(0.01)				—
Transition time τ_t	constant	3 years				—
Initial crack size a_0	normal	0.001m(0.18)				—
SWBM Ms	normal	$0.6M_{S,0}$(0.4)				—
VWBM M_w	extreme type I	Average arrival rate (ν_w) $=10^8$ in a 20 service life				—
Model error χ_u	normal	1.0 (0.05)				—
Model error χ_s	normal	1.0 (0.1)				—
Model error χ_w	normal	1.0 (0.25) for sagging condition				—

Note: * The value in parentheses is the coefficient of variance.
The values of yield stress with superscript 1,2 and 3 correspond to the values of deck, side shell and bottom grillages respectively.

TABLE 3
ULTIMATE STRENGTH RESULTS FOR ISSC-2000 HULL GIRDERS

Ship Name	Load Cond.	ISSC Mean (MNm)*	ISSC cov	Simplified Method (MNm)	Proposed Eq. (MNm)
Bulk Carrier	Sagging	1.52×10^4	0.07	1.53×10^4	1.53×10^4
	Hogging	1.86×10^4	0.04	1.72×10^4	1.70×10^4
Container Ship	Sagging	6.51×10^3	0.14	5.84×10^3	6.06×10^3
	Hogging	7.43×10^3	0.08	6.93×10^3	7.10×10^3
Double Hull VLCC	Sagging	2.24×10^4	0.11	1.98×10^4	2.23×10^4
	Hogging	2.91×10^4	0.04	2.76×10^4	2.68×10^4
Single Hull VLCC	Sagging	1.72×10^4	0.02	$1.46\times10^{4**}$	1.70×10^4
	Hogging	1.82×10^4	0.02	$1.79\times10^{4**}$	1.81×10^4
Frigate Model	Sagging	10.39	0.07	9.61	9.73
	Hogging	12.38	0.08	12.10	12.26

Note:* The results are obtained by averaging the values of all participators in ISSC VI.2 (2000).

**The external pressure is applied consistent with the analysis by Rutherford and Caldwell(1990)

Fig.3 shows the conditional reliability index as a time function under sagging condition.
Fig.4 illustrates the conditional reliability index as a time function under sagging condition with the initial crack size of 2mm. The time to unstable propagation of crack is significantly shortened and takes at the 10th, 12th, 13th and 20th service year for the bulk carrier, double hull VLCC, single hull VLCC and container ship respectively. The bulk carrier possesses the shortest duration and the single hull VLCC receives the highest reduction. It implies that the two types of single hull ships with less redundancy are vulnerable to sustain the fatigue damage.
Fig.5 demonstrates the conditional reliability index as a time function under sagging condition with the steady corrosion rate of 0.2mm/year. The single hull VLCC, bulk carrier, double hull VLCC and container ship reduce their the reliability index to 88.74%, 83.33%, 80.62% and 79.09% at 10th service year when compared with the initial value respectively. It indicates that the two types of single hull ships are also vulnerable to suffer corrosion wastage.

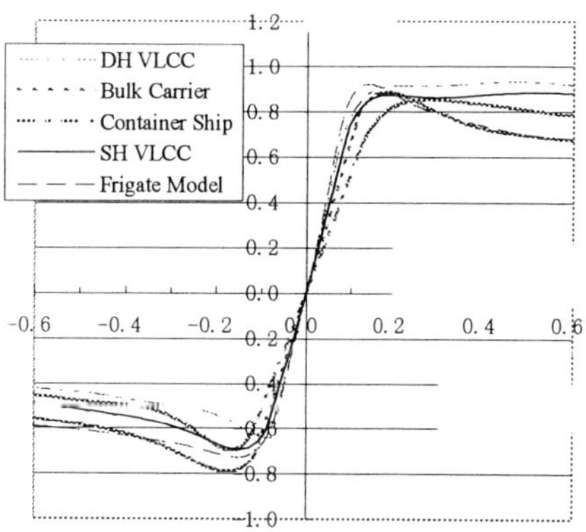

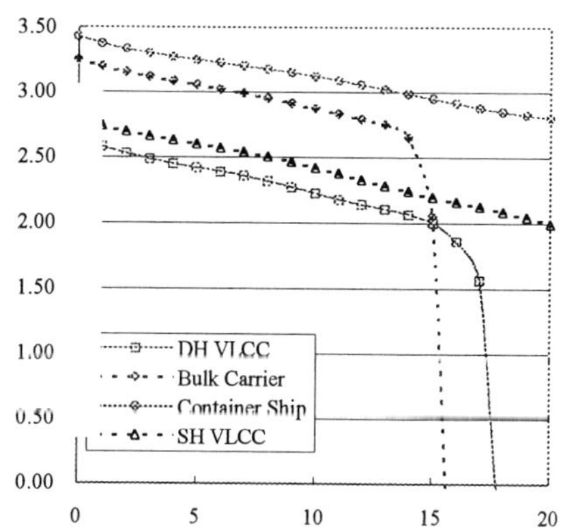

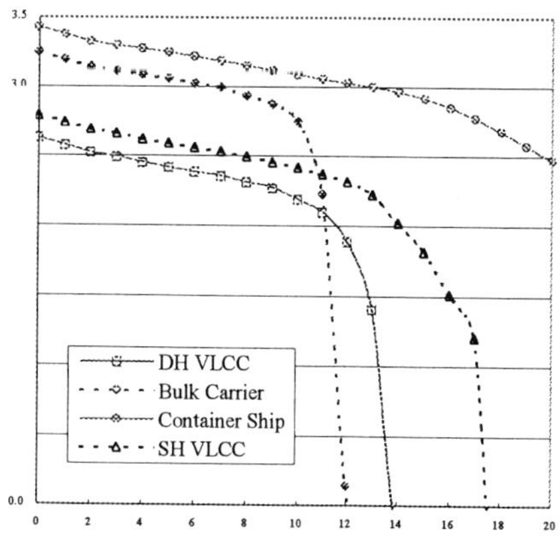

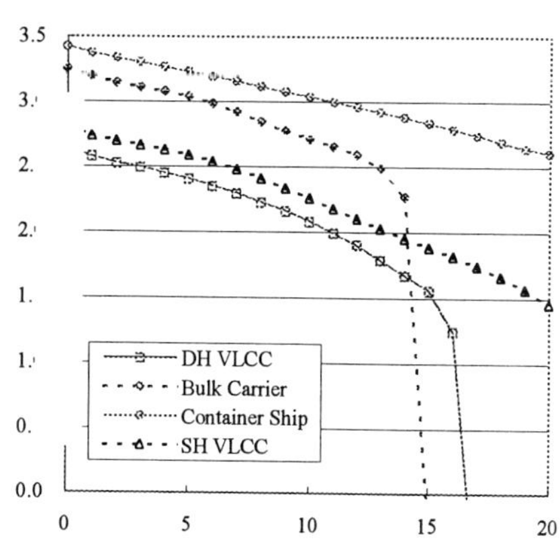

The results from the reliability assessment of different ship types might be used to explain why over 150 bulk carriers lost during past 20 years and the hull girder of "Energy Concentration" broke into two.

5 CONCLUSIONS

The ultimate strength and reliability analysis of ISSC-2000 hull girders were carried out. A simplified method combined with a formulas of effective width of plating by Guedes Soares(1988) and three distinct zones of stiffened panel (Rahman and Chowdhury,1996) was applied to evaluate the progressively collapse of the hull girders. An ultimate moment capacity equation was suggested for the future criteria development for hull girder collapse. It was demonstrated that the results obtained by the proposed simplified method and equation agree well with those from Yao et al (2000).
A time-variant reliability assessment method is applied to assess the reliability of the hull girders accounting for the time dependent degradation effects of corrosion and fatigue. It was found that the bulk carrier and single hull VLCC are weaker to sustain fatigue damage and corrosion wastage. However, the comparative study of different types of hull girders may not be generally interpreted, as the number of calculations is very limited.

References

Bai Y et al. (1993). Collapse Analysis of Ship Hulls. *Marine Structures* **6**, 485-507.

Bucher C G and Bourgound U A. (1990). A Fast and Efficient Response Surface Approach for Structural Reliability Problems. *Structural Safety* 7, 57-66.

Chen YK et al (1983). Ultimate Strength of Ship Structures. *SNAME Transactions* **91**, 149-168.

DnV.(2000). *Structural Design of Offshore Ships*. Offshore Standard OS-C102, Draft 23.6.

Ghose, DJ et al. (1995). *Residual Strength of Marine Structures*. Ship Structure Committee, SSC-381.

Guedes Soares C. (1988). Design Equation for the Compressive Strength of Unstiffened Plate Elements with Initial Imperfections. *Journal of Construction Steel Research* **9:4**, 287-310.

Guedes Soares C and Garbatov Y. (1999). Reliability of Corrosion Protected and Maintained Ship Hulls Subjected to Corrosion and Fatigue. *Journal of Ship Research* **43:2**, 65-78.

Hughes, O F.(1983). *Ship Structural Design*. New York : Wiley. 1983.

IACS rule.(1995). *IACS Requirement S11, Longitudinal Strength Standards*.

Paik J K et al. (1996). Ultimate Strength of Ship Hulls under Combined Vertical Bending, Horizontal Bending, and Shearing Forces. *SNAME Transactions* **104**, 31-59.

Rahman M K and Chowdhury M.(1996). Estimation of Ultimate Longitudinal Strength of Ships and Box Girders. *Journal of Ship Research* **40**, 244-257.

Rutherford S E and Caldwell J B. (1990). Ultimate Strength of Ships: A Case Study. *SNAME Transaction* **98**, 441-471

Sun H H and Chen T Y. (1997). Buckling Strength Analysis Of Ring-Stiffened Circular Cylindrical Shells Under Hydrostatic Pressure. *Proc. 7th ISOPE*, **4**, 361-366. Honolulu, USA.

Sun, H H and Bai Y. (2001) Time-variant reliability of FPSO's Hulls, Accepted for publication in SNAME Annual Conference and Transaction.

Xu X D, Cui, W. (2000).An Experimental and Theoretical Study on Ultimate Strength of A Box Girder. *Journal of Ship Mechanics* **4:5**, 36-43.

Yao T et al. (2000). Report of Special Task Committee VI.2, Proc. of 14th ISSC, Oct., 2000. Nagasaki, Japan.

Yao T and Nikolov P I (1991,1992). Progressive Collapse Analysis of A Ship's Hull under Longitudinal Bending (1st and 2nd Reports)". *J. SNAJ* **171**, 449-461 and **172**, 437-446.

Practical Design of Ships and Other Floating Structures
You-Sheng Wu, Wei-Cheng Cui and Guo-Jun Zhou (Eds)

AN ASSESSMENT OF THE ULTIMATE PLASTIC STRENGTH OF THE SHIP'S AGED HULLS

Gennadiy V. Egorov and Vitaly V. Kozlyakov

Marine Engineering Bureau, 65045, Odessa, Ukraine

ABSTRACT

Plastic fracture of hulls at sea and under cargo operations in practical situation has more disastrous natures than the theoretical pictures and occurs, as a rule, in one of the most weakened span with quick spreading of corrugation to the tensioned strake with the contravention of the hypothesis of flat sections. Maximum value of the ultimate moment is not stable and reduces to the nearly full release because of essential change of the structure after buckling and possible forming the viscous cracks, take-off longitudinals from the shell and etc. These particularities occurs very often in partly worn-out structures. The most important reason of nearly full release of the ultimate moment is a sharp reduction of the effective width of the compressed strake in the area of the plastic hinge. After nearly full release of the ultimate moment both parts of the hull functions as two pieces and have greater remaining deflection.

In this paper some particularities of ships hulls of ductile collapses are considered, in connection with the ductile hull's destruction of seagoing ships, river ships and floating dry docks. For stabilization and realization upper rate limits of ultimate hull girder strength are offered some structural advantange, ensuring high carrying capacity of compressed structures.

KEYWORDS

Ultimate hull girder strength, Initial imperfections, Service damages, Brittle and ductile fracture

1 INTRODUCTION

Problem of ensuring ultimate hull girder strength of seagoing cargo ship's became to be actual because of growing ship's dimensions and using high tensile steel (Kozlyakov (1995)). It is always interest for river and river-sea ships because of difficulties in ensuring stability of thin-web structures (Ershov et al.(1970), Kozlyakov et al. (2000), Papkovich (1941)). In 1958 Russian Maritime Register of Shipping introduced requirements to ensuring ultimate hull girder strength. Since 1974 in Rules of the Russian River Register these requirements were stated for the new hull and with service damages at the end a lifetime with reducing not only compressed, but tension-curved elements. Careful evaluation of the ultimate moment in these requirements is based on the hypothesis of flat sections and terms that stresses equal to yield point arises only in hard elements of one of the extreme strakes.

Smith (1977) had proposed procedures of the full ultimate moment determination based on the idea of Caldwell (1965). This procedure also used the hypothesis of flat sections using strength characteristics of each panel of the hull taking into account geometric and physical non-linear functioning in post buckling stage. So the ultimate moment is defined as the extreme of function relationship of bending moment and curvature. Kutt et al. (1985), Paik (1992), Ueda (1984) made programs, based on non-linearity schemes of finite element method. Careful test of accuracy of offered schemes was executed by Special Committee VI.2 ISSC-2000 (Yao (2000)). For practical purposes, aproximate method of Smith is recommended, which results in satisfactory comparison with experiments. In spite of these successes, process of ductile collapse during cargo operations on still water, as well as at sea stays far from clear. In this paper an attempt is made to explain some particularities of the hull ductile collapse process during cargo operations and sharp reduction of the ultimate moment under constant load with quick growing of deflection and to recommend some structural-technological and control actions for stability of the ultimate moment.

2 PARTICULARITIES OF THE SHIP HULL DUCTILE COLLAPSE

The most severe collapses are events of hull break-up of ships under cargo operations and floating docks. Very often break-up hulls of river ships, particularly non-self propelled (Kozlyakov et al. (2000)). But in connection with the growing of powers of cargo handling equipment more often break-up and large cargo ships, for example, tanker "Energy Concentration" in 1980 (Rutherford et al.(1990)) at hogging and OBO-Ship "Trade Daring" in 1994 at sagging. Arrows of the hogging and sagging have - 16 м and + 18 м accordingly.

For the clarification of main particularities of the process, let us consider the hull in the manner of the floating parallelepiped with symmetrical load (Fig.1a). Neglecting favourable influence of elastic foundation, one will obtain the value of bending moment in midship section and the ultimate load.

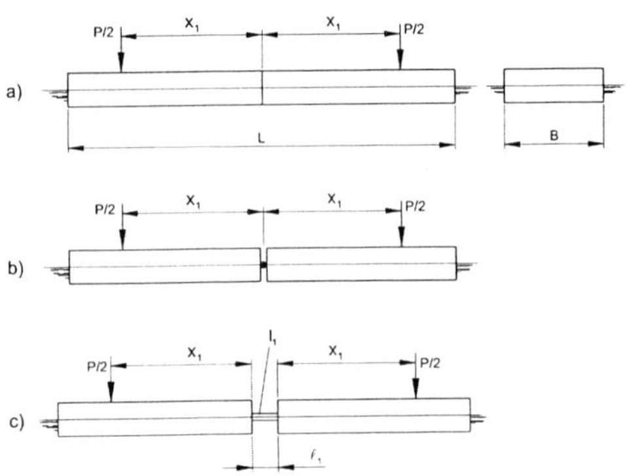

$x_1 > L/4$ – hogging,
$x_1 < L/4$ – sagging

$$M_{\otimes} = -\frac{PL}{8}\left(1-\frac{4x_1}{L}\right) \qquad (1)$$

$$|P_{ULT}| = \frac{8M_{ULT}}{L\left(1-\frac{4x_1}{L}\right)} \qquad (2)$$

Figure 1: Scheme of ships loading.

In accordance with traditional concept of plastic hinge value M_{ULT} stays constant and hull can carry additional load ΔP, as two-piece system (Fig.1b) with increment of trim angle each of piece

$$\Delta\psi = \frac{96\cdot\Delta M}{\rho g B L^3}; \qquad \Delta M = \frac{\Delta P}{8}\left(1-\frac{4x_1}{L}\right). \qquad (3)$$

Diagram of increment of maximum deflection with hinge amidships $\Delta w_{\otimes} = f(\Delta P)$ given on the Fig.2. It shows possibility of essential increasing load after arising plastic hinge, which absolutely disagree with actual data. Growing of deflection when $P = P_{ULT}$ is realized practically without increasing load, so as cargo operations stop and process of transition to the ultimate condition progressed very quickly because of nearly full release of the ultimate moment M_{ULT} in the plastic hinge $\Delta M = - M_{ULT}$. Value of deflection is also defined by formula (4), and $\Delta w = f_1(\Delta M)$ drawn by dashed line on the Fig.2. Quick growing of deflection is possible only under the sharp reduction of stiffness of the weakened section (Fig.1c). Really, from the condition of approximate equalities angles of rotation of weakened and hard parts of the hull will get value of reduction bending moment ΔM and increments of sagging $\Delta w_{\otimes}$

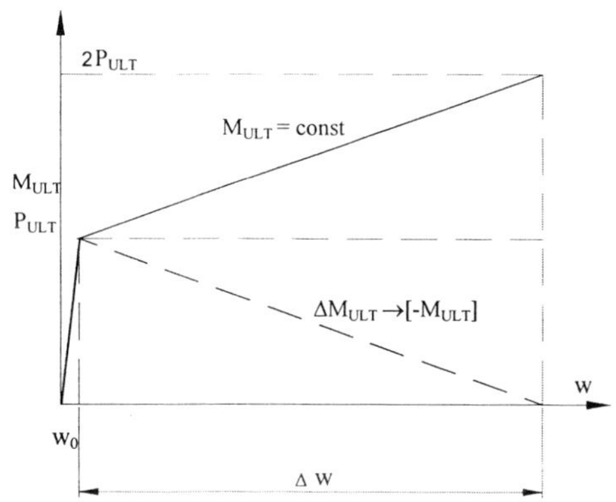

$$\Delta w_{\otimes} = \Delta\psi \cdot \frac{L}{2} = \frac{48 \cdot \Delta M}{\rho g B L^2} \quad \text{or} \quad \Delta M = \frac{\rho g B L^2 \Delta w}{48}. \tag{4}$$

When $\Delta M = M_{ULT}$, $\Delta P = P_{ULT}$

$$\Delta w_{\otimes} = \frac{48 \cdot M_{ULT}}{\rho g B L^2} = \frac{6 \cdot P_{ULT}\left(1 - \frac{4x_1}{L}\right)}{\rho g B L}. \tag{5}$$

$$\Delta M_{\otimes} = \frac{M_{\otimes}}{1 + \frac{192 E J_1}{\rho g B L^3 \ell_1}},$$

$$\Delta w_{\otimes} = \frac{48 M_{\otimes}}{\rho g B L^2 \left(1 + \frac{192 E J_1}{\rho g B L^3 \ell_1}\right)} \tag{6}$$

Figure 2: Load(moment)-deflection relationship.

When $M_{\otimes} = M_{ULT}$, $\Delta M \to M_{ULT}$ when $J_1 \to 0$. Thereby, in the ultimate condition occurs quick release of M_{ULT} and sharp reduction of actual stiffness of the weakened section. If release M_{ULT} occurs flash then t = 0, possible to define speed of vertical moving of ship's ends at the moment of grounding (limited depth). For the damage of the tanker "Energy Concentration" (LxBxDxd – 313x48.2x25x12) in 1980 at the depth of channel h = 20 m, M_{ULT} = 17800 MNm (Rutherford et al.(1990)) process of break-up lasted several seconds and after ship's ends contact with bottom of the channel hogging of the ship reach Δw_{max} = 16 m. From the formula (4) $\Delta M = 1.58 \cdot 10^7$ kNm $= 0.89 M_{ULT}$; $\Delta\varphi_{ST} = 5.86^0$

$$\Delta\varphi_D(t) = \Delta\varphi_{ST}\left(1 - \cos\frac{2\pi t}{\tau_K}\right), \tag{7}$$

where $\tau_K \cong 2.75\sqrt{d} = 2.75\sqrt{12} = 9.46$ sec – period of pitch. In accordance with dependency (7) contact with bottom of channel will occur through $t_1 = 2.19$ sec with vertical speed $w = 5.98$ m/sec.

The remaining deformation of forward and aft ends structures, correspond to striking loads at the speed near 6 m/sec. Thereby, all actual consequences, fixed in the process broke back of the tanker, quite agree with suggestion nearly full release of the initial ultimate moment and results of such simplified calculation. Formula (4) were used by authors for the evaluation ΔM in the area of the plastic hinge on measured values Δw_{max} at the test of many events of break up river ships (Kozlyakov (1999)) and floating docks. It was stated, that $\Delta M = (0.5 \div 0.9) M^{0}_{ULT}$, moreover faster and full release occurs in longitudinally framed hulls, were fracture is more catastrophic, than in transverse framed hulls.

Nature of the plastic fracture is like brittle, but brittle cracks quickly spreads from tensile strake to compressed, but in considered event pseudo-cracks in the manner of corrugation quickly spreads to tensile strake. In that and other events the deformation distribution does not correspond to the hypothesis of flat sections. One of the main reasons "explosive" nature of the process of transition from the whole hull to two-piece is a quick reduction of the efficient width of the compressed strake after the sharp reduction of its stiffness because of the loss of stability, arising the plastic deformation and take-off stiffeners from the sheeting. Known output on participation of strake with breadth up to 0.25L in bending can be explained only at the term of the constant thickness of flats of boxed girders (Papkovich (1939)). After the loss of stability compressed strake in the area of the narrow band in 1-2 spans occurs a sharp redistribution of the intensity of shear stresses to the proportional redistribution shear forces in the place where strake connected to the ship's sides (Fig.3). In the ultimate case, when equivalent thickness on this area $t_{EQ} \to 0$, compressed strake is divided up on two part and efficient width strake in weakened section $b_{EFF} \to 0$, as in pontoon floating docks. After the loss strake sharply increase stress in sides and corrugation in the manner of pseudo-cracks quickly moves to the tensile strake under the sharp reduction of actual stiffness of the weakened section.

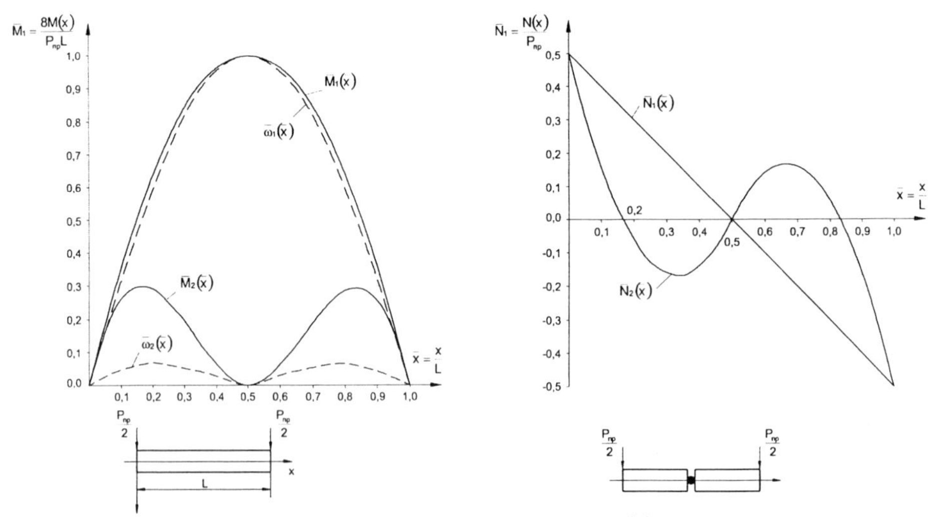

$$\bar{N}_1(x) = \frac{N_1(x)}{0.5P_{ULT}} = 2(x - 1.5x^2)$$

$$\bar{M}_1(x) = \frac{8M_1(x)}{P_{ULT}L} = \frac{M_1(x)}{M_{ULT}} = 4(x - x^2)$$

$$\bar{\omega}_1(x) = \frac{76.8 \cdot \omega_1(x) EI_0}{P_{ULT} \cdot L^3} = 6.4(0.4x + 0.6x^5 - x^4)$$

$$\bar{N}_2(x) = \frac{N_2(x)}{0.5P_{ULT}}$$

$$\bar{M}_2(x) = \frac{M_2(x)}{M_{ULT}}$$

$$\bar{\omega}_1(x) = \frac{76.8 \cdot \omega_2(x) EI_0}{P_{ULT} \cdot L^3}$$

Figure 3: Diagrams of $\bar{M}_1$, $\bar{N}_1$, $\bar{w}_1$ before and after arising plastic hinge.

On the Fig.3 given non-dimensional curves of bending moments $\overline{M}_1$ and $\overline{M}_2$, shear forces $\overline{N}_1$ and $\overline{N}_2$ and deflection $\overline{w}_1$ и $\overline{w}_2$ for example of floating parallelepiped for cases whole hull (index "1") and two-pieces (index "2"), as well as values of potential energy of bending U_1 and U_2

$$U_1 = \frac{1}{2}\int_0^L \frac{\overline{M}_1^2(x)}{EJ_0}dx = 0.27\frac{M_{ULT}^2 L}{EJ_0}; \quad U_2 = \int_0^{L/2} \frac{\overline{M}_2^2(x)}{EJ_0}dx = 0.019\frac{M_{ULT}^2 L}{EJ_0}. \tag{8}$$

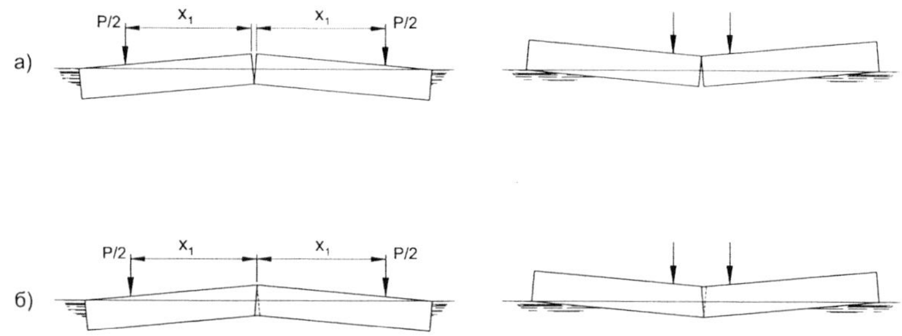

Figure 4: Brittle (a) and ductile (b) collapses.

When turning from the first condition to second bending moments decrease in 4 times, but energy of bending in 14.2 times. Thereby, quick spreading of corrugation - pseudo-cracks explained either as in brittle destruction, by releasing of the enormous internal energy and have catastrophic nature. For quick spreading a brittle crack it is necessary combination of critical temperature of material crack resistance and critical length of the crack, probably, for quick spreading of corrugation - pseudo-cracks it is necessary combination of the critical initial length of corrugation-dents and initial imperfects of structure in vertical and horizontal planes. In accordance with the noted analogy possible making a new section of mechanics of the fracture, as follows mechanics of the plastic fracture of floating thin web structures. Variability of ultimate moment's values is defined as degree of accuracy of used methods and programs, so and actual technical condition of hulls after building and in the process of operation. The most full study of accuracy of used methods was executed in the report (Yao (2000)). Also for hulls of four real ships had been explored sensitivity to changing of thickness, yield point, values initial deflection of plates and stiffeners in vertical and horizontal planes.

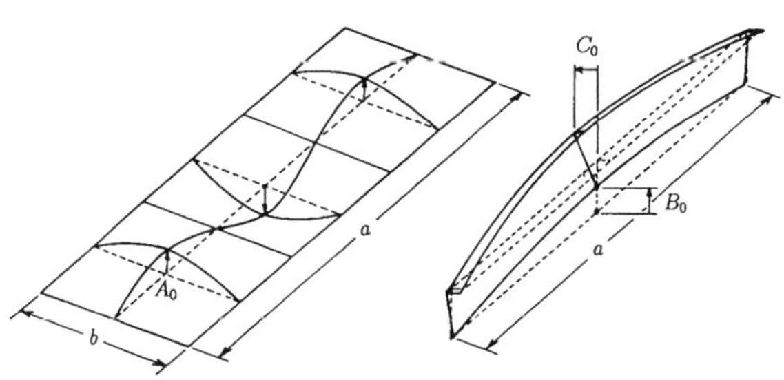

(a) Panel (b) Stiffener

Figure 5: Scheme of initial imperfections.

Statistical characteristics M_{ULT} at sagging and hogging (average value $\overline{M}_{ULT}$, standard $\sigma_{M_{ULT}}$, coefficient of variation $\vartheta_{M_{ULT}}$, and minimum value $M_{ULT}^{min} = \overline{M}_{ULT} - 3\sigma_{M_{ULT}}$) are given in the Table 1. In accordance with the Fig. 4 were accepted the following average and standards of imperfections value $\bar{A}_0 = 0.01t_p$; $\vartheta_{A_0} = 0.5$; $\bar{B}_0 = \bar{C}_0 = 0.0005a$; $\vartheta_{B_0} = \vartheta_{C_0} = 0.2$, where t_p, a - thickness of plates and length of stiffeners (span) accordingly. Values of $\bar{M}_{ULT}$ turn out to be sufficiently stable, $\vartheta_{M_{ULT}} = 0.027 \div 0.064$, under very moderate initial imperfections, corresponding to beginning of lifetime (after building). Greater values of initial imperfections different in different Classification Societies are allowed during operation. In particular, Rules of the Russian River Register are allowed $A_0 \leq 0.05b, B_0 \leq 0.07a$ or 100 mm, $C_0 \leq 0.2h_{ст}$ where $b, a, h_{ст}$ - distance between stiffeners, length and height of stiffener accordingly. When spreading corrugation or dent can reach $b_1 \leq 0.4B$.

TABLE 1

	Double Hull VLCC LxBxD=315x58x30.3		Container Ship LxBxD=230x32.2x21.5		Bulk Carrier LxBxD=285x50x26.7		Energy Concentration LxBxD=313x48.2x25.2	
	sag	hog	sag	hog	sag	hog	Sag	hog
$\bar{M}_{ULT}$, мкН	20387	22995	7227	7938	15304	19307	20314	23000
$\sigma_{M_{ULT}}$, мкН	903.9	1220.1	242.15	362.22	418.56	548.51	1100.35	1478
$\vartheta_{M_{ULT}}$, мкН	4.437	5.305	3.121	4.563	2.735	2.841	5.398	6.426
M_{ULT}^{min}, мкН	17647	19335	7004	6851	14048	17662	17083	18566
$\frac{M_{ULT}^{min}}{M_Y}$, мкН	0.74	0.80	1.07	0.95	0.92	1.16	0.84	0.91

For longitudinally framed large ships M_{ULT} various within limits $\cong (0.7 \div 1.16)M_Y$, where $M_Y = W_{min}(0) \cdot \sigma_Y$, $W_{min}(0)$ - minimum modulus of hull at the beginning of lifetime (without reduction). For transversally framed sea ships and restricted navigation ships M_{ULT} various within limits $M_{ULT} \cong (0{,}2 \div 0{,}8)M_Y$ (smaller value pertains to transversally framed river ships at the end of lifetime). Thereby, for the whole range of standard size cargo ships relative values of $\bar{M}_{ULT}$ various over the wide range $\bar{M}_{ULT} = M_{ULT}/M_Y = 0.2 \div 1.16$ and all determinate by structure features and it's technical condition. As far as the most bright feature break-up ships is value Δw (formula (4)), change within $\Delta w/L = 0.01 \div 0.06$, then for feature of process of transition hull in ultimate condition reasonable to build diagram dependency of the moment from deflection of ship under different combinations of parameters of initial imperfections A_0, B_0, C_0 and wear-out. Initial part of this graphics before beginning of loss of stability is built by traditional way up to $M = M_{ULT}$ with deflection w_0. When $w > w_0$ it is necessary to use non-linear method of final elements, though end maximum value Δw_{ULT} can be determined by the formula (4). Requirements to M_{ULT} and service damages must be unified on the IACS's level.

3 RECOMMENDATIONS ON ULTIMATE MOMENT STABILITY

Except tightening and unifications of requirements to initial and service imperfections of plates and ribs it is necessary to use constructive actions, raising stability of the system "plate-rib". For this in extreme strakes of the hull reasonable to ensure critical buckling stress in elements up to σ_Y (elastic stress $\sigma_E \geq 3\sigma_Y$).

On large sea ships this successfully solved by certain redistribution of the material from plates to stiffeners with the constant total areas of strakes. For ensuring stability under bent-torsional forms necessary to connect pairs of longitudinals by anti-torsional brackets in the middle of the span. For increasing stability of stiffeners and plates (for transverse framed system) reasonable to use efficient slope bars on webs of transverse elements, Reut (1962). These bars greatly raise stiffeners of transverses at the torsion. Welded seams must be double-sided and continuos, any holes in webs must be as small as possible and compensated by reinforcements. For ensuring high ultimate strength of ship's hulls necessary to ensure reliable regular supervision of structure's technical condition and control loads during cargo operations and in stormy conditions.

4 CONCLUSIONS

Energetic analogy had been stated between brittle and ductile type of fractures (collapses) of hulls. New factors determinate variation of ultimate moment for aging ships. Grounded necessity of unified requirements to permissible values of ultimate moments and characteristics of initial and service imperfections on the level of IACS.

References

Caldwell I.V. (1965). Ultimate Longitudinal Strength. *Trans. RINA* 107, 411 – 430.

Ershov N.F., Svechnikov O.I. (1970). *Ultimate Condition and Reliability of River Ship Structures*. Leningrad, Shipbuilding Publisher, (in Russian).

Kozlyakov V.V. (1995). On the Way of Improving the Strength Criteria in Design of Ship Structures. Proceedings of the 7th International Symposium on PRADS-95. Vol. 2, 2.1022-2.1034, Seoul, Korea.

Kozlyakov V.V. (1999). Some Problem Assessment of Ship Hulls Ultimate Plastic Strength. Proceedings International Conference "Strength and Exploitation Reliability of Ships". Vladivostok, Russia.

Kozlyakov V.V., Egorov G.V. (2000). An Analysis of River-Sea Going Ships Operation Experience and the Way of Improving Requirements to the Strength and Reliability of Their Structures. Proceedings of the IX Congress IMAM 2000, Ischia, Italy.

Kutt L.M. et al. (1985) Evaluation of the Longitudinal Ultimate Strength of Various Ship Hull Configurations. Trans. SNAME vol.93, 33 – 53.

Paik J.K. (1992). An Efficient Method for Ultimate Strength Analysis of Ship Structures – The Idealised Structural Unit Method (ALPS/ISUM Theoretical Manual). Pusan National University, Korea.

Papkovich P.F. (1939). *The Theory of Elastisity*. Leningrad, Oborongiz, (in Russian).

Papkovich P.F. (1941). *Ship Structural Mechanics*, **2**, Leningrad, Sudpromgiz, (in Russian).

Reut V.I., Kholopzev V.V. (1962). Some Methods of the Increasing of the Ship Grillages Elements Stability. NTO Sudprom. *Ship Structural Mechanics Issue*. No.42, Leningrad, Russia, (in Russian).

Rutherford S.E., Caldwell I.B. (1990). Ultimate Longitudinal Strength of Ships, a case study, *TSNAME* **98**.

Smith C. (1977). Influence of Local Compressive Failure on Ultimate Longitudinal Strength of a Ships Hull. Proceedings International Symposium PRADS-77. Tokyo, Japan, 75 – 79.

Ueda Y. et al. (1984). Plate and Stiffened Plate Units of the Idealised Structural Unit Method (ISUM) – Under in-plane Loading. *ISNAI*. **156**.

Yao T. et al. (2000). Special Task Committee V1.2. Ultimate Hull Girder Strength. *Proceedings of the 14th ISSC*. **2**. ELSVIER.

Practical Design of Ships and Other Floating Structures
You-Sheng Wu, Wei-Cheng Cui and Guo-Jun Zhou (Eds)

A NEW DESIGN MODEL FOR ULTIMATE AND BUCKLING STRENGTH ASSESSMENT OF STIFFENED PLATES

EIVIND STEEN, TOM K. ØSTVOLD AND SVERRE VALSGÅRD

Marine Advisory Services, Det Norske Veritas, Norway

ABSTRACT

A new computerised design model for the buckling strength assessment of stiffened panels is presented. The overall formulation is very general, and in principle, any type of stiffening arrangements of open or closed profile type, corrugations etc. can be analysed.
The model is based on an orthotropic version of Marguerre's non-linear plate theory. The stiffened panel is treated as an integrated unit, allowing for internal redistribution of membrane stresses between component plates, while preventing overall buckling and permanent deformations/sets.
By using non-linear plate theory, the strength model is more theoretical consistent than existing code formulations, which are mainly based on empirical curve fitting to a limited number of numerical and experimental results. Complicated items such as bi-axial loading combined with in-plane shear loads and non-linear mode interaction problems are dealt with in a sound physical framework, and empirical approximations are reduced to a minimum.
The model also provides a set of reduced anisotropic/orthotropic macro material coefficients that can be used in refined linear global FE analysis of ship hulls to reflect the increased membrane flexibility experienced by compressed stiffened panels. This area of application allows for redistribution of loads between gross elements such as stiffened panels, frames, girders and bulkheads and ensures a more realistic assessment of the nominal stress flow in a ship hull.
The presented model is planned to constitute the basis for a new DNV buckling procedure for stiffened panels.

KEYWORDS

Strength, Buckling, Postbuckling, Ultimate strength, Stiffened panels, Redundancy, Combined loads

1 INTRODUCTION

Existing rules, codes and guidelines for buckling design of stiffened panels in ship structures has been developed at a time where it was of main importance to have simple formulas and rule formats convenient for hand calculations. However, simple design rule formulas and codes, presently used by classification societies, consultants and designers, can hardly be expected to cover all the complex items necessary for the optimum and safe design of stiffened panels. Typical complex items are combined load situations, interaction between different buckling modes, effects from of out-of-flatness size and shape, residual stresses, heat affected zones and support conditions. These facts, combined with the present personal computer hardware and software capabilities, makes it possible to develop more modern and direct type of computerised buckling code formulations.

A computerised buckling code must be based on a recognised theoretical basis, and the non-linear thin-plate theories according to von Karman and Marguerre constitute such a set. Adopting these plate theories, special features such as elastic buckling and ultimate strength boundaries, deflections, stiffness properties and three-dimensional buckling mode and stress distributions may be visualised.

The present paper gives a brief introduction to a new computerised buckling code together with a very limited selection of verifications and applications. For a more comprehensive documentation see DNV (2000-seminar), DNV(2001).

2 PULS – COMPUTERISED BUCKLING MODEL FOR STIFFENED PANELS

Buckling of thin-walled stiffened plates is a non-linear phenomenon. However, from a global hull strength perspective, practical design procedures have to be based on linear elastic structural stress analyses with separate buckling checks of local elements. Such linearized design procedures are well established today. Obviously, the linear method will work satisfactory as long as the structure behaves linearly, i.e. as long as buckling failure of local elements is not accepted in any form. Though, to exclude all types of buckling failures of stiffened panels in a hull structure design is rather unrealistic, as it will lead to very conservative scantlings of plates and stiffeners. A design procedure should therefore accept some types of harmless buckling, typically elastic buckling of plate components between stiffeners. The consequence of accepting elastic buckling will be a change of the nominal stress flow in the hull and it is of paramount importance to account for this in the design of the surrounding structure.

In order to provide such a design procedure it is essential that the buckling model describe as closely as possible the real non-linear structural behaviour. To accomplish such a task, a new computerised buckling procedure called Panel Ultimate Limit State (PULS) is proposed. The PULS procedure is a simplified non-linear buckling model for assessing the strength of integrated hull elements. So far it is developed for integrated stiffened flat panels as typically found in ship hulls between frames and girders. It gives strength information at two levels:

i) Elastic buckling and design ultimate capacity of stiffened panels.
ii) Reduced stiffness properties of compressed and buckled panels.

The present PULS procedure is mainly constructed as an Ultimate Limit State design approach, i.e. the stiffened panels is to be designed to resist the most probable highest load during the design lifetime of the ship. The procedure is, however, based on some fundamental design principles, which eliminates certain types of buckling failures and allows others. The purpose behind these principles is to limit the probability for permanent buckles and to constrain the local panel flexibility to stay within reasonable limits. This is in full compliance with the design principles given in existing rules and guidelines of classification societies, e.g. DNV(2000), DNV CN30.1(1995).

2.1 Design Principles

A consistent design procedure must be founded on some basic design principles, and for the present PULS procedure three main principles apply

i) Elastic local buckling of the component plates in a panel cross-section is accepted.
For open profiles (as analysed as a case study here, Figure 2), local elastic buckling means buckling of plating between stiffeners, sideways/torsional buckling of stiffeners and stiffener web plate buckling. Acceptance of this type of buckling assumes the stiffeners to be strong, see principle iii).

ii) Permanent buckles are not accepted.
By ensuring the maximum membrane stresses within a panel to stay below the yield stress condition (von Mises), permanent sets and buckles are prevented. The maximum membrane stress locations are typical intersections between component plates in a cross-section, and are called hard corners. Maximum membrane stresses are the result of second order stresses from local buckling adding to the direct applied stresses.

iii) Overall buckling of the panel is not accepted.
This principle ensures the panel as a whole (stiffeners) to have sufficient out-of plane bending stiffness to avoid global (overall) buckling. Sufficient overall bending stiffness of the stiffeners ensures lateral support to the component plates, which is a requirement for accepting elastic buckling of local plate components (principle i)).

2.2 Theoretical Foundation

The present buckling model is based on an orthotropic version of Marguerre's non-linear plate theory, see e.g. Washizu (1975). In its original form, the orthotropic plate theory is not suitable for calculating the buckling strength of thin-walled stiffened panels, since local buckling effects of the component plates in a cross-section is not included. In order to cope with these local buckling modes, separately and in interaction with the overall panel mode, the general orthotropic plate theory is modified. This is done by introducing the concept of reduced orthotropic macro material coefficients.

The PULS procedure is based on a six-dimensional orthotropic macro material law. According to non-linear plate theory this macro material law takes the form of an incremental relation between the in-plane loads (N_1, N_2, N_3) and moments (M_1, M_2, M_3), and the corresponding strains $(\varepsilon_1, \varepsilon_2, \varepsilon_3)$ and curvatures $(\kappa_1, \kappa_2, \kappa_3)$ of a selected reference plane. In mathematical terms the orthotropic macro material law takes the following form

$$\begin{bmatrix} \Delta N_1 \\ \Delta N_2 \\ \Delta N_3 \\ \Delta M_1 \\ \Delta M_2 \\ \Delta M_3 \end{bmatrix} = \begin{bmatrix} C_{11} & C_{12} & C_{13} & Q_{11} & Q_{12} & Q_{13} \\ C_{21} & C_{22} & C_{23} & Q_{21} & Q_{22} & Q_{23} \\ C_{31} & C_{32} & C_{33} & Q_{31} & Q_{32} & Q_{33} \\ Q_{11} & Q_{21} & Q_{31} & D_{11} & D_{12} & D_{13} \\ Q_{12} & Q_{22} & Q_{32} & D_{21} & D_{22} & D_{23} \\ Q_{13} & Q_{23} & Q_{33} & D_{31} & D_{32} & D_{33} \end{bmatrix} \begin{bmatrix} \Delta\varepsilon_1 \\ \Delta\varepsilon_2 \\ \Delta\varepsilon_3 \\ \Delta\kappa_1 \\ \Delta\kappa_2 \\ \Delta\kappa_3 \end{bmatrix} \quad (1)$$

In Eqn.1 the symbol Δ indicates incremental quantities. Each of the coefficients in the stiffness matrix can be written as a sum of a linear and a non-linear part as follows:

$$\begin{aligned} C_{ij} &\equiv C_{ij}^{\ L} + C_{ij}^{\ N} \\ D_{ij} &\equiv D_{ij}^{\ L} + D_{ij}^{\ N} \\ Q_{ij} &\equiv Q_{ij}^{\ L} + Q_{ij}^{\ N} \end{aligned} \quad (2)$$

The linear part is given a superscript L symbolising linear fixed values. The non-linear part is symbolised with N as superscript. The latter part is the most interesting as it in addition to the dependency of the applied loads also depends on the level of geometrical imperfections, residual stresses, heat affected zones properties (e.g. in aluminium) etc. Most importantly, they depend on the level of compactness of the thin-walled cross-section.

In order to fully account for the simultaneous growth of both local and global buckling modes in a compressed stiffened panel, a full non-linear iterative solution should be implemented, Steen (1999). However, in order to obtain a more computer efficient and fast procedure, a simplified approach has been chosen here. The simplification amounts to use a fixed set of macro stiffness coefficients evaluated at first hard corner yield. This is a generalisation of the standard effective width approach used in most modern codes. The fixed macro stiffness values are then used as parameters in the overall nominal strength assessment, using the orthotropic buckling theory. It is emphasised that in the present design model we apply the macro tangent stiffness properties rather than the macro secant stiffness properties. This is done in order to cope with the possible severe and unstable interaction between local and overall panel buckling.

2.3 Limit States

Five limit state functions have been formulated in order to capture the possible collapse mechanisms in the panel. They are based on stress control against material yield in critical selected hard corner positions, and are typically related to plate and stiffener induced failure modes, as used in DNV/CN30.1(1995), in addition to a plate edge failure control criterion. The limit states are defined as five independent functions

$$f^{(i)}(N_1, N_2, N_3) > 0 \qquad i = 1,2,\ldots,5, \qquad \text{Acceptable} \tag{3}$$

At the limit $f^{(i)}(N_1, N_2, N_3) = 0$, each of the limit state functions describe a surface in load space (N_1, N_2, N_3). The inner envelope of all these surfaces will represent the ultimate design load capacity, Figure 1. The scaling of the loads in this his figure is with respect to the corresponding squash loads.

For a case by case evaluation of the strength, the distance from the applied load point (N_{10}, N_{20}, N_{30}) to the collapse surface (N_{1u}, N_{2u}, N_{3u}) is evaluated, using a proportional loading history.

The ratio between the applied load radius vector and the corresponding ultimate load radius vector is a direct measure of the safety margin, see Figure 1. In ship rule terminology this safety parameter is referenced to as the usage factor η (inverse of safety factor) and it follows that it is defined as

$$\eta = L_0 / L_u \tag{4}$$

where the radius vectors L_0 and L_u in load space are defined as

$$L_u = \sqrt{(N_{1u}^2 + N_{2u}^2 + N_{3u}^2)} \tag{5}$$

$$L_0 = \sqrt{(N_{10}^2 + N_{20}^2 + N_{30}^2)} \tag{6}$$

The present PULS procedure calculates the actual usage factor η, which is to be measured against the acceptable level given in the rules. This acceptance level will typically be related to the consequence of panel failure. The discussion of acceptance levels belongs to the overall safety philosophy and is a separate topic not discussed here.

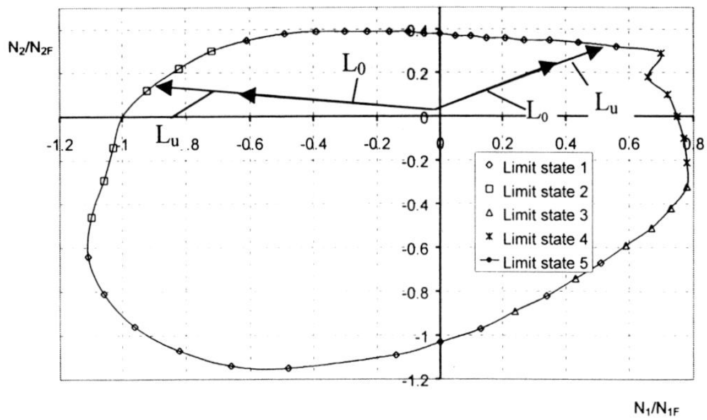

Figure 1: Inner envelope of the five PULS limit states

3 VERIFICATIONS OF PANEL STRENGTH

In order to have confidence in a new proposed model for buckling strength assessment, verifications against results obtained with more sophisticated analysis tools are necessary. For this purpose an extensive parameter study of stiffened panels subjected to bi-axial compression/tension loads were carried out using the recognised non-linear finite element code ADVANCE (1999), which is a subset of the well known ABAQUS(1998) code. The examples correspond to double bottom designs used in Bulk and LNG carriers. A full documentation is given in DNV(2000-seminar) from which the LNG carrier example in Figure 2 is extracted.

Figure 2 shows PULS 1.0 predictions together with ADVANCE/ABAQUS results for an out-of-flatness level corresponding to the tolerance levels given for ship structures, DNV/(1995-IS). The comparsions are resonable, but not perfect. One of the main problems encountered when doing non-linear buckling analysis with bi-axial loads, are associated with non-linear mode changes in certain regions of the load space. This effect is seen in the ABAQUS analyses as well as in the more simplified PULS code model and is the main reason for the irregular ultimate buckling boundary presented in Figure 2. The subject of mode change is very complicated and no attempt is made here to discuss it in detail. However, some improvements in the PULS code are needed in this respect.

Added in Figure 2 is also the bi-axial plate buckling criterion in DNV Ship Rules and DNV CN30.1-Chaper 3.4, stiffened panel strength criterion, even though they are not directly comparable. For this example (thick plating; 23 mm) it is noted that for dominating transverse compression, the DNV Ship rules formulas for bi-axial compression, gives optimistic strength estimates compared to both ADVANCE/ABAQUS and PULS results. This can be explained by the fact the DNV ship rules only calculate the strength between stiffeners, thus optimistically assumes that the stiffeners do not buckle. Note that the steel ship rules has an explicit requirement to stiffener moment of inertia for transversely compressed panels, which is not a part of the buckling assessment shown in Figure 2. This criterion is satisfied for the present example.

4 REDUCED PANEL STIFFNESS IN GLOBAL SHIP HULL ANALYSIS

Studying load-redistribution between gross elements in a ship hull due to buckling can only be solved in full by using non-linear analysis of a large part of the structure. This type of analysis is not dealt with in the present paper, as it requires a separate extensive study. However, by doing a sequence of linear elastic analysis with different levels of stiffness reductions, the consequence of an incrementally and continuously changing local panel flexibility can be simulated in a simplified manner. This is the method used here and results for an outer bottom panel in the double bottom of the LNG Carrier, are illustrated in Figure 2. As documented in DNV(2000-seminar), it is the membrane transverse and coupling stiffness coefficients C_{22} and C_{12} that by far will be most reduced due to elastic local buckling deformations of the plating between stiffeners. In the present example a minimum, an intermediate and a maximum stiffness reduction level of 20, 40 and 60 percent are shown.

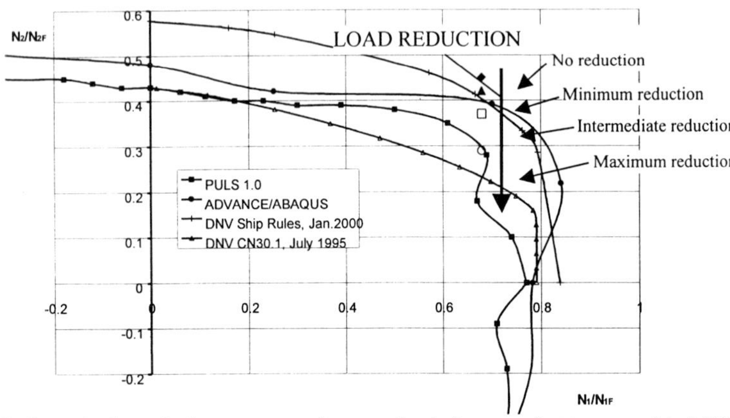

Figure 2: PULS Capacity boundaries curves, and stress check for outer bottom panel in LNG Carrier.

The global effect of these stiffness reductions is seen in Figure 2 illustrating a significant load reduction of the nominal transverse component N_2 (nominal stress σ_2) in the plating. By examining the stresses in the surrounding frame structure it was found that the stress change in the transverse girders were significant, while the stress change in the longitudinal girders was only marginal.

The aim with these studies was to show how linear FE models can be used for global hull redundancy assessment applying a reduced stiffness macro material technique. A main result from the double bottom study was the illustration of how loads redistribute between the plating and girders, as compared to what is normally calculated using the full stiffness standard linear approach. Clearly, due to the cellular construction with longitudinal and transverse frames, the double bottom construction has high torsional stiffness and significant strength reserves were documented. More documentation on this type of analysis, using linear reduced stiffness methods, can be found in Nissen-Lie, Steen & Østvold (1999) and Bakken, Østvold & Steen (2000).

It is concluded that in statically indeterminate plated hull structures, locally prone to plate buckling, global redistribution of forces takes place. The redistribution of forces can be significant and it is of equal importance to assess this effect as to assess the buckling capacity of individual stiffened panels. Further documentation, using sophisticated non-linear finite element programs considering larger parts of ship structures are needed, before firm conclusions on an acceptable design strength approach can be made.

5 CONCLUSIONS

A new computerised buckling code for flat stiffened panels has been described. The model is based on orthotropic plate theory, which makes the formulation applicable to all types of stiffening arrangements and materials. The results presented and discussed in this paper are for panels with welded open steel profiles.

Computerised buckling codes like the present PULS model has the obvious benefit of predicting more closely the real non-linear structural behaviour than existing rules and guidelines. This gives an improved basis for weight optimisations, together with a more consistent control of the safety margin against failure. It also provides additional valuable information, i.e. typically buckling mode shapes, elastic buckling and failure boundaries in load space, stress distributions and stiffness properties.

The present PULS code is based on some basic design principles in order to constrain the panel designs to have some minimum stiffness properties for efficient in-plane load transfer. Design principles for control of maximum stresses and prevention of permanent sets and buckles are also implemented. These design principles are consistent with the rules and guidelines given by classification societies today, even though in the latter they are not explicitly stated and not consistently included for complex load situations.

As a step in the direction of developing more consistent procedures for redundancy strength assessments of larger parts of hull structures, a reduced stiffness approach using coarse linear FE models, has been tested. This method has so far not been verified against large size non-linear analysis of ship hulls, but it is believed to capture the main physical effects. Future verification of this approach will rate its possible usefulness as a simple and practical approach for global hull strength assessment.

Acknowledgements

A special thanks goes to our colleague G. Holtsmark at DNV for his continuous support of the present work. Prof. J. Hellesland at the University of Oslo has also been an important inspirator and deserves honour for his interest and backing. Thanks also to DNV for sponsoring several internal R&D buckling and ultimate strength projects during the last years.

References

ABAQUS(1998), User's Manual, Version 5.8, Hibbit, Karlsson and & Sorensen, Inc.

ADVANCE (1999), SESAM User's Manual, ADVANCE/ABAQUS 5.8, 1999, DNV.

Bakken K, Østvold, T.K. and Steen, E. (2000). "Reduced stiffness modelling of double bottom in MRV LNG Carriers", DNV Report No.2000-0322, Høvik.

DNV CN30.1 (1995), Classification Notes 30.1, Buckling Strength Analysis, July 1995.

DNV (1995-IS), Rules for Classification of Ships, Instructions to Surveyors, I-B3.3

DNV (2000), Rules for Classification, Part 3, Chap.1, Hull Structural design, Steel, Jan. 2000.

DNV(2000-Seminar), Proceedings of DNV Seminar "Buckling and Ultimate Strength of Ship Structures", DNV- Høvik, September 2000.

DNV(2001), DNV Report no. 2001-0420, PULS 1.0 User's Manual.

Nissen-Lie, A., Steen, E. and Østvold, T.K. (1999)."Reduced stiffness modelling of double bottom in Bulk Carriers", DNV Report No. 99-0517, Høvik.

Steen, E. (1999). "Buckling of Stiffened Plates using a Shanley Model Approach". University of Oslo, Department of Mathematics, Mechanics Division, Report No. 99-1, ISBN 82-553-1182-3.

Steen E. and Østvold, T.K. " Buckling analysis of beam-columns using an incremental perturbation method", 13th Nordic seminar on Computational Mechanics, Univ. of Oslo, 20-21 Oct. 2000.

Washizu, K. (1975). "Variational methods in elasticity and plasticity", Pergamon Press, Second Edition, Bath, Great Britain.

ULTIMATE STRENGTH OF LONGITUDINALLY STIFFENED PANELS: MULTI-CRITERIA COMPARATIVE ANALYSIS

J.Y. Pradillon [1], T. Quesnel [1], C. Toderan [2] and Ph. Rigo [2]

[1] I.R.C.N (French Shipbuilding Research Institute), 1 rue de la Noë, BP 72108, 44321 Nantes Cedex 3, France
[2] Departement of Naval Architecture (ANAST), University of Liege, Chemin des Chevreuils n°1 (B52/3), B-4000 Liege, Belgium

ABSTRACT

Based on a recommendation of the ISSC 2000 Committee VI.2 (Yao et al. 2000) which has shown that the main factor to assess the ultimate bending moment of a hull girder is the quality of the "σ-ε" curves, we achieved an extensive comparison on the available methods providing the ultimate compressive strength and the average "σ-ε" relationship of longitudinally stiffened panels. This assessment mainly focuses on passenger ships (cruise ships, fast ferries) and naval vessels having the plate and column slenderness of standard steel stiffened panels produced in France. At the first stage, we compared 18 different methods to assess the ultimate strength of a hull girder. These methods may be classed into 3 types: empirical, intermediate (analytical or semi-analytical) and generic (e.g. ISUM and FEM). Only the methods able to rapidly provide "σ-ε" relationships are relevant for our purpose but others were also considered as valuable references. From this comparison, a set of "σ-ε" models were selected using a multi-criteria comparative analysis for advanced numerical tests. Then, the selected "σ-ε" models were compared on different stiffened panels. In addition, comparative tests on simple box girders were carried out to assess the effect of the "σ-ε" model capabilities (initial deflections, residual stress, lateral loads, ...) on the ultimate bending moment of a hull girder.

KEYWORDS

Ultimate strength, Axial compression, Longitudinally stiffened panel, Collapse analysis, Hull girder, Buckling, Ultimate bending moment, Multi-criteria analysis.

1 INTRODUCTION

One of the main IRCN's concerns deals with the development of a new shipyard-oriented tool which has at the design stage the capability to assess the ultimate bending moment of a hull girder. Our purpose is not to implement a high sophisticated approach asking for long and time-consuming numerical analysis but a method which gives a good compromise between accuracy and simplicity. In order to meet this requirement, a multi-criteria analysis was performed. To compare the "average stress

– average strain" curve ("σ-ε" curve) provided by different approaches, it was decided to review a series of models able to evaluate the ultimate bending moment of a hull-girder. All these models provide the panel's ultimate strength but only some of them give a "σ-ε" curve (these ones being the most relevant for our purpose). In fact, the quality of component behaviour is the predominant factor to assess the ultimate bending moment. However, in a first stage, all models have been considered to qualify all the different types of approach and to have more references.

The 18 methods listed in our study are briefly presented in Table 1. A method can be either an original method or a regrouping of very closed formulations proposed by different authors.

TABLE 1

INVENTORY OF METHODS TO ASSESS THE ULTIMATE STRENGTH OF A HULL GIRDER

N°	Name of Methods	References
1	Caldwell (Direct approach)	(Caldwell 1965)
2	Paik (Direct approach)	(Paik et al. 1995)
3	Smith (FEM for "σ-ε")	(Smith 1977, Dow et al. 1981)
4	Adamchak	(Adamchak 1984)
5	Hughes (SSD) without post collapse strength	(Hughes 1988 - Ch. 17)
6	Lloyd Register (Rutherford)	(Rutherford et al. 1990)
7	Bureau Veritas & Gordo-Soares	(Beghin et al. 1995, Gordo et al. 1996)
8	Yao (HULLST)	(Yao et al. 1991, 1992)
9	Rahman + Hughes	(Rhaman et al. 1996, Hughes 1988 – Ch. 14)
10	FEM (full non-linear analysis)	
11	ISUM	(Ueda et al. 1991, Paik et al. 1996)
12	I.A.C. (*) + FEM for "σ-ε"	
13	I.A.C. (*) + SPINE (Paik) for "σ-ε"	(Paik 1999)
14	I.A.C. (*) + Hughes for "σ-ε"	(Hughes 1988 – Ch. 14)
15	I.A.C. (*) + Imperial College for "σ-ε"	(Dowling et al. 1991)
16	I.A.C. (*) + Paik (empirical) for "σ-ε"	(Paik 1997)
17	I.A.C. (*) + ULSAP (Paik) for "σ-ε"	(ULSAP 2000)
18	I.A.C. (*) + Faulkner (ABS) for "σ-ε"	(Guedes Soares 1997)

(*) I.A.C.: Incremental Algorithm of Curvature for progressive collapse analysis of a hull girder (Smith method)

To perform this comparative review, criteria with associated weighting are used (see Table 2).

TABLE 2

MAIN CRITERIA FOR THE MULTI-CRITERIA COMPARATIVE ANALYSIS

Criteria	Weighting A	Weighting B
1. Considered approach Direct Approach, Progressive Collapse Analysis, Global analysis	10% [15 - 10 %(*)]	7% [9 - 5% (*)]
2. Quality of the geometrical modelling of the structure From simplified (box-girder) to detailed (fine FEM mesh)	15% [15 - 10 %(*)]	10.5% [13.5 - 7.5%(*)]
3. Quality of the "σ-ε" model Model capabilities: - considered failure modes (plate, web and column buckling, tripping, yielding) - interaction between basic components - initial imperfections (initial deflections, residual stress) - lateral pressure	50% [40 - 60 %(*)]	35% [45 - 25% (*)]
4. Facility of use (from the end-user point of view)	25% [30 - 20 %(*)]	17.5% [22.5 - 12.5%(*)]
5. Availability of the model (from the developer point of view)	Not used	30% [10 - 50 %(*)]

(*): Variants for sensitivity analysis of classification to the weighting selection

All the methods have been ranked twice. The first selection was performed with weighting A, **"σ-ε" model quality** and **facility of use** being the main criteria (Fig. 1). This group was then compared to the leading methods with weighting B. For each case, in order to assess influence of weighting selection, a range of variation for the five criteria was considered (variants in Table 2). For instance, Figure 1 shows results for weighting A.

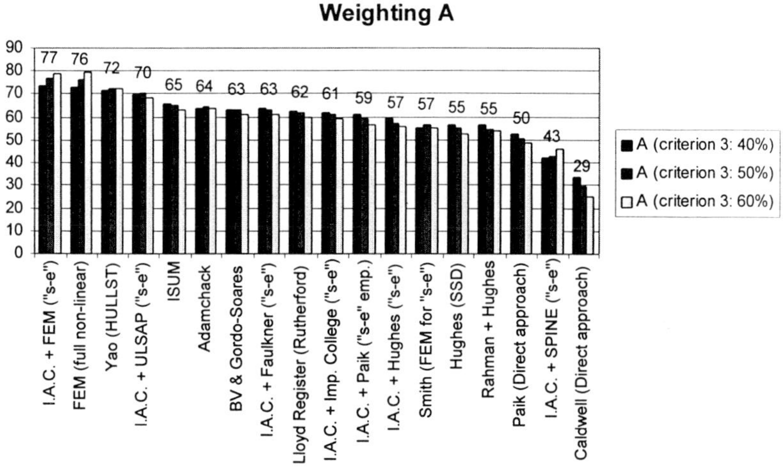

Figure 1: Classification of methods with weighting A

Based on these classifications, we have eliminated the direct approaches (1 and 2), the too "simple" progressive methods (5: no post-collapse strength, 9: too approximate geometrical modelling) and also method 13 (SPINE is only recommended for unstiffened plate). Moreover, the full FEM non-linear analysis does not suit for our purpose (too long analysis in design stage). We noticed that *BV-Gordo Soares* and *I.A.C. + Faulkner* methods have very similar formulations for the stiffened panel behaviour ("σ-ε" curves) and get close results. So, these two methods were grouped in method 7 (Table 1). Due to the availability criterion, methods 3 and 6, which have got average results, and also method 11 were not kept. At the end, it remains 5 methods of level 1 (good availability and well suited to our purpose): the *Yao (HULLST)* method which offers a good potential with a "σ-ε" numerical-analytical formulation, the *Adamchak, I.A.C. + Imp. College* and *I.A.C. + Hughes* methods with "σ-ε" analytical formulations and the *I.A.C. + Paik (empirical)* method with "σ-ε" empirical formulation. Three other methods (level 2), were also kept: the *BV-Gordo Soares* method which is consistent with our purpose but with restrained availability, the *I.A.C. + ULSAP* method which capacities must be verified and the *I.A.C. + FEM ("σ-ε")* method with "σ-ε" curves for references. Afterwards, more tests were completed for methods of level 1 than for methods of level 2.

2 DIRECT COMPARISON OF "σ-ε" ELEMENT MODELS

We have defined 10 test panels to compare the 8 remaining methods. Geometry of test panels corresponds to various stiffened panels either from ships manufactured in French shipyards (passenger ships, fast ferries, naval vessels) either from cargo ships. Dimensions, plate and columns slenderness for S235 are shown in table 3. We have considered 8 standard stiffened panels (with all identical stiffeners) and 2 mixed stiffened panels (with primary and secondary stiffeners). Each test panel was considered in two steel grade (S235 and S355 or other grade with yield stress higher than 235 MPa), with and without welding residual stress (σ_r = 0%, 7.5% and 15 % of σ_y), with and without initial plate and stiffener deflections (maximum plate imperfection from 4 to 15 mm. in accordance with quality standards of fabrication, maximum stiffener defection at 0.1% of the span length), and with or without lateral pressure panels (from 5.10^{-3} MPa for deck (classical rule pressure) to 0.13 MPa for bottom panels). For each panel, 10 tests were carried out.

TABLE 3

TEST PANEL DIMENSIONS AND SLENDERNESS FOR S235 STEEL

		Plate characteristics		Stiffener characteristics					S235 slenderness	
Designation	**a**	**b**	**t**	**Type**	h_w	t_w	b_f	t_f	**Plate β**	**Column λ**
P1 (deck of cargo ship)	4000	800	15	flat-bar	250	19			1.784	0.599
P2 (side of cargo ship)	3750	785	19	angle	500	12	150	23	1.382	0.197
P3 (side of cargo ship)	4000	800	15	angle	235	10	90	15	1.784	0.516
P4 (bottom of cargo ship)	4000	800	20	angle	383	12	100	17	1.338	0.317
P5 (bottom of cruise ship)	2750	700	25	angle (equivalent bulb-bar)	263.5	11	82.52	16.5	0.937	0.343
P6 (bottom of fast ferry)	1500	315	5	angle (equivalent bulb-bar)	111	6	29.33	9	2.107	0.370
P7 (deck of naval vessel)	2000	500.4	9.67	tee-bar	119.07	6.22	103.9	8.06	1.731	0.457
P8 (deck of naval vessel)	1500	700	6	tee-bar	200	6	100	10	3.903	0.195
P9 (deck of cruise ship)	2700	750	5	1 tee-bar (primary stiffener)	420	10	200	10	5.018	0.163
				8 angle-bars (eq. bulb) (secondary stiffener)	91	6	25.33	9	5.018	1.062
P10 (deck of fast ferry)	1500	300	4	1 tee-bar (primary stiffener)	300	5	80	8	2.509	0.128
				14 angle-bars (eq. bulb) (secondary stiffener)	54	4	23.67	6	2.509	0.819

a: span length (mm)
b: stiffener spacing (mm)
t: plate thickness (mm)
h_w: web height (mm)
t_w: web thickness (mm)
b_f: flange width (mm)
t_f: flange thickness (mm)

We can notice that common plate slenderness of passenger ship and fast ferry deck are greater than the standard values met in panels of cargo ship. On the contrary, stiffeners of P8 panel and primary stiffeners of P9 and P10 panels are very stocky. Their column slenderness are much smaller than the standard values for which most of the available formulations have been established and validated.

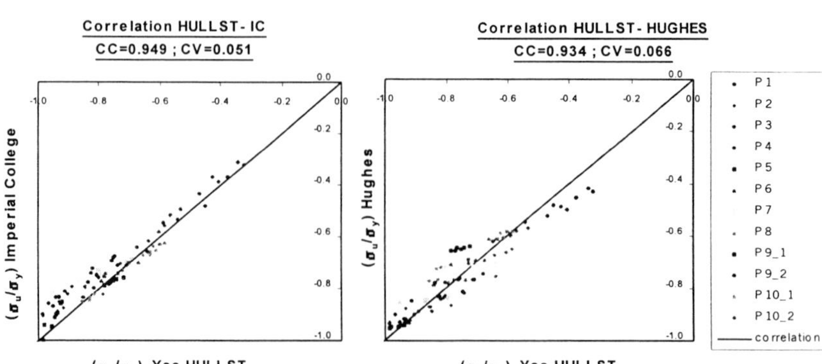

Figure 2: Correlation of ultimate stress for all test panels

The first decisive factor in quality of "σ-ε" model is the value of the ultimate stress associated to the critical collapse mode. We have considered *FEM*'s results as valuable reference unless panel geometry is perfect. Moreover, *Yao (HULLST)* based on a rational formulation also provides reliable results. So, we have used results from these two methods as references to compare the ultimate stress between all the models (Fig. 2). A synthesis of the comparison of the various "σ-ε" models is presented in Fig. 3.

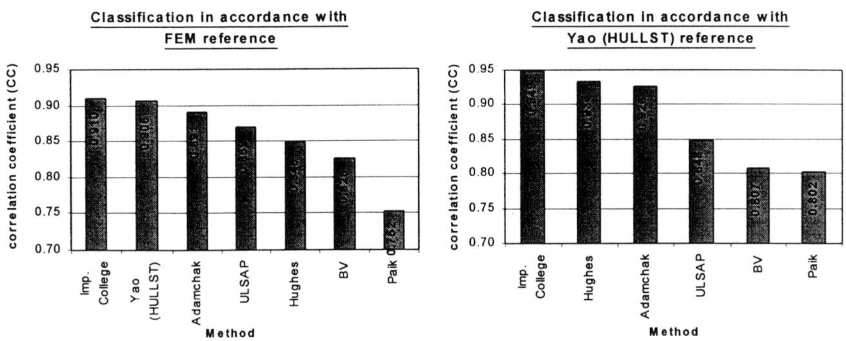

Figure 3: Synthesis of ultimate stress comparison

In fact, we can see that *BV* and *Paik (empirical)* models get the least accurate results. These models are yet valuable at design stage because they generally provide conservative values of ultimate stress. Moreover, they implicitly considered a basic imperfection, so they give an unique value for each panel (geometry and material combination). In fact, they are suited for rule design. The *Imperial College, Adamchak and Hughes* models get good correlation coefficients. These three models use analytical formulations, so they quickly provide "σ-ε" curves on the contrary to the two reference models. Furthermore, their formulations explicitly consider the level of some imperfections (residual stress, column deflection and lateral load).

Another important factor to establish the "σ-ε" relationship is the shape of the "σ-ε" curve. So, "σ-ε" curves of each test (panel, material and set of imperfections) have been plotted in order to compare their shapes. For instance, curves for one test of panel 1 are shown on Fig. 4.

The shapes are various and the differences are bigger in the post-collapse domain. In fact, *FEM* and *Yao (HULLST)* incrementally compute "σ-ε" curves, so they provide a smooth strength decay in post-collapse. Formulations of *Hughes-Rahman (as extended by Rahman for post-collapse)* and *Adamchak*

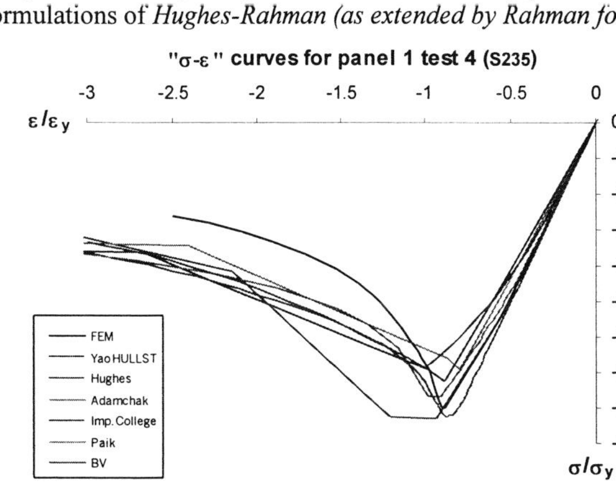

Figure 4: "σ-ε" curves of panel behaviour with 7 different models

present 3 separate domains in their load-shortening curves: a stable part in pre-collapse up to ultimate state, a plateau with practically no unloading and an unloading part. For these analytical formulations, the first part is based on an elastic-plastic buckling analysis and the third part is based on behaviour of a mechanism of collapse with plastic hinge. As *Imperial College* and *Paik (empirical)* models don't provide a shape for post-collapse domain, Rigo implemented an extent in post-collapse. This extent is based on an arbitrary calibrated shape just function of the ultimate state (σ_u, ε_u). For the *BV* model, the shape in post-collapse is also arbitrary but its calibration is based on FEM analysis. For all the tests, the more important difference noticed in shape concerns the plateau's length for the analytical formulations.

3 COMPARISON ON SIMPLE BOX-GIRDER

As our purpose is to assess ultimate hull girder bending moment of ships, it is important to compare bending moment provided with the different "σ-ε" element models. For that, we have used the previous "σ-ε" curves to assess ultimate bending moment of a symmetrical box girder (square of 6 stiffeners with associated plate on each side) with the same incremental algorithm of curvature. Fig. 5 shows moment-curvature curves associated with "σ-ε" element curves.

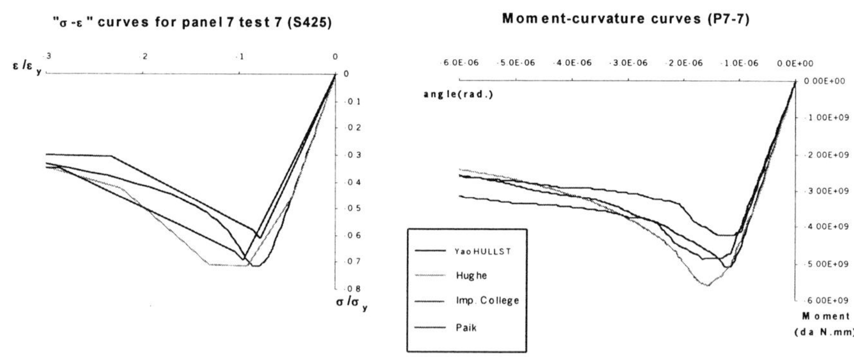

Figure 5: Moment-curvature relationships of a box-girder under longitudinal bending

The influence of "σ-ε" curve on ultimate bending moment is very significant and the shape of the "σ-ε" curve is often the most important factor: ultimate moment associated with *Hughes-Rahman* model is greater than the one from *Yao* model whereas ultimate stresses from each model are close. The excessive length of the plateau at ultimate strength in *Hughes-Rahman* model leads to an optimistic assessment of ultimate moment. The sensitivity analysis performed by ANAST (Rigo et al.) give more details about this remark.

4 CONCLUSIONS

In practice, it is very important to use well suited "σ-ε" model in order to assess with accuracy the ultimate bending moment of ships. But, most of the element models don't provide rational formulation for the post-collapse behaviour. For our purpose, it means that we can prefer a model with an explicit formulation for this post-collapse domain. To get a reliable and fast computing "σ-ε" model, an interesting way is to combine a good assessment of the ultimate stress from, for instance, *Imperial College* or *Hughes* models with an analytical post-collapse formulation as proposed by the *Adamchak* and *Rahman* models. Such mixed "σ-ε" relationship model is being developed in the IRCN's ultimate strength tool. However, this mixed model has to be calibrated with experimental results of actual stiffened panels. Particularly, this model must be able to represent the initial imperfections and to integrate the actual production's parameters of a specific shipyard (worker skill, welding type, etc...).

References

- Adamchak J.C. (1984). An Approximate Method for estimating the Collapse of a Ship's Hull in Preliminary Design. *Proc. Ship Structure Symposium' 84*, SNAME (Ed.), Arlington, US, 37-61.
- Beghin D., Jastrzebski T. and Taczala M. Result – a Computer Code for Evaluation of the Ultimate Longitudinal Strength of Hull Girder. *Proc. Int. Symp. PRADS'95*, Seoul, Korea, 832-834.
- Caldwell J.B. (1965). Ultimate Longitudinal Strength. *Trans. RINA* **107**, 411-430.
- Dow R.S., Hugill R.C., Clark J.D. and Smith C.S. (1981). Evaluation of Ultimate Ship Hull Strength. *Proc. Extreme Loads Response Symposium*, SNAME (Ed.), Arlington, US, 133-148.
- Dowling and al. (1991). Design of Flat Stiffened Plating: Phase 1 report. *CESLIC Report* **SP 9**, Dept. of Civil Engineering, Imperial College, London, UK
- Guedes Soares C. and Gordo J.M. (1997). Design Methods for Stiffened Plates Under Predominantly Uniaxial Compression. *Marine Structures* **10:6**, 465-497
- Gordo J.M. and Guedes Soares C. Approximate Method to Evaluate the Hull Girder Collapse Strength. *Marine Structures* **9:3-4**, 449-470.
- Hughes O.F. (1988). *Ship Structural Design*, SNAME, New Jersey, US
- Paik J.K. and Mansour A.E. (1995). A Simple Formulation for Predicting the Ultimate Strength of Ships, *Journal of Marine Science and Technology* **1**, 52-62.
- Paik J.K., Thayamballi A.K. and Che J.S. (1996). Ultimate Strength of Ship Hulls under Combined Vertical Bending, Horizontal Bending, and Shearing Forces. *Trans. SNAME* **104**, 31-59.
- Paik J.K. and Thayamballi A.K. (1997) An Empirical Formulation for Predicting the Ultimate Compressive Strength of Stiffened Panels. *Proc. 7th Int. Offshore and Polar Engineering Conf.*, Honolulu, USA, **4**, 328-338.
- Paik J.K. (1999). *SPINE, A Computer Program for Analysis of Elastic-plastic Large Deflection Behavior of Stiffened Panels by the Incremental Energy Method. User's Manual*, Pusan National University, Korea
- Rahman R.K. and Chowdhury M. (1996). Estimation of Ultimate Longitudinal Bending Moment of Ships and Box Girders. *Journal of Ship Research*, **40:3**, 244-257.
- Rigo P., Toderan C. and Yao T. (2001). Sensitivity Analysis on Ultimate Bending Moment. *Proc. 8th Int. Symp. PRADS*, Elsevier Science Ltd., Shanghai, China (to be published).
- Rutherford S.E. and Caldwell J.B. (1990). Ultimate Longitudinal Strength of Ships: A Case Study. *Trans. SNAME* **98**, 441-471.
- Smith C.S. (1977). Influence of Local Compressive Failure on Ultimate Longitudinal Strength of a Ship's Hull. *Proc. Int. Symp. PRADS'77*, Tokyo, Japan, 325-359.
- Ueda Y. and Rashed S.M.H. (1991). ISUM (Idealized Structural Unit Method) Applied to Marine Structures. *Trans. JWRI* **20:1**, 123-136.
- ULSAP (2000). *ALPS/ULSAP User's Manual*, Proteus Engineering, USA, 46p.
- Yao T. and Nikolov P.I. (1991). Progressive Collapse Analysis of a Ship's Hull under Longitudinal Bending. *J. Soc. Naval Arch. of Japan* **170**, 449-461. – (1992). 2nd Report, **172**, 437-446.
- Yao T. et al. (2000). Ultimate Hull Girder Strength: Report of Committee VI.2. *Proc. 14th International Ship and Offshore Structures Congress*, Ohtsubo H. and Sumi Y. (Eds.), Elsevier, **2**, 321-391.

Practical Design of Ships and Other Floating Structures
You-Sheng Wu, Wei-Cheng Cui and Guo-Jun Zhou (Eds)

ULTIMATE STRENGTH OF SUBMERSIBLE STRUCTURES

I. P. Pasqualino and S. F. Estefen

Department of Ocean Engineering, COPPE, Federal University of Rio de Janeiro
P.O. Box 68508, Rio de Janeiro, RJ, CEP 21945-970, Brazil

ABSTRACT

Numerical and experimental study on the behavior of stiffened cylindrical shells under hydrostatic pressure have been carried out. The structures analyzed are representative of the main body of submersible vehicles. Local and overall collapse modes were studied with the aid of experimental tests. Small scale models were constructed in order to evaluate their ultimate strengths and failure modes through experimental tests using an hyperbaric chamber. The models were specially designed and built to detect the local failure mode of the shell between ring frames or the overall failure mode involving ring frames and shell. The small scale models were first mapped in a specific facility, so that the out of roundness and eccentricity of the transverse sections could be evaluated. The material properties were also determined through uniaxial tension tests. The theoretical behavior of the structure was obtained through finite element analyses incorporating large displacements, plasticity and the arc-length control in the post-collapse regime. Correlation between numerical and experimental results associated with the structure ultimate strength showed to be very imperfection sensitive. The paper contributes to the design of submersible vehicles with emphasis on ring stiffened shells under hydrostatic pressure.

KEYWORDS

Submersible structure, Collapse, Post-collapse, Experimental test, Numerical analysis.

NOMENCLATURE

D_{max}	maximum external diameter;
D_{min}	minimum external diameter;
h	height of the stiffener web;
l	spacing between ring frames;
L_{tot}	model overall length;
R	internal radius of the cylindrical shell;
t	thickness of the cylindrical shell;
t_w	thickness of the stiffener web;
Δ_o	initial ovality;
λ	scale factor.

1 INTRODUCTION

The typical failure modes of stiffened cylindrical shells under hydrostatic pressure are the local collapse of the shell between ring frames and the overall collapse of shell and rings between rigid sections such as flat heads. The local collapse is an interaction between elastic-plastic buckling and axisymmetric membrane shell yielding at midbay. It may occur in one or more ring spacings. The overall collapse is associated with the ring frames that are weak to resist the combined effect of out of roundness bending and sideways tripping leading to its premature yielding.

The elastic-plastic instability of the shell between ring frames is characterised by the formation of inward and outward lobes. This mode of failure indicates that the rings have greater resistance to buckling than the shell between them. Axisymmetric shell yielding is characterised by longitudinal compressive strains at the extreme fibbers at both the outer surface of the shell between ring frames and the inner surface of the shell at the stiffeners. The hoop membrane strains of the shell at the ring frames are partially prevented due to the stiffeners rigidity. However, the longitudinal bending strains of the extreme fibbers at this region, tensioned at the outer surface and compressed at the inner, when added to membrane strains due to the axial load, may lead to premature yielding of the shell. The region of the shell at midbay is particularly critical because of the maximum hoop strains. In this case, the compressive bending strains of the shell outer surface are added to compressive membrane strains in longitudinal direction.

The overall collapse is normally verified between heavy stiffeners, such as large frames or bulkheads of submersibles. Its occurrence increases if the spacing between these heavy stiffeners is enlarged. It is characterised by the formation of one or more lobes involving shell and stiffeners. The out of roundness of the ring frames contributes to the bending of transverse sections which may collapse due to excessive plastic hoop strains. The bending of the transverse sections also contribute to the non orthogonality of the ring frames in relation to the shell, which may lead to sideways tripping of the stiffeners. Therefore, it may be concluded that the overall collapse is highly dependent of the initial imperfections magnitude.

The collapse pressure of stiffened cylindrical shells is mainly influenced by the structure geometry, the material, the boundary conditions, the residual stresses and the initial imperfections magnitude. The overall collapse is more affected by the out of roundness than the local collapse. Therefore, these structures are normally designed based on the local collapse pressure. Submersibles are designed in order to minimise the structural weight, a critical requirement to achieve good operational performance. Therefore, the optimised design of these structures must be carefully studied in order to preserve the structural safety.

This paper presents a numerical and experimental study of the collapse of ring stiffened cylindrical shells under hydrostatic pressure. The aim of the work is to evaluate the ultimate strength and the failure modes of such structures. Small scale models were defined through geometric similarity with the main body of a submarine prototype. These models were first evaluated with the aid of a finite element framework to determine its collapse pressure. Afterwards, they were instrumented and tested under hydrostatic pressure in a hyperbaric chamber.

2 LABORATORY TESTS

Laboratory tests were performed in order to identify the collapse behavior of ring stiffened cylindrical shells under hydrostatic pressure. Small scale models were conceived in order to study the local and the overall failures modes. Design, construction and instrumentation of the models are discussed below.

2.1 Model Design

The small scale models were defined through linear geometric similarity with the main body of a full scale submarine prototype. The ratios between radius and thickness (*R*/*t*) and between ring frames spacing and radius (*l*/*R*) were preserved.

Thin plates with thickness equal to 1.5 mm has been selected to construct the models, due to geometric limitations of the hyperbaric chamber. Therefore, a scale factor (λ) equal to 31.33, representing the ratio between prototype and model thickness, was employed to establish the model geometry (Table 1). In order to simplify the construction of the models, flat bar stiffeners were considered instead of the T-stiffeners of the prototype. The model ring stiffeners were designed with web thickness equal to 1.5 mm and the same inertia of the scaled T-stiffener.

TABLE 1
MAIN PARAMETERS OF MODEL GEOMETRY IN MM

R	t	l	h	t_w	L_{tot}
156.4	1.5	23.3	14.5	1.5	582.5

The models were designed to represent a ring framed cylindrical shell between a pair of transversal bulkheads. Twenty four equally spaced ring frames and a uniform cylindrical shell represented the submarine main body. Two robust lids were used to prevent water leak during the hyperbaric tests and to simulate the bulkheads.

2.2 Model Construction

Two small scale models (MST3 and MST4) have been manufactured to conduct the experimental tests. Based on Scott et al. (1987) proposal, a specially designed steel apparatus was developed to help with the manufacturing process. Considering the tolerances established, this facility can keep the longitudinal alignment between model bays as well as assure orthogonality between ring frames and shell during the welding process in hoop direction.

The cylindrical shell of each bay was obtained conforming a sheet strip over a die and TIG welding it longitudinally. The ring frames were directly cut from the steel sheets and positioned at its respective bay to be welded. Finally, the TIG welding process was employed for joining the model bays that compose each model. Specimens were cut in orthogonal directions of the steel sheet to perform uniaxial tensile tests. These experiments have provided the material properties to be used in the numerical analyses.

The initial geometric imperfections of the models were mapped with the aid of a special facility (Netto et al., 1991) designed to obtain radial displacements at measuring points of cylindrical shells. The outer surface of the models were evaluated at 40 equally spaced points in hoop direction and 101 transverse sections. These data were processed by a computer program (Netto, 1987) specially developed to generate the actual outer surface of a mapped cylindrical shell. The initial ovality (Δ_O) of the model sections were calculated with the aid of the following expression:

$$\Delta_O = \frac{D_{max} - D_{min}}{D_{max} + D_{min}}, \qquad (1)$$

where D_{max} and D_{min} are the maximum and minimum external diameters, respectively, of a particular section considered. The average initial ovality of the models MST3 and MST4 were equal to 0.0937% and 0.1443%, respectively.

2.3 Instrumentation and Test

The small scale models were provided with strain-gauges and sealed at its ends with the aid of robust circular lids to be tested in the hyperbaric chamber of the Submarine Technology Laboratory – COPPE. The tests were accomplished with the aid of a hydraulic pump which allows an increasing pressure at a low rate of 0.25 N/mm^2 per minute. The strain-gauges were installed on the inner and outer surfaces of the shell as well as on the nearest ring frame webs of the model mid section. They were positioned in both longitudinal and hoop directions spaced by 90 degrees.

Model MST3 was submitted to three consecutive hydrostatic tests designated as MST3-1, MST3-2 and MST3-3. In the first test the failure occurred close to the model end (section 7) at a collapse pressure of 4.31 N/mm^2. It was probably caused by the structural discontinuity between the shell and the rigid lid. The damaged region of the model (up to section 19) was cut off and the remained part was tested again. A similar behavior was observed for the model MST3-2 at a collapse pressure of 4.57 N/mm^2. Again the damaged region (sections 19 to 38) was cut off to the accomplishment of a new test. The end regions were reinforced by longitudinal stiffeners spaced by 45 degrees and welded on the ring frames. Despite of the reinforcement of the model MST3-3 the failure occurred again close to the end at a collapse pressure of 4.86 N/mm^2 (Figure 1).

Model MST4, nominally identical to MST3-1, was also reinforced at the end regions. In this case, the number of longitudinal stiffeners was doubled and the L cross-section stiffeners increased from 12.7 x 12.7 x 3.2 mm to 19 x 19 x 3.2 mm. The model shell was also reinforced at its ends with sheet strips of thickness equal to 1.5 mm. They were welded on the outer surface of the model shell between sections 2 to 17 and sections 85 to 100. The failure mechanism was then initiated at the central region of the model at section 55, involving the ring frames located from the sections 53 to 57 (Figure 2). It was registered a collapse pressure equal to 4.81 N/mm^2.

Figure1: Failure mode of the model MST3-3

Figure 2: Failure mode of the model MST4

3 NUMERICAL ANALYSIS

The theoretical behavior of the small scale models under hydrostatic pressure was obtained with the aid of the finite element method. The framework ANSYS release 5.3 was employed in the numerical analyses. The geometric and material properties of the manufactured models were carefully defined in order to simulate its collapse behavior.

The finite element mesh was generated with the nominal dimensions of the small scale models MST3 and MST4 presented in Table 1. The initial out of roundness was incorporated in the numerical model through the modification of the shell node coordinates in the radial direction, using the mapped geometry acquired before the experimental tests. Since the finite element mesh was discretized in the hoop direction with 80 elements, the unknown points were obtained through linear interpolations.

No symmetries were adopted to simplify the finite element model. A detail of the mesh refinement used can be observed in Figure 3. The four nodes square shell element STIF43 (DeSalvo, 1996) with six degree of freedom per node (three translations and three rotations) was used in the numerical analysis. Compatible two nodes beam elements BEAM3 were employed to simulate the rigidity of the L stiffeners at the model ends. The straight lines over the web frame edges represent these elements in Figure 3.

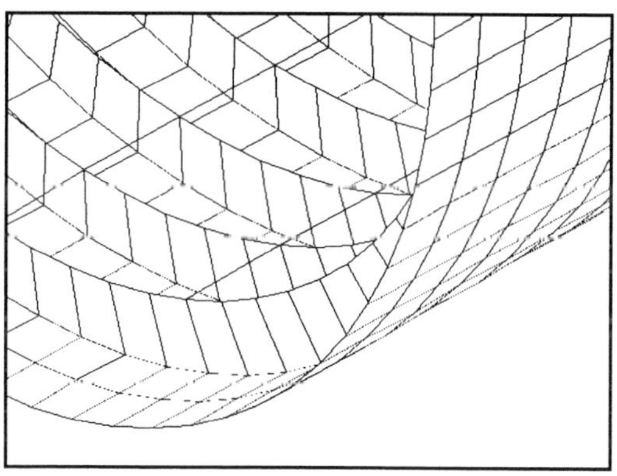

Figure 3: Detail of the mesh refinement used in the numerical model

Nonlinear capabilities such as large displacements and plasticity were included in the numerical model. The uniaxial tensile tests of the experimental study provided the necessary data to define the material properties. The stress-strain curve was inputted through a multi-linear approximation. The material tests have furnished proportionality and yield stresses equal to 254.3 and 337.7, respectively. The isotropic linear elastic constitutive model considered Young modulus equal to 205,000 N/mm^2 and Poisson ratio equal to 0.3. The elastic-plastic behavior was modeled through a potential flow law theory with isotropic hardening.

In this case, the hydrostatic load comprises a radial pressure applied on the shell outer surface plus an axial compressive force applied on the constrained edges of the model. The axial force was calculated considering the external pressure value and the lid area. The axial translations of the edge nodes were let free so that this force could be applied, while its remained degrees of freedom were made constrained. Also, some nodes at the model central section were constrained in axial direction to avoid rigid body movements. The arc-length control method was used to defined the load increments to overcome the limit load point as well as to configure the post-collapse load history.

4 RESULTS

The finite element mesh was adapted to simulate the four experimental tests carried out (MST3-1, MST3-2, MST3-3 and MST4). Not only the model length but also its end reinforcements were changed in relation to a base mesh. All the analyzed models verified the local collapse mode. The failures have initiated at the shell between frames with considerable deflections inward. The adjacent ring frames were also affected forming a localized lobe involving two or more stiffeners. The deformed shape of Figure 4 shows the localized collapse of the model MST4 which occurred close to the point observed in the experimental test. The failure of the models MST3-1, MST3-2, MST3-3 was initiated very close to the model extremities

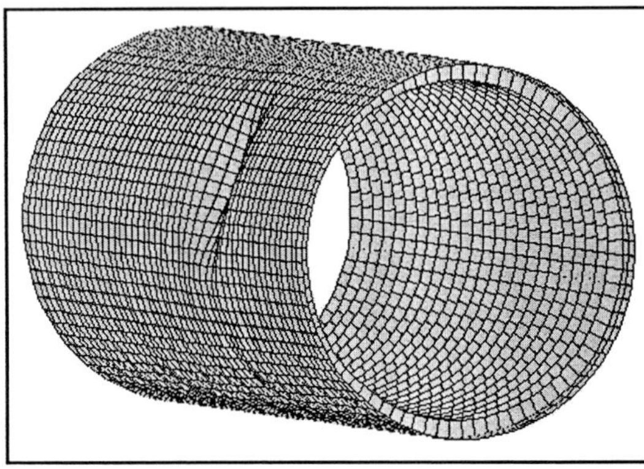

Figure 4: Deformed shape of the model MST4

The numerical and experimental results of the collapse pressure are compared in Table 2. The differences obtained may have been occasioned by the boundary conditions used to simulate the robust lids (clamping of the edge nodes with exception to the axial translations). This restriction has not allowed the longitudinal bending of the model, as experimentally verified in Figure 2. This effect is minimized to shorter models, resulting in decreasing differences between numerical and experimental results (see Table 2). Moreover, the numerical models have not considered the effects of residual stresses, material anisotropy and sideways imperfections of the ring frames

The post-collapse behavior was analyzed with the aid of Figures 5 and 6 which show the load history of the models MST3 and MST4, respectively. These graphs represent the applied pressure versus the edge axial displacements. It can be verified that the edge displacements further increase after the limit point and natural unloading of the models MST3-3 and MST4. The models MST3-1 and MST3-2, with lighter reinforcements at its edges, presented the opposite behavior.

TABLE 2

COMPARISON BETWEEN NUMERICAL AND EXPERIMENTAL RESULTS

Model	Collapse Pressure (N/mm^2)		Difference
	Experimental	Numerical	(%)
MST3-1	4.31	5.30	18.7
MST3-2	4.57	5.33	14.3
MST3-3	4.86	5.36	9.0
MST4	4.81	5.29	9.1

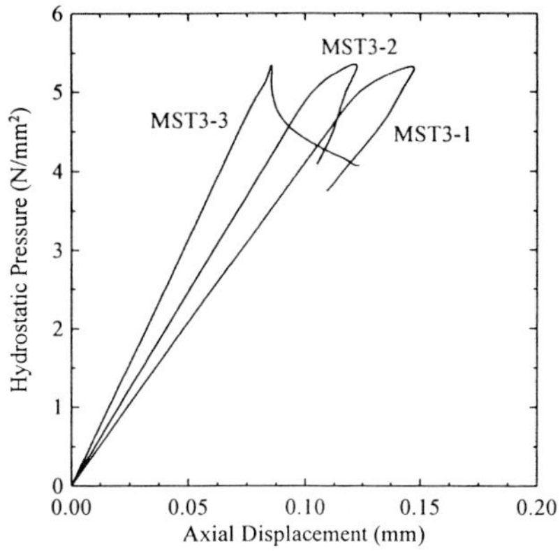

Figure 5: Load history of the Models MST3

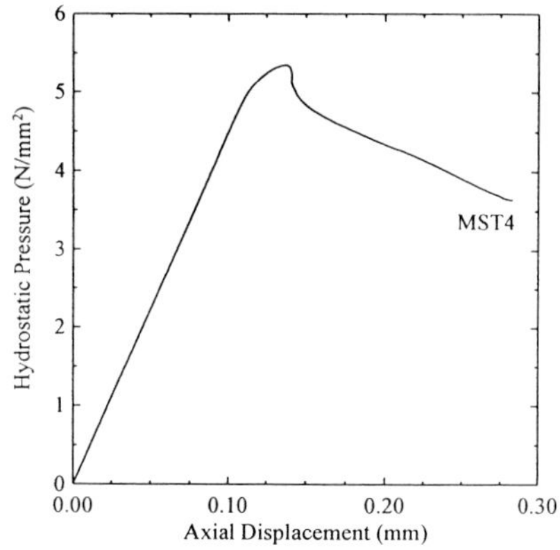

Figure 6: Load history of the model MST4

5 CONCLUDING REMARKS

The small scale models MST3-1, MST3-2 and MST3-3 have undergone local collapse of the shell close to its extremities, because of the severe structural discontinuity between the shell and the rigid lids. The small scale model MST4, correctly reinforced at its end, has undergone a local collapse of the shell at mid section, that also affected the adjacent ring frames. The finite element model has estimated larger collapse pressures than the experimental ones because of the simplification assumed to simulate the rigid lids. Nevertheless, a more sophisticated model including the rigid lids can overcome this problem. Qualitatively, the theoretical model represented adequately the failure modes verified in the experimental tests. Further improvements of the numerical model can generate a reliable tool to predict the collapse behavior of ring stiffened cylindrical shells. It can be used to the development of design equations based on limit states.

References

DeSalvo, G.J., Gorman, R.W. (1996). *Ansys User's Manual (Release 5.3).* Ansys, Inc., Houston, Pennsylvania, USA.

Netto, T.A. (1987). IGOR Computational Program to the Evaluation of Initial Geometric Imperfections of Small Scale Models of Semi Submersible Platform Columns. *I° Congresso Regional de Iniciação Científica e Tecnológica em Engenharia*, UFRJ, Rio de Janeiro, Brazil (in Portuguese).

Netto, T. A., Pasqualino, I. P. e Estefen, S. F. (1991). Development of Tests Facilities. *Buckle Propagation in Submarine Pipelines*, Phase I, Vol. I, Coppetec Report ET-170176, COPPE/UFRJ, Rio de Janeiro, Brazil (in Portuguese).

Scott, N.D., Harding, J.E. and Dowling, P.J. (1987). Fabrication of Small Scale Stiffened Cylindrical Shells. *Journal of Strain Analysis* **22:2**, 97-106.

Practical Design of Ships and Other Floating Structures
You-Sheng Wu, Wei-Cheng Cui and Guo-Jun Zhou (Eds)

A REPORT ON FATIGUE FAILURE OF A HIGHLY SKEWED FIXED PITCH PROPELLER

Hochung Kim, Keunjae Kim, Sungpyo Kim and Moonchan Kim

Daewoo Shipbuilding and Marine Engineering Co., Ltd
CPO Box 6208, Seoul, Korea

ABSTRACT

This paper reports a propeller blade damage case of a large ro-ro ship due to fatigue failure and describes the joint efforts followed by the builder, propeller maker, class and the owner to resolve the problem. Comprehensive investigations had been made and it was revealed that the fatigue failures were initiated by stern operation when stress at the trailing edge of a highly skewed blade can rise dramatically and the cracks were then continued to propagate during normal ahead operation. The major conclusions drawn is that sufficient fatigue margin must be reserved in case of propeller design for ro-ro ship operating frequently in astern mode during the port call.

KEYWORDS

Fatigue failure, Highly Skewed Fixed Pitch Propeller, FEM, Corrosion Fatigue Diagram

1 INTRODUCTION

A highly skewed propeller has been used more often in recent years and has been particularly successful in overcoming the possible risk of vibration problems arising from propeller cavitation excitation forces. However, for the propeller design with increased skew angle and especially intended for ro-ro ships making a frequent port call, special attention should be paid to the fatigue strength. When a propeller undergoes a transient manoeuvre, considerable changes occur in blade stress levels and this may lead propeller blades to fail by fatigue action.

Daewoo Shipbuilding and Marine Engineering Co., Ltd(DSME) has signed a contract with Wallenius Lines for the building of three 6,000 units pure car/truck carriers with the option of three vessels on January 19, 1993. The Vessel is a multi-purpose single screw ro-ro ship capable of carrying out 6,000 units pure car/trucks for worldwide service. The main dimensions are; Length overall(LOA) 199.0m, Length between perpendiculars(LBP) 190.5m, Breath, moulded(B) 32.26m, Depth, moulded up to upper deck(D) 32.98m and design draft, moulded(Td) 9.5m. The ship is equipped with convensional low speed diesel engine propulsion system composing of one set of B&W 8S60MC of MCR 22,420PS at 105.9 rpm together with a fixed pitch propeller and a Schilling rudder. The propeller has been designed jointly by the builder and the maker paying particular attention to fatigue strength under LR requirements. The builder was responsible for the hydrodynamic design while the propeller maker was responsible for the strength calculation. Figure 1 shows the original propeller drawing. The maker has carried out FEM calculations and reported that the propeller has the normal safety factor 2 against fatigue. LR has independently made FEM analysis in 1993 for the approval, and they found even somewhat lower stresses than calculated by the maker.

The ships have been constructed smoothly under good cooperation between the owner and the builder, and delivered successfully on Nov. 1, 1994 for the first ship with two months interval for the later ships. The ships have been proved to be very excellent in overall performances including speed and vibration characteristics through sea trials and have been in service with full satisfactions. As the result, repeat order was followed to build four more ships based on the same hydrodynamic design on hull form/propeller of the said series. Unfortunately, however, the first ship has lost part of the propeller blade tip on one blade after one and half years in services

2 GENERAL OBSERVATIONS

In January 1997, two blade tips of the propeller of the second ship in the series were reported to be bent due to ice impact. The spare propeller was then mounted, and the original propeller subsequently repaired. In April 1998 one blade tip was broken from the propeller of the first ship after about 4 years in service, of which the last 1.5 years were reported to be at 100% MCR. The propeller was replaced by the repaired propeller of the second ship. In July 1998 two blade tips of the propeller of the third ship were reported bent. These blades were repaired. In August 1998 one blade tip of the propeller of the second ship was broken.

3 INVESTIGATION ON DAMAGED PROPELLER

Comprehensive investigations were made to find possible cause of the propeller blade damage. The following works have been performed during the investigation.

3.1 Underwater Inspection

All propellers of the series vessels were checked very carefully by diver inspection. During the inspection of H4411 and H4412, it was noted that the tip of one of the propeller blades was missing. A photo of one of the damaged propellers is shown in Fig. 2.

3.2 Material Analysis

A part including the fracture surface of the broken propeller has been analysed by the Swedish National Testing and Research Institute with regard to chemical composition, hardness and fracture characteristics. In conclusion the material composition and hardness was found to be normal for a propeller nickel aluminum bronze. An examination of the fracture surface revealed that the probable cause for fracture is fatigue initiated at the trailing edge. The fatigue failures of the broken blades were initiated by astern operation. The cracks are then propagating during normal ahead operation and also at backing manoeuvres with low rpm. No macroscopic defects were found at the initiation point. No traces of cavitation erosion were evident on the propeller blade.

3.3 Strength Analysis by FEM

FE stress calculations were made in 1993 by the propeller maker for the 100% MCR condition. These calculations were approved by Lloyd's Register of Shipping. The results of stress analysis indicated that the calculated stresses are within the safety limit of a factor 2. The calculation was repeated in 1998 after the reporting of the two failures. Approximately the same results as in 1993 were found. Next, the propeller maker ordered the German research institute SVA in Potsdam to calculate both the hydrodynamic load, and the stresses. The stresses were calculated with the well known program ANSYS. Meanwhile, the shipowner ordered the Swedish research institute SSPA to make the same calculations. All results are summarized in Table 1. Various calculations indicated large differences as to the levels of stress; the stress calculations by SSPA show that the local stress level are extremely high in the region of trailing edge where the blades were broken, and the ANSYS calculation also predict higher stresses than the propeller maker own calculation. This may demonstrate the sensitivity of FE analysis depending on hydrodynamic loading, FE modeling and the method used. Whichever method is used between SSPA and SVA, the stress levels calculated based on today's knowledge are much higher than normally accepted merchant ships and the fatigue safety margin is less than 1.0.

3.4 Cause of Failure

The conclusion drawn from the above investigation is that the fatigue failure was attributable to three main features:

1. The stress level in trailing edge is too high and this is the cause of fatigue failure,
2. One cause of high stress level is due to the thin trailing edge which implies high stress originated from not only the external load but also the built-in residual stress from the casting procedure,
3. The fatigue failures of the broken blades were initiated by astern operation. The cracks are then propagating during the normal ahead operation and also at backing maneuvers with low rpm.

4 CORRECTIVE MEASURES

Based on the increased knowledge accumulated through the above investigation on the behavior of highly skewed fixed pitch propellers, the following two corrective measures were prepared:

4.1 Propeller Blade Modification

The first modification to change the contour line for the blade tip was proposed by the propeller maker based on their earlier experiences with success. Three more subsequent modification proposals were followed by SSPA and LR. The chosen trailing edge modification is shown in Fig. 3 and compared to the original design. This modification is aimed to enhance the blade strength properties in both the ahead and astern operating conditions.

4.2 New Propeller Design

A new propeller was designed in accordance with more enhanced class rule of today and additional requirement specified by the owner for sufficient safety margin against fatigue in the astern mode. The ship owner recognized that some compromise on propulsive efficiency and cavitation performances might be necessary in order to achieve the specified fatigue criteria. The most suitable means of achieving this was to increase the thickness at trailing edge and to change blade contour line by controlling the effective skew angle and chord distribution. The propeller geometries are shown in Fig. 4.

4.3 Model Tests

Towing tank and cavitation tests have been performed with the modified and new design propellers at SSPA. The self-propulsion tests showed that very small differences in predicted shaft power between all three propellers including the original propeller. The propeller cavitation for both modified and new propellers mainly appears as reasonably stable back sheet cavitation extending from approximately 0.8R to the tip. The cavitation characteristics on the new design propeller is very stable while that on the modified propeller is of a more bubbly character. The extent both in the circumferential and radial direction are somewhat larger for the modified propeller. The measured pressure pulses were of a moderate magnitude for all propellers. The modified propeller shows about 15% higher pressure pulse levels when compared to the original and new design propeller. In conclusion, all propellers are considered to be acceptable from a cavitation and vibration point of view.

4.4 Strength Analysis

The proposed LR modification to the original propeller design has been shown to give a satisfactory margin against fatigue failure in the ahead condition and against bending in the astern mode of operation. Comparing to the original design the blade stress in the ahead condition are reduced by about 38% and 18% for the mean and fluctuating components as shown in Table 2. In the astern condition the computed bollard astern stress reduces by 51%. As such, factors of safety of 1.87 and 1.38 are estimated against fatigue failure in the ahead condition and plastic failure in the astern condition respectively(see Figure 5). The Figure 6 compares the distribution of stress on blade at 100% MCR ahead condition and provides a graphic illustration of significant reduction of local stress level at trailing edge, 0.7R in the modification design

The results of FE analysis with the new design propeller confirmed its achievements of the design objectives – a reduction of normal stress level of 75% was predicted for a penalty on pressure pulses of 5% in comparison with the original design.

After consideration, the ship owner has decided in favor of the new design for the ships, They were approved by LR, manufactured and fitted to a complete series of these ships. Each of the ships has now successfully entered into service, and no fatigue problems have been encountered.

5 CONCLUSIONS

In case of propeller design for ro-ro ship where the full beam is carried right aft, forming large flat stern overhang above the propeller with small propeller-tip clearance, special attention should be focussed to the fatigue strength. More sophisticated procedure should be employed to ensure that the prepared design is fulfil the fatigue based stress criterion. Due allowance must also be made for operation in astern mode when stresses at the trailing edge of a highly skewed blade can rise dramatically.

Acknowledgements

The authors are grateful for the support and cooperation of Mr. Ulf Alexanderson and Mr. Per Croner from the Wallenius Lines over the various phases of work.

References

Calton J. S.(1984).Marine Propeller Blade Stress, Trans. I. Mar. E

TABLE 1

CALCULATED STRESSS LEVEL AT FULL SPEED AHEAD 21.4 KNOTS

Method	Stress		Power [kW]	RPM	Thrust [kN]	Remarks
	Mean [Mpa]	Fluct. [%]				
Propeller Maker(1993)	52	54	16030	105	-	T.E. 0.60R
Propeller Maker(1998)	53	48	16030	105	1410	T.E. 0.60R
SVA ANSYS	104	25	16030	105	156	T.E. 0.63R
SSPA	193	15	15907	111	1354	-

TABLE 2

MAXIMUM VON MIESES EFFECTIVE STRESS[MPA] AT T.E. AT 0.6R

Blade Design	Original	Modification	New Design
Normal Stress	132	86	56 at root 30 at T.E. 0.6R
Maximum Stress	155	119	97 at root
Minimum Stress	115	72	35 at root
Mean Stress	135	96	66 at root

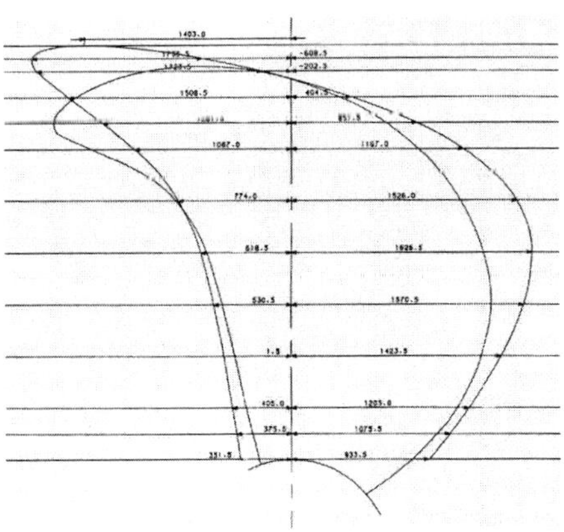

Blade Number	4
Diameter	8200 mm
Mean Pitch	5650 mm
Area Ratio	0.6
Skew Angle	39.5^{O}
Design Point	
- Structural	16380 kW x 105.0 rpm
- Hydrodynamic	13020 kW x 102.5 rpm
Service Speed	21.0 knots

Figure 1: Original propeller design

Figure 2: Damaged propeller blade and fracture surface

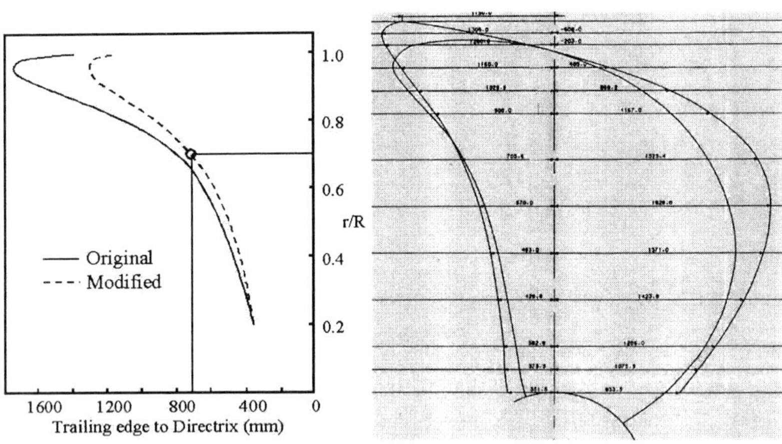

Figure 3: Comparison of trailing edge modification proposal and modified propeller

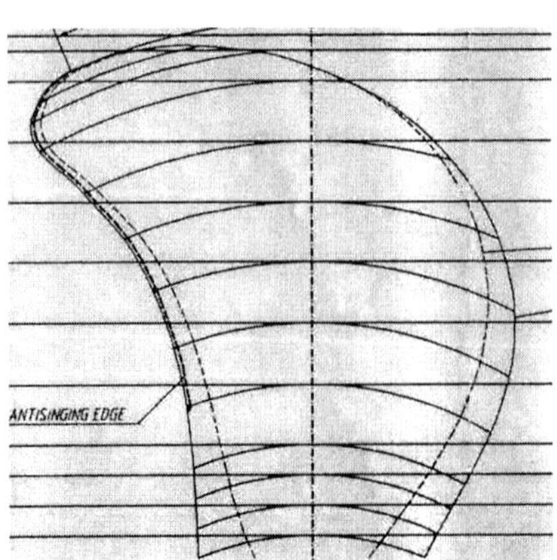

Blade Number	4
Diameter	8200 mm
Mean Pitch	5320 mm
Area Ratio	0.59
Skew Angle	29.7^{O}

Figure 4: New design propeller

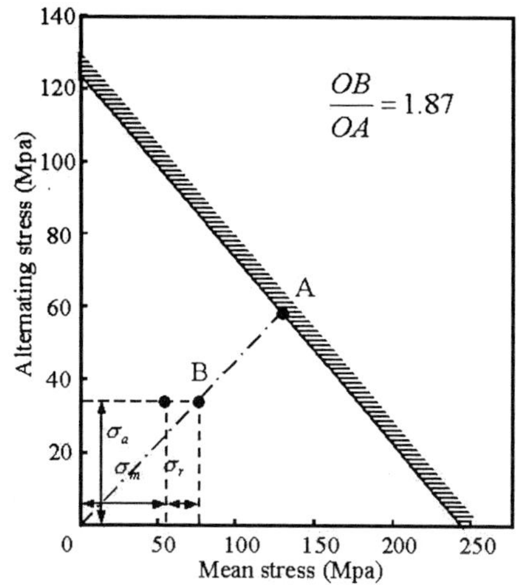

Figure 5: Fatigue assessment of the blade modification proposal in the ahead MCR condition

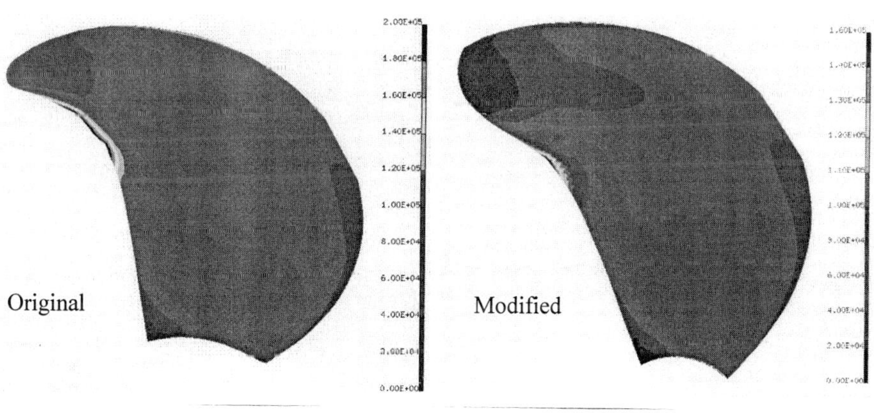

Figure 6: Stress analysis results with maximum load in the ahead MCR condition

Practical Design of Ships and Other Floating Structures
You-Sheng Wu, Wei-Cheng Cui and Guo-Jun Zhou (Eds)

FATIGUE ANALYSIS OF ALUMINIUM BOX-STIFFENER LAP JOINTS BY NOMINAL, STRUCTURAL AND NOTCH STRESS RANGE APPROACHES

Naiquan Ye[1], Torgeir Moan[1] and Bård Wathne Tveiten[2]

[1] Department of Marine Structures, Faculty of Marine Technology, Norwegian University of Science and Technology, Trondheim, Norway
[2] Halliburton AS, Structural and Marine Department, Stavanger, Norway

ABSTRACT

Fatigue behaviour of an aluminium box-stiffener lap joint is systematically investigated in this study. Fatigue analysis based on nominal stress range approach, two local approaches based on structural (hot spot) as well as notch stress respectively are applied. The structural stress range approach depends on how the structural stress concentration factors (SCFs) are obtained by means of several stress extrapolation methods. Different stress extrapolation methods are discussed. Stress analysis is performed by the finite element (FE) method. Twelve full-scale test specimens are included in the test program. The results show that the notch stress range approach is the most conservative one among the three approaches. The nominal stress range approach, however, gives the least conservative approach to the joint.

KEYWORDS

Fatigue, Box-stiffener, Lap joint, Aluminium, Nominal stress, Structural stress, Notch stress, SCF

1 INTRODUCTION

Fatigue is a highly relevant failure mode for welded structural components that have high stress concentration and are subjected to a large number of variable stress cycles. Traditionally, fatigue design is a stress-based procedure even though some strain-based and fracture mechanics methods have been developed. In this study, three stress-based fatigue analysis and design approaches are reviewed. Furthermore, the fatigue strength of a welded box-stiffener lap joint used in high speed light crafts (HSLCs) is systematically investigated by a nominal, a structural, and a notch stress range approaches.

The stress experienced in a weld joint is mainly determined by the following factors: global geometry, structural geometry and the weld. Correspondingly, three different definitions of stress are applied, namely: nominal stress, structural stress (that is also called "hot-spot" stress) and notch stress are defined by respectively taking the above factors into consideration. Figure 1 schematically illustrates these three stresses experienced in a single plate lap joint.

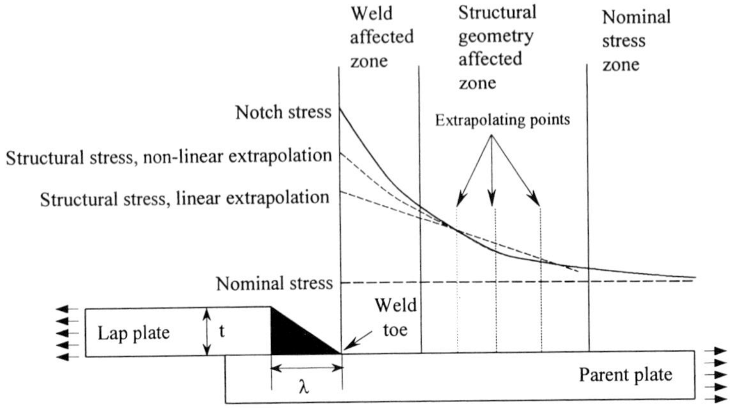

Figure 1: Stress components in a welded lap joint and structural stress extrapolation methods

1.1 Nominal Stress Range Approach

The nominal stress σ_n is defined such that only the effect of the global structure is taken into consideration. This stress can normally be estimated by applying elastic strength of materials (beam) theory, frame analysis, or coarse FE analysis. The stress-raising factors due to the structural geometry and the weld are implicitly included in the fatigue strength curves (SN-curves). This implies that a database which includes several typical structural details ("detail classes") is established in each design code. A characteristic stress level is assigned to each "detail class" referring to a fatigue life of 2×10^6 cycles. The fatigue analysis is then greatly simplified by using only the nominal stress. However, the "detail class" database can only be established by a significant amount of experimental work. Therefore, only a limited number of standard joint types is possible to include in the database. However, when new and efficient joints are proposed and produced, they are unable to be satisfied by the limited database. Thus, there is a need for a fatigue assessment method that can deal with these geometrical "irregularities". This situation becomes even more evident when aluminium extrusion fabricating technique is used to produce complicated cross sections and joints to optimise the structure design. Moreover, people have admitted long since before that fatigue is an extremely localised phenomenon which depends heavily on the local geometry of the joint and especially the weld. The embarrassment here is that it is difficult to find a comparable "detail class" in the design codes for those well-behaved new joints.

1.2 Structural Stress Range Approach

The structural stress is an attempt to advance further to account for the effect of the structural geometry on the fatigue strength of the joints. As stated above, the structural geometry of the joint plays an important role in the local stress level. In order to reflect its influence, a structural SCF K_g is introduced as:

$$K_g = \sigma_s / \sigma_n \tag{1}$$

where, σ_s and σ_n are the structural stress and nominal stress.

The effect of the structural geometry on the stress at the weld toe is then accounted for by this factor, but the notch effect of the weld is taken into account implicitly by the corresponding structural stress-based SN-curves, the "hot spot" fatigue strength curves. Compared to the nominal SN-curve database, much less experimental work is required for a structural stress SN-curve database. For instance, compared with the 22 nominal stress SN-curves, Eurocode 9 (1998) proposed only 6 structural stress SN-curves for all the joints (classified by the thickness of the stressed member). However, the structural stress method normally includes some kind of extrapolation procedure of the stress to the weld toe (hot spot). The challenge is how to exclude the stress-raising effect due to the weld (notch) and only include the local structural effect in the stress calculation. In order to describe the structural stress at the weld toe, various extrapolation methods have been proposed (Figure 1), either linear extrapolation (Hobbacher 1996, DNV 1997), quadratic extrapolation (Niemi 1994, Hobbacher 1996), or by a "singularity check" (Tveiten and Moan 2000).

Stress extrapolation methods are usually based on an assumption that the effect of the weld is localised within a distance close to the weld. This distance is normally expressed as a function of the main plate thickness. In addition, the extrapolation methods are quite joint dependent. In other words, the method is easily calibrated to a given joint type, but fails to get a right answer for a different configuration and dimension of the joint. Obviously, the assumption can not guarantee that the real influence of the weld on the structural stress has been removed. Furthermore, the joint-dependence narrows the application range of the methods because many joint types usually differ greatly from the proposed one.

This restriction is, however, avoided by Tveiten and Moan (2000). They proposed an analytical solution to obtain the structural stress at the weld toe by taking the weld as a singularity. Williams (1952) had proved that an exponential law as stated in equation (2) governs the stress near a singularity:

$$\sigma(r) = C_{1,}f(\theta)r^{\beta(\alpha)} + 0(\theta) \tag{2}$$

where r and θ are polar coordinates, C_1 is a constant, $f(\theta)$ is a dimensionless function of θ and $0(\theta)$ are higher order terms. It is seen that the stress close to the singularity ($r \to 0$) is governed dominantly by the singularity and varies log-linearly with the distance from the weld toe. Therefore, the influence of the weld on the stress at the weld toe can be separated through a log-log plot of stress against distance. On the basis of the asymptotic behaviour of stresses adjacent to a singularity, the following extrapolation procedure is proposed.

- to separate the zone solely influenced by the structural geometry as the distance from the end of the singularity-affected zone to the point where the structural stress and the nominal stress coincide.
- to fit a curve to the stress field influenced by the structural geometry by means of a polynomial of second order and extrapolate to the weld toe.

The proposed extrapolation method eludes the weakness of the extrapolation methods suggested by e.g. Niemi (1994) and Hobbacher (1996) in that the extrapolation method is independent of any structural dimension to determine the region influenced solely by the structural geometry and thus the location of the extrapolation points. The proposed method is instead dependent on a numerical stress singularity, which is possible to reproduce in all structural components normally found in ship structures independent of the actual geometrical dimensions of the detail. A drawback with the proposed extrapolation method is that it requires a rather detailed (time consuming) FE analysis to ensure that the local singularity at the notch (weld toe) has been properly accounted for.

1.3 Notch Stress Range Approach

With respect to stress calculation, the state-of-the-art notch stress range approach goes even further than the structural stress range approach. The stress referred to is the notch stress (total stress) at the weld toe that compromises all the possible stress-raising effects due to the structural geometry and the notch effects at the weld. Theoretically, only one SN-curve based on the plain material is needed for all the joints in practice. Actually, only 4 SN-curves were proposed in the DNV classification note (DNV 1997). A proper notch stress calculation is then a critical issue when this approach is applied.

Two alternatives to calculate the notch SCF are proposed in the classification notes (DNV 1997). One is direct FE analysis and the other is parametric formulae. Good experience on FE analysis would be beneficial in using the former technique to manage the mesh size and element type close to the weld toe. Besides, a suitable weld toe radius must be specified in the FE model otherwise an infinity will occur when the mesh size decreases to zero (DNV 1993). The latter, however, requires the relevant joint to match the geometry that serves as basis for the parametric SCF. A poor matching may result in an unacceptable error.

2 FATIGUE BEHAVIOUR OF THE BOX-STIFFENER LAP JOINT

2.1 Specimen Description

Lap joint is a widely used structural detail to join individual components. In aerospace industry, many lap joints are used to joint sheets either by adhesive or bolt. With the development of section fabrication technique, box-stiffener lap joints are used to join the pre-fabricated panels in the production of HSLCs and bridges. Figure 3 represents a typical box-stiffener lap joint.

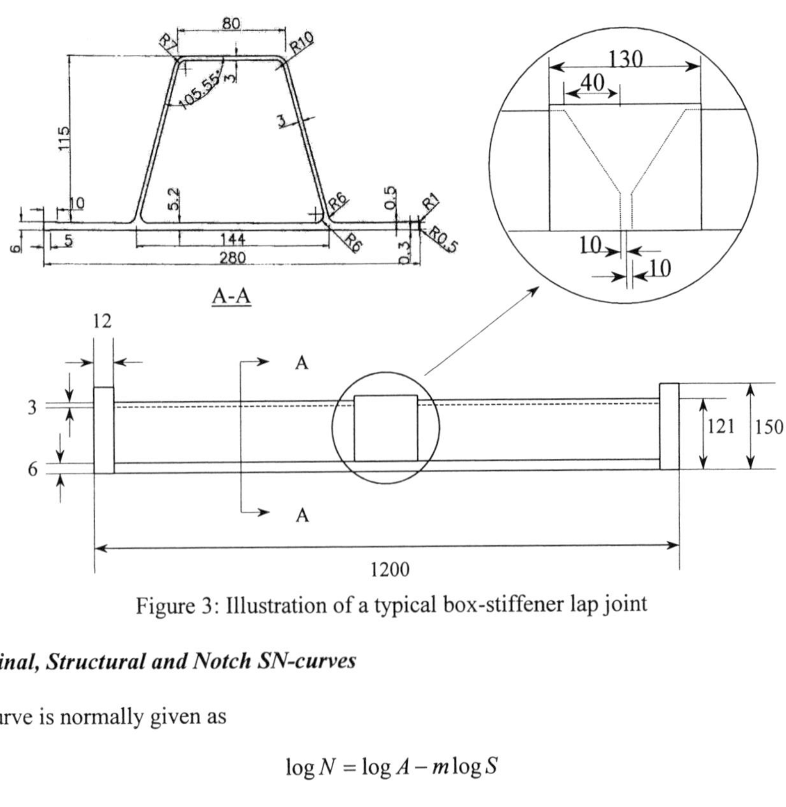

Figure 3: Illustration of a typical box-stiffener lap joint

2.2 Nominal, Structural and Notch SN-curves

An SN-curve is normally given as

$$\log N = \log A - m \log S \tag{3}$$

where m is the negative inverse slope of the SN-curve, $logA$ represents the intercept of the $logN$-axis by the SN-curve. N and S refer to the fatigue life and the corresponding stress range respectively.
A characteristic fatigue life is normally set to be 2×10^6 cycles and the corresponding stress range is correspondingly called a "detail class". It has been stated previously that the definitions of the stress in the three approaches for the fatigue assessment are different. With respect to the box-stiffener lap joint, Table 1 lists the characteristic SN-curves corresponding to each approach.

TABLE 1
SN-CURVES FOR THE BOX-STIFFENER LAP JOINT

Approach	Detail class[1)]	m	Classification
Nominal stress	25	3.2	Eurocode 9
Structural stress	44	3.2	Eurocode 9
Notch stress	55	4.32	DNV

1) Refers to stress range at $N=2\times10^6$ cycles.

It should be pointed out that among existing aluminium fatigue design codes such as ECCS, IIW, and BS8118, Eurocode 9 (1998) assigns the highest "detail class" to the load carrying fillet weld in the lap joints. Not every design code provides a possibility to use the structural stress as a design stress. Eurocode 9 (1998) supports this design concept by providing a series of structural SN-curves based on the thickness of the stressed member. However the DNV (1997) is the only design code that provides the alternative to use the notch SN-curves. The "detail class" in the DNV (1997) is not explicitly given but it can be directly derived by the given parameter $logA$ through equation (3). Theoretically, the three SN-curves should be equivalent to each other for one given joint type.

2.3 Stress Calculation

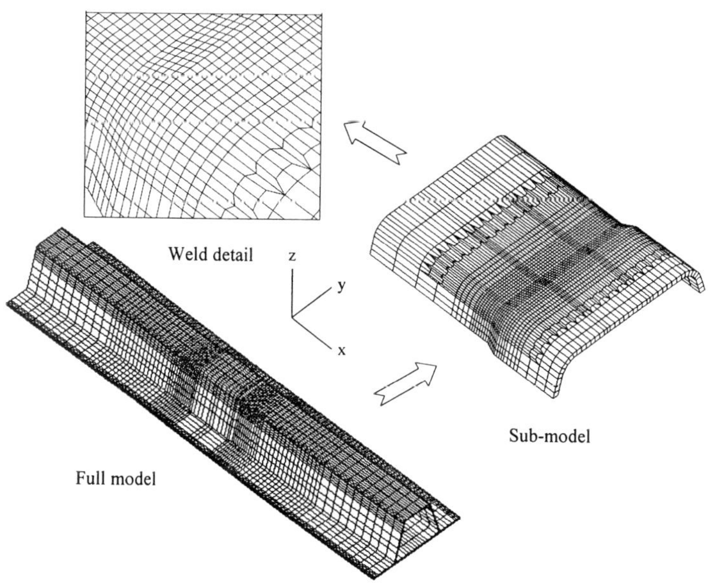

Figure 4: FE model of the box-stiffener lap joint

Three different levels of stress calculation were carried out to obtain the nominal stress, structural stress and notch stress. A FE model was established for this purpose. A sub-model was also built to

You-Sheng Wu, Wei-Cheng Cui and Guo-Jun Zhou (Eds)

FATIGUE STRENGTH ASSESSMENT OF CRUCIFORM JOINTS

W. Fricke and R. Wernicke

Dept. of Ship Structural Design and Analysis
Technical University Hamburg-Harburg, Hamburg, Germany

ABSTRACT

A large number of cruciform joints can be found in ship structures. In case of high dynamic loading, full or partial penetration welding is usually applied. The fatigue strength of this detail, with or without weld penetration, depends on several influence factors, including plate and weld throat thickness, width of the root gap as well as fabrication effects such as weld shape and misalignment. The fatigue strength assessment is reviewed on the basis of earlier and more recent investigations on small scale test specimens and some large scale tests. Different approaches are applied, which are able to consider relevant factors affecting the fatigue life. As misalignment is a stochastic variable not known during design, a probabilistic approach is applied which allows its rational consideration in fatigue design.

KEYWORDS

Welded joint, Cruciform joint, Fatigue, Fabrication effect, Misalignment, Local stress, Fracture mechanics

1 INTRODUCTION

The cruciform joint is one of the 'standard cases' of welded connections. Its fatigue strength was subject of several investigations in the past. Fatigue strength data were collected and re-analysed a. o. by Gurney & Maddox (1973) and Olivier & Ritter (1979). An interesting feature of this joint is that in case of incomplete root penetration, different crack initiation sites exist, i. e. at the weld toes, leading to plate fracture, and at the ends of the non-welded root gap, leading to a fracture through the weld (see Fig. 1). The question, where the first crack will appear, depends mainly on the geometrical configuration, in particular on the amount of weld penetration and throat thickness, and on the type of loading.

The danger of root cracking, which can only be detected after the whole weld has been separated, has resulted in preferring full or at least partial penetration welding for structures which are subjected to cyclic loading. This is supported by corresponding guidelines and partly stringent fatigue strength requirements for joints only with fillet welds.

On the other hand, these fillet-welded cruciform joints without weld penetration are typical for large plated structures such as ships and mobile units. Partial or full penetration welding would require additional preparation of plate edges which is expensive and, therefore, normally limited to areas where it is absolutely required. Wherever possible, weld reinforcement would be preferred over plate edge preparation by many shipyards.

Therefore, the fatigue strength assessment of cruciform joints without and with partial or complete weld penetration is of great interest for practical application. For this reason, cruciform joints have been further investigated in two research projects, in the European project FatHTS and a national R&D project. After a short review of the current fatigue design procedures for cruciform joints, the results of the R&D projects will be described, showing the effects of several influence factors on fatigue strength, based on small scale tests. These are supplemented by some large scale tests showing the behaviour of cruciform joints in complex structures. Finally, a probabilistic consideration of misalignment is presented, offering a rational assessment of this important, but uncertain influence factor.

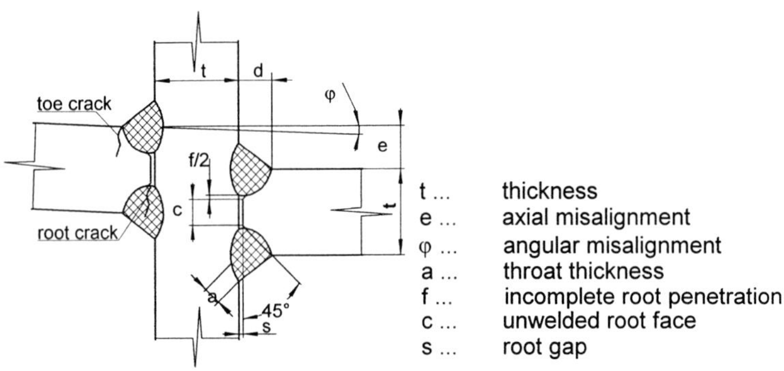

Figure 1: Geometry of a Cruciform Joint

2 FATIGUE STRENGTH ASSESSMENT

The fatigue strength of cruciform joints can be evaluated using any of the available approaches. One possibility is offered by the *nominal stress approach*, which is used in many codes and guidelines. In accordance with the two possible crack initiation points, two nominal stresses are assessed:

a) the nominal stress in the interrupted plate, for which detail category FAT 63 is specified by IIW (1996), defining the fatigue strength reference value of the design S-N curve at 2 million cycles. FAT 63 is slightly higher than the corresponding class F2 of BS 5400 (1980).

b) the nominal stress in the weld throat, i. e. the stress in the plate multiplied by 2 x throat thickness and divided by the plate thickness, which is classified by IIW (1996) as FAT 45, corresponding to class W of BS 5400.

In case of full penetration, only (a) is applicable, but with an increased fatigue class (FAT 71 or F, respectively). The design curves incorporate an allowance for increased stress due to misalignment, which amounts up to 45% according to IIW (1996).

The *hot spot stress approach* allows only the assessment of the crack initiation at the weld toes, i. e. crack initiation site (a). IIW (1996) classifies 45° fillet welds into FAT 100. However, some codes

such as Germanischer Lloyd (2000) propose a reduced fatigue class for fillet welds carrying the whole load of the attached plate, i. e. also for cruciform joints without weld penetration. The difference between the fatigue classes of the nominal and the hot spot stress approach is due to the fact that in the latter any stress magnification due to misalignment has to be considered in the applied stress.
A more refined approach is the *effective notch stress approach*, which allows the consideration of specific geometry and load parameters. According to Radaj (1990), the effective notch stress at weld toes and roots in steel structures can be analysed using a fictitious radius of 1 mm. IIW (1996) defines a corresponding fatigue class FAT 225 for these notches. Radaj (1990) and Anthes at al. (1993) derived parametric formulae for the notch stress at cruciform joints. Possible stress magnification due to misalignment has to be considered in the applied stress here as well.

Finally, the *fracture mechanics approach* is mentioned, which has been applied particularly to the assessment of cracks initiating from the notch root. The root gap is considered as initial crack and the crack propagation through the weld is computed using a crack propagation law such as that of Paris and Erdogan. The geometry function required to account for the effect of the weld toe and root can be found in several handbooks, such as IIW (1990).

3 FATIGUE TESTS WITH SMALL SCALE SPECIMENS

In the projects mentioned, additional fatigue tests have been performed using small scale specimens of cruciform joints. The main objective was to investigate fabrication effects, in particular axial misalignment, as well as the effect of steel strength in view of applying higher and very high strength steel with yield point up to 690 MPa. Cruciform joints with almost full penetration welds were considered, so only toe cracking was expected.
Axial misalignment creates secondary bending stresses under axial (membrane) loading. As shown by several authors before (a. o. Maddox, 1985; Berge and Myrhe, 1977), the increased stress range $\Delta\sigma$ can be calculated from the membrane stress range $\Delta\sigma_M$ with the following formula for unrestrained cruciform joints with constant thickness t for both fixed and pinned ends:

$$\Delta\sigma = \Delta\sigma_M (1 + 3\, e/t) \qquad (1)$$

Additional external bending stresses of the cruciform joint is not affected by axial misalignment e.
The test specimens were fabricated by two different shipyards (series A, L) from plates 300 x 600 mm with different thickness and cut into 60 mm wide specimens. Three levels of axial misalignment were realised: S - small: $e/t \approx 0.1$; U - usual: $e/t \approx 0.3$; L - large: $e/t \approx 0.5$
Fig. 2 shows typical sections of the joints. Several of the L-series had to be re-welded with additional passes to avoid root cracking. In total 15 test series, each with 11 - 28 specimens, were investigated. Table 1 gives an overview together with the series designation formed by four letters or numbers, indicating the manufacturer (A, L), the plate thickness in mm, the steel strength (M = mild steel; H = higher tensile steel with nominal yield strength 355 - 390 MPa; V = very high tensile steel with 690 MPa) and the level of misalignment as given above.

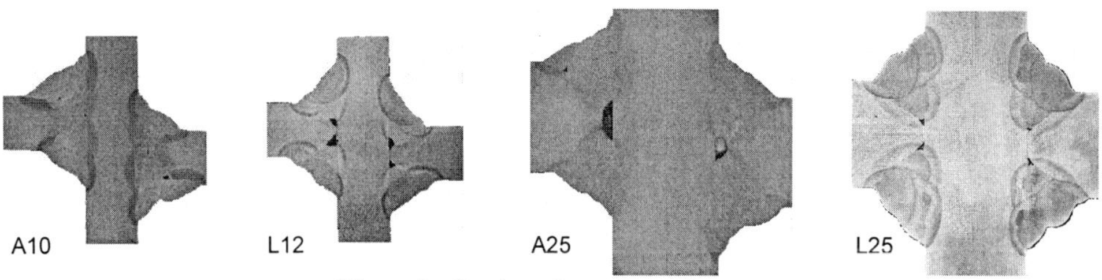

Figure 2: Sections through specimens

The load-controlled fatigue tests were carried out with pulsating axial load (R = 0) in air at frequencies between 20 and 30 Hz. The load level corresponded in most cases to a membrane stress of 140 MPa, while 80 and 100 MPa were chosen for series A10H-L. The number of cycles until failure, which was defined as fracture of the specimen, was statistically analysed. Almost all failures were due to cracking from the weld toe, except for a few specimens of A25-series showing fabrication-related weld defects.

TABLE 1

AVERAGE GEOMETRICAL VALUES OF SPECIMENS (SEE ALSO FIG. 1)

Series	R_{eH} [MPa]	t [mm]	e [mm]	φ [°]	e / t	c + f [mm]	a [mm]	Number of specimens
A10H-S	355	10.6	1.3	0.1	0.13	1.2	5.2	11
A10H-L	355	11.6	6.2	0.3	0.53	1.7	5.3	11
A25H-S	355	26.0	3.6	1.2	0.14	7.7	7.3	11
A25M-U	235	25.5	6.6	0.4	0.26	4.5	6.8	11
A25H-L	355	26.0	9.1	1.7	0.35	6.7	8.2	11
L12M-U	235	12.5	3.1	0.2	0.25	4.6	5.9	14
L12H-S	390	12.2	2.8	0.2	0.23	5.8	4.3	14
L12H-U	390	12.2	4.3	0.4	0.35	5.6	6.4	27
L12H-L	390	12.2	7.4	0.2	0.61	5.9	6.0	14
L12V-S	690	11.8	1.3	0.2	0.11	5.6	7.1	14
L12V-U	690	11.8	1.3	0.3	0.33	-	6.9	14
L25H-S	390	24.8	4.2	0.2	0.17	8.0	9.7	14
L25H-U	390	24.7	7.5	0.2	0.30	9.2	10.0	28
L25H-L	390	24.7	9.5	0.9	0.39	8.4	10.0	14
L25V-S	690	24.9	2.1	0.3	0.09	7.4	10.3	13

Fig. 3 shows the relationship between cycles to failure of the series (the asterisk stands for different materials and misalignments) and misalignment e/t together with the scatter band for 10 and 90% survival probability P_S. It can be seen that axial misalignment significantly reduces the fatigue life. The results are in good agreement with the curves plotted in Fig. 3 which consider the reduction in fatigue life under the assumption of the stress increase acc. to eq. (1) and an S-N curve with a slope exponent m = 3. Furthermore the tests verify that steel strength has absolutely no effect on the fatigue behaviour.

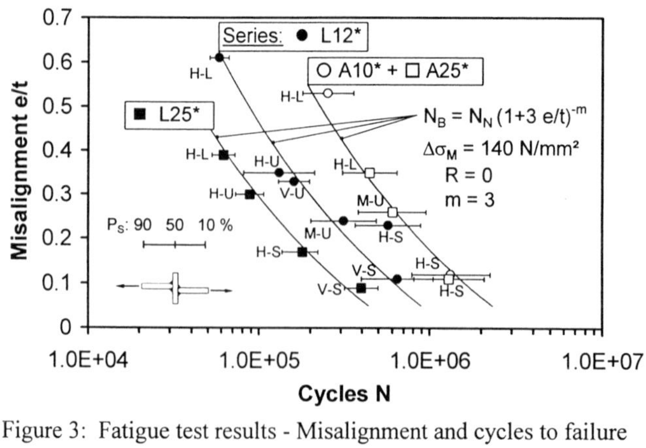

Figure 3: Fatigue test results - Misalignment and cycles to failure

However, large differences can be observed between the series from different manufacturers (A, L). The plate thickness effect is only significant in the L-series, where it corresponds to the usual thickness correction exponent 0.25 (Maddox, 1991). No thickness effect was observed in the A-series. Although all test series are within the scatter band found in the evaluation of cruciform joints by Olivier and Ritter (1979), additional investigations were performed in order to clarify these differences.

At first, the *notch stress approach* was applied allowing the actual weld shape to be considered. The notch stress was calculated with 2D finite element models, using average weld shape parameters for the A- and L-series. The measured notch radius was considered in addition to the fictitious radius r_f = 1 mm which accounts for the micro-supporting effects to be taken in the worst case of zero weld toe radius. The resulting effective notch factors K_f for axial loading and bending were then used to derive a characteristic nominal fatigue strength value $\Delta\sigma_{90}$ (fatigue strength at 2 million cycles for 90% survival probability) taking the corresponding value $\Delta\sigma_R$ = 270 MPa for effective notch stresses as proposed by Radaj (1990).

Fig. 4 shows the formula and resulting curves in relation to the misalignment. The factor k in the formula considers the stress reduction due to the distance between the specimen centre and the weld toe location. The symbols in Fig. 4 represent the corresponding test results, i. e. for 90% survival probability and shifted to a fatigue life of 2 million cycles.

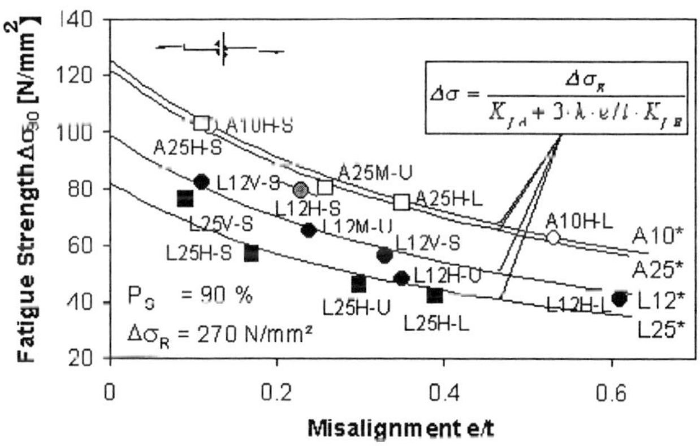

Figure 4: Fatigue strength assessment with the notch stress approach

It can be seen that the curves based on the *notch stress approach* agree quite well with the test results. As both plate thickness effect and local weld geometry are considered, it can be concluded that they cancel each other for the L series, i. e. the thicker specimens had a better weld shape. This was actually the case as Fig. 2 shows for the critical weld toes.

Figure 5: Crack surfaces of specimens from A-series (left) and L-series (right)

However, the remaining differences between the A- and L-series cannot be explained by notch effects. Therefore, refined investigations based on *fracture mechanics* were performed. Some test specimens were used to observe more thoroughly the crack propagation as well as the initiation phase. By investigating the crack surfaces and the size and shape of the starting crack, it was found that in the A-series the cracks started at one point and propagated with elliptical front through the thickness. The L-series in contrary, which were mostly re-welded with additional passes, showed a quite different behaviour. Due to small defects in a depth 0.03 mm below the toe, several initial cracks grew together and formed a long and shallow crack propagating through the thickness. Fig. 5 shows examples of the different crack surfaces.

Crack propagation analyses based on IIW (1990) as well as on Bell et al. (1987) and Seeger et al. (1991) regarding the coalescence of cracks showed a good agreement between measurement and calculation including the effect of secondary bending. More details are described by Wernicke (2001). These findings underline the fact that the scatter in fatigue life largely depends on the quality of fabrication which may differ considerably between the manufacturers.

4 LARGE SCALE FATIGUE TESTS

In real structures the situation is often not as simple as in the small scale specimens, owing to rather complex geometry and loading. This was investigated on three models of frame corners in a scale close to reality (Fig. 6). Two misaligned connections of flanges were investigated in one symmetrical model. The frame corners consisted of 3850 mm long branches, made of 12 mm thick plates, while the flanges were fabricated from flat bars 300 x 20 mm. The vertical flange is interrupted and connected with a full penetration weld. The built-in misalignment was 7 mm, i. e. 35% of the flange thickness.

The load was applied by a diagonally acting hydraulic cylinder creating mainly bending and shear in the branches. The tests were carried out at load levels between 300 and 380 kN. The stress state in the vicinity of the connection was measured by strain gauges. For comparison, finite element calculations were performed with a 3D model of the structure.

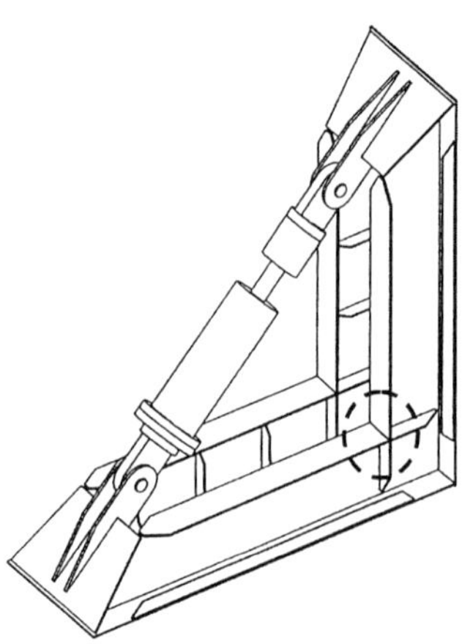

Figure 6: Large scale specimen

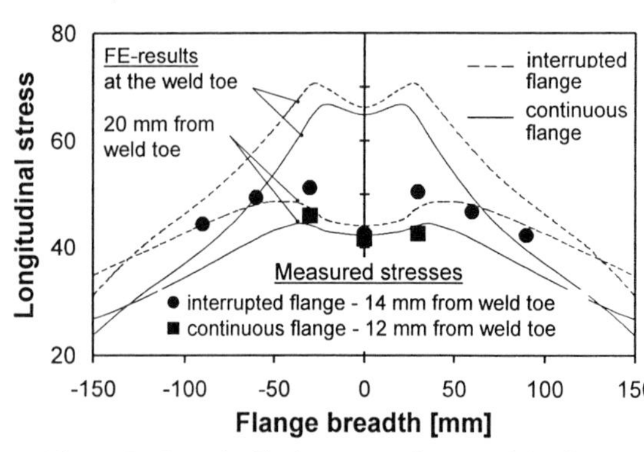

Figure 7: Longitudinal stress at the top of the flange

Fig. 7 shows the comparison between the measured and computed stress distribution in both flanges close to the welded connection. Generally a good agreement is observed. The interrupted flange clearly shows higher stresses, although the difference is much smaller than in unrestrained specimens having the same misalignment, where the bending stress would even be larger than the axial stress. In the large scale test models, high restraining is caused by the web and the continuous flange. Furthermore, a stress reduction towards the edges can be observed which is due to the reduced effective breadth. Therefore, the highest stress occurs close to the web, however, in a distance of approx. 30 mm away

from the web. The stress increase towards the weld toe is very steep. The hot spot stress at the weld toe has been calculated by applying quadratic stress extrapolation over three points (0.5, 1.5 and 2.5 x plate thickness away from the hot spot).
During the fatigue tests the cracks appeared at the weld toes of the interrupted flanges. These cracks were detected when they reached a length of 1 – 2 mm. Subsequently several small cracks coalesced to a shallow surface crack. Failure was defined for a crack length of approx. 50 mm. The fatigue lives were above the hot spot design S-N curve defined by IIW (1996) and GL (2000).

In summary the large scale tests have shown that rather complex stress distributions occur in 3D structures. The assumption of a stress magnification factor as given e. g. by IIW (1996) might be too conservative. However, the fatigue strength of misaligned joints can be well assessed by the hot spot stress approach in such structures, if both misalignment and effective breadth are taken into account.

5 PROBABILISTIC CONSIDERATION OF MISALIGNMENT

The S-N approach offers a relatively simple way to consider the uncertainties on the strength and load side. The strength is characterized by the S-N curve, i.e.:

$$N = K \cdot \Delta\sigma^{-m} \tag{2}$$

The high scatter in fatigue life is usually considered by assuming K to be a stochastic variable, while m is set to 3.0. But also the load side, which is represented by $\Delta\sigma$ and the number n_{max} of applied load cycles, may contain high uncertainties, especially if variable amplitude loading occurs. In this case a damage accumulation law such as Miner's rule is needed, creating additional model uncertainties.

As shown above, misalignment magnifies the stress, e. g. acc. to eq. (1). If misalignment e/t is a stochastic variable, $\Delta\sigma$ will also be stochastic and include the uncertainties.

Observations in practice indicate that the frequency distribution of axial misalignment may be approximated by an exponential distribution (Berge and Myrhe, 1977). If we assume that the "permissible misalignment" e_p/t is a characteristic value, which is exceeded in 5% of all cases, the frequency distribution, illustrated in Fig. 8, can be written as

$$f\left(\frac{e}{t}\right) = \frac{3}{e_p/t} \exp\left(-\frac{3e/t}{e_p/t}\right) \tag{3}$$

Using the stochastic variables K and e/t, if appropriate together with others, the probability of failure P_f can be calculated for a given fabrication quality. The same can be done for an enhanced quality, characterized by a reduced "permissible misalignment", allowing higher stresses if the original failure probability shall be maintained. This can be of great interest if high tensile steel is applied.

In the following, the effects are illustrated by a relatively simple example, i. e. a cruciform joint with full penetration welding and with an axial misalignment being distributed acc. to eq. (3). The loading is assumed to be purely axial with $n_{max} = 2 \cdot 10^6$ cycles and a constant stress range $\Delta\sigma_M = 71$ MPa.

In the nominal stress approach, the probability of failure for this case is assumed to be 2,3% (IIW, 1996), because this detail is related to FAT 71 with a corresponding design S-N curve located two standard deviations below the mean curve. In the probabilistic analysis, the probability of failure is computed for various "permissible misalignments e_p/t", applying the hot spot S-N curve acc. to FAT

100 and assuming for this a standard deviation 0,064 in logarithmic scale (related to stress range). The mean S-N curve becomes then in logarithmic scale:

$$\log N = \log K_{50} - m \cdot (\log \Delta\sigma + 0.128) \tag{4}$$

By inserting the data of the design hot spot S-N curve, we obtain a mean value E(log K) = 12.685 and a variation coefficient COV(log K) = 0.015135. (log K) is assumed to be normally distributed.

Failure occurs, if the number of load cycles n_{max} is greater than the endurable number N acc. to the S-N curve, i. e. $n_{max} > N$. Inserting eq. (1) and (2) in logarithmic form into this condition, we obtain the following limit state function with two independent stochastic variables (log K) and e/t:

$$\log K - \log n_{max} - m \cdot \log[(1 + 3e/t)\Delta\sigma_M] < 0 \tag{5}$$

The numerical evaluation was performed using the FORM/SORM method by applying the STRUREL software (RCP Consult, 1998). The probabilities of failure plotted in Fig. 9 were obtained for different characteristic values of e_p/t. One can see that $P_f = 2{,}3\%$ as assumed in the nominal stress approach is just kept for a "permissible misalignment" $e_p/t = 0.22$.

If the fabrication quality is enhanced, the probability of failure will reduce as shown in Fig. 9. If, on the other hand the failure probability is maintained, the load may be increased. Further calculations showed that a reduction of the misalignment by 50% allows the stress range to be increased by approx. 20%.

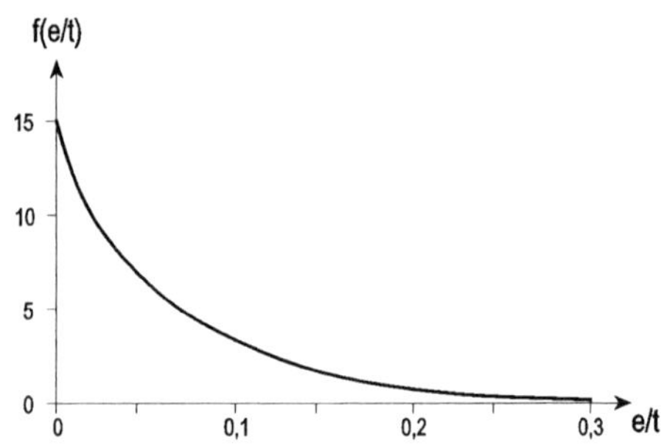

Figure 8: Exponential distribution for "permissible misalignment" $e_p/t = 0.2$

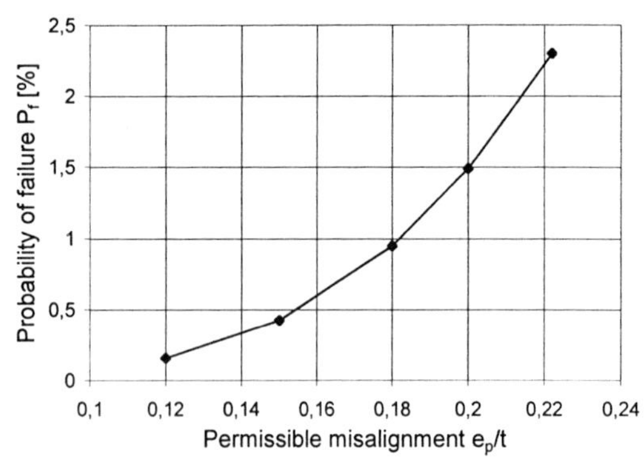

Figure 9: Probability of failure vs. permissible misalignment

6 CONCLUSIONS

Cruciform joints are typical elements in ship structures. Their fatigue strength is influenced to a large extent by the amount of weld root penetration and other fabrication effects such as misalignment. From the investigations described, the following conclusions can be drawn:

- The tests confirm that steel strength has no (beneficial) effect on the fatigue behaviour.
- Axial misalignment affects the fatigue strength due to secondary bending, as already shown by other investigators.
- As axial misalignment is an unknown quantity during design, it should be treated in a probabilistic manner using characteristic values for fabrication quality.
- In complex 3D joints, the effect of axial misalignment might be fairly small, which can be verified by hot spot stress analyses.

- The large scatter in fatigue lives is highly affected by fabrication quality in terms of weld shape and small defects close to the weld toe.

Acknowledgement

The investigations were partly performed within the joint research project "Life Cycle Design" (Project E3), funded by the German Federal Ministry for Education and Research (BMB+F), while the major part was carried out in the years 1996 - 2000 within the European research project FatHTS, funded by the European Union under the Industrial and Materials Technology Programme Brite Euram III. The partners of this project were: ÅF/Industritechnik (S), Bureau Veritas (F), Chalmers University of Technology (S), Chantiers de l'Atlantique (F), Fincantieri (I), Germanischer Lloyd (D), Instituto Superior Técnico (P), Lisnave (P), Odense Steel Shipyard (DK), RINA (I), Royal Schelde Shipyard (NL), Technische Universität Hamburg-Harburg (D) and VTT (FI). The EU-project was coordinated by TNO. An overview of the project was given by Janssen (2000).

References

Anthes, R.J.; Köttgen, V.B. and Seeger, T. (1993). Notch Factors of Butt and Cruciform Joints (in German). Schweißen und Schneiden 45, pp. 685 – 688.

Bell, R.; Vosikovsky, O.; Burns, D.; Mohaupt, U. (1987). A Fracture Mechanics Model for Life Prediction of Welded Plate Joints. Proc. of 3rd Int. Offshore Conf. Steel in Marine Structures, Delft.

Berge, S.; Myhre, H. (1977). Fatigue Strength of Misaligned Cruciform and Butt Joints. Norwegian Maritime Research; No. 1/77

BS 5400 (1980). Code of Practice for Fatigue. British Standard Institution.

Germanischer Lloyd (2000). Rules for Classification and Construction, I – Ship Technology, Part 1: Seagoing Ships, Chapter 1 Hull Structures. Self-Publishing, Hamburg.

Gurney, T.R. and Maddox, S.J. (1973). A Re-Analysis of Fatigue Data for Welded Joints in Steel. Weld. Res. Int., Vol. 3, No. 4, pp. 1 – 54.

IIW (1990). IIW Guidance on Assessment of the Fitness for Purpose of Welded Structures - Draft for Development. FORCE Institutes, Copenhagen.

IIW (1996). Fatigue Design of Welded Joints and Components. Ed. A. Hobbacher, International Institute of Welding, Abington Publishing, Cambridge (UK).

Janssen, G.T.M. (2000). Fatigue Based Design Rules for the Application of High Tensile Steel in Ships. Proc. of the 7th Int. Marine Design Conference, Kyongju.

Maddox, S.J. (1985). Fitness-for-purpose assessment of misalignment in transverse butt welds subject to fatigue loading. IIW-Document XIII-1180-85, International Institute of Welding (IIW)

Maddox, S.J. (1991). Fatigue Strength of Welded Structures. Abington Publishing, Cambridge

Olivier, R. and Ritter, W. (1979). Catalogue of S-N Curves of Welded Joints in Structural Steels – Vol. 1 – 5. Report 56, Deutscher Verband für Schweißtechnik (DVS).

Radaj, D. (1990). Design and Analysis of Fatigue-Resistant Welded Structures. Abington Publishing, Cambridge (UK).

RCP Consult (1998). STRUREL – A Structural Reliability Analysis Program System, User Theoretical Manual. RCP GmbH & RCP ApS.

Seeger, T.; Olivier, R.; Köttgen, V. (1991). Fatigue Strength Analysis of Welded Joints on Basis of Local Stresses (in German). DVS-Report 133, DVS-Verlag, Düsseldorf.

Wernicke, R. (2001). Assessment of cyclic loaded Cruciform Joints with Geometrical Imperfections (in German). Doctoral Thesis at TU Hamburg-Harburg (to be published).

Practical Design of Ships and Other Floating Structures
You-Sheng Wu, Wei-Cheng Cui and Guo-Jun Zhou (Eds)

FATIGUE STRENGTH ASSESSMENT OF HULL DETAILS FOR AN FPSO

S. Berge[1], A. Johansen[2] and Lars G. Bjørheim[3]

[1] Department of Marine Structures, Norwegian University of Science and Technology
[2] MARINTEK, Trondheim, Norway
[3] Statoil, Stavanger, Norway

ABSTRACT

The aim of the work reported herein was to make an assessment of the fatigue strength of typical connections between longitudinal stiffeners and transverse frames/bulkheads along the ship side of a Floating Production, Storage and Offloading vessel (FPSO). Two identical models with full size details of the stiffener/frame connections were tested. Experimental stress analysis was carried out on the models. The sections were fatigue tested to failure with loads representing the hydrostatic pressure from wave action on the ship side. Finite element analysis was carried out to investigate the effect of boundary conditions being different for the sections in the tests compared to in a ship side. The finite element analysis of the detail geometry was verified against the experimental stress data. The results were analysed using the DNV fatigue design procedure based on finite element analysis. A major uncertainty was identified to be related to the definition of "fatigue life" for ship structural details.

KEYWORDS

Ship structures, FPSO, Ship design, Fatigue design, Fatigue strength, Finite element analysis, Experimental stress analysis, Fatigue testing.

1 INTRODUCTION

Until some few years ago, there was no explicit fatigue design of cargo ships. Based on service experience, ships were designed against ultimate loads with the implicit assumption that fatigue would – at worst – be a maintenance problem.

During the 1980's new grades of steel with improved weldability and increased tensile strength were developed for ship structures. Thus, the allowable stress level increased and scantlings were decreased to minimise fabrication costs and weight of structures. As is well known, fatigue strength of welded joints is relatively insensitive to the tensile strength of the steel, and it turned out that implicit fatigue design based on static strength criteria was not sufficient to cater for the increased stress level.

Another development was the advent of floating production, storage and offloading (FPSO) ships for oil and gas. These structures are moored offshore and operated in a stationary mode for the production life of an oil and gas field, typically up to 20 years. Docking for inspection and repair is very costly for an FPSO, and design is thus aimed at uninterrupted operation. Fatigue design should thus be based on a safe life approach.

For these reasons, during the latter half of the 1990's several major classification societies have issued new or revised rules and recommendations for fatigue design of ships, ISSC (1997). Various methods are prescribed for assessment of load and load effects, from simplified design wave approaches to full spectral analysis. For strength assessment, a range of schemes are applied, based on different formulations and families of SN curves for welded joints.

A comparative study has shown that for a rather simple weld detail, subjected to nominal ship girder stress only, the variability in fatigue life prediction using different rules and recommendations, was a factor of 10, ISSC (2000).

The purpose of this paper is to present the main results from a study on fatigue strength of a ship hull detail, with full scale fatigue test results, and finite element analyses.

2 EXPERIMENTAL STRESS ANALYSIS

2.1 Test Models

Two test models were built to the same dimensions and specifications. The essential feature of the test models was the transition between longitudinal stiffener and transverse frame, Figure1. The global geometry of the test models is shown in Figure 2. The models were fabricated according to shipbuilding specifications.

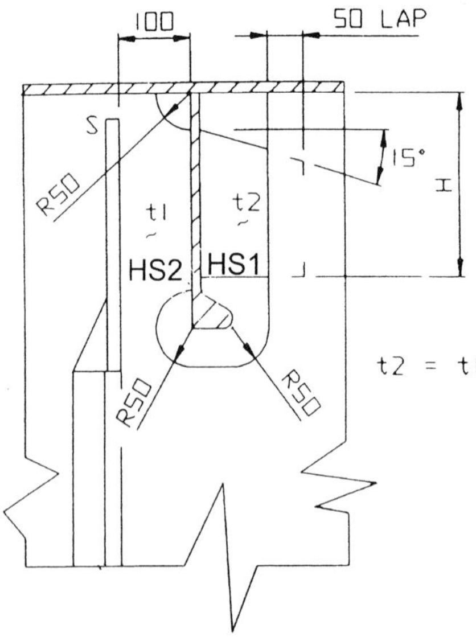

Figure 1: Dimensions of stiffener/frame transition
The figure is oriented with ship side shown as top horizontal plate. Location of Hot Spot 1 (stiffener to lug weld) and Hot Spot 2 (stiffener to web) are marked as HS1 and HS2 respectively.

Dimensions, cf. Figures 1 and 2:

- Shipside: 6400 x 800 x 16 mm (L x B x T)
- Frames: 1300 x 800 x12 mm (H x B x T)
- Longitudinal stiffener: HP 320 x 14
- Stiffeners on frames: HP 260 x 12
- Lug: 12 mm

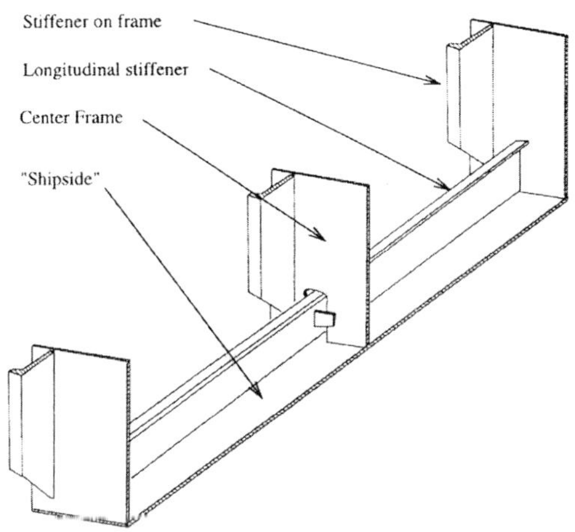

Figure 2: Global geometry of test model (not to scale)

2.2 Test Set-up

The models were welded to a steel plate which was bolted to the strong floor. The actual load on a ship side is hydrostatic pressure. As an approximation, two equal point loads at mid-span were applied, Figure 3. The effect on hot spot stress due to the approximate load application and other effects, was assessed by finite element analysis.

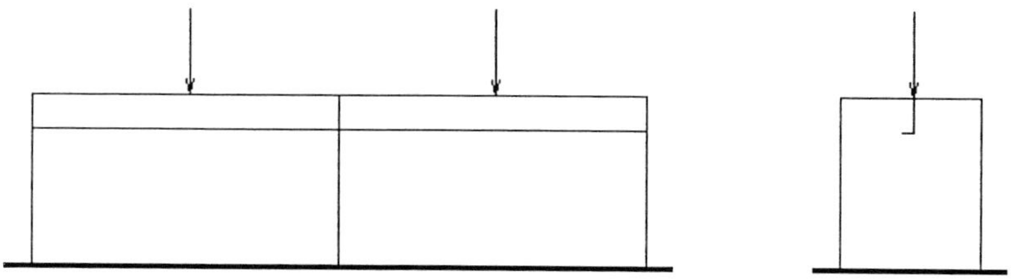

Figure 3: Loading arrangement (not to scale).

Strain gauges were applied to measure strain at different locations, on top of the stiffener bulb (nominal stress), and at various locations on the stiffener/frame transition (hot spot stress locations). Static stress was measured with an applied load of 200 kN (2 x 100 kN point load, cf. Figure 3).

3 FINITE ELEMENT ANALYSIS - GENERAL

Finite element analyses were carried out using the computer program ABAQUS, with a linear elastic material model. Two different geometries were modelled, the geometry of the laboratory test, and the geometry of a full size ship side. The purpose of the former model was to compare FEA with experimental stress analysis. The purpose of the latter was to investigate the effect on hot spot stress due to boundary conditions, which were different in the laboratory tests compared to in a ship side. Herein results from the former model is reported only.

3.1 Laboratory Model – Computed vs. Measured Stress

The purpose of this model was to simulate the test conditions and to verify the hot spot stress calculations against experimental data.
The stress analysis yielded a large volume of data. Only the most significant results are presented here. The test model was modelled with 9356 4-noded shell elements. Elements with thickness variation were used to model the stiffener bulbs. Eccentricity of lug was modelled. Welds were not modelled. A refined mesh was used in the instrumented areas. Two point loads, 100 kN each, were applied to the side shell at mid-span, directly above the longitudinal stiffeners. This corresponds to the load applied in the tests, Figure 3. The model was fixed in all translational and rotational degrees of freedom at the end of the three transverse frames, as for the experimental model.
The stress analyses produced a large volume of data. Only the most significant results for the fatigue tests are presented herein. As a general comment, computed and measured stress in non-welded regions of the model, e. g. at top of the stiffener (bulb), were in very good agreement. For welded components (lug, web, etc.) subjected to more complicated loading, the agreement was somewhat poorer. This is typical, due to the limitations of alignment and fabrication tolerances in welding fabrication, and the associated secondary bending stresses. The welds were not modelled in the FE analysis, also contributing to errors in the analysis.
The computed stress distributions for HS1 and HS2 are shown in Figures 4 and 5, along with the measured stresses.
At HS1 (Figure 4) the gradient of stress towards the weld toe and the bending component due to eccentricity is reasonably well represented by the FE analysis. The analysis appears to underestimate the peak stress by approximately 20%, essentially due to an underestimation of the bending component in the lug.
At HS2 (Figure 5) the bending component towards the weld (right hand side of graph) is clearly underestimated. The reason could be the underestimation of bending stiffness at the joint due to the omission of the weld fillet in the FE model. The gradient towards the hot spot is reasonably well represented.
In the same graph (Figure 5) is shown a large stress concentration at 90° angle. However, this stress is compressive, and at a location with no weld. Fatigue crack growth from this point is therefore quite unlikely, as was verified in the fatigue tests.

4 FATIGUE TESTS – CRACK INITIATION AND GROWTH

The model was mounted as shown in Figures 2 and 3. The load range was 40 - 440 kN, i.e. a load range ΔP = 400 kN with a mean load of 240 kN, and an R-ratio $P_{min}/P_{max} = 0.091$. The cyclic rate of loading was 0.5 Hz. The environment was laboratory air at ambient temperature.

The hot spot regions were inspected regularly for crack detection and crack size measurements, by spraying with white spirit. With this method, cracks become easily visible and can be detected when the crack depth is less than 1 mm.

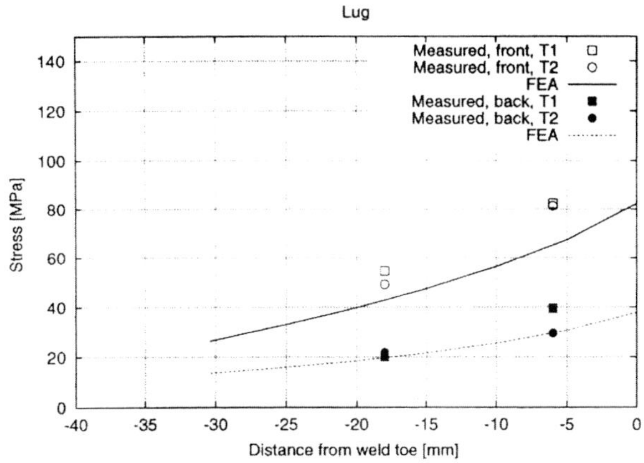

Figure 4: Measured stress in lug, Hot Spot 1 (cf. Figure 1).
Note bending component due to the eccentricity of the lug.

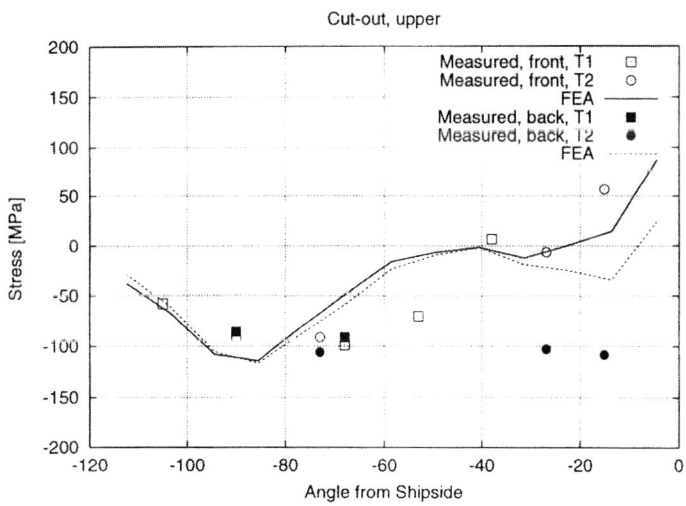

Figure 5: Measured stress in Hot Spot 2 (cf. Figure 1).

4.1 Crack Initiation and Growth

Crack initiation was observed at two locations, HS1 and HS2, cf. Figure 1. In both cases the crack was initiated at the weld toe and grew through the fillet of the weld causing gradual detachment from the transverse frame of the lug (HS1) and stiffener web (HS2) respectively.

In Model 1, the first crack initiation was observed at HS1. After some time, crack growth at HS1 appeared to slow down, while the crack at HS2 continued to grow and to cause detachment of the entire weld. In Model 2, crack initiation and growth at HS1 was observed only. Life to initiation and life to failure was not significantly different for the two models.

In Tables 2 and 3 the observed crack lengths are shown, for different stages of the test. For the initial stage, the surface crack length is given, Figure 6. For the through weld stage, the length of cracked weld is shown. The crack tended to grow at different rates at the respective sides of the weld. The length of the longest crack is given.

The test was terminated when the crack had grown through the full cross section of the weld. At this stage the model was still fully load carrying with respect to the fatigue loading. Vertical load through the severed weld was apparently transmitted by shear, due to the roughness of the fatigue crack surface.

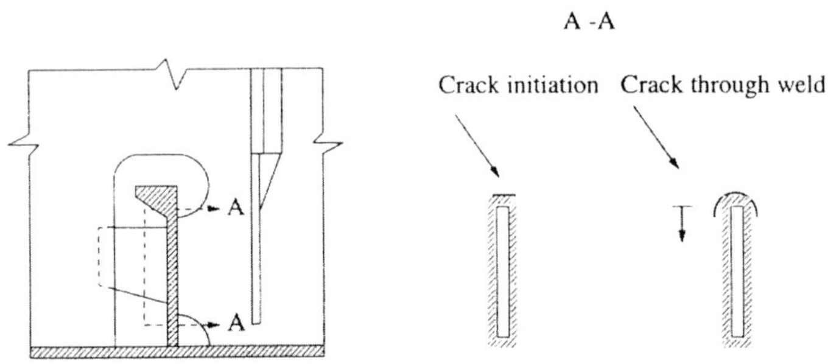

Figure 6. Observation of crack initiation and through weld crack.

Measured and computed stress was largest at HS1, and SN data is plotted in terms of stress at this location. Stress range at HS1 was assessed by strain gauge measurements and from FE analysis. The stress component normal to the weld was taken at distances t/2 and 3t/2 from the weld toe and extrapolated to the location of the weld toe, DNV (1998). The lug at the weld was subjected to significant secondary bending, stress range was taken at the side of the lug with maximum stress. The extrapolated measured stress and the stress calculated by FE are shown in Table 1.

TABLE 1
STRESS RANGE AT HOT SPOT HS1 (WELD TOE) FOR A LOAD RANGE $\Delta P = 400$ KN

Model 1, measured	200 Mpa
Model 2, measured	190 Mpa
FE analysis	154 Mpa

5 DISCUSSIONS

In Figures 7 and 8 the results are plotted into the SN design curves given by DNV (1998). For welded joints in air Curve I applies. The hot spot stress was multiplied by the factor $F_W = 1.5$ to take care of the weld geometry factor.
Observations:

- Life to *crack initiation* in the tests was significantly less than the design life. At this stage the cracks were innocuous surface cracks with depth of the order of one millimetre. These cracks would be of no concern for strength or integrity of a ship. With normal inspection procedures cracks at this stage would not be detected.
- Life to the stage where *through weld* cracking was initiated (cf. Figure 6) is close to the design curve. Fatigue life at this stage was non-conservative in relation to a "mean life SN curve" (design

curve plus two standard deviations). However, also at this stage a crack would be innocuous. A crack would be detectable by magnetic particle inspection and similar methods, but not by visual inspection.

- Life to *complete cracking* of the weld (end of test) indicates a safety factor of the order of 2 on measured hot spot stress range, corresponding to a factor of 10 on fatigue life. This implies that at this stage the fatigue life was conservative in relation to the mean life criterion underlying the DNV design curves. At this stage the crack would be detectable by visual inspection, provided the joint had been cleaned. In the constant amplitude tests, the joint was fully load-carrying even at this stage. For a ship structure subjected to randomised loading, this may not be true.
- The crack growth stage constituted more than 90% of the fatigue life. Throughout the entire crack growth stage the joint was fully functioning as a load carrying member, albeit in constant amplitude loading at a level considerably less than the ultimate state load. There is a need for establishing criteria for "end of fatigue life" of ship details. This issue was not addressed in the present study.

Legend (Figures 7 and 8):

O - First detection of surface crack, initiation
Δ - Through weld cracking
□ - End of test, weld fully cracked
Open symbols: Measured hot spot stress (DNV extrapolation x 1.5)
Closed symbols: Hot spot stress calculated by FE (DNV extrapolation x 1.5)

TABLE 2
CRACK DEVELOPMENT, MODEL 1

Cycles	Hot Spot 1 Surface crack length (mm)	Hot Spot 2 Surface crack length (mm)	Comments
122.300	Surface crack 2 mm	No crack	First detection of crack
420.000	Through weld crack 10 mm	Surface crack 7 mm	
638.000	Through weld crack 20 mm	Through weld crack 50 mm	Crack growth at HS 2 is considerably faster
762.000	Through weld crack 22 mm	Through weld crack 65 mm	Crack growth at HS 2 is dominating. No further growth of crack at HS 1.
1.422.000	No crack growth	Weld fully cracked	End of test

TABLE 3
CRACK DEVELOPMENT, MODEL 2

Cycles	Hot Spot 1 Surface crack length (mm)	Hot Spot 2 Surface crack length (mm)	Comments
72.000	Surface crack 8 mm	No crack	First detection of crack
404.000	Through weld crack 10 mm	No crack	
2.250.000	Weld fully detatched	No crack	End of test

EVALUATION OF SIMPLIFIED PREDICTION METHOD OF STRESS RESPONSE FUNCTION FROM THE VIEWPOINT OF FATIGUE STRENGTH ANALYSIS OF A SHIP

T. Fukasawa[1], K. Hashimoto[2] and Y. Tomita[2]

[1] Division of Mechanical Engineering, Kanazawa Institute of Technology, Nonoichi, Ishikawa 921-8501 Japan,
[2] Department of Naval Architecture and Ocean Engineering, Osaka University, Suita, Osaka 565-0871 Japan

ABSTRACT

Similar Response Function Method is proposed as a simplified prediction method of the frequency response function of stress fluctuation caused by wave pressure. With the use of the method, computational efforts in FEM analysis to calculate the stress response function can extraordinary be reduced. The accuracy of the method is evaluated from the viewpoint of fatigue strength analysis of a ship. Fatigue crack growth analyses are carried out, where the storm model is used as the time history of wave elevation and the occurrence probability of wave encounter angle is taken into account. Comparing the results calculated by the simplified method and the exact method, the accuracy and the applicability of Similar Response Function Method in fatigue strength design are discussed.

KEYWORDS

Fatigue Strength Design, Stress Response Function, Crack Growth Analysis, Simplified Prediction Method, Similar Response Function Method, Wave Pressure, Storm Model, Sailing Model

1 INTRODUCTION

Fatigue strength design is currently carried out by applying the linear cumulative damage law, the Miner rule, on the basis of crack initiation life. However, the Miner rule is unsatisfactory, because the actual fatigue damage is accumulated nonlinearly during random stress cycles; in other words, the fatigue damage strongly depends on the order of stress occurrence. To overcome this shortcoming, fatigue strength should be evaluated by means of crack growth analysis. Tomita (2000) has proposed a fatigue strength evaluation method in which the fatigue crack growth is analyzed according to the Paris law. A storm model is introduced as the time history of stress occurrence, and the probability of occurrence of wave encounter angle is taken into account according to the sea condition.

In order to analyze the fatigue crack growth by means of the Paris law, it is necessary to know the stress time history of a certain structural member of a ship in waves. The stress response function can be available to generate the stress time history. However, in order to calculate the frequency response function of stress fluctuation caused by wave pressure (the stress response function), enormous numbers of FEM calculations are necessary. That is, since not only the pressure magnitude but also the pressure distribution pattern along ship's hull vary in time according to wave condition, FEM analysis should be conducted in every wave condition to obtain the stress response function in this case. To reduce this computational effort in FEM analysis, Fukasawa (2000) have proposed a simplified method called as the Similar Response Function Method (abbreviated as SRFM hereafter), by which the frequency response function of stress fluctuation caused by wave pressure can easily be estimated with the use of pressure response functions with much less number of FEM calculations.

In the present paper, firstly the frequency response function of stress fluctuation caused by wave pressure is calculated by SRFM as well as the exact method. The crack growth analyses are then conducted according to the Paris law with the use of the storm model. Comparing the results of the exact method and of SRFM, the accuracy and the applicability of SRFM are discussed from the viewpoint of the fatigue strength analysis of a ship.

2 FATIGUE STRENGTH EVALUATION PROCEDURE

Firstly, the sea condition which ocean-going ships encounter is categorized into two groups; a calm sea condition and 6 kinds of storm conditions, which are specified by the wave height. The North Pacific route is chosen as the sailing route in this paper. Considering 20 years life time, the occurrence probability of calm sea condition is taken to be 1991/2084, while that of storm sea conditions is 93/2084 in total (Tomita, 2000). Occurrence of wave is assumed to be time independent in the calm sea condition. Time history of wave elevation is determined with the use of exponentially distributed random numbers, and the wave period is assumed to be constant in the calm sea condition. In the storm condition, on the other hand, occurrence of wave is assumed to be time dependent and the storm model (Tomita, 1992) is incorporated as a time history of wave elevation. The occurrence probability of wave period in the storm condition is determined according to the wave height, which is given in the wave scatter diagram provided by Ship & Ocean Foundation (1992), and the wave period is determined with the use of random numbers. The wave encounter angle is also determined using random numbers according to the occurrence probability given in weather reports (Tomita, 2000).

Time histories of working stress can be calculated by means of the stress response function. Once the wave height and the wave period are determined both in the calm sea and the storm conditions and the wave encounter angle is fixed as mentioned above, the nominal stress can be obtained from the stress response function. Assuming the stress concentration factor, the fatigue strength is then evaluated according to the Paris law, instead of the Miner rule, in this paper. With the use of this method, the amount of crack growth under a certain working stress and the amount of crack length at any given moment can be calculated taking account of the effect of the order of stress occurrence.

3 STRUCTURAL MODEL

In FEM analysis, the structural model shown in Figure 1 is used in the present paper, which is a sub-structure model of a 210,000 DWT bulk carrier. The stress evaluation points and the definition of stress directions are also shown in the figure. MSC.NASTRAN is used for the solver in the structural analysis. Since the external forces are not balanced themselves, counterbalance forces are added to avoid the stress concentration at the constraint points. The counterbalance forces can be obtained so as to balance the reaction forces at each constraint point which are come out in FEM calculation when the external forces are only applied.

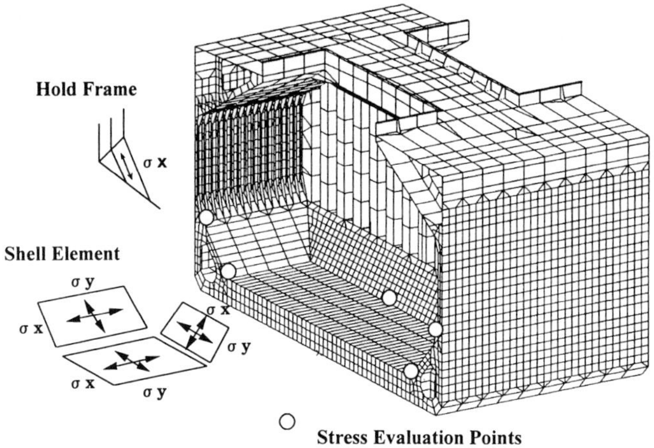

Figure 1: Finite element model of bulk carrier to be analyzed

4 CALCULATION OF STRESS RESPONSE FUNCTION

4.1 Exact Method

In the exact method, FEM calculations are conducted at 8 different instants during 1 wave encounter period, applying the wave pressure distribution of each instant. Interpolating the obtained stresses of 8 instants as a sinusoidal function in time, the stress fluctuation can be calculated. Since the external forces are not balanced themselves in each stress analysis, the counterbalance forces are applied in the calculation. The total number of FEM calculations in this method is 2688; that is, 7 (wave encounter angles) x 24 (wavelengths) x 8 (time instants) x 2 (counterbalance force calculations). This method can be considered to be the strictest one at present.

4.2 Similar Response Function Method (SRFM)

In SRFM, the stress response function is estimated by means of the pressure response functions. Assuming that the pressure of the point nearby the stress calculation point has much effect on the stress concerned, rather than the pressure of the point far from the stress calculation point, the pressure influence function, $\alpha(\chi, \lambda)$, is defined as follows.

$$\alpha(\chi, \lambda) = \sum_{i}^{M} \left\{ p_i(\lambda) \Big/ (\ell_i)^k \right\} \tag{1}$$

where $p_i(\lambda)$ stands for the pressure response function at i-th pressure point, and M is the number of pressure points. ℓ_i is the distance between i-th pressure point and the stress point. In Eqn. 1, it is assumed that the pressure affects on the stress in inverse proportion to the k-th power of the distance between pressure point and stress point. Then, the pressure influence function is normalized as

$$\hat{\alpha}(\chi, \lambda) = \alpha(\chi, \lambda) / \alpha(\chi, \hat{\lambda}) \tag{2}$$

where $\hat{\lambda}$ is the specified wavelength, which is determined for each wave encounter angle as follows:

1. The first and second highest peak of the normalized pressure influence function $\hat{a}(\chi, \lambda)$ is searched for. The wavelength of the first highest peak is denoted as λ_{1st} and that of the second peak is λ_{2nd}.
2. The fluctuation component of pressure is calculated by using the Min-Max Pressure Method (Fukasawa, 2000) at the wavelength of λ_{1st} and λ_{2nd}.
3. FEM analysis is carried out with the use of the obtained fluctuation component of pressure at the wavelength of λ_{1st} and λ_{2nd}, and the fluctuation component of stress is calculated.
4. Comparing the stress magnitude at the wavelength of λ_{1st} and λ_{2nd}, the wavelength of which the stress magnitude is larger than the other is adopted as the specified wavelength $\hat{\lambda}$.

Multiplying the normalized pressure influence function $\hat{a}(\chi, \lambda)$ by the stress magnitude at the specified wavelength $\hat{\lambda}$ obtained above, the stress response function is estimated at each wave encounter angle χ.

Although the number of pressure points is arbitrary in principle in SRFM, such 5 points are adopted in this paper; that is, the points at still water line and bilge corner in weather and lee sides and the point at bottom centerline. It should be noted here that the stress response function of every stress point at a certain wave encounter angle can be estimated by the SRFM by only 4 times FEM calculations, including the counterbalance force calculation.

5 FATIGUE CRACK GROWTH ANALYSIS

Fatigue crack growth rate is given by the Paris-Elber equation, that is,

$$da/dN = C(\Delta Keff)^m \tag{3}$$

where $\Delta Keff$ is the effective stress intensity factor range, N is the number of stress cycle, a is the crack depth or crack length, and C and m are the material constants, which are assumed to be $C = 1.45\times10^{-11}$ and $m = 2.75$ in this paper. Crack growth life can be calculated by integrating Eqn. 3. Effective stress intensity factor range is defined by

$$\Delta Keff = \Delta K - Kop \tag{4}$$

where Kop is the stress intensity factor at the crack opening point, which can be obtained from the storm model fatigue test results (Kawata, 1996). It is assumed that a unique surface crack with a semi-circle shape initiates at weld toe and grow to a thickness-through crack in this paper. Approximate equation for the stress intensity factor of a surface crack in a ship structural member can be given by the modified Raju-Newman equation, that is,

$$\Delta K = Ks \times \sigma\sqrt{\pi a} \times Fa(a,b,t) \times MK(a,b,L,t) \tag{5}$$

where $2a$ and b are the length and depth of surface crack, σ is the nominal stress, t is the plate thickness, and L is the weld leg length. Fa is the finite plate correction factor and MK is the weld bead correction factor, for which the values of the reference (Ship Research Panel 219, 1996) are used. Ks is the stress concentration factor of the structural member, which is assumed to be $Ks = 4.5$ in this paper. The initial crack size is set to be 0.4mm in length and 0.2mm in depth.

6 COMPARISON OF STRESS RESPONSE FUNCTION

Firstly, the stress response functions of slant plate at lower stool are shown in Figure 2. The wave height is taken as 5m. In the figure, "linear", "square", and "cubic" are the cases where the k-value in Eqn. 1 is taken as 1, 2, and 3, respectively. In the case of "linear (k=1)", there seems a large false peak in the stress response function. This peak corresponds to the rolling resonance peak in the pressure response function. As the rolling effect is significant in the pressure at the side shell, it can be recognized from the figure that the pressure at the bottom affects much on the stress at lower stool. The false peak is getting to disappear with the increase of k-value.

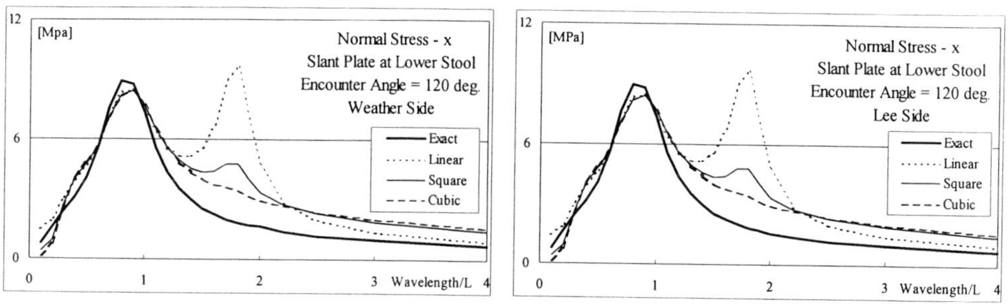

Figure 2: Stress response functions (Slant Plate at Lower Stool)

Secondly, the estimated stress response functions of slant plate at bilge knuckle are shown in Figure 3. The stress response function estimated by SRFM agrees with the exact one satisfyingly. The estimated results are slightly affected by the k-value only in the shorter wavelength range.

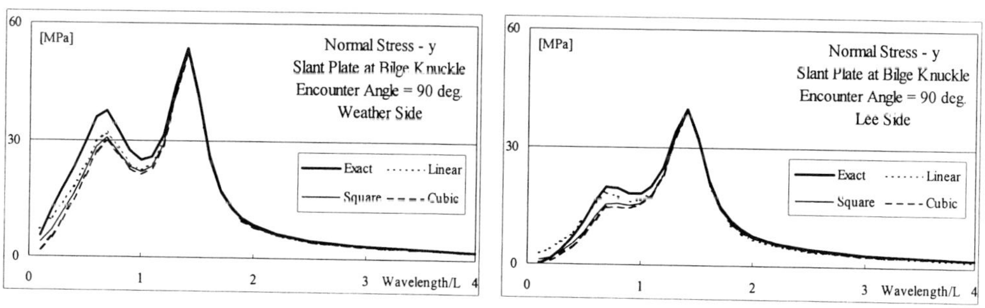

Figure 3: Stress response functions (Slant Plate at Bilge Knuckle)

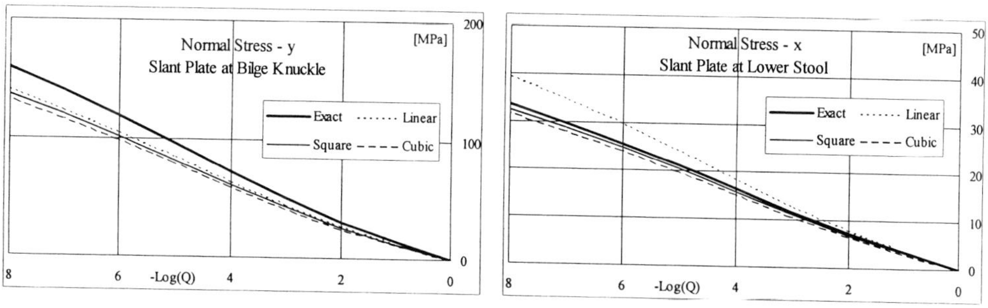

Figure 4: Long-term probability of occurrence of stress fluctuation

In Figure 4, the long-term probability of occurrence of stress fluctuations of the slant plate at lower stool and the slant plate at bilge knuckle are shown. Although the results of SRFM differ from the exact one according to the k-value, the case of "square (k=2)" may be the best one in the accuracy, and is used in the following crack growth analysis.

7 COMPARISON OF CRACK GROWTH

Crack growths are computed for 12 loading patterns. The plate thickness is assumed to be 12mm in every stress point, and simulations are conducted for 20 years time span; however, in case where a surface crack comes to a thickness-through crack in 20 years, the simulation is stopped at the moment. Growths of crack length are shown in Figures 5 and 6. In Figure 5, the crack does not come to a thickness-through crack in 20 years, while the crack grows through the plate thickness within 20 years in Figure 6. It can be seen from the figures that the amounts of crack growth estimated by SRFM are lower than that by the exact method on the whole. This means that the fatigue life of structural members is overestimated by SRFM. However, there are some scatters of the amount of crack growth according to the loading pattern, and the results by SRFM and the exact one are somewhat overlapped. The crack growth behavior largely depends on the number, the order and the size of storm, and there exists a loading pattern which makes cracks grow much faster or much slower. The estimation error in SRFM is inside the scatter range caused by the difference of loading pattern.

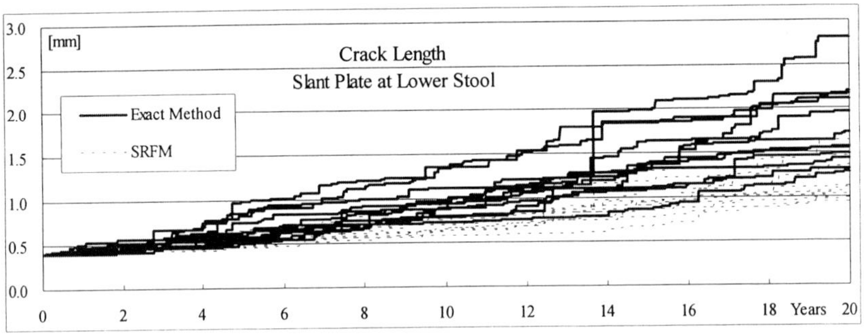

Figure 5: Crack growth curve (Slant Plate at Lower Stool)

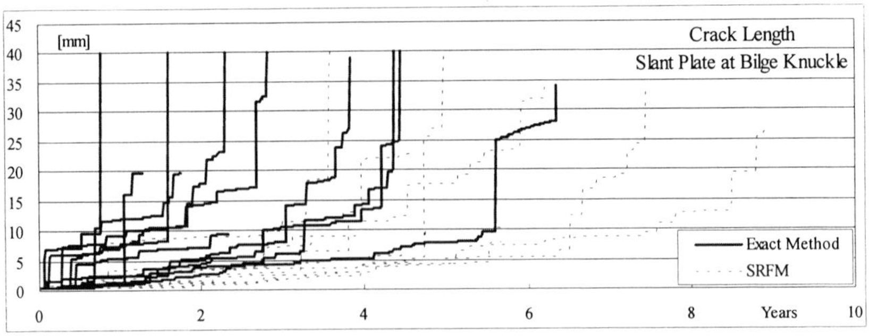

Figure 6: Crack growth curve (Slant Plate at Bilge Knuckle)

The crack length/depth and the fatigue life averaged over 12 loading patterns are listed in Table1, where the fatigue life is defined as the year which is necessary to increase the surface crack to the thickness-through crack. The difference of crack length obtained by SRFM and the exact method is at most 2.5mm in 20 years, while the difference of the depth is at most 1mm. In the case of thickness-through crack, the difference of fatigue life between these two methods is about 2-3years.

TABLE 1
AVERAGE CRACK LENGTH/DEPTH AND AVERAGE FATIGUE LIFE

Crack Growth in 20 years	Exact Method		SRFM	
	Length [mm]	Depth [mm]	Length [mm]	Depth [mm]
Lower Part of Hold Frame	7.86	2.78	5.41	1.75
Slant Plate at Lower Stool	1.80	0.52	1.16	0.36
Inner Bottom at Lower Stool	0.49	0.21	0.46	0.21

Fatigue Life	Exact Method		SRFM	
	Length [mm]	Fatigue Life	Length [mm]	Fatigue Life
Slant Plate at Bilge Knuckle	22.73	3.1 years	20.23	5.0 years
Inner Bottom at Bilge Knuckle	18.74	5.8 years	26.79	8.5 years

8 CONCLUDING REMARKS

SRFM was proposed as a handy method to estimate the frequency response function of stress cause by wave pressure, with which the number of FEM calculation is only 4 times to obtain the stress response function at a certain wave encounter angle, including the counterbalance force calculations. The fatigue life of structural members are a little bit overestimated by SRFM, however, the estimation error of SRFM is inside the scatter range caused by the difference of loading pattern. The averaged error of SRFM in estimated crack length is at most 2.5mm in 20-year life of a ship, while the error in the depth is at most 1mm. As for the thickness-through crack initiation, the difference of the fatigue life obtained by these two methods is about 2-3years. These results show that SRFM can be used in fatigue strength design

References

Fukasawa T., Morikawa M., Fukuoka T., Shibazaki K. and Ito A. (2000). On the Approximation of Stress Response Function and Long-Term Probability of Occurrence of Stress caused by Wave Pressure. *J. Soc. Naval Architects of Japan* **187,** 241-251. (in Japanese)

Fukasawa T., Morikawa M., Fukuoka T., Shibazaki K. and Ito A. and Tomita Y. (2000). On the Simplified Method to Predict the Long-term Probability of Occurrence of Stress caused by Wave Pressure acting on a Ship. *Int. Conf. on Ship and Shipping Research (NAV2000)* **2,** 7.4.1-7.4.10.

Kawata N. (1996). *Experimental Study on Fatigue Crack Growth Behavior under Storm Model, Graduation thesis*, Dept. of Naval Arch. and Ocean Eng., Osaka University, JAPAN (in Japanese)

Ship & Ocean Foundation. (1992). *Winds and Waves of the North Pacific Ocean (1974-1988)*, Ship & Ocean Foundation, JAPAN

Ship Research Panel 219. (1996). *Study on Practical Use for Fatigue Crack Growth Analysis*, Shipping Research Association of Japan, JAPAN (in Japanese)

Tomita Y., Kawabe H. and Fukuoka T. (1992). Statistical Characteristics of Long-Term Wave-Induced Load for Fatigue Strength Analysis for Ships. *Practical Design of Ships and Mobile Units (PRADS)* **2,** 2,792-2,805.

Tomita Y., Hashimoto K., Osawa N., and Terai K. (2000). Fatigue Strength Evaluation of Ship Structural Members Based on Fatigue Crack Growth Analysis. *Int. Conf. on Ship and Shipping Research (NAV2000)* **2,** 10.4.1-10.4.11.

Practical Design of Ships and Other Floating Structures
You-Sheng Wu, Wei-Cheng Cui and Guo-Jun Zhou (Eds)

COMBINATION OF FATIGUE DAMAGES PRODUCED BY SEVERAL WAVE-INDUCED LOADS BASED ON CORRELATION COEFFICIENT METHOD

H. Kawabe[1] and K. Shibazaki[2]

[1]National Maritime Research Institute, 6-38-1 Shinkawa, Mitaka, Tokyo 181-0004, Japan
[2]NKK Co, 1-banchi, Kumozukokanmachi, Tsu, Mie, 514-0393, Japan

ABSTRACT

This report is concerned with the combination method for ship hull structural fatigue damages produced by several wave-induced loads based on correlation coefficient method. Several loads, such as dynamic water pressures, cargo inertia forces and hull girder bending moments, act on ship's hull structural members at the same time with different amplitudes and phases. The correlation coefficient method is proposed as one method to combine these extreme loads in the long-term distributions. Authors extend the correlation method to combine not only extreme values but also the whole shape of wave-induced long-term distribution. If properties of the most severe short-term wave conditions which dominates the long-term distribution of wave-induced load are almost same condition in each wave-induced loads, the long-term distributions can be combined using correlation coefficient. Based on S-N curves $\sigma = C N^n$, long-term distribution of wave-induced stresses by the loads and Miner's rule, fatigue damages in each stress components can be calculated by following equation.

$$\{ D_{12} \}^{-2n} = \{ D_1 \}^{-2n} + \{ D_2 \}^{-2n} + 2 \rho \{ D_1 \}^{-n} \{ D_2 \}^{-n}$$

Where D_1 and D_2 are fatigue damages obtained from each stress components, D_{12} is combined fatigue damage caused by two stress components and ρ is the correlation factor between loads. Some numerical examples are demonstrated.

KEYWORDS

Fatigue, Miner's Rule, S-N curve, Wave-induced load, Long-term distribution, Correlation coefficient, Combination of load, Design wave condition,

1 INTRODUCTION

Several loads, such as dynamic water pressures, cargo inertia forces and hull girder bending moments, act on ship's hull structural members at the same time with different amplitudes and phases. A standard calculation method of a long-term distribution of wave-induced stress combined loading is conducted as following procedures. First, we calculate the stress response amplitude function in regular waves by using strip method and FEM structural analysis. Then, combining with wave spectrum, we can get the short-term response within the scope of linear theory. The long-term probability distribution of wave-induced stress is calculated in account of operating condition and long-term probability wave distribution. The greatest advantage of these calculation methods is that the response amplitude function of wave-induced stress on the ship structure could be calculated while the several kinds of loads with various phase differences act on the structure. However, these methods require great number of calculations, manpower and costs. As another method of calculating the long-term distribution of wave induced stress on the ship structure where several kinds of load act at the same time, the 216th Research Panel organized in the Shipbuilding Research Association of Japan (SR216, 1994) proposed three simple methods, such as, the peak value ratio method, the correlation coefficient method and the design wave method. These methods calculate each long-term distributions of wave-induced stress components respectively. Then, the extreme values of each stress components are combined by using the above methods. SR 216 reported that there is not a significant difference among these three combination methods.

In this report, authors extend the correlation coefficient method to combine not only the extreme values but also the whole shape of long-term distribution of wave-induced stress. Authors examined which factor of wave condition decide the whole shape of the long-term distribution, such as significant wave height, average wave period and relative angle between ship course and wave direction. To investigate the effect of wave condition and operating condition on the long-term distribution, the long-term distribution is decomposed into every these factors and a contribution rate analysis method for these factors in each load levels in the long-term distribution was introduced. By using the method, authors clarify the governing wave condition and operating condition over long-term distribution and the most severe wave condition to decide the extreme value of wave-induced load. If the governing wave conditions in each stress components have same characters, the long-term distributions can be combined by using the correlation coefficient method. Based on S-N curves $\sigma = C N^n$, long-term distribution of wave-induced stresses by the loads and Miner's rule, fatigue damages in each stress components can be combined by using the stress correlation coefficient. For numerical examples, the combination of long-term distribution of wave-induced stresses and fatigue damages in a 210,000 DWT bulk carrier structure are demonstrated.

2 GOVERNING WAVE CONDITION OVER LONG-TERM DISTRIBUTION OF WAVE-INDUCED LOAD

Under short-term irregular wave condition, characterized by the significant wave height H and the mean wave period T, and sailing ahead relative angle between ship course and wave direction θ, a spectrum of wave-induced loads shows narrow band characteristics. A short-term probability distribution of wave-induced loads can be approximated by the Rayleigh probability function with a parameter R(H, T, θ). The long-term distribution of wave-induced load consists of a large number of short-term conditions. The equation of the long-term distribution function Q(σ) is as follows.

$$Q(\sigma) = \int Q\{\sigma \mid R(H,T,\theta)\}\, p\{R(H,T,\theta)\}\, dR(H,T,\theta) \tag{1}$$

where $Q\{\sigma \mid R(H,T,\theta)\}$ is short-term probability Rayleigh distribution. $p\{R(H,T,\theta)\}$ is the long-term probability density function of parameter R(H,T, θ).

From above formulation of long-term distribution, the equation implicitly has a information of

short-term wave conditions (H,T) and operating conditions (θ).
To observe the effect of these conditions on the long-term probability distribution, the distribution is discretized in each wave component and wave direction and is decomposed into following three type of distributions.

$$\begin{aligned} Q(\sigma) &= \sum_{i=1}^{L} Q_H\{\sigma \mid R(H_i)\}\, p_H\{R(H_i)\}\, \Delta R_H(H_i) \\ &= \sum_{j=1}^{M} Q_T\{\sigma \mid R(T_j)\}\, p_T\{R(T_j)\}\, \Delta R_T(T_j) \\ &= \sum_{k=1}^{N} Q_\theta\{\sigma \mid R(\theta_k)\}\, p(\theta_k)\, \Delta\theta_k \end{aligned} \tag{2}$$

where $Q_H\{\sigma \mid R(H_i)\}p_H\{R(H_i)\}\Delta R_H(H_i)$ denotes the decomposed long-term distribution into significant wave height. $Q_T\{\sigma \mid R(T_j)\}p_T\{R(T_j)\}\Delta R_T(T_j)$ and $Q_\theta\{\sigma \mid R(\theta_k)\}p_T(\theta_k)\,\theta_k$,is component of long-term distribution divided into mean wave period and wave direction.

The contribution rate of the short-term wave condition or operating condition to the long-term distribution with load level σ can be obtained by dividing the above equation with Q(σ).

$$\begin{aligned} 1 &= \sum_{i=1}^{L} \frac{Q_H\{\sigma \mid R(H_i)\}\, p_H\{R(H_i)\}\, \Delta R_H(H_i)}{Q(\sigma)} \\ &= \sum_{j=1}^{M} \frac{Q_T\{\sigma \mid R(T_j)\}\, p_T\{R(T_j)\}\, \Delta R_T(T_j)}{Q(\sigma)} \\ &= \sum_{k=1}^{N} \frac{Q_\theta\{\sigma \mid R(\theta_k)\}\, p(\theta_k)\, \Delta\theta_k}{Q(\sigma)} \end{aligned} \tag{3}$$

The right hand side terms in the first equation gives the contribution rate of significant wave height H to the long-term distribution at the load level σ. The second and third equations mean the contribution rate of mean wave period and operating condition to the long-term distribution.

By using the method, we can examine the governing wave condition that a specific mean wave period, T_m, and wave direction, θ_m plays a key role over the long-term distribution. Also we can pinpoint the most severe short-term wave condition that the maximum wave height, H_m, with mean period T_m in the supposed long-term wave statistics decides the extreme value of wave-induced load.

3 RELATION BETWEEN LONG-TERM DISTRIBUTION PARAMETER AND THE MOST SEVERE WAVE CONDITION

The long-term distribution of wave-induced load can be described by Weibull distribution function.
The Weibull distribution parameters can be estimated from the governing wave condition and the most severe wave condition. So that the parameters are function of (H_m, T_m, θ_m). For example, the relation between the Weibull shape parameter and the mean wave period of governing wave condition are shown in Table 1.

$$Q(\sigma) = \exp\{-(\frac{\sigma}{\alpha_\sigma})^{\beta_\sigma}\} \tag{4}$$

TABLE 1

WEIBULL SHAPE PARAMETER β_σ FOR ALL HEADINGS LONG-TERM DISTRIBUTION

governing mean wave period T_m	North Atlantic		North Pacific		Japan-Persian Gulf route	
	β_σ	H_m	β_σ	H_m	β_σ	H_m
6.5	0.6935	9.5	0.7310	10.5	0.7948	10.5
7.5	0.8261	11.5	0.7942	11.5	0.8392	10.5
8.5	0.8646	14.5	0.8143	12.5	0.8440	10.5
9.5	0.8930	16.5	0.8132	13.5	0.8224	10.5
10.5	0.8993	17.5	0.8223	13.5	0.8322	9.5
11.5	0.8933	16.5	0.8465	12.5	0.8227	8.5
12.5	0.8727	15.5	0.8178	12.5	0.7904	7.5
13.5	0.7527	13.5	0.8539	11.5	-	-

The most probable extreme value in the long-term distribution with probability of exceedance 10^{-8} ,σ_{-8}, can be approximated by the extreme value in the most sever short-term distribution with probability of exceedance 10^{-3}. Therefore, the Weibull scale parameter can be calculated by the following equation.

$$\begin{aligned} \sigma_{-8}(H_m, T_m, \theta_m) &= R_{max}(H_m, T_m, \theta_m)\sqrt{2\ln(1000)} \\ \alpha_\sigma &= \sigma_{-8}(H_m, T_m, \theta_m)\{8\ln(10)\}^{-\frac{1}{\beta_{\sigma 1}(T_m)}} \end{aligned} \tag{5}$$

where, $R_{max}(H_m,T_m,\theta_m)$ is the Rayleigh parameter in the most severe wave condition.

4 COMBINATION OF LONG-TERM DISTRIBUTION AND FATIGUE DAMAGE BASED ON CORRELATION COEFFICIENT

Usually, to evaluate the strength of a ship hull structure, the wave-induced stresses are estimated from a long-term distribution assuming the all heading operating condition which is uniformly sailing toward all directions between ship course and wave direction. If the long-term distributions of each stress components, σ_1 and σ_2, have the same governing wave condition and the most severe wave condition, the two distributions can be combined into one long-term distribution using correlation coefficient ρ.

$$\begin{aligned} \sigma_{12,-8} &= \sqrt{\sigma_{1,-8}^2 + \sigma_{2,-8}^2 + 2\rho\sigma_{1,-8}\sigma_{2,-8}} \\ \beta_{\sigma 12} &= \beta_{\sigma 1} = \beta_{\sigma 2} \\ \alpha_{\sigma 12} &= \sqrt{\alpha_{\sigma 1}^2 + \alpha_{\sigma 2}^2 + 2\rho\,\alpha_{\sigma 1}\alpha_{\sigma 2}} \end{aligned} \tag{6}$$

The correlation coefficient method was initially introduced by Fukuda (1978) to seek the extreme value of combined stress in long-term distribution. Applying the governing wave condition and the most severe wave condition, authors extend the correlation coefficient method to combine not only the extreme stress values but also the whole shape of wave-induced long-term distribution.

Based on S-N curves $\sigma = C\ N^{n,}$ long-term distribution of wave-induced stresses by the loads and Miner's rule, fatigue damages can be calculated by the following equation.

$$D = \int_0^\infty \frac{N_0\, p(\sigma)}{N(\sigma)}\, d\sigma = N_0\, C^{\frac{1}{n}}\, \Gamma\left(1 - \frac{1}{n\beta_\sigma}\right) \alpha_\sigma^{-\frac{1}{n}} \tag{7}$$

where $\Gamma(\ )$ is a Gamma function.

If each stress component has the same governing wave condition and the most severe wave condition, the fatigue damage components can be combined by substituting equation (7) into (6).

$$\{D_{12}\}^{-2n} = \{D_1\}^{-2n} + \{D_2\}^{-2n} + 2\rho\{D_1\}^{-n}\{D_2\}^{-n} \tag{8}$$

where D_1 and D_2 are fatigue damages obtained from each stress component, D_{12} is combined fatigue damage caused by two stress components

5 NUMERICAL EXAMPLES

For numerical examples of combination of long-term distributions and fatigue damages, authors examine double bottom structure stress of a 280,000 DWT bulk carrier (L*B*d = 250m*50m*26.3m). Assumed conditions are as follows; (1) wave condition : wave statistics in the North Atlantic reported by Hogben (1986) and ISSC wave spectra. (2) operating condition : All Headings.

Figure 1 shows contribution rate of mean wave periods and wave directions at bottom plate stress at hopper produced by wave pressure. The contribution rates around lower stress levels are widely distributed over each wave periods and wave directions. The contribution rate of a specific wave period and a wave direction increases with increasing the stress level in the long-term distribution.

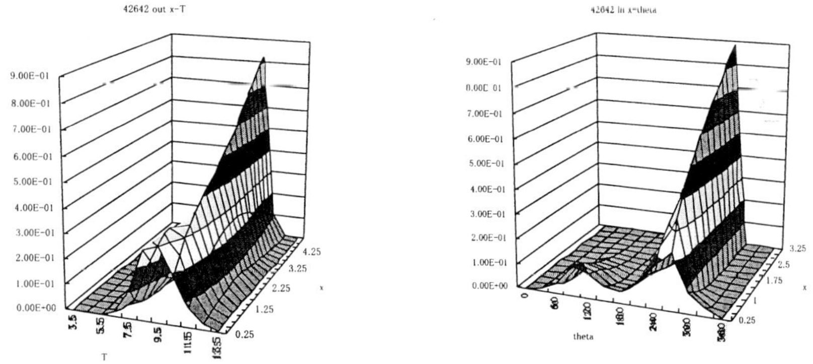

Figure 1 : Contribution rate of mean wave period (left) and wave direction (right) on inner bottom plate stress at hopper produced by wave pressure

From these results, we can pinpoint the governing wave conditions and the most sever wave condition in the long-term distribution. Table 2 shows the governing wave conditions in each load. As the governing wave conditions for wave pressure and internal load are the same condition, the long-term distributions can be combined using a correlation coefficient. Figure 2 shows the combined long-term distribution based on the correlation coefficient method and the exact distribution which is calculated using stress response amplitude function considering phase difference between each loads. Table 3 shows correlation coefficients for extreme value with $Q=10^{-8}$ in the long-term distribution and fatigue damages. These values are calculated from equation (6) and (8). From these results, if the governing wave conditions on long-term distribution have the same properties, we can combine not only the whole shape of the distributions but also the fatigue damages.

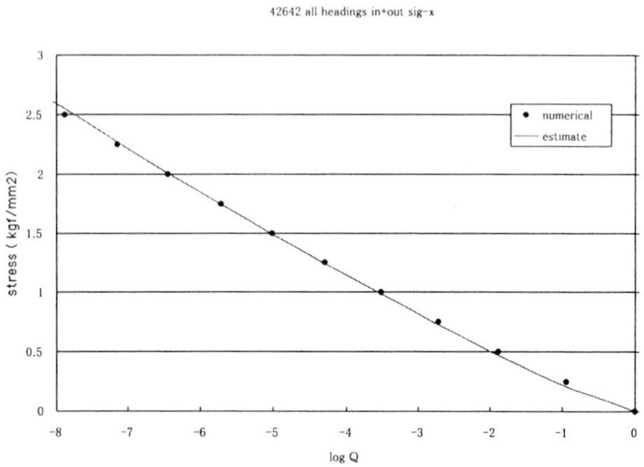

Figure 2 : Comparison of long-term distribution of wave-induced stress on inner bottom plate at hopper produced by wave pressure

TABLE 2
THE GOVERNING WAVE CONDITION

condition	Inner bottom plate at hopper	
	wave pressure	inner load
T_m (sec)	10.5	10.5
θ_m (deg)	270	270
H_m (m)	17.5	17.5

6 CONCLUSIONS

To examine which factor gives influence by load level in the long-term distribution, such as significant wave height, mean wave period of supposed wave condition and wave direction, the long-term distribution is decomposed into every factor and a contribution rate analysis method in the long-term distribution is introduced. The governing wave condition and the most severe wave condition on a long-term distribution is clarified by using contribution rate analysis. Authors extend the correlation method to combine not only the extreme values but also the whole shape of long-term distribution of wave-induced stress. Moreover, if the governing wave conditions on long-term distribution have the same properties, we can combine the fatigue damages.

TABLE 3

CORRELATION COEFFICIENTS FOR EXTREME VALUE $Q=10^{-8}$ AND FATIGUE DAMAGES

position	嘆Hold frame σ_x		嘴 Inner bottom plate at center line lower stool σ_x	
Extreme value at $Q = 10^{-8}$ (kgf/mm^2)	$\sigma_{x\text{-in}}$ = 12.727	ρ = -0.811	$\sigma_{x\text{-in}}$ = 3.348	ρ = -0.677
	$\sigma_{x\text{-out}}$ = 16.946		$\sigma_{x\text{-out}}$ = 4.860	
	$\sigma_{x\text{-in+out}}$ = 9.952		$\sigma_{x\text{-in+out}}$ = 3.576	
Damage ratio	$D_{x\text{-in}}$ = 0.1734	ρ = -0.762	$D_{x\text{-in}}$ = 2.9921E-3	ρ = -0.701
	$D_{x\text{-out}}$ = 0.7177		$D_{x\text{-out}}$ = 1.4963E-2	
	$D_{x\text{-in+out}}$ = 0.1763		$D_{x\text{-in+out}}$ = 4.9578E-3	
	嵺Inner bottom plate at hopper σ_x		嵺Inner bottom plate at hopper σ_y	
Extreme value at $Q = 10^{-8}$ (kgf/mm^2)	$\sigma_{x\text{-in}}$ = 2.192	ρ = -0.699	$\sigma_{y\text{-in}}$ = 7.172	ρ = -0.819
	$\sigma_{x\text{-out}}$ = 3.526		$\sigma_{y\text{-out}}$ = 11.701	
	$\sigma_{x\text{-in+out}}$ = 2.542		$\sigma_{y\text{-in+out}}$ = 7.132	
Damage ratio	$D_{x\text{-in}}$ = 5.8746E-4	ρ = -0.641	$D_{y\text{-in}}$ = 2.537E-2	ρ = -0.768
	$D_{x\text{-out}}$ = 4.0630E-3		$D_{y\text{-out}}$ = 20.455E-2	
	$D_{x\text{-in+out}}$ = 1.7134E-3		$D_{y\text{-in+out}}$ = 5.696E-2	

References

Fukuda, J., Shinkai, A. and Moriyama, A. (1978). Long-Term Prediction of the Longitudinal Stresses Induced on a Ship Hull in Sea Waves. *Journal of the Society of Naval Architects of Japan* **144**, 230-239.

Report of the 216th Research Panel organized in the Shipbuilding Research Association of Japan. (1996).

Hogben, N., Dacumha, N. M. C. and Oliver, G. F. (1986). *Global Wave Statistics*. Unwin Brothers Limited, UK.

Practical Design of Ships and Other Floating Structures
You-Sheng Wu, Wei-Cheng Cui and Guo-Jun Zhou (Eds)

FATIGUE ANALYSIS OF AN AGED JACK-UP PLATFORM STRUCTURE REFITTED TO CANTILEVER-BEAM TYPE

Wu Nie, Yu-Wu Sun and Li-Ping Sun

Department of Naval Architecture and Ocean Engineering
Harbin Engineering University, Harbin, China

ABSTRACT

The fatigue problem of a jack-up platform refitted from slot type to cantilever-beam type, Bohai 10 jack-up platform, is discussed in this paper. According to the problems met in service and the particularity of the platform, the fatigue life is analyzed and calculated from a new aspect, which can provide new information for safe service and reasonable refitment of this kind of aged platform structures.

The fatigue analyses on platforms are mostly focused on tubular joints, but also on new fatigue hot points appearing in the platform due to the refitment mentioned above. To deal with this problem, fatigue analyses different from previous tubular joint fatigue are carried out based on the calculation model of Bohai 10 jack-up platform, which has been refitted and operated over 12 years.

Firstly, comparing the region where crack occurred in service and the finite element analysis results, one point at the 120-degree angular welding line in yoke house is defined as fatigue hot point. Then cyclic loading that leads to fatigue is obtained. Through stress analyses and comparative studies, it is shown that wave loads have little effect on the fatigue accumulated damage and can be ignored, cyclic bit pressure is the main reason that causes fatigue in the special case discussed in this paper. By statistical analysis the distribution of the cyclic bit load ranges is normal. The factors such as welding residual stress and stress concentration and mean stress are taken into account. The stress concentration factor is obtained through finite element analysis. At last, the accumulative damage is calculated by using P-S-N curves and the remained fatigue life of this platform is predicted.

Many jack-up platforms are refitted from cantilever-beam type to slot type and most of them are aged. The fatigue life of this kind of platforms is estimated from a different point of view in this paper. The approach presented in this paper has a practical meaning to the costly platforms refitting and it also can be extended idea to fatigue analyses of aged platform structures.

KEYWORDS

Jack-up platform, Slot type, Cantilever-beam type, Fatigue life, Yoke house, Aged platform structures, Cyclic bit pressure

1 INTRODUCTION

In the exploitation of oil under sea, various kinds of offshore platforms are indispensable, especially the jack-up platforms. In the past most of jack-up platforms are designed as slot type, because the moving area of its oil derrick is limited due to the structure form, it can drill few wells during each jacking up operation and has low efficiency. Cantilever-beam type Jack-up platform can deal with the problem well because it has a more advantage structure form, which can enlarge the moving area of the oil derrick and so the number of wells to be drilled increases. This brings great economic benefit to offshore oil drilling company, so there are more and many cantilever-beam platforms in the world. But it is well known that it costs much financially to construct a new platform, so the method of refitting a slot type jack-up platform into cantilever-beam type is chosen first. Presently there are recordings of cantilever-beam type jack-up platforms in the world and most of them, for example, Bohai 8 and Bohai 10 jack-up platforms, are refitted from slot type [1].

Although the refitment mentioned above improves the drilling efficiency, it brings some structural problems. The structures of the refitted platform and the loads they carried changed. If the resistance change of the platform caused by structural refitment cannot match the change of load, some problems will gradually appear during the service of the platform. The problems met in service of Bohai 10 platform are summarized as follows:

(1) serious deformation of the machinery deck
(2) cracks initiated and propagated in the yoke house

To deal with these problems, analysis and calculation of static strength, fatigue strength and buckling strength are carried out based on the calculation model of Bohai 10 jack-up platform. The content of this paper is focused on the second problem and the fatigue accumulative damage is analyzed, fatigue life is calculated from a new point of view.

2 FATIGUE OF PLATFORM AND FATIGUE HOTSPOT ANALYSIS

A great deal of work has been done on the fatigue problem of platforms [2], but most of the analyses are focused on tubular joints of the platform leg caused by wave load. For the jack-up platform refitted from slot type to cantilever-beam type, Bohai 10, a new fatigue problem in its main part has been found after inspection in dock.

2.1 Analysis of the Reason that Causes Cracks in the Yoke House Area

For offshore structures, wave loads are the main reason that causes fatigue, but after comparative study, this is not true to the fatigue problem of the refitted jack-up platform analyzed in this paper. Jack-up platform is generally in drilling state and the main hull is all above water. Based on this state, the finite element model is established and loads such as wind, wave and current are applied according to operating manual. Through analysis, it shows that wave load has little effect on the stress variation in the perpendicular 120^0 welding line in yoke house and cannot form cyclic load to cause fatigue damage. It is reported that this kind of cracks only appeared after the platform refitted, so it is doubt that refitment maybe the reason. Fig.1 shows that the cantilever-beam extends far backward and the distance between the oil derrick on the beam and the stern is more than 10 meters in drilling. FEA results show that the perpendicular load at there can lead tensile stress in the top part of the welding line and the amplitude variation of bit pressure can cause relative bigger cycling load, which can induce fatigue damage as time goes by.

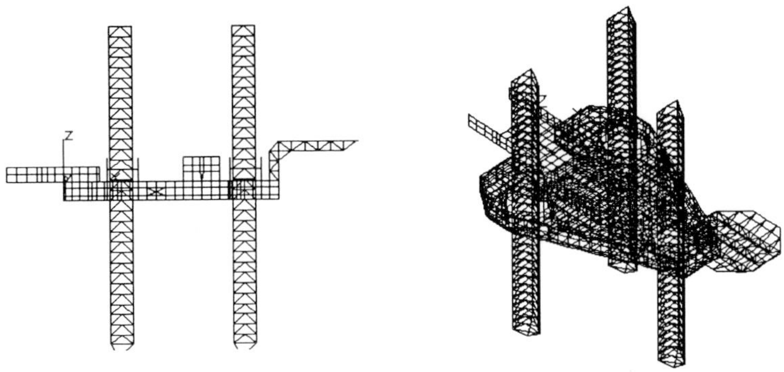

Figure 1: Bohai 10 Element Model

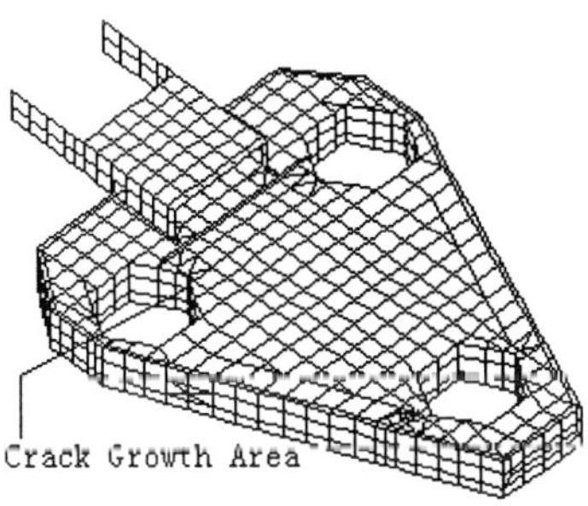

Figure 2:Fatigue Calculation Point

2.2 Determination of the Fatigue Calculation Area

After pointing out the reason that caused fatigue damage in the platform, varying loads are applied to the drilling area on the cantilever-beam to find the area where stress range is the biggest. After FEM analysis, the node that has the biggest stress range is found at the top corner of the inside perpendicular 120^0 welding line in the stern yoke house (shown in Fig.2). This is consistent with the case found during an annual inspection in dock, so this area is taken as the fatigue calculation hotspot to analyze the fatigue accumulation damage of the platform and to estimate its fatigue life.

3 FATIGUE CYCLING LOAD

A random distributed fatigue load model is obtained in this paper, in which the influence of welding residual stress, mean stress and stress concentration is taken into account.

3.1 The Source of Cyclic Load

Presently tricone bit is commonly used in drilling [3, 4]. During the drilling process, periodic perpendicular vibration happens when single and double cones of the bit touch the well bottom alternatively. The intense vibration of the bit can make the bit pressure varied in a great range. Its value can be zero or several times of the average bit pressure and is distributed randomly. The varying of bit pressure is transferred from the hole toe to the main hull of the platform and cause cyclic stress at the hotspot.

3.2 Cyclic Bit Pressure at Pit Bottom

3.2.1 Analysis of a record

Fig.3 is a bit pressure variation curve of a tricone bit recorded in a site [5]. It shows that it is a stochastic process and needs statistical analysis. Before analysis, the load ranges are handled by modified rain-flow method. Then the sampling data are divided into groups according to the value of the range and histograms are plotted.

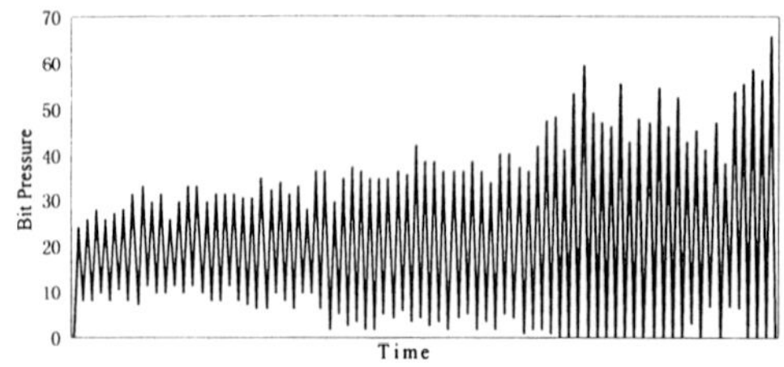

Figure 3: Bit Pressure-Time Curve

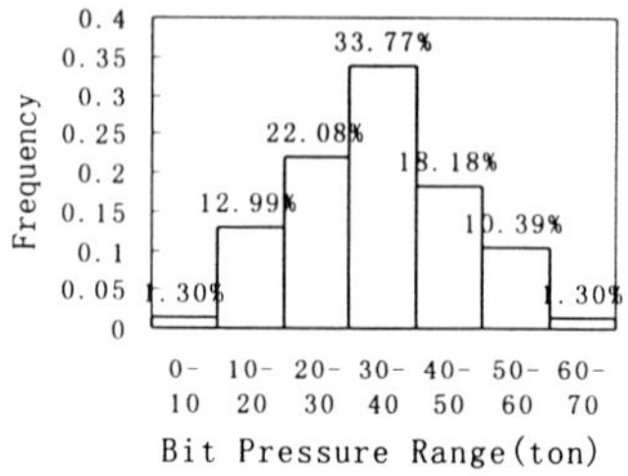

Figure 4: Histogram

The histogram shows that the distribution of bit pressure ranges approximates to normal distribution. Now probability paper is used to further determine its distribution type and function. Taking exceeding accumulative frequency as abscissa and bit pressure range as as ordinate, then an approximate straight line is obtained on the normal distribution probability paper. So it is reasonable to have the opinion that the bit pressure range can be considered as a normal distraction.

$$f(x)=\frac{1}{\sigma\sqrt{2\pi}}e^{-\frac{(x-\mu)^2}{2\sigma^2}} \tag{1}$$

From Fig.5, the mean value μ and standard deviation σ can be obtained.

$$\mu = x_{50} = 34.5\,(\text{ton})$$

$$\sigma = x_{84.1} - \mu = 11.5\,(\text{ton})$$

Then the distribution function of bit pressure frequency range is

$$f_r(x)=\frac{1}{11.5\sqrt{2\pi}}e^{-\frac{(x-34.5)^2}{2\times 11.5^2}} \tag{2}$$

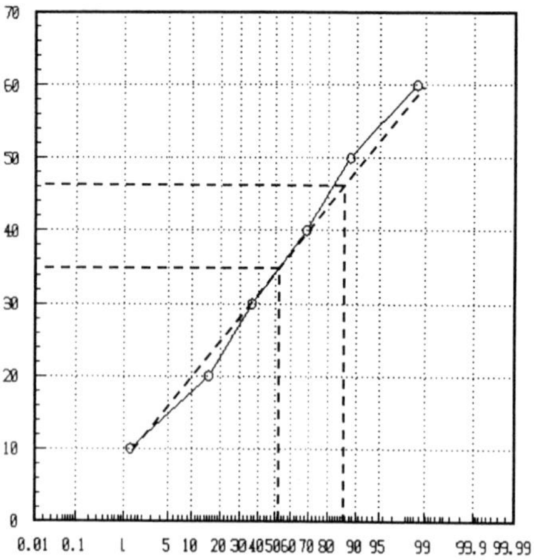

Figure 5: Normal Distribution Probability Paper

3.2.2 Survey data

The survey data show that bit pressure range is to a great extent in proportion to mean bit pressure. So it is assumed that the distribution of bit pressure range of tricone bit has the same type, which is normal distribution with different parameters. There are some information about Bohai 10 jack-up platform in Table.1 and Table.2.

According to the survey data and the hypothesis above, distribution parameters of Bohai 10 are determined as follows.

So the distribution function of calculated bit pressure amplitude range is

$$f_c(x)=\frac{1}{8\sqrt{2\pi}}e^{-\frac{(x-24)^2}{2\times 8^2}} \tag{3}$$

TABLE 1

DRILLING PARAMETERS OF BOHAI 10

Bit type	Tricone bit
Mean bit pressure (ton)	15
Rotational frequency(r/min)	50
Working time (day/year)	365×80%≈290

TABLE 2

BIT PRESSURE PARAMETERS

Mean value of B.P* (ton)	Proportion coefficient	Mean value of B.P.R** (ton)	Standard deviation of B.P.R** (ton)
21.6 (record)	1	34.5	11.5
15.0 (Bohai 10)	0.694	24.0	8.0

B.P*: Bit Pressure; B.P.R**: Bit Pressure Range

3.3 Cyclic Stress in Hotspot

After having load input, the distribution function of hot spot cycling stress can be obtained, by taking account of some parameters.

3.3.1 Transmissibility between pit bottom bit pressure and structure load

When calculating node stress through finite element method, the pressure is applied on the cantilever-beam. Because of friction and other reasons, there is a loss of the bit pressure when it transfers from pit bottom to the main hull. It is difficult to calculate the loss of bit pressure theoretically. Analysis of site record shows that the transmissibility is relevant to mean bit pressure and rotational frequency of the bit. So the transmissibility is obtained by statistical analysis and by interpolation of the site measuring data. The transmission rate is determined as

$$r = 0.6$$

3.3.2 Relation between perpendicular load and hotspot stress

A liner elastic structural finite element model is assumed and adopted in this paper. The relation between the perpendicular cycling load N(ton) transferred form pit bottom and hotspot cycling stress range S is

$$S = -6.95N \times 10^{-2} \text{(MPa)}$$

3.3.3 Frequency of cyclic stress in hotspot

The investigation data show that bit pressure changes three times while the bit rotates one circuit. So the frequency of the fatigue stress range in hotspot can be represented as

$$f_s = 3f_b = 150 \text{(r/min)}$$

where

f_s: cycling stress frequency $\quad$ f_b: bit rotation frequency

3.3.4 Stress concentration factor (SCF)

The fatigue life of a detail is governed by the notch stress range. For components other than smooth specimens the notch stress is obtained by multiplying the nominal stress by SCF.

The relation between the notch stress range to be used together with the S-N-curve and the notch stress range is

$$\Delta\sigma = K \cdot \Delta\sigma_{no\ min\ al} \tag{4}$$

All stress amplifications have to be considered when evaluating notch stress. The resulting SCF used for calculation is derived as

$$K = K_g \cdot K_w \cdot K_{te} \cdot K_{t\alpha} \cdot K_n \tag{5}$$

where

K_g = SCF due to the gross geometry of the detail considered

K_w = SCF due to the weld geometry. K_w =1.5 if not stated otherwise

K_{te} = additional SCF due to eccentricity tolerance (normally used for plate connections only)

$K_{t\alpha}$ = additional SCF due to angular mismatch (normally used for plate connections only)

K_n = additional SCF for unsymmetrical stiffeners on laterally loaded plates, applicable when the and nominal stress is derived from simple beam analysis.

The stresses of the hotspot in yoke house are derived from finite element analysis. As the structure of the platform is very huge and it is composed of too many details so that it is impossible to make mesh fine enough everywhere. Mostly geometry stress concentration exists in structure discontinuities. The local structure of yoke house is taken out to compute geometry concentration factor using finite element method. Finite element analysis software NASTRAN is adopted in this paper and geometry concentration factor is obtained as

$$K_g = 3.65$$

Other concentration factors are chosen according to the rule, DNV Classification Notes (1993). The resulting SCF is

$$K \approx 5.5$$

3.3.5 Welding Residual Stress

The influence of transverse welding residual stress is also considered in this paper. According to survey data, [6, 7], the residual stress , which is compressive , is derived as

$$\sigma_y = \sigma_y' + \sigma_y'' = 200\text{MPa} \tag{6}$$

where

σ_y': welding residual stress due to longitudinal constriction of plastic deformation in weld and its nearby area

σ_y'':welding residual stress due to asynchronism of transverse constriction in plastic deformation area

σ_y : combined transverse welding residual stress

3.3.6 Influence of Mean Stress

Using Goodman formula, the equivalent cycling stress range of the calculation node is derived as

$$\Delta\sigma_{st} = \Delta\sigma \Big/ \left(1 - \frac{\sigma_m}{\sigma_{ult}}\right) \approx 4.0\Delta\sigma \tag{7}$$

where

$\Delta\sigma_{st}$: maximum cyclic stress range of standard cycling curve of the material at given cycles

σ_{ult} : ultimate strength of the material; $\Delta\sigma$: cycling stress amplitude of the node
σ_m : mean value of cyclic stress of the node

4 FATIGUE DAMAGE CALCULATION AND FATIGUE LIFE ESTIMATION

4.1 P-S-N Curves

According to DNV Rules [8, 9], the S-N curves for welded joints with a 97.6% probability of survival is given as

$$\log N = \log \bar{a} - m \log \Delta\sigma \quad (8)$$

where

m : negative inverse of S-N — $\Delta\sigma$: stress range

$\log \bar{a}$: intercept of $\log N$ -axis by S-N curve — $\log \bar{a} = \log a - 2\log s$

a : constant relating to mean S-N curve — $\log s = 0.2$

s: standard deviation of log N — $s = 0.2$

The parameters are taken as following (shown in Table 3).

TABLE 3
PARAMETERS OF S-N CURVE

Parameter	Bi-liner S-N curve		One-slope S-N curve
	$N \le 10^7$	$N \ge 10^7$	
$\log \bar{a}$	12.65	12.64	12.76
m	3.0	5.0	3.0

4.2 Accumulated fatigue damage calculation and fatigue life estimation

According to Miner liner accumulated damage theory [10], a FORTRAN program is developed to calculate the accumulated fatigue damage of the hotspot. It is supposed that the structure damaged when accumulated fatigue damage value $\Delta D = 1$. The results are shown in Table.4.

TABLE 4
CALCULATION RESULT

	Cycles/year	Fatigue damage/year	Fatigue life (year)
bi-liner S-N curve	6.3072×10^7	0.072	14
one-slope S-N curve	6.3072×10^7	0.084	12

One-slope S-N curve is more pessimistic than bi-liner S-N curve, so the calculated fatigue life is lower. Bohai 10 platform was built in 1980 and refitted in 1988. It has worked over 12 years up to now. Crack and its propagation are found in the weld of the yoke house. This is close to the calculated result. Bohai 10 is an aged jack-up platform and it is approaching its fatigue life limit. It should be maintained carefully to prolong its service time.

Bohai 10 is an aged jack-up platform and it is approaching its fatigue life limit. It should be maintained carefully to prolong its service time.

5 CONCLUSIONS

Many jack-up platforms in the world today are refitted from slot type to cantilever-beam type, like Bohai 10. Some of them are in later stage of service. But the fatigue problems caused by refitment have not been widely realized. Quantitative fatigue analysis of this kind of platforms from a different point of view is presented in this paper. The fatigue damage and fatigue life caused by bit pressure loading different from others can be estimated through the approach presented here.

Reference

[1] Marine Company. (1988). *Handbook of Bohai10 – a mobile offshore jack-up platform*, authorized by ABS on January 4th.
[2] Mohri,M. Uemura. (1996). Fatigue strength of welding joints under variable amplitude load, *Ishikawajina Harima Engng Rev*, V.36 n4, 292-297.
[3] France Petroleum Institute. (1995). *Drilling data handbook*(Sixth Edition), Beijing, Geology Publisher. (in Chinese)
[4] 《Ocean drilling handbook》 editor, (1996). *Ocean drilling handbook*, China Ocean Petroleum Company. (in Chinese)
[5] Zhao Guozhen, Gong Weian. (1988). *Elementary drilling mechanics*, Beijing, Petroleum Industry Publisher. (in Chinese)
[6] D.Lardaii. (1994). *Fatigue strength of welded structures*, Beijing, Mechanical Industry Publisher, (in Chinese)
[7] Tian Xitang. (1997). *Welded structures*, Beijing, Mechanical Industry Publisher.
[8] Classification Notes. (1993). *Note No. 30.2*, Norway, Det Norske Veritas.
[9] Draft DNVC Classification Note, *Fatigue assessment of ship structures*, Client DNVC/JTC Report NO.93-0432. (in Chinese)
[10] Hu Yuren, Chen Bozhen. (1996). *Fatigue Reliability Analysis of Ship and Offshore Structure*, People's Marathon Letournean Communication Publisher. (in Chinese)

Practical Design of Ships and Other Floating Structures
You-Sheng Wu, Wei-Cheng Cui and Guo-Jun Zhou (Eds)

ANALYSIS OF THREE-DIMENSIONAL CRACKS IN SHIP STRUCTURES SUBJECTED TO AN ARBITRARY LOADING BY NUMERICAL WEIGHT FUNCTION METHOD

Y. Sumi

Department of Naval Architecture and Ocean Engineering
Yokohama National University, Yokohama 240-8501, Japan

ABSTRACT

In the present paper a numerical weight function method for the stress intensity analysis of cracks in ship structures is proposed by applying the Maxwell-Betti's reciprocal theorem. In addition to an original problem, the auxiliary problems are introduced, which are subjected to three kinds of force couples acting on the crack surfaces near the limiting periphery of a crack front line. In the present formulation, a weight function can be approximated by using a finite element solution combined with an analytical expression near the condensation point, where hyper-singularities occur. A general-purpose finite-element codes can be used for the numerical calculation. This method has successfully been applied to through-the-thickness cracks in three-dimensional plate structures and also to a surface crack in a detail of a double hull structure. The weight function method is extremely useful for the fatigue strength assessment of cracks in ship structural details, because a quite large number of the same structural details subjected to wide varieties of loading exist in a ship structure.

KEYWORDS

Crack analysis, Weight function, Computational method, Ship structures

1 INTRODUCTION

A new computational method is proposed for the calculation of stress intensity factors of cracks in plate structures and ship structural details by using weight functions. Traditionally weight functions are defined in two ways; the one is defined as the hyper-singular term of the eigen-function expansion of the displacement field of a cracked body (Bueckner, 1970), and the other, the variation of displacement field with respect to a virtual extension of a crack tip (Rice, 1972). Recently, Sumi (2001) has proposed a new numerical weight function method. The original problem and the auxiliary problems subjected to three kinds of force-couples acting on the crack surfaces near the limiting periphery of a crack front line are first defined. Then the numerical weight functions are derived for the three modes of stress intensity factors by applying the Maxwell-Betti's reciprocal theorem to the original and auxiliary problems.

This approach is a generalization of the weight function method, which was presented by Paris *et al.* (1976) for a two-dimensional crack problem by using a finite element method with the special treatment near a crack tip. In the present formulation, weight functions of two- and three-dimensional cracks are approximated by finite element solutions combined with analytical expressions near the condensation point, where hyper-singularities exist (Bueckner 1970, 1987). The proposed method does not require any special numerical treatment for the hyper-singularities at the condensation point so that general-purpose finite-element codes can be used to obtain the three-modes of weight functions of the stress intensity factors at any point of an arbitrary three-dimensional crack front-line. Having obtained the weight function for the given geometry of a structure, stress intensity factors for an arbitrary loading can easily be calculated by an integration of the inner product of the weight function and the applied load. This method will be successfully applied to through-the-thickness cracks in three-dimensional plate structures and also to a surface crack in a detail of a double hull structure. The weight function method could be extremely useful for the fatigue strength assessment of cracks in ship structural details, because a quite large number of the same structural details subjected to wide varieties of loading exist in a ship hull.

2 WEIGHT FUNCTIONS FOR A THREE-DIMENSIONAL CRACK

2.1 Three-Dimensional Crack Problem

Figure 1 shows an arbitrary three-dimensional crack in an elastic body, whose crack surfaces are assumed to be smooth and their crack front-line forms a smooth curve. We shall consider the stress intensity factor at point P on the crack front line, and introduce a coordinate system P-$x_1x_2x_3$, where the x_2-axis directs normal to the tangential plane of the crack surface, and the x_3-axis directs to the tangent of the crack front-line at the origin P. The governing equation and the boundary conditions are expressed by

$$\begin{aligned} &\sigma_{ij,j} + f_i = 0 && \text{in V,} \\ &\sigma_{ij} n_j = t_i && \text{on S}_t, \\ &u_i = v_i && \text{on S}_u, \end{aligned} \tag{1}$$

where a body force f_i, surface traction t_i, and surface displacement v_i are prescribed in the elastic domain V, on the traction boundary S_t, and on the displacement boundary S_u, respectively. The unit external normal vector is denoted by n_i. This original problem will be denoted by Problem-O.

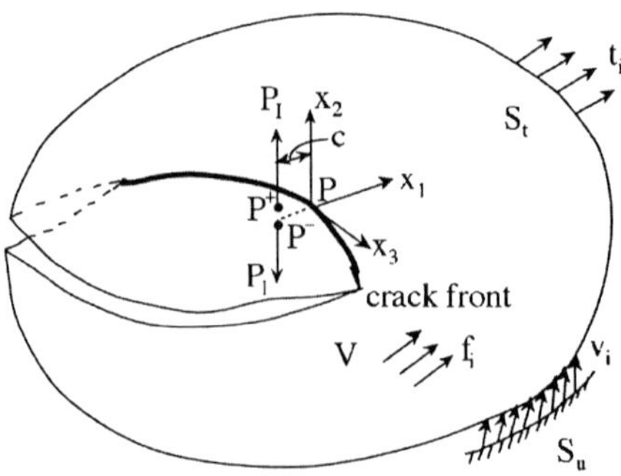

Figure 1: A three-dimensional crack problem

2.2 Auxiliary Problems

Let us introduce auxiliary problems (Problem-λ, λ = I, II, III) which correspond to the three modes of stress singularities ahead of the crack tip,

$$\begin{aligned} &\sigma_{ij,j} = 0 && \text{in V,} \\ &\sigma_{ij} n_j = P_i^{(\lambda)} && \text{on } S_t, \\ &u_i = v_i && \text{on } S_u,\ \lambda = \text{I, II, III,} \end{aligned} \tag{2}$$

where the three loading conditions, $P_i^{(\lambda)}$, represent the three concentrated force-couples of magnitude, P_λ (λ = I, II, III), acting in the x_2-, x_1-, and x_3-directions, respectively, at the points $P^{\pm}$ on the both sides of the crack surfaces at a distance *c* from the point, P (see Figure.1). The solution of Eqn. 2 will be denoted by u_i^λ, and σ_{ij}^λ (λ = I, II, III), for the displacements and stresses, respectively.

2.3 Determination of Weight Functions by Using Maxwell-Betti's Reciprocal Theorem

Let us consider the crack opening displacement of Mode-I of Problem-O at point P^+, which is given by

$$u_2 = \frac{4\sqrt{2}}{\sqrt{\pi}E'} K_I \sqrt{c} \ , \tag{3}$$

where K_I is the Mode-I stress intensity factor, and

$$E' = \begin{cases} E & \text{for plane stress} \\ E/(1-\nu^2) & \text{for plane strain} \end{cases} \tag{4}$$

in which *E* and ν are Young's modulus and Poisson's ratio, respectively.

Having applied the Maxwell-Betti's reciprocal theorem to Problem-O and Problem-I, we have

$$u_2 P_I = \int_V u_i^I f_i dV + \int_{S_t} u_i^I t_i dS \ . \tag{5}$$

Substitution of Eqn. 3 into Eqn. 5 leads to the expression of Mode-I stress intensity factor;

$$K_I = \frac{\sqrt{\pi}E'}{4\sqrt{2}} \frac{1}{P_I \sqrt{c}} \left\{ \int_V u_i^I f_i dV + \int_{S_t} u_i^I t_i dS \right\} . \tag{6}$$

Since Eqn. 3 exactly holds for the limiting case (c→0), we calculate the limit of the right hand side of Eqn. 6. Then, we can obtain the weight function of Mode-I stress intensity factor given by

$$h_i^I(\text{P,Q}) \equiv \lim_{c\to 0} \frac{\sqrt{\pi}E'}{4\sqrt{2}} \frac{1}{P_I \sqrt{c}} u_i^I(\text{P,Q}) \ . \tag{7}$$

This function is a kind of a universal function, which depends only upon the geometry of the body and displacement boundary condition. Having taken the limit c→0, it has a hyper-singularity at the condensation point P. The stress intensity factor can be expressed by

$$K_I(\mathrm{P}) = \int_V h_i^I(\mathrm{P,Q}) f_i(\mathrm{Q}) dV(\mathrm{Q}) + \int_{S_t} h_i^I(\mathrm{P,Q}) t_i(\mathrm{Q}) dS(\mathrm{Q}) \quad . \tag{8}$$

Similarly, the weight functions of Mode-II and Mode-III stress intensity factors are obtained in the following form;

$$h_i^\lambda(\mathrm{P,Q}) \equiv \lim_{c\to 0} \frac{\sqrt{\pi}E'}{4\sqrt{2}} \frac{1}{P_\lambda \sqrt{c}} u_i^\lambda(\mathrm{P,Q}) \qquad \lambda = \mathrm{I, II, III}\ , \tag{9}$$

so that the general expression of the stress intensity factors are expressed by

$$K_\lambda(\mathrm{P}) = \int_V h_i^\lambda(\mathrm{P,Q}) f_i(\mathrm{Q}) dV(\mathrm{Q}) + \int_{S_t} h_i^\lambda(\mathrm{P,Q}) t_i(\mathrm{Q}) dS(\mathrm{Q}) \qquad \lambda = \mathrm{I, II, III}\ . \tag{10}$$

3 STRESS INTENSITY FACTORS OF CRACKS IN THREE-DIMENSIONAL STRUCTURES

3.1 Finite Element Calculation Procedure

When we calculate the weight functions for arbitrary geometry by the finite element method, we have the numerical difficulty to apply the concentrated force-couple with the limit c→0. In order to bypass this drawback, we shall take the following calculation procedure;

(a) to take a small but finite value of the distance c from the condensation point, P, and to apply the force-couple P_λ (λ = I, II, III), acting at the points $\mathrm{P}^\pm$ on the both sides of the crack surfaces,

(b) to use the finite element mesh which is in conformity with the application points of the force-couples in the vicinity of the condensation point,

(c) to use the analytical weight functions of a flat circular crack having the same radius of curvature (Bueckner, 1987) in the vicinity of the condensation point, and to use the finite element solutions for the rest of the domain, and

(d) to use a general-purpose finite element code with the conventional wedge shaped singular elements in the vicinity of the crack front-line.

3.2 Simply Supported I-Beam with Through-the-Thickness Crack

We shall consider a simply supported I-beam subjected to 3-point bending with a T-shaped penetrating crack at the middle span, whose lengths are $2a_1$ in the flange and a_2 in the web as illustrated in Figure 2. A typical mesh pattern used for the analysis is given in Figure 3, in which the wedge shaped singular elements with the element size of 2% of a_1 and a_2 are employed at the crack tip.

In a plane problem, an analytical expression of Mode-I crack opening displacement which correspond to the weight function by Eqn.7, is asymptotically represented by

$$u_2^I(x,0) = \frac{4P_I\sqrt{c}}{\pi E} \frac{1}{\sqrt{x}} \quad , \tag{11}$$

where x is the distance from the crack tip. In order to define the weight function of the problem, this analytical expression is used in the region $r/a < 0.15$, where r is the radial distance from the crack tip. The finite-element solution of the weight function is applied in the rest of the domain.

In Table 1, the results of the stress intensity factors by the proposed method are compared with those obtained by the direct displacement method. They show good agreement with each other so that the validity of the method has been confirmed.

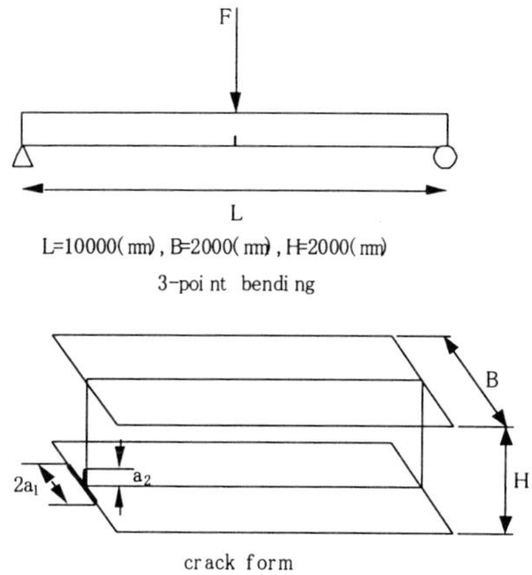

Figure 2: Simply supported cracked I-beam under 3-point bending

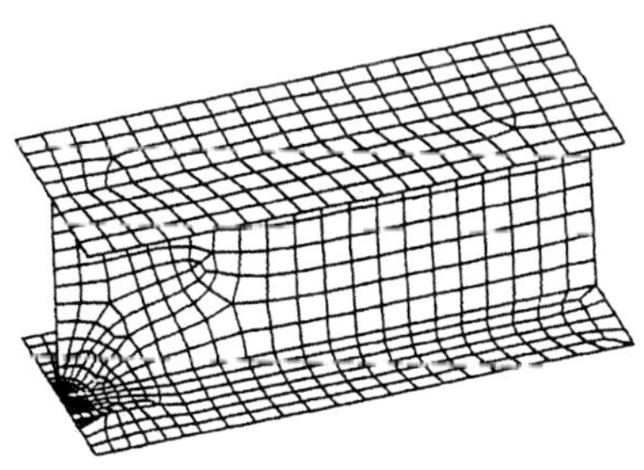

Figure 3: Finite element mesh subdivision

TABLE 1

STRESS INTENSITY FACTORS OF SIMPLY SUPPORTED I-BEAM UNDER 3-POINT BENDING

a_1(mm)	K_{Ia_1} (MPa · mm$^{1/2}$) weight function method	displ. method	a_2(mm)	K_{Ia_2} (MPa · mm$^{1/2}$) weight function method	displ. method
20	745	731	10	779	772
50	1188	-	25	1228	-
100	1690	1659	50	1724	1708
150	2084	-	75	2096	-
200	2426	2383	100	2413	2396

3.3 A Surface Crack in a Critical Detail of a Double Hull Structure

In order to investigate a surface crack in a critical detail of a double hull structure, a semi-circular crack of radius a at a bilge knuckle is analyzed by the present method. The numerical weight functions are evaluated by solving the auxiliary problem by using the finite element method with $c/a = 0.04$, where the analytical expression of the weight function (Bueckner, 1987) is applied in the vicinity of the condensation point.

Figure 4 shows the structural model used in SR245-project, in which the model was used for fatigue crack propagation tests at the bilge knuckle (SR245, 2000). The finite element model used for the crack analysis is illustrated in Figure 5, where solid hexahedron elements are used only in the region near the crack. The degenerated wedge-shaped singular elements having the wedge angle of 22.5° are allocated along with the crack-front line. The radial size of the elements are taken as 2% of the crack radius at the crack tip. The stress intensity factors at the surface and at the deepest point are compared in Table 2, where the stress concentration due to the structural discontinuity results in a relatively high stress intensity factor at the surface.

TABLE 2

STRESS INTENSITY FACTOR OF THE SURFACE CRACK IN BILGE KNUCKLE

a(mm)	b(mm)	K_{Ia}(MPa · mm$^{1/2}$)	K_{Ib}(MPa · mm$^{1/2}$)
5.0	5.0	1320	955

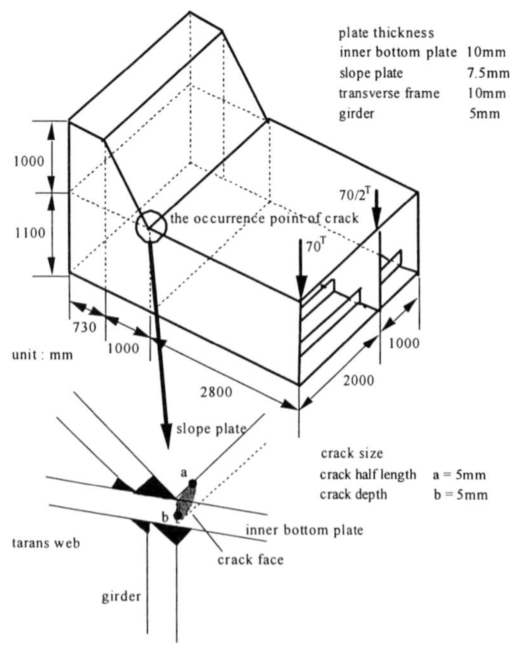

Figure 4: A semi-circular surface crack in a bilge-knuckle

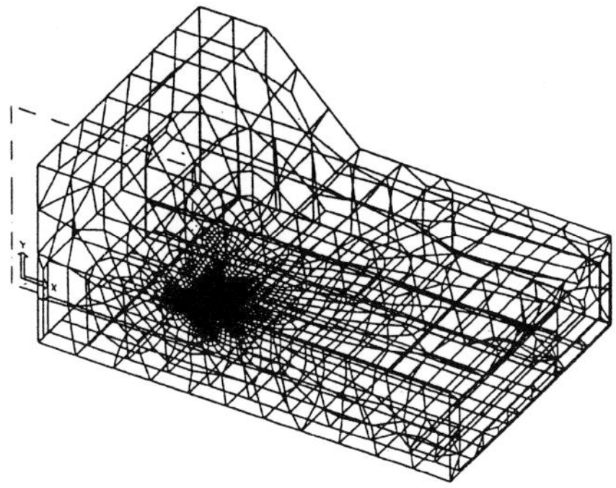

Figure 5: Finite element model of bilge knuckle with a surface crack

4 CONCLUSIONS

In the present paper, a new combined analytical and finite-element weight function method has been proposed for three-dimensional crack problems. Having compared the results with those obtained by the direct crack tip opening, the accuracy of the method has been confirmed. Since stress intensity factors are calculated by the simple integration of the inner product of a weight function and a loading vector, the proposed method is especially useful for cracks in same structural details with a variety of loading conditions.

Acknowledgement

This work was supported by Grant-in-Aid of Scientific Research B-2-10555338 from the Ministry of Education, Science and Culture to Yokohama National University, and in part by SR245-project, Shipbuilding Research Association of Japan. The author is grateful to Mr. T. Kashiuchi, Mr. K. Ito, and Mr. A. Okura for their assistance in numerical calculations.

References

Bueckner H.F. (1970). A novel principle for the computation of stress intensity factors. ZAMM **46**, 529-545.

Bueckner H.F. (1987). Weight functions and fundamental fields for the penny-shaped and the half plane crack in three-space. *Int. J. Solids Structures* **23-1**, 57-93.

Paris P.C., McMeeking R.M., and Tada H. (1976). The weight function method for determining stress intensity factors. In: *ASTM STP* **601**, 471-489

Rice J.R. (1972). Some remarks on elastic crack-tip stress fields. *Int. J. Solids Structures* **8**, 751-758.

SR245 (2000). Research on the structural lives of double hull tankers. Shipbuilding Research Association of Japan. (in Japanese)

Sumi Y. (2001). Numerical weight function method for arbitrary three-dimensional cracks. In: *Proceedings of the 10th International Conference of Fracture* (in press).

Practical Design of Ships and Other Floating Structures
You-Sheng Wu, Wei-Cheng Cui and Guo-Jun Zhou (Eds)

EFFECT OF MEAN STRESS CHANGES ON THE FATIGUE STRENGTH OF SPECTRUM LOADED WELDS

Gary B. Marquis, Timo P.J. Mikkola

VTT Manufacturing Technology, P.O. Box 1705, FIN-02044 VTT, Finland

ABSTRACT

Fatigue tests have been performed on high and very high strength steel I-beam type structures using constant amplitude loading and three spectra considered as representative for ships. Both "usual" quality, representing normal shipyard fabrication practices, and improved ground welds were tested. Special attention was given to the effect of mean stress of the spectra. Material strength did not significantly influence fatigue strength but grinding improved fatigue strength by 30% at $2x10^6$ cycles to failure. Comparison of the spectrum loading test data showed that the most suitable analysis hypothesis is that that nominally compressive portions of the loading cycles were non-damaging.

KEYWORDS

Welded structures, Spectrum fatigue, High strength steel, Improved welds, Mean stress

1 INTRODUCTION

Welded structures generally contain high residual stresses which significantly influence their fatigue behaviour. Even when the nominal loading on a structure is compressive, the presence of residual stresses may result in local tensile stresses near the weld toe where failure is likely to occur. Consequently, the fatigue life of welded structures is normally assessed on the basis of stress range regardless of the nominal applied stress ratio. Some test data suggests that stress ratio does influence fatigue strength of welded structures and this has led some committees responsible for developing fatigue design recommendations for welded structures to include correction factors for fatigue loading which allow for higher stress amplitudes when significant potions of the fatigue cycle are compressive [1-3].

Damage accumulation rules found in many design codes assume that the damage caused by a single stress cycle is independent of the other stress cycles in the spectrum. However, the fatigue damage caused by small stress range cycles is different in spectrum loading from that in a constant amplitude (CA) loading [4]. Typically ship structural details are loaded by wide banded stress spectra due to wave loading. Some types of ships which are also subject to cargo loading which is characterised by block-wise variations in mean stress as the operation alternates between fully loaded and ballast conditions. These slowly changing mean stresses are superimposed on the wave induced loading. In highly stressed regions this may cause reversed plastic deformation and affects the local mean and residual stresses. The damaging effect of small fatigue cycles in the presence of large mean stress variation has not received significant attention in the open literature. Matsuoka and Takahashi [5] tested longitudinal non-load carrying fillet weld specimens using spectrum loading with superimposed constant tensile, constant compressive and block-wise varying tensile / compressive mean stresses. They found that the compressive mean stress cycles as part of a block-wise varying mean stress spectrum contributed fatigue damage for cracks up to 1 mm in depth, but they did not contribute to the propagation of cracks longer than 1 mm.

Several researchers have studied the effects of spectrum loading where cycles in the spectrum have stress ratios of either R=0 or R=-1 [6-11]. It is generally found that spectrum loading with R=0 spectrum is less damaging than spectrum loading with an R=-1 spectrum. Even though R=0 will produce a spectrum with higher mean stresses, Gurney [7] notes that the maximum stress is much less for the R=-1 spectrum as compared to R=0 with the result that lower compressive residual stresses will be built up at the crack tip. In a Gaussian distributed load spectra, for example, the largest stress cycle will be several times larger than the characteristic stress level with the result that plastic deformation at the weld toe may occur. Dahle [10] and Borgren et al. [11] report tests for longitudinal non-load carrying fillet weld specimens using spectrum loading with superimposed tensile mean stress in addition to R=0 and R=-1 spectra. While the number of data is not large, it was found that the spectrum with a tensile mean stress was most damaging and the R=0 spectrum was least damaging and the R=-1 spectrum had about 25% longer life than the tensile mean stress spectrum.

2 EXPERIMENTAL PROGRAMME

2.1 Test Specimens

The I-beam type specimens shown in Fig. 1 were fabricated from 10 mm thick HTS 390 and VHTS 690 steel plate. Measured yield strengths of the two materials were 445 MPa and 712 MPa, respectively. Weld quality of the gusset attachment was defined as ‘usual’, i.e., corresponding to present shipyard fabrication standards and ‘round’ requiring full penetration welding and grinding of the gusset to obtain a smooth transition between gusset and I-beam flange.

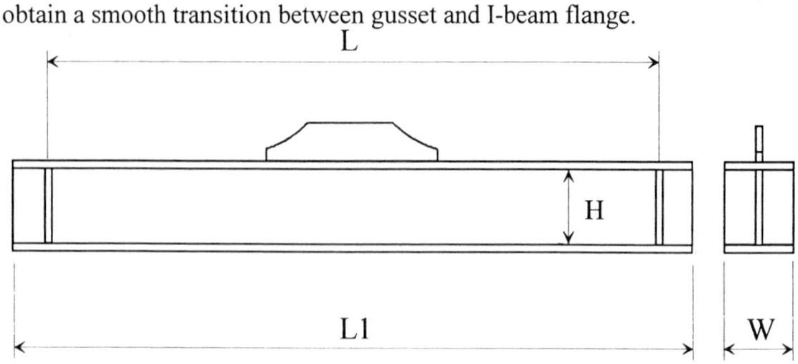

Figure 1. I-beam specimens. W = H = 100 mm, L = 850 mm and L1 = 1000 mm.

2.2 Load Spectrum

Load controlled three-point-bending was applied to the I-beam type specimens. Loading was both CA and spectrum loading using three alternate spectra with varying mean stress combinations. Blocks 12 and 13 of the WASH spectrum, Wave Action Standard History [12], were chosen for testing. The final spectrum is defined by the stress ranges in the original WASH spectrum with either superimposed block-wise varying mean stress, constant mean stress, or near-zero mean stress. Characteristics of the spectra are presented in TABLE 1. The spectrum with block-wise varying mean stress is illustrated in Fig 2.

TABLE 1
LOAD SPECTRA CHARACTERISTICS

Parameter	Changing mean spectrum	Tensile mean spectrum	Near-zero mean spectrum
load sequence	WASH block 12 and 13	WASH block 12 and 13	WASH block 12 and 13
superimposed mean stress	Block-wise variation between σ_{max} / 2 and $-\sigma_{max}$ / 2	Constant $\sigma_{mean} = \sigma_{max} / 2$	Constant $\sigma_{mean} = \sigma_{max} / 10$
maximum stress range	$\Delta\sigma_{max} = 2\sigma_{max}$	$\Delta\sigma_{max} = \sigma_{max}$	$\Delta\sigma_{max} = 1.67\sigma_{max}$
minimum applied stress	$-\sigma_{max}$	0	$-0.67\sigma_{max}$
equivalent stress	20.8% of $\Delta\sigma_{max}$	41.6% of $\Delta\sigma_{max}$	41.6% of $\Delta\sigma_{max}$
omission level	13% of $\Delta\sigma_{max}$	26% of $\Delta\sigma_{max}$	26% of $\Delta\sigma_{max}$
number of cycles	46 444	23 222	23 222
largest cycle occurrence	twice in spectrum	once in spectrum	once in spectrum

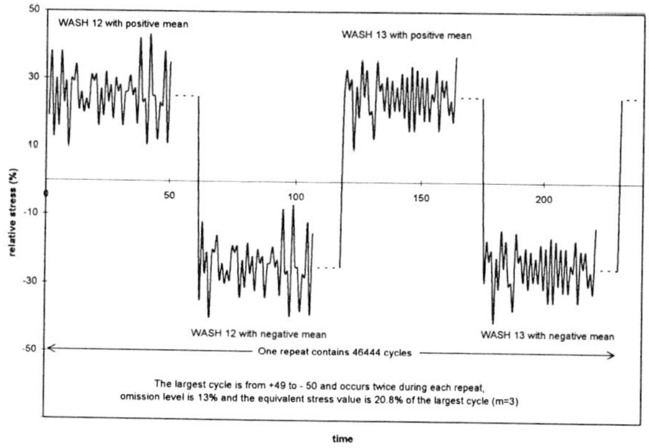

Figure 2. Block-wise changing mean stress spectrum.

2.3 Fatigue Results

Data for both the spectrum and CA fatigue tests are shown graphically in Fig. 3. Failure in all cases was defined as the number of cycles when the crack penetrated through the 10 mm thick flange. This corresponded to a surface crack length (2c) of about 35 mm.

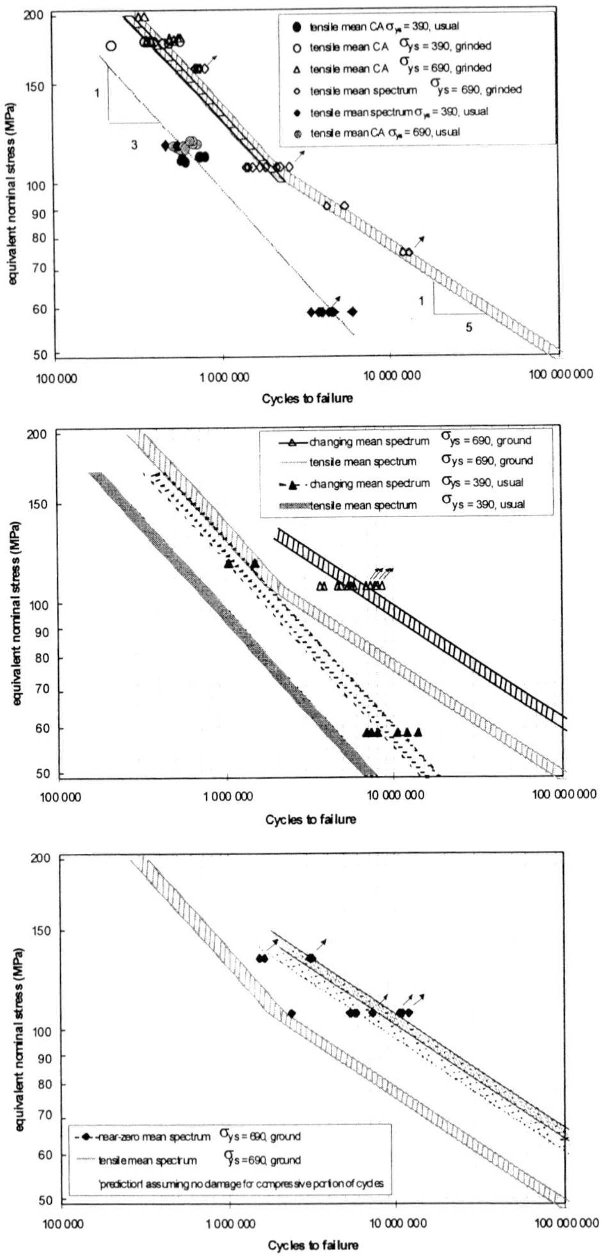

Figure 3 a) Constant tensile mean stress for "usual" quality and ground welds for both steel grades, b) effect of constant tensile vs. block-wise changing mean stress for "usual" quality $\sigma_{ys} = 390$ MPa and ground $\sigma_{ys} = 690$ MPa welds, c) near-zero mean spectrum for ground $\sigma_{ys} = 690$ MPa welds.

3 DISCUSSION

3.1 Effect of Material Strength

In the current investigation the yield strength of the base material did not greatly affect the fatigue life of either the usual quality or the ground welds. In both cases the higher strength material had only a few percent greater fatigue strength. This, however, was not statistically significant based on the number of specimens tested. This lack of dependence on material strength is not surprising for usual quality welds [13] but is contrary to some published data for improved welds. Weld improvement techniques such as grinding normally show greater fatigue strength improvement as material strength increases [14,15]. Figure 3a shows CA and tensile mean stress spectrum fatigue data collected for usual quality welds of both materials. Usual quality welds fabricated with VHTS 690 steel had a slightly higher mean strength than did welds fabricated from HTS 390 steel but when considering the number of data points, the greater strength is not statistically significant.

Figure 3a also shows the CA and tensile mean stress spectrum fatigue data collected for ground welds produced from both material grades. Ground VHTS 690 welds had a slightly higher mean strength than did welds fabricated from HTS 390 steel but the strength improvement is not statistically significant.

One possible explanation for this observed lack of influence on material strength is the presence of sub-surface welding defects. Finite element analysis showed that the most highly stressed location for the ground welds was along the gusset as compared to at the weld toe for the usual quality welds. A large number of failures, however, were observed at the region of the weld toe which had been ground away. Sub-surface welding flaws in this region propagated causing failure. Full penetration welds were specified but not completely achieved in the structures tested so that improved welding processes would be expected to improve the fatigue strength of the improved welds fabricated from the very high strength steel.

Both CA and spectrum loading data for usual quality welds tends to fall along a single S-N line with slope of 3. This is often observed for welded structures where equivalent stress range is used to compute stress [4]. It should be noted, however, that the degree of conservatism of this approach depends on the load spectra [7,16]. For ground specimens tested with spectrum loading there is a tendency for S-N curves to alter slope at about $2x10^6$ cycles to failure. This reduced slope is not observed for CA test data because of the fatigue limit effect which means that specimens will either fail prior to $2x10^6$ cycles or will not fail at all within a reasonable test interval. This illustrates one advantage of spectrum loading which provides the means to investigate long but finite fatigue lives that are common for many structures. Spectrum loading also provides a means for observing the damaging effect of small cycles in the presence of larger load events. Large cycles in a spectrum have been shown to affect the residual stress distribution near the weld toe [17,18].

3.2 Effect of Grinding

Figure 3a shows that the usual quality welds tended to follow a straight line S-N curve down to at least a stress level corresponding to $4x10^6$ cycles to failure. Other studies have shown straight line S-N curves for gusset attachments out to fatigue lives as long as $1x10^8$ cycles [16]. On the other hand the ground welds show a tendency for the S-N curve to change slope at about $2x10^6$ cycles to failure. The number of data points, however, is small and further testing is need to confirm this observation. The reduced S-N slope indicates either crack growth threshold effect for the large number of small cycles in the spectrum or the presence of a crack initiation period for the welded joint.

As compared to the usual quality welded specimens, the ground welds showed an improved fatigue strength of about 30% at $2x10^6$ cycles to failure. At longer lives the degree of improvement is even greater due to the difference in S-N curve slopes. Similar or even higher degrees of improvement have been reported by other researchers [14,15,17,18].

3.3 Effect of Mean Stress

The effect of block-wise changing mean stress can be assessed by referring to Fig. 3b. This figure shows the mean life for usual quality HTS390 and ground HTS690 welds for both constant tensile mean and block-wise changing mean spectrum. If all cycles in the spectra were equally damaging regardless of mean stress level, the changing mean stress data would be expected to lie on top of the constant mean stress data. For both specimen types, however, the fatigue life of specimens subject to block-wise changing mean stress was about double the fatigue life of specimens subject to the constant tensile mean stress spectrum. For the usual quality HTS 390 specimens the mean fatigue life for the block-wise changing mean stress spectrum was 2.2 times greater than for the constant tensile mean stress spectrum while for ground VHTS690 specimens the comparable ratio was about 3.3. In the block-wise changing mean stress spectrum the number of cycles with compressive mean stress was equal to the number with tensile mean stress so the easiest explanation for Fig 3b is that the cycles with compressive mean stress contributed no fatigue damage. This assumption is consistent with observations made by Matsuoka and Takahashi [5] who found that the compressive mean stress cycles as part of a block-wise varying mean stress spectrum did not contribute to the propagation of cracks longer than 1 mm for longitudinal non-load carrying stiffeners.

It is interesting to note that for the ground VHTS 690 specimens the ratio in fatigue life was even greater that a factor 2. This may mean that occasional compressive overloads are actually beneficial. This effect is not observed for the usual quality welds. This is observation will require further study.

The case of near-zero constant mean stress is illustrated in Fig. 3c. Results are consistent with the assumption that the compressive cycles do not contribute to fatigue damage. These tests were conducted using a spectrum where a portion of many cycles was compressive. If all cycles in the spectra were equally damaging regardless of mean stress level, the near-zero mean stress spectrum data would be expected to lie on top of the constant tensile stress spectrum data. In fact, all specimens had fatigue lives greater than that predicted in based on this assumption. It is also possible to hypothesise that only the portion of the fatigue cycles where the applied stress is tensile contributed to fatigue damage. The equivalent stress for the "effective" spectrum obtained by truncating all computed damage from the applied spectrum at zero load is 74% of the equivalent stress for the entire spectrum assuming that the full stress range is damaging. Consequently the fatigue lives are expected to be significantly longer. Figure 2c shows the expected mean life based on the assumption of a truncated effective spectrum. This predicted mean life overlaps the mean life computed from the data so in this case it is more accurate to assume that only the tensile portions of cycles are damaging.

4 CONCLUSIONS

Fatigue tests have been performed on high and very high strength steel I-beam type structures. In addition to constant amplitude loading, spectrum loading with both constant tensile, alternating tension / compression, and near-zero mean stresses were used. Both "usual" quality, representing normal shipyard fabrication practices, and improved ground welds were tested. Special attention was given to the effect of mean stress of the spectra. For both "usual" quality HTS welds and VHTS ground welds the spectrum loading with alternating tension and compressive mean stresses resulted in about twice the fatigue life as compared to spectrum loading with constant tensile mean stress. This indicates that the best analysis assumption is that the portion of the spectrum with compressive cycles was non-damaging.

As compared to the "usual" quality HTS 390 welded specimens, the 690 VHTS full penetration and ground welds showed and improved fatigue strength of about 30% in stress at $2x10^6$ cycles to failure. At longer lives the degree of improvement appears to be even greater due to the apparent difference in $\Delta\sigma$-N_f slope. Both for usual quality and ground welds and under both constant amplitude and spectrum loading, material strength was found to not strongly influence fatigue strength. However, it should be noted in this context that the fabrication procedures, especially for the ground welds were not optimal.

Acknowledgements

Specimens were manufactured at Fincantieri SpA Trieste, Italy. Spectrum tests were performed at VTT Manufacturing Technology and were partially funded by the European Commission as part of the project "FatHTS - Fatigue based design rules for the application of high tensile steels in ships" contract BE95-1937. Risto Laakso and Heikki Laukkanen are acknowledged for their assistance in performing the tests. Some CA results were performed at the Lappeenranta University of Technology, Laboratory of Steel Structures directed by Prof. Teuvo Partanen.

References

1. GL: *Rules for Classification and Construction. I-Ship Technology, Part 1 - Seagoing Ships, Chapter 1 - Hull Structures, section 20 - Fatigue. Hamburg*. Germanischer Lloyd, 1996, 13 p.
2. DNV: *Fatigue Assessment of Ship Struictures, Class*. Notes No. 30.7, 1998, 134 p.
3. Hobbacher A.(1996). Recommendations for the Fatigue Design of Welded Joints and Components. IIW Document XIII-1539-96 / XV-845-96, 138 p.
4. Marquis, G. (1996). Long Life Spectrum Fatigue of Carbon and Stainless Steel Welds. *Fatigue Fract Eng Mater Struct* **19:6**, 739-753.
5. Matsuoka K and Takahashi I. (2000). Effect of Loading History on Fatigue Crack Initiation Life of Welded Joint, IIW Document XIII-1833.
6. Blom A. (1989). Fatigue Strength of Welded Joints Subjected to Spectrum Loading. Scand J of Metall. **18**, 181-4.
7. Gurney TR. (1992). Comparative Fatigue Tests on Fillet Welded Joints Under Various Types of Variable Amplitude Loading. *ETSU WN* 6014 50.
8. Gurney TR. (1989). Cumulative Damage of Welded Joints: II Test results, *Joining & Mat* **2:8**, 390-5.
9. Gurney TR. (1993). Comparative Fatigue Tests on Fillet Welded Joints Under Various Types of Variable Amplitude Loading in Air. *OMAE V*. III-B, 537-542.
10. Dahle T. (1993). Spectrum Fatigue of Welded Specimens in Relation to the Linear Damage Rule. In: Blom AF(ed.) Fatigue Under Spectrum Loading and in Corrosive Environments, *EMAS*. 133-147.
11. Bogren J, Lopez Martinez L and Brunnberg M.(1991). The Influence of Various Spectrum Parameters on Fatigue Life and Residual Stress Relaxation. *The Aeronautical Research Inst. of Sweden*, Stockholm FFA TN 1991-44, 31.
12. TSG, *User Manual and Documentation for WASH (Version W, 1.02) Time History*. Technical Software Consultants Ltd., Milton Keynes, UK 1989, 72.
13. Maddox S. (1991). Fatigue Strength of Welded Structures, Abington, Cambridge, 198.

14. Haagensen PJ, Dragen A, Slind T and Orjasaeter O. (1987). Prediction of the Improvement in Fatigue Life of Welded joints Due to Grinding, TIG Dressing, Weld Shape Control and Shot Peening, *3rd Int Conf on Steel in Marine Structures*, Delft, Elsevier.
15. Huther I, Lieurade HP, Soissi R, Nussbaumer A, Chabrolin B and Janosch J.J. (1996). Analysis of Results on Improved Welded Joints, *Welding in the World.* **37:5**, 242-266.
16. Dahle T. (1994). Long Life Spectrum Fatigue Tests of Welded Joints, *Int J Fat.* **16:2**, 130-136.
17. Lopez Martinez L, Blom AF, Trogen H and Dahle T. (1997). Fatigue Behaviour of Steels with Strength Levels Between 350 and 900 MPa - Influence of Post Weld Treatment Under Spectrum Loading, Welded High Strength Steel Structures, A.Blom (Ed.), *EMAS*, London, 361-376.
18. Lopez Martinez L, Lin R, Wang D, and Blom AF. (1997). Investigation of Residual Stresses in as-Welded and TIG Dressed Specimens Subjected to Static/Spectrum Loading, Welded High Strength Steel Structures, A.Blom (Ed.), *EMAS*, London, 377-390.

Practical Design of Ships and Other Floating Structures
You-Sheng Wu, Wei-Cheng Cui and Guo-Jun Zhou (Eds)

A NEW LOOK AT THE EFFECT OF BANDWIDTH AND NON-NORMALITY ON FATIGUE DAMAGE

Lei Yu, P. K. Das, and Nigel D. P. Barltrop

Department of Naval Architecture and Ocean Engineering,
University of Glasgow, UK

ABSTRACT

A new attempt is made in this paper to quantify the effect of bandwidth and non-normality in fatigue damage analysis. For want of real time history of stress response, a series of non-Gaussian and homogeneous random fields are generated first. FFT technique is used to guarantee a large sample size and high efficiency. The stress range is then counted by rain-flow method and its moment estimates are obtained statistically. Finally, a factor is calculated to correct narrow-band and normality assumption. In this process a mathematical family of distribution is employed to study the variation of the correction factor with kurtosis, skewness and the moments of stress range. It is shown that fatigue damage evaluated through conventional method may be either conservative or rather unconservative. According to the result of simulation, the upper bound of the correction factor is found.

KEYWORDS

Random field generation, Fast Fourier transform (FFT), Fatigue analysis

1 INTRODUCTION

The widely used spectral fatigue analysis in ocean engineering is based on the assumption that the stress response is normal and narrow-banded (DNV 1998). Effective as it is in many cases, the assumption becomes vulnerable when the structure-fluid system is non-linear and the randomness of structural parameters needs to be taken into account. Wirsching et al. (1980) are among the first to study the effect of bandwidth in fatigue damage. In their work a generated Gaussian field is used as stress response. The wide band fatigue damage is calculated by rain-flow counting and a correction factor is obtained for narrow band assumption. This sets the very frame for succeeding researches. Later Lutes et al. (1984) extended the idea to account for the effect of non-normality. In the paper of Winsterstein(1985), an effort is made in analytical approximation. However, none of them gave any details of generated stress history, such as sample size, peak rate, and zero-crossing rate. All these properties play a decisive part in the final results. In addition, the effect of skewness in the early studies is totally ignored. In the present study, a long stress history is generated through spectrum representation and FFT. The parameters of random field generator are chosen in such a way that the

where ν_R is the rain flow cycle rate, S_R is the rain-flow stress range, and D_{RS} is fatigue damage of per unit σ^m. $\bar{S}_R$ in Eq. (12) is the stress range counted from stress history standardised by

$$\bar{X} = \frac{X - X_m}{\sigma} \tag{13}$$

where X is stress history, X_m is the average stress level. Apparently the definition of D_S and D_{RS} is very helpful in random field generation. Now the correction factor can be calculated by

$$\chi = \frac{D_{RS}}{D_S} \tag{14}$$

Because of the non-Gaussianness of stress distribution and the complexity of rain-flow counting method, the analytical analysis of Eq. (12) is impossible. So we will apply rain-flow counting to a generated random field and calculate the unbiased estimate of $E[\bar{S}_R^m]$ by

$$\hat{E}[\bar{S}_R^m] = \frac{1}{\hat{N}_R}\sum_{i=1}^{\hat{N}_R} \bar{S}_{Ri}^m \tag{15}$$

where $\bar{S}_{Ri}^m$ is the series of rain-flow stress range $\hat{N}_R$ is the corresponding number of rain-flow cycle. According to large number law $\hat{E}[\bar{S}_R^m]$ has an asymptotic normal distribution. The unbiased estimate of its variance is

$$\hat{V}ar[\hat{E}[\bar{S}_R^m]] = \frac{1}{\hat{N}_R - 1}\left[\frac{1}{\hat{N}_R}\sum_{i=1}^{\bar{N}_R} \bar{S}_{Ri}^{2m} - (\hat{E}[\bar{S}_R^m])^2\right] \tag{16}$$

In Eq. (16) we assume that $\bar{S}_{Ri}^m$ and $\bar{S}_{Rj}^m$ ($i \neq j$) are statistically independent, since rain-flow counting tends to weaken the correlation of the original stress series. From Eq. (15) and (16) we can estimate the coefficient of variation of χ by

$$\hat{C}ov[\chi] = \frac{\sqrt{\hat{V}ar[\hat{E}[\bar{S}_R^m]]}}{\hat{E}[\bar{S}_R^m]} \tag{17}$$

5 EFFECT OF BANDWIDTH

To study the effect of bandwidth Gaussian field is generated. The spectral density used is

$$S(\omega) = \frac{1}{4}\sigma^2 b^3 \omega^2 e^{-b|\omega|} \tag{18}$$

where σ^2 is the standard deviation of stress history, b is correlation distance. According to Wiener-Khinchine relation, the corresponding autocorrelation function is

$$R(\tau) = \sigma^2 \frac{b^4(b^2 - 3\tau^2)}{(b^2 + \tau^2)^3} \tag{19}$$

To reduce the peak and trough imperfection the time step is chosen as Δt = 1/64sec. The results of rain-flow counting is shown in Table 1, where rain-flow cycle rate is estimated by

$$\hat{\nu}_R = \hat{N}_R / T \tag{20}$$

TABLE 1
RESULTS OF RAIN-FLOW COUNTING

Period T (sec)	Rain-flow cycle rate	Peak rate	Zero-crossing rate
64	0.7969	0.8750	0.5313
256	0.8398	0.8594	0.5469
1024	0.8633	0.8662	0.5459
4096	0.8628	0.8650	0.5513
16384	0.8588	0.8592	0.5491
65536	0.8588	0.8590	0.5490
262144	0.8587	0.8588	0.5502
Theoretical value: $n_0 = 0.8717$, $\nu_0 = 0.5513$			

To obtain reliable estimate the simulation period must be long enough to guarantee the convergence of statistical properties. In this example, T = 262144secs can give well-converged results. This period will be used hereafter. Another thing interesting is that the rain-flow cycle rate ν_R is very close to peak rate n_0. This is also observed by Wirching(1977). Using the moment estimates by rain-flow method, we can calculate damage D_{RS} by Eq.(12), and then obtain the correction factor by Eq. (14). The result is shown in Fig. 1. Apparently, the result from narrow band assumption is rather conservative. The higher the order m, the more conservative it will be.

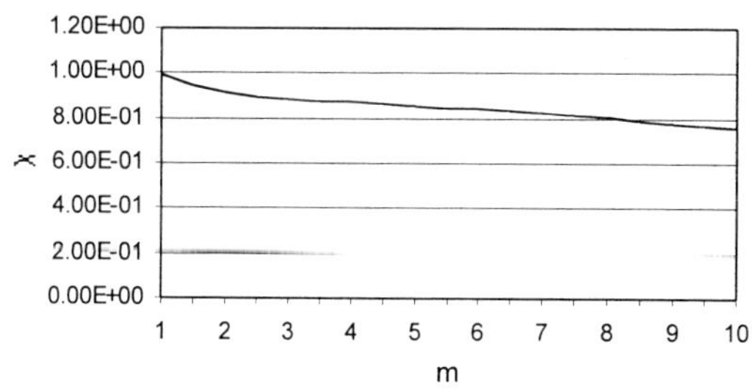

Figure 1: Correction factor of bandwidth

6 EFFECT OF NON-NORMALITY

To generate non-Gaussian random field, the S_U family of distribution proposed by Johnson (1972) is used. The corresponding ZMNL transform is

$$U = \gamma + \delta \sinh^{-1} X \tag{21}$$

where γ and δ are shape parameters, U is standard normal variable, and X is stress. The first four moments of S_U family are

$$\mu_1'(X) = -\omega^2 \sinh \Omega \tag{22}$$

$$\mu_2(X) = E[(X - \frac{1}{2}(\omega - 1)(\omega \cosh 2\Omega + 1) \tag{23}$$

$$\mu_3(X) = -\frac{1}{4}\sqrt{\omega}(\omega - 1)^2 \{\omega(\omega + 2)\sinh 3\Omega + 3\sinh \Omega\} \tag{24}$$

$$\begin{aligned} \mu_4(X) = \frac{1}{8}(\omega - 1)^2 \{\omega^2(\omega^4 + 2\omega^3 + 3\omega^2 - 3)\cosh 4\Omega \\ + 4\omega^2(\omega + 2)\cosh 2\Omega + 3(2\omega + 1)\} \end{aligned} \tag{25}$$

where μ'_1 is moment of origin, μ_2, μ_3, and μ_4 are central moments and

$$\begin{aligned} \omega &= \exp(\delta^{-2}) \\ \Omega &= \gamma / \delta \end{aligned} \tag{26}$$

Thus we obtain the ZMNL transformation of standardised S_U by

$$\overline{X} = \frac{X - \mu_1'}{\sqrt{u_2}} \tag{27}$$

The skewness and kurtosis of S_U family can be calculated by

$$\begin{aligned} \alpha_3 &= \frac{\mu_3}{\mu_2^{3/2}} \\ \alpha_4 &= \frac{\mu_4}{\mu_2 \mu_2} \end{aligned} \tag{28}$$

They are import parameters to measure the degree of non-normality. For standard normal distribution, we have $\alpha_3 = 0$, $\alpha_4 = 3$. In Jonson's family a pair of possible values (α_3, α_4) correspond to only one distribution. When Ω is large enough the distribution will be asymptotically a lognormal one. A series of random fields of S_U distribution are generated in which process parameter γ is used to control skewness and parameter δ to control kurtosis. The correction factor of non-normality against $\beta_1 = \alpha_3^2$, $\beta_2 = \alpha_4$, and m is shown in Fig. 2.

According to Fig.2, both β_1 and β_2 have significant effect on fatigue damage. At point $\beta_2 = 3$, the standardised S_U distribution reduces to normal distribution and the correction factor only reflects the effect of bandwidth. This is the starting point for all the curves in Fig.3, since those cases where $\beta_2 < 3$ are exclusively related to S_B family which has little value to be studied. It is very interesting to find that the correction factor is strictly bounded by two curves, one corresponding to symmetric distribution ($\gamma = 0$), the other corresponding to lognormal distribution. They offer the upper and the lower limit of correction factor as β_2 varies from 3 to 10. For the same kurtosis β_2, when γ increase and so is the skewness, the correction factor will decrease. Furthermore, when $m = 1$, non-Gaussianness will enhance the effect of bandwidth, and make D_S more conservative. However, when $m \geq 1.5$, non-Gaussianness tends to offset the effect of bandwidth at the very beginning and then outweigh it and make D_S a risky estimate. It should be stressed that when we change the sign of γ and hence the sign of skewness, the correction factor will not change. This is simply due to the fact that D_S and D_{RS} are only subject to stress range but not average stress.

7 CONCLUSIONS

Although, in the study of early researchers, there are attempts to fit the correction factor either as an empirical function of β_2 and m, or of m and effective bandwidth, no formula has been proposed to include all these explanatory parameters at the same time. This is partly because of the complexity involved, partly because of the diversity of response spectrum used in different areas. Obviously it is futile to try to obtain a panacea. However, if we focus on a specific problem with known response spectrum, pragmatic formula of correction factor can easily be established by the proposed method. Several important points in simulation are summarised as follows:

1. The accuracy of moment estimates of stress range by rainflow counting is decided by both sample length and time step.

2. Though Nyquist condition can prevent overlap in frequency domain, it is not enough to prevent the imperfection of peaks and troughs in the generated stress time series. To reduce the imperfection a smaller time step is needed, which will give rise to higher resolution in frequency domain.

3. The rain-flow cycle rate is very close to theoretical peak rate. From practical point of view, the latter can be used in place of the former.

References

Det Norske Veritas. (1998). *Fatigue assessment of ship structures*, Classification Note No. 30.7.
Johnson, N. I. and Kotz S. (1970). *Distributions in statistics: continuous univariate distributions*, New York: Houghton Mifflin.

Liu B. & Munson D. C. (1982). Generation of a random sequence having a jointly specified marginal distribution and autocovariance. *IEEE Transaction on Acoustics, Speech, and Signal Processing*. **ASSP 30**:**6**, 973-983.

Lutes, L. D., et al. (1984). Stochastic fatigue damage Accumulation, *Journal of Structural Engineering*, **110**:**11**, 2585-2601.

Shinozuka M. & Deodatis G. (1991). Simulation of stochastic process by spectral representation. *Applied Mechanics Reviews*, **44**:**4**, 191-204.

Spanos B. & Zeldin B. (1998). Monte Carlo treament of random fields: A broad perspective. *Applied Mechanics Review*, **4**:**5**, 219-237.

Wirsching, P. H., & Mohsen Shehata A. (1977). Fatigue under wide band random stresses using the rain-flow method. *Journal of Engineering Materials and Technology*, **99**:**3**, 205-211.

Wirsching, P. H., ASCE, A. M., and Light, M. C. (1980). Fatigue under wide band random stresses. *Journal of the Structural Division*, ASCE, **106**:**ST7**, 1593-1607.

Winterstein, S. R. (1985). Non-normal response and fatigue damage. *Journal of Engineering Mechanics*. **111**:**10**, 1291-1295.

Yamazaki F. & Shinozuka M. (1988). Digital generation of non-Gaussian stochastic fields. *Journal of Engineering Mechanics*, **114**:**7**, 1183-1197.

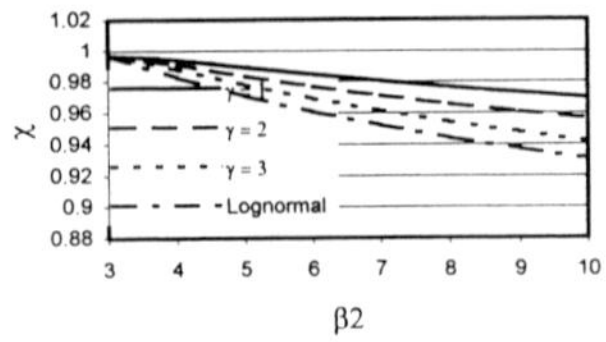

Figure 3(a) m = 1.0

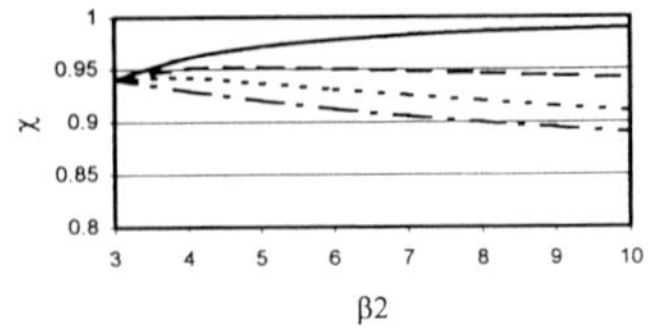

Figure 3(b) m = 1.5

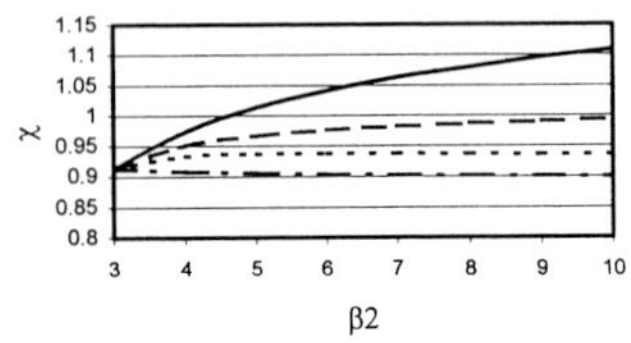

Figure 3(c) m = 2.0

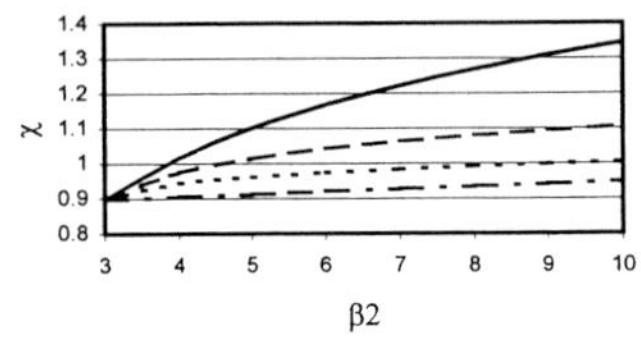

Figure 3(d) m = 2.5

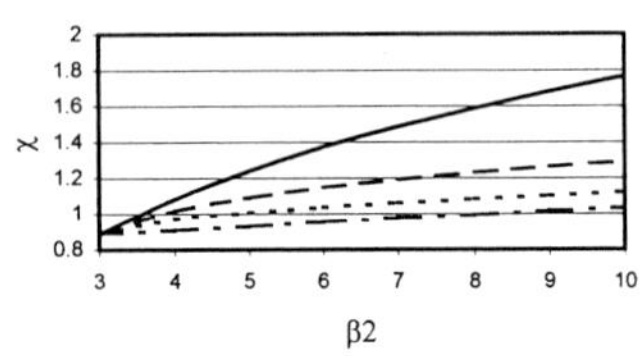

Figure 3(e) m = 3.0

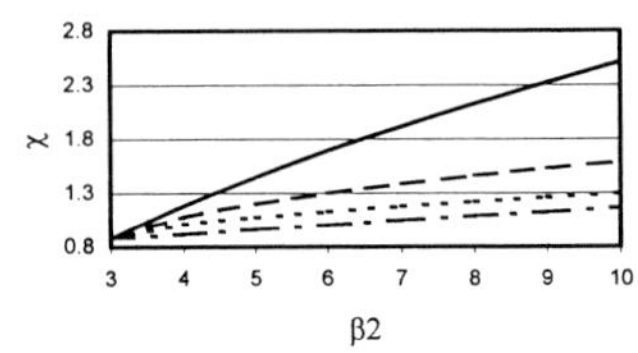

Figure 3(f) m = 3.5

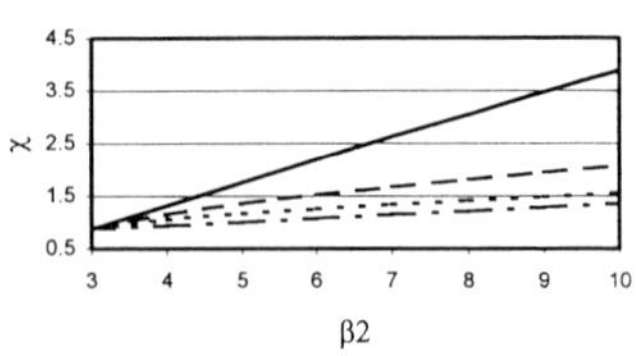

Figure 3(g) m = 4.0

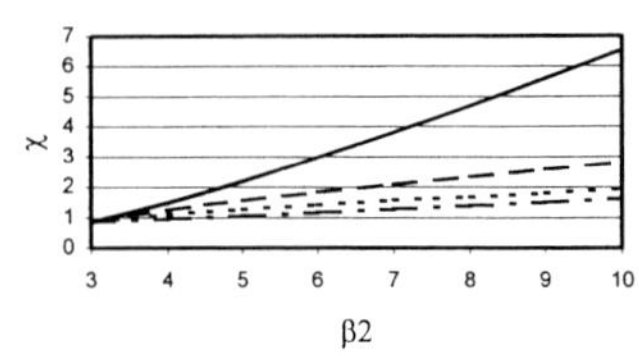

Figure 3(h) m = 4.5

Figure 2: Correction factor of non-normality

Practical Design of Ships and Other Floating Structures
You-Sheng Wu, Wei-Cheng Cui and Guo-Jun Zhou (Eds)

AN EXPERIMENTAL INVESTIGATION ON FATIGUE BEHAVIOR OF INVERTED ANGLE AND T-TYPE SIDE LONGITUDINALS IN TANKERS

Jinsoo Park [1], Kuk Bin Kim [1], Wha Soo Kim [2], and Doe Hyun Kim [2]

[1] Structure Research Department and [2] Hull Initial Design Department
Hyundai Heavy Industries Co., Ltd., 1, Cheonha-Dong, Dong-Ku, Ulsan, Korea

ABSTRACT

Results of a comparative experimental study on fatigue performance of two types of side longitudinal stiffeners in tankers are presented. One is an inverted angle stiffener supported by a soft-toed web stiffener with a soft-toed backing bracket. The other is a T-type stiffener supported by a flat bar web stiffener without a backing bracket, as an alternative to the angle type stiffener. Six full-scale structural models (three for each type) were tested under fully-reversed constant amplitude cyclic loading, resulting in fatigue lives ranging between about 5×10^4 and 2×10^6 cycles. Crack growth data as observed on the surface are presented and discussed. Finite element analyses were also conducted to evaluate the two types of longitudinals based on calculated hot spot stresses. Overall, the T-type longitudinal stiffener showed superior fatigue performance.

KEYWORDS

Side longitudinal, Tanker, Fatigue strength, Inverted angle stiffener, T-type stiffener, Full-scale fatigue test, Finite element analysis, Hot spot stress, Crack initiation, Crack growth

1 INTRODUCTION

Long-term operability with less maintenance cost for ship structures is a great concern of shipbuilders as well as ship owners in line with economical ship design. Among numerous aspects of design and construction, which influence the long-term operability of a ship, adequate fatigue performance of hull structures is one of the most critical items. Design of structural details for extending fatigue lives is always in connection with productivity and it is usually very hard to fulfill both fatigue-enhanced structure and moderate shipbuilding cost. In the late 1980s and early 1990s, a large number of serious fatigue damages were reported in the side longitudinals of single hull VLCCs, where the section of the side longitudinals was unsymmetrical. Most of these damages occurred at the intersection between the face plate of the longitudinal and the heel of the transverse web stiffener (Fig. 1). Side shell area between the load and ballast waterlines is most frequently subjected to fatigue damage due to cyclic wave loading (Hansen and Winterstein, 1995). In consideration of applied cyclic loading and fatigue damages experienced, various designs of structural details such as soft-toed web stiffeners with either a soft-toed bracket or a reversed radius at the heel of the web stiffener have been proposed to reduce the risk of fatigue cracking during service.

In this research, fatigue tests of full-scale structural models of a VLCC have been carried out to investigate comparative fatigue strength between two selective types of side longitudinals and their connections to the web frame. One is an inverted angle stiffener supported by a soft-toed web stiffener with a soft-toed backing bracket (Fig. 2), where the backing bracket is often demanded to reinforce fatigue strength. From now, this structure is briefly called "Angle-Type" longitudinal. The other is a T-type stiffener supported by an ordinary flat bar web stiffener without a backing bracket (Fig. 3), as an alternative to the angle type stiffener to reduce construction cost. From now, this structure is called "T-Type" longitudinal in short. The fatigue tests were aimed at investigating the rational structural design of side longitudinals considering fatigue-enhanced structure as well as cost-effective production. Finite element structural analyses were also conducted to evaluate the two types of side longitudinals based on calculated stresses.

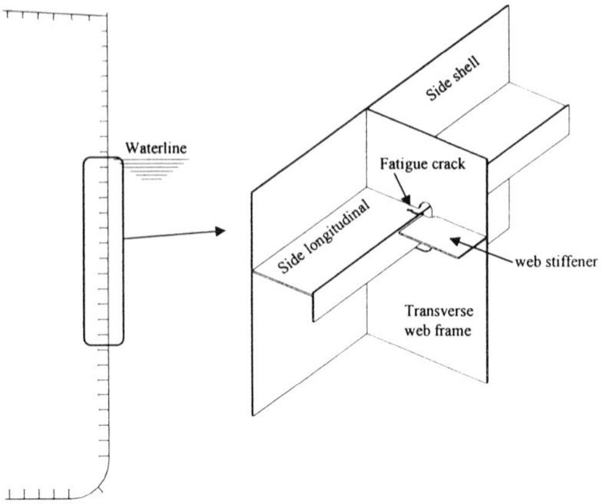

Figure 1: Typical fatigue cracks in side longitudinals

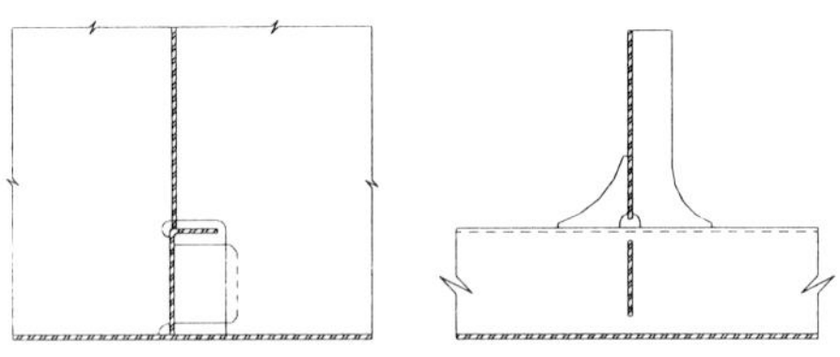

Figure 2: Angle-Type longitudinal

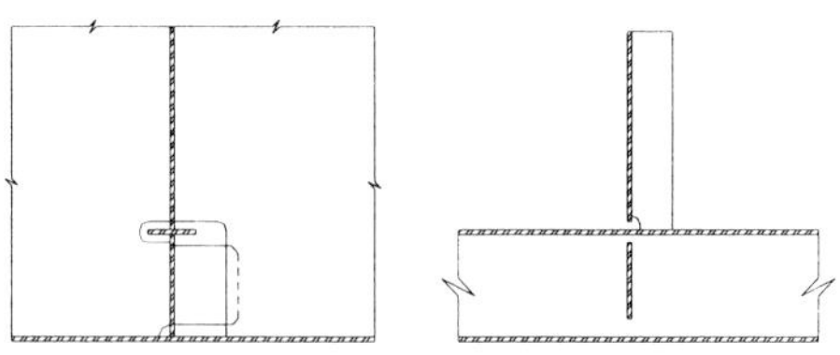

Figure 3: T-Type longitudinal

2 FINITE ELEMENT ANALYSIS

Finite element analyses were conducted to compare fatigue strengths of the Angle-Type and T-Type side longitudinals based on calculated hot spot stresses. A side shell structure of a double hull tanker was chosen as a subject structure and a finite element model for each type was made using shell elements (Fig. 4). The size of the inverted angle is 450×125×11.5/18mm and the T-type longitudinal has the equivalent vertical stiffness by moving the flange part of the inverted angle horizontally to make a T-shape. The sizes of the web stiffener and bracket were decided by the yard practices. The analysis models range three web frames, including three side longitudinals, a longitudinal bulkhead, and a side shell (Fig. 4). Stress concentration areas in the center longitudinal were very finely meshed to the size of plate thickness in order to calculate hot spot stresses. As loading, a uniform side pressure of 3.2-meter water head was applied for both cases on the outer side shell to evaluate comparative fatigue strength. Symmetry boundary conditions were imposed at the model boundaries. MSC/NASTRAN and PATRAN were used for the analysis.

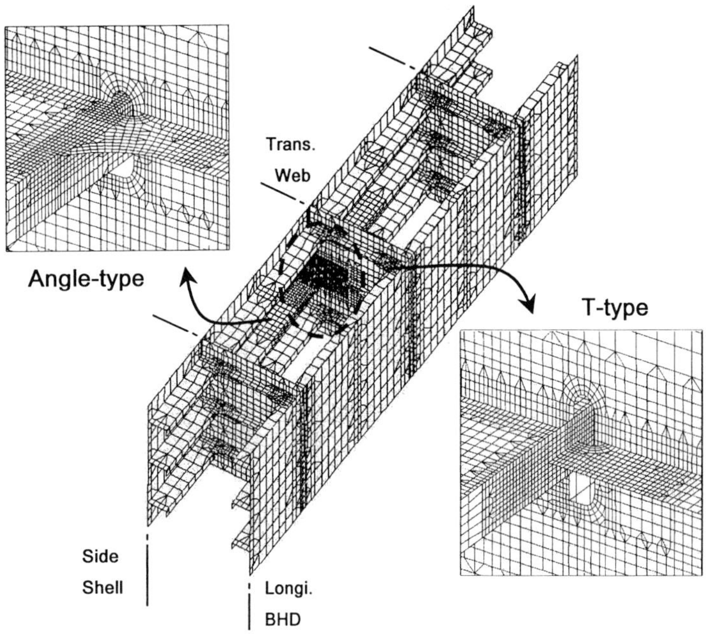

Figure 4: Finite element model of the double hull side shell structure

For the typical stress concentrated locations, hot spot stresses were calculated as in TABLE 1 using the procedure widely used in fatigue evaluation, i.e., the linear extrapolation using the stresses of the two neighboring thickness(t)-sized elements at the distances t/2 and 3t/2 from the hot spot point. Locations A-1 and T-1 in TABLE 1 are the areas where fatigue cracks usually occur. Comparing these two locations, hot spot stress of the Angle-Type longitudinal is about 30% larger than that of the T-type one. Stress at location T-3 was calculated to be a maximum of 89.0 MPa; however, fatigue damages at this area usually have been much less compared to location T-1.

TABLE 1
CALCULATED HOT SPOT STRESSES FOR SIDE PRESSURE LOADING (MPA)

Type	Angle-Type		T-Type			
Location	A-1	A-2	T-1	T-2	T-3	T-4
Hot Spot Stress	66.4	51.4	50.1	33.3	89.0	28.4

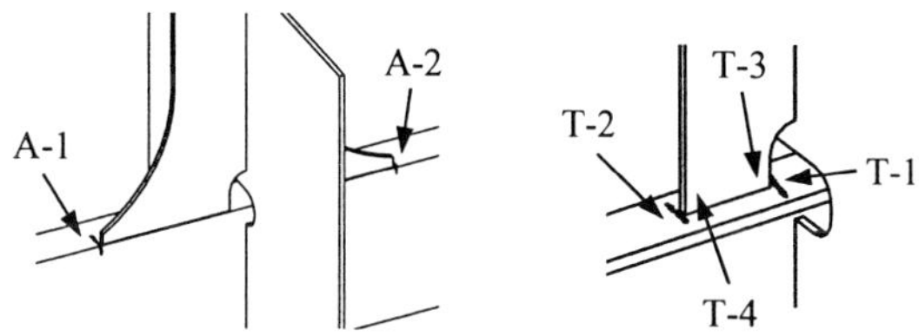

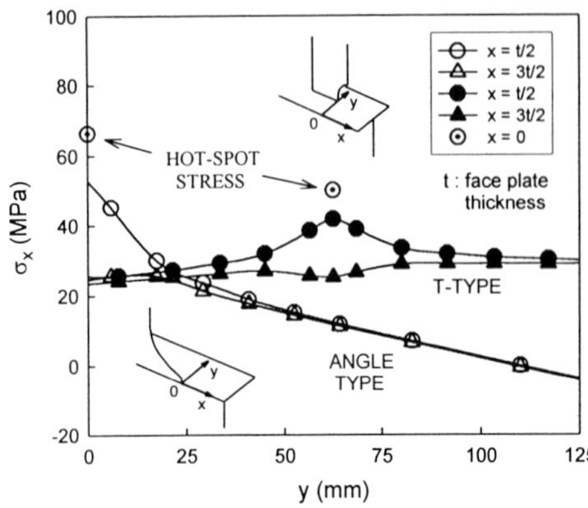

Figure 5: Stress distributions at A-1 and T-1

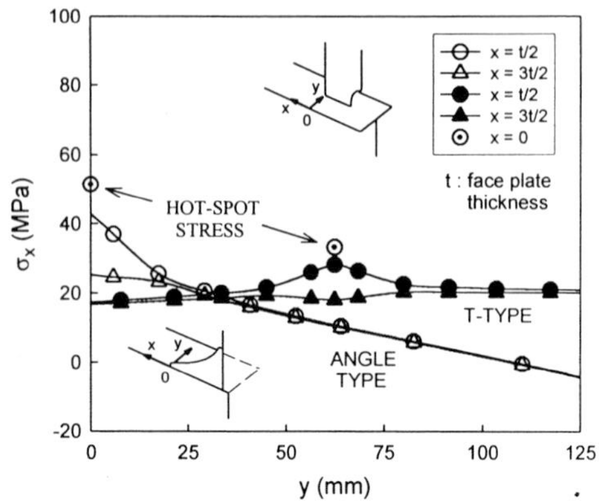

Figure 6: Stress distributions at A-2 and T-2

Figures 5 and 6 compare longitudinal (x-dir.) stress distributions on the face plate of Angle-Type and T-Type (See TABLE 1 for locations A-1, A-2, T-1, and T-2), where the terms x and y are distances from the connection and face plate (or flange) edge as illustrated in the figures, respectively. The term t indicates the plate thickness. It can be seen that stress gradients are steeper in Angle-Type, resulting in larger hot spot stresses.

3 FATIGUE TESTS

In order to investigate fatigue strengths of the Angle-Type and T-Type side longitudinals, fatigue tests were conducted using six (three for each type) full-scale single hull side shell structural models. The model size was determined from a VLCC built in the yard and the main dimensions are as follows:

- model size: 4400×1600×1200mm
- side shell thick.: 16mm
- side longi.: 300×90×11/16 I.A. (Angle-Type)
- side longi.: 300×11+90×16 built-up (T-Type)
- vert. stiffener thick.: 11mm
- weld leg length: 6mm

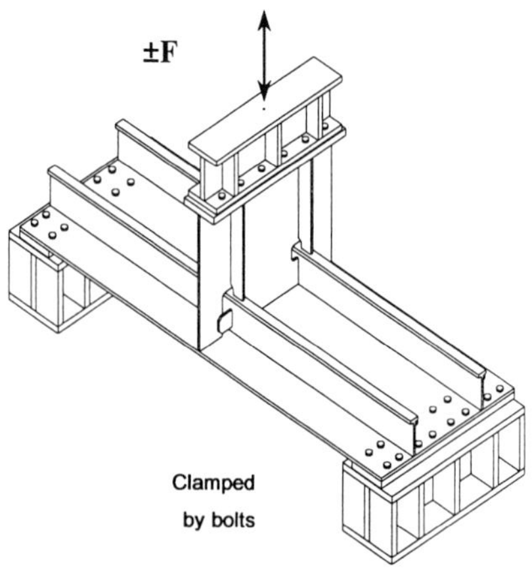

Figure 7: A schematic of the test model

Figure 7 shows a schematic of the test model, where the both ends are clamped by bolts to the test floor. Load-controlled cyclic loading is applied using a 25-ton hydraulic actuator to an auxiliary structure installed on the top of the web frame in order to apply an equivalent load to side pressure loading. The models include two longitudinals, ranging a half-to-half web frame space, and were made from mild steel according to the standard welding procedure. Three different magnitudes of fully-reversed cyclic load (±180kN, ±150kN, and ±120kN) were applied sinusoidally at about 7Hz for each type of models, aiming at fatigue lives between 10^5 and 2×10^6 cycles. Strain gages were installed at the stress concentrated areas to measure hot spot stresses. Figures 8 and 9 show a view of the test model with a different direction.

Figure 8: An overall view of the test model

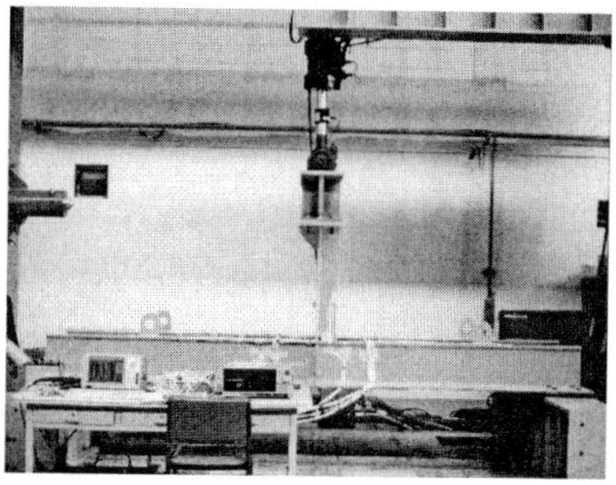

Figure 9: A side view of the test model

For the test models, finite element analyses were also done for a load of 100kN and calculated hot spot stresses are listed in TABLE 2. The overall trend is similar with that of the double hull side shell structure mentioned previously; however, the ratios of the stress levels changed more or less. In order to confirm the equivalence of the concentrated test loading with the uniform side pressure loading, stress distributions on the face plate of the longitudinal were checked as shown in Figs. 10 and 11, where the overall distributions are very similar with those of the previous analysis results for the pressure loading of Figs. 5 and 6. TABLE 3 shows comparison of the hot spot stresses measured from the strain gages and calculated from the analysis for locations A-1, A-2, T-1, and T-2 in the two (L:left and R:right) longitudinals. The ratio of the measured stresses to the calculated stresses ranges about 0.8~1.45. In TABLE 3, L1 indicates the location 1 (See TABLE 1) of the left-side longitudinal in the test model and R2 the location 2 of the right-side longitudinal.

TABLE 2

CALCULATED HOT SPOT STRESSES FOR 100KN IN THE TEST MODEL (MPA)

Type	Angle-Type		T-Type			
Location	A-1	A-2	T-1	T-2	T-3	T-4
Hot Spot Stress	83.7	77.1	74.4	50.5	86.3	8.2

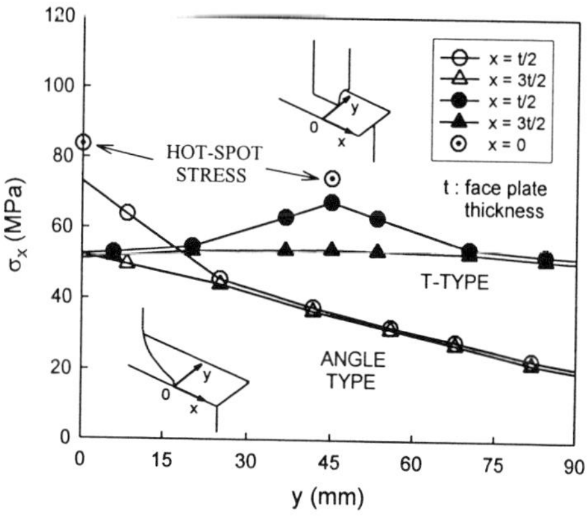

Figure 10: Stress distributions at A-1 and T-1 in the test model (See TABLE 1)

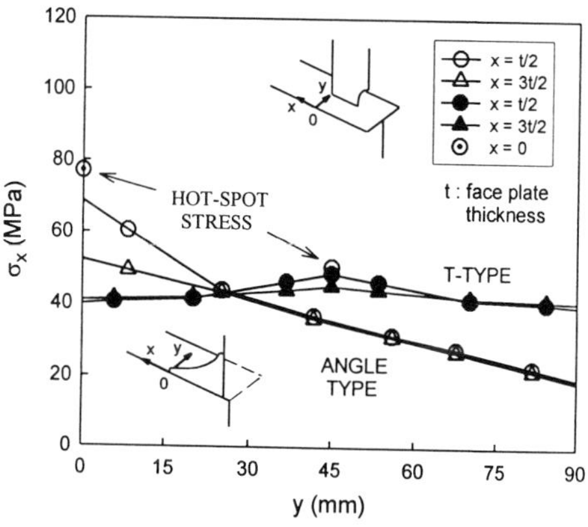

Figure 11: Stress distributions at A-2 and T-2 in the test model (See TABLE 1)

TABLE 3
HOT SPOT STRESSES MEASURED FROM THE TEST MODEL AND CALCULATED FROM FEA

MODEL TYPE		Hot spot stress (MPa) from measurement / FEA for 100kN			
		L1	L2	R1	R2
Angle Type	Model 1	83.0 / 83.7	73.8 / 77.1	81.6 / 83.7	75.7 / 77.1
	Model 2	68.7 / 83.7	75.0 / 77.1	72.5 / 83.7	75.0 / 77.1
	Model 3	74.3 / 83.7	70.2 / 77.1	72.3 / 83.7	68.0 / 77.1
T-Type	Model 4	79.7 / 74.4	69.1 / 50.5	78.1 / 74.4	67.9 / 50.5
	Model 5	76.8 / 74.4	73.8 / 50.5	69.9 / 74.4	73.6 / 50.5
	Model 6	70.7 / 74.4	73.7 / 50.5	74.8 / 74.4	67.6 / 50.5

The fatigue tests were periodically stopped to examine crack initiation and to measure the surface crack length by the penetrant testing. All of the cracks occurred at the weld toes, i.e., hot spot locations. In most cases, after the first crack initiated at one weld toe, one to three more cracks also initiated at other weld toes and grew together with the first one. From the crack growth data of Figs. 12 to 15, it can be clearly seen that crack growth rates in Angle-Type are higher than those in T-Type, and cracks of T-Type grew steadily with almost the same growth rate. With lower load (120kN), the difference of the crack initiation and growth patterns between the two types is clearer (Fig. 12); T-Type showed much superior fatigue performance. The tests continued until a leading surface crack propagated across the face plate. Figure 15 is the results for leading cracks of the six test models. After testing, several cracked sections were examined to see the crack shape. Two typical crack shapes are given in Figs. 16 (Angle-Type) and 17 (T-Type), where the crack shapes look semi-elliptical very much.

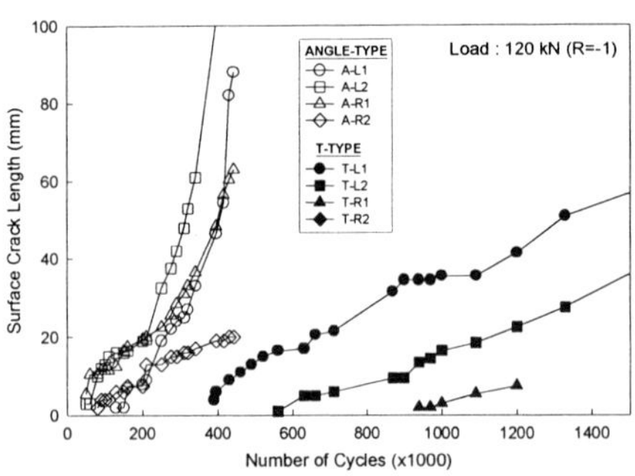

Figure 12: Crack length vs. cycles (120kN)

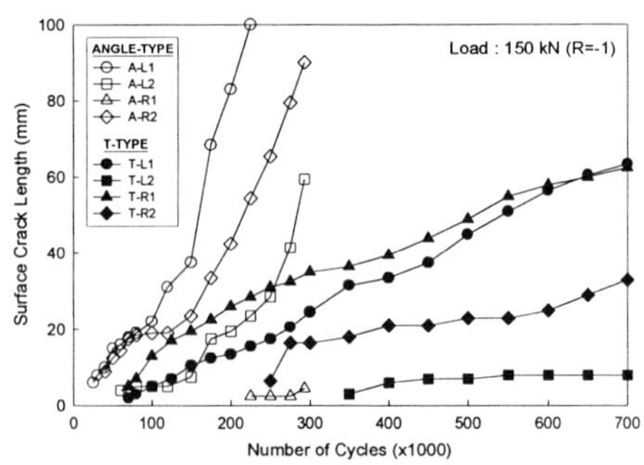

Figure 13: Crack length vs. cycles (150kN)

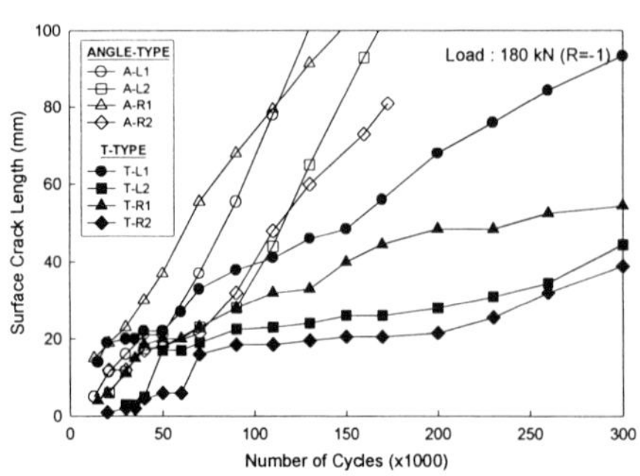

Figure 14: Crack length vs. cycles (180kN)

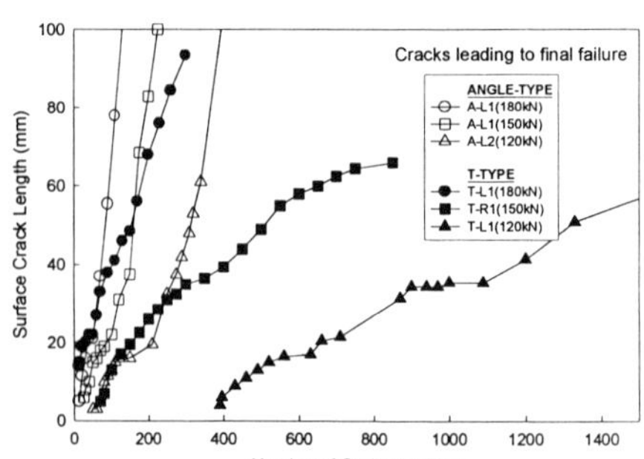

Figure 15: Crack length vs. cycles (Leading cracks)

Figure 16: Crack shape in Angle-Type

Figure 17: Crack shape in T-Type

In this testing, final failure of the test model was regarded as 50mm surface crack length. Data of the measured hot spot stresses vs. fatigue lives for all of the cracks initiated are plotted in Fig. 18, where the broken and solid lines are best-fitted lines to the data of Angle-Type and T-Type, respectively. Data for 15mm crack length are also given in the figure as information of initiation lives. It can be seen that the T-Type structure endured about 2~4 times more lives than the Angle-Type one for the criterion of 50mm crack length. Overall, T-Type shows better fatigue performance than Angle-Type with lower stresses. Figure 19 shows the results for the hot spot stresses calculated from the structural analysis, where the difference of the fatigue lives of the two types reduced more or less; however, T-type is still slightly superior to Angle-Type for 50mm cracks.

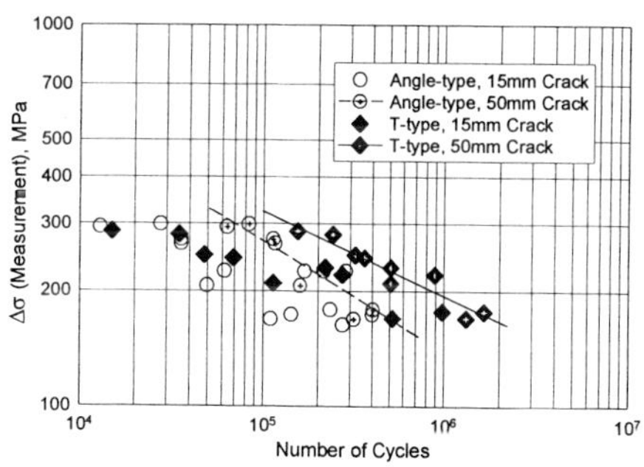

Figure 18: Measured stresses vs. lives

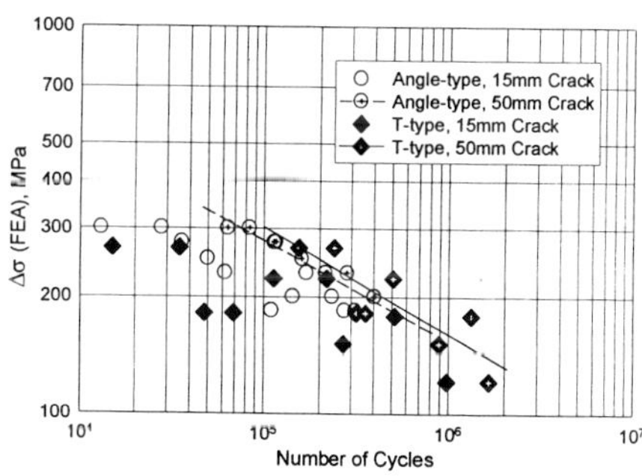

Figure 19: Calculated stresses vs. lives

4 CONCLUSIONS

Fatigue strength of two selective types of side longitudinals were compared from full-scale fatigue testing: inverted angle-type longitudinal supported by a soft-toed web stiffener with a soft-toed backing bracket (Angle-Type) and T-type longitudinal supported by a flat bar web stiffener without a backing bracket (T-Type). T-Type was found to be superior to Angle-Type with regard to fatigue performance under the same loading in the test cycle range between 5×10^4 and 2×10^6 cycles.

References

Hansen P.F. and Winterstein S.R. (1995). Fatigue damage in the side shells of ships. *Marin Structures* **8,** 631-655

Det Norske Veritas. (1998). *Fatigue Assessment of Ship Structures*, DNV CN No.30.7

Practical Design of Ships and Other Floating Structures
You-Sheng Wu, Wei-Cheng Cui and Guo-Jun Zhou (Eds)

FATIGUE BEHAVIOUR OF DIFFERENT BRACKET CONNECTIONS

H. Paetzold [1], O. Doerk [1] and H. Kierkegaard [2]

[1] Department of Ship Structural Design and Analysis, Technical University Hamburg-Harburg, Lämmersieth 90, 22305 Hamburg, Germany
[2] Odense Steel Shipyard LTD, P.O. Box 176, 5100 Odense C, Denmark

ABSTRACT

The paper presents fatigue tests with large scale models and analysis results of a European R&D project, called FatHTS. Tested were frame corners with three different bracket details: with symmetrical flange, with symmetrical flange and additional bracket nose modification, and with one-sided flange with sniped ends. In order to be able to make a comparison, the hot spot strain was determined by strain gauge measurements at all corners of each model. The FE results lead to higher hot spot strains and also the strain increases earlier in front of the hot spot. This is due to the different weld geometry in the drawing and in reality. In most cases the crack initiation did not occur as expected at the hot spot. The crack started at the root and propagated through the weld. Additional stress analyses revealed the cause of the unexpected crack behaviour. This was due to the fact that the weld carries the full load. The fatigue tests showed that it is difficult to predict the root failure in complex structures. The hot spot concept does not cover the metallurgical notch which occurs during fabrication. For toe failure good agreement is obtained between prediction and tests if the failure is defined as through-thickness crack.

KEYWORDS

Welded structure, Large scale fatigue test, Hot spot stress, Fabrication effect, Root crack

1 INTRODUCTION

Modern ocean-going vessels are large all-welded structures of individual design. Improved knowledge of load sequences and extended means for stress analysis and fatigue life prediction have led to a quick progress towards a rational assessment of the fatigue strength of ship structures, which is reflected by pertinent rules of Ship Classification Societies. Fatigue tests with structural components assist in validating calculation models. They are especially important in case of unconventional structural design, new materials or fabrication processes.

Within the European research project FatHTS which was presented by Janssen (2000), an analysis of possible application of high tensile steel was carried out for different ship types in cooperation with a number of shipyards. Focal point was to raise the permitted design stress in order to increase the economy of present and future ship types. However, the fatigue strength of welded steel structures is largely independent of the material strength. Therefore, proof of sufficient fatigue strength is of great importance in structures made of high strength steel. For this reason frame corners were tested because they are a typical structural details of ships. The following describes the experiences gained from the large scale fatigue tests and the application of the hot spot stress approach and also describes the expected and unexpected cracks which occurred during the tests.

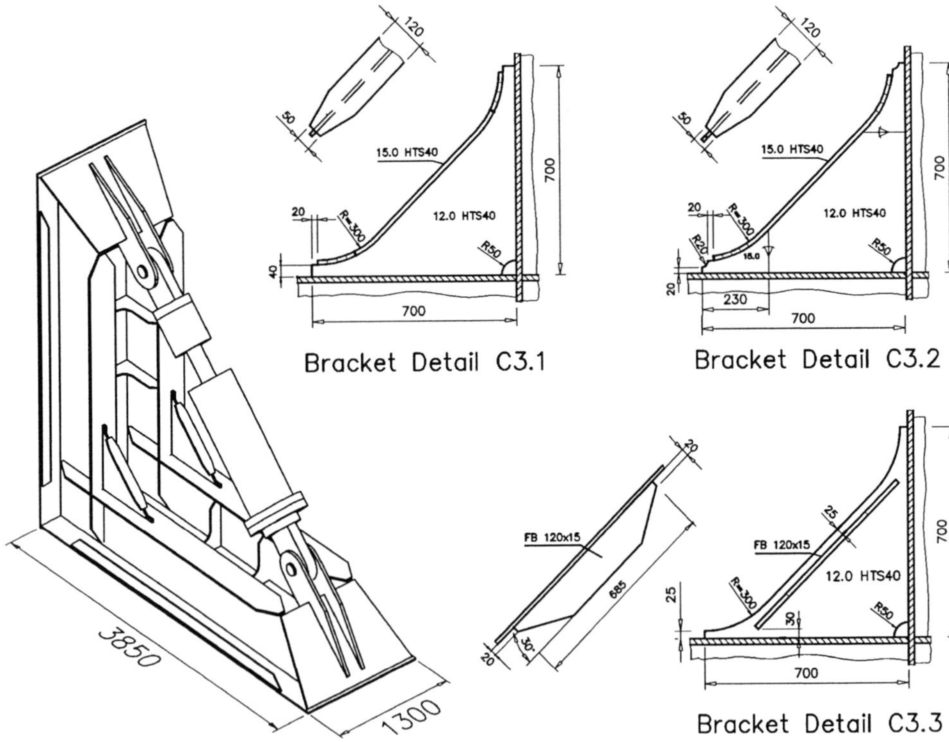

Figure 1: Test arrangement and detail design of the different bracket types

2 DESCRIPTION OF THE MODELS INVESTIGATED

The double symmetric design of the modelss leads to four critical details at each model. Fig. 1 shows the principal test arrangement. The models are loaded via a diagonally acting hydraulic cylinder. The load creates mainly bending and shear stresses in the I-beams below the brackets together with a relatively small axial force. Altogether three different bracket types, called C3.1, C3.2 and C3.3 were tested. To enable an easy comparison of the fatigue behaviour of the different bracket types an identical construction of the test rig has been used for all models. Due to this, the models differ only in the type of the bracket. The models consist of high tensile steel (WELDOX 420) and have been fabricated by Odense Steel Shipyard. Static tests and fatigue tests have been carried out at the Technical University Hamburg-Harburg.

Fig. 1 shows the different bracket types in detail. While the flange is located symmetrically on the free edge of brackets C3.1 and C3.2, the flange is arranged at the side of the bracket C3.3 in a non-symmetric way. Types C3.1 and C3.2 differ only in the shape of the bracket nose. In order to reduce stress concentration, the nose of type C3.2 is softened with the help of a radial cut-out and additionally the thickness of the bracket is increased from 12 to 15 mm in an area of 230 mm from the bracket nose.

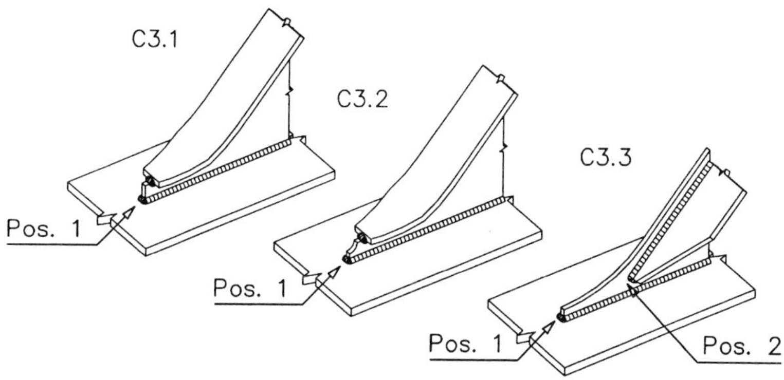

Figure 2: Hot spot positions at the different bracket types

3 RESULTS OF STRESS ANALYSIS AND MEASUREMENT

In order to avoid large plastic deformations of the models the static pre-tests have been carried out with a cylinder force of 100 kN resulting in tensile stress in the bracket. The load was the same for all models. After some load cycles stabilising the material behaviour, the strains have been measured by the application of strain gauges. Fig. 2 shows the hot spots of the different bracket types. The measured strains at the hot spots of the models have been compared with the results of finite element calculations. Fig. 3 shows the finite element mesh of the bracket type C3.1. With respect to the symmetric shape only one quarter of the specimen has been modelled. According to the hot spot concept described by Petershagen et al. (1991) the finite element meshes consist of 20 node solid elements with a reduced integration scheme. Welds at hot spots have been implemented in the finite element models with their nominal dimensions. A quadratic extrapolation of the surface stresses perpendicular to the weld, at reference points located 2.5t, 1.5t and 0.5t away from the weld toe (t is the plate thickness) has been used for the calculation of the hot spot stress at the weld toe. The same procedure has been used for the calculation of hot spot strains. A comparative study, where the influence of different modelling techniques and extrapolation procedures has been investigated, is presented in ISSC (2000).

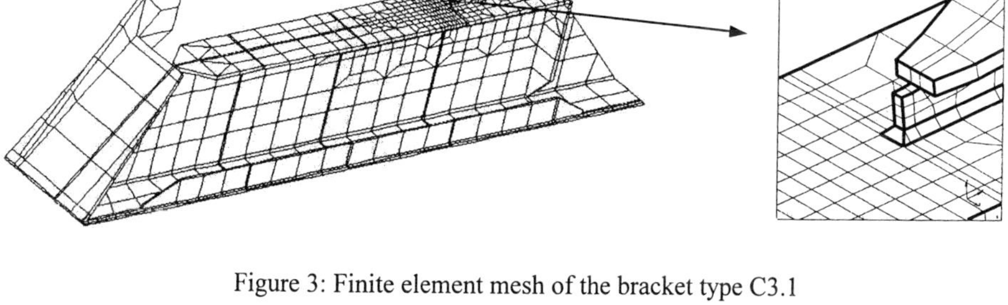

Figure 3: Finite element mesh of the bracket type C3.1

In general the finite element calculations led to higher stresses respectively strains at the hot spot locations than the measurements and also a higher stress respectively strain increase in front of the hot spot. Since a good agreement between the finite element results and the measured values was observed in far distance from the welds, the influence of the welds in the finite element models and in reality was investigated in more detail. Measurement of the weld dimensions at the different models showed that the weld dimensions are in reality much bigger than the nominal weld dimensions in the drawing. Additionally a different weld shape was observed in reality. Fig. 4 shows the measured and calculated longitudinal strains in front of the bracket nose of the C3.1 bracket. When the weld is implemented in the finite element model with its actual dimensions, a very good agreement with the measured values is obtained.

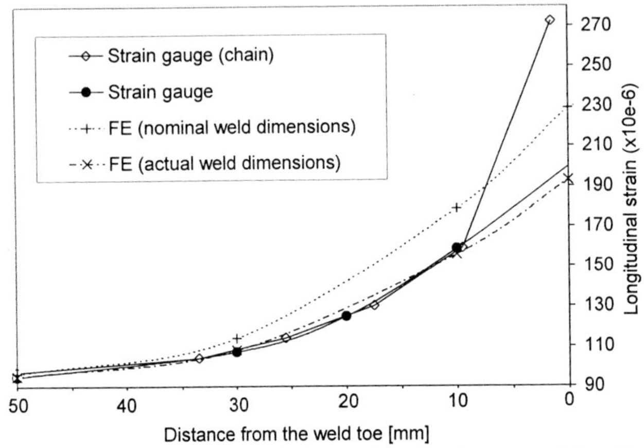

Figure 4: Measured and calculated strains in front of the bracket nose of the C3.1 bracket type

Based on the nominal weld dimensions it is possible to compare the fatigue behaviour of the different bracket types with the help of the hot spot approach. The stresses at the hot spots of the different bracket types, obtained by finite element calculations, are shown in Table 1.

TABLE 1

CALCULATED STRESSES AT THE HOT SPOTS OF THE DIFFERENT BRACKET TYPES FOR 100 KN

Bracket type	Stress 2.5t away from weld toe [MPa]	Stress 1.5t away from weld toe [MPa]	Stress 0.5t away from weld toe [MPa]	Hot spot stress [MPa]
C3.1	18.97	22.27	38.63	**51.70**
C3.2	18.89	22.26	37.70	**49.95**
C3.3 Pos.1	19.00	20.44	28.47	**34.96**
C3.3 Pos.2	58.72	70.67	90.34	**103.07**

The lowest hot spot stress occurs at bracket type C3.2 for which the highest fatigue life of the three analysed bracket types would be expected. The hot spot stress of bracket type C3.1 is only slightly higher, but the fabrication of this bracket seems to be much easier because there is no radial cut-out and the plate thickness is not increased in the nose area. Due to their similar shape these bracket types have the same hot spot location at the weld toe on the flange in front of the bracket nose. Bracket type C3.3 has the highest hot spot stress. However, the most critical hot spot is located at the weld toe in front of the side-attached flange. Additional bending due to the eccentricity is responsible for this. On the other hand this bracket type has the lowest fabrication costs of the investigated bracket types because the flange does not have to be fitted to the shape of the bracket plate.

4 RESULTS OF THE FATIGUE TESTS

The fatigue tests have been carried out under pulsating load with constant amplitude ($R = 0$). In total three models have been tested, one of each bracket type. During the tests unexpected cracks have been observed. After a very small number of load cycles root cracking occurred at the C3.1 model. Fig. 5 shows the root cracks in a transverse section, close to the bracket nose starting at the weld root.

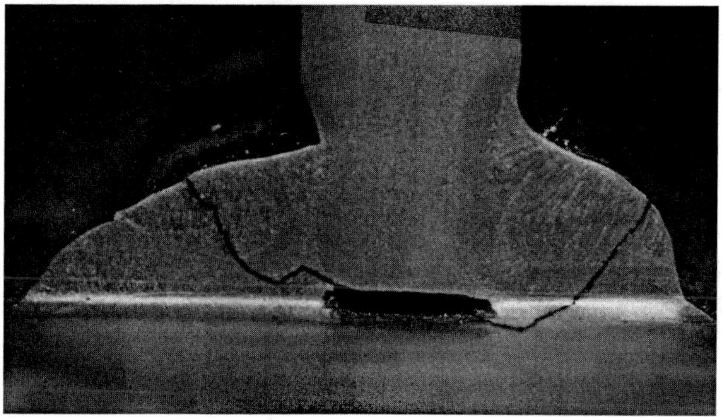

Figure 5: Root cracks at the C3.1 model after the fatigue test

Due to this the fillet welds at the bracket nose of this model have been replaced by full penetration welds and afterwards the model has been tested again. In the following, this model is called C3.11. As expected the cracks started in the additional test at the weld toe on the flange.

The fatigue test of the C3.2 model showed two simultaneous crack initiations at the weld toe on the flange and at the weld root.

Although the hot spot stress at the weld toe in front of the side-attached flange is much bigger than the hot spot stress in front of the bracket nose, the fatigue test of the C3.3 model showed a crack initiation at the weld toe on the flange in front of the bracket nose (Pos.1).

4.1 Fatigue Strength Assessment

In order to enable a comparison between the measured and predicted fatigue behaviour based on the hot spot concept the visible crack lengths and the appertaining number of load cycles N_{TEST} for the different bracket types, when the test was stopped, as well as the calculated number of load cycles until failure N_{CAL} are presented in Table 2. According to Hobbacher (1996) an S-N curve with a negative slope of $m = 3$ and a reference point $\Delta\sigma_R = 100$ MPa for the stress range at $N = 2 \cdot 10^6$ load cycles has been used for calculating N_{CAL}. The theoretical hot spot stress ranges $\Delta\sigma_{HSCAL}$ have been calculated using the values in Table 1. It should be noticed that all models except the C3.11 and the C3.3 model showed a crack behaviour which is not covered by the hot spot concept, which explains the non-conservative life prediction.

Although the C3.3 model shows the highest calculated stress concentration, its fatigue behaviour during the test was good. The observed crack occurred at Pos.1, although the stress at Pos.2 is about three times higher. The test of the C3.11 model where the crack started at the predicted weld toes showed a good agreement with the results obtained by the hot spot stress concept when the failure is defined as a through-thickness crack. In contrast to small scale models a reasonable definition of the failure criteria has big influence on the fatigue strength assessment. If the fatigue behaviour is influenced or determined by root cracking (Models C3.1 and C3.2) the measured fatigue life is significantly lower then the calculated fatigue life.

TABLE 2
RESULTS OF THE FATIGUE TESTS FOR THE DIFFERENT BRACKET TYPE

Bracket type	C3.1	C3.11	C3.2	C3.3
Weld type at bracket nose	Fillet weld	Full penetration weld	Fillet weld	Fillet weld
Test load [kN]	560	547	528	222
$\Delta\sigma_{HSCAL}$ [MPa]	289.5	282.8	263.7	228.8 (Pos.2)
N_{CAL}	82413	88429	109024	166945
N_{TEST}	4000	192100	61500	900000
Measured crack lengths [mm]	Through weld throat thickness	123, 94, 65, 41 (at 4 bracket noses)	54, 42, 28, 22 (at 4 bracket noses)	36 (at 1 bracket nose)
Crack start location	Weld root at the bracket nose	Weld toe at the bracket nose (Pos.1)	Simultaniously at weld root and weld toe at bracket nose	Weld toe at the bracket nose (Pos.1)

4.2 Weld Root Crack Assessment

It is principally not possible to analyse cracks starting at the weld root with the hot spot concept. Normally the gap at the weld root of fillet welds is not implemented in a coarse finite element model for a fatigue calculation based on the hot spot concept. For this reason realistic stresses at the weld root can not be extracted from those models. Additionally the element size at the hot spot, which is about plate thickness, seems to be too large to catch local effects of the gap at the weld root. Besides this, the actual shape of the gap is usually not known.

Generally it is possible to cover cracks starting at the weld root with the notch stress concept according to Radaj (1990), but the very high modelling and calculation costs makes an application of this concept not practicable, with the exception of cases where plane strain or plane stress conditions can be assumed.

Nevertheless it is an important task to cover this crack behaviour because in many cases the service life of a structure where this type of crack behaviour occurs is much lower than in cases where a crack starting at the weld toe is responsible for the fatigue strength of a structure. A simple concept which covers the root crack problem in complex 3D structures is desirable.

5 CONCLUSIONS

The following conclusions can be drawn from the described large scale fatigue tests:

- Good agreement is obtained between prediction and tests for toe failure if the failure is defined as a crack having propagated through the plate.
- Tests showed that cracks may occur at locations not predicted by the hot spot approach
- Especially in welds carrying high loads cracks may appear at imperfections and/or root gaps.

- The following weld details play an important part:
  - Full penetration or fillet weld with gap

- One pass of weld or more
- Welding sequence
- Fabrication quality

These metallurgical and geometrical effects are not covered by hot spot stress calculations.

- It is difficult to predict the root crack in welded complex structures e.g. at the toes of brackets.

Acknowledgement

The investigations have been carried out within the European research project FatHTS in the years 1996 - 2000. The European Union under the Industrial and Materials Technology Programme Brite Euram III funded this project. The partners were: ÅF/Industritechnik (S), Bureau Veritas (F), Chalmers University of Technology (S), Chantiers de l'Atlantique (F), Fincantieri (I), Germanischer Lloyd (D), Instituto Superior Técnico (P), Lisnave (P), Odense Steel Shipyard (DK), RINA (I), Royal Schelde Shipyard (NL), Technische Universität Hamburg-Harburg (D) and VTT (FI). The project was coordinated by TNO.

References

Hobbacher A. (1996). *Fatigue Design of Welded Joints and Components*, Abington Publishing, Cambridge (UK).

ISSC (2000). Finite Element Comparative Study of Ship Structural Detail. Proceedings of the 14th International Ship and Offshore Structures Congress, Volume 1, 176-182.

Janssen G.T.M. (2000). Fatigue Based Design Rules for the Application of High Tensile Steel in Ships. Proceedings of the Seventh International Marine Design Conference, Kyongju, 317-328.

Petershagen H., Fricke W., Massel T. (1991). Application of the Local Approach to the Fatigue Strength Assessment of Welded Structures in Ships. IIW-Document XIII-1409-91.

Radaj D. (1990). *Design and Analysis of Fatigue-Resistent Welded Structures,* Abington Publishing, Cambridge (UK).

1 INTRODUCTION

Four large-scale tests have been carried out in the framework of the Brite/Euram project "**Fat**igue based design rules for the application of **H**igh **T**ensile **S**teels in ships (**FatHTS**)" [1, 2].

The specimens represent a corner connection of ship structure and are fabricated by Royal Schelde in Flushing, the Netherlands. The specimens are 3.23 m high, 3.23 m long and 1.2 m wide (Figure 1).The material used is a high tensile steel, with the following specifications:

yield strength:	390 MPa
tensile strength:	510 - 560 MPa
elongation A5:	20 %
impact energy:	27 J, transversal, at T = -40 °C

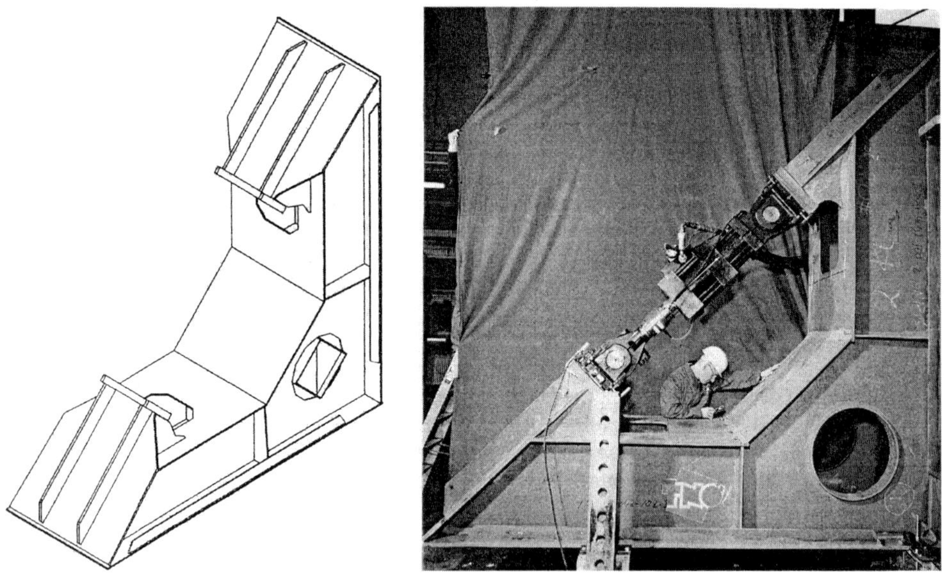

Figure 1: Test specimen (left 3D drawing; right laboratory set up at TNO Building and Construction Research)

The specimens were loaded by a hydraulic jack between the two legs of the specimen.
Four specimens were tested with the following geometrical specifications (Figure 2) and test conditions (specimens mentioned in order of testing):

- specimen 2CA a welded detail with open mouse holes tested with a constant amplitude load
- specimen 3CA a welded detail with filled mouse holes tested with a constant amplitude load
- specimen 1CA a bent plate with open mouse holes tested with a constant amplitude load
- specimen 2VA a welded detail with open mouse holes tested with a variable amplitude load

The nominal cross-section is given in Figure 3. The knuckle detail is at the top side in the 1100 mm wide plate.

The fatigue load applied to the specimens varied between tension and compression (R = -1). The constant amplitude load varied between Fmax = +200kN to Fmin = –200kN (ΔF = 400 kN) at a frequency of 3 HZ. The basis of the variable amplitude loading was the sequence "C-spectr.s14"

supplied by VTT (Finland). This sequence was modified to 46422 load levels (peaks and troughs). The maximum load level in this sequence was +500 kN and the minimum load level was –500kN. A rainflow counting on these load levels was carried out. An equivalent load range ΔF_{equi} was calculated from the variable load ranges assuming Miners rule and a slope of the S-N curve m = 3.
The resulting ΔF_{equi} for the whole spectrum is 412 kN, which is close to the constant amplitude fatigue load range of 400 kN.

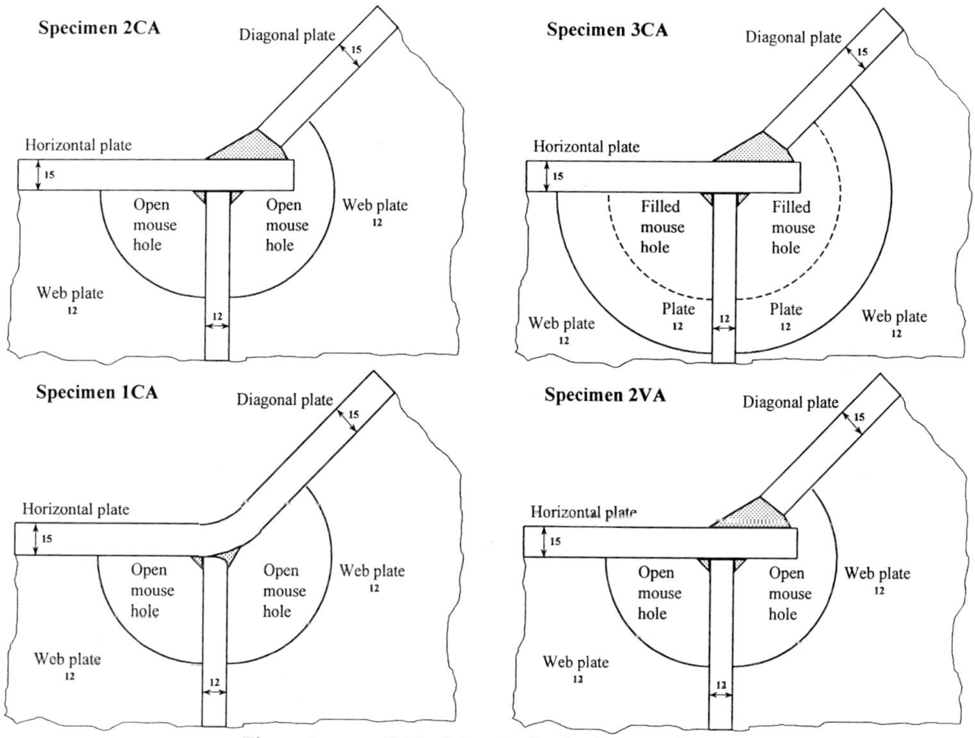

Figure 2: Critical details for the four specimens

2 TEST RESULTS

2.1 Specimen 2CA

Specimen 2CA has a welded joint with open mouse holes (Figure 2).

The specimen was statically loaded before the fatigue test with a load range of 400 kN (between -200 kN and +200 kN). The results across the horizontal plate shows a concentration at the web location (see Figure 4). To the weld toe the measured strain range increased to a relatively high value (see Figure 5).

The specimen was subjected to a fatigue load between –200 kN and +200 kN. The first fatigue crack, located at the weld in the open mouse hole, was discovered after 125700 cycles. Further cracks initiated at all the mouse hole locations and at the weld toes in the butt welds. All the mouse hole cracks grow through the plate (for further crack development see Figure 6 ; cracks are alphabetically numbered in order of initiation).

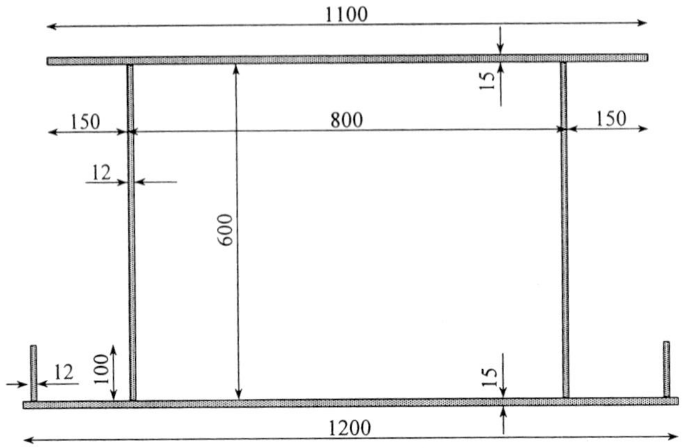

Nominal cross-section

Figure 3: Cross-section at knuckle detail

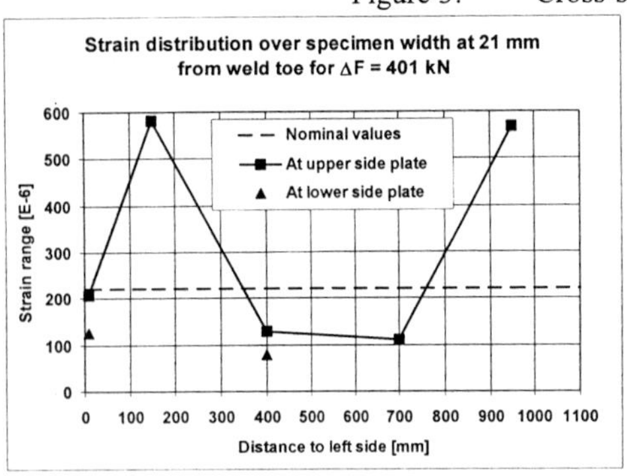

Figure 4: Strain distribution in horizontal plate

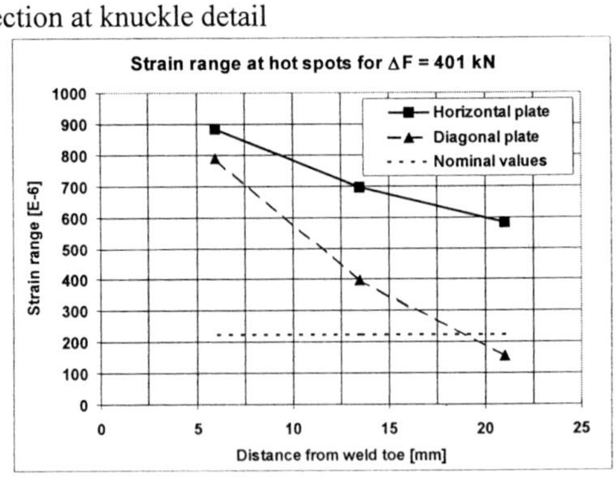

Figure 5: Strain distribution near weld toe at web locations

The test was stopped after 3.77 million cycles, when the specimen was cracked over the whole width of the specimen (crack C-J in Figure 6). At the left side of the specimen the main crack was initiated at the edge of the mouse hole in the diagonal plate and at the right side initiation of the main crack took place at the weld toe in the horizontal plate.

2.2 Specimen 3CA

Specimen 3CA has a welded joint with closed mouse holes with a collar plate (Figure 2). The strain measurements showed lower values compared to specimen 2CA.

The specimen was loaded with a fatigue load between -200 kN and + 200 kN.

The first crack, located at the weld in the inserted plate at the filled mouse hole, was discovered after 552900 cycles. Further cracks initiated at all these locations and at the weld toe of the butt weld. All the mouse hole cracks grow through the diagonal plate.

The test was stopped after 3.81 million cycles, when the specimen was cracked over the whole width of the specimen. The total crack was at the weld toe in the horizontal plate.

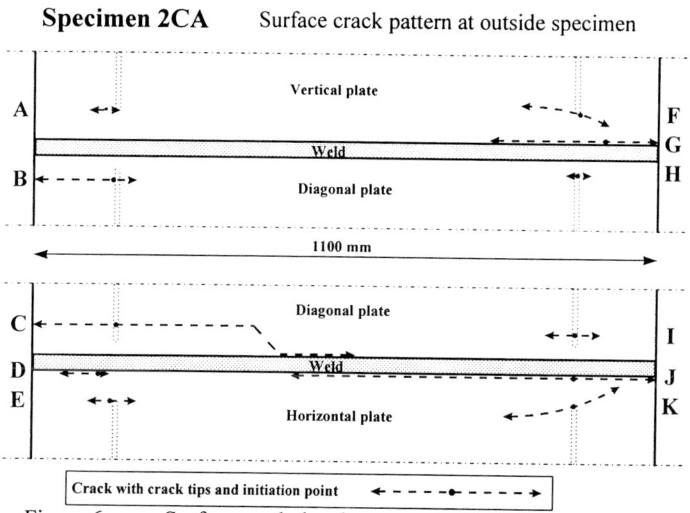

Figure 6: Surface crack development at outside of specimen 2CA

2.3 Specimen 1CA

Specimen 1CA has a bent plate with an open mouse holes (Figure 2). The strain measurements showed values between specimen 2CA and 3CA.

The specimen was loaded with a fatigue load between -200 kN and + 200 kN.

The first crack, located at the weld in the open mouse hole, was discovered after 203000 cycles. Further cracks initiated at all these locations. All the mouse hole cracks grow through the diagonal plate

The test was stopped after 9.07 million cycles, when the specimen was cracked over the whole width of the specimen. The total crack was in the diagonal plate at the lower side. Due to the absence of the weld no crack was initiated in the bent area.

2.4 Specimen 2VA

Specimen 2VA has a welded joint with open mouse holes (Figure 2). The geometry is the same as for specimen 2CA.

The specimen was loaded with a variable amplitude fatigue load (see above).

The first crack, located at the weld in the house hole at the lower detail, was discovered after 2500 cycles. Further cracks initiated at the other lower mouse hole and at the weld toes of the butt weld. All the two mouse hole cracks grow through the diagonal plate.

The test was stopped after 1.66 million cycles, when the specimen was cracked over the whole width of the specimen. The total crack was at the weld toe in the vertical plate.

3 EVALUATION AND CONCLUSIONS

The test results are summarised in TABLE 1.

TABLE 1
SUMMARY OF TEST RESULTS

Specimen	2 CA	3 CA	1 CA	2 VA 1)
Knuckle detail	welded	welded	bent plate	welded
Mouse hole	open	closed	open	open
Load range	400 kN	400 kN	400 kN	412 kN
Total number of cycles	$3.77 * 10^6$	$3.81 * 10^6$	$9.07 * 10^6$	$1.66 * 10^6$
Calculated nominal strain range	$222 * 10^{-6}$	$222 * 10^{-6}$	$222 * 10^{-6}$	$222 * 10^{-6}$
max. strain range in hor. or vert. plate at 21 mm from weld toe	581	505	500	649
max. strain range in horizontal plate at 6 mm from weld toe	884	668	776	832
extrapolated strain range in horizontal plate	993	737	961	912
max. strain range in diagonal plate at 21 mm from weld toe	155	306	52	179
max. strain range in diagonal plate at 6 mm from weld toe	790	626	724	767
extrapolated strain range in diagonal plate	1019	745	1153	1003
approximate number of cycles at outside surface crack length of 100 mm	$550*10^3$	$950*10^3$	$1800*10^3$	$< 350*10^3$
approximate number of cycles at outside surface crack length of 250 mm	$2000*10^3$	$2400*10^3$	$4500*10^3$	$630*10^3$

1) Variable amplitude load on 2VA with equivalent load range of 412 kN
Measured strain ranges for 2VA are at load range of 400 kN

From the constant amplitude tests it can be concluded that the fatigue life of a bent plate detail is more than two times the life of a welded detail.
Closing the mouse hole with a plate does not have a beneficial effect on the fatigue life (compare specimen 2CA with 3CA).
The fatigue life of the variable amplitude loaded specimen (2VA) is only 44% of the life of the constant amplitude loaded specimen (2CA). From the small increase in equivalent load range one would expect a decrease in life to 92%. So, the fatigue life is approximately half of the fatigue life predicted from the constant amplitude test.

The crack growth curves of all the four specimens show a long period of stable crack growth (Figure 7).

The main conclusions and recommendations from the tests and the evaluations of the results are:

a) The specimens show an early crack initiation at the mouse hole location and a long period of stable crack growth. So, in practice there is a long period in which a growing crack can be discovered by inspections.
b) The specimen with the mouse hole filled with an eccentric collar plate connected with fillet weld has the same fatigue life as a specimen with open mouse holes.

c) The specimen with a bent plate detail has a 2.4 times longer fatigue life than a specimen with a welded detail. This longer life is due to the absence of the weld toe (a local stress raiser) in the bent detail.
d) The specimen loaded with a variable amplitude loading has only 50% of the lifetime of the comparable specimen loaded with a constant amplitude loading. One reason for this difference could be unstable crack growth during a high load in the random sequence. Another reason could be crack closure effects in these R = -1.0 tests.
e) A comparison with S-N curves from design guides show that designing a large scale geometry with stress concentrations using the hot spot stress method leads to a very conservative estimation of the fatigue life (5% of the experimental life).
f) The influence of variable amplitude in R = -1 load situations has to be investigated further.
g) The lifetime prediction of large scale geometries with stress concentrations and redundant load paths has to be improved by additional testing and theoretical investigations.

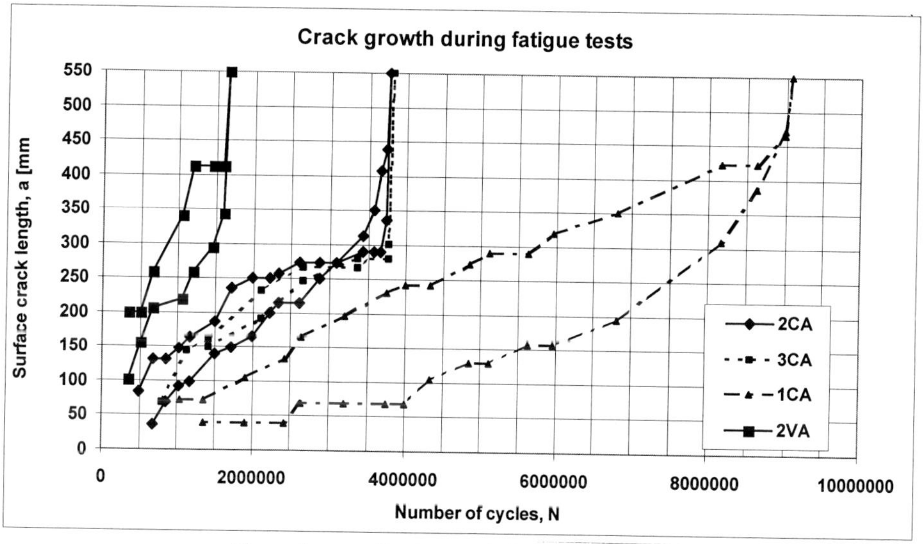

Figure 7: Crack growth during the fatigue tests

References

[1] Janssen G.T.M. (2000). Fatigue based design rules for the application of high tensile steel in ships. *Poc. 7th Int. Marine Design Conference*, 317-328, May 21-24, Kyongju, Korea.

[2] Dijkstra, O.D., Janssen, G.T.M. and Ludolphy. J.W.L. (1999) Fatigue tests on large scale knuckle C2 specimens. TNO report 1999-CON-BiS/R5023/1

Practical Design of Ships and Other Floating Structures
You-Sheng Wu, Wei-Cheng Cui and Guo-Jun Zhou (Eds)

THE PRE-FABRICATED HULL DETAILS FOR APPLICATION IN DESIGN AND REPAIR

S. V. Petinov

Department of Structural Mechanics of Ships, Marine Technical University St.Petersburg, 190008 St.Petersburg, Russia

ABSTRACT

Many of structural details such as bracket connections, shell stiffener-transverse web assemblies, etc., are essentially susceptible to fatigue due to superposition of the stress concentration and the weld endings specific by the feasible weld imperfections and high residual welding stress.

In certain applications the fatigue performance of designed and repaired details may be improved in conjunction with the current practice of repair by implementation of the fatigue designed pre-fabricated details. The principle of the pre-fabricated details is to provide the parent material in the areas of stress concentration and to avoid the coincidence of high stress concentration and weld ending in one location. Transfer of the extensive stress flow via the parent material in modified elements offers essentially better control of fatigue reliability of hull details in design and in service by reducing effects of uncertainties attributed to the production and to the database of fatigue performance of typified welded joints. It also provides application of the local strain approach for evaluation of the necessary configuration of critical locations.

KEYWORDS

Stress concentration, Fatigue properties, Fatigue design of hull details, Pre-fabricated details, Critical locations, Weld endings, Local strain approach.

1 GENERAL

The frequent structural details such as bracket connections, shell stiffener-transverse web assemblies, etc., are generally prone to fatigue due to superposition of the stress concentration and the weld endings specific by the unsteady weld profile, feasible weld imperfections and high residual welding stress. Fatigue strength assessment of such locations based on the present rules reveals unsteady depending on the analyst experience and influenced by uncertainties of the current methodology and database. In conjunction, these factors can not safely provide the necessary fatigue properties of hull details.

The conventional and preferred practice of repair of the affected structural details is to crop and part renew the damaged elements without considerable alteration of the initial detail design and without assessing the fatigue life which may be attained. The repair may suggest introducing additional elements to provide an auxiliary load transfer, etc., however, the coincidence of the weld endings and stress concentrations remains usually unchanged in renewed detail. Fig.1 illustrates the typical of structural detail coincidence of stress concentration and weld endings.

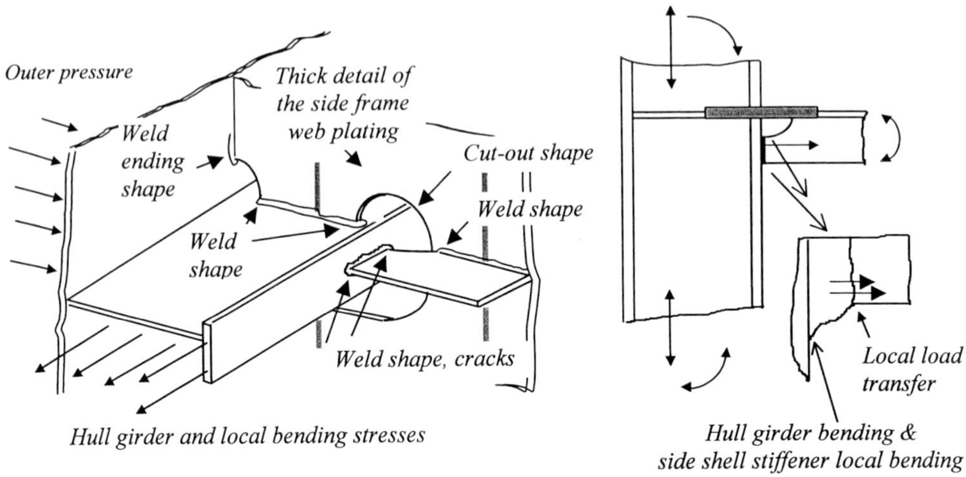

Figure 1: Example of feasible critical locations and fatigue affecting parameters: The side longitudinal-side frame web connection

In certain applications the fatigue performance of designed and repaired details may be improved in conjunction with the current practice of repair by implementation of the fatigue-designed pre-fabricated details.

The principal idea of pre-fabricated details is to provide parent material in the areas of stress concentration and to avoid the coincidence of high stress concentration and weld ending in one location. Transfer of the extensive stress flow via the parent material in modified elements offers essentially better control of fatigue reliability of hull details in design and in service by reducing effects of uncertainties attributed to the production and to the database of fatigue performance of typified welded joints. The parent material in areas with stress concentration provides favorable conditions for successful application of the local strain method, which is physically more correct, than the stress-life approach.

It also makes feasible application of different methods of fabrication of critical elements: the cast and pressed elements optimized for the stress flow may be applied, especially when aluminium alloys are the hull material.

2 EXAMPLES OF MODIFIED DETAILS

Application of the suggested technology may be illustrated referring to the example of repair of the deck beam web given in (IACS, 1995). The damaged, cracked web at the slot (Area 3, example 7) has to be repaired by cropping and part renewing the web; also, the lug between web and deck longitudinal is recommended, etc. The Guidelines indicate the possible causes of damage: local stress concentration due to shear deformation of the web, deficient welds.

A version of repair (and design) may be suggested considering the above principles. The repaired detail (to be renewed) may be designed for the necessary residual fatigue life and fabricated of several elements. Essential features of the design are the introduction of a thick web plate element, which includes the "heading" part of the slot – to increase the shear stiffness and to reduce the stress concentration in it at the slot (as shown in Fig.1 with solid gray lines). Another is the estimation of the curvature of the slot so that together with the local increased web thickness it would provide the stress concentration factor value not greater than required by fatigue analysis. The procedure of evaluation of the allowable stress concentration factor using the local strain approach is given in (Ma, et al, 1998).

As an example, a transverse web – shell stiffener connection was analyzed according to the data given by Debbink (1997). The shear mode and local load transfer were considered using local FE model designed according to the basic structure and assuming local thick detail of the web (1.5 of the basic thickness). Comparative analysis shows, that the stress concentration factor at the slot may be decreased by 1.4, extending the detail fatigue life 3.8 times.
When the stiffness of a detail is sufficient, and fatigue damage may be caused by the combined effects of weld ending and high stress concentration, a possible solution might be achieved by transferring the weld ending away from the stress concentration area. The problem of providing the stress "flow" with necessary elements may be solved by applying inserted pressed or cast details, the shape of which can be respectively optimized considering the modes of deformation.
Fig.2 shows finite-element model of a bracket detail in the web and shell stiffener connection. Such details are characterized by combined high stress concentration and the weld ending in one location at the bracket ending. When the longitudinal loading is considered, the stress concentration factor due to combined effects of the bracket shape and the weld geometry may be of around 2.5 – 4.0 depending upon the weld metal profile. Inserted one-piece detail (indicated with solid gray lines) provides the smooth transition of the stiffener flange into the connection detail and continuous material at the stress concentration locus. Approximate optimization of shape of the inserted detail (Fig.2) allows decreasing the stress concentration factor to 1.5 level.
It is essential, the detail geometry may be identical in production in many locations, with identical fatigue performance under identical loading conditions. Furthermore, the detail shape can be found based on required fatigue life, e.g., through evaluation of the allowable stress concentration factor.

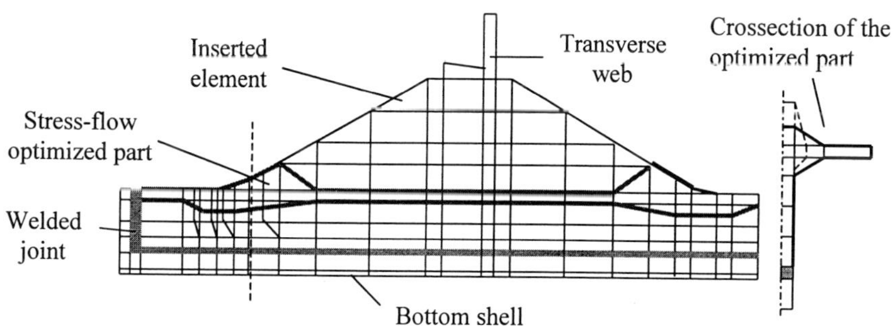

Figure 2: Shell stiffener bracket connection FE model

3 PRINCIPLES OF FATIGUE DESIGN

With respect to the fatigue design of pre-fabricated details it is appropriate to explain why the local strain approach should be preferred. Firstly, the basic components of the stress-life approach should be discussed. Despite certain differentiation, the basic principles of fatigue model of hull details are common in the rules. Analyzing these principles from the fatigue mechanics standpoint, the following deficiencies of the S-N approach should be indicated:

- the stress amplitude controls the fatigue process indirectly: in the initiation phase – via development of the inhomogeneous microplastic and plastic deformation, and in the crack growth phase – in the complicated form, via the effective crack opening conditions and transfer of fracture through the material structure

- the stress-life criterion corresponds to the design principle according which the stress (nominal) in structural members has to be less than the elasticity limit stress. Respectively, the design S-N curves are related to the intermediate and high-cycle ranges of fatigue; the low-cycle range ($N < 10^4$) is excluded from the database. Meanwhile, when the linear damage summation rule is applied, the S-N diagrams are automatically extrapolated into the low-cycle range, which is not consistent with the elastic-plastic behavior of material within this range
- the correspondence of the S-N failure criterion, including the crack origination and crack propagation until failure of a test-piece, and of the damage state in actual structure is not considered
- the latter is essential for the purposes of fatigue analyses and design of hull details: the initiation phase in hull structural detail develops under control of local conditions, whereas the crack propagation progressively reflects the changes of a specimen stiffness in the course of crack growth
- the loading conditions in laboratory tests are usually the load-range or the maximum-minimum load control, whereas in hull structures the loading conditions are those of displacement control due to the redundancy of most structural details
- application of the hot-spot technique or of a method based on multiplying the elastic stress concentration factors assume elastic material deformation, whereas the fatigue process may be associated solely with microplastic and plastic deformation, even at stress amplitudes less than fatigue limit stress.

The local strain method based on the strain-life criterion is free from the mentioned deficiencies, although it forms also an approximate model of fatigue through common uncertainties (e.g. uncertain loading history, material structure and microplastic deformation random inhomogeneity, etc.). It operates with elastic-plastic behavior of material and is supported by the necessary experimentally obtained database. The advantage of the local strain concept application may be explained by the following fundamental reasons:

- The material failure criterion relates fatigue life of material to the local strain range, which controls almost directly the fatigue process in the crack initiation phase and during the early crack growth under influence of the local conditions
- The fatigue analysis of welded details when the strain-life criterion is applied, would need mostly in parent material and heat-affected zone material failure criteria (Testin et al, 1987), since cracks typically originate in either of materials at critical locations
- The strain-life criterion is obtained in tests under the strain range or displacement range control, which approximately corresponds to the loading conditions of most details in hull structure and provides approximately the identity of fatigue damage in laboratory specimen and at the critical location in hull detail
- Since the approach uses basics of the elastic-plastic analysis, it enables mechanically correct assessment of notch conditions for fatigue damage.

When applying the inelastic strain range – life criterion, an analyst would need in the local strain at the critical location. This may be obtained using one of several approaches: a method based on the Neuber heuristic rule, a method which employs empirical conditional cyclic (load range-strain range) diagram, FE elastic-plastic cyclic analysis. These methods operate with experimentally obtained generalized cyclic diagrams, and usually a stabilized diagram is applied in analyses (Petinov, 1990; Ellyin, 1997, etc.). When the strain-life criterion is applied, the FE modeling would need in higher resolution of the stress field at critical location. The hot-spot stress concept requires the FE mesh design in which the finite element dimensions are equal to the structural detail, flange, shell, etc., thickness (Niemi and Partanen, 1998). The local strain concept assumes the element dimensions (and element type) compatible with the stress gradients at the critical location. The concept also allows application of the vast stress concentration data collected in handbooks, e.g. Peterson (1973).

4 THE LOCAL STRAIN APPROACH-BASED FATIGUE ANALYSIS

To demonstrate efficiency of the local strain approach (LSA), an example of the case study given in the ISSC-2000 documents (Committee, 2000) may be used. The Committee experts analyzed fatigue behavior of a welded attachment at the longitudinal coaming of a Panamax containership.

Since the purpose of the present demonstration is to display the elements of the approach, the loading history of the ship analyzed may be approximated by appropriate selection of the long-term Weibull nominal stress distribution. Assuming the world-wide operation profile, the shape parameter for the long-term stress distribution may be taken equal to unity (in case the only North Atlantic region should be considered, the shape parameter may be approximated as 1.1 – 1.2 since the ship is large and fast). The scale parameter may be varied in the range: $A_S = 6 - 9$, MPa, and the corresponding maximum stress ranges (related to the probability to occur in the range of 10^{-8} ... 2×10^{-8}) result in, respectively, $S_{max} = 212 - 319$, MPa (matching the range of the data presented in the Committee report).

For simplicity of the analysis the feasible effects of the SW load conditions may be neglected. However, the local strain fatigue model as seen from the below readily allows considering effects of the SW loading component in assessing the local strain: to do it, one needs in detailed information on the operation profile characteristics[1].

The material (HT36 steel) fatigue characteristics should be given in the form of strain-life criterion parameters and the cyclic stress-strain diagram. The simplest criterion form is given by the Coffin-Tavernelli (1962) equation:

$$\Delta\varepsilon = CN^{-\alpha} + 2\sigma_{-1}/E \qquad (1)$$

the parameters of which for the above steel were obtained experimentally (Petinov, 1976): C = 0.400 and $\alpha = 0.65(4)$; the fatigue limit stress amplitude for the heat-affected zone (HAZ) material[2], rolling skin removed, is around 90 MPa. To consider effects of the random loading, the Gassner-Seeger correction has to be assumed, here - equal to 0.8 (Heuler & Seeger, 1986). Respectively, the fatigue limit stress amplitude is reduced to 72 MPa.

The local strain at the weld toe may be estimated approximately, using a conditional stabilized stress-strain diagram, a version of the Ramberg-Osgood formula derived based on testing of hull steel specimens with smooth stress concentrations, $K_t \leq 4$ (Petinov, 1990):

$$\Delta\varepsilon = (K_t \Delta\sigma_n / E)(1 + M_s(r,t) F(\sigma_m)(AK_t \Delta\sigma_n / \sigma_c - 1)^{\beta}) \qquad (2)$$

where $\Delta\sigma_n$ is the nominal (or any reference) stress range, A and β are empirical parameters, which for the mentioned steel are: $A = 1.25$, and $\beta = 1.20$; $M_s(r,t)$ is the correction of the plastic part of the total strain to account for effects of the notch size, of the stress state with respect to a provisional structural parameter (a),

$$M_s(r,t) = M_0 \Big/ \left(1 + (a/t)^s (r/t)^q\right)$$

where M_0, a, s and q are empirical parameters, $M_0 = 0.155$, $a = 0.24$ (mm), $s = 0.30$ and $q = 1.50$ for a low-alloy steel (HT36) and $M_0 = 0.107$, $F(\sigma_m)$ is the correction for the influence of the mean stress on the plastic strain range, neglected in the present analysis,

---

[1] The influence of the SW loads on the local strain decreases after application of relatively high variable loads due to relaxation of the mean stress

[2] The crack initiation is assumed in the heat-affected zone at the weld toe

$$F(\sigma_m)=1+B\sigma_m/(K_t\Delta\sigma_n)$$

where σ_m is the nominal mean stress, and B is an empirical parameter, $B = 2.50$.

The fatigue life is calculated using linear damage summation-based formula (Petinov, 1990):

$$N=(CE/\sigma_{-1})^{1/\alpha}/\int p(\sigma)(E\Delta\varepsilon/\sigma_{-1}-2K_t/K_f)^{1/\alpha}\,d\sigma \quad (3)$$

where K_f is the fatigue notch factor, applied in the local strain format to approximately modify the failure criterion when the high-cycle fatigue is considered. The Peterson's formula (Peterson, 1973) may be used in case of stress concentrations with relatively small notch root radii, when fatigue properties of welded structural details should be assessed

$$K_f=1+(K_t-1)/(1+g/r) \quad (4)$$

where r is the notch root (weld toe) radius, g is the so-called material structural parameter, a value of the order of highly stressed zone size at a notch root. Extension of Peterson's data (Peterson, 1973) on ship hull steels of 235 - 390 grades allows to define the structural parameter as

$$g=0.38(350/\sigma_b)^{1.16} \quad (5)$$

with σ_b - the ultimate strength, MPa.

It was suggested that non-sailing time was about 0.3 of the total service life (20 years), the SW load conditions were not considered in fatigue life calculations. The 1-mm weld toe radius was assumed in (4). The stress concentration factor values (theoretical SCF, related to elastic material behavior, weld shape and loading mode – longitudinal tension-compression) were varied in the range 1.3 – 2.3 (the smaller figure may correspond to smooth transition, the bigger – to the bulging weld material). This range encircles the Committee data, however, the actual weld profile-based FE model and respective analysis would be desirable. The results, fatigue life, yrs, in dependence on K_t and A_S are given in Table in the below.

Comparing the present results and those given in Table 1 (Committee, 2000), one may assume the obvious correspondence: the present results cover the range of fatigue lives calculated by the Committee experts.

However, the LSA provides strict correspondence of varied parameters and fatigue lives, in contrast to the Committee experts' findings, where the hot-spot stress-based stress concentration factors estimated occurred in the range 1.63 – 2.15 with respective scatter of fatigue lives.
Analysis also indicates that fatigue life reduction is feasible at stress concentration $K_t > 1.7$ and severe sea conditions ($A_S > 8$ MPa). Insignificant modification of the weld profile at the critical location, e.g. by the weld toe dressing and respective increasing the weld toe radius may provide the necessary fatigue life of the component.

To conclude, it should be noted, that the strain-life fatigue model can be applied in the format of the allowable stress concentration factor and the joint geometry assessment (Ma, et al, 1998), which may provide the necessary information for the detail design and fabrication procedure.

TABLE 1
FATIGUE LIFE, OBTAINED USING THE LOCAL STRAIN METHOD, YEARS

Scale parameter, A_s , MPa	Stress concentration factor, K_t					
	1.3	1.5	1.7	1.9	2.1	2.3
6	>20	>20	>20	>20	>20	20.4
7	>20	>20	>20	24.7	12.5	7.0
8	>20	>20	19.8	9.6	5.1	3.0
9	>20	19.5	8.7	4.3	2.3	1.6

References

Committee III-2 "Fatigue and Fracture" (2000) *Report, 14 International Ship and Offshore Structures Congress (ISSC-2000).* Nagasaki, Japan, October

Coffin L.F. and Tavernelli J.F. (1962) Experimental Support for Generalized Equation Predicting Low-Cycle Fatigue. *Trans. ASME*, **Ser. D, 4,** 533-538

Debbink M. (1997) Hull Structural Design for a 40,000 DWT Double-Hull Products Carrier. *Prevention of Fracture in Ship Structure. Symposium and Workshop.* SSC Report PFSS95, Washington, pp. 247-263

Ellyin F. (1997) *Fatigue Damage, Crack Growth and Life Prediction,* Chapman & Hall, London

Heuler P. and Seeger T.A. (1986) Criterion for Omission of Variable Amplitude Loading Histories. *Intern. Journal of Fatigue*, **8: 4**, 225-230

Ma T., Petinov S.V. and Thayamballi A.K. (1998) Toward the Automated Determination of Geometry for a Target Fatigue Life. *The Eighth International Offshore and Polar Engineering Symposium ISOPE-98,* Montreal, Canada, May 24-29

Niemi E. and Partanen T. (1998) Fatigue Analysis and Resistance of Welded Steel Components based on the Hot Spot Stress Approach. *International Shipbuilding Conference, St. Petersburg, Proceedings ISC'98,* **Sec. C**, 137-144

Peterson R.E. (1973) *Stress Concentration Factors. A Handbook.* J.Wiley & Sons, New York

Petinov S.V. (1976) Crack Initiation Period of Fatigue and Strain Criterion-based Prediction of Structure Fatigue Life. *Report SK/R-35, Division of Ship Structures,* NTH, Trondheim

Petinov S.V. (1990) *Basics of Engineering Analysis of Fatigue of Ship Structures,* Sudostroenie Pubs, Leningrad

Testin R.A., et al (1987) Predicting the Fatigue Resistance of Steel Weldments. *Welding Res. Suppl. to the Welding Journal,* Apr., 93-98 S

IACS (1995) *Bulk Carriers: Guidelines for Surveys, Assessment and Repair of Hull Structure.* Witherby & Co. Ltd., London

Practical Design of Ships and Other Floating Structures
You-Sheng Wu, Wei-Cheng Cui and Guo-Jun Zhou (Eds)

FATIGUE STRENGTH OF LOAD-CARRYING BOX FILLET WELDMENT IN SHIP STRUCTURE

Wha Soo Kim [1], Doe Hyun Kim [1], Sang Gab Lee [2], and Yoon Ki Lee [2]

[1] Hull Initial Design Department, Hyundai Heavy Industries Co., Ltd.
1, Cheonha-Dong, Dong-Ku, Ulsan, Korea
[2] Division of Marine Systems Engineering, Korea Maritime University
1, Dongsam-Dong, Yeongdo-Ku, Busan, Korea

ABSTRACT

In case of load-carrying box fillet weldments such as connection of a longitudinal stiffener and a web stiffener in ship structure, welding residual stresses at the weld toe on the longitudinal stiffener are tensile while those on the web stiffener are compressive. The distribution of welding residual stresses and the re-distribution of residual stresses by a pre-load history in the load-carrying box fillet weldment are studied in experimental and analytical ways. The study shows that welding residual stresses near weld toes decrease by a pre-load. The bigger the pre-load becomes, the more the residual stresses decrease. Fatigue tests were carried out to investigate the effect of such a welding residual stress distribution. Three types of crack were observed depending on applied stress ranges due to the effect of the residual stress. Additional fatigue tests were also carried out to investigate the effect of shaken-down residual stresses and tensile mean stresses. The fatigue strength decreases drastically by the tensile mean stress and this decrement is predominant at lower cyclic stress range. However, there is no significant difference of the fatigue strength depending on the conditions of pre-load itself.

KEYWORDS

Load-carrying box fillet weldment, Pre-load, Tensile mean stress, Re-distribution of residual stress, Shaken-down residual stress, Sectioning method, Thermo-elasto-plastic analysis

1 INTRODUCTION

It has been known that there in general exist tensile residual stresses up to yielding point of the material around welded joints. Many research works were carried out to investigate effects of the welding residual stresses, and such residual stresses were known to reduce fatigue strength qualitatively. Most of these research works were, however, concentrated on the investigation of the effects of the initial welding residual stress (IWRS). Only a few research works were performed on the effect of the re-distributed residual stress (RDRS) caused by the actual service condition (Ki et al., 1997). Static loads on ship structure induced either by water pressure before service or by cargo pressure during first laden voyage cause relatively higher stress history at welded joints compared with cyclic loads induced by wave during the service. Due to these high static loads, IWRS at welded joints, where fatigue strength is concerned (in most cases, where stress concentration occurs), is shaken-down in a great extent by the elasto-plastic deformation behavior of the material. Ship structural members are, then, exposed to cyclic loads during the service. Therefore, it is reasonable to assess the fatigue strength of ship structure considering the effect of shaken-down residual stresses.

It was reported that the effect of cyclic stress ratio (minimum stress/maximum stress) was minor on the fatigue strength of structure under the condition of IWRS (Maddox, 1982). This means the mean stress effect on the fatigue strength is minor as far as they are exposed to cyclic loads under as-welded conditions. However, if the ship structure with RDRS from a fairly large static pre-load is exposed to cyclic loads induced by wave, the effect of a certain level of the mean stress associated with static loads of cargo or ballasting would not be minor.

In this paper, IWRS and RDRS by the static pre-load for the load-carrying box fillet weldment, which is a typical weld joint in ship structure, are evaluated using FE analysis and measured by an ordinary sectioning method. Fatigue tests are performed to evaluate the effect of RDRS on the fatigue strength of the weldments using the as-welded specimens and statically pre-loaded ones. Also the effects of the tensile mean stress (TMS) on the fatigue strength of pre-loaded specimens are investigated.

2 STATIC PRE-LOAD APPLIED IN SHIP STRUCTURE

Scantlings of main ship structure are determined in general by the requirements of Classification Societies' rule, and the values of allowable stresses for design static loads are approximately in the range of 50~70% of yield points of materials on nominal stress basis. In ship structure, the stress concentration is inevitable due to structural discontinuity or structural geometry. Considering the stress concentration, it is expected that there are large re-distributions of residual stresses by the design static pre-load at these areas. In most cases, the fatigue damage occurs at these stress-concentrated points. It is, therefore, important to verify the fatigue characteristics related to the RDRS to obtain more reliable fatigue strength of the actual ship structure.

3 DISTRIBUTION OF RESIDUAL STRESS

Details of the specimen of the load-carrying box fillet joint are illustrated in Figure 1. The specimens were fabricated in accordance with the actual shipbuilding workmanship and practice. The material of the specimen was ship structural mild steel of grade A and its major chemical composition and mechanical properties are listed in Table 1. Although the actual yielding stress of the material is about 300 MPa, Classification Societies define the design yield stress (σ_o) as 235 MPa for ship structural mild steel. In this regard, hereinafter, the design yield stress is defined as 235 MPa according to the Classification Societies' specification.

Distributions of the residual stresses normal to the weld toe at the specimens ($\sigma_{x,res}$) under following three types of statically pre-loaded cases, which are listed in Table 2, were evaluated by FE analysis and measured by a sectioning method.

- Case 1: As-welded condition (No pre-load)
- Case 2: Pre-load inducing $0.5\sigma_o$ of nominal stress at weld toe
- Case 3: Pre-load inducing $0.85\sigma_o$ of nominal stress at weld toe

TABLE 1

MAJOR CHEMICAL COMPOSITION (%) AND MECHANICAL PROPERTIES OF MILD STEEL

C	Si	Mn	P
0.13~0.17	0.15~0.18	0.46~0.65	0.012~0.019
	Yield Stress (MPa)	Ultimate Strength (MPa)	Elongation (%)
From Mill Sheet	299~336	441~468	28~30
From Tensile Test	290~299	427~457	34~36

TABLE 2

STATIC PRE-LOAD CONDITIONS

Case	1	2	3
Pre-load (nominal stress)	As-welded	$0.5\sigma_o$ (117.5 MPa)	$0.85\sigma_o$ (199.75 MPa)
Load pattern	0	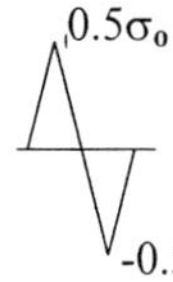 $0.5\sigma_o$, $-0.5\sigma_o$	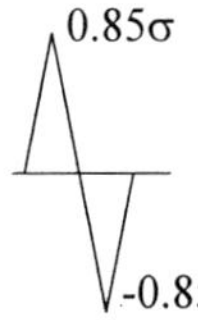 0.85σ, $-0.85\sigma_o$

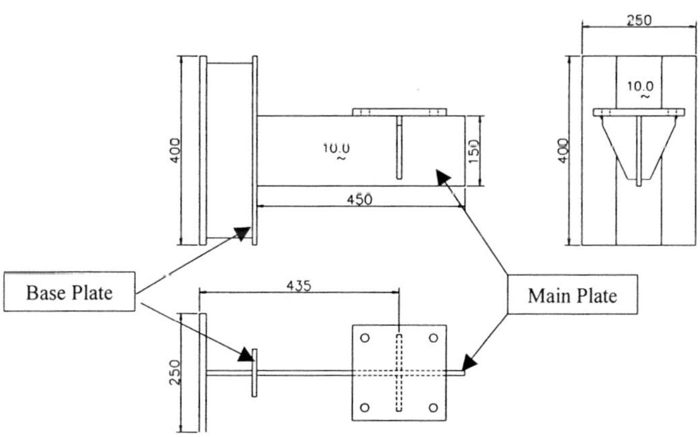

Unit: mm

Figure 1: Shape and size of specimen of load-carrying box fillet weldment

FE analysis with the general-purpose program of ABAQUS was performed to simulate residual stress distributions of the specimen. The FE model is shown in Figure 2. Half of the specimen was used in the model imposing a symmetric boundary condition. The used element type was 8-noded solid and the element size around weld toe was about 2.5×2.5×2.5 mm. The main plate in the model was connected to the base plate by means of weld bead only and the flank angle of weld bead was assumed to be 45°. Von-Mises yield criterion with isotropic hardening rule was applied. The distribution of the temperature was calculated by a transient thermal conduction analysis with respect to time ignoring heat convection and radiation to the air. Using the calculated distributions of the temperature in respect to time, a thermo-elasto-plastic analysis was carried out to simulate the distribution of IWRS. After obtaining the distribution of IWRS, static pre-loads described in Table 2 were applied to simulate RDRS by an elasto-plastic analysis. The force was applied to the loading position as shown in Figure 2.

IWRS and RDRS by pre-loads were measured by using an ordinary sectioning method. As shown in Figure 3, two-dimensional strain gauges with a gauge length of 1 mm were bonded on the main plate. Then, the released strains were measured and converted into the residual stresses by sectioning the main plate around the strain gauges into small cubes.

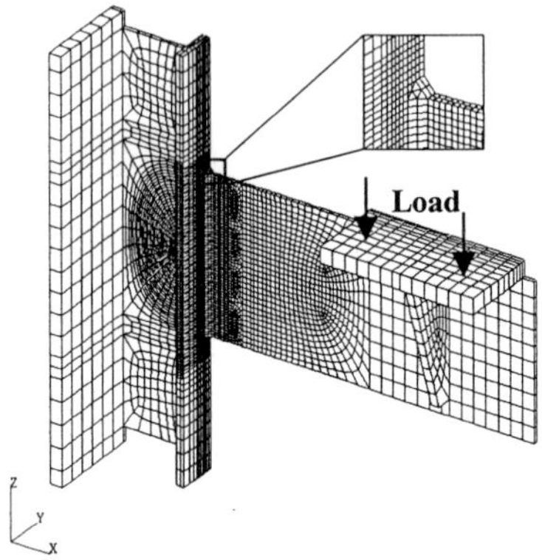

Figure 2: FE model

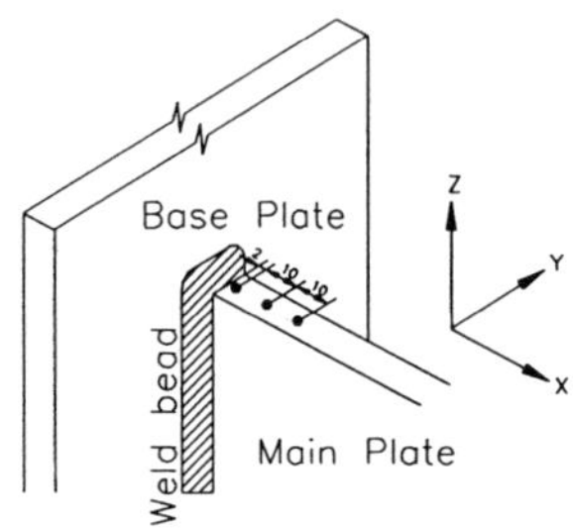

Unit: mm

Figure 3: Locations of strain gauges

Figure 4 shows the results of the FE analysis and the measurements for $\sigma_{x,res}$ distribution in longitudinal direction from the weld toe. According to the results of the FE analysis, IWRS near the weld toe of the base plate ('Position B' in Figure 5) reaches almost the tensile yield stress of the material or more. However, IWRS at the main plate edge ('Position A' in Figure 5) is compressive according to the results of both the measurement and the FE analysis while its magnitude is somewhat different. Both tensile IWRS at 'Position B' and compressive IWRS at 'Position A' is shaken-down by applied pre-loads

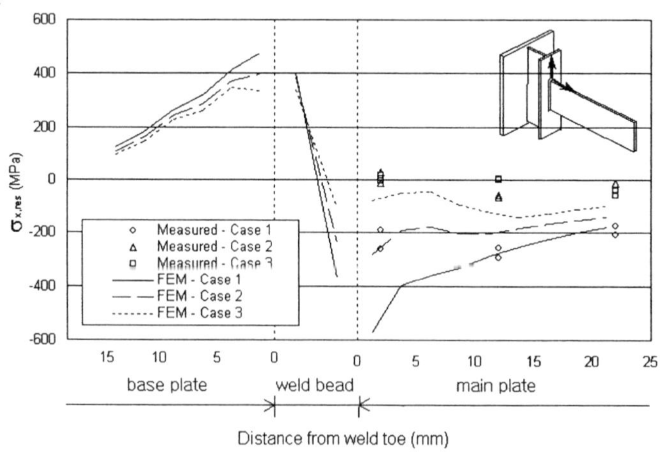

Figure 4: Residual stress distribution at surface

4 FATIGUE TESTS

Fatigue tests were carried out under load-controlled axial loading with the fully reversed constant amplitude at room temperature and in air. Test frequency was in the range of 5 to 8 Hz. Fatigue tests were carried out until around 5×10^6 cycles of loading and stopped unless a fatigue crack was initiated. The fatigue life of the specimen, Nf, is defined as the number of load cycles when a crack length reaches about 20 mm. Fatigue tests under following five conditions were carried out to evaluate the effect of both RDRS by pre-loads and TMS by static loads.

- Condition 1 : Pre-load = 0, Mean stress = 0
- Condition 2 : Pre-load = $0.5\sigma_o$, Mean stress = 0
- Condition 3 : Pre-load = $0.5\sigma_o$, Mean stress = 0
- Condition 4 : Pre-load = $0.85\sigma_o$, Mean stress = $0.5\sigma_o$
- Condition 5 : Pre-load = $0.85\sigma_o$, Mean stress = $0.85\sigma_o$

Three types of fatigue cracks were initiated and propagated as shown in Figure 5. Fatigue test results under the condition 1, 2 and 3 are shown in Figure 6, which represent the relation between nominal stress range of ΔS at 'Position A' and fatigue life of Nf. The test results for the crack occurred at weld root were discarded. Some fatigue test results showed mixed type of crack initiation at two or three points simultaneously or progressively. In these cases, the results were based on the major crack. It is expected that the crack would occur at the point where the value of applied stress is the highest due to the stress concentration. Figure 7 illustrates the stress distribution at the specimen when unit load is applied. 'Position A' is the most vulnerable to crack. It was observed, however, that the crack occurred at 'Position B' under the condition 1 (as-welded condition) when the cyclic load of relatively lower stress range was applied. This is due to the effect of IWRS. IWRS at 'Position A' are compressive and this makes it hard to initiate the crack at lower stress range. Instead, the crack occurred first at 'Position B' where IWRS was tensile, although the applied stress values were lower than those at 'Position A'. A pre-load reduces both compressive IWRS at 'Position A' and tensile IWRS at 'Position B'. The fatigue test results under the condition 3 show that the most cracks occurred at 'Position A' when the high pre-load applied, as expected, because the effect of IWRS would be reduced.

Fatigue test results for the cracks occurred at 'Position A' under each conditions are plotted in Figure 8. This figure represents the relation between nominal stress range of ΔS at 'Position A' and fatigue life of Nf. There is no significant difference of the fatigue strength by the different conditions of pre-load. This seems to be mainly due to the compressive residual stress. However, there are significant differences of fatigue strength by TMS and the differences are predominant if the stress range becomes lower. For reference only, mean S-N curve of 'G' class in BS5400 (1980), that is known to be applicable to evaluate the fatigue strength of this type weldment, is also drawn in the figure. There is significant difference of the fatigue strength between the test results and the BS5400 'G' curve

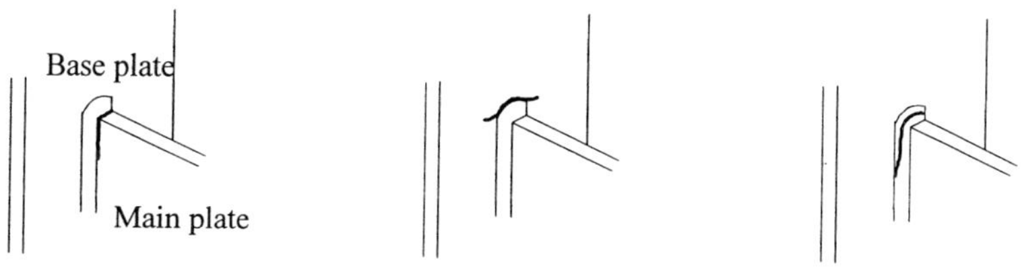

Crack at weld toe of main plate ('Position A') Crack at weld toe of base plate ('Position B') Crack from weld

Figure 5: Crack initiated points during fatigue tests

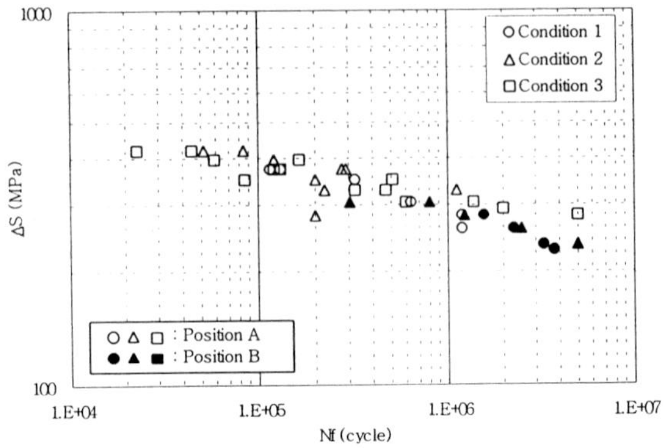

Figure 6: Fatigue test results under the condition 1, 2 and 3

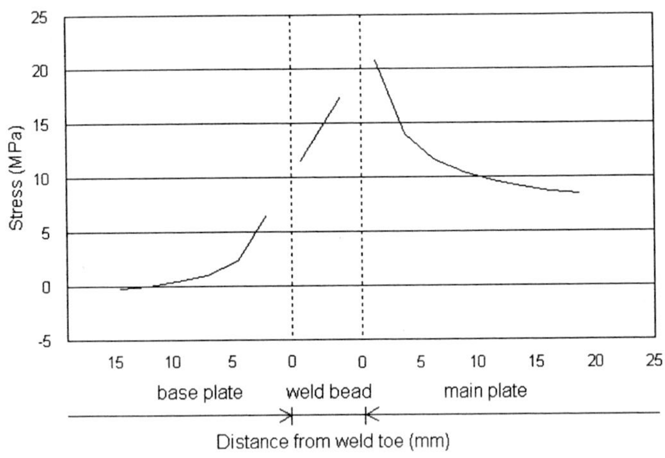

Figure 7: Stress distribution when unit load of 1 kN is applied

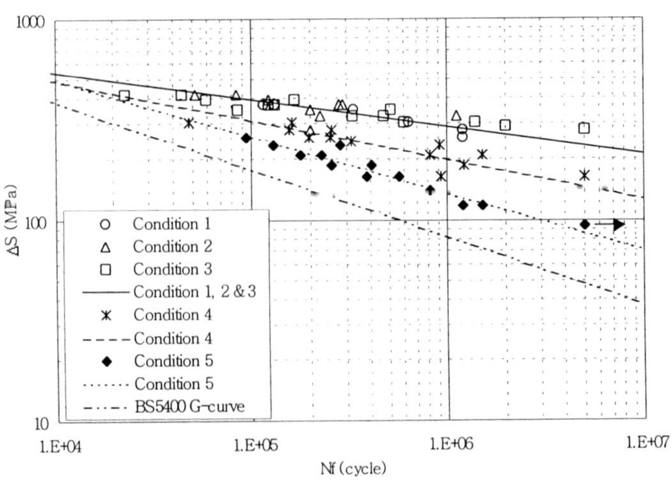

Figure 8: ΔS-Nf curves for 'Position A'

5 CONCLUDING REMARKS

The fatigue strength were investigated with the specimens of load-carrying box fillet weldment. To investigate the re-distribution of the initial welding residual stress (IWRS) due to static loading on ship structure, the simulation by FE analysis and the measurement by a sectioning method were carried out applying the pre-loads on specimens. IWRS at the weld toe is shaken-down by the pre-load. The bigger the pre-load becomes, the more the residual stress decreases, according to the results of both the simulation and the measurement.

To investigate the effect of the pre-load and the tensile mean stress (TMS) on fatigue strength, fatigue tests were carried. It has been found that there are significant differences of fatigue strength depending on the existence of TMS when a pre-load has shaken down IWRS. The fatigue strength decreases drastically by TMS and this decrement is predominant at lower cyclic stress range. However, there is no significant difference on the condition of pre-load itself when IWRS is compressive.

The present widely used practice for the assessment of fatigue strength was derived from the research results for other industries such as steel bridges and offshore structures. It should be noted that, not only structural details of geometry and material but also loading patterns in ship structures are different from those of other structures. Therefore further research works are strongly recommended to establish the more reliable S-N diagram for ship structural design.

Acknowledgment

A joint industry project to assess the fatigue capacity of 'Floating Production, Storage and Offloading units' was established by DNV under the name of 'FPSO JIP – Fatigue Capacity'. The objective of this project is to provide data to obtain a reliable design basis that can be used to ensure sufficient fatigue capacity of FPSO units. This is to avoid costly maintenance during in-service life. Eighteen companies including oil companies, classification societies, engineering companies, a governmental organization and shipbuilders joined the JIP. Hyundai Heavy Industries Co., Ltd., as a participant in the JIP, has performed the fatigue tests with small specimens of typical fillet weld joints in ship structure to investigate the effect of mean and residual stress on fatigue strength, which is one of the main tasks in the JIP. Authors express the appreciation to all participants of the JIP for their permission of the publication of fatigue test results.

References

British Standards Institution. (1980). *Steel, Concrete and Composite Bridges, Part 10. Code of Practice for Fatigue.*

Kim W. S., Tomita Y., Hashimoto K. and Osawa N. (1997). Effects of Static Load on Fatigue Strength of Ship Structure. *7th International Offshore and Polar Engineering Conference (ISOPE-97)* **5**, 565-571.

Maddox S. J. (1982). Influence of Tensile Residual Stresses on the Fatigue Behavior of Welded Joints in Steel. *Residual Stress Effects in Fatigue, ASTM STP* **776**, 63-96.

Practical Design of Ships and Other Floating Structures
You-Sheng Wu, Wei-Cheng Cui and Guo-Jun Zhou (Eds)

A VBAR MODEL TO IDENTIFY THE DYNAMIC CHARACTERISTICS OF MARINE STRUCTURES

C. F. Hung; Y. T. Peng and W. J. Ko

Department of Naval Architecture and Ocean Engineering,
National Taiwan University, Taipei, Taiwan., China

ABSTRACT

In this paper, an algorithm to identify the dynamic characteristics (natural frequency, damping ratio, and relative mode shapes) of marine structures from measured displacement, velocity, or acceleration of multiple channels has been developed.

A Vector type of Backward Auto-Regressive model (VBAR) is adopted to describe the relationship between the measured output response in different time steps and channels. A backward type of time-discrete state equation consisted of measured dynamic response is established to determine the physical vibration modes of structures.

The numerical simulation of identification of dynamic characteristics for a 200t ship structure is selected as study cases. A 3D FE-model for hull structure is established, the results of modal analysis are used as reference basis, and the displacements on multiple measured points calculated by transient dynamic analysis are taken as output vector. The natural frequencies, damping ratios, and relative mode shapes between measure points are identified from the measured out put vectors. The results show that the physical modes and spurious modes with noise condition can be separated clearly by the proposed method, and the identified mode parameters are consistent with reference results.

KEYWORDS

Vibration, Ship structures, Backward vector auto-regressive model, System identification, FEM

1 INTRODUCTION

Estimation of vibration mode in design stage to avoid the possible resonance due to periodic loading is the fundamental topic of vibration problem. Since 1950, many researches to examine the dynamical characteristics of marine structures have been carried out, and many simplified approaches to predicate the natural frequency and the valuable design guide have been proposed, e.g. DnV(1985), NK(1984), BV(1987) and Noonan(1990). The FEM enable the engineer to consider every detail structure parts to establish a fine model for a high accuracy level of dynamic analysis, e.g. Carlsen(1977) and Gepner & Ahrens(1985). Because of the complicacies of marine structures, there are a lot of uncertainties existed in theoretical and mathematical model for description of vibration problems. The measurement of vibration behaviors for new developed structure is necessary to examine the effects of the uncertainties on the vibration of structures.

The natural frequency, damping ratio, and mode shape, are the dynamic characteristics of structure, which governs the vibration behaviors of structures. The conventional vibration testing focus on the natural frequencies and response levels. It is not easy to identify the local or global modes, or double frequency for different modes. The modal parameter identification described in Ewin(1986) and Juang(1994) provide a lot of methods to find out the mode parameters from measured input force and output dynamic response. Because the vibration behaviors of marine structure are very complicate, and the background noise is relative high, the separation of system modes and ghost modes is not easy. Cooper(1992) examined the backward predication error model proposed by Kumarsen & Tufts(1982) for modal parameter identification, the results showed the advantage of being able to distinguish system modes from spurious modes. In this paper we proposed an algorithm to use vector type of backward auto-regressive model for identification of the dynamic characteristics of marine structures. We establish a backward state equation from measured output vector, the system modes are determined in this state equation of output vector, and the relative mode shapes are determined in a reduced state equation.

2 EQUATION OF MOTION

The equation of motion for vibrating structure is a second-order differential equation.

$$M\ddot{z}(t) + E\dot{z}(t) + Kz(t) = Fu(t) \tag{1}$$

Where M, E, and K are the mass, damping, and stiffness matrices, respectively; z(t) is the displacement vector; u(t) is the input force vector; and the matrix F specifies the distribution of the input force component related to the measured degree of freedom on model.

An output equation to define the relationship between measured response y(t) and the model response can be expressed as:

$$y(t) = c_a\ddot{z}(t) + c_v\dot{z}(t) + c_d z(t) \tag{2}$$

Where c_d , c_v and c_a are the output influence matrices for displacement, velocity and acceleration, respectively.

The state vector of motion consisted of displacement and velocity is defined as:

$$x(t) = \begin{Bmatrix} z(t) \\ \dot{z}(t) \end{Bmatrix} \tag{3}$$

The equations of motion and the output equation can be transferred into a first-order State equation of motion and state output equation, which are expressed in terms of state vector.

$$\dot{x}(t) = Ax(t) + Bu(t) \tag{4}$$

$$y(t) = Cx(t) + Du(t) \tag{5}$$

Where A, B, C and D are 2n×2n system matrix, 2n×p input matrix, p×2n output influence matrix and p×p transmission matrix, respectively. And they are defined as

$$A = \begin{bmatrix} 0 & I \\ -M^{-1}K & -M^{-1}E \end{bmatrix}, \qquad B = \begin{bmatrix} 0 \\ M^{-1} \end{bmatrix} \qquad (6),(7)$$

$$C = [c_d - c_a M^{-1} K \quad c_v - c_a M^{-1} E] \,, \quad D = c_a M^{-1} \tag{8), (9}$$

3 THE VECTOR BACKWARD AUTO-REGRESSIVE MODEL

For a structure with p-measured output channels, the output vector at current time step t can be expressed as summation of the output vector in next q time steps.

$$y_t = \sum_{i=1}^{q} a_i y_{t+i} + \varepsilon_t \tag{10}$$

Where t is the integer time index. ε_t is a white noise sequence, a_i is the backward auto-regressive (BAR) parameter matrix and the output vector y_t is a time series vector. The BAR parameter matrix carries the pivotal dynamical characteristic of structural system. Following linear matrix equation can be established from equation (10).

$$y = aW + \varepsilon \tag{11}$$

Where y is the output matrix assembled from the time step 1 to N-q. W, ε and a are the data matrix, error matrix and BAR parameter matrix, respectively. These matrices can be expressed as

$$y = [\, y_1 \;\; y_2 \;\; \ldots \;\; y_{N-q} \,]_{p \times N-q} \tag{12}$$

$$W = \begin{bmatrix} y_2 & y_3 & \cdots & y_{N-q+1} \\ y_3 & y_4 & \cdots & y_{N-q+2} \\ \vdots & \vdots & \ddots & \vdots \\ y_{q+1} & y_{q+2} & \cdots & y_N \end{bmatrix}_{pq \times (N-q)} \tag{13}$$

$$a = [\, a_1 \;\; a_2 \;\; \ldots \;\; a_q \,]_{p \times pq} \tag{14}$$

$$\varepsilon = [\, \varepsilon_1 \;\; \varepsilon_2 \;\; \ldots \;\; \varepsilon_{N-q} \,]_{p \times N-q} \tag{15}$$

Where N is the number of total time steps, and pq is the production of channel number p and order q The BAR parameter matrix can be computed by least squares method as

$$a = yW^{+} \tag{16}$$

Where W^{+} is the Moore-Penrose pseudo-inverse of W.
The state vector of output is defined as

$$x_t = \begin{bmatrix} y_t \\ y_{t+1} \\ \vdots \\ y_{t+q-1} \end{bmatrix}_{pq \times 1} \tag{17}$$

The VBAR model can be transformed into a discrete backward state space model as

$$x_t = A_d x_{t+1} + B_d \varepsilon_t \tag{18}$$

$$y_t = C_d x_t \;\; + D_d \varepsilon_t \tag{19}$$

Where A_d, B_d, C_d and D_d are discrete state system matrix, input matrix, output influence matrix and transmission matrix, respectively. And they are expressed as

$$A_d = \begin{bmatrix} a_1 & \cdots & a_{q-1} & a_q \\ I & \cdots & 0 & 0 \\ \vdots & \ddots & \vdots & \vdots \\ 0 & \cdots & I & 0 \end{bmatrix}_{pq\times pq}, \qquad B_d = \begin{bmatrix} I \\ 0 \\ \vdots \\ 0 \end{bmatrix}_{pq\times p} \tag{20), (21}$$

$$C_d = [\, a\, A_d^{-1} \,]_{p\times pq}\,, \quad D_d = [\, I \; \square a\, A_d^{-1}\, B_d \,]_{p\times p} \tag{22),(23}$$

Where the subscript in right side of equations indicates the dimension of corresponding matrix. The main advantage of this approach is that the VBAR model has the ability to automatically separate the physical system mode from non-system mode induced by the numerical and measured noise.

3.1 *Reduction of discrete state space model*

A n-DOFs dynamic system has n physical mode, and the dimension of state equation pq in equation (18) is much higher than 2n. A reduced technique is introduced to obtain the reduced-order system from eq. (18) and (19), which can describe adequately the n-DOFs dynamic system.

The system matrix A_d in equation (18) and (20) can be decomposed as

$$A_d = \Psi \Gamma \Psi^{-1} \tag{24}$$

Where Ψ is the eigenvector matrix consisting of pq eigen-vectors, and Γ is the eigenvalue matrix which is a diagonal matrix consisted of pq eigenvalues.

Only 2n of the pq eigen-values are the system roots, and the 2n roots are n pairs of conjugate complex variables, the system roots located outside a unit circle, and can be appeared repeatedly in pq root. These three key properties of system root play an essential role in extraction of system modes and also in reduction of the discrete backward state space model. The eigen- decomposition method is employed to reduce the size of backward system from pq to 2n. Compared with the separated system and spurious roots, the eigenvector and eigenvalue matrice can be partitioned as

$$\Psi = \begin{bmatrix} \Psi_{rr} & \Psi_{re} \\ \Psi_{er} & \Psi_{ee} \end{bmatrix}, \qquad \Gamma = \begin{bmatrix} \Gamma_r & 0 \\ 0 & \Gamma_e \end{bmatrix} \tag{25), (26}$$

Where subscripts r and e indicate the "retained" mode with regard to the 2n system modes and "excluded" modes with regard to the pq-2n spurious modes, respectively. Hereafter, the matrices C_d and $\Psi^{-1}B_d$ are also partitioned as

$$C_d = [C_r \quad C_e]\,, \quad \Psi^{-1}B_d = \begin{bmatrix} B_r \\ B_e \end{bmatrix} \tag{27), (28}$$

The reduced-order discrete backward state space model is then expressed as

$$x_t = A_r x_{t+1} + B_r \varepsilon_t \tag{29}$$

$$y_t = C_r x_t + D_r \varepsilon_t \tag{30}$$

Where A_r, B_r, C_r and D_r are the 2n×2n reduced state system matrix, 2n×p reduced input matrix, p×2n reduced output matrix, and the p×p reduced transmission matrix, respectively. And

$$A_r = \Psi_{rr}\Gamma_r\Psi_{rr}^{-1}, \qquad B_r = \Psi_{rr}B_r \tag{31), (32}$$
$$C_r = C_r + C_e\Psi_{er}\Psi_{rr}^{-1}, \qquad D_r = D_d \tag{33), (34}$$

The reduced discrete backward model can be transform into forward state space model as

$$x_{t+1} = \overline{A}_r x_t + \overline{B}_r \varepsilon_t \tag{35}$$
$$y_t = \overline{C}_r x_t + \overline{D}_r \varepsilon_t \tag{36}$$

Where $\overline{A}_d$, $\overline{B}_d$, $\overline{C}_d$ and $\overline{D}_d$ are the matrices in forward state.

$$\overline{A}_r = A_r^{-1}, \qquad \overline{B}_r = -A_r^{-1}B_r \tag{37), (38}$$
$$\overline{C}_r = C_r A_r^{-1}, \quad \overline{D}_r = D_r - C_r A_r^{-1}B_r \tag{39), (40}$$

3.2 *Transformation Matrix*

In experimental modal analysis, in general, the available instruments are limited. In this paper, the displacements, velocity and acceleration are available to be measured as output information in the free response or impulse response testing. The state system matrix, consisting of the matrices of mass, damping and stiffness, plays a crucial role in extracting the normal and complex modes. That is to say, normal modes and complex modes can be derived from the eigen-analysis of matrices $M^{-1}K$ and the system matrix A in equation (6), respectively. The pair $<\overline{A}_r, \overline{C}_r>$ in equation (37) and (38), which is reduced from the state equation of measured output vector, can be converted into a pair of continuous state $<\overline{A}, \overline{C}>$ with the following transformation by zero-order-hold technique

$$\overline{A} = \ln(\overline{A}_r)/\Delta t \quad ; \quad \overline{C} = \overline{C}_r \tag{41), (42}$$

Where Δt is the time interval of sampling. The pair $<\overline{A}, \overline{C}>$ is not the same form as the pair <A,C> in equation (6), but they have the same dynamic properties: the natural frequency and mode shape. The complex modes can be extracted from $<\overline{A}, \overline{C}>$. For extraction of normal modes the pair $<\overline{A}, \overline{C}>$ should be transferred in the form of the pair <A,C >. A nonsingular transformation matrix T is introduced to define the relationship between these two pairs.

$$T = \begin{bmatrix} c_d\overline{C} + c_v\overline{C}\,\overline{A}^{-1} + c_a\overline{C}\,\overline{A}^{-2} \\ c_d\overline{C}\,\overline{A} + c_v\overline{C} + c_a\overline{C}\,\overline{A}^{-1} \end{bmatrix} \tag{43}$$

And then the identified pair <A,C > are:

$$A = T\overline{A}T^{-1}, \quad C = \overline{C}T^{-1} \tag{44), (45}$$

4 NUMERICAL EXAMPLE

In this paper, the system identification for a 200 tons ship is carried out to demonstrate the feasibility and validity of the proposed method. The natural frequencies and mode shapes are calculated by 3D FEM modal analysis, and taken as reference basis. The natural frequencies calculated by FEM are listed in the second column of Table 1. To simulate the impact test on board, the transient analysis by direction integration is then performed. Figure 1 show the FEM model and node number of measured point, an impulse force applied on node 19083, the free response in X-, Y- and Z- direction of node 17160, 17156, 8486, 19099, 11787, 9751, 5251, 4251, 19520, 18101 and 20436 are measured as output vector. Each channel has 2048 points and with sample rate of 1000 Hz. Figure 2 shows the Power Spectrum Density (PSD) of response on node 19099, which can be used to observe the distribution of natural frequencies.

Two case of noise are taken into consideration, case 1 is noise free, in case 2 each output channel contains random noise, whose peak value is 10% of maximum response.
The identified modal parameters are listed in Table 1 and 2. The results show that the identified natural frequency and mode shape are identical to the results from FEM modal analysis for noise free case. And the identified results have also good agreement with reference mode parameters in case with noise. Some modes may be missed due the disturbance of noise modes.

5 CONCLUSION

In this paper, a VBAR model is proposed to identify the model parameters form output data, which can be displacement, velocity or acceleration. The results show that the presented method is feasible, the physical modes and spurious modes with noise condition can be separated clearly, and the identified mode parameters are consistent with reference results.

Acknowledgements

The authors would like to acknowledge that this work was supported by National Science Council under the grant NSC89-2611-E-002-007.

TABLE 1

THE NATURAL FREQUENCY AND IDENTIFIED DAMPING RATIO

Item	FEM modal analysis	Identified by VBAR			
		Noise free		With noise	
Para-meter / Mode	Frequency (Hz)	Frequency (Hz)	Damping ratio (%)	Frequency (Hz)	Damping ratio (%)
1	7.7613	7.7590	2.9091	7.7582	2.9115
2	12.2507	12.2416	4.5884	12.2410	4.6787
3	13.1489	13.1377	4.9239	13.1392	4.9732
4	14.0953	14.0817	5.2766	----	-----
5	17.5700	17.5434	6.5727	17.8234	8.2087
6	18.0363	18.0076	6.7457	----	-----
7	23.7293	23.6948	8.9158	22.4654	6.0375

TABLE 2

THE RELATIVE MODE SHAPES IN Y AND Z DIRECTION

Mode / Node	1		2		3		4		5	
	Y	Z	Y	Z	Y	Z	Y	Z	Y	Z
19099	0.1909	0.1211	0.0572	-0.1924	0.3534	-0.3352	0.0585	0.0195	-0.0964	0.4255
11787	0.1118	-0.0685	-0.0130	0.0480	0.0384	-0.0812	-0.1268	-0.0267	-0.0959	0.0956
9751	0.0501	-0.1233	-0.0235	0.0270	-0.0150	-0.0215	-0.1261	-0.0344	-0.0958	0.1553
5251	-0.1402	0.0113	-0.0085	-0.0620	0.1653	0.2462	0.0668	0.0317	-0.0978	0.0466
4251	-0.1114	-0.1514	-0.0050	-0.3770	0.0685	0.3801	0.1762	0.0695	-0.1011	-0.0892
19520	0.5536	0.9148	0.0276	0.7438	-0.6792	0.6860	0.5398	-0.0806	-0.1300	-0.7089

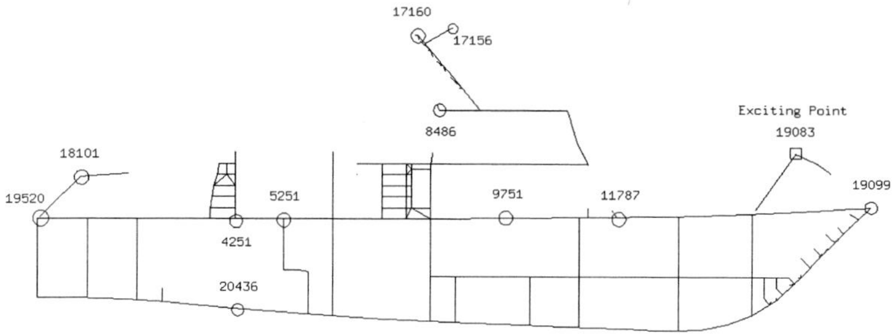

Figure 1: The FEM model and measured point of ship structure

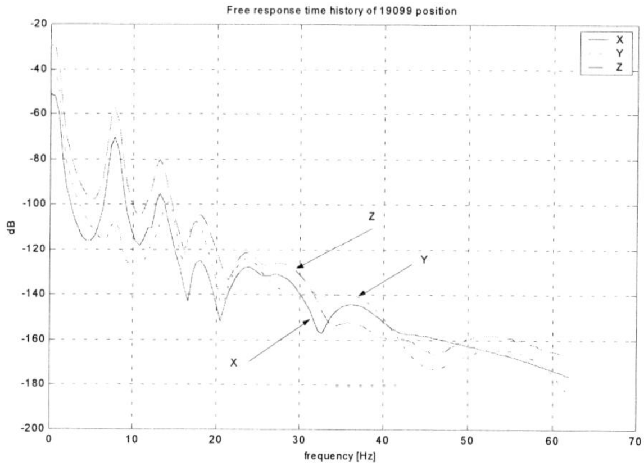

Figure 2: The PSD of free response in X, Y and Z direction on node 19099

References

BV (1987). *Building and Operation of Vibration free Propulsion Plants and Ships*, Bureau Veritas.
Carlsen L.A. (1977). *A Parametric Study of Global Hull Vibration and Superstructure Vibrations – Analysis by Means of the Finite Element Methods*, Trans. RINA.
Cooper J.E. (1992). *The Use of Backwards Model for Structural Parameters Identification*, Mechanical System and Signal Processing 6, 217-228.
Dnv (1985). *Vibration Control of Ships*, VERITEC, A.S
Ewins D.J. (1986) *Modal Test: Theory and Practice*, Research Studies Press LTD., England.
Gepner W. and Ahrens J. (1985). *FEM-Calculation of Aft-Superstructure Vibration and Check with Experiment*, Schiffbauforschung, 24, Vol. 24, pp. 166-168.
Juang J.N. (1994) *Applied System Identification*, Prentice Hall PTR, Englewood Cliffs, NJ.
Kumaresan R and. Tufts D.W. (1982). *Estimating the Parameters of Exponentially Damped Sinusoids and Pole-Zero Modeling in Noise*, IEEE Transactions on Acoustics, Speech, and Signal Processing 30, 833-840.
NK (1984). *Guide to Ship Vibration*, Nippon Keiji Kyokei.
Noonan D. (1990). *Ship Vibration Design Guide*, SSC-350, Ship Structure Committee, UAS.

You-Sheng Wu, Wei-Cheng Cui and Guo-Jun Zhou (Eds)

VIBRATION ANALYSIS METHOD OF SHIP STRUCTURES IN THE MEDIUM FREQUENCY DOMAIN

F. Besnier[1] G. Bechepay[1] Y. Mavrakakis[1] and M. Ferry[2]

[1] Institut de Recherches de la Construction Navale, Nantes, France
[2] MFRDC, Orvault, France

ABSTRACT

Frequency response of weakly damped structures, like ships, reveals three frequency domains: a low frequency domain with well separated modes and peaked response, a high frequency domain with high spectral density and regular response and a medium intermediate frequency band with high but non constant modal density and highly coupled modes. Modal method is well mastered and widely used in the low frequency domain. Energy methods - like statistical energy analysis (SEA) - cover the high frequency band. Until now, no method seems to be emerging in the medium frequency domain. Increasing severity of vibration levels required by ship owners, especially in the case of passenger ships, implies more accurate vibration level prediction. First, second and in some cases third harmonic of the propeller excitation have to be taken into account, broadening the frequency band to be studied in numerical simulation. The corresponding number of modes to be calculated is increasing raising the question of validity of modal methods. This paper presents the application of displacement but non-modal methods initially developed at ONERA in France, to ship structures. They have the advantage of using stiffness and mass matrices built from a standard finite element model as used for static or modal analysis. Two methods will be described. In both methods the frequency domain is analysed by frequency bands. Validation of the methods will be presented on test cases and application to a ship model will be compared to traditional methods, in term of accuracy and performances. The extension and limit of application of such methods will also be discussed.

KEYWORDS

Structural Vibration, Medium Frequency, Non-Modal Method, Energy Operator, Reduced Model.

1 INTRODUCTION

Evaluation of vibration level of ships, especially passenger ships, is of prime importance due to severe ship owner requirements. Typical levels for vertical velocities of 1.5 $mm.s^{-1}$ are now required for cabins and public premises in a broad frequency band covering first and second harmonic of the propeller induced excitation (typically 7-20Hz). Since mid seventies, modal methods applied to finite element models have been developed and widely used for the prediction of dynamic behavior of ships. The effect of surrounding fluid is taken into account by semi-infinite fluid elements or boundary equations, via added mass terms [Zienckiewicz & Newton 1969, Orsero & Armand 1979].

Though modern finite element software and computers allows for eigenmode analysis on large models, it is necessary to use reduction techniques such as static condensation [Guyan 1965] or improved methods [O'Callahan,1990] to filter out local modes. Large passenger's ships are becoming more flexible, due to the presence of numerous windows and openings and inner open spaces (atrium, theatre, restaurants, etc.). Even after condensation, the modal density is high and the number of needed modes is increasing rapidly. In the frequency band of the propeller excitation, the modal density is high and the frequency response does not always exhibit marked peaks, even if resonant response still exists. This domain can then be referred to as medium frequency domain (MF), and classical low frequency methods can hardly be applied, to achieve acceptable accuracy.

2 MEDIUM FREQUENCY DOMAIN PROBLEMATIC

The increasing complexity of large passenger ship structure and associated finite element models produces a shift of the modal spectrum towards lower values. Usual propeller induced excitation enter therefore into the medium frequency domain, with high but irregular modal density and highly coupled modes as can be seen on figure 1.

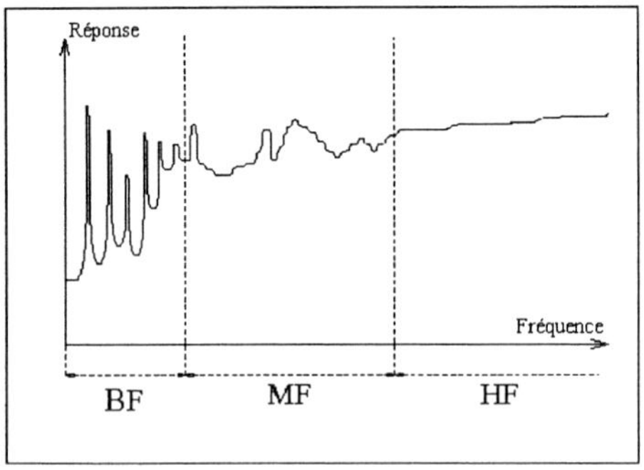

Figure 1: Spectral response of a complex structure

This domain may start about 15 Hz. The number of modes needed for the Ritz-Galerkin projection becomes important and the presence of very close frequencies makes modal extraction numerically difficult and costly. Finer meshes are also needed to capture higher frequencies, involving numerous local modes in the frequency band of interest. The MF frequency domain cannot clearly be handled using the methods of the low frequency domain on finer meshes and on a larger modal basis. On the other side, energy methods used in the high frequency domain (Statistic Energy Analysis or Energy Finite Element Formulation) cannot be easily adapted here due to the existence of resonance in the MF domain. In the sequel, after a reminder of the classical modal formulation, two displacement methods will be presented for the solution of the dynamic response in the MF domain.

3 DISPLACEMENT METHODS

3.1 Modal Frequency Method (Ritz-Galerkin)

The general dynamic problem to be solved is the following:

$$\left\{-\omega^2 M + i\omega C + K\right\}u = F(\omega) \qquad (1)$$

K, M, C being the stiffness, mass and damping matrixes built using the finite element formulation, *u* the displacement and *F* the vector of excitation forces. This equation is projected on a modal basis *X* built from a given number of eigenmodes of the associated conservative system: $MX=KX\Lambda$. Then, with $u=X\eta$ and after a premultiplication by X^T, relation 1 becomes:

$$\left\{-\omega^2[m] + i\omega[c] + [k]\right\}\eta = X^T F \qquad (2)$$

Thanks to the orthogonal properties of the modes and assuming a modal damping $c_i=2\xi_i c_{cr}$, relation (2) is a set of uncoupled equations with *[m], [c]* and *[k]* diagonal. Then the accuracy of the response computed here is depending on the quality of the modal basis *X*. Modal truncation correction is usually performed improving the solution for low frequency, but it is necessary to compute modes up to a frequency higher than the highest frequency component contained in the excitation. Typically, for a large passenger ship model of 300,000 d.o.f., the 370th frequency appears at 6.8Hz.When condensed on 10,000 translational dof, 370 modes are contained up to 22Hz. The second harmonic frequency is 18.4Hz. Using the same method for the response to the third harmonic (27.6Hz) is clearly intractable. Moreover, the accuracy of this modal approach, on a condensed model will be discussed in the sequel.

3.2 Non Modal Medium Frequency Method

As mentioned previously, modal methods are not adapted when modal density is high. The number of modes to be computed would be high. For example, considering a 300,000 d.o.f. model of a large passenger ship, condensed on 30,000 translational degrees of freedom (which is needed according to IRCN experience), the number of modes needed to cover the [0-30Hz] excitation band is about 600, including modes up to 35Hz for accuracy reasons. Modes in the upper part of this band are very sensitive to the mass distribution. But if the mass distribution of the steel part of the structure and heavy equipments is well known and can be accurately modelized, this is not the case for outfitting and lighter equipments. This non structural masses may represent more than half of the steel structure weight and they are entered in the model by regional modification of the density. This uncertainty on mass distribution leads to uncertainty on the eigenfrequencies and modes, mainly in the upper part of the frequency band raising the question of the validity of the modal basis. Hull girder modes are not sensitive to this uncertainty, which justifies modal methods in the low frequency domain.

Non modal displacement methods have been developed in France at ONERA, in Christian Soize team. These methods are based on finite element models and matrixes (stiffness, mass and damping) developed for low frequency domain and are therefore good candidates to replace, classical modal methods in the MF domain.

The first method studied here [Soize 1982] can be used for the response analysis of a structure to random or deterministic excitation. This method can be considered as a multiple scale one and is based on a shift in the frequency domain and a sampling technique [Liu & al 1991]. The problem, defined on a narrow band $B_n=[\Omega_n - \Delta\omega/2,\ \Omega_n + \Delta\omega/2]$, with $\Delta\omega << \Omega_n$, is first shifted to a low frequency band $B_0=[-\Delta\omega/2,\ +\Delta\omega/2]$, centred on the null frequency, and then this new problem is solved in the time domain by numerical integration. In other words, MF algorithm methodology is based upon introduction of two time scales: a short time scale associated to the central frequency of the band B_n and a long time scale associated to the bandwidth of B_n. The analysis domain is partitioned in a finite number of narrow bands B_n and MF analysis is performed separately for each B_n.

Considering Eqn. 1, the matrixes M, C and K as well as the excitation forces F are linearized assuming that they are independent of the frequency ω on the band B_n. Then the frequency content of the response u is shifted by Ω_n, which leads to the function $u_0(t)$ in the time domain:

$$u_0(t) = u(t)e^{-i\,\Omega_n t} \tag{3}$$

where $u_0(t)$ is a square integrable time signal whose Fourier transform $U_0(\omega)$ is such that $U_0(\omega)=0$ for all ω *outside* B_0. Then applying Shannon's theorem and taking the Fourier transform of the result we obtain, on the band B_n, an exact relation between the response in the frequency domain and the time domain function $u_0(t)$:

$$u(\omega) = \frac{2\pi}{\Delta\omega}\sum_{m\in Z} u_0\left(\frac{2m\pi}{\Delta\omega}\right) e^{-i\,m\frac{2\pi(\omega-\Omega_n)}{\Delta\omega}} \tag{4}$$

Hence, solving Eqn. 1 on the band B_n reduces to the evaluation of u_0 for sampling points. For that, Eqn. 1 is written using Eqn. 3 as:

$$M\,\ddot{u}_0 + \widetilde{C}\,\dot{u}_0 + \widetilde{K}\,u_0 = F_0(t) \tag{5}$$

where the pseudo stiffness and damping matrixes are given by:

$$\widetilde{C} = C + 2\,i\,\Omega_n M \quad \text{and} \quad \widetilde{K} = K + i\Omega_n C - \Omega^2 M \tag{6}$$

and where the excitation $F_0(t)$ is the Fourier transform of the unit function $F_0(\omega)=F(\Omega_n)$ on B_0 and $F_0(\omega)=0$ outside B_0.

$$F_0(t) = \frac{1}{\pi\,t}\sin\left(\frac{\Delta\omega}{2}t\right)F(\Omega_n) \tag{7}$$

Because of the form of the sampling function $F_0(t)$ the infinite summation of the exact relation 4 may be truncated between $m_i=-3$ and ending value m_f damping dependent, usually lower than +10. Therefore Eqn. 5 is solved in the time domain with initial conditions $u_0(t_i) = \dot{u}_0(t_i) = 0$ by an implicit, unconditionally stable Newmark integration scheme.

The main restriction of the method is that the bandwidth should not be greater than $0.1\,\Omega_n$. Then, applied to ship vibrations, the bandwidth $\Delta\omega$ corresponding to the propeller excitation will range from 0.7 to 3Hz. Nevertheless, it is not necessary, as in the modal approach, to extend the frequency band, and the computer cost of the method for one frequency band is then essentially governed by the Newmark step, which implies a factorization of the pseudo stiffness. Moreover, the bandwidth increasing with the frequency so that the number of "narrow" bands is finally limited in the MF domain.

A second restriction of the method is due to the spatial distribution of forces $F(\Omega_n)$ in Eqn. 5. This means that a new analysis is needed for each effort distribution. In the case of pressure fluctuation response, this is not a real problem, one case being usually studied. But in the general case, a reduction of the model in the MF domain [Soize 1998], based on the representation of the energy in a narrow band is of interest and actually under evaluation for ship. Such a reduced model is obtained by a Ritz-Galerkin projection of Eqn. 1 on the subspace spanned by the eigenvectors X of the energy operator

$GE_nGX=GX\Lambda$ corresponding to the highest eigenvalues. Where G is the real matrix resulting of the discretization of the sesquilinear form $\int u\overline{\delta u}dx$ and E_n the positive-definite symmetric matrix such that:

$$E_n = \frac{1}{\pi}\int_{B_n}\omega^2 \operatorname{Re}\{\overline{T}(\omega)\, M\, T(\omega)\}d\omega \tag{8}$$

with $T(\omega)$ being the symmetric complex inverse of the dynamic stiffness matrix.

$$T(\omega) = \{-\omega^2 M + i\omega C + K\}^{-1} \tag{9}$$

Following [Soize 1998] the dominant subspace is built by a subspace iteration method in which the matrix $T(\omega)$ is not directly computed but where the MF method describe above is used to evaluate the products E_nGX when needed.

4 EVALUATION OF THE METHODS

4.1 Basic Test Case

In order to test the method, a first small size test case has been defined, built to present a modal density similar to that of a typical passenger ship. Starting from a bar of constant section (1D model, 250 dof), random stiffness perturbations have been introduced in order to obtain a target spectrum, derived from a passenger ship, in the [0-25Hz] band. There are 22 modes in the [19-21Hz] band. A harmonic force is applied at the free end of the bar. A reference solution was obtained by modal superposition using a complete modal space of 250 modes.

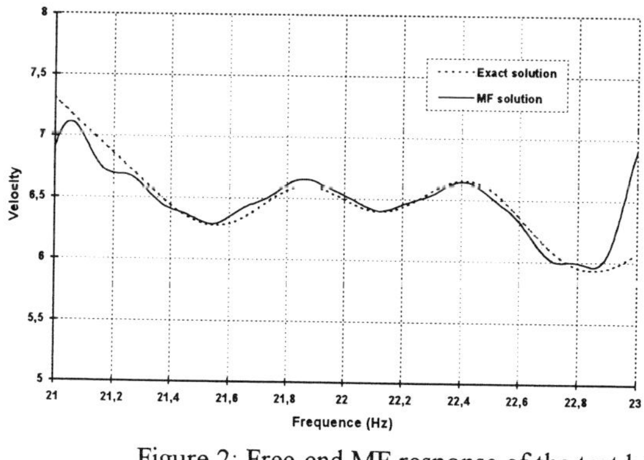

Figure 2: Free-end MF response of the test bar

Velocities at the free end of the bar are compared. The MF method gives very good results, with slightly increasing error at the extremities of the band (Fig. 2). The two extreme values have to be discarded. CPU time is in favour of the MF method in that case by several orders of magnitude.

4.2 Industrial Validation

A model of a small ship has been used for further evaluation of the MF method. The ship model comprises 10,788 shell and beam elements leading to 20,172 dof (figure 3).
There are 22 modes in the 27-35 Hz band. The excitation is a nodal harmonic vertical force applied above the propeller. Several cases have been compared:

- Modal superposition on a model condensed on 3,200 dof, using a modal basis of 100 modes, up to 48 Hz
- Modal superposition on a model condensed on 3,200 dof, using a modal basis of 200 modes, up to 72 Hz
- MF Soize method on a condensed model
- Direct frequency response
- Direct frequency response on a condensed model

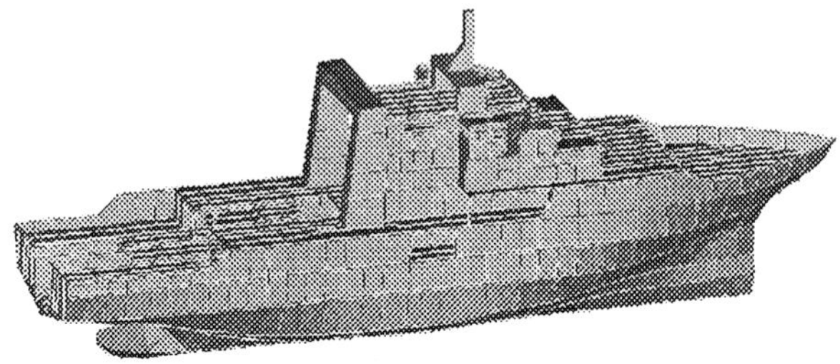

Figure 3: Finite element model of ship

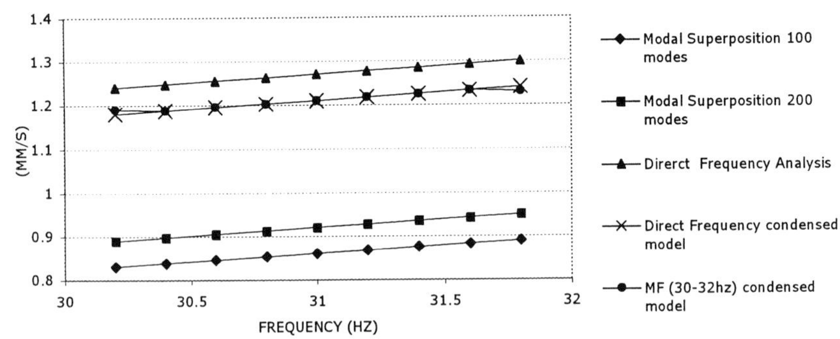

Figure 4: Response (velocity) comparison in the 30-32Hz band

The direct frequency response is obtained by solving for each frequency ω equation (1) and can be considered as a reference solution, no approximation being done on the initial model. In the 10-13 Hz band, splitted into 3 bands of 1Hz to comply with the $0.1\Omega_n$ criteria, the difference between MSM and MF method is lower than 3.7%. The comparison is more interesting in the 30-32Hz band, where the modal density is higher. The figure 4 shows the results at the excited node (velocity), the summarized in table 1.

This table shows clearly the error induced by modal superposition method, even with a modal basis more than 2 times the excitation frequency. Direct frequency and MF method give the same accuracy Moreover, there is no restriction on damping in any of those 2 methods, which is interesting in medium frequency domain. Direct frequency method can be very fast if the solver is well optimized. Nevertheless, the frequency step must be small enough to capture possible resonance, leading to a great number of factorization. In the MF formulation, the maximum bandwidth is increasing with the central frequency, which favours the method when frequency is increasing.

TABLE 1
VELOCITY COMPARISON

Method	Direct Frequency	Direct Frequency Condensed Model	MF (Soize) Condensed	Modal 200 modes	Modal 100 modes
Error	0%	4%	4%	27%	32%

5 CONCLUSIONS

Dynamic response in medium frequency domain, with high modal density, cannot be accurately obtained by traditional modal superposition methods. The displacement MF method presented here is a suitable tool in and can follow after the previous one on the same finite element models. The model reduction method has not yet been tested on an industrial case. Further work is needed to improve the robustness and performances of the algorithm. Nevertheless, this method, as mentioned by its author [Soize 1998b] can be a very powerful approach for the representation of structural-acoustic systems with structural complexity (associated in ship structure to outfitting) and modal synthesis methods in the MF domain. This reduced model could also be used in a non parametric approach of random uncertainties [Soize 1999]. This problem is of prime importance in medium frequency domain for passenger ship structures: initial stresses, geometrical imperfections are ignored in the models and outfitting, besides the uncertainty in the mass distribution, act as random resonant systems attached to the main structure. There is a need for the construction of probabilistic models in order to compute bounds for the dynamic response in the medium frequency domain.

Acknowledgements

This publication is a result of a work sponsored by the French Ministry of Economy & Industry and Chantiers de l'Atlantique.

References

Guyan R.J. (1965). Reduction of Stiffness and Mass Matrices. *AIAA Jnl*, **3:2**, 380.
Liu W.K., Zhang Y. & Ramirez M.R. (1991). Multiple Scale Finite Element Methods, *International Journal for Numerical Methods in Engineering*, **32**, 969-990.
O'Callahan J. (1990). Comparison of Reduced Model Concepts, *8th Int. Modal Analysis Conf.*, Orlando, Florida.
Orsero P., Armand J.L. (1978). A Numerical Determination of the Entrianed Water in Ship Vibrations, *International Journal for Numerical Methods in Engineering*, **13.**
Soize C. (1982). Medium Frequency Linear Vibrations of Anisotropic Elastic Structures, *La Recherche Aérospatiale (English edition)*, **5**, 65-87.
Soize C. (1998a). Reduced Models in the Medium Frequency Range for General Dissipative Structural-Dynamics Systems, *Eur.J. Mech. A/Solids*, **17:4**, 657-685.
Soize C. (1998b). Reduced Models in the Medium Frequency Range for General Dissipative External Structural-Acoustic Systems, *J. Acoust. Soc. Am.*, **103:4**, 3393-3406.
Soize C. (2000). A NonParametric Model for Random Uncertainties for Reduced Matrix Models in Structural Dynamics, *Probabilistic Engineering Mechanics*, **15**, 277-294.
Zienckiewicz O.C., Newton R.E.(1969). Coupled Vibration of a Structure Submerged in a Compressible Fluid, *Proc. Symposium on Finite Element Technique, ISD-ISSC*, Stuttgart.

Practical Design of Ships and Other Floating Structures
You-Sheng Wu, Wei-Cheng Cui and Guo-Jun Zhou (Eds)

A NEW METHOD FOR DETERMINING ACOUSTIC ADDED MASS AND DAMPING COEFFICIENTS OF FLUID-STRUCTURE INTERACTION

Q. Zhou[1•], W.Zhang[1] and P.F.Joseph[2]

[1]Huazhong University of Science and Technology,
N0.101, 339 Jie Fang Road, Wuhan 430033, China
[2]ISVR, University of Southampton, Highfield,
Southampton, SO17 1BJ, UK

ABSTRACT

According to the technique of potential decomposition, an uncoupled boundary element method has been presented to solve the problem of fluid-structure interaction. Hydrodynamic forces exerted on structural wetted surfaces are expressed as matrices of added mass and damping coefficients respectively, which are independent of unknown structure variables. Symmetrical properties of added mass and damping coefficients are given. The method presented is validated by comparing the numerical results with theoretical results.

KEYWORDS

Acoustics, Noise, Sound, Numerical method, BEM, FEM, Fluid-structure interaction, Added mass, Damping.

1 INTRODUCTION

Computational approach for computing added mass of fluid-structure interaction problem have been reported by DeRunts & Geers (1978), Everstine (1991). Both obtain the solution by solving Laplace's equation in the low frequency regime. Their results are therefore only valid for incompressible fluid without damping effects. The computation of acoustic added mass and damping coefficients for compressible fluid without the low frequency limitation is still an unsolved problem, as pointed out by Ali (1995). In this paper, an uncoupled boundary element method is presented for the calculation of the acoustic added mass and damping coefficients without the limitation of low frequency.

The crux of solving fluid-structure interaction problems is how to uncouple the problem. In this paper a new boundary element method is presented to achieve this. The method is based on the linearity of the governing equations. The inertial and damping forces of fluid are dependent on the acceleration and velocity on structure surface separately, but their added mass and damping coefficients, are independent of the acceleration and velocity for a given frequency and a given structure surface. Nevertheless, the situation is not as simple as a rigid body moving in fluids since the structure surface is generally elastic. So one must use superposition principle of linear equation to decompose velocity potential of fluid into a sum of element potentials. This approach is called the potential decomposition technique, a method widely used in ship-wave interaction problems presented by Salvesen (1970). Also one can expect that the added mass will be very different from conventional mass since there is something like mutual radiant impedance. So the acoustic added mass and damping coefficients are two-dimension tensors.

Once the effective acoustic added mass and damping coefficients are known, the acoustic problem of fluid-structure interaction can be easily solved by conventional FEM. This solving procedure is different from those approaches presented by Seybert (1993), Everstine (1997) in that the mass and damping matrices of fluid and structure can be simply added together to obtain the solution of fluid-structure interaction problem. And also by the method presented herein the acoustic eigenvalue analysis is possible for submerged structures.

2 GOVERNING EQUATIONS AND SPATIAL VELOCITY POTENTIAL

The motion of a fluid particle is assumed to exist long enough that all transient effects have decayed and the acoustic velocity potential can be written as $\Phi = \mathrm{Re}\{\phi\, e^{-i\omega t}\}$, where $\boldsymbol{i} = \sqrt{-1}$, t is time variable. Let $\mathrm{Re}\{W\, e^{-i\omega t}\}$ be the normal displacement at any point on the wetted surface, then the inward normal velocity of any point is

$$U_n = \mathrm{Re}\left\{ \boldsymbol{i}\omega\{W\, e^{-i\omega t}\}\right\} \tag{1}$$

Under these conditions, the governing equation can be written as:

$$\begin{cases} \nabla^2\phi + k^2\phi = 0 \\ \lim\limits_{r\to\infty}|r\phi| < \infty \\ \lim\limits_{r\to\infty} r(\dfrac{\partial\phi}{\partial r} - \boldsymbol{i}k\phi) = 0 \\ \dfrac{\partial\phi}{\partial n_e}\Big|_{P\in S_o} = \boldsymbol{i}\omega W(P) \end{cases} \tag{2}$$

where ω is circular frequency, k is the wave-number, $\vec{n}_e$ is the unit inward normal from the wetted surface of structures, r is the distance between field point $P(x, y, z)$ and source point $Q(\xi, \eta, \zeta)$. Obviously the linear superposition principle can be used to decompose the velocity potential into a sum of element potentials, $\phi(x, y, z)$ is called spatial velocity potential depending only on the spatial variables.

3 DECOMPOSITION OF VELOCITY POTENTIAL AND UNCOUPLING OF THE PROBLEM

The last of Eqn. 2 means that the normal derivative for ϕ at any point of wetted surface must be equal to the amplitude of normal velocity of the elastic body at the same point. In order to carry out numerical calculation of the sound field due to the elastic shell vibration, we can discretize the shell into finite triangular elements. Let $W_j = \mathrm{Re}\{\overline{W}_j\, e^{-i\omega t}\}$ denote the normal displacement of average-piston vibration of element j, where $\overline{W}_j$ is average displacement amplitude of element j. Normal displacement amplitude W at any point of the wetted surface can be expressed as

$$W(P_r) = \sum_{j=1}^{N_{eo}} \delta_{rj} \cdot \overline{W_j} \tag{3}$$

where δ_{rj} is Kronecker δ, N_{e0} is the number of elements on wetted surface.

On the basis of linear superposition, the velocity potential ϕ can be decomposed as

$$\phi = \sum_{j=1}^{N_{eo}} \phi_j \cdot (-i\omega \overline{W_j}) \tag{4}$$

The corresponding governing equations become

$$\begin{cases} \nabla^2 \phi_j + k^2 \phi_j = 0 \\ \lim\limits_{r \to \infty} \left| r\phi_j \right| < \infty \\ \lim\limits_{r \to \infty} r(\dfrac{\partial \phi_j}{\partial r} - ik\phi_j) = 0 \\ \left. \dfrac{\partial \phi_j}{\partial n_e} \right|_{P_r \in S_0} = -\delta_{rj} \end{cases} \tag{5}$$

where ϕ_j is the velocity potential due to element j which vibrates with velocity of unit amplitude when all other elements are kept stationary. It is called normalized velocity potential. Thus the problem of solving velocity potentials, ϕ, *is* transferred to solving N_{e0} numbers of velocity potential, ϕ_j. Because of this step, the solution of ϕ_j does not depend on the unknown displacement $\overline{W_j}$ on the wetted surface. It depends on the shape of the wetted surface and the vibrating frequency of element *j*. A solution that is valid for all time on a given structural surface and at a fixed frequency can therefore be obtained.

4 SOLUTION FOR SOURCE STRENGTH

The velocity potential at any point *P(x, y, z)* of the sound field can be expressed as

$$\phi_j(P) = \iint_{S_0} \sigma_j(Q) G(P,Q) dS(Q) \tag{6}$$

where G is a Green function, $\sigma_j = \sigma_j^c + i\sigma_j^s$ is the source strength at the source point $Q(\xi, \eta, \zeta)$ of the wetted surface when the element j has velocity of unit amplitude.

The Eqn. 6 shows that the potential ϕ_j is a superposition of the elementary solutions $G(P,Q)$ weighted by the source strength $\sigma_j(Q)$. The unknown source strength σ_j can be determined from the surface boundary condition of Eqn. 5. When element j vibrates with velocity of unit amplitude, suppose the source strength of any element t, σ_{tj}, is a constant and rewrite source strength of element *t* as

$$\sigma_{tj} = \sigma_{tj}^c + \boldsymbol{i}\sigma_{tj}^s \tag{7}$$

Then Eqn. 6 can be rewritten as

$$\phi_j(P) = \sum_{t=1}^{N_{e0}} \left[\iint_{S_t} G(P, Q_t) dS(Q_t) \right] \sigma_{tj} \tag{8}$$

where S_t denotes that the integral is taken on element t, $dS(Q_t)$ expresses infinitesimal area on element t. Define influential coefficients of source strength, I_{rt}, as

$$I_{rt} = I_{rt}^c + \boldsymbol{i}\, I_{rt}^s = (\vec{N}_r \cdot \nabla) \iint_{S_t} G(P, Q_t) dS(Q_t) \Big|_{P=P_r \in S_r} \tag{9}$$

where S_r is rth element on the wetted surface S_0, $\vec{N}_r$ is unit outward normal vector of element r, the differential operator ∇ is applied to $P(x, y, z)$. From surface boundary condition of Eqn. 5, we can obtain a set of $2N_{e0}$ linear algebraic equations for unknowns σ_{tj} $(t=1, 2, 3, \cdots N_{e0})$

$$\begin{cases} \sum_{t=1}^{N_{e0}} (I_{rt}^c \sigma_{tj}^c - I_{rt}^s \sigma_{tj}^s) = \delta_{rj} \\ \sum_{t=1}^{N_{e0}} (I_{rt}^s \sigma_{tj}^c + I_{rt}^c \sigma_{tj}^s) = 0 \end{cases} \tag{10}$$

where $j=1, 2, 3, \cdots N_{e0}$. By solving this equations, we can obtain the source strength of any element t, σ_{tj}, on the wetted surface.

5 SOLUTION OF VELOCITY POTENTIAL AT ANY POINT ON THE WETTED SURFACE

We define influential coefficient of velocity potential as

$$A_{kt} = A_{kt}^c + \boldsymbol{i}A_{kt}^s = \iint_{S_t} G(P_k, Q_t) dS(Q_t) \tag{11}$$

which is the velocity potential of element k contributed by unit source strength of element t. The velocity potential at any point P_k on the wetted surface can then be expressed as

$$\phi_j(P_k) = \sum_{t=1}^{N_{e0}} A_{kt}\sigma_{tj} = \sum_{t=1}^{N_{e0}} \left[(A_{kt}^c \sigma_{tj}^c - A_{kt}^s \sigma_{tj}^s) + \boldsymbol{i}(A_{kt}^s \sigma_{tj}^c + A_{kt}^c \sigma_{tj}^s) \right] \tag{12}$$

6 SOLUTION OF ADDED MASS AND DAMPING COEFFICIENT

If added mass $\overline{m_{kj}}$ and damping coefficient $\overline{n_{kj}}$ are defined by

$$\overline{m_{kj}} = -\rho \sum_{t=1}^{N_{e0}} (A_{kt}^c \sigma_{tj}^c - A_{kt}^s \sigma_{tj}^s) \tag{13}$$

$$\overline{n_{kj}} = -\rho\omega\sum_{t=1}^{N_{ek}}(A^s_{kt}\sigma^c_{tj} + A^c_{kt}\sigma^s_{tj}) \tag{14}$$

where ρ is the density of the medium, then the pressure at the point P_k on the element S_k can be expressed as

$$p(P_k) = \sum_{j=1}^{N_{e}}(-\overline{m_{kj}}\ddot{W}_j - \overline{n_{kj}}\dot{W}_j) \tag{15}$$

here $W_j = \text{Re}\{\overline{W_j}e^{-i\omega t}\}$ is the normal displacement of element j , $\dot{W}_j$ and $\ddot{W}_j$ are normal velocity and acceleration of element j respectively. From Eqn. 15, the pressure exerted on element k consists of two parts: one is directly proportional to the vibration acceleration of element j, its coefficient $\overline{m_{kj}}$ is called added mass; another is directly proportional to the vibration velocity of element j, its coefficient $\overline{n_{kj}}$ is called the damping coefficient. The meanings of $\overline{m_{kj}}$ and $\overline{n_{kj}}$ are added mass and damping coefficient respectively per unit area on element k produced by the motion of element j.

It can be proved that if the element scale is much smaller than wavelength, the product of added mass and element area has symmetrical property,

$$S_k \cdot \overline{m_{kj}} = S_j \cdot \overline{m_{jk}} \tag{16}$$

where S_k and S_j are the element areas. The meaning of Eqn. 16 is that the added mass on element k produced by the motion of element j is equal to the added mass on element j caused by the motion of element k. The same property is true for damping coefficient,

$$S_k \cdot \overline{n_{kj}} = S_j \cdot \overline{n_{jk}} \tag{17}$$

Eqn. 16 and 17 are a consequence of the reciprocal theorem.

7 CALCULATION FOR INFLUENTIAL COEFFICIENT OF SOURCE STRENGTH I_{rt}

Assume the Green function as the form

$$G(P,Q) = -\frac{1}{4\pi}\cdot\frac{e^{ikr}}{r} \tag{18}$$

Thus Eqn. 9 becomes

$$I_{rt} = \frac{-1}{4\pi}(\bar{N}_r \cdot \nabla)\iint_{S_t}\frac{e^{ikr}}{r}dS = I^c_{rt} + \boldsymbol{i}\, I^s_{rt} \tag{19}$$

This leads to the calculation of vector $\mathbf{C}_{rt}$

$$\mathbf{C}_{rt} = \nabla\iint_{S_t}\frac{e^{ikr}}{r}dS = C^x_{rt}\vec{i} + C^y_{rt}\vec{j} + C^z_{rt}\vec{k} \tag{20}$$

If $r \neq t$, there is no singular point for integrand inside element S_t. From Eqn. 20, we have

$$C_{rt}^{xR} = -\iint_{S_t} \frac{x-\xi}{r^3}(\cos kr + kr\sin kr)d\xi d\eta \tag{21}$$

$$C_{rt}^{xI} = \iint_{S_t} \frac{x-\xi}{r^3}(kr\cos kr - \sin kr)d\xi d\eta \tag{22}$$

$$C_{rt}^{yR} = -\iint_{S_t} \frac{y-\eta}{r^3}(\cos kr + kr\sin kr)d\xi d\eta \tag{23}$$

$$C_{rt}^{yI} = \iint_{S_t} \frac{y-\eta}{r^3}(kr\cos kr - \sin kr)d\xi d\eta \tag{24}$$

$$C_{rt}^{zR} = -\iint_{S_t} \frac{z}{r^3}(\cos kr + kr\sin kr)d\xi d\eta \tag{25}$$

$$C_{rt}^{zI} = \iint_{S_t} \frac{z}{r^3}(kr\cos kr - \sin kr)d\xi d\eta \tag{26}$$

To sum the above expressions, we see that

$$I_{rt}^{c} = -\frac{1}{4\pi}(N_r^x C_{rt}^{xR} + N_r^y \cdot C_{rt}^{yR} + N_r^z C_{rt}^{zR}) \tag{27}$$

$$I_{rt}^{s} = -\frac{1}{4\pi}(N_r^x C_{rt}^{xI} + N_r^y \cdot C_{rt}^{yI} + N_r^z C_{rt}^{zI}) \tag{28}$$

where $\vec{N}_r = N_r^x \vec{i} + N_r^y \vec{j} + N_r^z \vec{k}$ is outward normal unit vector on r th element. If $r=t$, there is a singular point for the integrand of Eqn. 19 inside element S_t. It can be proved that

$$I_{rt}^{c} = \frac{1}{2} \tag{29}$$

$$I_{rt}^{s} = 0 \tag{30}$$

8 CALCULATION OF PRESSURE AT ANY POINT IN THE SOUND FIELD

The pressure at any point $P_k = (x_k, y_k, z_k)$ in the sound field is given by

$$p(P_k) = -\rho\frac{\partial}{\partial t}\Phi = |\overline{p}|\cos(\omega t + \alpha) \tag{31}$$

here the notation $\overline{p}$ expresses the amplitude of pressure, that is

$$\overline{p} = \boldsymbol{i}\rho\omega\phi = \sum_{j=1}^{N_{e_0}} \rho\omega^2\phi_j\overline{W_j} \tag{32}$$

Similar to Eqn. 12, we have

$$\phi_j^c(P_k) = \sum_{t=1}^{N_{e_0}} (A_{kt}^c \sigma_{tj}^c - A_{kt}^s \sigma_{tj}^s) \tag{33}$$

$$\phi_j^s(P_k) = \sum_{t=1}^{N_{e_0}} (A_{kt}^s \sigma_{tj}^c + A_{kt}^c \sigma_{tj}^s) \tag{34}$$

Suppose $\overline{W_j} = \overline{W_j}^R - \boldsymbol{i}\overline{W_j}^I$, then

$$\bar{p}^c(P_k) = \rho\omega^2 \sum_{j=1}^{N_{e_0}} \left(\phi_j^c(P_k)\cdot\overline{W_j}^R + \phi_j^s(P_k)\overline{W_j}^I\right) \tag{35}$$

$$\bar{p}^s(P_k) = \rho\omega^2 \sum_{j=1}^{N_{e_0}} \left(\phi_j^s(P_k)\cdot\overline{W_j}^R - \phi_j^c(P_k)\overline{W_j}^I\right) \tag{36}$$

$$|\bar{p}| = \sqrt{(\bar{p}^c)^2 + (\bar{p}^s)^2}\ , \quad \alpha = \tan^{-1}(-\bar{p}^s/\bar{p}^c)\ . \tag{37}$$

9 NUMERICAL EXAMPLE AND VALIDATION OF THE METHOD

In order to investigate the validity of the method, a vibrating spherical crown in a rigid sphere is considered. Suppose a spherical crown has a piston vibration, as shown in Figure 1. Amplitude of vibration velocity is given by

$$u_r(\theta) = \begin{cases} u_0 & 0 \le \theta \le \theta_0 \\ 0 & \theta_0 < \theta \le \pi \end{cases} \tag{38}$$

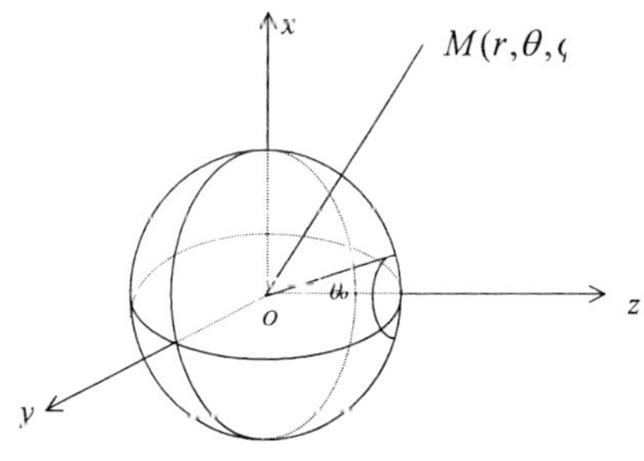

Figure 1: A spherical crown

where θ is the latitude of a spherical coordinate system (r,θ,φ), having the pole through the center of the "piston", and θ_0 is spherical crown angle.

The theoretical solution of the sphere can be found from Morse (1948). The numerical method presented herein can also be used to deal with the same problem. Figure 2 is the net division of the hemi-sphere. Supposing that there are $N_{e\theta_0}$ elements inside the spherical crown, the sum of their area is S_Δ. Then the radius of equivalent circle is $a_p = \sqrt{S_\Delta/3\pi}$. Corresponding spherical crown angle is $\theta_0 = 2\sin^{-1}(a_p/2a)$. The spherical radius a is set equal 1.0 m. The field points for comparison are located at $r = 2.0a$. The hemi-sphere is divided into 324 triangle elements. The hemispherical crown is divided into 18 elements. The scale of element ΔB is 0.175 m.

Theoretical and numerical solution can be compared by dimensionless pressure of Eqn. 37. Figures 3, 5 and 7 are the dimensionless amplitude of sound pressure varies via the latitude θ, when the spherical crown vibrates at different frequencies. Solid lines are theoretical results and dash lines express numerical results. The figures show that the numerical solutions give good agreement with the theoretical results.

Figures 4, 6 and 8 are the phase angles of the sound pressure varies with latitude θ at different frequencies. In these figures, the solid lines are theoretical calculations and small circles are numerical results. From these figures, we see that at low frequency, the phase angles show good agreement, but at high frequency there are differences.

Figure 9 and Figure 10 are the pressure comparison at the front point ($\theta = 0^\circ$) and the rear point ($\theta = 180^\circ$) respectively. Abscissa is $f = \omega/2\pi$ *(Hz)* and ordinate is $|\bar{p}|/\rho c u_0$. The solid lines are theoretical results and the dash lines express numerical results. From these figures, it can be seen that if $f > 1000\,Hz$, the difference of rear point pressure begins to be obvious. The differences are caused by the element scales. Figure 11 is that dimensionless pressure of the rear point varies with the element scale $\lambda/\Delta B$. Here λ is wavelength. Figure 11 implies that if $\lambda/\Delta B \geq 9.0$, which is almost the same as $\Delta B \leq \lambda/3\pi$ from Schenck (1968), numerical error is small.

Figure 12 is the results that verify the symmetrical properties expressed by Eqn. 19 and 20. Abscissa is the element scale $\lambda/\Delta B$. Ordinate indicates the relative error of added mass or damping coefficient from symmetry requirements. It is illustrated that if $\lambda/\Delta B \geq 9.0$, then the relative error is less than 5%. It means that the symmetry property is valid.

Figure 13 and 14 are plots of dimensionless added mass $\frac{\bar{m}_{kj}}{\rho c/\omega}$ and damping coefficient $\frac{\bar{n}_{kj}}{\rho c}$ versus frequency f for different elements. Abscissa is frequency f. Ordinate are nondimensional added mass and damping coefficient. From these figures it can be seen that dimensionless added mass and damping coefficients are much larger when $k = j$ than $k \neq j$. Figure 13 shows that added mass $\bar{m}_{168,1}$ and damping coefficient $\bar{n}_{168,1}$ change substantially with frequency. Here element *168* has 90° angle with element *1* on the sphere.

10 CONCLUSIONS

According to the technique of potential decomposition, an uncoupled boundary element method has been presented. This new method treats fluid load as added mass and damping coefficient matrices while the present approaches form complex matrices by discretization of the Helmholtz surface integral equation. It gives a convenient way for engineers to apply present FEM software packs (such as MSC NASTRAN) to solve acoustic problem of fluid-structure interaction by simply combining added mass matrices with structure mass matrices, and by adding damping coefficient matrices to the structure damping matrices.

From the theoretical deduction, a FORTRAN program has been developed. The computational results prove the validity of the method, and the limitations of the method are provided.

Acknowledgment

The first author gratefully acknowledges the benefit of discussions with professor Guo Ri-Xiu during the preparation of this paper.

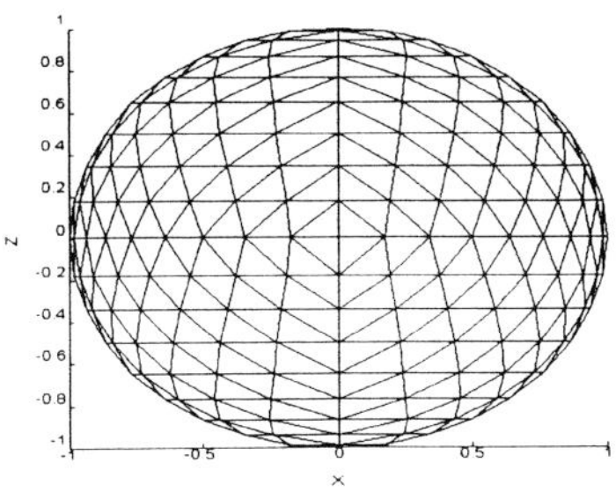

Figure 2: Net division of a hemi-sphere

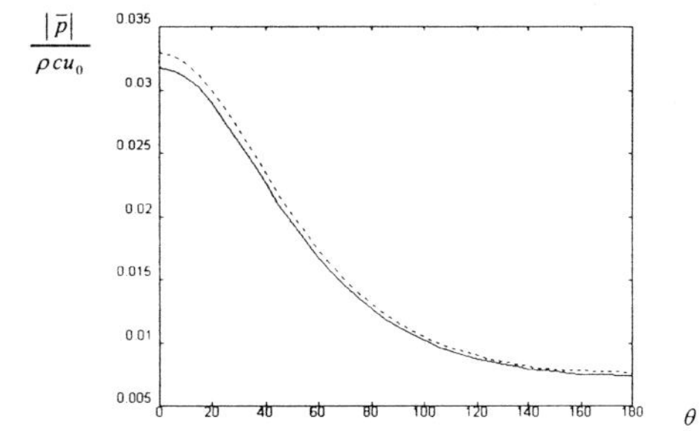

Figure 3: Comparison of theoretical and numerical pressure $\frac{|\bar{p}|}{\rho c u_0}$ versus latitude θ for ω=628 1/s , ---- numeral, —— theory

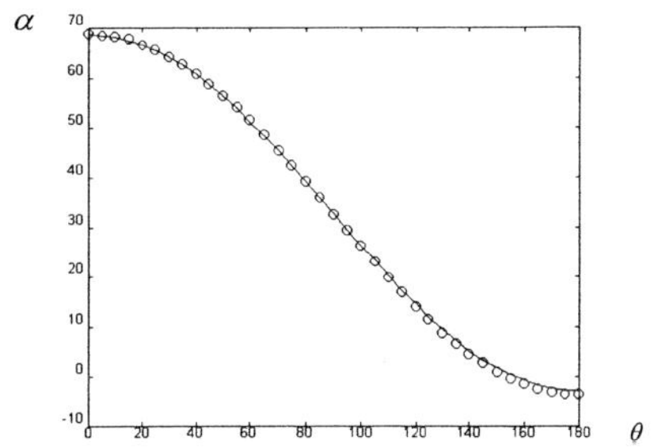

Figure 4: Comparison between theory and numeral of pressure phase α versus latitude θ for ω=628 1/s , --o-- numeral, —— theory

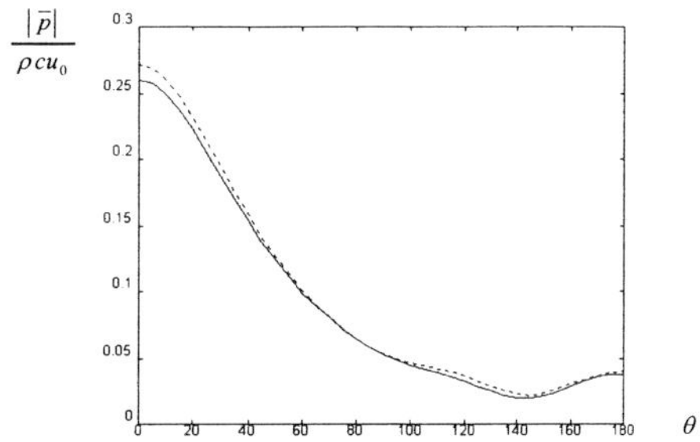

Figure 5: Comparison of theoretical and numerical pressure $\frac{|\bar{p}|}{\rho c u_0}$ versus latitude θ for ω=4000 1/s , ---- numeral, —— theory

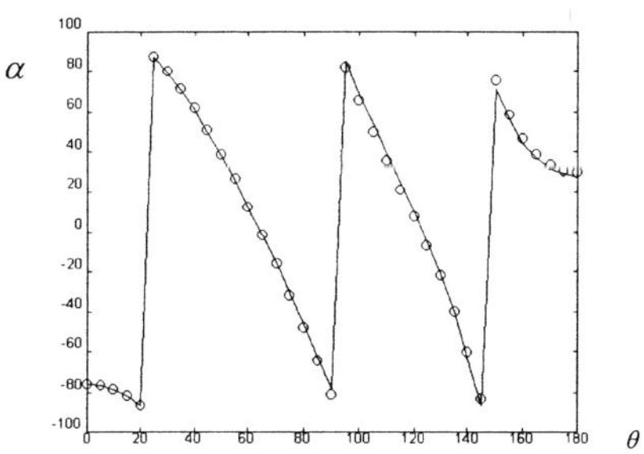

Figure 6: Comparison between theory and numeral of pressure phase α versus latitude θ for ω=4000 1/s , --o-- numeral, —— theory

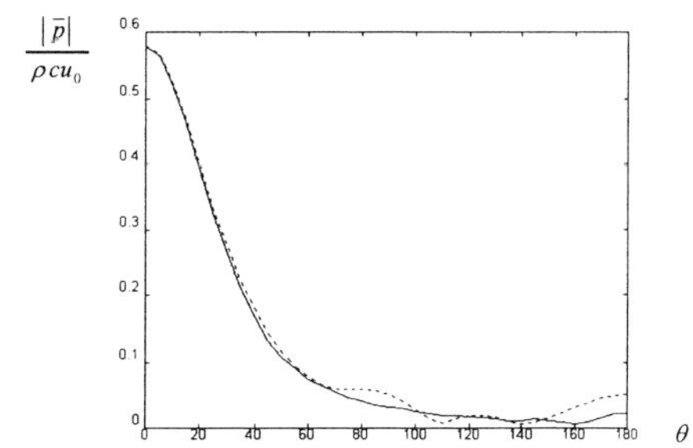

Figure 7: Comparison of theoretical and numerical pressure $\frac{|\bar{p}|}{\rho c u_0}$ versus latitude θ for ω=8676 1/s , ----numeral, —— theory

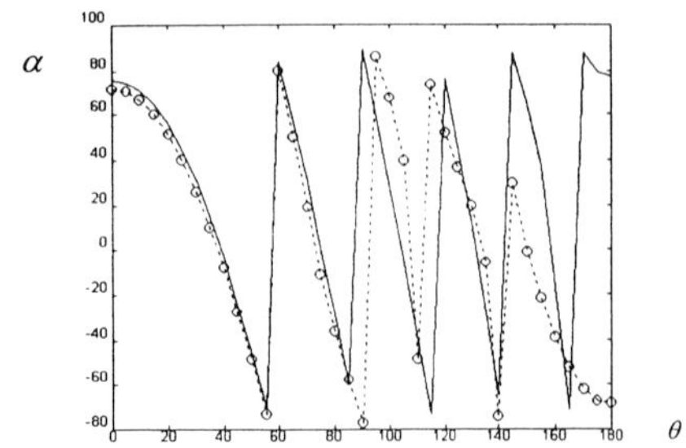

Figure 8: Comparison between theory and numeral of pressure phase α versus latitude θ for ω=8676 1/s , -o- numeral, —— theory

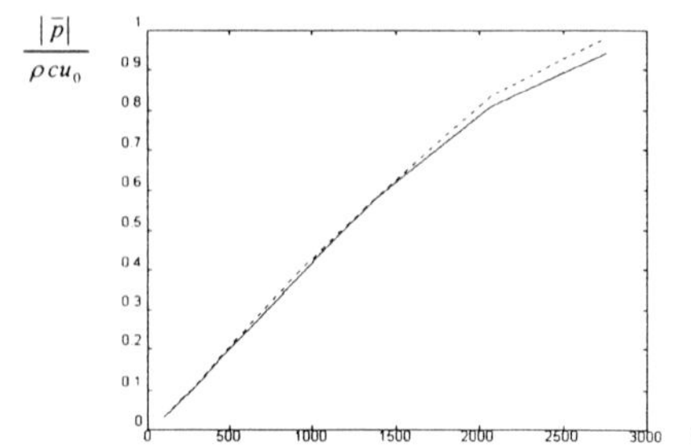

Figure 9: Comparison of theoretical and numerical pressure $\frac{|\bar{p}|}{\rho c u_0}$ at the front point θ =0 versus frequency f, ---- numeral, —— theory

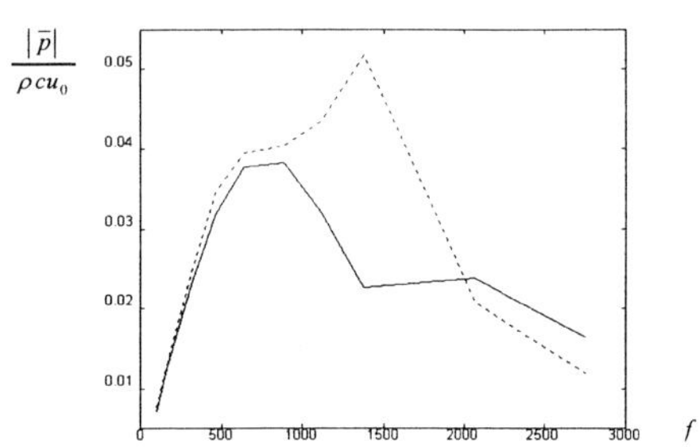

Figure 10: Comparison of theoretical and numerical pressure $\frac{|\bar{p}|}{\rho c u_0}$ at the rear point θ =180 versus frequency f, ----numeral, —— theory

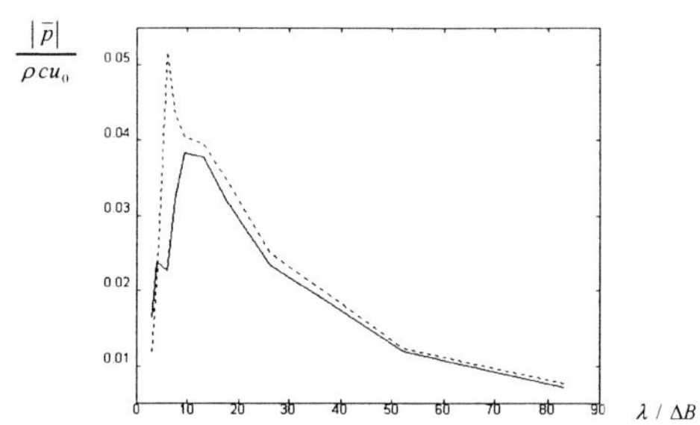

Figure 11: Pressure $\frac{|\bar{p}|}{\rho c u_0}$ of the rear point varies with element scale $\lambda/\Delta B$, ----numeral, —— theory

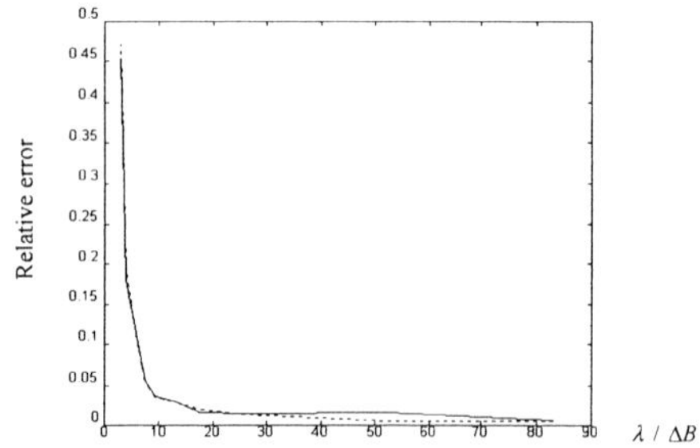

Figure 12: Relative error of added mass and damping coefficient versus element scale $\lambda/\Delta B$, ----- damping coefficient, —— added mass

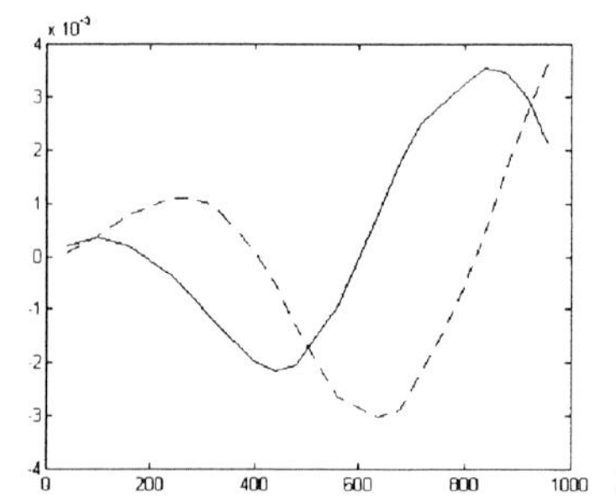

Figure 13: Plots of nondimensional added mass $\frac{\bar{m}_{kj}}{\rho c/\omega}$ and damping coefficient $\frac{\bar{n}_{kj}}{\rho c}$ versus frequency f for k=168, j=1, —— added mass, ----damping coefficient

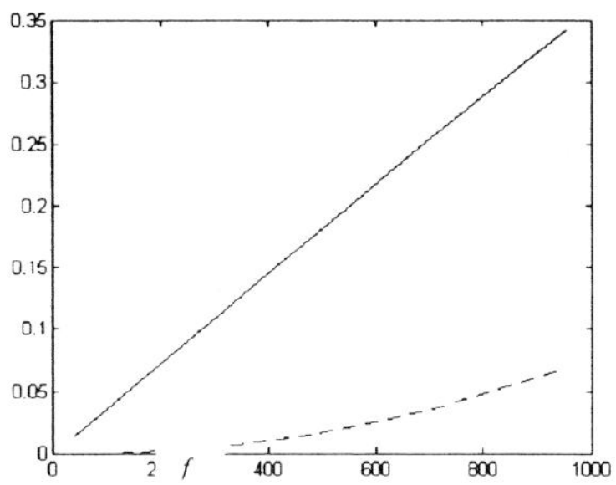

Figure 14: Plots of nondimensional added mass $\frac{\bar{m}_{kj}}{\rho c/\omega}$ and damping coefficient $\frac{\bar{n}_{kj}}{\rho c}$ versus frequency f for k=1, j=1, added mass, ----damping coefficient

References

Ali A., C. Rajakumar and S. M. Yunus (1995). Advances in acoustic eigenvalue analysis using boundary element method. *Computers & Structures* **56:5**, 837-847.

DeRuntz J. A. and Geers T. L. (1978). Added mass computation by the boundary element method. *Int. J. Numer. Meth. in Engrg* **12** ,531-549.

Everstine G. C. (1991). Prediction of low frequency vibrational frequencies of submerged structures. *Journal of Vibration and Acoustics* **113**, 187-191.

Everstine G. C. (1997). Finite element formulations of structural acoustics problems. *Computers & Structures* **65:3**, 307-321.

Morse P.M. (1948). *Vibration and sound*, 2nd ed. McGraw Book Company, Inc., New York.

Salvesen N., Tuck E. O., Faltinsen O. (1970). Ship motion and sea loads. *Trans. SNAME* **78**.

Schenck H. A. (1968). Improved integral formulation for acoustic radiation problems. *the Journal of the Acoustical Society of America* **44**, 41-58.

Seybert A. F., Wu T.W., Li W.L. (1993). Coupled FEM/BEM for fluid-structure interaction using Ritz vectors and eigenvectors. *Journal of Vibration, Acoustics, Stress, and Reliability in Design* **115:2**, 152-158.

• The present address of first author: ISVR, University of Southampton, Highfield, Southampton, SO17 1BJ, UK

Practical Design of Ships and Other Floating Structures
You-Sheng Wu, Wei-Cheng Cui and Guo-Jun Zhou (Eds)

VIBRATION PREDICTION OF RECTANGULAR TANK STRUCTURES

Yu Takeda

Industrial Machine & Plant Development Center
Ishikawajima-Harima Heavy Industries Co., Ltd.
1, Shin-Nakahara-Cho, Isogo-Ku, Yokohama, 235-8501,Japan

ABSTRACT

Heavy vibration of tank structures in way of engine room or aft region is sometimes caused by the decrease of natural frequency of tank wall. On the basis of investigations, it is clarified that the natural frequency of fundamental mode of rectangular tank may be reduced than the designed frequency due to the increase of added mass induced by hydro-elastic interactions. This paper describes the mechanism to cause this hydro-elastic interaction, in view of narrow shape of tank, effect of multi tiers, effect of coupling between panel and stiffeners, and effect of thermal stress. Also vibration prediction method is proposed for practical design purpose and its accuracy is verified by FE analysis.

KEYWORDS

Natural frequency, Hydroelastic interaction, Rectangular tank, Breath mode, Fundamental mode, Added mass, Multi tiers, Coupling mode between panel and stiffener, Narrow tank, Thermal stress

1 INTRODUCTION

Recently heavy vibration of tank structures in way of engine room or aft region is sometimes caused by the decrease of main engine revolutions or by the reduction of main engine's cylinder numbers or by the increased power of main engine per cylinder. Such vibration in the tank structure results in leakage of fuel oil or water due to the structural failures of tank integrity, therefore, tank vibration is an important problem.

Fig.1 shows the structural damages caused by this vibration in way of vertical stiffener at bracket toe attached on 2nd deck in drinking water tank of a VLCC, Nippon Kaiji Kyokai (1998).

These vibrations are induced by the resonance of the natural frequency of the tank structures with the exciting frequency of main engine or of the propeller blade, however, that phenomenon is quite curious, since the tank structure had been designed so as to avoid the resonance with these exciting forces. Therefore, natural frequency of the tank structure in actual condition may be reduced than the designed frequency.

In order to clarify the reason of this phoenomenon, several investigations were carried out and then the mechanism of this resonance can be explained. Furthermore, on the basis of these investigations, practical prediction method of the natural frequency of tank panel structures is proposed.

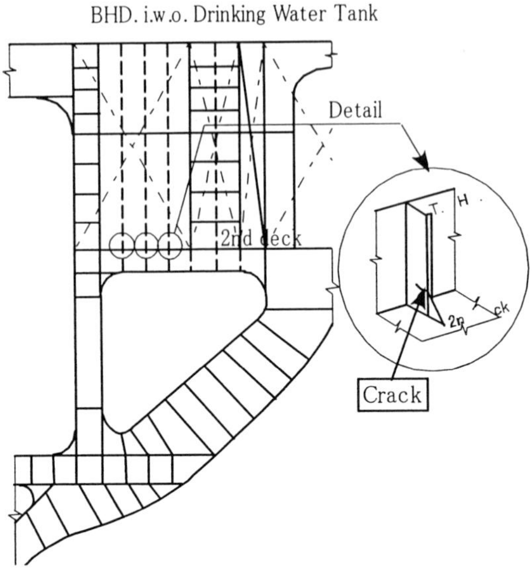

Figure 1: Example of structural failure

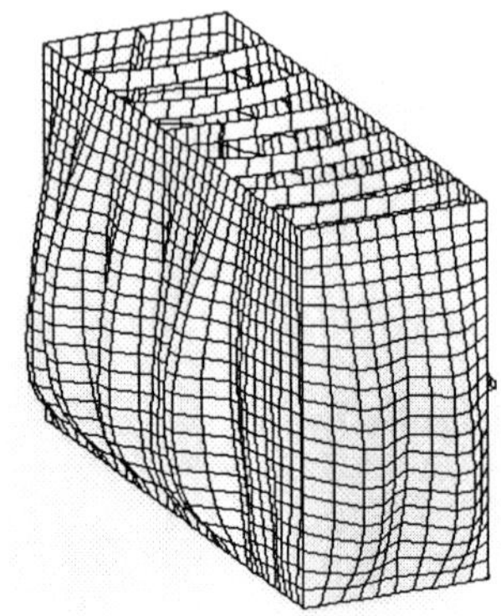

Figure 2: Calculated result of "Breath Mode"

2 CLASSIFICATION OF DAMAGE CAUSES

Regarding the causes of vibration damage in rectangular tanks, investigations were carried out to clarify the mechanism of vibration. While details of investigation can be referred to the relevant papers, overview of the vibration causes is summarized in this paper. In case that the tank wall vibrates heavily, many of the cases are brought by the resonance between the exciting frequency and the reduced natural frequency of fundamental vibration mode of the tank wall. The reduction of the natural frequency of fundamental mode is caused by increase of added mass based on the hydro-elastic interactions. This interaction is induced by the following factors:

2.1 Narrow Shape of Tank

When the rectangular tank is filled with fluid and the distance of a pair of opposite side walls is short, movement of these walls interferes each other through the fluid, which results in the rise of fluid pressure, Kito (1959). This rise of pressure causes the increase of added mass of fluid and then it reduces the natural frequency of the tank wall. There may be two types of vibration mode of the opposite side wall, i.e. in-phase mode and out-phase mode, however, the fundamental mode is out-phase because the added mass of out-phase mode is always larger than that of in-phase. Therefore, out-phase mode of vibration in side wall appears as fundamental mode as shown in Fig.2. This mode is recently called "Breath Mode" by Bae, Yasuzawa and Kagawa (2000). In this manner, the natural frequency of tank wall should be predicted considering the change of added mass according to the tank size.

2.2 Effect of Multi Tiers

Tanks in engine room tend to have tall and slender shape, as a result of the restriction of space. These tall tanks are reinforced by horizontal stringers and the tank becomes multi-tiered tanks. The tank is rigidly supported by the stringers but not tightened, therefore, the fluid is able to move freely from a tier to another tier. In this multi-tiered tanks, oscillation of upper and lower walls causes the increase of added mass of "Breath Mode".

Hence the fundamental vibration mode of multi-tiered tanks becomes in-phase mode between upper and lower wall and out-phase mode of opposite walls as illustrated in Fig.3. One example of calculation shows that the added mass coefficient, which means the percentage of effective tank volume acting as added mass. In case of the two tiered tanks vibrating in "Breath Mode", added mass coefficient increases to about 3 times of that of single tier tank as shown in Fig.4, Takeda (1999).

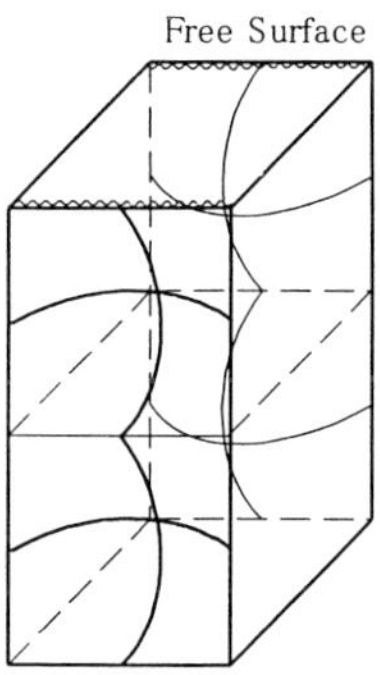

Figure 3: "Breath Mode" in double-tired tank

No. of Tiers LxBxH	1 1x1x1	2 1x1x2	2 1x1x2
Vib. Mode	Out-phase : Opposite Walls	Out-phase : Opposite Walls Out-phase : Upp&Low. Walls	Out-phase : Opposite Walls In-phase : Upp&Low. Walls
Figure of Vib. Mode	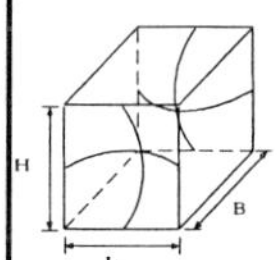	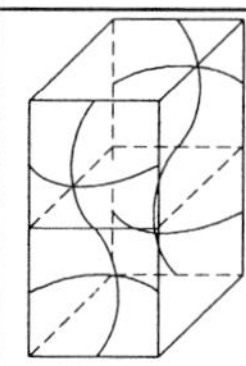	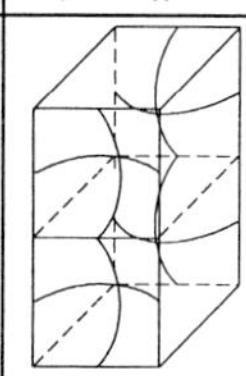
Added Mass Coeff. M	0.16	0.16	0.49

Figure 4: Added mass coefficient of tank

2.3 Effect of Interference between Adjacent Structures

Increase of added mass due to the hydro-elastic interaction between adjacent structures may appear not only in upper and lower tank wall but also in the interference between side wall and bottom plate or in the interference between side wall and adjacent side wall.

2.4 Effect of Coupling between Panel and Stiffener

Fundamental vibration mode of stiffened wall is usually considered to be a panel vibration mode "p" or stiffened plate vibration mode "s" as illustrated in Fig.5. However, panel and stiffener coupling mode of vibration "sp" would appear in tank wall contacting with fluid as fundamental vibration mode, Takeda and Niwa (2000). This phenomenon is caused by the increase of added mass in "sp" mode due to the tank dimension. In case that the tank dimension, L in length, B in breadth and H in height, satisfies the following condition, "sp" mode would appear as fundamental mode.

$$B/L < 0.5 \text{ or } H/L > 0.5 \tag{1}$$

If such vibration mode occurs, it may easily cause structural failure, since the bending stress of plate becomes high at the connection with stiffener due to the clamped edge.

2.5 Effect of Thermal Stress

Although this factor is not classified as hydro-elastic interaction, it sometimes causes serious damage due to the reduction of natural frequency. In case of fuel oil tank, in particular heavy fuel oil tank, the tank is always heated in order to reduce the viscosity of fuel oil. Whereas the panels of the fuel oil settling tank will extend due to heated oil, surrounding structure of the tank, such as bulkheads or flats, are in normal temperature. Thus the extension of tank structure is restricted by flats or bulkheads, therefore, thermal stress of the heated tank generates compressive stress in tank panel.

It is known that the compressive stress reduces the natural frequency of tank panel. In the failure analysis of a heavy fuel oil settling tank of a VLCC, it is reported that the natural frequency of a panel under thermal stress of 0.7 times of critical buckling stress is reduced to about half of the original frequency, Takeda (1999).

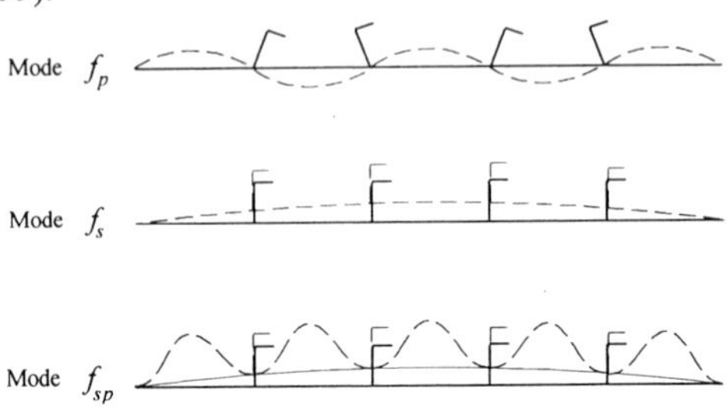

Figure 5: Typical vibration mode of stiffened wall

3 NATURAL FREQUENCY OF COUPLED VIBRATION

3.1 Prediction Method

Let us assume a rectangular tank whose length is L, breadth is B, height is H, consist of two opposite walls reinforced by stiffeners of number n-1 as represented in Fig.6, then the strain energy of stiffened wall is obtained in the following equation, Takeda and Niwa (2000):

$$V = \frac{1}{2}\iint D\left\{\left(\frac{\partial^2 w}{\partial x^2}\right)^2 + \left(\frac{\partial^2 w}{\partial y^2}\right)^2 + 2\nu\frac{\partial^2 w}{\partial x^2}\cdot\frac{\partial^2 w}{\partial y^2} + 2(1-\nu)\left(\frac{\partial^2 w}{\partial x \partial y}\right)^2\right\}dxdy + \sum_{i=1}^{n-1}\frac{1}{2}\int_0^H \kappa\ \alpha EI\left(\frac{\partial^2 w}{\partial x^2}\right)^2\Bigg|_{y=\frac{L}{n}i} dx \qquad (2)$$

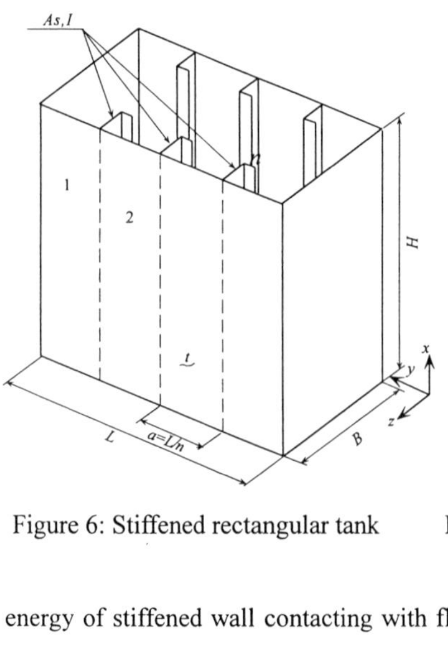

Figure 6: Stiffened rectangular tank

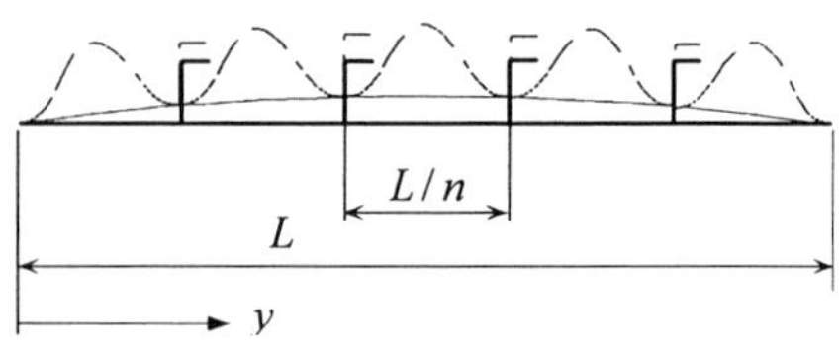

Figure 7: Coupling mode of panel and stiffener

Kinetic energy of stiffened wall contacting with fluid can be expressed as a sum of kinetic energy of stiffened wall and kinetic energy of fluid in the following form:

$$T=\frac{1}{2}\rho_m t\iint\left(\frac{\partial w}{\partial t}\right)^2 dxdy+\sum_{i=1}^{n-1}\frac{1}{2}\int_0^H \rho_m A_S\left(\frac{\partial w}{\partial t}\right)^2\Bigg|_{y=\frac{L}{n}i} dx+\frac{1}{2}\rho_w\iint\phi|_{Z=0}\cdot\frac{\partial\phi}{\partial z}\Bigg|_{z=0} dxdy \tag{3}$$

By assuming that the vibration mode of stiffened wall is represented by the coupling of large wall and small panel divided by stiffeners as illustrated in Fig. 7, vibration mode is given in the form:

$$w=\sin\left(\frac{\pi}{H}x\right)\cdot\left\{\lambda\sin\left(\frac{\pi}{L}y\right)+\sin^2\left(\frac{\pi}{L/n}y\right)\right\}\cdot\cos\omega t \tag{4}$$

where λ indicates amplitude ratio of large wall against small panel

With the aid of Layleigh's principle, circular frequency of stiffened wall can be derived in the following form:

$$\omega^2=\frac{k_{2p}+2k_{sp}\lambda+k_s\lambda^2}{\left(m_{2p}+\mu m_{2p}^*\right)+2\left(m_{sp}-\mu m_{sp}^*\right)\lambda+\left(m_s+\mu m_s^*\right)\lambda^2}\times\frac{\pi^4 D}{\rho_m H^4 t} \tag{5}$$

In Eq.(5), suffixes "s", "2p" and "sp" indicate the vibration mode of stiffened wall, mode of independent panel and coupling mode of panel and stiffener, respectively. Coefficients m, k are non-dimensional modal mass and modal stiffness, respectively, and suffix * represents mass associated with fluid. These coefficients are listed in Table 1, according to vibration modes. In order to minimize ω in Eq.(5), $\partial u/\partial\lambda=0$ must be satisfied, as long as the first term of right side equation in Eq.(5) is represented by u. Hence minimum ω can be obtained by solving the following equation and:

$$\left\{\left(m_s+\mu m_s^*\right)\left(m_{2p}+\mu m_{2p}^*\right)-\left(m_{sp}-\mu m_{sp}^*\right)^2\right\}u^2-\left\{k_s\left(m_{2p}+\mu m_{2p}^*\right)+k_p\left(m_s+\mu m_s^*\right)-2k_{sp}\left(m_{sp}-\mu m_{sp}^*\right)\right\}u+\left(k_s\cdot k_{p2}-k_{sp}{}^2\right)=0 \tag{6}$$

Putting positive minimum root of Eq.(6) as $u_{\min}$, natural frequency of the stiffened panel contacting with fluid can be expressed in the following equation:

$$f_{sp}^*=\frac{1}{2\pi}\sqrt{u_{\min}\times\frac{\pi^4 D}{\rho_m H^4 t}} \tag{7}$$

TABLE 1
MODAL MASS AND STIFFNESS COEFFICIENT

	k	m	m^*
s	$\left\{1+\left(\frac{H}{L}\right)^2\right\}^2+\frac{n\kappa\alpha EI}{DL}$	$1+\frac{nA_s}{Lt}$	$\frac{4}{\pi^4}\sum_i\sum_j\varepsilon_i^2\gamma_j^2\frac{\coth(Bn_{ij}/2)}{Bn_{ij}}$
sp	$\frac{2}{\pi}\cdot\frac{4n^2}{4n^2-1}\left\{1+\left(\frac{H}{L}\right)^2\right\}^2$	$\frac{2}{\pi}\cdot\frac{4n^2}{4n^2-1}$	$\frac{2}{\pi^3}\sum_j\gamma_{0,j}^2\frac{\coth(Bn_{0,j}/2)}{Bn_{0,j}}+\frac{2}{\pi^3}\sum_j\gamma_{2n,j}\delta_j\frac{\coth(Bn_{2n,j}/2)}{Bn_{2n,j}}$
$2p$	$\frac{3}{4}+2n^2\left(\frac{H}{L}\right)^2+4n^4\left(\frac{H}{L}\right)^4$	$\frac{3}{4}$	$\frac{1}{4\pi^2}\sum_j 2\gamma_{0,j}^2\frac{\coth(Bn_{0,j}/2)}{Bn_{0,j}}+\frac{1}{4\pi^2}\sum_j\delta_j^2\frac{\coth(Bn_{2n,j}/2)}{Bn_{2n,j}}$

3.2 Accuracy of Prediction Method

In order to confirm the accuracy of this prediction method, natural frequency calculation of the rectangular tank is carried out by FEM as illustrated in Fig.8. The tank size is 4m in length, 3m in breadth and 3m in height, of which opposite walls are reinforced by 3 vertical stiffeners. The tank is filled with water up to the top but the water has free surface. In Fig.8, a pair of side walls and a bottom plate are rigid wall represented by cross-hatching. The water is modeled as virtual mass.

Calculated fundamental mode by FE analysis is shown in Fig.9 and comparison of the natural frequency by FE analysis and the present method is listed in Table 2. According to Fig.9, fundamental mode of this tank is "Breath Mode", namely the opposite walls vibrate in out-phase mode and the coupling mode "sp" appears. While natural frequency of 9.10Hz is obtained by FEM, 8.66Hz is calculated by the present method, which means 5% error.

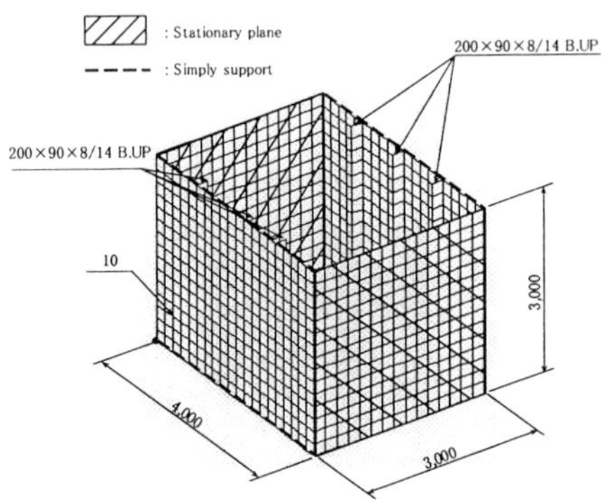

Figure 8: FEM model of rectangular tank

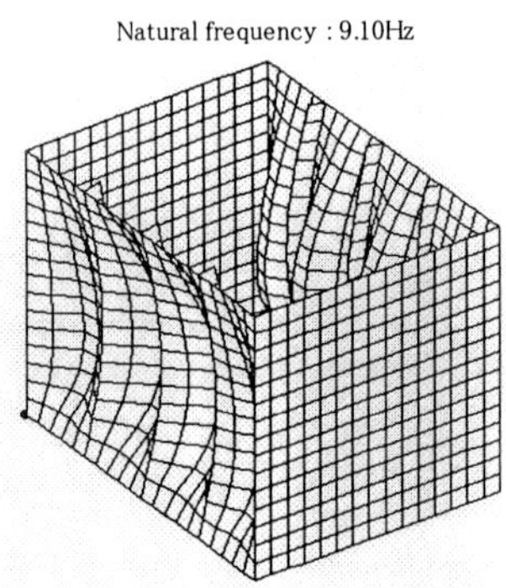

Figure 9: Calculated fundamental mode (Natural frequency : 9.10Hz)

TABLE 2
COMPARISON OF ESTIMATED NATURAL FREQUENCY

Unit : Hz

Mode	FEM (A)	Present method (B)	Error (B/A)
f_{sp}	9.1	8.66	0.95

4 CONCLUDING REMARKS

1) On the basis of several investigations, it is clarified that the natural frequency of fundamental mode of rectangular tank structure may be reduced than the designed frequency due to the following reasons:

- Narrow shape of tank
- Effect of multi tiers
- Effect of interference between adjacent structures
- Effect of coupling between panel and stiffeners
- Effect of thermal stress

2) Vibration prediction using energy method is proposed for practical design purpose and its accuracy is confirmed by FE analysis.

Nomenclature

f : natural frequency
L : tank length
B : tank breadth
H : tank height
t : plate thickness of wall
a : stiffener space $a = L/n$
n : number of panel
ρ_m : mass density of plate and stiffeners
ρ_w : mass density of fluid
EI : bending rigidity of stiffener
ν : Poisson's ratio
D : bending rigidity of plate
$D = Et^3/12(1-\nu^2)$
κ : reduction of bending rigidity due to stiffener shape
α : reduction of bending rigidity due to shearing deformation
A_s : sectional area of stiffener
μ : ratio between equivalent mass of fluid and equivalent mass of tank wall
$\mu = 4\,\rho_w B/\rho_m t$
w : deformation shape of tank wall
ϕ : velocity potential of fluid
ω : circular frequency

$$\varepsilon = \begin{cases} 2 & \text{for } i = 0 \\ 4 & \text{for } i \neq 0 \end{cases}$$

$$Bn_{ij} = \frac{\pi B}{L}\sqrt{i^2 + \left(\frac{L}{H} j\right)^2}$$

$$\gamma_{ij} = \frac{1}{(i^2-1)(j^2-1)}$$

$$\delta_j = \frac{1}{j^2-1}$$

References

Bae S. Y., Yasuzawa Y. and Kagawa K. (2000), Breath Mode and Pressure Distribution in Fluid-Structure Coupled Vibration. *Trans. of The West-Japan Society of Naval Architects*, 99, 211-223. (in Japanese)

Kito F. (1959), On Vibration of a Rectangular Tank Filled with Water. *Journal of the Society of Naval Architects of Japan*, 107, 75-82. (in Japanese)

Nippon Kaiji Kyokai (1998), Casualty Review.

Takeda Y. (1999), Natural Frequency of Multi-Tiered Tanks Subjected to Thermal Stress. *Proc. of TEAM99 in Keelun*, 41-50.

Takeda Y. and Niwa F. (2000), Fundamental Vibration Modes of Stiffened Tank Plate in Contact with Fluid. *Journal of the Society of Naval Architects of Japan*, 188, 569-577. (in Japanese)

Practical Design of Ships and Other Floating Structures
You-Sheng Wu, Wei-Cheng Cui and Guo-Jun Zhou (Eds)

INFLUENCE OF JOURNAL BEARING MODELLING METHOD ON SHAFT LINE ALIGNMENT AND WHIRLING VIBRATIONS

L. Murawski

Institute of Fluid Flow Machinery, Polish Academy of Sciences,
14 Fiszera, 80-952 Gdańsk, Poland
Ship Design and Research Centre,
8 Rzeczypospolitej, 80-369 Gdańsk, Poland

ABSTRACT

Calculation correctness of journal bearings stiff characteristics is one of the most important data during static and dynamic analysis of a marine power transmission system. Especially shaft line alignment and whirling vibration analysis are sensitive to proper boundary conditions determination. Nowadays it is still difficult to find out realistic data of the all shaft line foundations characteristics. Three elements are analysed in the paper: local hull stiffness, stiffness of intermediate bearing frame and lube oil stiff characteristics.

Stern part of a container ship is modelled and analysed by Patran – Nastran software (finite element method). There is calculated static and dynamic stiffness of a stern tube bearing and stiffness of an intermediate bearing foundation. In the next part of the paper there are shown analysis of intermediate bearings frame stiffness. Natural vibration analysis show that construction detuning from excitation forces is very good, then only static stiff calculations have been done. Specialised software (based on finite difference method) has been made by author to determining lube oil stiff characteristics of journal bearings. There are analysed stern and intermediate bearings. Calculated static and dynamic pressure distribution determined horizontal and vertical stiff distribution along bearing length. Analysis has been done as a function of shaft line revolutions and bearings deflections.

Several analysis of shaft line alignment and whirling vibrations of typical marine power transmission system have been done. Influence of different boundary conditions modelling method as well as shaft line's operation conditions on calculations results have been analysed. In author's opinion, stiffness characteristics should be taken into account during static and dynamic calculations of the shaft line alignment and lateral vibrations. Stern tube bearings should be modelled as a continuous support, intermediate bearings' and main engine bearings' support could be modelled as points.

KEYWORDS

Shaft line alignment, Whirling vibrations, Stern tube bearing, Intermediate bearing, Journal bearing, Boundary conditions, Stiffness characteristics.

1 STIFFNESS OF A BEARINGS FOUNDATION

Stern part of a container ship 2000 TEU has been modelled (fig. 1) and analysed. The model contained 134708 degrees of freedom. Added water has been taken into account. Excitation forces are placed at the: stern (k8) and bow (k11) edge of a stern tube bearing, stern collision bulkhead (k16) and intermediate bearing (k20). Static stiffness is shown in the table 1 in each of above points. Dynamic stiff characteristics of the bearing's foundation are shown in fig. 2.

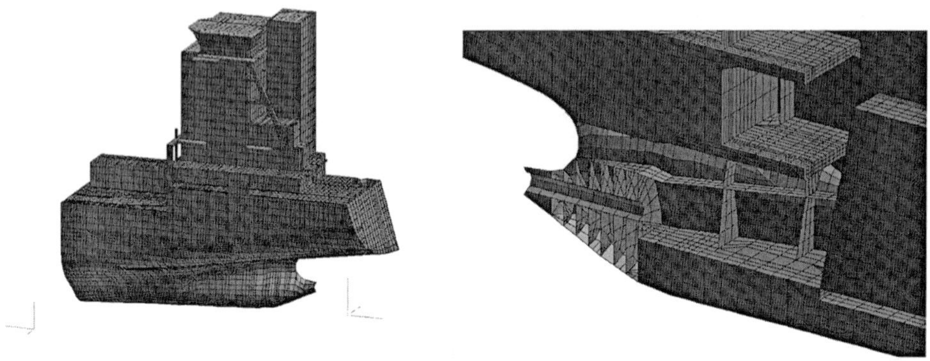

Figure 1: Model of a container's stern part

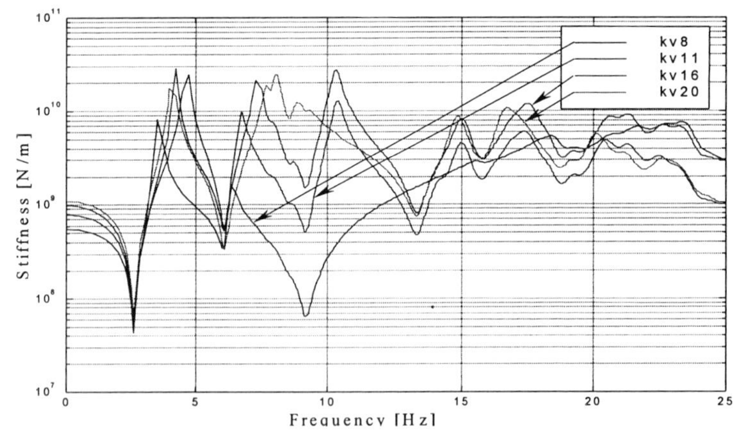

Figure 2: Vertical dynamic stiffness of container hull

TABLE 1

STATIC STIFFNESS OF CONTAINER HULL

	Stiffness [N/m]	
	Vertical	***Horizontal***
k8	5.55×10^8	3.44×10^8
k11	7.91×10^8	6.01×10^8
k16	9.96×10^8	1.02×10^9
k20	1.10×10^9	1.11×10^9

2 STIFFNESS OF AN INTERMEDIATE BEARING FRAME

FEM Model of a typical intermediate bearing frame (fig. 3) contains 195253 degrees of freedom. Three type size of bearing's frame have been analysed. The lowest natural frequency (bearing type ϕ600/ϕ620) is equal 132 Hz. Typical marine excitation frequency is less than 30 Hz. Detuning is so high that only static calculations has been done. Static stiffness is shown in the table 2 for each of the analysed bearing frames.

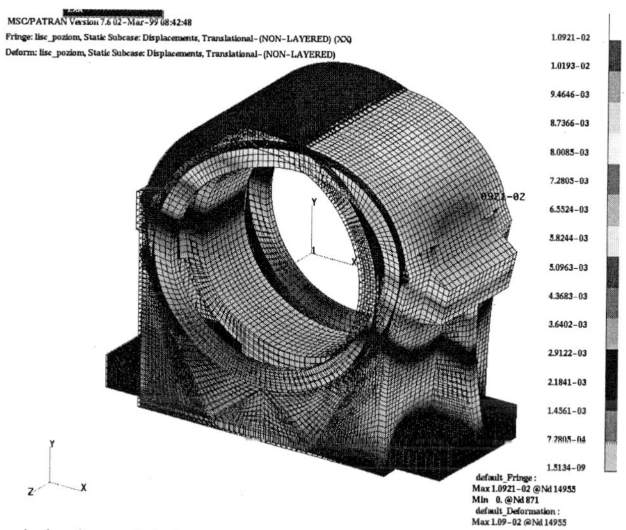

Figure 3: Static horizontal deformation of the intermediate bearing frame type ϕ520/ϕ540

TABLE 2

STATIC STIFFNESS OF INTERMEDIATE BEARING FRAMES

Bearing's type	**Stiffness [N/m]**	
	Vertical	***Horizontal***
ϕ430/ϕ460	$2.98x10^9$	$2.01x10^9$
ϕ520/ϕ540	$3.35x10^9$	$2.31x10^9$
ϕ600/ϕ620	$3.84x10^9$	$2.57x10^9$

3 FILM OIL CHARACTERISTICS

Action of a power transmission system of the container ship (the same as in the previous chapters) has been analysed. Oil film characteristics (stiff and damping) of a stern tube bearing and intermediate bearing has been modelled. Stern tube bearing's dynamic pressure distribution in nominal speed (95 rpm) and design shaft line alignment (+1.75 mm) is shown in fig. 4 (assumptions: 0 m - stern edge, 0 deg. - top point). Vertical stiffness distribution along stern tube bearing as a shaft line revolutions function is shown on fig 5. Also stiffness – shaft line alignment (bearing's relative vertical shift) dependence is shown on fig 5. Dynamic characteristics have been calculated for the first propeller's blade (5 blade) harmonic excitation frequency. Absolute values of the stern tube bearing's oil film static and dynamic stiffness and damping are shown in the table 3. The same values for the intermediate bearing are shown in the table 4.

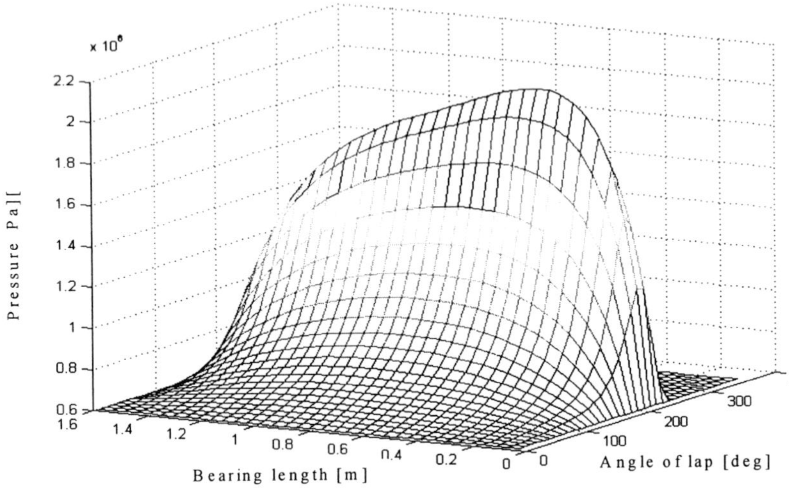

Figure 4: Stern tube bearing's dynamic pressure distribution in nominal speed

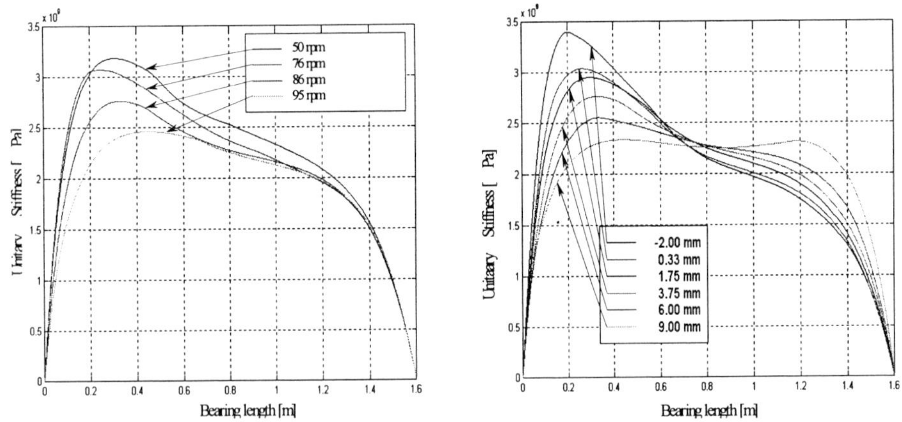

Figure 5: Stern tube bearing's vertical stiffness distribution

TABLE 3

CHARACTERISTICS OF STERN TUBE BEARING'S OIL FILM

Revolutions [rpm]	Static stiffness k_v [N/m]	Dynamic stiffness k_{vd} [N/m]	Damping c_v [Ns/m]
30.0	6.197×10^9	3.109×10^{10}	9.509×10^8
50.0	3.677×10^9	2.507×10^{10}	5.377×10^8
76.0	3.513×10^9	2.460×10^{10}	4.711×10^8
86.0	3.285×10^9	2.412×10^{10}	4.189×10^8
95.0	3.102×10^9	2.325×10^{10}	4.050×10^8

TABLE 4

CHARACTERISTICS OF INTERMEDIATE BEARING'S OIL FILM

Revolutions [rpm]	Static stiffness k_v [N/m]	Dynamic stiffness k_{vd} [N/m]	Damping c_v [Ns/m]
30.0	1.375×10^9	7.803×10^9	4.911×10^8
76.0	1.286×10^9	9.318×10^9	2.019×10^8
95.0	1.448×10^9	1.072×10^{10}	1.865×10^8

4 ANALYSIS OF SHAFT LINE ALIGNMENT AND WHIRLING VIBRATIONS

Specialised software (based on finite element method) has been made by author for a shaft line alignment and whirling vibration calculations. Power transmission system of the container ship (the same as in the previous chapters) has been analysed. The model and example of shaft line's deformations is shown on fig 6.

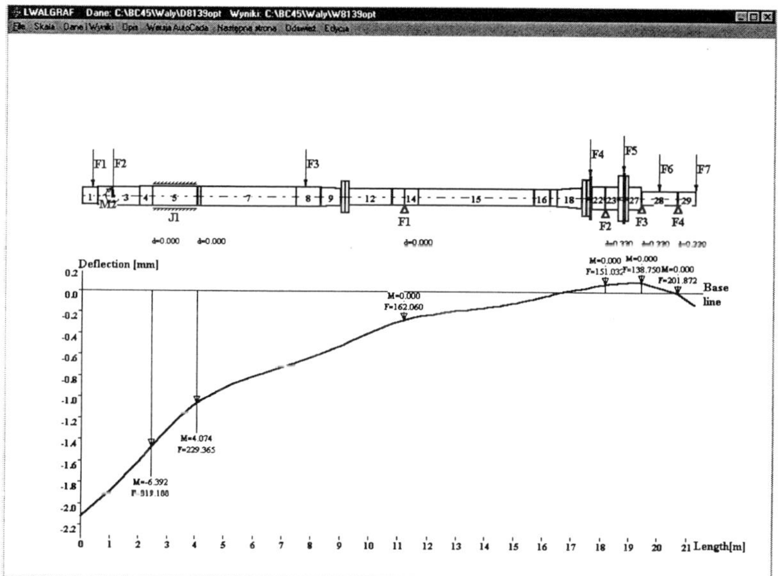

Figure 6: FEM model of the container's shaft line

Influence of different boundary conditions (especially stiff and damping characteristics of propulsion's bearings) has been determined. The following seven variants of calculations have been done: 1 – classical model – bearings modelled as pointwise ideal stiff support; 2 – shaft line's support elasticity, determined in the previous chapters is used, stern tube bearing is modelled as a continuous support; 3 – all bearings are modelled as a continuous support, rest assumptions the same as 2; 4 – calculations are limited to the vertical plane, rest assumptions the same as 3; 5 – main engine bearings are modelled as pointwise ideal stiff support (producers recommendation), rest assumptions the same as 3; 6 – simplified, rectangular stiff distribution in bearings' models, rest assumptions the same as 3; 7 – all bearing's stiffness are multiplied by 2.0, rest assumptions the same as 3. Crankshaft loading (by shear forces and bending moments) and shaft line vertical absolute deformation for above mentioned variants are shown in fig. 7.

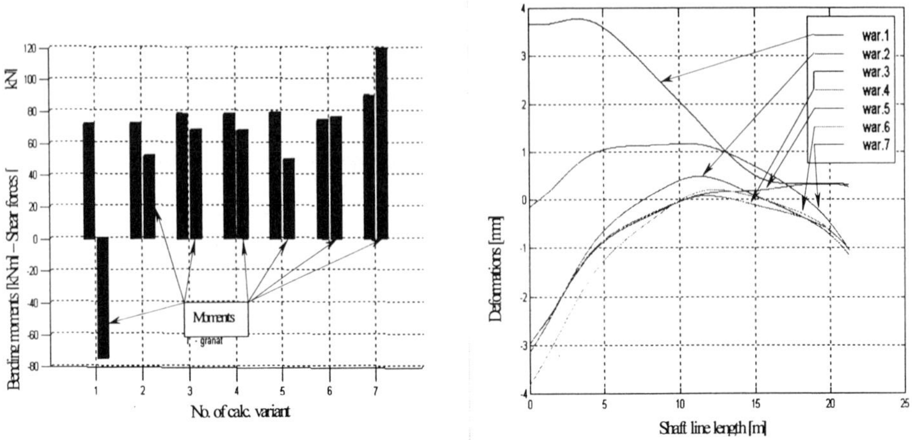

Figure 7: Crankshaft loading and shaft line vertical deformations in different assumptions

In the next step shaft line alignment and whirling vibrations in different operation conditions have been analysed. The model described in point 3 is used. Variants from hot main engine conditions in nominal speed to zero speed cold conditions are calculated. Bearings' vertical, static and dynamic reactions are shown in fig. 8. No. 1 and 2 mark stern tube bearing (stern and aft edge); no. 3 – intermediate bearing; no. 4, 5 and 6 first main engine's bearings.

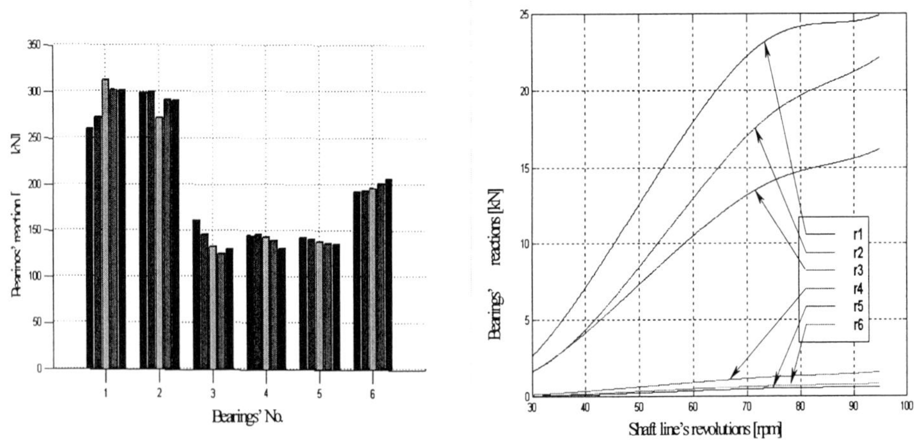

Figure 8: Static and dynamic bearing's reaction

5 SUMMARY

Numerical analysis has been verified by measurements. Three different experimental investigation methods of propulsion shaft alignment are used and compared. The first method (standard method used in the shipyard) is based on the direst determination of bearing reaction by measuring pressure on a hydraulic jack. The second one is based on the identification of shafting alignment by measuring bending stress in the two points of an intermediate shaft. The last attempt at the direct extensometric measurement of intermediate bearing reaction by determining the strain in the bearing foundation pads.

Measurements and numerical analysis (with bearings stiff characteristics taken into consideration) have got good correlation in cold main engine condition. Also the influence of an engine thermal conditions in the bearing reaction level was confirmed. The influence of ship ballast condition in the shafting alignment, not taken into account so far, seems not to be negligible. The influence of the blade components of the propeller-induced hydrodynamic forces on the power transmission system in question is small. However accounting for the constant components of hydrodynamic forces is reasonable.

Shaft line alignment analysis could be confined to one, vertical plane for most of the ships. Multiplane calculations have to be done only if horizontal hydrodynamic forces are not negligible. During shaft line alignment analysis correct distribution of the oil film pressure should be taken into consideration as well as bearing's reaction distribution and crankshaft loading. The same level of Sommerfeld number should be reached in each propulsion bearing. Not only do static shaft alignment have to be analysed but also dynamic parameters (whirling vibrations) should be calculated. Especially problem of detaching shaft neck from bearing pipe should be analysed.

In author's opinion, stiffness and damping characteristics should be taken into account during static and dynamic analysis of the shaft line alignment and whirling vibrations. It is essential while shaft line alignment is performed in the dock or on the slipway. Stern tube bearings should be modelled as a continuous support, intermediate bearings' and main engine bearings' support could be modelled as points. Hull's local stiffness, intermediate bearing's frame and oil film stiffness are in the same order of magnitude, therefore all this components should be taken into account. Foundation stiffness might be determined by modelling only stern part of the ship's hull even without deckhouse and chimney (to the upper deck). Natural frequencies of the intermediate bearing's frame are high, therefore only static analysis could be done.

References

(1995). Shaft Alignment for Direct Coupled Low-Speed Diesel Propulsion Plants. *MAN B&W Diesel A/S*, Copenhagen, Denmark.

Cowper B., DaCosta A., Bobyn S. (1999). Shaft Alignment Using Strain Gages: Case Studies. *Marine Technology*, **36:2**, 77 91.

Murawski L. (1997). Analysis of Measurement Methods and Influence of Propulsion Plant Working Parameters on Ship Shafting Alignment. *Polish Maritime Research* **4:2** and **3**, 19-22 and 15-20.

Pressicaud J.P. (1986). Correlation Between Theory and Reality in Alignment of Line Shafting. *Bureau Veritas*, Paris, France.

Volcy G.C. (1976). Forced Vibrations of The Hull and Rational Alignment of The Propeller Shaft. *Bureau Veritas*, Paris, France.

Volcy G.C., Ville R. (1981). Actual Tail Shaft Behaviour on Oil Film Taking Account of Propeller Forces and Moments. *Bureau Veritas*, Paris, France.

Practical Design of Ships and Other Floating Structures
You-Sheng Wu, Wei-Cheng Cui and Guo-Jun Zhou (Eds)

EXPERIMENTAL STUDIES ON RESISTANCE REDUCTION AND VIBRATION REDUCTION BY BUBBLY LAYER

Wen-Cai Dong, Fan Wu, Yun-Xiang Zhu and Ri-xiu Guo

Dept.of Naval Architecture and Ocean Eng'g, Naval university of Eng'g
Wuhan 430033, China.

ABSTRACT

There are two parts in this paper. The first part presents the experimental study on flow in Air-liquid two phase mixed boundary layer of flat plate. The flow field is visualized by light sheet in low turbulence water tunnel. The horizontal velocity and thickness of the boundary layer at the outermost edge are measured by Laser Doppler Velocity technique. The influence on flow field by plate orientation, free stream velocity and form of air injection are studied. The results show that the thickness of the boundary layer increases as air flow rate increases, the velocity at the outermost edge of the boundary layer decreases due to air injection. The second part presents the experimental study on vibration reduction of ship hull model by bubbly layer. The experiment was carried out in a circulating flow water tank. The effect of bubbly layer on free vibration, on response of forced vibration and on damping of hull model are measured under several selected variations of model draft, air injecting pore size, air flow rate, and water flow velocity. The results show that the response of forced vibration of ship hull model is effectively reduced by bubbly layer, the maximum reduction is about 50%.

KEYWORDS

Resistance reduction, Vibration reduction, Bubbly layer, Air injection, Air-liquid two phase mixed boundary layer, LDV technique

1 INTRODUCTION

A thin bubbly layer, formed by injecting air between ship's bottom and water, can affect ship resistance, vibration response and noise of ship structures. To study these effects of bubbly layer and to investigate their mechanism, problems of ship structure-bubbly layer-liquid interactions were highlighted by researchers in the field of ship mechanics in last two decades. The main researches carried out by the authors in this field in recent years are presented in Zhu & Guo (1994, 1996a, 1996b) and Dong et al (1999, 2000a, 2000b)

Experimental study on resistance reduction of flat plat by bubbly layer (Dong & Guo 1999) indicates that the local frictional resistance is effectively reduced by bubbly layer, the maximum reduction is about 80%. In order to investigate the flow characteristics of bubbly layer, which is a mixed flow of air and liquid, especially to investigate the mechanism of resistance reduction by bubbly layer, the authors carried out an experimental study on flow in air-liquid mixed boundary layer of flat plate.

It is reported that the bubbly layer can reduce vibration response of ship's afterbody, which is observed during a test of noise control of ship propeller by bubbly layer (Zhao et al 1990), but Zhao et al did not study the phenomena of vibration reduction further. In order to investigate the effect of bubbly layer on vibration reduction of ship hull, the authors carried out an experimental study on vibration reduction of ship model by bubbly layer.

In this paper, the experimental studies on resistance reduction and on vibration reduction by bubbly layer are presented separately.

2 EXPERIMENT ON FLOW IN AIR-LIQUID TWO PHASE MIXED BOUNDARY LAYER

2.1 Experimental Setup

The experiment was conducted in the low turbulence water tunnel (400mm×400mm ×6000mm) at State Key Laboratory on Turbulence Research of Peking University. The free stream velocity ranges from 10mm/s to 1300mm/s; the turbulence is less than 0.3%.

The test plate is made of perspex, with oval-shaped leading edge and beveled trailing edge in longitudinal section. The dimensions of the test plate are 1200mm ×380mm ×10mm. A porous panel (70mm×130mm) is placed at 410mm from the leading edge. The sketch of the setup is shown in Fig.1. The free stream velocity U is measured at section in front of the leading edge by 1080mm.

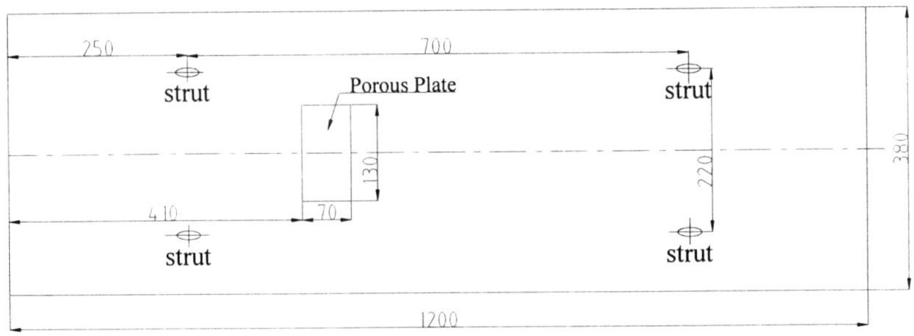

Fig.1 Experimental Model

The porous panel is a stainless steel plate, on which pores are scribed by means of laser. The pores are oval in shape, spaced at 1.5mm center to center, their nominal diameters have 3 variations, 0.11mm, 0.2mm, 0.5mm respectively.

The flow is visualized by light sheets, which are projected to the lower surface of the test plate from the sides near the bottom of the tunnel. The recorder lens and camera lens are placed just under the lower surface of the test plate. The flow velocity at test section is measured by Laser Doppler Velocity technique. The laser is perpendicularly projected to the central longitudinal plane of the tunnel. The airflow rate Q is measured by a LZDH-15 metal rotameter, its maximum range is $10m^3/h$.

2.2 Results and Discussions

2.2.1 Flow visualization

The influences on the flow field of air –liquid two phase mixed boundary layer by 3 variations of free stream velocity, 3 variations of pore size, different plate orientation (horizontal, trim 5^0, heel 7.5^0), and by variation of air flow rate are studied by flow visualization technique. The typical flow field photographs for 0.5mm pore size are shown in fig.2 ~ fig.5.

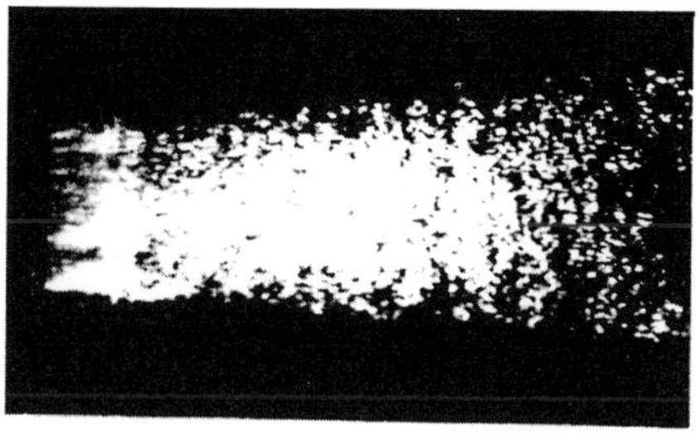

Fig. 2: Q=1m^3/h ,U=0.84 m/s

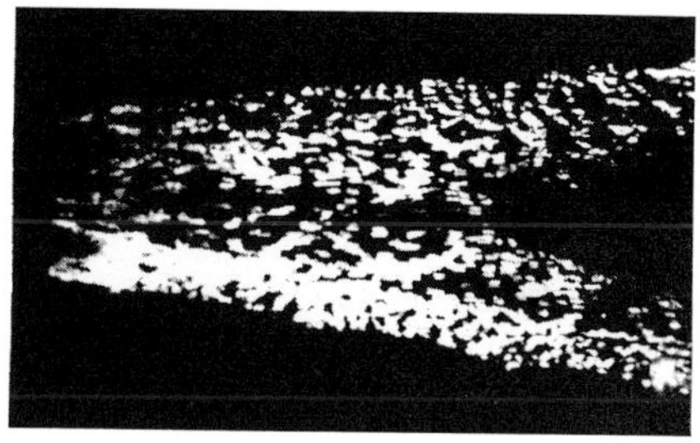

Fig.3 Q=4 m^3/h, U=1.29 m/s

Fig.4 Q=1 m^3/h, U=1.29 m/s, heel

Fig.5 Q=4 m^3/h, U=1.29 m/s, heel

The flow visualizations show that: (1) When plate is placed horizontally, the air-liquid two phase mixed flow starts from the air injection section, extends longitudinally, diffuses transversely along the downstream. The diffuse range in transverse direction increases as airflow rate increases, but decreases as free stream velocity increases. (2) The air bubbles are ellipsoidal approximately, their diameters range from 0.3mm to 10mm. The bubble size increases as airflow rate increases, decreases as free stream velocity increases, but is little affected by pore size, when free stream velocity exceeds a certain value. (3) When plate is at trim condition and when the free stream velocity is less than 0.8m/s, the mixed flow behaves as crescent or triangular air sheet, air leaks out from both sides of the test plate near the air injection section. The length of air sheet decreases as airflow rate increases. When free stream velocity exceeds a certain value, the flow pattern for test plate at trim condition is very similar to that for plate at horizontal condition, but the length of mixed boundary layer for plate at trim condition is somewhat shorter than that for plate at horizontal condition. (4) When plate is at heel condition, the mixed flow is biased towards rising along heel direction, further more, when airflow rate exceeds a certain value, vortices are formed, which start from air injection section, and diffuse away downstream.

2.2.2 Velocity of flow field and thickness of mixed boundary layer

The horizontal velocity and thickness of mixed boundary layer δ at the outermost edge are measured for 3 variations of free stream velocity (0.5m/s, 0.8m/s, 1.3m/s), 2 variations of air injection pore size (0.11mm, 0.5mm) and different plate orientation. The thickness of mixed boundary layer given in fig.6 and fig.7 corresponds to plate in horizontal position and with 0.11mm air injection pores; x, in these figures, is measured from the leading edge of the plate.

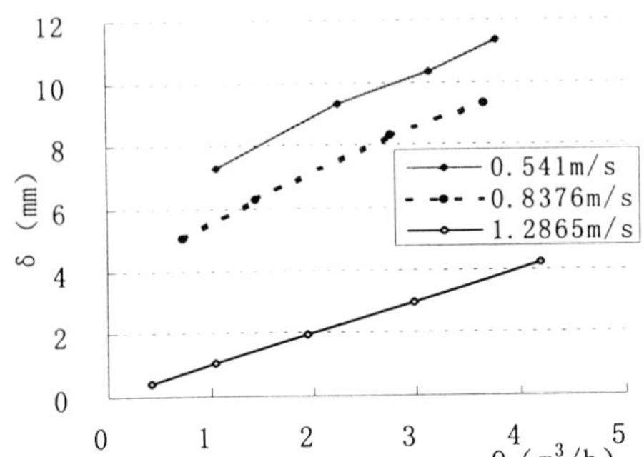

Fig.6 δ at x=450mm

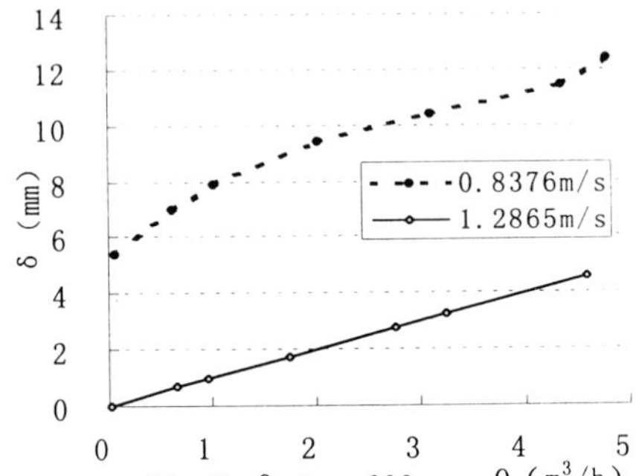

Fig.7 δ at x=600mm

The results show that the horizontal velocity at the outermost edge of mixed boundary layer decreases due to air injection. The average decrease is about 8%; the thickness of the mixed boundary layer increases as airflow rate increases, decrease as free stream velocity increases.

In air liquid mixed boundary layer, the reduction of skin friction is related to the inhibition of turbulence in liquid phase. The factors governing the variation of turbulence in liquid phase are as follows: (1) the wake flow generated due to change of bubbles shape, (2) change of momentum due to partial volume of inflow liquid being displaced by air bubbles, (3) bubble resistance arising from the difference in velocities between air and liquid phases, (4)dissipation of turbulent energy due to change of volume or shape of bubbles. The first and second factor promotes turbulence, the third and fourth factor suppresses turbulence. When the bubble shapes change very small, the first factor can be neglected. When bubbles sizes are significantly smaller than the characteristic length of turbulence, the turbulence is suppressed. The main factor to the inhibition of turbulence is the bubble resistance. The reduction of horizontal velocity at the outermost edge of mixed boundary layer and the increment in thickness of boundary layer reveal that the flow field structure of the boundary layer of flat plate is changed by air injection.

3 EXPERIMENT ON VIBRATION REDUCTION OF SHIP MODEL BY BUBBLY LAYER

3.1 Experimental Model

The model is made of Aluminum plate. It has the following main dimensions: length 320cm, breadth 47cm; depth 31cm, plate thickness 0.12cm, rib 20×20×1.2mm, rib spacing 15cm. The bow compartment of the model, 50cm in length, is water tight for air injection. A porous panel is placed at the bottom of the bow compartment. There are two different designs of the panel, one is of pores with diameter D 1.2mm, the other, pore diameter 0.5mm. The instrumentation setup and the location of measurement points are shown in fig.8.

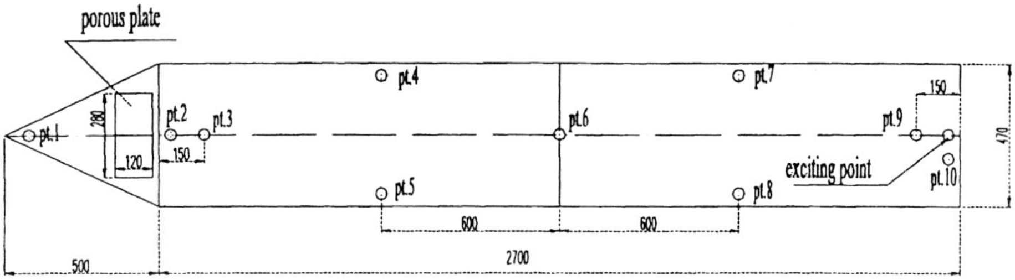

Fig. 8: Measurement system and location of measurement points

3.2 Experimental Procedures

The experiment was carried out in a circulating flow water tank. The natural frequency of the model is measured under 2 variations of flow velocity(1.0m/s,1.5m/s), 2 variations of drafts(light draft, loaded draft) and 5 variations of air flow rate (1.0m^3/h, 1.5m^3/h, 2.0 m^3/h, 3.0 m^3/h, 5.0 m^3/h). Impose on the model a periodic force with frequency equal to the resonance frequency of the model, on which different bubbly layer is applied. The corresponding dynamic responses are measured. The periodic exciting force is unloaded suddenly when the model is in steady state of resonance. Then the damping process is recorded.

3.3 Results of the Experiment and Analysis

3.3.1 The effect of bubbly layer on natural frequency of the model

The stern of the model is hammered in loaded draft state. The natural frequency is measured under two variations of water flow velocity and three variations of air flow rates. The results are shown in table 1.

TABLE 1
NATURAL FREQUENCY OF THE MODEL

Serial number	U (m/s)	Q (m^3/h)	Mode 1 (HZ)	Mode 2 (HZ)
1	0	0	170.0	375
2	1	0	172.5	373.75
3	1	1	171.25	368.75
4	1	3	170.00	372.50
5	1	5	172.5	362.5
6	1.5	0	173.75	375.00
7	1.5	1	172.50	373.75
8	1.5	3	172.50	375.00
9	1.5	5	177.50	381.25

From table 1, we can see that the change of natural frequency under different flow velocity and air flow rate are not significant. Therefore, it can be concluded that the effect of bubbly layer on natural frequency of the model can be neglected.

3.3.2 The effect of bubbly layer on responses of forced vibration of the model

When exciting force is imposed with frequency equal to the first natural frequency of the model, the accelerations (effective value) at the measurement points 1, 3, 6, 10, under different loading conditions are shown in figure 9. When the exciting force with frequency equal to 2nd natural frequency, the acceleration responses are shown in figure 10. From these figures, we can see that the bubbly layer has significant reduction effect on the dynamic response of the model under periodic exciting force. The maximum reduction of vibration response is about 50%.

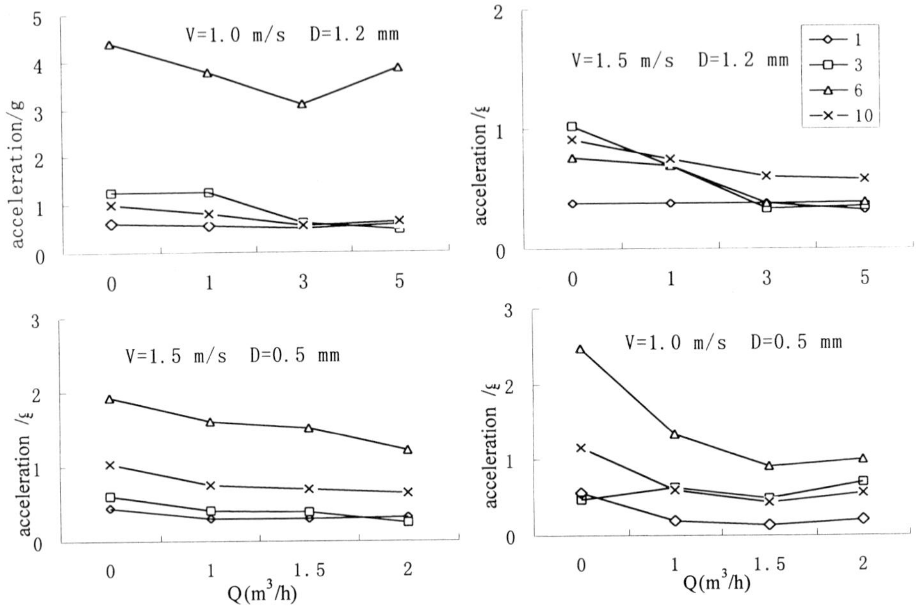

Fig. 9: Acceleration response at measurement points (1,3,6,10) under the mode 1

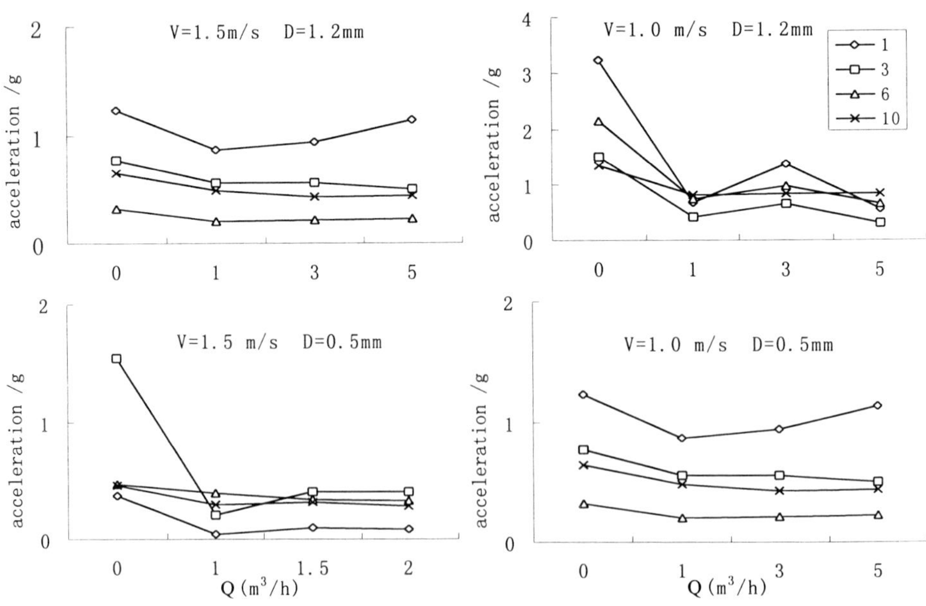

Fig. 10: Acceleration response at measurement points (1,3,6,10) under the mode 2

3.3.3 *The effect of bubbly layer on damping*

The damping process of the vibration responses at the measurement points is recorded when the exciting force with resonance frequency of the model is unloaded suddenly. Based on the linear damping theory, the equivalent viscous damping at measurement points 6 and 10 are calculated, as shown in table 2. From table 2, we can see that the bubbly layer can increase the damping of the model vibration in some extent.

TABLE 2

EQUIVALENT VISCOUS DAMPING

Serial number	Draft	U (m/s)	Q (m^3/h)	Equivalent viscous damping			
				Mode 1		Mode 2	
				pt. 6	pt.10	pt.6	pt.10
1	Light draft	0	0	0.0233	0.0224	0.0116	0.0129
2	Light draft	1	0	0.0156	0.0151	0.0133	0.0138
3	Light draft	1	1	0.0224	0.0180	0.016	0.0254
4	Light draft	1.5	0	0.0181	0.0173	0.0122	0.0127
5	Light draft	1.5	1	0.0262	0.0261	0.0276	0.01346
6	Loaded draft	0	0	0.0194	0.0192	0.0120	0.0133
7	Loaded draft	1	0	0.0217	0.0233	0.0124	0.0138
8	Loaded draft	1	1	0.0403	0.0415	0.0203	0.0236
9	Loaded draft	1.5	0	0.0177	0.0181	0.0151	0.0171
10	Loaded draft	1.5	1	0.0196	0.0211	0.0325	0.0302

4 CONCLUSIONS

(1) The reduction of skin friction of flat plate by bubbly layer is due to the inhibition of turbulence in the liquid phase. (2) The bubbly layer has very small effect on free vibration of ship model ,but it can effectively reduce the vibration response of the model under periodic exciting force. The maximum reduction of vibration response is about 50%. (3) The bubbly layer has an effect on increasing the damping of the model vibration in some extent. (4) it is worthwhile to carry out further study on the application of bubbly layer to resistance reduction of high speed crafts and flat bottom barges and to vibration reduction of ship structure.

References

Dong W.C. & Guo R.X. (1999). Experimental Study on the Skin Friction Reduction of Flat Plate by Formation of Air Cavity. *Proc., International Workshop on Ship Hydrodynamics*, 21-27. Wuhan.

Dong W.C., Guo R.X. and Liu X.W. (2000a).Air Injection on the Bottom of Stepped Planing Craft and Its Effect on Resistance. *Journal of Ship Mechanics* **4:3**, 36-42.

Dong W.C., Guo R.X. and Liu X.W. (2000b). Effects of Different Air Injection Technique on Resistance of Stepped Planing Boat. *Shipbuilding of China* **41:2**, 8-14.(in Chinese).

Zhao X.R., Zhang W.G., Li D.R., Hao Y.W. and Wang H.B.(1990). The Experimental Research of Air –Bubble Curtains for Reduction the Propeller Noise and the Stern Hull Vibration. *Shipbuilding of China, Supplement*, 178-186.(in Chinese)

Zhu Y.X. and Guo R.X.(1994). Basic Equation of a Simplified Theory of Air-Liquid Two Phase Flow. *Proceedings, Chinese Symposium on Hydrodynamics* 621-625, Ocean Press.(in Chinese)

Zhu Y.X. and Guo R.X.(1996a). Vibration of Elastic Spherical Shell in Bubbly Layer. *Journal of Vibration Engineering* **9:3**, 237-243. (in Chinese).

Zhu Y.X. and Guo R.X.(1996b). Hydroelastic Analysis of Structure-Bubbly Layer-Liquid Interaction Problems, *Journal of Solid Mechanics* **17:2**, 157-162.(in Chinese)

Practical Design of Ships and Other Floating Structures
You-Sheng Wu, Wei-Cheng Cui and Guo-Jun Zhou (Eds)

APPLICATION OF HIGHER ORDER BALANCER TO CONTROL THE SUPERSTRUCTURE VIBRATION OF A CONTAINER SHIP

Soo-Mok Lee, Won-Hyun Kim and Kyoon-Yang Chung

Maritime Research Institute, Hyundai Heavy Industries Co., Ltd.
1 Cheonha-Dong, Dong-ku, Ulsan 682-792, Korea

ABSTRACT

This paper presents the results of analysis, measurement and countermeasure to solve the vibration problem predicted in the superstructure of a 4,800 TEU container carrier. The main reason of vibration was due to the resonance between the deckhouse longitudinal vibration mode and the 4th order main engine unbalance moment. Detailed analysis based on 3-dimensional FE model was performed for the vibration prediction and onboard measurement and exciter test were done for confirmation and identification. The 4th order mechanical balancer was installed to compensate engine excitation and reduce deckhouse vibration. As consequences it was found effective in reducing the vibration level to the allowable limit but resultant vibration was still not negligible seemingly due to the shaft thrust excitation. It was further considered to determine the optimal phase of the balancer control, which gives minimal deckhouse vibration through phase vector analysis method.

KEYWORDS

4,800 TEU container carrier, Deckhouse vibration, Main engine unbalance moment, Finite element m odel, Exciter test, 4th Order balancer, Phase vector analysis

1 INTRODUCTION

It is often the general mechanism of ship structural vibration that the global or substructural vibration mode is resonant with main excitation sources of the ship as main engine and propeller. Of these substructures the deckhouse is of primary concern being normal accommodation space for crew's work and rest and its vibration is subject to evaluation under ISO 6954 standard [1] which has recently been revised. This paper deals with prediction and solution of the deckhouse vibration of a 4,800 TEU container ship due to main engine excitation. Some excessive vibration exceeding the allowable limit in main operation range was predicted through the analysis and an appropriate measure to reduce the vibration was required.

As is often the case in the vibration problem, the vibration mode, excitation and solution were all different from existing cases which were worth to note and discuss herein. Natural frequency of the deckhouse was lower than any conventional one which gave the possibility of resonance with lower

frequency excitation. Fourth order main engine inertial moment which seldom becomes a magnitude to be a vibration source was the main source of excitation. And higher order mechanical balancer was newly introduced to control the vibration in actual application.
Detailed vibration analyses with 3-dimensional FE model were performed for prediction and followed by various onboard tests and measurements for experimental confirmation. Phase vector analysis was applied in the determination of optimal phase of the balancer. Main particulars of the ship are presented in Table 1 for reference.

TABLE 1
MAIN PARTICULARS OF THE SHIP

Ship Type	4,800 TEU container carrier
L.O.A.	294.1 m
L.B.P.	283.2 m
Bmld	32.2 m
Dmld	21.8 m
Design Draft	12.0 m
Main Engine	HYUNDAI-B&W 7K98MC
MCR	54,460 BHP x 94.0 RPM
NCR	47,000 BHP x 89.5 RPM
Propeller	F.P.P
No. of Blade	5
Diameter	8.4 m

2 VIBRATION ANALYSIS

2.1 Analysis Model

Finite element model for vibration analysis included whole ship structure of aft end, engine room, deck house, hold and fore end with simpler main engine model. Both sides of symmetric ship structure were modeled with the size of 5476 nodes and 14332 elements with plates and beams. The model is shown in Figure 1.

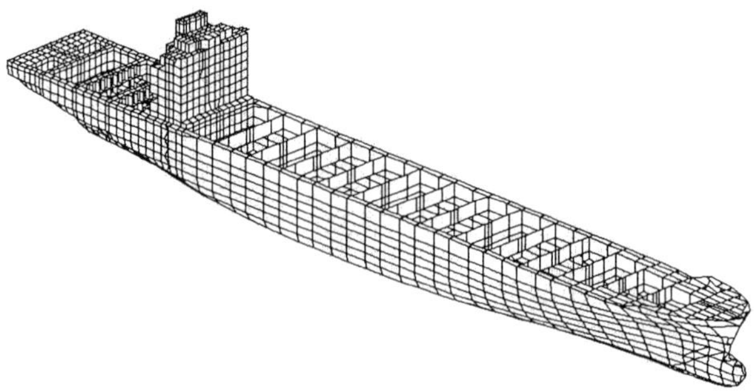

Figure 1 : Model for vibration analysis

2.2 Excitation

Excitations in global ship vibration come from the main engine and propeller where main engine excitation shows multiple of engine revolution speed and the propeller excitation has blade passing frequency and its multiples. In this case, the excitation of concern was main engine unbalance moment coming from the inertial forces of engine moving masses. Usually the first and second order components are significant which sometimes needs counterbalancing or compensation but fourth order component was specially significant in this 7-cylindered engine with large bore and super-long stroke. The basic vibration concern in the design stage was the resonance of deckhouse longitudinal vibration mode and 4th order engine unbalance moment.

Another 4th order excitation comes from the thrust force by the shaft axial vibration. Shaft thrust force does not have uniform or simple variation with respect to engine speed whereas the engine unbalance moment as well as the balancer force have exact proportionality to the square of the engine speed. Usually the shaft thrust is not clearly predictable as the engine unbalance moment so that it is more practical way to determine from the experimental method in onboard measurement.

The location and direction of major excitations related with deckhouse vibration are illustrated in Figure 2.

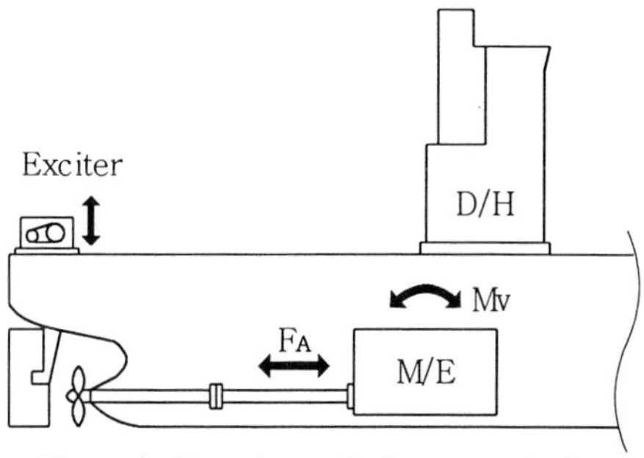

Figure 2 : Locations of relevant excitations

2.3 Mode and Response Analysis

Natural frequencies and mode shapes were extracted from the normal mode analysis of structural model. Lanczos algorithm and virtual mass capability of MSC/NASTRAN were used to solve the eigenvalue problem and fluid-structure interaction problem due to the added mass effect of surrounding water respectively [2].

Vibration response was predicted with assumed excitations. Response of navigation bridge deck usually chosen as the reference point of the vibration evaluation was predicted to exceed the allowable limit due to the resonance between deckhouse longitudinal mode and the fourth order engine excitation In Figure 3 the expected responses of deckhouse by the main engine unbalance moment M_{4V} and shaft axial force F_{4A} are presented separately in a range of engine speed. It was expected that the amplitudes of vibration by the engine and the shaft are nearly in same level and both exceed the allowable limit of 4 mm/s in main operation range.

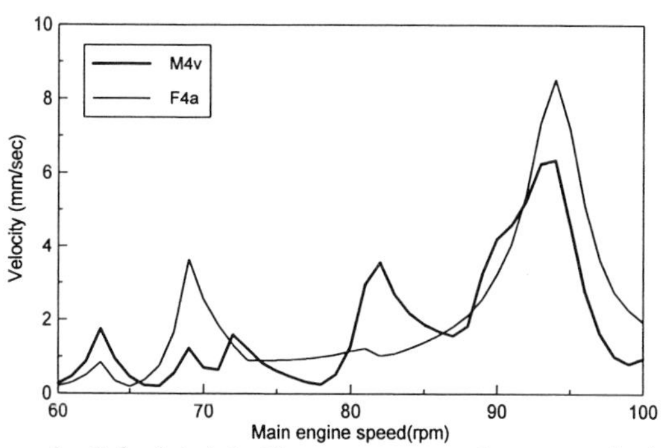

Figure 3 : Calculated deckhouse responses by two excitations

3 VIBRATION MEASUREMENT

3.1 *Tests and Measurements*

Exciter tests were performed to identify the vibration characteristics experimentally. As shown in Figure 2, a counter-rotating mechanical exciter of 15 ton capacity was installed and operated in the upper deck aft end for vertical excitation while vibration measurements of the major response points were being done to find out the vibration characteristics of each substructures including mode, resonance range, mobility and phase information, etc. Vibration measurement during the sea trial test was made in ballast condition for full speed range. It was found out that the 4th order component vibration is most dominant as expected in the analysis.

3.2 *Comparison with Analysis Results*

Generally good coincidences were found between analysis and measurement results. The transfer frequency response function between the aft end vertical excitation and deckhouse longitudinal response is shown in Figure 4. It showed good coincidence below 5 Hz where for hull girder beam mode governs and above 7.5 Hz for non-resonant frequency range. Intermediate frequency zone between 5-7.5 Hz showed some discrepancies mainly due to the estimation error in deckhouse stiffness and other factors. Another comparison in actual response of deckhouse for the 4th order component is presented in Figure 5 where the level and trend of the vibration are very similar each other but show some differences in local resonant peaks.

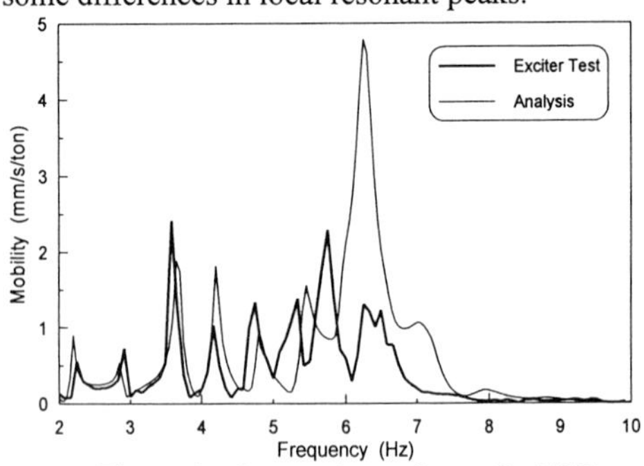

Figure 4 : Comparison of transfer FRF between aft end and deck house

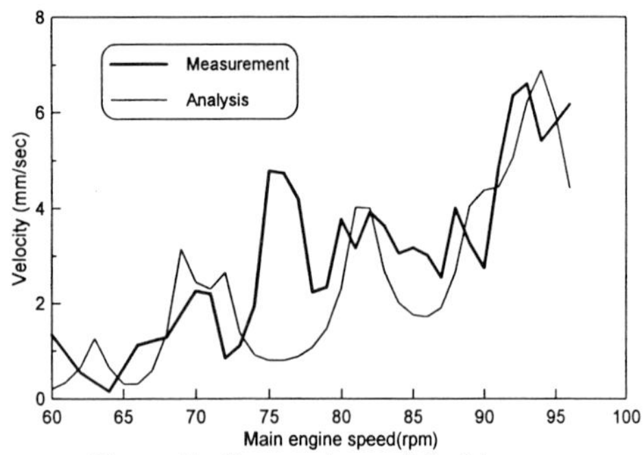

Figure 5 : Comparison of deckhouse vibration (4th order component)

4 BALANCER CONTROL

4.1 *Balancer Installation*

Mechanical balancer is installed in fore and aft of the main engine to compensate the engine-generated 4th order unbalance moment. Each set of balancer consists of two counter-rotating eccentricities to generate unidirectional force and is designed to rotate 4 times the speed of main engine shaft rotation. The size and phase of the eccentricities are set to exactly compensate the engine excitation. In order to observe the effect and to find out the optimal phases of the balancer, both the balancer operating and non-operating conditions are tested with successive measurements.

4.2 *Effect of Balancer Operation*

The vibration responses measured before and after the balancer running are compared. In the Figure 6 the relevant vibration magnitude w.r.t. engine speed are shown and responses in complex plane with phase information are presented for two cases. It was found out that the balancer operation made distinct effect in reducing the vibration in the speed range over the 90 rpm but not clear effect below that range. In the complex plane representation the real and imaginary parts of the response are shown for each rpm in Figure 7. One could find considerable difference for even similar level of vibrations which is contributed by the balancer operation.

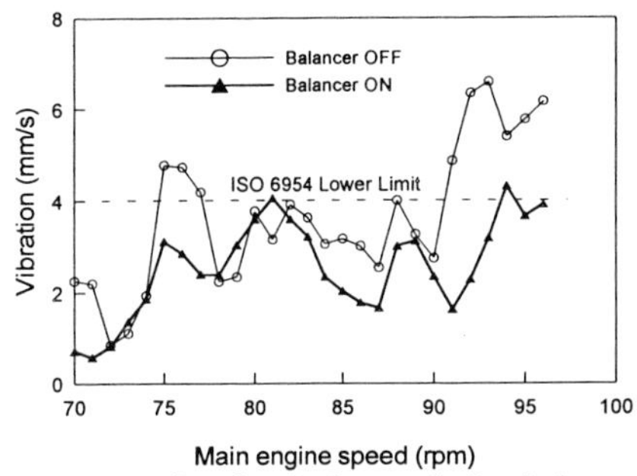

Figure 6 : Vibration before and after balancer running

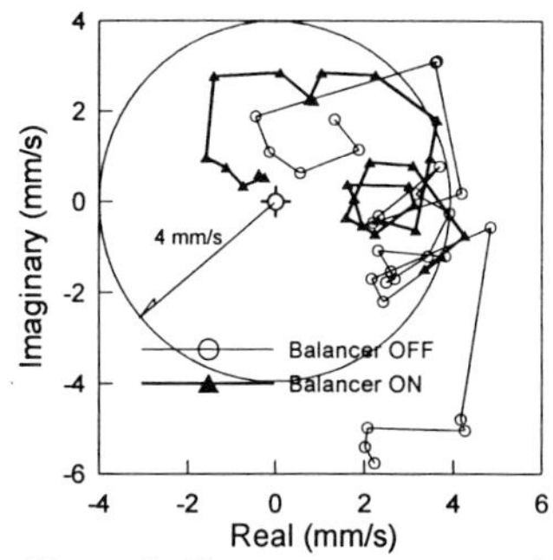

Figure 7 : Response representation in complex plane

The reason why the 4th order vibration is not completely removed or drastically reduced in most of the engine speed is in the presence of other excitations as shaft thrust mentioned previously. Ideally, it could be interpreted that the response under the balancer operating is the response purely due to the shaft axial thrust force. As a whole the response under the balancer operating is within the allowable limit of 4 mm/s.

4.3 Phase Vector Analysis

It is also required to eliminate or reduce the shaft thrust excitation but there is no availability to directly control it in practical way. On this account it is required to reduce two inherent excitation sources of main engine and shaft thrust by a single set of balancer simultaneously. But this should not be excitation-based but response-based strategy, that is, the balancer phase should be determined which gives minimum vibration in deckhouse area [3].

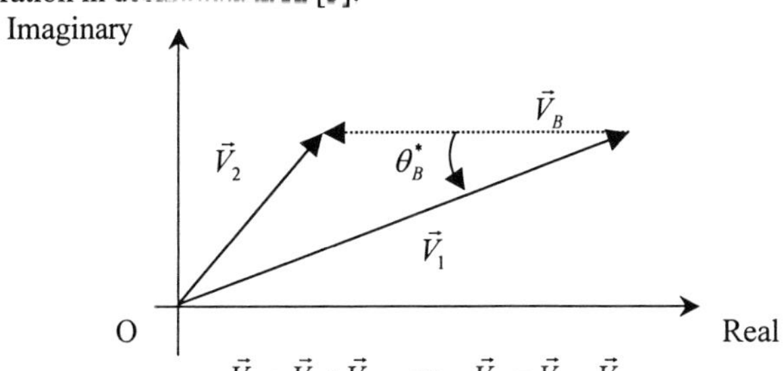

$$\vec{V}_2 = \vec{V}_1 + \vec{V}_B \quad \text{or} \quad \vec{V}_B = \vec{V}_2 - \vec{V}_1$$

Figure 8 : Vector relation between vibration response

Figure 8 shows a typical vector diagram of vibration responses where $\vec{V}_1$ and $\vec{V}_2$ are vibration response vectors before and after the balancer operating, and $\vec{V}_B$ is their difference and equivalent to the response by the excitation of balancer itself. In this case the response could representatively be the navi. bridge deck response in a certain engine speed.

As can be understood from the figure it is when $\vec{V}_B$ is in line with $\vec{V}_1$ in opposite direction that the resultant $\vec{V}_2$ gets minimum magnitude. In that case the phase of the $\vec{V}_B$ should be adjusted by θ_B^*.

Since $\vec{V}_1$ and $\vec{V}_2$ vectors have all different values in each different rpm as shown in Figure 6 and Figure 7, $\vec{V}_B$ and $\theta^{\bullet}_B$ have similar variation w.r.t. engine speed accordingly. In Figure 9 $\vec{V}_B$ and $\theta^{\bullet}_B$, the magnitude of net effect due the balancer and required phase of the balancer for minimal vibration are extracted from the vectorial operation or complex value treatment.

4.4 Response with Optimally Phased Balancer

It is shown that by the characteristics of structural dynamics the magnitude has some considerable variation for given speed range but the required phase has rather small variation seemingly between 0 to 90 degrees. Since the phase of the balancer practically needs to be set to a certain fixed value for all the running, the best optimal phase setting may be chosen as the mid line through the variation.

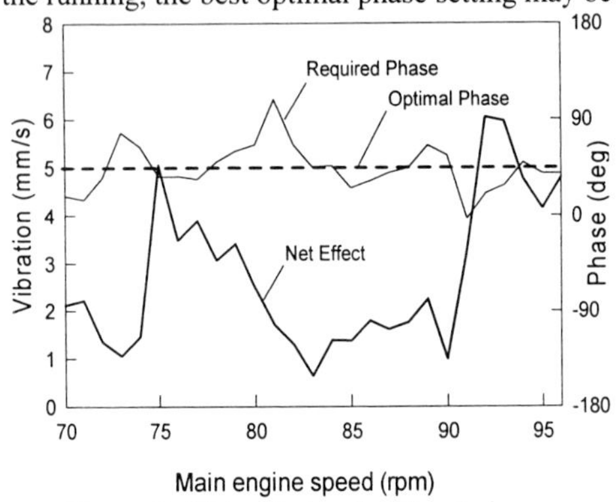

Figure 9 : Net effect and optimal phase of balancer

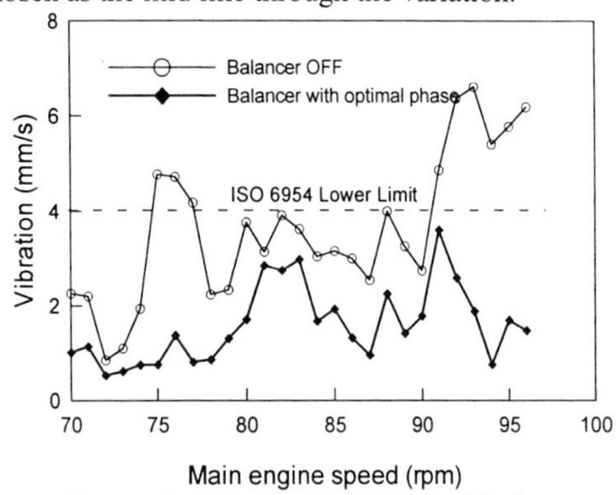

Figure 10 : Simulated response with the optimally phased balancer

It is simulated for the response under the balancer with optimal phase setting. As shown in Figure 10 the response under the balancer with optimal phase setting is much reduced than that with original phase in most of the speed ranges especially in main operating range over 90 rpm. In reality the response under original phase was within the allowable limit, which needed no further phase adjustment, but the strategy and method for optimal phase setting could be applied to other more serious cases.

5 CONCLUSIONS

Deckhouse vibration of a 4,800 TEU container carrier was accessed by FE analysis and onboard measurement. Excessive vibration was predicted due to the resonance between deckhouse longitudinal mode and 4th order main engine unbalance moment. Higher order mechanical balancer was installed to reduce vibration and found to meet satisfactorily relevant allowable limit in operation. For more optimal operation of the balancer, phase vector analysis of the vibration response was attempted. As conclusions good coincidence between measurement and analysis and establishment of optimal control strategy for balancer operation were both encouraging to reliable vibration prediction and informative future application.

References

[1] ISO 6954. (1984). Mechanical vibration and shock - Guidelines for the overall evaluation of vibration in merchant ships

[2] MSC. (2000). MSC/NASTRAN User's Manual Version 70.0

[3] D.S.Cho, S.M.Lee and K.Y.Chung. (1999). Ship Vibration Control using a force- adjustable mechanical actuator, *Journal of Vibration and Control* **5:5**

Practical Design of Ships and Other Floating Structures
You-Sheng Wu, Wei-Cheng Cui and Guo-Jun Zhou (Eds)

NONLINEAR DYNAMICS OF TOWED UNDERWATER VEHICLES
– NUMERICAL MODELLING AND EXPERIMENTAL VALIDATION –

G.F. Clauss and M. Vannahme

Institute of Naval Architecture and Ocean Engineering
Technical University Berlin, Germany

ABSTRACT

Controllable towed underwater vehicles are used as carrier systems in a wide range of maritime applications. The quality and accuracy of the registered multimedia seabed and subsoil data depends on the dynamic behaviour and stability of the towed sensor carrier as well as on the transmission of wave induced motions along the submarine umbilical. This paper presents a comprehensive numerical and experimental analysis of the dynamic behaviour of towed systems, taking into account the non-linear characteristics of drag and lift coefficients as well as the non-linear dynamic cable force. It is proved that ship motions are inducing substantial cable oscillations resulting in significant excitation of the towed body. A computer program is developed to determine the cable curvature for stationary conditions as well as the dynamic response of the towed body due to ship oscillations. The numerical model is validated by towing tests, performed at the TU Berlin in one of the world largest cavitation and circulation tunnels with free surface.

KEYWORDS

Towed underwater vehicle, Nonlinear dynamics, Nonlinear cable forces, Orbital oscillations, Sensor carrier, Depressor wings, Numerical simulation, Experimental validation

1 INDRODUCTION

Vertically tethered systems or towed bodies – applied in a wide spectrum of offshore operations – are characterised by hydroelastic effects. In general, such systems consist of the components research or support vessel, tether, umbilical or towing cable, and a suspended load or a towed sensor carrier (Figure 1). Tethers connecting the ship with the subsea unit allow the precise deployment of the suspended equipment like drill strings, subsea manifolds, BOP stacks or templates at a preselected

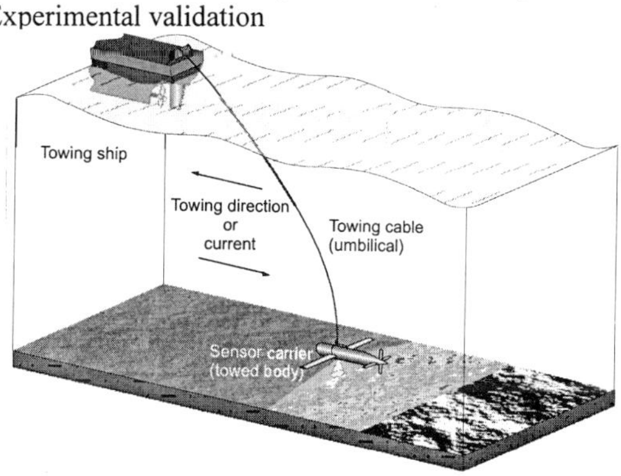

Figure 1: Typical components of a towing system

location on the seabed. Umbilicals enable a ROV or towed body to stay for a long time at the underwater work site and allows real-time communication.
The quality and suitability of the recorded multimedia seabed and subsoil data for further analysis depends on the dynamic behaviour and stability of the towed sensor carrier as well as on the transmission of wave induced motions along the submarine umbilical. The tether or umbilical couples the motion of the floating structure with the oscillations of the underwater system. Under certain conditions resonance is observed, with substantially increased motion and acceleration amplitudes. Due to high inertia and drag forces the deep dived module cannot follow the umbilical oscillations, with the consequence of cable slackening (Clauss and Hoog (2001)). In this case the cable tension decreases to zero, and the motions of the ship and the submerged structure are decoupled for a short while. During this time the upper cable end and the towed body are oscillating independently with different amplitude and phase. When the cable becomes taut again, large snap loads and erratic motions of the submerged body occur resulting in severe damage of the umbilical with its internal electrical and optical conductors. The resulting degradation reduces the life span of the umbilical and endangers the recovery of the suspended or towed subsea module.
A validated analysis model is required, based on theoretical investigations of the non-linear cable dynamics, using the method of finite elements. A dynamic time-domain simulation of marine operations of vertically tethered or towed systems allows the evaluation of occurring load and motion amplitudes during specific missions in advance, defining operational limitations due to weather conditions giving rise to snap loads. Furthermore, designers are enabled to minimise motions and to limit tensions by choosing optimal system parameters.
The dynamic analysis of deep dived towed bodies comprises the determination of the dynamic behaviour of vertically suspended loads. In addition, it considers the hydrodynamic forces and moments due to stationary towing speed as well as hydroelastic effects. Ablow and Schechter (1983), and Milinazzo et al. (1987) use the lumped-mass-method to analyse the three-dimensional dynamic behaviour of submerged towed bodies and to determine the trajectory during a given ship manoeuvre.
Wu and Chwang (1997) describe the three-dimensional dynamic behaviour of a towing system subdivided into a primary and a secondary cable. They show that the heave and pitch motions of the sensor carrier attached to the secondary cable is significantly damped compared to the oscillations of the depressor body at the lower end of the primary cable. Buckham et al. (1999) develop a numerical model – based on the lumped-mass-method – to calculate the three-dimensional dynamic behaviour, i.e. the velocity and depth profile as well as the horizontal trajectory of a towed body during a 180°-manoevre, considering the influence of active control surfaces at the tail plane.
Clauss et al. (1998) determine experimentally and numerically the seaway dependent limitations of offshore pipeline operations. The applied numerical formulation – based on the finite element method – is split in a simplified stationary approach by solving only the stationary components of the differential equations. In a second step the oscillation around the stationary equilibrium – caused by forced excitation at the upper pipeline end – is determined.
This paper presents a comprehensive numerical and experimental analysis of the dynamic behaviour of towed systems, based on the numerical model for pipelaying analysis of Clauss et al. (1998), taking into account the non-linear characteristics of drag and lift coefficients of the sensor carrier as well as non-linear components of dynamic forces due to cable curvature. The analysis proves that ship motions are substantially transmitted along the submerged cable, and excite oscillations of the towed body. A numerical model is developed to determine both, the cable curvature for stationary conditions as well as the dynamic response of the towed body due to ship oscillations. The numerical model is validated by towing tests, performed at the TU Berlin in one of the world largest cavitation and circulation tunnel with free surface. The motions of the towed model are recorded by video cameras and evaluated by digital video post–processing. Cable forces are measured at the upper suspension point (winch) and at the lower end (body attachment).

2 NUMERICAL ANALYSIS

2.1 Coordinate Systems

For numerical analysis the cable is subdivided in small segments of length ds. The motion of each cable segment is described by the vector $\underline{r}(s,t)$, depending on the cable coordinate s and time t. The cable curve can be split in a stationary and a dynamic component, i.e.

$$\underline{r}(s,t) = \underline{r}_{st}(s) + \underline{u}(s,t)\,. \tag{1}$$

Two coordinate systems (Figure 2) are introduced: The first system with the components X, Y, Z travels with constant towing speed U. X points forward in towing direction, Y points to port and Z points vertically upwards. The second, body fixed coordinate system is located at the respective cable position s, distorted against the first coordinate system by the inclination angle $\varphi(s)$ and the deviation angle $\psi(s)$. The transformation is performed, using the transformation matrix $\underline{\underline{A}}$ as follows:

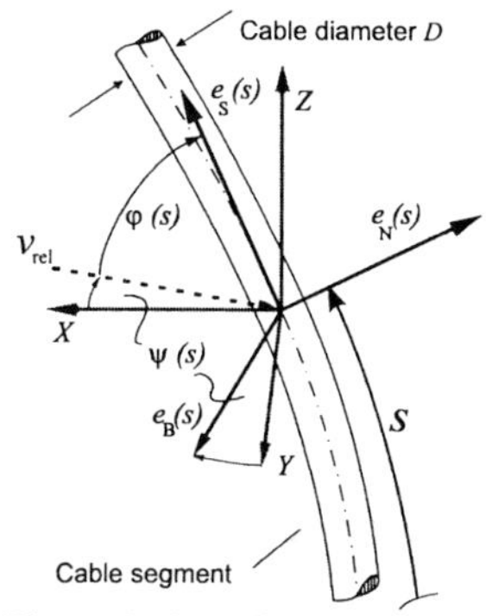

Figure 2: Coordinate systems

$$\begin{Bmatrix} \underline{e}_S(s) \\ \underline{e}_B(s) \\ \underline{e}_N(s) \end{Bmatrix} = \underline{\underline{A}} \cdot \begin{Bmatrix} \underline{e}_X \\ \underline{e}_Y \\ \underline{e}_Z \end{Bmatrix}, \text{ where } \underline{\underline{A}} = \begin{vmatrix} \cos\varphi(s)\cos\psi(s) & \cos\varphi(s)\sin\psi(s) & \sin\varphi(s) \\ -\sin\psi(s) & \cos\psi(s) & 0 \\ -\sin\varphi(s)\cos\psi(s) & -\sin\varphi(s)\sin\psi(s) & \cos\varphi(s) \end{vmatrix}. \tag{2}$$

The local derivatives are therefore:

$$\underline{e}'_S(s) = \varphi' \cdot \underline{e}_N + \psi' \cos\varphi \cdot \underline{e}_B\,; \quad \underline{e}'_N(s) = -\varphi' \cdot \underline{e}_S - \psi' \sin\varphi \cdot \underline{e}_B\,; \quad \underline{e}'_B(s) = \psi'(\sin\varphi \cdot \underline{e}_N - \cos\psi \cdot \underline{e}_S)\,. \tag{3}$$

2.2 Hydrodynamic Cable Load

The line load due to viscous drag $\underline{q}_D$ is a superposition of the tangential and normal drag force which are evaluated with empirical tangential and normal drag coefficients, c_S and c_D, respectively, acting at each finite cable element in the plane opened up by the cable tangent $\underline{e}_S$ and the relative free flow direction $\underline{v}_{rel}$:

$$\begin{Bmatrix} q_{DS} \\ q_{DB} \\ q_{DN} \end{Bmatrix} = \frac{\rho D}{2} \begin{bmatrix} \pi \cdot c_S \cdot v_{rel,S} & 0 & 0 \\ 0 & c_D\sqrt{v^2{}_{rel,B} + v^2{}_{rel,N}} & 0 \\ 0 & 0 & c_D\sqrt{v^2{}_{rel,B} + v^2{}_{rel,N}} \end{bmatrix} \begin{Bmatrix} v_{rel,S} \\ v_{rel,B} \\ v_{rel,N} \end{Bmatrix}.$$

The relative free flow velocity $\underline{v}_{rel}$ follows from two components, the steady towing speed U and the cable oscillation $\dot{\underline{r}}$:

$$\underline{v}_{rel} = -U\underline{e}_X - \dot{\underline{r}}\,,$$

with: $v_{rel,S} = -U\cos\varphi\cos\psi - \dot{u}_S\,;\quad v_{rel,B} = U\sin\psi - \dot{u}_B\,;\quad v_{rel,N} = U\sin\varphi\cos\psi - \dot{u}_N\,.$ (4)

In addition, the ribbed surface of armored umbilicals causes a transverse force $\underline{q}_{\Delta L}$ normal to the relative free flow direction $\underline{v}_{rel}$. This asymmetrical load is taken into account by introducing the empirical transverse force coefficient $c_{\Delta L}$, i.e.

$$\begin{Bmatrix} q_{\Delta LS} \\ q_{\Delta LB} \\ q_{\Delta LN} \end{Bmatrix} = c_{\Delta L} \frac{\rho D}{2} v_{rel,S} \begin{Bmatrix} 0 \\ v_{rel,N} \\ -v_{rel,B} \end{Bmatrix} = c_{\Delta L} \frac{\rho D}{2} v_{rel,S} \begin{bmatrix} 0 & 0 & 0 \\ 0 & 0 & 1 \\ 0 & -1 & 0 \end{bmatrix} \begin{Bmatrix} v_{rel,S} \\ v_{rel,B} \\ v_{rel,N} \end{Bmatrix}.$$

The inertia force $\underline{q}_I$ is derived from the general Morison-equation (with $c_A = 1$):

$$\underline{q}_I = -c_A\rho\frac{\pi D^2}{4}\underline{\ddot{u}}_{normal} = -c_A\rho\frac{\pi D^2}{4}\begin{bmatrix}0&0&0\\0&1&0\\0&0&1\end{bmatrix}\begin{Bmatrix}\ddot{u}_S\\\ddot{u}_B\\\ddot{u}_N\end{Bmatrix}.$$

In conclusion, the hydrodynamic load follows from the sum of all components:

$$\underline{q}_0 = \underline{q}_I + \underline{q}_D + \underline{q}_{\Delta L} = -\underline{\underline{m}}_{hy}\cdot\underline{\ddot{r}} + \underline{\underline{b}}\cdot(-U\underline{e}_X - \underline{\dot{r}}), \quad (5)$$

where $\underline{\underline{m}}_{hy}$ denotes the hydrodynamic mass tensor and $\underline{\underline{b}}$ the hydrodynamic drag and lift tensor, i.e.

$$[m]_{hy} = c_A\rho\frac{\pi D^2}{4}\begin{bmatrix}0&0&0\\0&1&0\\0&0&1\end{bmatrix};\ [b] = \frac{\rho D}{2}\begin{bmatrix}\pi\cdot c_S|v_{rel,S}| & 0 & 0\\ 0 & c_D\sqrt{v^2_{rel,B}+v^2_{rel,N}} & c_{\Delta L}v_{rel,S}\\ 0 & -c_{\Delta L}v_{rel,S} & c_D\sqrt{v^2_{rel,B}+v^2_{rel,N}}\end{bmatrix}.$$

The velocities $\underline{\dot{u}}$ as well as the accelerations $\underline{\ddot{u}}$ of the incremental cable displacements vanish for stationary conditions, leading to the relative stationary free flow velocity of $v_{rel} = -U\underline{e}_X\cdot\underline{e}_S$, $-U\underline{e}_X\cdot\underline{e}_B$, $-U\underline{e}_X\cdot\underline{e}_N$ (see Eqn. (4)). The corresponding hydrodynamic load follows from Eqn. (5), and – for stationary conditions – results in:

$$\begin{Bmatrix}q_{stS}\\q_{stB}\\q_{stN}\end{Bmatrix} = \frac{\rho DU^2}{2}\begin{Bmatrix}-\pi\cdot c_S\cos^2\varphi\cos^2\psi\\ c_D\sqrt{1-\cos^2\varphi\cos^2\psi}\,\sin\psi - c_{\Delta L}\cos\varphi\sin\varphi\cos^2\psi\\ c_D\sqrt{1-\cos^2\varphi\cos^2\psi}\,\sin\varphi\cos\psi + c_{\Delta L}\cos\varphi\cos\psi\sin\psi\end{Bmatrix} \quad (6)$$

2.3 Cable Elongation

Cable elongation ε comprises a stationary and a non-steady component: $\varepsilon = \varepsilon_{st} + \varepsilon_{dy}$, with: $\varepsilon_{st} = F_{st}\cdot EA$. At stationary conditions the finite cable element length is $|d\underline{r}_{st} = ds|$, at non-steady conditions it is elongated to $|d\underline{r}_{st} + d\underline{u}| = |\underline{r}'_{st} + \underline{u}'|ds = |\underline{e}_s + \underline{u}'|ds$ (with: $\underline{r}'_{st} = \underline{e}_s$).

$$\varepsilon_{dy} = \frac{|d\underline{r}| - |d\underline{r}_{st}|}{|d\underline{r}_{st}|} = \frac{|\underline{e}_S + \underline{u}'|ds - ds}{ds} = |\underline{e}_s + \underline{u}'| - 1 \Rightarrow \sqrt{\underline{e}^2_s + 2\underline{e}_s\underline{u}' + \underline{u}'^2} - 1 \approx \underline{e}_s\underline{u}' + \frac{1}{2}\underline{u}'^2.$$

2.4 Cable Equation

By definition, bending and torsion moments are not transferred via an ideal cable: Therefore the remaining force at the free margins of a cable element is the axial force only. Thus the equation of motion for a cable element of submerged weight $\underline{w}$ is:

$$m_{cab}\underline{\ddot{r}} - (F\underline{r}')' = \underline{w} + \underline{q}_0.$$

Introducing the hydrodynamic load vector $\underline{q}_0$ from Eqn. (5) results in:

$$\underline{\underline{m}}\cdot\underline{\ddot{r}} + \underline{\underline{b}}\cdot\underline{\dot{r}} - (F\underline{r}')' = \underline{w} + \underline{q}, \quad (7)$$

with: $\underline{\underline{m}} = m_{cab}[E] + \underline{\underline{m}}_{hy}$ and the stationary line load $\underline{q} = \underline{q}_{st} + \underline{q}_{dy}$.

2.5 Evaluation of the Stationary Cable Curve

Considering the towing system at constant towing speed, excluding any external excitation at the upper cable end, the non-steady terms $\underline{\dot{r}}$, $\underline{\ddot{r}}$ vanish. With $\underline{q} = \underline{q}_{st}$ Eqn. (7) reduces to $-(F_{st}\underline{r}'_{st})' = -w\underline{e}_z + \underline{q}_{st}$, leaving only the stationary components of the line load and the axial cable force. With $\underline{r}'_{st} = \underline{e}_s$ and Eqn. (3) the above equation is transformed into the body fixed coordinate system:

$$-(F_{st}\underline{e}_S)' = -w(\sin\varphi\underline{e}_S + \cos\varphi\underline{e}_N) + \underline{q}_{st} \quad (8)$$

$$\Rightarrow -F'_{st}\underline{e}_S - F_{st}(\varphi'\underline{e}_N + \psi'\cos\varphi\underline{e}_B) = -w(\sin\varphi\underline{e}_S + \cos\varphi\underline{e}_N) + \underline{q}_{st}\,.$$

Splitting Eqn. (8) into the three directional components S,B,N and introducing the stationary hydrodynamic load from Eqn. (6), the characteristics of the stationary cable curve are derived from:

$$F'_{st}(s) = w\sin\varphi(s) + \frac{1}{2}\pi c_S \rho D U^2 \cdot \cos^2\varphi(s)\cdot\cos^2\psi(s)\,,$$

$$\varphi'(s) = \frac{w\cdot\cos\varphi(s) - \frac{1}{2}\rho D U^2 (c_D \sin\varphi\cos\psi\sqrt{1-\cos^2\varphi\cos^2\psi} + c_{\Delta L}\cos\varphi\cos\psi\sin\psi)}{F_{st}(s)}\,,$$

$$\psi'(s) = \frac{\frac{1}{2}\rho D U^2(-c_D \sin\psi\sqrt{1-\cos^2\varphi\cos^2\psi} + c_{\Delta L}\cos\varphi\sin\psi\cos^2\psi)}{F_{st}(s)\cdot\cos\varphi(s)}\,.$$

Integration starts at the towed body suspension point, where the axial cable tension is equivalent to the sum of body weight, drag and hydrodynamic depressor force. Consequently the direction of the resulting force vector corresponds to the inclination and deviation angle of the first finite element. Since the integration starts at the towed end, all cable nodes have to be shifted down by the vertical stationary displacement of the last integrated node point, in order to adjust the winch point at water surface level.

2.6 Solution for Non-steady Conditions

To determine the displacements of each cable node – caused by external excitations at the upper cable end – only the non-steady components of Eqn. (7) are considered:

$$\underline{\underline{m}}\cdot\ddot{u} + \underline{\underline{b}}\cdot\dot{u} - \underline{F}'_{dy} = \underline{q}_{dy} \qquad \text{with} \quad \begin{cases} \underline{F}_{dy} = F_{st}\underline{u}' + F_{dy}(\underline{e}_S + \underline{u}') \\ \underline{q}_{dy} = -U(\underline{\underline{b}} - \underline{\underline{b}}_{st})\cdot e_X\,. \end{cases}$$

Applying the Galerkin-procedure, the equation for each cable segment is multiplied with a kinematical possible displacement $\delta\underline{u}$ and integrated over the whole cable length:

$$\int_L \left(\delta\underline{u}\cdot\underline{\underline{m}}\cdot\ddot{u} + \delta\underline{u}\cdot\underline{\underline{b}}\cdot\dot{u} - \delta\underline{u}\cdot\underline{F}'_{dy}\right)\cdot ds = \int_L \delta\underline{u}\cdot\underline{q}_{dy}\,ds\,.$$

Partial integration results in:

$$\int_L \left(\delta\underline{u}\cdot\underline{\underline{m}}\cdot\ddot{u} + \delta\underline{u}\cdot\underline{\underline{b}}\cdot\dot{u} - \delta\underline{u}'\cdot\underline{F}_{dy}\right)\cdot ds = \int_L \delta\underline{u}\cdot\underline{q}_{dy}\,ds + \left[\delta\underline{u}\cdot\underline{F}_{dy}\right]_0^L\,. \tag{9}$$

Non-steady cable force oscillations are expressed by displacement quantities, according to the law of elasticity:

$$\underline{F}_{dy} = (F_{st} + F_{dy})(\underline{e}_S + \underline{u}') - F_{st}\underline{e}_S = F_{st}\underline{u}' + F_{dy}(\underline{e}_S + \underline{u}')\,; \quad \text{where} \quad F_{dy} = EA(\underline{u}'\cdot\underline{e}_S + \underline{u}'^2/2)\,.$$

Note, that the dynamic cable force is split into a linear and a non-linear component:

$$\underline{F}_{dy} = \underline{\underline{F}}\cdot\underline{u}' + \underline{F}_{nl}$$

with: $\underline{\underline{F}} = ((F_{st} + F_{dy})\underline{\underline{E}} + EA\underline{e}_S \otimes \underline{e}_S)$; $\quad F_{dy} = EA(\underline{u}'\cdot\underline{e}_S + \underline{u}'^2/2)$; $\quad \underline{F}_{nl} = EA\cdot\underline{u}'^2/2\cdot\underline{e}_S$, (10)

transforming Eqn. (9) into

$$\int_L \left(\delta\underline{u}\cdot\underline{\underline{m}}\cdot\ddot{u} + \delta\underline{u}\cdot\underline{\underline{b}}\cdot\dot{u} + \delta\underline{u}'\cdot\underline{\underline{F}}\cdot\underline{u}'\right)\cdot ds = \int_L \left(\delta\underline{u}\cdot\underline{q}_{dy} - \delta\underline{u}'\cdot\underline{F}_{nl}\right)\cdot ds + \left[\delta\underline{u}\cdot\underline{F}_{dy}\right]_0^L\,. \tag{11}$$

The sum of the virtual displacements with respect to tangential and normal components of the stationary cable curve are

$$\int_L \left(\delta\{u\}^T[m]\{\ddot{u}\} + \delta\{u\}^T[b]\{\dot{u}\} + \delta\{u'\}^T[F]\{u'\}\right)\cdot ds = \int_L \left(\delta\{u\}^T\{q\}dy - \delta\{u'\}^T\{F\}_{nl}\right)\cdot ds + \left[\delta\underline{u}\cdot\underline{F}_{dy}\right]_0^L\,, \tag{12}$$

$\{u\}^T = \{u_S \quad u_B \quad u_N\}$ denotes the components of the non-steady displacement, and its derivative is:

$$\{u'\} = \begin{Bmatrix} u'_S \\ u'_B \\ u'_N \end{Bmatrix} + \begin{Bmatrix} \psi' \sin\varphi \\ -\varphi' \\ \psi' \cos\varphi \end{Bmatrix} \times \begin{Bmatrix} u_S \\ u_B \\ u_N \end{Bmatrix} = \{u\}' + [\kappa] \cdot \{u\} \quad \text{with} \quad [\kappa] = \begin{bmatrix} 0 & d_N & -d_B \\ -d_N & 0 & d_S \\ d_B & -d_S & 0 \end{bmatrix}.$$

From Eqn. (10) we obtain the forces

$$[F] = \begin{bmatrix} EA + F_{st} + F_{dy} & 0 & 0 \\ 0 & F_{st} + F_{dy} & 0 \\ 0 & 0 & F_{st} + F_{dy} \end{bmatrix} \quad ; \quad \{F\}_{nl} = \frac{1}{2} EA\{u'\}^T \{u'\} \begin{Bmatrix} 1 \\ 0 \\ 0 \end{Bmatrix}.$$

$[m]$, $[b]$, $\{q\}_{dy}$ in Eqn. (12) are expressed as

$$[m] = m_{cab}[E] + [m]_{hy}; \quad [m]_{hy} = c_A \rho \frac{\pi D^2}{4} \begin{bmatrix} 0 & 0 & 0 \\ 0 & 1 & 0 \\ 0 & 0 & 1 \end{bmatrix}; \; [b] = \frac{\rho D}{2} \begin{bmatrix} \pi c_S |v_{rel,S}| & 0 & 0 \\ 0 & c_D |v_{rel,normal}| & c_{\Delta L} v_{rel,S} \\ 0 & -c_{\Delta L} v_{rel,S} & c_D |v_{rel,normal}| \end{bmatrix}$$

$$\text{with:} \quad \begin{cases} v_{rel,S} = -U \cos\varphi \cos\psi - \dot{u}_S \\ |v_{rel,quer}| = \sqrt{(U \sin\psi - \dot{u}_B)^2 + (U \sin\varphi \cos\psi - \dot{u}_N)^2} \end{cases}; \qquad \{q\}_{dy} = -U([b] - [b]_{st}) \begin{pmatrix} 1 \\ 0 \\ 0 \end{pmatrix}.$$

Since this formulation is valid for the entire cable, it will be applied to an individual finite element at the position $s_i < s < s_{i+1}$. In this case the degrees of freedom are just the translatory motions at each end. Between the two end points (nodes) the displacement is interpolated linearly. A normalized coordinate ξ is introduced to integrate along the finite element. With consideration to the paper limitations the explicit integration formulation over the finite element cannot be explained in detail.

2.7 Towed body

The dynamic behaviour of the towed body is described by rigid body equations of motion, corresponding to a conventional airplane configuration, i.e. all translatory and rotatory degrees of freedom are considered. The motions of the first cable node are identical with the translatory motion of the rigid body. Due to the additional rotatory motion of the towed body the number of system degrees of freedom increases by three, and the equation of motion of the rigid body results in:

$$\underline{\underline{m}}_{bdy} \cdot \ddot{\underline{u}}_{bdy} + \underline{\underline{b}}_{bdy} \cdot \dot{\underline{u}}_{bdy} + \underline{\underline{c}}_{bdy} \cdot \underline{u}_{bdy} = \sum \begin{Bmatrix} F_a \\ M_a \end{Bmatrix}. \tag{13}$$

The external force F_a is equivalent to the cable force at the towed body suspension point. The displacement vector $\underline{u}_{bdy}$ relates to the coordinate system XYZ:

$$\underline{u}_{bdy}^{\;T} = \{u_X \quad u_Y \quad u_Z \quad \phi \quad \theta \quad \gamma\}.$$

The mass-matrix considers that the towed body is symmetric along the XZ-plane. Horizontal and vertical distances between the cable suspension point and centre gravity are x_g and z_g.

$$\underline{\underline{m}}_{bdy} = \begin{bmatrix} m + m_{a11} & 0 & 0 & 0 & m \cdot z_g & 0 \\ 0 & m + m_{a22} & 0 & -m \cdot zg & 0 & m \cdot x_g \\ 0 & 0 & m + m_{a33} & 0 & -m \cdot x_g - m_{a35} & 0 \\ 0 & -m \cdot z_g & 0 & \Theta_{xx} + m \cdot z_g^2 + m_{a44} & 0 & -\Theta_{xz} - m(x_g \cdot z_g) \\ m \cdot z_g & 0 & -m \cdot x_g - m_{a53} & 0 & \Theta_{yy} + m(x_g^2 + z_g^2) + m_{a55} & 0 \\ 0 & m \cdot x_g & 0 & -\Theta_{xz} - m(x_g \cdot z_g) & 0 & \Theta_{zz} + m \cdot x_g^2 + m_{a66} \end{bmatrix}.$$

The damping matrix of the towed body has the following hydrodynamic drag and lift components. Note the non-linear drag and the additional lift force with larger angles of incidence

$$\underline{\underline{b}}_{bdy} = \frac{\rho}{2} A_{bdy} \begin{bmatrix} -\left(C_{dX} + \frac{\partial C_{dx}}{\partial \alpha}\theta^2\right)\cdot|\dot{u}_X| & 0 & 0 & 0 & 0 & 0 \\ 0 & -C_{dY}\cdot|\dot{u}_Y| & 0 & 0 & 0 & 0 \\ \left(C_{L0} + \frac{\partial C_L}{\partial \alpha}(\theta - \alpha')\right)\cdot|\underline{v}_{rel}| & 0 & -C_{dZ}\cdot|\dot{u}_Z| & 0 & 0 & 0 \\ 0 & 0 & 0 & 0 & 0 & 0 \\ 0 & 0 & 0 & 0 & -C_{d\theta}\cdot|\dot{\theta}|\cdot l_\theta & 0 \\ 0 & 0 & 0 & 0 & 0 & -C_{d\gamma}\cdot|\dot{\gamma}|\cdot l_\gamma \end{bmatrix}$$

$$\text{with:} \quad \alpha' = \arctan\left(\frac{\dot{u}_Z}{\dot{u}_X}\right).$$

The stiffness-matrix of the towed body considers the restoring moments due to the stabilising tailplane around the pitch and yaw axes:

$$\underline{\underline{c}}_{bdy,55} = \underline{\underline{c}}_{bdy,66} = \frac{\rho}{2} A_{bdy} \frac{\partial C_M}{\partial \alpha} \cdot l \cdot v_{rel}^{\;2}.$$

3 EXPERIMENTAL VALIDATION

Model towing tests are performed at the TU Berlin in one of the world largest cavitation and circulation tunnels to validate the non-linear numerical model (Figure 3). The stationary cable curve as well as the orbital oscillating motion of the towed body due to external disturbances at the upper cable end are determined using video postprocessing. Underwater video cameras are attached to a transverse adjustable camera mounting device. The X and Z coordinates of the lenses are known with respect to the water surface and the horizontal winch position. The experimental procedure commences by recording a reference frame to calibrate the dimensions of the obtained video picture in the plane of the submerged towing system. After a constant flow velocity is reached, external excitations are introduced at the upper cable end using a linear oscillator device. Resulting oscillations of the model sensor carrier are recorded. Finally, the recorded reference coordinate system is transformed into a transparent digital bitmap-file and superimposed with the digitised video frames of the towed system. In addition to motion measurements, cable forces at the towed body suspension point and at the upper winch point as well as the horizontal oscillation amplitudes are registered.

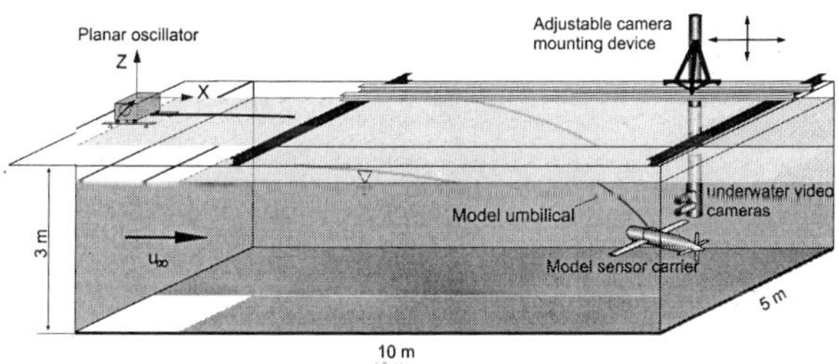

Figure 3: Sketch of the working section of the TU Berlin cavitation and circulation tunnel

4 RESULTS

Figure 4 and 5 presents stationary cable curves as well as towed body motions and forces comparing experimental and numerical results. Comparison between measured and calculated orbital motion of the towed body illustrate good agreement. Note, that shape variations of the orbital trajectory due to increased excitation frequency and towing speed is observed. The numerical simulations illustrate that non-linear effects are dominating the motion behaviour if the sensor carrier is equipped with depressor wings. In this case the towed body motion becomes substantially flatter, and the associated cable force exhibits high peaks – three times as high as without depressor wings. Furthermore, the dynamic analysis proves that ship motions induce substantial cable oscillations resulting in significant excitations of the towed body with a transfer factor of about one.

5 CONCLUSIONS

The paper introduces a numerical model to analyse the dynamic response of towed submarine sensor carriers due to wave-induced oscillations at the winch of the towing ship. The evaluation proves that ship motions are substantially transmitted along the towing cable or umbilical resulting in significant excitations of the towed body. Experimental validation of the calculated stationary cable curves, orbital oscillations of the towed body around its stationary origin as well as axial forces at the upper and lower cable ends show good agreement between measured and calculated data. Even super-harmonic cable force oscillations at the towed body suspension point of twice the frequency of the external excitation are determined. It can be stated, that the tense configuration of the stationary cable curve acts like a virtual system-stiffness, since the non-linear cable force component considers the incremental displacement of each finite element as well as cable curvature. This significant non-linear effect has never been considered before.

The dynamic analysis reveals that the shape of the orbital trajectory of the towed body caused by wave-induced excitations depends on the geometric and therefore hydrodynamic configuration of the submarine body. The numerical model which considers the complete rigid body equations of motions accurately predicts the effect of depressor wings considering the non-linear increase of drag force due to a higher angle of incidence or velocity. As a consequence a significant increase in axial cable force is observed.

References

Ablow C.M.; Schechter S. (1983). Numerical Simulation of Undersea Cable Dynamics, in Ocean Engineering 1983, Vol. 10, No. 6, pp. 443-457.

Buckham B.; Nahon M.; Seto M. (1999). Three-Dimensional Dynamics Simulation of a Towed Underwater Vehicle, in Proc. of the 18th Int. Offshore Mechanics and Arctic Engineering Symp., July 1-6, 1999, St. John's, Newfoundland, Canada, Paper-No. 99-3068.

Clauss G.F.; Hoog S. (2001). Design and Hydroelastic Optimization of the Mobile Docker for GEOSTAR 2 – Seafloor Stations, in Proc. of the 11th Int. Offshore and Polar Engineering Conf., Stavanger, Norway, June 17-22, 2001, Paper No: 2001-JSC-367.

Clauss G.F.; Saroukh A.; Weede H. (1998). Prediction of Limiting Seastates for Pipelaying Operations, in Proc. of the 17th Int. Offshore Mechanics and Arctic Engineering Symp., July 5-9, 1998, Lisbon, Portugal, Vol. IV, Paper-No. 98-611.

Milinazzo F.; Wilkie M.; Latchman S.A. (1987). An Efficient Algorithm for Simulating the Dynamics of Towed Cable Systems, in Ocean Engineering 1987, Vol. 14, No. 6, pp. 513-526.

Wu J.; Chwang A.T. (1997). 3-D Simulation of a Two-Part Underwater Towed System, in Proc. of the 7th Int. Offshore and Polar Engineering Conf., Honolulu, USA, May 1997, Vol. II, pp. 44-50.

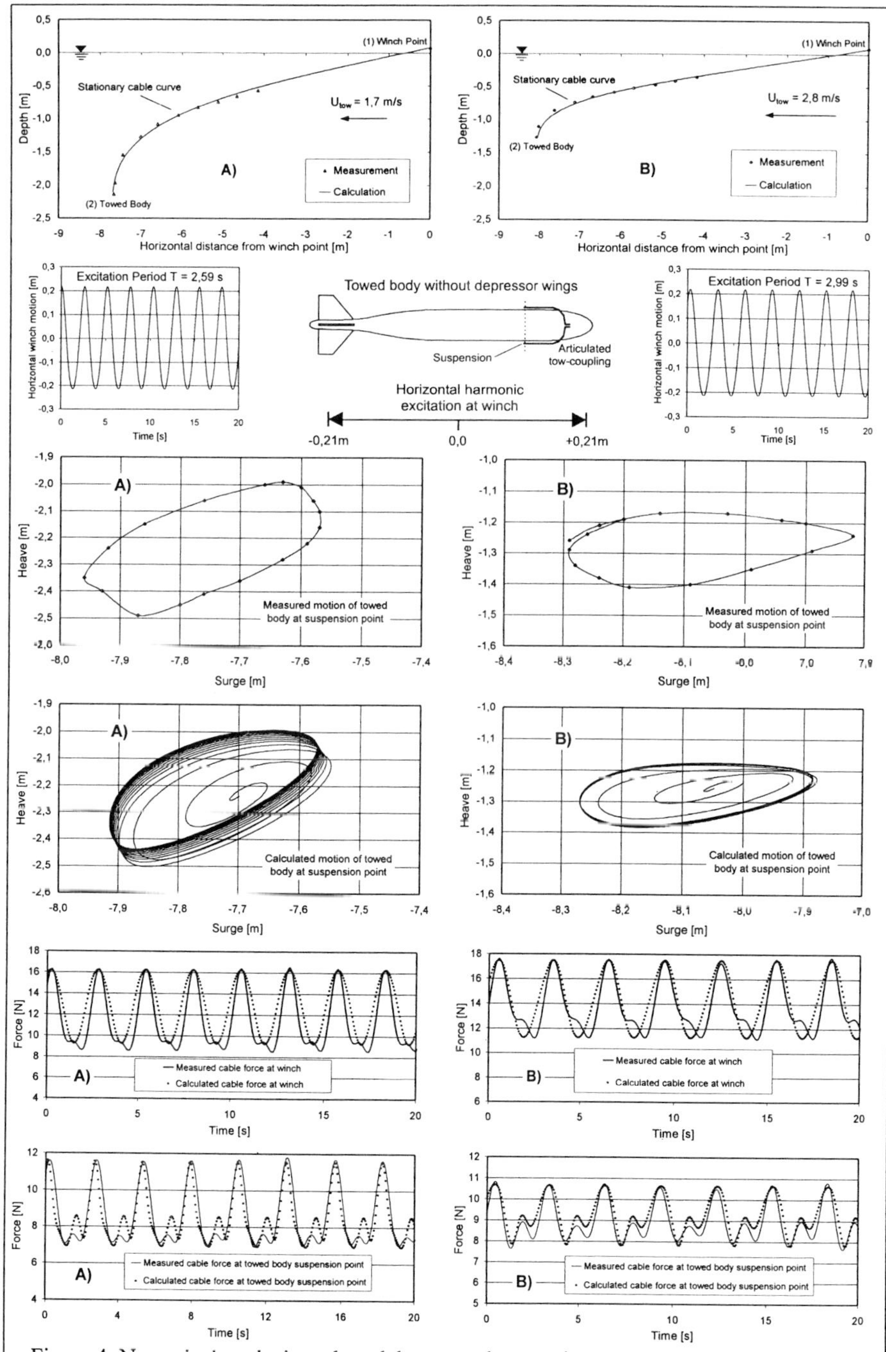

Figure 4: Numerical analysis and model test results – stationary cable curve and dynamic behaviour of towed body without depressor wings:

A) U_{tow} = 1,7 m/s and T = 2,59 s; **B)** U_{tow} = 2,8 m/s and T = 2,99 s

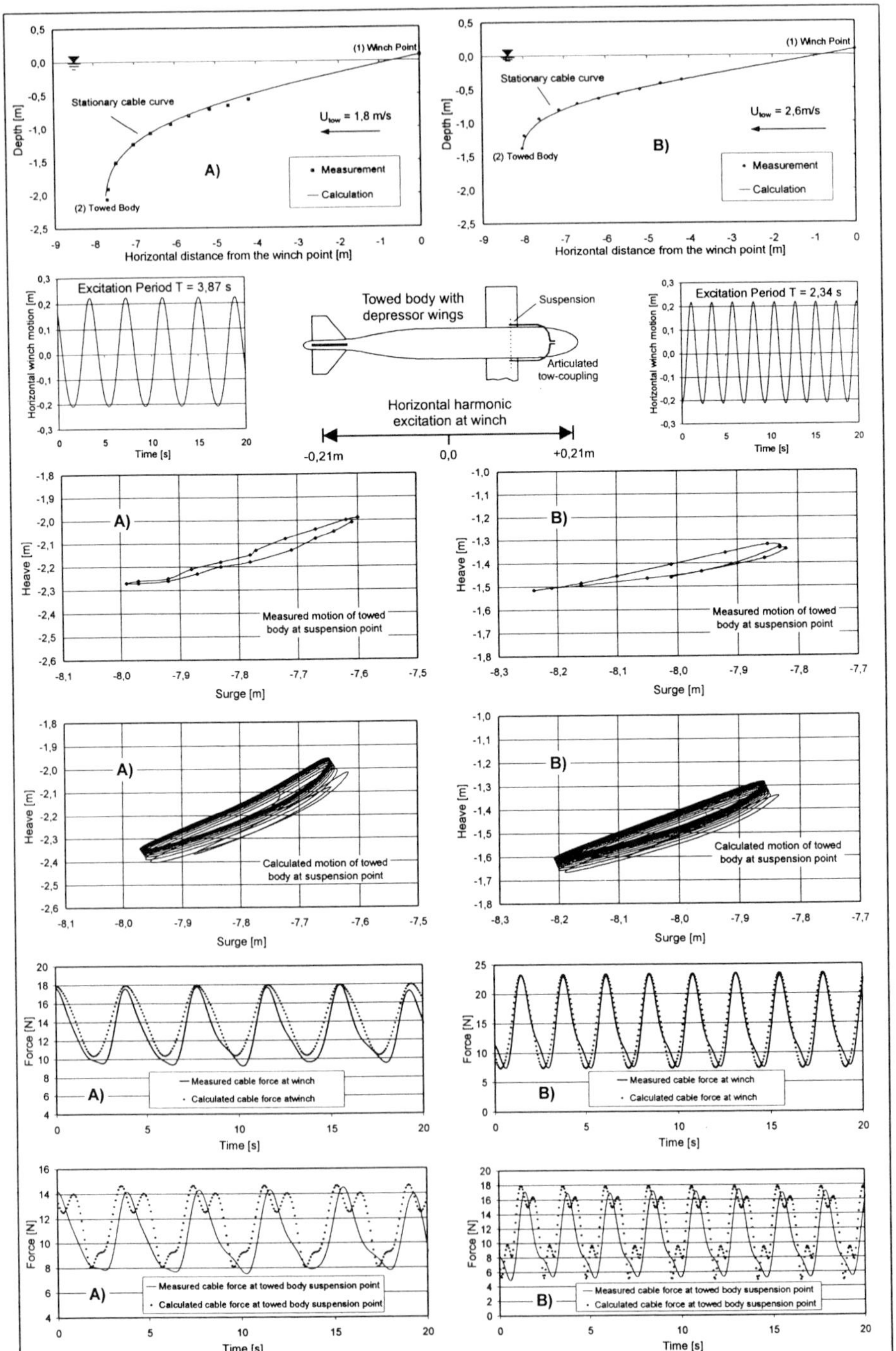

Figure 5: Numerical analysis and model test results – stationary cable curve and dynamic behaviour of towed body with depressor wings :

A) U_{tow} = 1,8 m/s and T = 3,87 s; **B)** U_{tow} = 2,6 m/s and T = 2,34 s

Practical Design of Ships and Other Floating Structures
You-Sheng Wu, Wei-Cheng Cui and Guo-Jun Zhou (Eds)

VORTEX-INDUCED VIBRATION OF TWO DIMENSIONAL WING-SPRING COUPLED SYSTEM

Zhi-Xing Yu, Ying-Zhong Liu and Guo-Ping Miao

Shanghai Jiao Tong University,
Shanghai, China, 200030

ABSTRACT

The fluid-solid coupled motion of a two dimensional wing-spring system is numerically simulated using Navier-Stokes Equations in the paper. The simplified spring system, which has two freedoms, i.e., the sway and the roll substitute the recovering forces and moments due to deformation of the solid. In order to stir up the vortex-induced vibration, the attack angle of the wing is initialized with a large value. The coupled vibrations caused by the periodical vortices and fluid dynamics is then simulated after successfully checking with samples.

KEYWORDS

Wing flow, Vortex-induced vibration, Fluid-solid coupling, Vortex shedding

1 INTRODUCTION

There exist many components of naval ships and ocean constructions suffering viscous flow, which forms vortex shedding and further damaging vibrations on structures. A great deal literatures, such as Yu(1999), Yu & Liu(2000), Blevinx(1983) etc, concentrated on fluid induced vibration on a cylinder, however, fewer researches are on the induced vibration of wing. Ohmi & Coutanceay(1991) made experiments on wing forced rotation and displacement at large attack angle; Morikawa & Gronig(1995) studied the formation and the structure of vortex in this case. Piperno(1997) computed 2D non-viscous fluid-structure coupled aeroelasticity from numerical view point.

In this paper, the viscous flow around 2D wing is simulated with Navier-Stokes Equations with consideration and fixed wing or elastic supported wing. The result shows that flow around the simplified wing-spring system will produce the similar phenomena of cadence and resonance under certain conditions.

1.1 Mathematical Modal and Numerical Method

Because the contour of wing is a complex curve and the computational grid in physical space is not uniform, the description of the problem is under general curvilinear coordinate. Moving grid should be considered due to the movable wing. Given characteristic constants such as the length of the wing L, the inflow speed U, the fluid density ρ and the kinetic viscosity ν, the non-dimensional imcompressible Navier-Stokes equations can be written as:

$$\frac{\partial}{\partial \xi^j}(J^{-1}\frac{\partial \xi^j}{\partial x^i}u_i)=0 \tag{1}$$

$$J^{-1}\frac{\partial u_l}{\partial \tau}+\frac{\partial}{\partial \xi^j}[J^{-1}\frac{\partial \xi^j}{\partial x^l}(u_l-v_l)u_i]=-\frac{\partial}{\partial \xi^j}(J^{-1}\frac{\partial \xi^j}{\partial x^i}p)+\frac{\partial}{\partial \xi^j}(\upsilon J^{-1}g^{jl}\frac{\partial u_i}{\partial \xi^l}) \tag{2}$$

where, (x_i,t) and (ξ,τ) are coordinates in physical and calculation space respectively; v_l is the grid velocity. Eq.(1) and (2) are written with contra-variant variables:

$$\frac{\partial U^j}{\partial \xi^j}=0 \tag{3}$$

$$J^{-1}\frac{\partial \vec{S}_i U^i}{\partial \tau}=\vec{C}+\vec{D}+\vec{R} \tag{4}$$

where

$$\vec{C}=-\frac{\partial}{\partial \xi^j}[(U^j-V^j)\vec{u}] \qquad \vec{D}=\frac{\partial}{\partial \xi^j}(\upsilon\vec{S}^j\cdot\nabla\vec{u}) \qquad \vec{R}=-\frac{\partial}{\partial \xi^j}(\vec{S}^j p)$$

are convective part, diffusive part and pressure part respectively.

By using staggered grid, Eq.(3) and (4) are discretized by finite volume method. In each time step the momentum equation is advanced by Adams-Bashforth/Crank-Nicolson scheme(Wang, 1991); $\nabla\vec{u}$ the diffusive part is expressed by velocity of element boundary with Gauss theorem; the projection method is used to couple the pressure and velocity. The final linear algebra equation is solved by GMRES(Saad and Schultz, 1986) with ILU preconditioner(Lin and Chen, 1996). The algorithm is of second order in both time and space globally. The fixed grid is generated by algebraic method(Thompson etc,1985) and then smoothed by partial difference equation; as for movable grid, only the former is utilized to speed up the grid generation. The outer boundary is set with a buffer to damp the reflection of error (Yu,1999).

The control equation of 2D wing with linear spring support is

$$[M]\{\frac{\partial^2 Y(t)}{\partial t^2}\}+[C]\{\frac{\partial Y(t)}{\partial t}\}+[K]\{Y(t)\}=\{F(t)\} \tag{5}$$

where, $Y=\{y(t),\theta(t)\}^T, F=\{F_y,F_\theta\}^T$ are displacement vector and fluid dynamic vector;

$$[M]=\begin{bmatrix} M_y & S_\theta \\ S_\theta & I_\theta \end{bmatrix},\ [C]=\begin{bmatrix} C_y & 0 \\ 0 & C_\theta \end{bmatrix},\ K=\begin{bmatrix} K_y & 0 \\ 0 & K_\theta \end{bmatrix}$$

respectively, the system mass, the spring damp and the elasticity. Eq.(5) is integrated by the traditional Newmark-β method(Bathe & Wilson, 1976).

2 COMPUTATIONAL RESULTS AND ANALYSIS

2.1 Flow around Fixed Wing

Calculation of the fixed wing flow is to check the present method and the corresponding program and to seek the suitable attack angle when periodical vortex shedding is produced.

In the paper lots of comparison with current literatures on flow around fixed wing are carried. Fig.1 shows one of them, which is of wing NACA0015 at Re=35000. Fig.1(i) is the experiment result of Ohmi et al(1991); Fig.1(ii) is the numerical result of Morikawa & Gronig(1995) with uniform acceleration to nondimensional time 0.598; Fig.1(iii) is the result of Nair & Sengupia(1998); Fig.1(a,b) is the present result. It can be seen that the present result give clearly vortex layout of the main vortex, the second vortex at leading edge and the third vortex at trailing edge.

2.2 Forced Vibration of the Wing

Wing ACA0012 is selected to rotate with angle

$$\alpha = -(\alpha_0 + \Delta\alpha \sin(2kt))$$

where $2k = 0.15$, $\alpha_0 = 15^\circ$, $\Delta\alpha = 10^\circ$. The center of the rotation is at the chord center. Fig.2 give the force revolution with time at Re=1000. With large attack angle, the vortices have more complex structure that interacts with each other making irregular force fluctuation. When it is smaller, the irregularity vanishes gradually.

2.3 Vortex Induced Vibration

The simplified wing-spring coupled system is shown in Fig.3 that has two freedoms of up-down displacement and rotation around the elastic center. The system can be described with Eq.(5), where the coefficients are shown in the following table(*c* is the chord length):

TABLE 1
COEFFICIENS DIFINITION IN EQ(6)

$b = c/2$,	$S_\theta = m_y b x_\theta$,	$I_\theta = m_y b^2 \gamma_\theta^2$,	$k_y = m_y \omega_y^2$
$k_\theta = I_\theta \omega_\theta^2$,	$\omega_y = \omega_\theta$,	$c_y = 2\xi_y \omega_y m_y$,	$c_\theta = 2\xi_\theta \omega_\theta I_\theta$

According to the motion theory(Blevinx, 1983), as for two-freedom system, the interaction between them will become maximum at $\omega_y / \omega_\theta \approx 1$, thus in the present paper, the two are made equal. These values are chosen from(Piperno, 1997):

TABLE 2
DATA FOR COEFFICIENTS

m_y	c	ω_θ^2	x_θ	γ_θ	ξ_y	ξ_θ	a_y
1.0kg	1.0m	$100\,s^{-1}$	1.8	1.865	0.01	0.01	0

Fig.4 and Fig.5 show the time evolution of displacement of the two freedoms with nature frequencies of $1.25\,s^{-1}$ and $1.05\,s^{-1}$ at Re=1000. In order to get the periodical vortex shedding of the wing, the attack angle is chosen to be –30 degree and maintains the same angle and vertical position until stable vortex separation obtained. The computing experience suggests the non-dimensional time should be about 53.0. From Fig.4, the 'beat' phenomenon appears so that vortex shedding frequency and system nature frequency coexist, while the vertical translation is not distinct. When $\omega = 1.05$ (Fig.5), it is found that the system vibration will advance to a resonance state of high amplitude from beat state as time increases in both freedoms. The damp restricts the amplitude to a certain limit that may be a kind of 'lock' phenomenon. However, because of the complexity of the vortex flow, the frequency of the vortex can hardly be foreseen. So the behavior of the moments is not the most representative.

3 CONCLUSIONS

Calculations in the present paper have been fully checked with literatures to make the program credible. The following conclusions can be drawn:

(1) In order to obtain vortex vibration, the attack angle should be large enough. There will

appear complex vortex structure under large attack angle at high Reynold number, which interacts with each other making irregular force evolutions.

(2) Under large attack angle, 'beat' phenomenon emerges similarly to the flow around cylinder; at a certain nature frequency, resonance with large amplitude will be produced leading to a similar 'lock' phenomenon.

(3) Because the periodical vortex shedding is related to a certain large attack angle and type of the wing, there is a possibility to control the vortex induced vibration through changing the shape of the wing.

References

Zhixing Yu. (1999). Hydroelastic Calculations of Viscous flow, *PhD Thesis(in Chinese) of Shanghi Jiao Tong University..*

Zhixing Yu and Yingzhong Liu (2000). Hydroelastic Calculation of Viscous Flow Around a Cylinder-Spring Aystem. *Proceedings of ISOPE-2000, I, Seattle.*

Bathe, K.J. and Wilson, E.L. (1976). Numerical Methods in Finite Element Analysis, Prentice-Hall Inc.

Thomas. (1982). Composite Three-Dimensional Grids Generated by Elliptic System, *AIAA J.* **20:9**.

Ohmi K., Coutanceay M. et al. (1991). Further Experiments on Vortex Formation Around an Oscillating and Translating Airfoil at Large Incidences. *J. Fluid Mech.* **225**, 607-630.

Morikawa K. and Gronig H. (1995). Formation and Structure of Vortex Systems Around a Translating and Oscillating Airfoil. Z, Flugwiss Weltrqumforsch. **19**, 391-396.

Nair M.T. and Sengupta T.K. (1998). Orthogonal Grid Generation for Navier-Stokes Computations. *Int. J. Numer. Meth. fluids.* **28**, 215-224.

Blevinx R.D. (1983). Fluid Induced Vibration (in Chinese), Translated by Shushan Wu, et al, Publishing House of Ministry of Machine Building.

Piperno, S. (1997). Explicit/Implicit Fluid/Structure Staggered Procedure with a Structural Predictor and Fluid Subcycling for 2D Inviscid Aeroelastic Simulations. *Int. J Numer. Meth. Fluids.* **25**, 1207-1226.

Xianfu Wang, et al. (1991). Computational Ship Fluid Dynamics(in Chinese). Shanghai Jiao Tong University Press.

Van Leer B.V. (1977). Towards the Ultimate Conservative Difference Scheme. V. A new approach to numerical convection. *J Comp. Phys.* **23**, 32-78.

Saad Y. and Schultz M.H. (1986). GRRES: A Generalized Minimal Residual Algorithm for Solving non-Symmetric Linear System. *SIAM J. Sci. Stat. Comput.* **7**, 856-869.

Lin H.W. and Chen L.D. (1996). Application of the Krylov Subspace Method to Numerical Heat Transfer. *Numer. Heat Trans. Part A.* **30**, 246-270.

Thompson J.F., Warsi Z.U.A. and Mastin, C.W. (1985). Numerical Grid Generation. North-Holland.

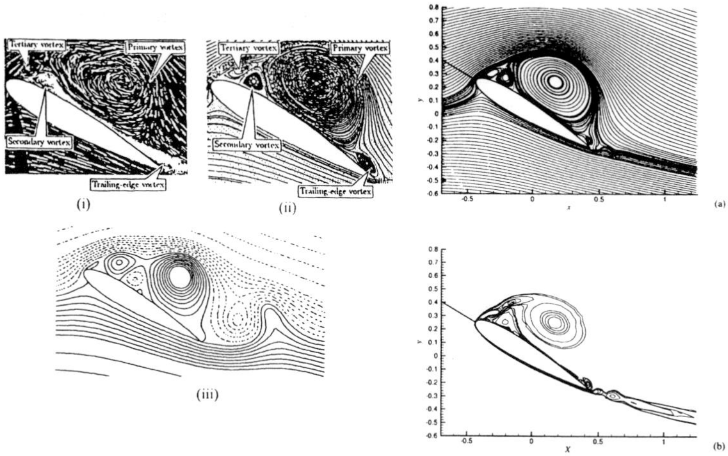

Figure 1: Comparison of streamline and vortex of wing NACA0015, Re=35000, $\alpha = 30°, t \approx 2.9$

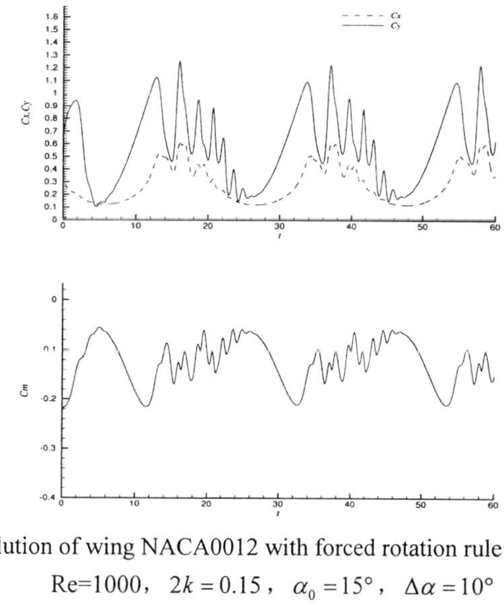

Figure 2: Forces evolution of wing NACA0012 with forced rotation rule: $\alpha = -(\alpha_0 + \Delta\alpha \sin(2kt))$, Re=1000, $2k = 0.15$, $\alpha_0 = 15°$, $\Delta\alpha = 10°$

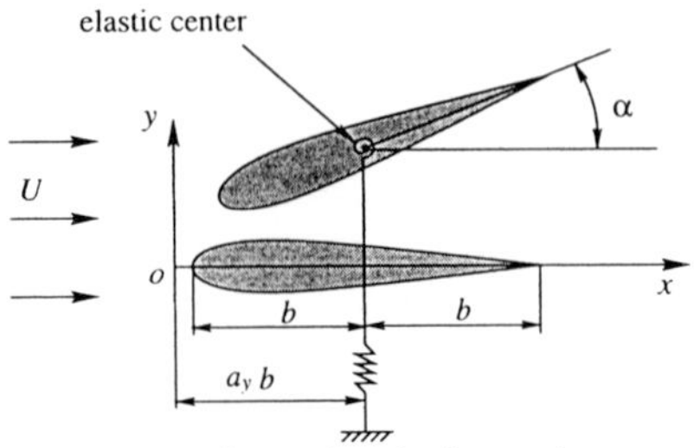

Figure 3: Two dimensional wing-spring system

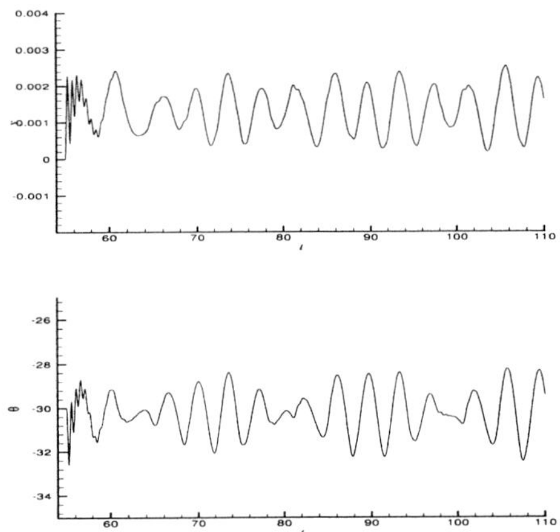

Figure 4: Displacements evolution of the win-spring system,
$\omega = 1.25s^{-1}$, Re=1000, $\alpha = 30°$

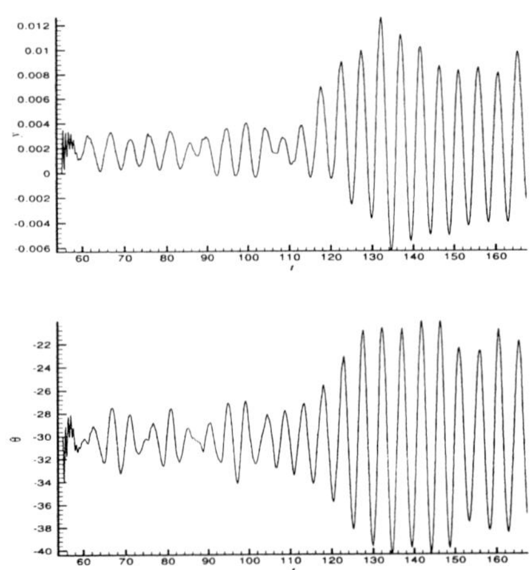

Figure 5: Displacements evolution of the win-spring system,
$\omega = 1.05s^{-1}$, Re=1000, $\alpha = 30°$

Practical Design of Ships and Other Floating Structures
You-Sheng Wu, Wei-Cheng Cui and Guo-Jun Zhou (Eds)

FIRE RISK ANALYSIS AND ITS APPLICATION TO SHIPS

M. Dogliani and A. Vergine

Innovation Research & Products Section
RINA–Registro Italiano Navale Group, Genova Italy

ABSTRACT

An innovative fire safety regulatory approach concerning the fire safety on board ships is about to be introduced by the International Maritime Organisation (IMO). It will be possible to use alternative, performance-based fire safety design and arrangements in lieu of traditional prescriptive designs.
After providing a summary of the background of this new approach, this paper briefly reviews the state of the art in the fire engineering science applied to ships and pinpoints gaps to be filled by future R&D. For the sake of illustration, an example of application is provided.

KEYWORDS

Fire, Safety, Risk Analysis, Fire simulation, Formal Safety Assessment

1 INTRODUCTION

Fire safety provisions for ships are specified in the SOLAS International Convention, chapter II.2: in order to be accepted by the competent authority the design has to fulfil its requirements. An advantage of this framework is that there is in general no need to carry out complex calculations and experimental activities during the design; however the drawback is that design solutions which significantly deviates from requirements cannot be accepted.
This aspect is being addressed within the revision of SOLAS II.2 chapter with the aim of opening the way to performance based design. One aspect of particular relevance is that the verification of compliance of alternative design will be addressed by a specific regulation (Part F, Regulation 17) indicating that engineering analyses are needed and will have to be carried out according to specified principles spelled out in a guidance document recently issued by IMO (2001).
When entered into force (1st July 2002), this new regulation has the potential to strongly influence fire safety design leading to the introduction of advanced fire engineering and fire risk analysis methods and tools into ship design.
The Guidance, IMO (2001), whose goal is to specify the methodology to be used in the Fire Risk Assessment of alternative design and arrangements, is based on the following main concepts:

- a comparative analysis is to be carried out where the alternative design is assessed, in performance terms, against an acceptable design;
- the assessment focuses on the deviations of the alternative design from fire safety regulations;
- additional fire risk mitigation measures must be considered in order to balance these deviations;

- reference is made to approaches and fire models applied in the civil building sector.

The main steps of the assessment are as follows:

1. definition of the problem;
2. identification of desired deviations from requirements;
3. definition of performance parameters to be used for comparative analysis;
4. definition of acceptance criteria;
5. definition of the method of analysis (i.e. type and extent of the analyses);
6. specification of the design fires to be considered;
7. execution of the analysis.

Depending on the case under analysis the performance parameters on which the comparative assessment is to be based are related to human vulnerability (effects of heat, smoke, toxicity) or on damage to ship systems (e.g. impact of fire on the structure, the escape system, etc.).
This implies that fire consequence models are needed which are able to quantify these parameters taking into account effects such as:

- ignition sources and their characteristics;
- rate of heat release of combustible material;
- proximity, amount and distribution of combustible material;
- ventilation characteristics;
- position and characteristics of sprinklers;
- fire growth rate and its modifications due to sprinkler and ventilation.

2 STATE OF THE ART

In broad terms, when carrying out a FRA, the following consequence modelling are to be considered:

- the fire and its development;
- subsequent smoke movement through the ship;
- explosions;
- mustering and evacuation.

A number of computer codes exist which model escape of people from complex structures, however Mustering and evacuation are dealt with in this paper; on this issue see e.g. Dogliani (2001). Because of the structure of a passenger ship, consequence models must handle different geometry, including:

- interconnected rooms;
- large atria;
- long corridors;
- multiple decks;
- stairwells.

Onboard a ship, the conduction of heat through metal bulkheads and decks will be particularly important. This could lead to the ignition of fires in adjacent compartments, even when there is no breach in the original fire compartment. No single model is likely to be able to handle all these aspects efficiently, but it should be practical to ensure that a combination of computer codes and engineering correlations are available which together cover all likely aspects. The selection of tools will need to take into account practicalities such as the time and cost required to perform calculations, as well as the reliability of the predictions under the conditions of interest.

2.1 Fire Models

Computer codes modelling fires can be divided into two categories, namely zone models and field models. Zone models simplify the problem by considering a limited number of spatial zones, whilst field models use a fine three-dimensional mesh to examine the regions of interest in more detail. Zone models are faster running and easier to use than field models, but care must be taken that their underlying assumptions are valid in the particular situation of interest. Perhaps the most used software model is CFAST, Peacock et al. (2000) which was also used in this paper.
Whilst more expensive to use, field models (CFD based) are more flexible than zone models and able in principle to handle more complex situations. However, the complexity of their input and their long run times makes it impractical to use CFD codes routinely at the present time.
The best way forward is then to define a series of standard fire scenarios, and investigate them by appropriate available tools (correlations and zone and field computer codes).

2.2 Fire Risk Assessment

Compared with other, more mature, engineering fileds, limited literature exist to date on fire risk analysis techniques, particularly in the maritime sector. Available documentation pertains to R&D papers, thematic conferences and publications on one side, and codes and standards on the other. Institutions such as the NIST (National Institute of Standards and Technology), the NFPA (National Fire Protection Association), the USCG (United States Coast Guard) and the FSEG (Fire Safety Engineering Group) at the Greenwich University provide on-line information and publications, generally addressing various industrial fields.
Issues of SNAME (Society of Naval Architects and Marine Engineers) Transactions provide relevant, though limited, information on what is currently going on in the field of maritime fire safety, both regarding research studies and technical innovations. Codes and Standards over a broad spectrum of interest on marine, railways, building and offshore fire protection were consulted. Also, standards (such as the European Standard prEN50126 for railways) that cover aspects related with risk analysis or, more generally, safety assessment guidelines for transportation systems can be considered. It is noted that the most recent activities in the fire safety field have concentrated on the development of performance codes and other forms of performance prediction, creating a Performance-Based approach to fire safety analysis, see e.g. Delichatsios and Karydas (1996), O'Hara (1998), Bukowski (1996). Every draft performance evaluation system is risk based; it is therefore clear that risk is the metric of choice for the performance-based safety regulations of the future. It is recognised that, within the shipbuilding industry, this shift towards performance based codes and design has not yet progressed to the extent recognisable in other industrial fields, such as civil engineering, railways and metro systems and offshore structures.

3 CASE STUDY – A FIRE RISK ANALYSIS FOR PASSENGER CABINS

The analysis is conducted through the following main steps:

1. Problem definition
2. Construction of the Event Tree (ET)
3. Probabilistic quantification
4. Consequences assessment
5. Risk evaluation.

3.1 Problem Definition

The fire safety objective is, according to SOLAS' philosophy, the containment of the fire in a main vertical zone. Accordingly, the analysis was applied to a portion of a vertical zone above the embarkation deck of a cruise ship compliant to 1992 Amendments of SOLAS (see Figure 1); it includes two rows of cabins (all equal, the only difference being the presence of a window in the cabins of one row) and small lockers separated by a corridor.

Cabin description

Passenger cabin, 15-m^2 area, containing a trash can, an upholstered chair, a bed, a table, a toilet and provided with low flame spread coverings as per SOLAS II-2, Reg. 34.3.

Fire Load

The data relative to the fire load and necessary for the fire simulation are summarised in Table 1. They are relevant to the t-squared fire growth model and classified according to BSI (1994). The item called "Others" encompasses toilet, TV set, refrigerator, luggage and coverings. In view of a simplified analysis these items were grouped in a single, "dummy" object. Information on the layout and fire load for a typical cabin was obtained from an internal investigation and from NFPA 72E (1987).

Ventilation

The cabin is equipped with a self-closing door as per SOLAS (Chapter II-2/12, 1992 Amendments) and a window. The door is not airtight, so it was assigned a leakage even when closed. Mechanical ventilation supplies air into the cabins with a flow rate of 74 l/s. In order to take into account the natural ventilation (including leakage through the door), the following geometric characteristics of the openings were assumed:

- door open: 0.6 m × 2.1 m; door closed: 0.01 m × 2.1 m; door grill: 0.2 m × 0.4 m, sill 0.2 m;
- window open: 1.2 m × 1.8 m, 0.7 m from the floor; window closed: 0.01 m × 2.1 m.

Sprinklers

The sprinkler characteristics (temperature intervention and specific flow rate) are consistent to SOLAS II-2, Reg. 12.3. (Ref. /8/); its performances are: flow rate: 5 liter/m^2/min = 1.5 kg/s (spray density 8.3 E-05 m/s); activation temperature 79°C; Response Time Index (RTI) of 278 $(m \cdot s)^{1/2}$ was assumed.

Smoke Detectors

The cabin is equipped with an optical detector, whose actual response time should be evaluated by assuming its activation if the temperature increases by 30° above ambient temperature, and Response Time Index (RTI) = 50 $(m \cdot s)^{1/2}$ as per NFPA (1993); if the detection signal is not cleared within 120 s, an alarm is activated in the affected zone.

Bulkheads

SOLAS II-2, Reg. 26 (1992 Amendments) prescribes:

- B-0 class between cabins (accommodations of moderate fire risks);
- B-15 class between cabins and corridors.

HRR (heat release rate) and fire growth

At first, the only presence of an upholstered chair was assumed to screen the most relevant and critical event sequences. This allows to apply the HRR reference curve obtained by laboratory tests with the furniture cone calorimeter. The model characteristic values are:

- fire growth rate: $\alpha = 0.1055$ kJ/s^3, i.e. is "fast" according to Sprague (1996) and between "fast" and "ultra-fast" according to BSI (1994);
- Peak value: t = 140 s, $\dot{Q} = 2.1$ MW;
- Transition to decay at t = 160 s;
- End combustion at t = 500 s;
- Total heat developed = 372 MJ.

The fire growth model of the upholstered chair was derived from the reference test for the sprinkler performance verification as per IMO Resolution A800(19). It was classed as intermediate between "fast" and "ultra-fast" as per BSI (1994). The chair was assumed to be in a corner of the room.

3.2 Event Tree Analysis

The Event Tree Analysis technique is a mean to identify the possible outcomes and their probabilities of occurrence resulting from a main hazard (Initiating Event – EI) sequence of events, relevant to failure or success of successive items of equipment or mitigating actions. In the present case, see Figure 2 and Table 1, an ignition was adopted as initiating event and the following events were included:

- door status (open or closed);
- window status (open/broken or closed);
- automatic suppression: local equipment (sprinkler) is able to extinguish fire.

3.3 Severity Criteria

The following criteria to estimate the severity of the consequences were adopted:

I. the FRI (full room involvement) temperature was assumed to be 500 °C. By comparing this value to the corresponding temperatures of the 8 scenarios, the likelihood of FRI is assessed;
II. the likelihood of the fire propagation in the nearby cabins only depends on the FRI occurrence; if FRI is not reached, the fire will self-terminate before ignition in the adjacent cabin.

The main limitations are:

- the 500°C corresponds to flashover conditions: sometimes, even if rarely, FRI is reached before;
- the likelihood of fire spread should take into account the walls' characteristics (strength, thermal conductivity, etc.) and layout (e.g. air gap between the walls of two adjacent cabins).

3.4 Simulation Results

The simulation for the eight scenarios led to the following results, see Table 2:

1. where either the door or the window is open (scenario # 1, 2, 3, 4, 5 and 6) the fire is fuel-controlled (its growth is limited by the amount of the fire load); two cases can be distinguished:
   sprinkler operating correctly (# 1, 3, 5): the sprinkler is instrumental in reducing the fire heat release rate, and consequently the likelihood of FRI, see Figure 3;
   sprinkler unavailable (# 2, 4, 6): fire heat release rate, thus FRI's likelihood, increases;
2. where the door is closed and no window is present (# 7 and 8) the fire is ventilation-controlled (its growth is limited by the amount of oxygen); the heat release rate, thus the maximum temperature, is low enough as to make the FRI occurrence unlikely.

3.5 Risk Matrix

According to the class of probability and to the simulation results, the overall ranking of the eight scenarios is as shown in the risk matrix of Table 3. It is to be noted that, as the probabilities are conditional to fire occurrence, it is not possible to obtain a complete risk evaluation; therefore, likelihood and severity were kept apart.
For comparative purposes, the same analysis was carried out by assuming the cabin door not self-closing (thus a probability of door open was increased to 0.5). It can be shown that, in this case, the probability of the severe outcomes # 2 and 4 would increase by a factor of 50.

3.6 Detailed Consequence Analysis

From the preliminary analyses, the most relevant scenario in terms of potential full room involvement has been investigated considering a most case scenario in which all objects burn simultaneously. Fire simulations were carried out with the main goals of quantifying the most severe scenario through more realistic fire dynamics, to be compared against the ship barrier fire performance. This comparison has been developed in terms of fire severity with reference to the standard time-temperature curve adopted in the shipping field for fire test;
For scenario # 4, natural ventilation through the door (assumed open) was considered and the sprinkler action was assumed to fail.

3.7 Calculation of HRR for the Whole Cabin

The Heat Release Rate (HRR) curves of each object were combined to obtain the overall HRR associated to the cabin. This was based on the conservation of energy and mass: the burned mass and the energy developed at any time equals the sum of the energies and burned masses of the various items. More specifically the global values were been obtained by the following relations:

$$Hc_{tot} = \frac{\sum m_i Hc_i}{\sum m_i}$$

$$\dot{Q}_{tot}(t) = \sum_i \dot{Q}_i(t)$$

$$\dot{m}_{tot} = \frac{\dot{Q}_{tot}}{Hc_{tot}}$$

where:

Hc = heat of combustion; $\dot{Q}$ = heat release rate; $\dot{m}$ = mass lost by the fuel.

3.8 Results and Comments

The results are the time histories of the overall heat release rate (Fig. 4) and theupper layer temperature (Fig. 5). Inspection of Fig. 5 shows that fire room temperatures rise much faster than the standard (SOLAS) time-temperature curve, remain above for a short period of time and eventually drop below it. In practical terms it is possible to correlate the temperature-time histories with the barrier's capacity of sustaining a fire: if a fire barrier is able to withstand the SOLAS fire test, then is able to sustain a fire whose temperature vs. time curve has the same or lower integral area: this is the so-called Ingberg's criterion which leads, for the present example an endurance of respectively 440 s for B10 class and 700 s for B15 class.

The analysis has basically highlighted the following points:

1. the fire growth model is of paramount importance to properly account for variations in materials as well as effects of the sprinkler; the fire load gives indications of the heat energy stored, thus of the likelihood of FRI and fire spread;
2. in ships' cabins, the HRR is driven by the ventilation, which tends to smooth down the peak temperatures; an assumption of free ventilation (fuel controlled fire) would be misleading,
3. the barriers' material plays an important role in the overall phenomenon: the relative high thermal conduction properties of ship's cabins reduces the temperature increase inside the cabin.

4 CONCLUSIONS

Fire risk analysis applicability to ship design has been proved to be feasible with a reasonable accouracy by using models and tools available from other industrial sectors.

However, in view of the huge scope for FRA application in the near future, as called by new SOLAS II-2 / Regulation 17, the following R&D issues are recommended:

a. develop fire growth models which are suitable to ship applications (i.e. low air intake, steel walls, forced ventilation, both small and very large rooms to be covered)
b. upgrade existing fire consequence models in order for them to better deal with ship features
c. define standard fire scenarios to be considered at the design stage and when carrying out a FRA
d. carry out a few significant prototype applications of the whole methodology to serve as a demonstrator of the approach.

References

IMO Fire Protection Subcommittee. FP44/WP.4, January 2001.

Dogliani M. (2001). An Overview of Present and Under-development IMO's Requirements concerning Evacuation from Ships. *International Conference of Pedestrian and Evacuation Dynamics – PED*. Gerhard Mercator University, Duisburg, Germany.

Peacock R.D., Reneke P.A., Jones W.W., Bukowski R.W., Forney G.P. (2000). *A User's Guide for FAST: Engineering Tools for Estimating Fire Growth and Smoke Transport*. National Institute of Standards and Technology, Special Publication 921, 2000 Edition.

Bukowski, R. W. (1996). Risk and Performance Standards. *13th Meeting of the UJNR Panel on Fire Research and Safety*, Gaithersburg, MD.

Delichatsios, M. A. and Karydas, D. M. (1996). A Framework for Fire Risk Assessment of Buildings Based on Performance Based Engineering Analysis. NIST Annual Conference of Fire Research, Gaithersburg, MD.

O'Hara, M. A. (1998). Future Fire Codes. *Civil Engineering-ASCE* **68:5**, 48-51.

BSI Standards 94/340 DC (draft). (1994). The Application of Fire Safety Engineering Principles to Fire Safety in Buildings.

NFPA 72E. (1987). Standard on Automatic Fire Detectors National Fire Protection Association.

NFPA - National Fire Protection Association. (1993). The SFPE Handbook of Fire Protection Engineering.

C.M. Sprague, P.E., B.L. Dolph. (1996). Theoretical Basis of the Ship Fire Safety Engineering Methodology United States Coast Guard R&D Center.

TABLE 1

ET NODES PROBABILITY

Event	Probability	Rationale
Door open	0.01	Engineering judgement: the door is self-closing
Door closed	0.99	see above
Presence of a Window	0.5	Half the cabins (8 out of 16) of the zone are provided with a window
Sprinkler working	0.9	Expert judgement/historical data
Sprinkler not working	0.1	see above

TABLE 2

EVENT TREE ANALYSIS RESULTS

Scenario	Door	Window	Sprinkler	Probability	Class
1	Open	Yes	Success	4.5 E-03	Unlikely
2	Open	Yes	Failure	5.0 E-04	Unlikely
3	Open	No	Success	4.5 E-03	Unlikely
4	Open	No	Failure	5.0 E-04	Unlikely
5	Closed	Yes	Success	4.5 E-01	Likely
6	Closed	Yes	Failure	5.0 E-02	Possible
7	Closed	No	Success	4.5 E-01	Likely
8	Closed	No	Failure	5.0 E-02	Possible

TABLE 3

RISK MATRIX

	Low	Medium	High	Very High
Likely	7 5	5		
Possible	8 6		6 2	4
Unlikely		1 3	2	4

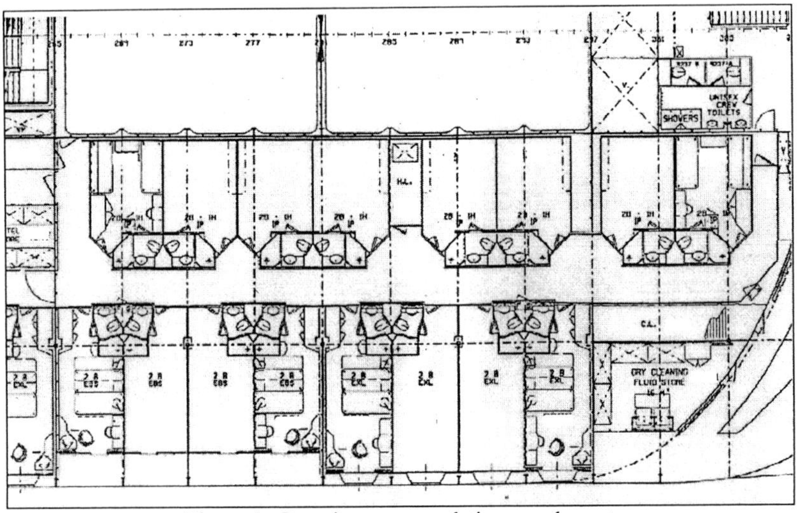

Figure 1: Generic accommodation area layout

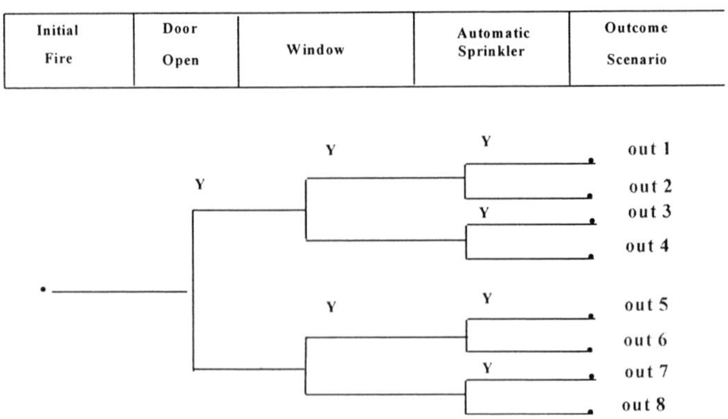

Figure 2: ET structure

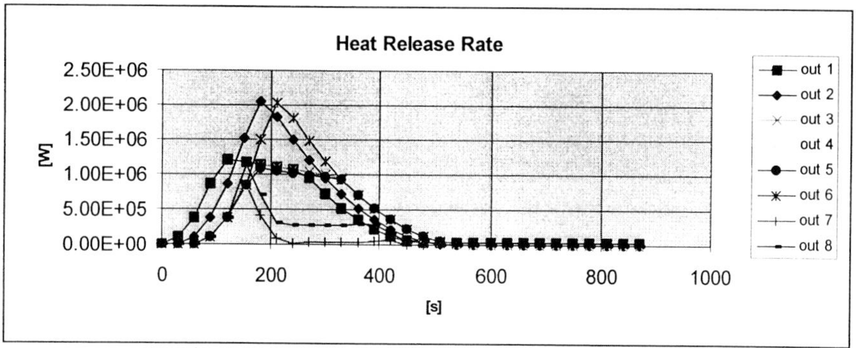

Figure 3: Heat release rate outcome for each scenario

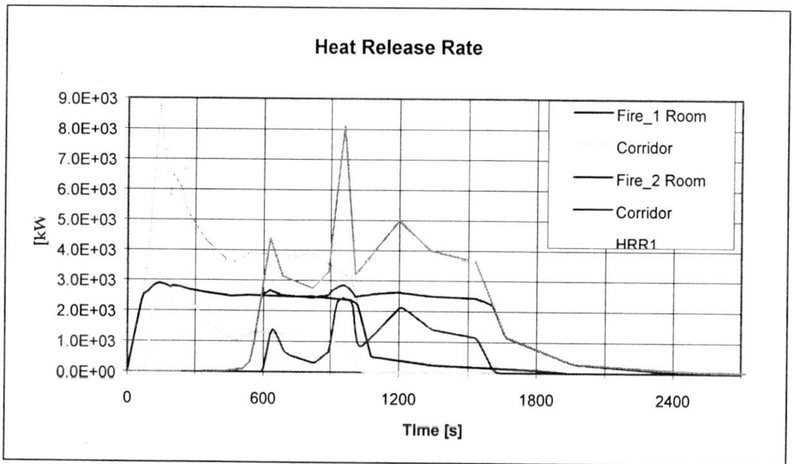

Figure 4: Applied Heat Release Rate for outcome 4

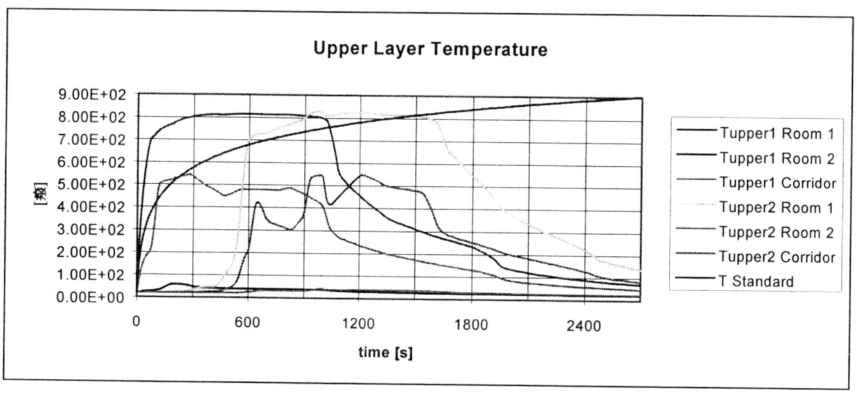

Figure 5: Temperature in the upper zone

Practical Design of Ships and Other Floating Structures
You-Sheng Wu, Wei-Cheng Cui and Guo-Jun Zhou (Eds)

THE CHARACTERISTIC ANALYSIS OF MARINE FIRE SPREAD PHENOMENA WITH MULTI-EQUATIONS SYSTEM FOR FIRE SAFETY DESIGN

Nobuyoshi Fukuchi and Changhong Hu

Faculty of Engineering, Kyushu University
6-10-1 Hakozaki, Higashi-ku, Fukuoka, 812-8581, Japan

ABSTRACT

For the preventive design of marine fire, it is necessary to be grasped entirely the fire spread phenomena by an analysis taking thoroughly account of the governing factors and the characteristics of compartment fire. The turbulent heat diffusion field system consisted of the momentum and heat transportation are applied in the two dimensional calculation on multi-linked compartments, while the gas balance equations with oxygen consumption and gas generation are solved by means of zone model. In order to investigate the threshold conditions of fire spread, some numerical simulations of fire accident in the multi-linked compartments are carried out on the variety of thickness of heat insulation

KEYWORDS

Fire safety design, Fire spread phenomena, Turbulent diffusion of heat and composed gas, Multi-linked compartments, The pseudo-field mathematical model, The thickness of heat insulation

1 INTRODUCTION

It is necessary to establish a fire safety plan that is secured sufficient time for the evacuation of the personnel onboard and the crew to engage in fire fighting activities, in the case of spontaneous outbreak of fire in a ship. For that reason, the fire spreading phenomena in multi-compartments has to be simulated numerically using a suitable mathematical model by considering a distinctive feature of the turbulent diffusion of heat and composed gas.

A pseudo-field mathematical model with multi-equations system expressing the fire spread phenomena is proposed herein by considering a characteristic of the turbulent diffusion of heat and composed gas in closures. In order to investigate the relations between the combustion characteristics and the variety of governing factors such as fire load and thickness of heat insulation, some numerical simulations of fire accident in the multi-linked compartments were carried out applying of the proposed model. From the results of the numerical calculation, it can be clarified on the threshold conditions of fire spread and the validity of pseudo-field mathematical model.

2 THE GOVERNING EQUATIONS OF FIRE SPREAD PHENOMENA

2.1 *The Heat Generation of Combustibles*

The reaction rate $r(t)$ of combustion on fire may be generally expressed by Arrhenius' formula as follows;

$$r(t) = M(t)\xi c(t)e^{-Ea/RT(t)} \tag{1}$$

where t is time, M is the burning weight, E_a is the activation energy, R is the gas constant, ξ is the reaction constant and T, c denote the absolute temperature and the oxygen concentration around combustibles respectively. The heat generation of combustibles can be obtained as $hr(t)$ to multiply the reaction rate $r(t)$ by the calorific value per unit weight h.

2.2 *The Mass Balance Equation of Composite Gases*

Accompanying the combustion of inflammable articles, oxygen is consumed and carbon-dioxide, carbon-monoxide, hydro-oxide are created in proportion to the reaction rate $r(t)$. Consequently, the mass balance of composite gases in each division with ventilation are given the following equations.

$$\frac{dm_X^i}{dt} = \sum_j \frac{v_{in}^j}{V^j} m_X^j - \sum_j \frac{v_{out}^j}{V^i} m_X^i + \sum_k^N L_X r \tag{2}$$

where m_X^i is the mass of a composite gas $\left(X = O_2, N_2, CO_2, CO, H_2O\right)$, V^i is the volume of objective division (i), V^j is the volume of adjacent division (j) and v_{in}^j, v_{out}^j are the changing quantities by flow in or out between divisions (i) and (j) respectively. Further, L_X is the proportional coefficient of creation (positive) or consumption (negative) to combustion rate.

2.3 *The Momentum and Heat Transport Equations in the Internal Space*

Let u_i, p and q be the flow velocity, pressure and gas temperature respectively in Cartesian coordinate ($\left(x_1, x_2, x_3\right); x_3$ is the vertical axis). The momentum transport equation in an anisotropic turbulent flow subjected to gravitational effects can be expressed as folloes;

$$\frac{\partial \bar{u}_i}{\partial t} + \frac{\partial}{\partial x_j}\left(\bar{u}_i \bar{u}_j\right) = -\frac{1}{\rho}\frac{\partial \bar{p}}{\partial x_i} + \frac{\partial}{\partial x_j}\left(\nu \frac{\partial \bar{u}_i}{\partial x_j} - \overline{u_i' u_j'}\right) + \beta g\left(\bar{\theta} - \bar{\theta}_0\right)\delta_{i3} \tag{3}$$

Hence, ν is the coefficient of kinetic viscosity, g is the gravitational acceleration, β is the expansion rate of interior gas, ρ is the density, $\bar{q}_0$ is the average temperature of the internal space, respectively. The symbol δ_{ij} is Kronecker's delta and $\overline{u_i' u_j'}$ denotes Reynolds stress. The symbols $(\bar{\ })$, $(')$ denote respectively the averaging operator upon some dependent variables and the fluctuated variables.

The heat transfer equation, in the field with generating heat $\bar{q}_{fk}$ of combustibles at $x_i = x_{fk}$ can be written as;

$$\frac{\partial \bar{\theta}}{\partial t} + \frac{\partial}{\partial x_j}\left(\bar{u}_j \bar{\theta}\right) = \frac{\partial}{\partial x_j}\left(k\frac{\partial \bar{\theta}}{\partial x_j} - \overline{u_i' \theta'}\right) + \sum_k^N \bar{q}_{fk}\delta\left(x_i - x_{fk}\right) + \bar{q}_c \delta\left(x_i - x_w\right) \tag{4}$$

where k is the thermal diffusion coefficient, $\bar{q}_c$ is the transfer heat flux from the wall located at $x_i = x_w$. Hence $\delta\left(x_i\right)$ represents the delta function and $\overline{u_i' \theta'}$ is the turbulent heat flux.

From standard eddy viscosity model, the Reynolds stress and the turbulent heat flux can be expressed as follows;

$$\overline{u_i'u_j'} = \frac{2}{3}k\delta_{ij} - \nu_t\left(\frac{\partial \bar{u}_i}{\partial x_j} + \frac{\partial \bar{u}_j}{\partial x_i}\right), \quad \overline{u_i'\theta'} = -k_t \frac{\partial \bar{\theta}}{\partial x_i} \tag{5}$$

where k is the turbulent kinetic energy and κ_t is the turbulent thermal diffusivity. The eddy viscosity ν_t is calculated by the formula; $\nu_t = C_u k^2 / \varepsilon$ (herein $C_u = 0.09$), where ε is the dissipation rate of kinetic eddy viscosity. The transportation equations on k and ε are as following equations.

$$\frac{\partial k}{\partial t} + \frac{\partial}{\partial x_j}\left(k\bar{u}_j\right) = D_k + P_k - \varepsilon + G_k \tag{6}$$

$$\frac{\partial \varepsilon}{\partial t} + \frac{\partial}{\partial x_j}\left(\varepsilon\bar{u}_j\right) = D_\varepsilon + \frac{\varepsilon}{k}\left(C_{\varepsilon 1}P_k - C_{\varepsilon 2}\varepsilon + C_{\varepsilon 3}G_k\right) \tag{7}$$

where $\sigma_k = 1.0, \sigma_\varepsilon = 1.3, C_{\varepsilon 1} = 1.44, C_{\varepsilon 2} = 1.92, C_{\varepsilon 3} = 0.7$ are used in this paper. Furthermore,

$$P_k = -\overline{u_i'u_j'}\frac{\partial \bar{u}_i}{\partial x_j} = \frac{1}{2}\nu_t\left(\frac{\partial \bar{u}_i}{\partial x_j} + \frac{\partial \bar{u}_j}{\partial x_i}\right)^2, \quad D_k = \frac{\partial}{\partial x_j}\left(\frac{\nu_t}{\sigma_k}\frac{\partial k}{\partial x_j}\right), \quad D_\varepsilon = \frac{\partial}{\partial x_j}\left(\frac{\nu_t}{\sigma_\varepsilon}\frac{\partial \varepsilon}{\partial x_j}\right), \quad G_k = g\beta\overline{u_i'\theta'}\delta_{i3}$$

By introducing the turbulent Plandtl number Pr_t, the thermal diffusivity can be calculated by the formula;

$$k_t = \nu_t / \mathrm{Pr}_t \tag{8}$$

The boundary conditions for the velocity are given as formulae $\bar{u}_t = u_T y^+$ (the viscous sublayer; $y^+ < 10$) and $\bar{u}_t = u_T / K \ln(Ey^+)$ (the log law zone; $y^+ \geq 10$), where $\bar{u}_t$ is the velocity in the tangent direction to the wall, K is Karman's constant(=0.42), E=09.7 , y^+ is the dimensionless wall coordinate, and u_τ is the friction velocity.

The thermal boundary conditions at the walls are given like as $\theta = \Theta_w$ or $-\lambda\,\partial\theta/\partial n = q_w$,where Θ_w is given temperature, q_w is given heat flux and λ is the heat conductivity of the wall.

2.4 The Heat Conduction Equations of Surrounding Walls

Taking account of heat convection and heat radiation, the heat conduction equation of the surrounding walls, floor and ceiling of each division is written as;

$$c_I\rho_I\frac{\partial \theta_w}{\partial t} = \frac{\partial}{\partial x_j}\left(\lambda_j\frac{\partial \theta_w}{\partial x_j}\right) + q_c\delta(x_i - x_w) + q_r\delta(x_i - x_w) \tag{9}$$

Where θ_w, c_I, ρ_I denote the temperature, specific heat, specific weight of surrounding walls, respectively. q_c is the quantity of heat convection per area element a_i through the thermal boundary layer.

$$q_c = \alpha a_i\left(\theta_w\big|_{xi=xw} - \bar{\theta}\big|_{xj=xw}\right) \tag{10}$$

Further, q_r is the heat radiation per area element a_i against element a_j and it can be expressed by following.

$$q_r = C_b\varepsilon_i\varepsilon_j\left(T_i^4 - T_j^4\right)\varphi_{ij}a_i \tag{11}$$

C_b is the radiative constant of black body, ε_i is the emissivity of body i and φ_{if} is the angle factor of area element a_i for a finite area a_j.

3 NUMERICAL SOLUTION

3.1 Method of Numerical Calculation

A finite volume method is applied here to the numerical analysis of turbulent heat diffusion on two dimensional models of multi-linked cabins like as shown in Figure 1. In order to determine the state values of the turbulence field from the governing equations, we need various interdependent state values which make, if necessary, to conduct iterative calculations between the equations.

3.2 Heat Transfer in Vertical Linked Compartments with an Opening

The analysis deals with vertical linked enclosed spaces, which are surrounded with thermal insulation walls and connected by an opening, in the case that the flow field is the state of turbulence as Rayleigh number $Ra = 1.5\times10^{12}$. Here, the following dimensionless quantities, time $t^* = t/t_0$, $\theta^* = (\overline{\theta} - \Theta_B)/(\Theta_F - \Theta_B)$, velocity $u^* = \overline{u_i}/U_0$ are used for the displaying the results. Where, $t_0 = H^2/\nu\sqrt{Gr} = \sqrt{H/g\beta(\Theta_F - \Theta_B)}$, $U_0 = H/t_0$, Θ_B is reference temperature, Θ_F is the temperature of heat source.

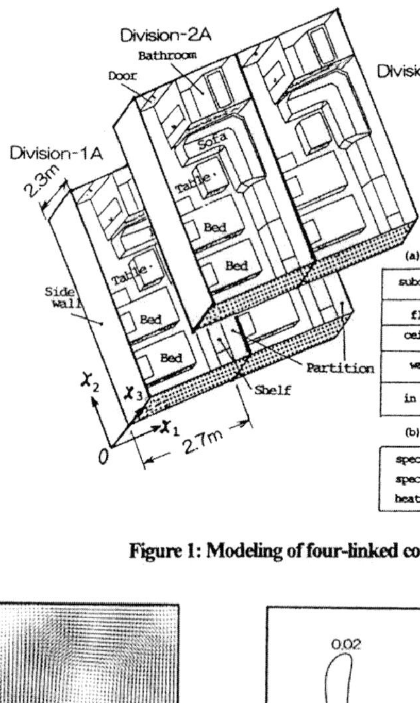

(a) Inflammable materials

subdivision	main material	wood equivalent distribution
floor	carpet	3.09 kg/m^2
ceiling	lining (vinyl-cloth)	1.00 kg/m^2
wall	lining (vinyl-cloth)	1.34 kg/m^2
in space	funitures (desk, bed etc)	106 kg

(b) Thermal insulation

specific weight	:	ρ_s = 180 kg/m^3
specific heat	:	C_s = 0.84 kJ/kg·K
heat conductivity	:	λ = 0.060 W/m·K

Figure 1: Modeling of four-linked compartments

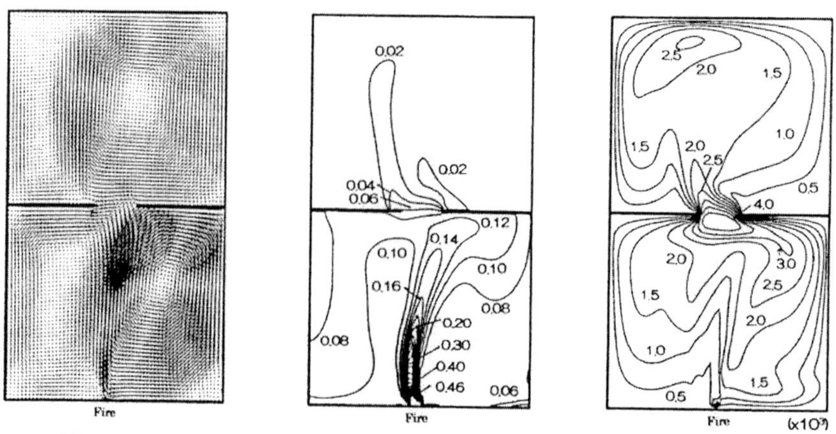

(a) Velocity (b) Gas Temperature : θ^* (c) Eddy Diffusivity : ν_t/ν

Figure 2: The state variables of vertical linked compartments with an opening

The state variables in two enclosures at $t^* = 0.07 \times 10^3$ are shown in Figure 2. The airflow rises from the fire source, separates at the ceiling, forms circulation in the room. Part of the circulation flow goes through the opening to the upper compartment, and the flow in the upper space circulates by hauling effect of lower space circulation at the opening. From the temperature distribution, it is found that the heat transfer comes mainly by the advection effect.

4 NUMERICAL SIMULATION OF FIRE SPREAD PHENOMENA

The analysis of fire spread phenomena on the models of four-linked cabins is carried out numerically in the conditions that the fire loads of inflammable materials and the thermal properties of heat insulation are shown in Figure 1. The structure of cabins are assumed to be made of aluminum-alloy, in which the point of structural rupture is estimated to be over 250°C. The dimensionless thickness of insulation $\delta = d\sqrt{\xi}/\sqrt{\lambda/c_I\rho_I}$ is used hereafter, where d is the thickness of insulation, λ, c_I, ρ_I are respectively the heat conductivity, specific heat, specific weight of furnishing insulation.

4.1 Fire Spread State Arising from Structural Ruptures

By simulating numerically the fire spread state arising from structural ruptures in the case of $\delta = 28$, Figure 3 shows the variations of averaged gas temperature and oxygen concentration as time proceeds in the cabin spaces, according to fire spread history initiated from cabin-1A as shown in Figure 4. Furthermore, the temperature distribution of the cabin spaces are shown in Figure 5.

By these results, a compartment fire is clarified to be "the oxygen-depended combustion" that is balanced consuming oxygen with supplying air. And the rupture of ceiling in a fire room tends to occur rapid fire spread for active heat convection in vertical direction. According to develop burning state, it has a tendency to form gradually the thermal layer with high temperature. It can be seen that the present pseudo-field mathematical model can be used aptly to predict the fire spread phenomenon.

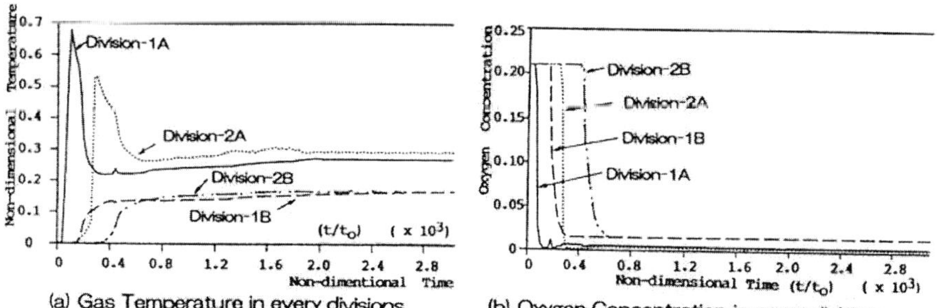

Figure 3: The variation of gas temperature and oxygen concentration in four-linked compartments

4.2 The Possibility of Fire Spread in Multi-linked Cabins

After the numerical simulations of fire accident of four-linked cabins were executed in the case of varying the insulation thickness for investigating the possibility of fire spread, the propagation times to neighboring divisions are obtained as shown in Figure 6. The insulation thickness of occurring the structural rupture can be estimated considerably to be $\delta < 42$, while the possibility of fire propagation may be pretty low in the case of $\delta > 55$.

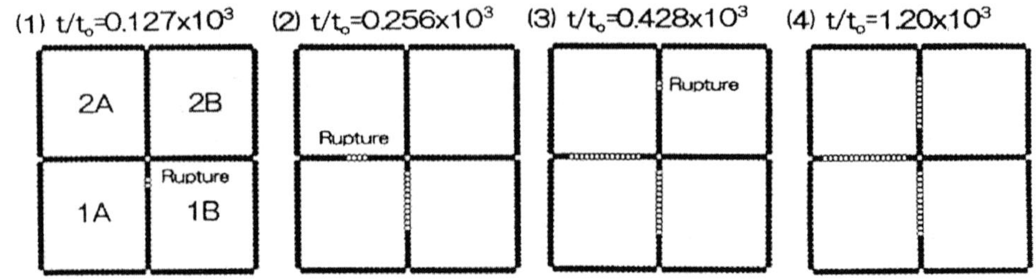

Figure 4: A rupture history of partitions of four-linked cabins initiated from Division-1A

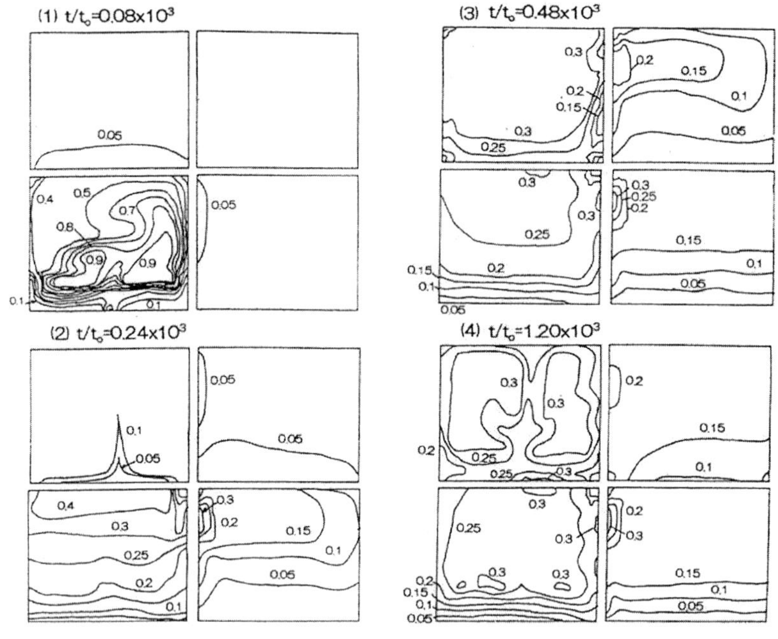

Figure 5: The variation of isotherm in four-linked compartments as time proceeds

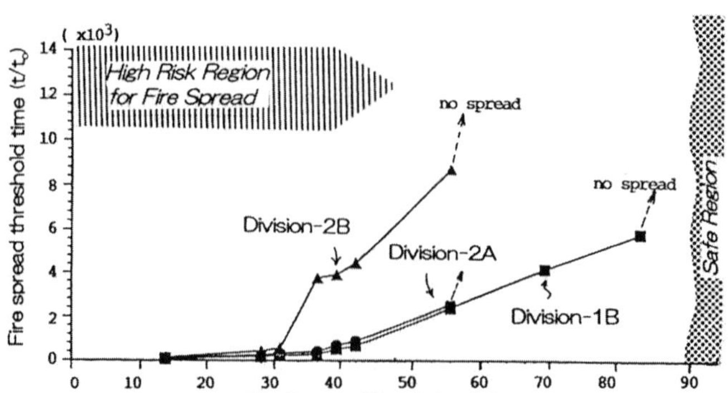

Figure 6: The fire spread threshold times of four-linked cabins for different thickness of heat insulation

5 CONCLUSIONS

Some numerical simulations of fire accident in the multi-linked cabins are carried out applying of the proposed mathematical model by considering a characteristic of the turbulent diffusion of heat and composed gas in closures. From the results of the numerical calculation, it can be clarified the characteristics of fire spread phenomena, the threshold conditions of fire spread and the validity of the present pseudo-field model.

References

N. Fukuchi and S. Koth. (1993). The Threshold Conditions and Characteristic Analysis of the Marine Fire Spread Phenomena (Part 1) The Fire Spread through Ruptured Partitions and Structural Members, *Jour. of Society of Naval Architects of Japan*, **174**, 833-842.

W. P. Jones and B. E. Launder. (1975). The Prediction of Laminarization with a Two Equations Model of Turbulence, *Jour. of Heat Mass Transfer*. **15**, 301-314.

S. V. Patankar. (1980). Numerical Heat Transfer and Fluid Flow. McGraw-Hill, New York.

C. Hu and N. Fukuchi. (1997). The Characteristic Analysis of Turbulent Diffusion on the Functional Design of Marine Systems (Part 1) Turbulent Heat Diffusion by Natural Convection, *Jour. of Society of Naval Architects of Japan*. **181**, 143-150.

Practical Design of Ships and Other Floating Structures
You-Sheng Wu, Wei-Cheng Cui and Guo-Jun Zhou (Eds)

APPLICATION OF COMPUTATIONAL FLUID DYNAMICS IN THE FIRE SAFETY DESIGN OF MARINE SYSTEMS

Changhong Hu and Nobuyoshi Fukuchi

Graduate School of Engineering, Kyushu University
6-10-1 Hakozaki, Higashi-Ku, Fukuoka 812-8581, Japan

ABSTRACT

The purpose of this research is to apply the computational fluid dynamics (CFD) method in the fire safety design of marine systems. This paper presents fire field model for the case of a fire within a compartment, to numerically simulate a three-dimensional, turbulent buoyancy-driven flow. The flow is considered as a low Mach number compressible flow, and the set of approximate control equations are used. The numerical solution is performed based on the large eddy simulation approach. As an example, a fire simulation is made on a passenger ship cabin with furniture; the predicted temperature variation in the room is compared to measured data of an experiment.

KEYWORDS

Fire safety design, Computational fluid dynamics, Fire field model, Low Mach number approximation, Large eddy simulation, Compartment fire

1 INTRODUCTION

Predicting the spread of smoke flow from a marine fire is still one of the challenges to the functional design of safe and comfortable environments for marine systems. In the design process associate with control of smoke spread or fire–extinguishing system, it is often required to know the heat transfer and flow motion induced by fires in compartments. This knowledge can be obtained by experimental observations and theoretical models. As fire experiments are generally expensive and dangerous, it is difficult to obtain sufficient data by conducting many experiments. For the theoretical models, the zone model is popular method but it can only give limited data we needed. In this paper, a field model, computational fluid dynamics method is proposed to predict the thermal convection from a compartment fire.

A distinguishing feature of the problem is that the temperature difference as well as the density variation is very large while the smoke flow induced by the heat release is much slower than the propagation of acoustic waves. Therefore, this flow should be considered as a low Mach number compressible flow. The generally used Boussinesq approximation, in which the density is assumed constant except for the buoyancy term, is limited to very small temperature difference, and is not an appropriate model for the problem (Gray and Giorgini, 1976). In this paper, a set of approximate equations will be used to solve the low Mach number flow, and the numerical solution will be conducted by a large eddy simulation (LES) method in which the Smagorinsky SGS model (Smagorinsky, 1963) is used. At this stage of research the simulation of combustion is not done, the heat release of combustible materials used for the computation is taken from an experimental database. As an example, a 3-D numerical simulation of a turbulent buoyant flow induced by a compartment fire is made. The compartment, which is a common type of cabin with furniture of a passenger ship, is connected through a door to a corridor. The geometry of the room is chosen so that comparison with available experimental results is possible. Two types of cabin fires are simulated and comparisons with the experimental results are made.

2 MATHEMATICAL FORMULATION

The governing equations are derived by applying a spatial, low-pass filter to the Navier-Stokes equations and energy equation for a compressible, constant-viscosity fluid. Based on Favre filtering, the turbulent field variable *f*, can be decomposed as follows.

$$f = \tilde{f} + f'' \tag{1}$$

Where $\tilde{f} = \overline{\rho f} / \bar{\rho}$ is the resolved (grid-scale or GS) part, and f'' the unresolved (subgrid-scale or SGS) part. Then the governing equations solved in the LES approach can be written as follows.

$$\frac{\partial \bar{\rho}}{\partial t} + \frac{\partial (\bar{\rho}\tilde{u}_i)}{\partial x_i} = 0 \tag{2}$$

$$\frac{\partial \bar{\rho}\tilde{u}_i}{\partial t} + \frac{\partial}{\partial x_j}(\bar{\rho}\tilde{u}_i\tilde{u}_j) = -\frac{\partial \bar{p}}{\partial x_i} + \frac{\partial}{\partial x_j}(\phi_{ij} - \tau_{ij}) - \bar{\rho} g \delta_{i3} \tag{3}$$

$$\frac{\partial \bar{\rho}\tilde{T}}{\partial t} + \frac{\partial}{\partial x_i}(\bar{\rho}\tilde{u}_i\tilde{T}) = \frac{\partial}{\partial x_i}(\theta_i - \sigma_i) + Q \tag{4}$$

$$\bar{p} = \bar{\rho} R \tilde{T} \tag{5}$$

Where $\phi_{ij} = 2\mu(\bar{S}_{ij} - 1/3\delta_{ij}\bar{S}_{kk})$, $\theta_i = k\partial\tilde{T}/\partial x_i$, $\bar{S}_{ij} = (\partial\tilde{u}_i/\partial x_j + \partial\tilde{u}_j/\partial x_i)/2$, $\tilde{u}_i$ is the velocity component in the *i*-th direction, $\tilde{T}$ the temperature, $\bar{p}$ the pressure, $k = \mu / P_r$, μ the viscosity and Pr the Prandtl number. τ_{ij} and σ_i are the SGS Reynolds stress and the SGS heat flux, and can be expressed as follows.

$$\tau_{ij} = \bar{\rho}\left(\widetilde{u_i u_j} - \tilde{u}_i\tilde{u}_j\right) \tag{6}$$

$$\sigma_i = \bar{\rho}\left(\widetilde{u_i T} - \tilde{u}_i\tilde{T}\right) \tag{7}$$

From Erlebacher et al. (1992), by using the classical concepts of SGS viscosity and diffusivity, the SGS Reynolds stress and the SGS heat flux can be expressed as the follows.

$$\tau_{ij} = -2\mu_e\left(\bar{S}_{ij} - \frac{1}{3}\delta_{ij}\bar{S}_{kk}\right) \tag{8}$$

$$\sigma_i = -\frac{\mu_e}{\Pr_t}\frac{\partial\tilde{T}}{\partial x_i} \tag{9}$$

Where, $\mu_e = \bar{\rho}(C_S\Delta)^2|\bar{S}_{ij}|$ is the SGS eddy viscosity, $|\bar{S}_{ij}| = [2\bar{S}_{ij}\bar{S}_{ij}]^{1/2}$, $\Pr_t$ the turbulent Prandtl number, C_s the Smagorinsky model constant, $\Delta = (\Delta x_1 \Delta x_2 \Delta x_3)^{1/3}$ the length scale of the model.

For the constants, $C_s = 0.12$, $Pr_t = 0.5$ are used.

This set of above compressible equations generally contains vorticity, entropy and acoustic modes, in different frequencies. For the low Mach number problem like flow motion induced by fires, the acoustic modes are in frequency bands much higher than the others. Numerical solution of such low Mach number flow using these compressible equations is impossible because it requires extremely small time step to track the acoustic fluctuations. Here, a set of approximate equations, which are derived from the equations (2)-(5) with the acoustic wave filtered out, is presented to express the low Mach number flow.

Rewriting equations (2)-(5) in non-dimensional form, defining $\varepsilon = \gamma M^2$, (M is the Mach number, γ the specific heat ratio) and assuming $\varepsilon << 1$, the dependent variables may be expanded as power series in ε. Collecting the zero-order terms in ε results in lowest-order equations, and they are written in dimensional form as follows.

$$\frac{\partial \bar{\rho}}{\partial t} + \frac{\partial(\bar{\rho}\tilde{u}_i)}{\partial x_i} = 0 \tag{10}$$

$$\frac{\partial \tilde{u}_i}{\partial t} + \tilde{u}_j \frac{\partial \tilde{u}_i}{\partial x_i} = \frac{1}{\bar{\rho}}\left\{ -\frac{\partial \bar{p}_D}{\partial x_i} + \frac{\partial \varphi_{ij}}{\partial x_j} + \bar{\rho} g \delta_{i3} \right\} \tag{11}$$

$$\frac{\partial \tilde{T}}{\partial t} + \tilde{u}_i \frac{\partial \tilde{T}}{\partial x_i} = \frac{1}{\bar{\rho}}\left\{ \frac{\partial}{\partial x_i}\left[(k + k_e)\frac{\partial \tilde{T}}{\partial x_i} \right] + \bar{Q} \right\} \tag{12}$$

$$\bar{p}_T = \bar{\rho} R \tilde{T} \tag{13}$$

Where $\varphi_{ij} = 2(\mu + \mu_e)(\bar{S}_{ij} - \delta_{ij}\bar{S}_{kk}/3)$, $k_e = \mu_e / \mathrm{Pr}_t$. It can be seen that in the approximate equations, the total pressure is decomposed in to two parts.

$$p = p_T + p_D \tag{14}$$

Where, p_T is the thermodynamic pressure, and it is a function only in the time. Therefore, this pressure is considered as the spatial mean pressure. In the low-Mach-number approximation, it assumes that the propagation speed of acoustic wave (the sound speed) is infinity, so that the pressure variation induced by the temperature change is supposed to spread through all the flow fields instantly. On the other hand, p_D is the dynamic pressure related to the fluid motion. It is a function in both space and time. Under the zero-order approximation there is no interference between p_T and p_D.

The approximate equations are solved using a finite difference method which has been developed for solution of incompressible flows. In the method a second-order Adams-Bashforth time stepping scheme is used to advance the equations in time. The pressure and velocity are solved by a fractional step method (Chorin, 1967) as follows.

$$\frac{\hat{\tilde{u}}_i - \tilde{u}_i^n}{\Delta t} = \frac{3}{2}H_i^n - \frac{1}{2}H_i^{n-1} \tag{15}$$

$$\frac{\partial}{\partial x_i}\left(\frac{1}{\bar{\rho}} \frac{\partial \bar{p}_D^{n+1}}{\partial x_i} \right) = \frac{1}{\Delta t}\left(\frac{\partial \hat{\tilde{u}}_i}{\partial x_i} - \frac{\partial \tilde{u}_i^{n+1}}{\partial x_i} \right) \tag{16}$$

$$\frac{\tilde{u}_i^{n+1} - \hat{\tilde{u}}_i^n}{\Delta t} = -\frac{1}{\bar{\rho}} \frac{\partial \bar{p}_D^{n+1}}{\partial x_i} \tag{17}$$

Where H_i represents the viscous, convective and buoyant components of the momentum equation. The important difference from the computation with incompressible equations is that in the pressure equation (16) the divergence of velocity $(\partial \tilde{u}_i / \partial x_i)^{n+1}$ is not zero. It can be obtained from (10), (12)

TABLE 1
HEAT RELEASE OF COMBUSTIBLES

Name of combustible	Mass (kg)	Heat release (MJ)
Single Bed	6.24	152.9
Double Bed	11.78	288.6
Desk	6.30	102.1
Locker	9.70	189.2
Table	3.00	40.5
Chair	1.05	22.5
Wall paper	7.72	91.1

and (13) by assuming the thermodynamic pressure p_T is a constant, as follows.

$$\frac{\partial \tilde{u}_i^{n+1}}{\partial x_i} = \frac{1}{\overline{\rho}\widetilde{T}}\left\{\frac{\partial}{\partial x_i}\left[(k+k_e)\frac{\partial \widetilde{T}}{\partial x_i}\right]+\overline{Q}\right\} \tag{18}$$

3 RESULTS

Compartment fires in two types of cabins of a passenger ship are selected for the numerical simulation using the method described in the previous section. The calculation conditions are the same as the full-scale fire experiment without fire–extinguishing systems (JSRS Report, 1998). These two types of cabins are a first-class cabin and a second-class cabin, which are shown in Figure 1.

The computation model includes the cabin and part of corridor; they are connected through an open door. The grid number is $72\times40\times40$. In the figure 'Temperature (1)' and 'Temperature (2)' indicate two columns of measuring points for comparison of temperature. As an example, the time variation of the temperature at the measuring points for the first-class cabin is shown in Figure 2.
At the current stage of the research, the combustion itself is not simulated, instead, the heat release used for the computation is taken from the experiment data of combustible materials. For example, the measured heat releases of combustibles for the first-class cabin are shown in Table 1. These values are used as the heat source for the energy equation in the computation.

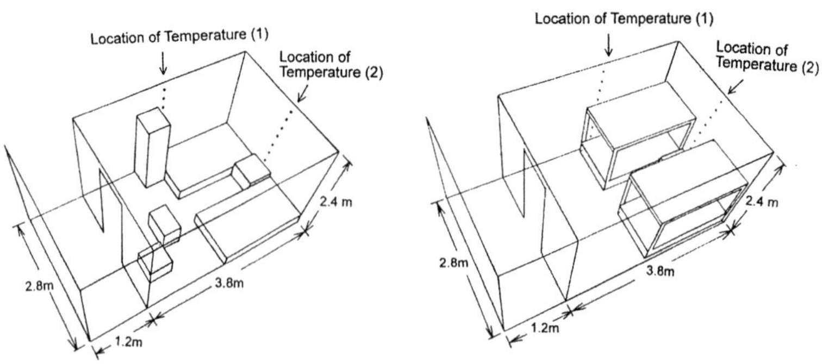

First-Class Cabin   Second-Class Cabin

Figure 1 : Geometries of two types of cabins with furniture

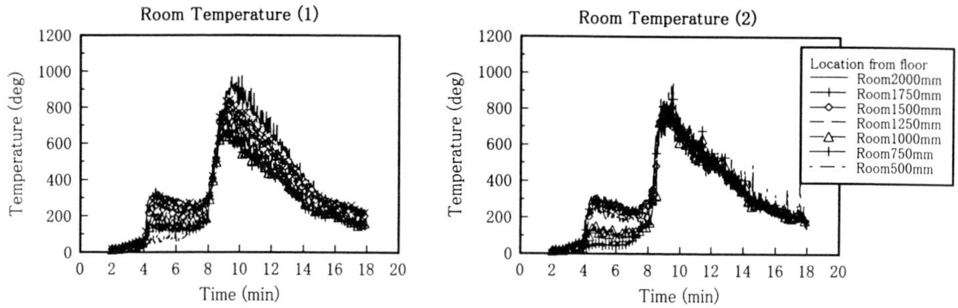

Figure 3 : Predicted time variation of the temperature at the measuring points for the first-class cabin

The fire process in the computation is set the same as that observed in the experiment [7]: for the first-class cabin, the fire starting from the left-side bed, spreading to the wall paper (at t=4 min), flash-over occurring after the fire spreading into the desk and the neighboring bed (at t=11 min). The calculated temperature variation at the measuring points for the first-class cabin is shown in Figure 3. Comparing to the experiments, the trends of the variation are the same. The different point is also found that for the experimental data, the temperatures at different measured points are almost the same after the flashover. It is considered that in the experiment, the flames may envelope the measuring points as the thermocouples are installed near the wall. The measured temperature is, therefore, different from the gas temperature computed at the time.

The calculated isothermal lines on the walls and the isothermal surface of at $t = 11\,\text{min}$ for the first-class cabin are shown in Figure 4. It is found the maximum temperature arise occurs at the ceiling of the room. The maximum temperature arises at the measured points for the two cases are shown in Figure 5, with the comparison of measured data in the experiment. It can be seen that the maximum temperature in the second-class cabin is higher than that in the first-class cabin, for that the double beds in the second-class cabin have larger heat releases than the single beds in the first-class cabin.

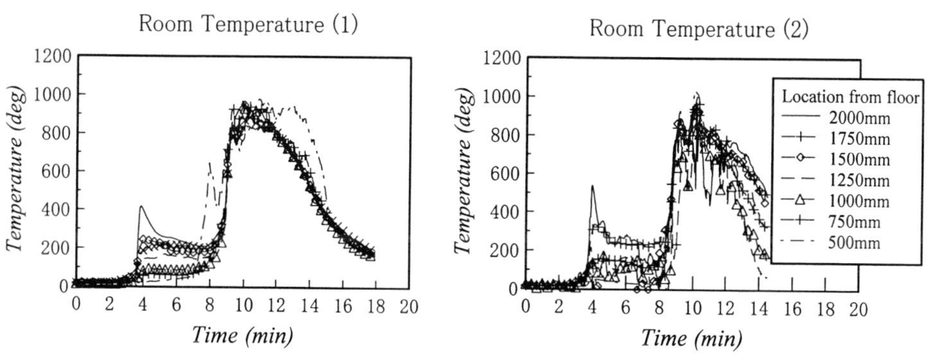

Figure 2 : Experimental time variation of the temperature at the measuring points for the first-class cabin

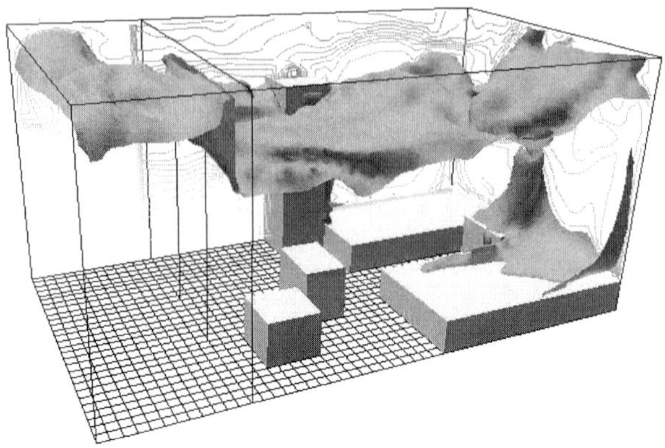

Figure 4 : Calculated isothermal lines on the walls and the isothermal surface of $T = 700^{o}C$ for the first-class cabin

From these results, it can be said that the present CFD method provides a powerful tool to predict the heat transfer and fluid motion induced by a fire. However, as the present calculation does not consider the effect of flame, the prediction of the temperature distribution is not satisfied occasionally. This is the weakest point of the present computation. We consider to improve it in the future research by introducing a kind of flame model into the computation.

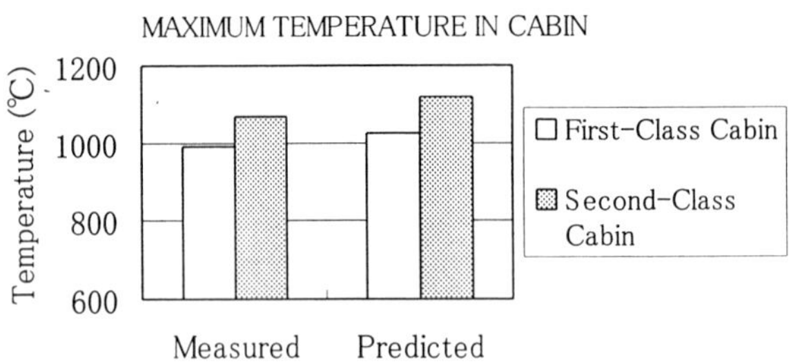

Figure 5 : Comparison of predicted and measured maximum temperature arise in the cabins

4 CONCLUSIONS

In this paper, we present some early work concerned with the application of Computational Fluid Dynamics method in the fire safety design of marine systems. By simply specifying a scenario based on known fire characteristics and experimental observations, the fire induced turbulent buoyant flow in a cabin of a passenger ship has been numerically simulated with a field model. The comparison of air temperature in the cabin between the predict results and experimental data has shown a fairly good agreement.

Reference

Chorin, A. J. (1967). A Numerical Method for Solving Incompressible Viscous Flow Problems. *J. Comp. Phys.* **2,** 12-26.

Erlebacher G., Hussaini M. Y., Speziale C. G. and Zang T. A. (1992). Toward the Large-Eddy Simulation of Compressible Turbulent Flows. *J. Fluid Mech* **238**, 155-185.

Gray D. D. and Giorgini A. (1976). The Validity of the Boussinesq Approximation for Liquids and Gases. *Int. J. Heat Mass Transfer* **19**, 545-551.

Smagorinsky, J. (1963). General Circulation Experiments with the Primitive Equations. *Mon. Weather Rev.* **91**, 99-164.

The Japanese Shipbuilding Research Society, RR73 Standard Section, (1998): *A 1997 Report: Investigation on the Fire-Protection of the Ship.*

Practical Design of Ships and Other Floating Structures
You-Sheng Wu, Wei-Cheng Cui and Guo-Jun Zhou (Eds)

AN EXAMINATION OF SOME STRUCTURAL LIMIT STATES FOR HYDROCARBON EXPLOSIONS

P. A. Frieze[1], R. B. Corr[2], R. O. Snell[2] and V. H. Y. Tam[2]

[1]P A F A Consulting Engineers, Hofer House, 185 Uxbridge Road,
Hampton, Middx, TW12 1BN, UK
[2]BP Exploration, Chertsey Road, Sunbury on Thames,
Middx, TW16 7LN, UK

ABSTRACT

The paper examines how limit states may be applied in the assessment of the consequences of hydrocarbon explosions. A conditional structural reliability analysis approach is used to account for both the likelihood of the explosion occurring and exceedance of a structural limit state. The structural limit state is expressed in terms of deflection limits. Typical offshore topsides explosion events lie in the range 10^{-3} to 10^{-5} per annum. Typical overpressure versus probability of exceedance curves are used to determine possible loads for a simply supported beam. Annual probabilities of failure are determined for a range of hydrocarbon explosion probabilities accounting for uncertainties in the loading, response and strength. The effect of structural system strength enhancement is simply accounted for. For a target probability of failure of 10^{-5}, load and resistance partial factors are determined. These are reported and discussed. Some unexpected values are realised which are reviewed and explained.

KEYWORDS

Explosions, Limit states, Probabilities of failure, Deflection limits, Offshore, Overpressures, Triangular pulse loading, Partial load and resistance factors.

1 INTRODUCTION

The present approach in the UK to the design against blast loads is described in the Interim Guidance Note (IGN), Steel Construction Institute (1993). Despite the accidental nature of gas explosions, the offshore industry has tended to follow the use of a 'normal' design approach in dealing with such events. This involves identifying a design level for an accident and considering it usually under two conditions, the event itself and a post-accident or residual strength assessment. The acceptability or damage sustained in resisting the event is not a normal consideration and assessments are usually conducted on a deterministic basis. The purpose of this paper is to identify some of the issues surrounding the determination of hydrocarbon explosion loads, the interpretation of structural response, the possible consequences of the explosion, and the reliability of the predictions within the context of the limit state philosophy.

1.1 Background

Gas Explosions are comparatively low frequency accidental events with the potential for serious consequences either through direct effects or as a result of escalation. Within the range of possible events there are a wide range of consequences. Figure 1 presents a plot relating the frequency of a typical topsides hydrocarbon explosion to overpressure. The overpressure v frequency plot is typical of plots due to many other types of accidental load and extreme environmental load. As the frequency reduces, the overpressure increases approaching a finite limit. Typical design events are in the range of 10^{-3} to 10^{-5}. However, adopting one case as the basis of design tends to mask the performance standards required of essential safety systems to control lesser events with a high degree of reliability and avoid escalation.

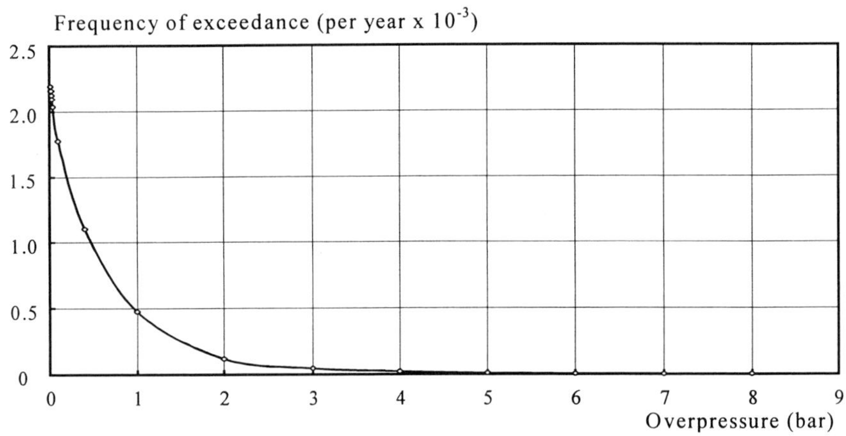

Figure 1: Typical explosion frequency of exceedance versus overpressure curve

Currently, for the design event, load and resistance factors of unity are adopted assuming the 'design scenario' to be inherently conservative despite the relatively large uncertainty associated with hydrocarbon explosion events. Normally, it would be expected in a partial factor limit state design approach that the load and resistance factors would have values that reflected the degree of uncertainty associated with the loads and resistance respectively.

2 DESIGN LIMIT STATES

2.1 Selected Limit States

ISO 19902 (2000) defines four categories of limit states of which three are relevant to the present application:

a) an ultimate limit state corresponding to the maximum resistance to an applied action
b) a serviceability limit state governing normal functional use
c) a fatigue limit state that corresponds to the accumulated effect of repeated actions
d) an accidental limit state corresponding to an event where damage occurs. The intention of this limit state is to ensure that the structure can tolerate the damage due to specified accidental

incidents and can subsequently maintain structural integrity for a sufficient period under specified environmental conditions to enable evacuation to take place.

The basis for the selected design limit states is discussed by Corr & Tam (1998). It is an overall approach to dealing with the risks associated with gas explosion events and includes an initiating event, the release of gas and its composition, the size of the gas cloud, ignition, the resulting overpressure-time history curve, structural response, and consequences. Three levels of events are identified, of which the relevant factors are:

a) High probability 10^{-3} Pa

- Safety Critical Systems should remain intact and functional.
- Primary and secondary steelwork should remain within elastic limits and, in particular, no permanent damage to primary steelwork. Local damage to tertiary steelwork is acceptable.

b) Intermediate probability 10^{-4} Pa

- Minor damage to primary steelwork is acceptable whilst secondary steelwork can undergo permanent deformation.

c) Low probability 10^{-5} Pa

- Decks and primary steelwork may be severely distorted but the structure should have a greater than 90% probability of withstanding collapse.

3 DYNAMIC EFFECTS

3.1 General

Explosions produce significant dynamic effects that increase as the severity of the event itself increases. In addition to the direct impact of blast loads on topsides structure and equipment, indirect loading occurs as a consequence of equipment and structure deforming and failing as well as motions and vibrations, a result of the aftermath of the main event. Although in general, the concern would be for all manifestations involving structural response, namely, beams, columns, and stiffened plates and webs, the last of which are also prime components of blast walls, the prime concern here is with beams.

3.2 Strain-Rate Effects in Steel

To offset some of the dynamic loading effects, steel undergoes strain-rate effects, which tend to enhance the basic properties and improve the resistance characteristics. Thus strain-rate as well as ductile effects is important because these mechanisms help absorb energy in a hydrocarbon explosion and improve the structural performance in resisting intense transient loads. Strain-rate effects considered in the present work are addressed in P A F A (1999). This provided a basis for selecting a rate enhancement factor given a strain rate and also enabled the corresponding uncertainties (bias and COV - coefficient of variation) to be quantified for use in the reliability assessment described below.

3.3 Yield Stress Values of Analysis

The basis for the high probability value is that the primary and secondary steel should remain elastic but tertiary steel can suffer some deformations. By using the mean value of yield stress in the assessment, some yielding will occur based on the above ratio (unity) but it will be minimal. For the intermediate level, adopting a ratio of 3 allows maximum deflections to reach this value but the

permanent value will be less than this. Permanent deflections less than 3 times the elastic value are relatively minor in keeping with the limit state requirement. For low probability events, severe distortions can occur without collapse, for which 10 seems appropriate.
One of the factors affecting this choice of deformation limits was that the ratios between adjacent numbers should be similar, ie, 1:3 should be approximately equal to 3:10. This is necessary to ensure that consistency with the ratios of probabilities of occurrence is maintained otherwise distortion of the final outcome will result.

For a beam considered by Biggs (1964) with a span of 6.1 m, the above ratios correspond to deflections of 13.8, 41.4 and 138 mm or span divided by 442, 147 and 44.2. Solutions have been obtained for the Biggs' SDOF model for an idealised triangular pressure time history for ratios of R_m/F_1 equal to 1.6, 1.0, 0.8 and 0.4, where R_m is the plastic moment capacity of the single degree-of-freedom (SDOF) system and F_1 the magnitude of the triangular loading. The maximum velocities corresponding to each of these deflection ratios were determined and are tabulated immediately below. Given the beam size, the strain-rates consistent with these deflections and corresponding velocities have been calculated and are also listed. From P A F A (1999), the maximum strain-rate effect on yield was determined and is tabulated together with the uncertainty in that estimate.

Limit state probability	Maximum deflection ratio (elastic limit = 1.0)	Maximum velocities (m/s)	Strain-rate (s^{-1})	Maximum strain-rate enhancement	Coefficient of variation
High	1	0.785	0.0561	1.213	0.031
Intermediate	3	1.75	0.1252	1.256	0.040
Low	10	3.41	0.2440	1.297	0.050

From the tabulation, it is seen that the effect of the selected deflection ratios has led to yield stress increases ranging from 1.213 to 1.297. These are maximum values. More representative values will be smaller than these so, for this study, enhancement factors of 1.182, 1.216 and 1.247 were adopted. These values approximate to the maximum velocity less one standard deviation.

3.4 SDOF Systems - Load Characterisation

One approach to determining the response of a structural element is to idealise the structural system as a SDOF and to idealise the applied load as a simple triangular pressure time trace. The solution of the equations for the idealised form is expressed in the classic Biggs' graph which expresses the key parameters in non-dimensional form. However, the Biggs' idealisation hides a number of uncertainties.

Some of these issues can be addressed directly by modification of the Biggs approach but other effects such as the development of membrane action cannot be readily incorporated. In addition, measured data from full-scale tests depart significantly from the idealised version as can be appreciated from the measured pressure-time history plot shown in Figure 2.

4 PARTIAL FACTOR DETERMINATION

4.1 Structural Reliability Analysis (SRA) Approach

In the present circumstances, structural failure is assumed to occur following an explosion. Failure will not occur unless the explosion takes place. Structural failure is therefore conditional on the explosion having taken place. Formulating this as a conditional SRA approach:

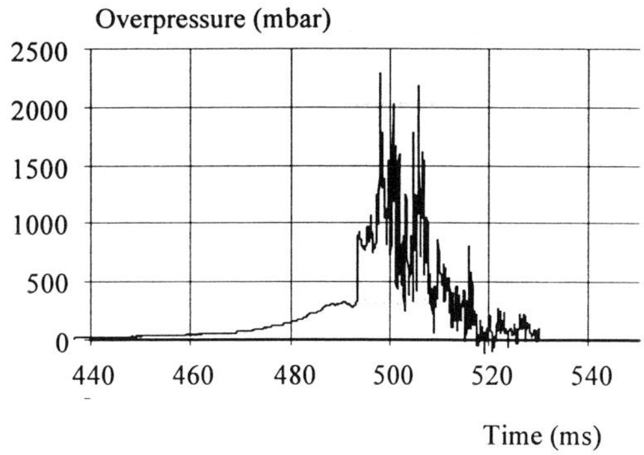

Figure 2: Sample measured pressure-time history

$$P(A \text{ and } B) = P(B)\, P(A|B) \quad (1)$$

where P(B) is the probability of an explosion occurring, P(A|B) is the probability of structural failure occurring given an explosion has taken place, and P(A and B) is the probability that both structural failure and an explosion occur. The probability of an explosion occurring has already been determined for three limit state levels as described above. For structural failure, the normal (Gaussian) version of the level 1 reliability index (β) procedure will be used. It has the form

$$\beta = (R_m - Q_m)/(\sigma_R^2 + \sigma_Q^2)^{0.5} \quad (2)$$

where R and Q refer to resistance and loading, 'm' indicates mean value and σ is standard deviation.

4.2 Limit State Equation

The proposed limit state equation is

$$R_m - Q_m = X_s\, X_m\, R_c - Q_m \quad (3)$$

where X_s is a factor to account for system strength, X_m is the modelling uncertainty parameter associated with R_c, ie, a measure of the accuracy of component strength equation under consideration, and R_c is the mean value of component strength. Component strength and the mean and nominal value of loading Q_{nom} are given by

$$R_c = \gamma_m R_d \qquad Q_{nom} = Q_d / \gamma_e \qquad Q_m = X_Q\, Q_{nom} \quad (4)$$

where R_d is the component design strength equation, γ_m is the partial resistance factor, Q_d is the specified design accidental (explosion) load magnitude, γ_e is the partial load factor and X_Q is the loading bias.

In the present application, although component selection has not been limited in any way, most consideration has been given to beam-type behaviour as in, for example, the problems extracted from Biggs. This theme will be continued and the statistical descriptions, as discussed in the next section, developed accordingly. For beam elements, strength will be a function of yield stress, strain-rate and geometry parameters. Of these, geometry usually plays a minor role in affecting failure probability.

Accordingly, only yield stress and strain-rate will be considered to influence R_d. For design, $R_d \geq Q_d$, whilst for analysis, $R_d = Q_d$.

4.3 Input Variables

The input variables are listed in Table 1 together with their distribution parameters that characterise their uncertainty, ie, bias and COV. The basis for their selection is described in PAFA (1999).

TABLE 1
LIMIT STATE VARIABLES AND THEIR DISTRIBUTION PARAMETERS

Variable	Bias	COV
System strength X_s	1.5, 2.0	0.15/mean
Modelling uncertainty parameter X_m	1.0	0.10
Yield stress σ_Y	1.12	0.04
Strain-rate $\dot{\varepsilon}$	1.182, 1.216, 1.247	0.031, 0.040, 0.050
Load bias X_Q	1, 1.5, 2	0.0
Explosion load magnitude Q_{nom}	1.0	0.15, 0.30, 0.45

4.4 Limit State Reliabilities

For partial load and resistance factors of unity, Table 2 summarises the derived annual reliabilities.

TABLE 2
PROBABILITIES OF FAILURE PER ANNUM FOR EXPLOSION LIMIT STATES

Strength bias	Load bias	High probability (10^{-3} Pa)	Inter. Probability (10^{-4} Pa)	Low probability (10^{-5} Pa)
1.5	1.0	7.44×10^{-6}	1.74×10^{-6}	3.61×10^{-7}
	1.5	1.15×10^{-4}	1.36×10^{-5}	1.64×10^{-6}
	2.0	5.14×10^{-4}	4.65×10^{-5}	4.38×10^{-6}
2.0	1.0	8.20×10^{-7}	2.13×10^{-7}	5.69×10^{-8}
	1.5	1.41×10^{-5}	2.12×10^{-6}	3.40×10^{-7}
	2.0	1.08×10^{-4}	1.15×10^{-5}	1.31×10^{-6}

In can be seen from Table 2 that, for the present range of input parameters, overall probabilities of failure, ie, when accounting for the frequency of occurrence of the explosions themselves, are relatively small. This is primarily a result of determining a level of loading that basically satisfies the design equation, namely, $R_d = Q_d$. This is appropriate in the case of design where account is taken of the occurrence of explosion events during the design phase. For re-assessment, of course, no such equality can be presumed to exist. However, no relevant information regarding typical topsides member sizes and strengths relative to explosion loading is to hand so this aspect cannot be examined further at this point in time.

4.5 Partial Load and Resistance Factors

As a first-pass to determining load factors, resistance factors were introduced as listed in Table 3, approximately in keeping with the design limit states. The corresponding load factors were derived for a target reliability of 10^{-5}. The objective was to minimise the difference between the target and the achieved probabilities across all nine combinations, ie, high, intermediate and low limit states combined with load biases of 1.0, 1.5 and 2.0. The load factors were constrained to be equal for each limit state category and a minimum load factor of 0.5 was introduced. The factors were derived for system strength biases of 1.5 and 2.0. The resulting partial load factors are listed in Table 3.

TABLE 3

TENTATIVE LOAD AND RESISTANCE FACTORS

Limit state	Resistance factor	Load Factors	
		System strength bias	
		1.5	2.0
High probability (10^{-3} Pa)	1.1	1.81	1.39
Inter. Probability (10^{-4} Pa)	1.0	1.40	1.05
Low probability (10^{-5} Pa)	0.9	0.50	0.50

Load factors less than unity are generally not acceptable. In deriving typical probabilities of failure, a spread of some four orders of magnitude was obtained. Such a spread makes it difficult to determine an appropriate target reliability and necessarily leads to a wide spread in the load factors calculated to achieve the selected target. One contributing factor to the spread in reliabilities appears to be the chosen limits on deflections for each of the limit states. The range of these (1, 3, 10) is probably too small bearing in mind that the difference in frequency between each of the events covered by the limit states is one order of magnitude (10^{-3}, 10^{-4}, 10^{-5}).

5 CONCLUSIONS

A conditional probability reliability analysis procedure was used to determine typical probabilities of failure and load and resistance factors for three hydrocarbon explosion limit states. For input parameters typical of the design of topsides structures subjected to such explosions, the range of probabilities of failure across the three limit states was found to be 5.7×10^{-8} to 5.1×10^{-4} per annum. These are judged to be low bearing in mind that the 'failure mode' corresponding to the least reliable values is only first yield.

References

Biggs J.M. (1964). *Introduction to Structural Dynamics*, McGraw-Hill, USA.

Corr R.B. and Tam V.H.Y. (1998). Development of a Limit State Approach for Design against Gas Explosions. *Proceedings Conference on Fire and Explosion Engineering*, ERA Technology, Leatherhead, UK.

ISO. (2000). *ISO 19902 Petroleum and Natural Gas Industries - Offshore Structures - Part 2: Fixed Steel Structures*, ISO (Draft).

P A F A Consulting Engineers. (1999). *A Reliability-Based Methodology for Determining Response-Based Load Factors for Design of Topsides against Hydrocarbon Explosion Loads*, P A F A, Hampton, UK.

Steel Construction Institute. (1993). *Interim Guidance Note on The Design and Protection of Topsides Structures against Explosions and Fire*, SCI, Ascot, UK.

Practical Design of Ships and Other Floating Structures
You-Sheng Wu, Wei-Cheng Cui and Guo-Jun Zhou (Eds)

DESIGN AGAINST MINOR IMPACTS

Marie Lützen and Preben Terndrup Pedersen

Department of Mechanical Engineering, Maritime Engineering
Technical University of Denmark
DK-2800 Kgs. Lyngby, Denmark

ABSTRACT

A procedure for design and for prediction of damages to ship bows and side shells of vessels during minor impacts is presented. The procedure is illustrated by practical examples.

Many impacts happen during manoeuvring in harbour areas. If the vessel is not completely controlled, it may have severe impacts with quays or obstacles. Similarly, when a vessel collides with floating objects energy is released for structural deformation of the vessel. Floating objects may be cargo lost by other vessels such as containers and barrels or nature-made objects such as logs or floating ice bits.

Two application examples are presented. The first example deals with an impact with a fender system during a harbour manoeuvre. Contact forces and energy absorbed by the system during the manoeuvre are calculated and the maximum safe berthing velocity for the vessel is estimated. Secondly, collision with floating objects is illustrated by an example where a bulk carrier impacts small icebergs.

KEYWORDS

Collision, Minor Impacts, Fender Systems, Floating Objects.

1 INTRODUCTION

Insurance company statistics shows that large sums are paid every year for repair of minor contact damages. Minor ship impacts may arise at sea during collision with floating objects or by contact with obstacles during manoeuvring in harbour areas.

When a ship collides with a floating object energy is released for structural deformation of the vessel. Floating objects can be cargo lost from other ships such as containers and barrels or nature-made objects such as logs or floating ice bits. Loss of deck cargo may happen in bad weather due to insufficient lashing combined with large motion or green water on deck. The risk of colliding with a floating object depends on factors such as the size and the geometrical form of the object. Weather conditions and traffic density will influence the possibility of detection of the object.

Most minor impacts occur during berthing manoeuvres in the harbour, or when the vessel is alongside or passing narrow canals or locks. During harbour manoeuvres the vessel is exposed to many forces, which may arise from own propulsion, mooring lines, wind, current, swell, etc. To avoid unintended contact, the control options available for the navigator during berthing of the vessel are e.g. rudder, thrusters, winches, anchors, and occasionally tug assistance. In such a complex situation it is not surprising that the frequency of severe impacts to a large extent depends on the experience and the education of the crew.

Since impacts cannot be avoided, it is necessary to design ships so that a certain robustness of the hull structure is ensured. The purpose of this paper is to present a procedure for design against minor impacts against floating objects and berthing forces.

2 COLLISION THEORY

The analysis procedure is divided into an external dynamics analysis, e.g. prediction of global motions of the ship and the struck object, and an internal mechanics analysis, which is the prediction of the resulting deformation of the structure.

The external dynamics can be solved by numerical solution of the equations of motion or by an integrated approach where the conservation of energy, momentum and angular momentum during the impact is used to derive analytical expressions for the dissipated energy. The present paper is based on a method of the latter kind developed by Pedersen and Zhang (1998). The energy loss for dissipation by structural deformation is estimated and expressed in closed-form solutions. The procedure is based on a rigid body mechanism, where it is assumed that there is negligible strain energy for deformation outside the contact region, and that the contact region is local and small. The collision can then be considered instantaneous as each body is assumed to exert an impulsive force on the other at the point of contact. The model includes friction between the impacting surfaces so that situations with glancing blows can be identified. Three degrees of freedom are considered: surge, sway and yaw. The interaction between the ship, the object and the surrounding water is approximated by simple added mass coefficients, which are assumed to remain constant during the collision.

To illustrate the assumptions behind the method for prediction of energy released for structural deformation we shall consider a ship, which sails with a forward speed of v_{ax} and a speed of v_{ay} in the sway direction. The vessel collides with a floating or fixed object. An *xyz*-coordinate system is fixed to the sea bottom, see Figure 1. The origin of the *xyz*-system is placed so that the midship section is in the *yz*-plane at the moment $t=0$. The origin of a $\xi\eta$-system is located at the impact point *c*, the ξ-direction is normal to the impact surface, the angle between the *x*-axis and the η-axis is α, and the angle between the *x*- and *l*-axis is β.

The equations of motion for the striking vessel, due to the impact force component F_ξ in the ξ-direction and F_η in the η-direction, can be expressed as

$$
\begin{aligned}
M_a(1+m_{ax})\dot{v}_{ax} &= -F_\xi \sin\alpha - F_\eta \cos\alpha \\
M_a(1+m_{ay})\dot{v}_{ay} &= -F_\xi \cos\alpha - F_\eta \sin\alpha \\
M_a R_a(1+j_a)\dot{\omega}_a &= F_\xi\left(y_c \sin\alpha - (x_c - x_a)\cos\alpha\right) + F_\eta\left(y_c \cos\alpha - (x_c - x_a)\sin\alpha\right)
\end{aligned}
$$

Here M_a is the mass of the striking vessel, $\dot{v}_{ax}$, $\dot{v}_{ay}$ and $\dot{\omega}_a$ denote the accelerations of the ship during the collision of the striking vessel in the *x*- and *y*-direction and around the centre of gravity. The radius of the ship mass moment of inertia around the centre of gravity is R_a, the coordinate of the centre of gravity is $(x_a, 0)$ and the coordinate of the impact point is (x_c, y_c). The parameters m_{ax} and m_{ay} are the added mass coefficients for the surge and the sway motions, and j_a is the added mass moment coefficient for the rotation around the centre of gravity. The motions of the floating object can be expressed in a similar way. The equations of motion for the vessel and the object are used to determine the relative accelerations $\ddot{\xi}_a - \ddot{\xi}_b$ and $\ddot{\eta}_a - \ddot{\eta}_b$ at the collision point c as functions of the impact forces F_ξ and F_η. By integration of the impact forces in the time domain the impact impulses are found as functions of the initial velocities. The energy released for structural deformation is determined from integration of the impact forces in the domain of the relative motions, see Pedersen and Zhang (1998). The method is developed for ship-ship collisions but may as well be used for ship-object collisions and for vessels impacting a quay.

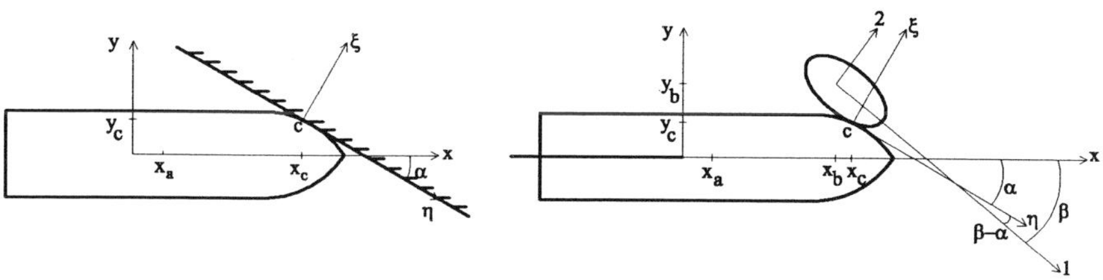

Figure 1: Berthing vessel, ship – object collision.

When the energy released for structural deformation during the impact is known, the next step is to apply the internal mechanics analysis to prediction of the resulting structural deformation. Structural elements such as plates or beams may withstand a pressure above the limit for elastic deformation without unacceptably large permanent deformation after unloading. The permanent deformation, which may be accepted, depends on the design and the type of the vessel. Hughes (1981) has as a design criterion suggested that a permanent deformation of the plate thickness is acceptable for merchant ships. If the vessel has been exposed to larger deformation, ship repair may be required. As the velocity of a vessel is relatively small, dynamic effects can be neglected.

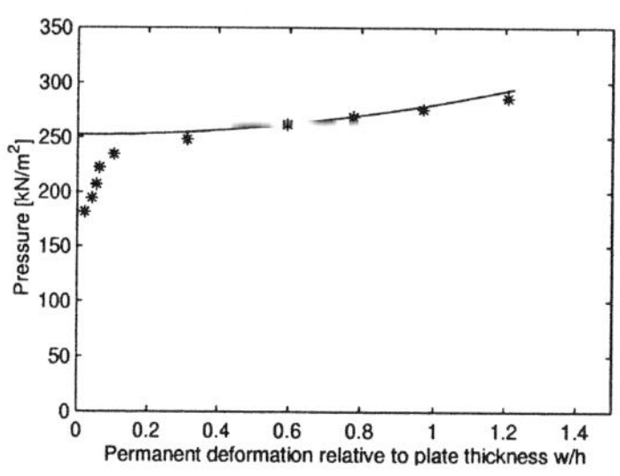

Figure 2: Experimental (*), (Clarkson, (1963)) and analytical results (-) for plate under uniform pressure.

The prediction of the structural deformation is based on a balance between the external work from loads on the structure and the internal energy stored or dissipated in the structure. Due to page limitation, the theory (Jones, 1989) is illustrated by an example. A flat plate grillage under uniform pressure is considered. The plate is stiffened both longitudinally and transversely. As the transverse stiffeners are deep, it is assumed that these are fully clamped ends. The spacing between transverse stiffeners is 1016 mm, the plate thickness 6.2 mm and the longitudinal stiffening consists of T-stiffeners (61.1 x 44.4 x 6.2/8.7) with a spacing of 508 mm. The yield strength of the steel is 323 MPa. The permanent deformation versus the pressure at the plate is shown in Figure 2, together with experimental results (Clarkson, 1963). The energy absorbed in the plate is calculated by an integration procedure.

3 APPLICATION EXAMPLES

Two application examples will be presented. The first example deals with a fender system impact during a harbour manoeuvre. Contact forces and energy absorbed by the system during the manoeuvre are calculated, and the maximum safe berthing velocity for the vessel is estimated. Secondly a collision with a floating object will be illustrated by an example where an ice-strengthened bulk carrier impacts a small iceberg. The plastic deformation of the bow is given as a function of the mass of the iceberg and the velocity of the striking vessel. For both examples the energy absorbed by elastic deformation of the ship structure is negligible.

3.1 Harbour Manoeuvre

The manoeuvre of berthing a vessel is so difficult that the frequency of accidents is very high. The increase in the size of vessels has made berthing manoeuvres even more difficult. Special problems are related to large vessels, which are not able to enter some harbours and therefore must berth in locations exposed to waves, swell and currents.

3.1.1 Berthing velocities

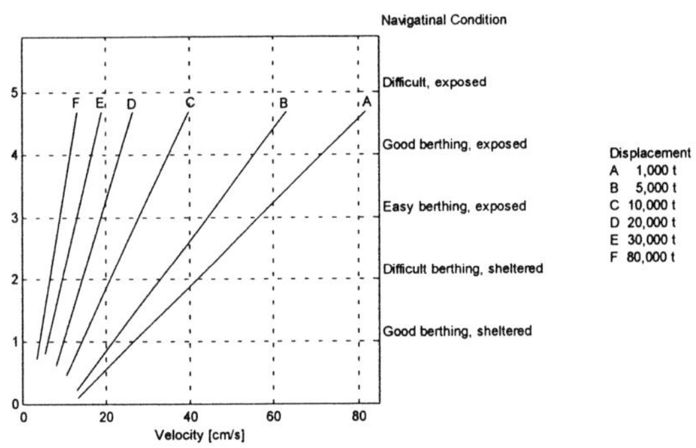

Figure 3: Berthing velocity as a function of navigation conditions for tankers of various displacements, data from Pianc (1983).

In the berthing procedure the vessel is moved slowly towards the quay by tugs, thrusters or mooring winches. During this procedure, the wind may push the vessel towards the quay with a speed above what is desirable, or a strong current may cause the vessel to drift in an undesirable direction. Berthing forces are a function of the vessel momentum and the berthing velocity, where the impact velocity depends on the navigational condition, the difficulties of berthing and the location of the quay. Figure 3 shows an empirical relation between navigation conditions and velocity for tankers of various displacements. The diagram is based on field data analysed by Pianc (1983).

3.1.2 The Fender System

The type and the size of a fender system in a specific harbour are mainly determined with respect to the following considerations:
The fender system must be capable of absorbing the loss in kinetic energy during the procedure of berthing the vessel. If berthing the quay is difficult, the requirements of energy absorption are more strict, as the velocity of the vessel may be increased

- If only small movements of the vessel are allowed during the loading/discharging procedure, a fender with small deformation for large energy absorption is required
- The size of the reaction force and the contact area of the fender must be considered together with the design strength of the weakest part of the side shell

From these considerations it can be concluded that the design of fender systems is highly dependent on the vessels expected to berth at a particular quay.

Berthing a tanker of 150 m in length

Main particulars of the vessel:			Scantlings:		
Length (L)	150.0	m	Frame spacing	3000	mm
Breadth (B)	23.4	m	Plate thickness	18	mm
Draught (T)	9.7	m	Spacing between longitudinals	600	mm
Displacement (Δ)	27721	t	Longitudinals	180 x 80 x 12/12}	
Yield strength	235	MPa			

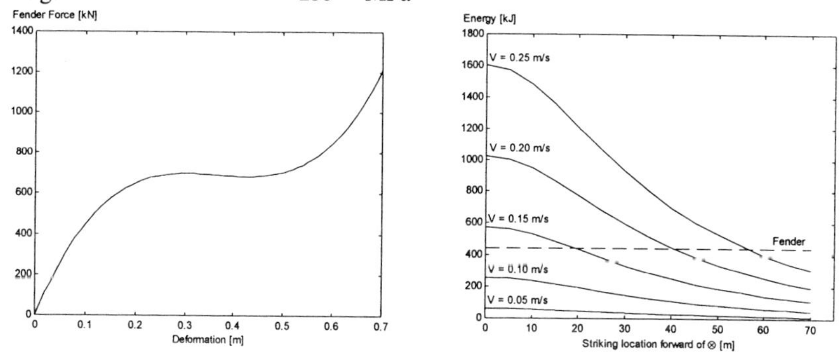

Figure 4: Left: Deformation diagram for fender. Right: Energy to be absorbed by the fender as a function of the collision location forward of amidships for various velocities.

The added mass coefficients are assumed to be 0.05 for surge, 0.85 for sway and 0.21 for yaw motion. The coefficient of friction between the fender and the side of the vessel is taken to be 0.6. It is assumed that the vessel is berthed as parallel to the quay as possible, but that only one fender is in contact with the side of the vessel. The design berthing velocity for a tanker of approx. 30,000 t is of the size 0.1 to 0.2 m/s. (See Figure 3). The smallest velocity is for berthing at sheltered quay and good berthing, the largest velocity for difficult berthing at exposed quays.

The loss in kinetic energy or the energy to be absorbed by the fender as a function of the collision location forward of amidships is shown in Figure 4 (right). The velocity is taken to be 0.05 m/s to 0.15 m/s. It is assumed that the quay is equipped with a fender system of the buckling type (Derucher, 1983) the deformation diagram for this fender type is shown in Figure 4 (left).

According to Agerschou et al. (1983), the design height and length of a fender are based on the size and the type of vessels, which will berth at a particular quay. For tankers of 10,000 to 100,000 DWT the ratio between the fender height and the longitudinal spacing should not be less than about two. For this example the smallest height is chosen. According to Agerschou et al. (1983), this will give a length equal to the frame spacing. The limit strength P_c for the side shell is calculated and compared to the fender reaction force. The absorption of energy by the fender, before the force reaches the limit strength of the side shell, may now be calculated by integration. The limit strength is calculated to 1117 kN, which for the chosen fender system gives an energy absorption of 441 kJ for a travelled distance of 0.68 m. From Figure 4, where both the energy to be absorbed and the maximum allowable energy absorbed by the fender are shown, it is seen that at a berthing velocity of 0.13 m/s the fender can absorb all energy regardless of the striking location. A velocity of 0.13 m/s relates to a berthing condition referred to as easy berthing at exposed quays, see Figure 3.

3.2 Collision with a Floating Object

A collision between a vessel and floating objects will be illustrated by an example where a bulk carrier impacts a small iceberg. The struck vessel is a 150,000 DWT bulk carrier of the high ice class (DNV ice class 1B). The scantlings are found in Pedersen at al. (1993).

The main particulars of the vessel:

Length (L)	274.0	m
Breadth (B)	47.0	m
Draught (T)	15.96	m
Displacement (Δ)	174,850	t
Service speed (V)	15	knots

The floating object is in this example a cylindrical iceberg. The added mass coefficients for the cylinder are assumed to be one for both the surge and the sway motions, whereas the coefficient for yaw is zero. The radius of inertia may be taken to be $R = r/\sqrt{2}$ where r is the cylinder radius. The object is assumed to be rigid, which means that the loss in kinetic energy during the collision must be transformed into deformation of the bow of the vessel. The density of ice and water is 920 kg/m^3 and 1025 kg/m^3 respectively, and the height above the water level of the iceberg or the cylinder is in all cases assumed to be 0.3 m, which means that the object can hardly be seen visually or by radar. The coefficient of friction between ice and steel can according to Bruun (1989) be taken to be 0.2. The highest local loads will occur in a head-on collision. But it is not necessarily this situation, which will induce the highest stresses or cause damage, as the bow is the strongest part of the vessel. Higher stresses or larger damages may occur by glancing impacts on the side of the bow. Therefore, a situation like this is analysed for the side.

The striking location is assumed to be 7 m aft of the forward perpendicular where the breadth of the vessel is 6 m. The form of the bow is such that the angle between the centre line and the side of the vessel is 15°. The bow region is longitudinally stiffened. The frame spacing is 3200 mm; the plate thickness 33 mm and the longitudinal stiffeners are T-stiffeners (400 x 100 x 19/19) with a spacing of 600 mm. The yield strength of the steel is 315 MPa.

The energy absorption due to crushing of the side shell as a function of the ratio between the permanent deformation and the plate thickness w/h is shown in Figure 5. Figure 5 also shows the loss in kinetic energy during the collision calculated by use of the described external dynamics analysis. It is assumed that the vessel has a velocity up to its service speed of 15 knots, and a permanent deformation of the size of the plate thickness is accepted. Figure 5 shows that a collision with an object of 10 t is acceptable without reducing the velocity of the vessel. If the velocity is reduced to 13 knots a collision with a 20 t object is acceptable.

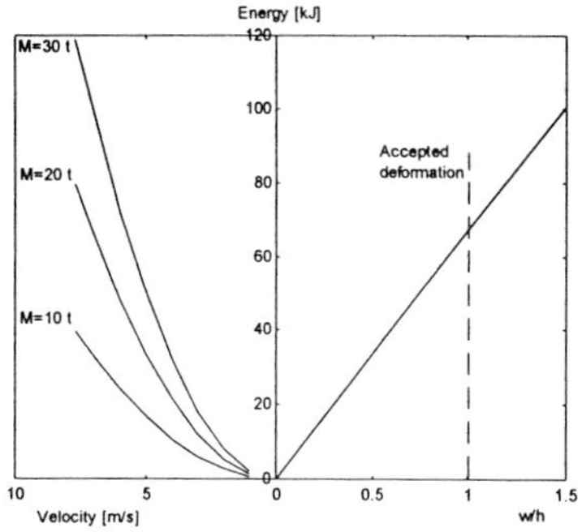

Figure 5: Left: Energy to be absorbed as a function of the velocity of the bulk carrier for various masses of the iceberg. Right: Absorbed energy in the side shell versus the ratio between permanent deformation and plate thickness.

4 CONCLUSIONS

This paper has briefly described a procedure for rational design of ship hulls against minor impacts. The procedure has been illustrated by two examples. In the first example an impact with a fender system during berthing of a vessel has been analysed. In this case, it is required that the fender must absorb all energy for a given set of scantlings. The maximum allowable berthing velocity for a tanker impacting a specific fender system has been determined. The second example presents a model for prediction of minor damages to a vessel during collision with a floating object. In this case the object is considered rigid, which means that the loss in kinetic energy during the collision must be transformed into deformation of the vessel. Permanent deformations of the size of the plate thickness are in this case accepted.

References

Agerschou, H., Lundgren, H., Sørensen, T., Ernst, T., Korsgaard, J., Schmidt, L. and Chi, W. (1983), *Planning and Design of Ports and Marine Terminals*, A Wiley - Interscience Publication

Bruun, P. (1989), *Port Engineering, volume 1*, Library of Congress Catalog – in Publication Data, 4th edition

Clarkson, J. (1963), Test of Flat Plated Grillage under Uniform Pressure, *Transactions, Royal Institution of Naval Architects*, 105

Derucher, K. N. (1983), Evaluation of Pier Fendering Systems, *The Dock and Harbour Authority*

Hughes, O. F. (1981), Design of Laterally Loaded Plating - Uniform Pressure Loads, *Journal of Ship Research*, 25(2)

Jones, N. (1989), *Structural Impact*, Cambridge University Press

Pedersen, P. T., Valsgaard, S., Olsen, D. and Spangenberg, S. (1993), Ship Impacts: Bow Collision, *Int. J. of Impact Engineering*, 13(2): 163--187

Pedersen, P. T. and Zhang, S. (1998), On Impact Mechanics in Ship Collision, *Marine Structures*, 11:429—449

Pianc (1983), Report of the International Commision for Improving the Design of Fender Systems, *Permanent International Association of Navigation Congress (Pianc)*, Supplement to Bulletin (45)

EXPERIMENTAL STUDY ON THE BUFFER BOW STRUCTURES

H. Endo[1], Y. Yamada[1]1 and O. Kitamura[2]

[1]National Maritime Research Institute, 6-38-1 Shinkawa, Mitaka, Tokyo, 181-0004, Japan
[2]Mitsubishi Heavy Industries, Ltd., Nagasaki R&D Center, 5-717-1 Fukahori, Nagasaki, 851-0392, Japan

ABSTRACT

In order to prevent a disastrous damage induced by collision accident which might cause cargo oil spill from a struck oil tanker, the adoption of double hull system in side hull of oil tanker has been recognized as an effective countermeasures. However, when considering that ocean-going vessels are increasing not only in size but also in speed, the threat of disastrous collision accident should be further mitigated also on the responsibility of striking ships. Thus RR76 panel of Shipbuilding Research Association of Japan and NMRI (SRI by old name) have started, as a joint work, to investigate into the effectiveness of buffer bow structure since 1998.

NMRI has conducted a series of crush test using three kinds of simplified scale models of the buffer bow. The test results have been compared with those obtained by FEA simulation and a simple analysis method which was developed to evaluate the crashworthiness of buffer bow structure. The performance of the buffer bow is discussed focusing on the collapse mechanism and the P/ δ characteristics. Then the recommendation for the practical design of buffer bow structure is presented.

KEYWORDS

Collision, Buffer bow, Soft bow, Oil spill, Crush test, FEA simulation, Collapse mechanism, Conical shell, Folding.

1 INTRODUCTION

Ship-ship collision might cause serious damage to the struck ship and also threaten the safety of life onboard. When oil tanker is collided by large ship, disastrous oil spill might be induced and marine environment nearby might be seriously contaminated. In order to prevent such oil spill from a struck tanker, the double hull(D/H) system in side hull has been recognized as an effective countermeasures.

However, it is the fact that tanker oil spill has been still induced by collision accidents like the case of oil tanker "Evoikos" which was collided and caused disastrous sea pollution off Singapore in 1997. Then further improvement to minimize the probability of oil spill accident has been required.

The buffer bow concept has been proposed by ASIS (1997-1998) in order to further decrease the oil spill probability as the next-step countermeasures. Buffer bow is expected to lessen the threat of the damage on the side hull of struck ship by absorbing the striking energy.

Although buffer bow (soft bow) concept was proposed by Cheung (1969) in the 1960's for high speed vessels, only a few research activities, Lee (1983), have been conducted so far. ASIS (The Association for Structural Improvement of Shipbuilding Industry) has attempted the application of buffer bow to large cargo ships in their research project between 1991 to 1998. ASIS has shown the preliminary design of prototype buffer bow and conducted FEM simulation analysis to evaluate its effectiveness, ASIS (1997-1998) and Kitamura et al.(1998).

In order to introduce the buffer bow concept in the international regulations in the near future, RR76 panel of Shipbuilding Research Association of Japan and NMRI have started, as a joint work, to study the necessary items for the buffer bow structure since 1998, JSRA (2001) and Kitamura (2000).

In the previous researches on ship-ship collision, the attention have been paid to the side structure of collided ship. A lot of experimental and analytical works have been dedicated to the crashworthiness of side hull structure of collided ships especially in relation to nuclear powered ships and radioactive material carriers, JSRA (1971), JSRA(1978-1980) and JSRA(2000).

As far as the bow structure damage is concerned, only a few researches have been carried out , Cheung (1969), Lee (1983), JSRA (1978-1980) and Ohnishi et al. (1983). Ohtsubo and Suzuki (1994) developed a simplified analysis method to estimate collapse strength of bow. Then they expanded their simplified method to be applicable to the collapse of bulbous bow, JSRA (2000) and JSRA (2001). Lehmann et al. (1995) developed an analytical method to follow the collapse mechanism of bulbous bow and they demonstrated the effectiveness of their method by applying it to the ship-ship collision accidents which had occurred recently. Lehmann (1995) treated the collapse of bulbous bow based on the progressive collapse of conical shell.

Bulbous bow is regarded as the threat to struck ship as most of large ships have a bulbous bow and it usually contacts the struck ship first then might penetrate the side shell. It is considered that structure after collision bulkhead may be rarely broken in head-on-collision accidents because large ships have large sectional area at the root of bulbous bow. Then the bulbous bow is regarded as the key structure part for buffer bow design.

In this research the attention has been focused on the bulbous bow part. For Simplicity the bulbous bow part was modeled as a symmetric conical shell shape then three kinds of scale models were constructed and the collapse mechanism and energy absorption capacity of them were experimentally and analytically studied through the crush test. The models were designed as about one fifth the size of Suezmax. class tanker which has a transverse stiffening system. In order to study the size effect, the scale models are designed different in ring frame space and shell thickness.

Simplified analysis method has been developed to evaluate the crashworthiness of buffer bow by jointing the method by Lehmann (1995) and the method by Ohtsubo (1994). Its accuracy was verified by comparing the results with the test and FEM calculation as well.

2 EXPERIMENT

2.1 Scale Model of Buffer Bow

In order to investigate the basic characteristics of buffer bow, three kinds of conical buffer bow model were designed and the crush tests were conducted. As the buffer bow structure the transverse stiffening system "ring frame" was adopted as it is expected to be advantageous in reducing the axial crushing strength. Each model was designed so that ring frames space(FS) and shell plate thickness (t) are different. The particulars of each scale model are shown in Figure 1 and Table 1.

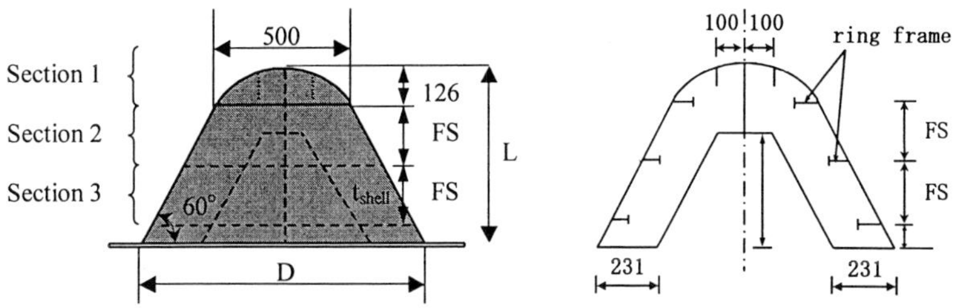

Figure 1: Particulars of the models

TABLE 1
PRINCIPAL DIMENSION OF THE MODELS

Model	L	D	t_{shell}	t_{stiff}	FS	N_{rf}
BC-A	626	1077	4.24	3.08	200	3
BC-B	751	1222	4.24	3.08	250	3
BC-C	626	1077	1.55	1.55	100	5

t : shell plate thickness, FS : frame space (in mm)
N_{rf} : number of ring fame

2.2 Test Method

A series of crush tests were conducted with buffer bow models. The models were fixed with the bow top faced upwards and was compressed quasi-statically by a rigid board from top of the bow as if the bow collide into the rigid side shell in right angle. Displacement of rigid plate, reaction force and strains were measured.

2.3 Experimental Results

2.3.1 Collapse mode

In the collapse process all of the models collapsed progressively frame space by frame space from top of the bow. The collapse mode observed was spherically concentric folding as shown in Figure 8. The folding span of Model BC-C was about the same as its frame space(=100mm) whereas the folding span of BC-A and BC-B were larger than that of model BC-C but smaller than the frame space. In both cases of BC-A and BC-B, the folding span was about 100 to 140mm which corresponds to 1 to 2 folding wrinkles within one frame space. The first folding occurred in Section 1. Folding span of BC-A and BC-B is equivalent to the interval which the peak of reaction force appears as shown in Figures 2 , 4 and 6. It was found that the folding span corresponds to the frame space in the case of small frame space whereas it does not correspond to the frame space in the case of larger frame space.

The folding span is considered to be also affected by the rigidity longitudinal girders.

2.3.2 Time history of reaction force

Relations between reaction force(P) and displacement(δ) for each model are shown in Figures 2,4 and 6. It is found that the time history of reaction force shows periodical ups and downs. Observing the test and P- δ curve carefully, it is found that the first peak appears at Section 1, both the second and the third peaks at Section 2 and the fourth at Section 3.

In the case of BC-C, no obvious peak force was observed in the test whereas the peak is expected to appear in FEA results. However, the collapse mode that spherical wrinkle repeats with ring frame span is quite similar between the test and FEA. The discrepancy in the characteristics of peak force is regarded due to initial deflection and residual stress caused by welding fabrication. Since the shell thickness of Model BC-C (t=1.6mm) is much thinner than others (t=4.5mm), it is supposed to be highly vulnerable to welding heat input. It should be noted that we had better use steel plate with thickness of 4 mm or greater as the material for scale models.

The cyclic fluctuation of the reaction force is regarded as highly related to the folding mechanism. There are two stages in one cycle of folding. After small buckling occurs the reaction force goes down with developing wrinkle. Then the reaction force goes up after the folding is completed and the surface of folded plate touches the adjacent plate until the next folding occurs.

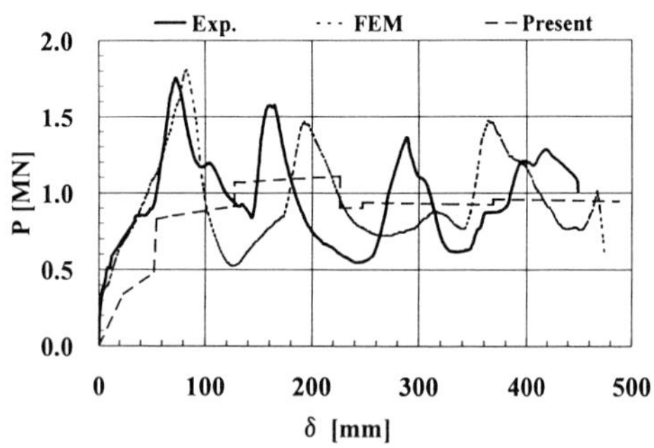

Figure 2: P/ δ relation for model BC-A

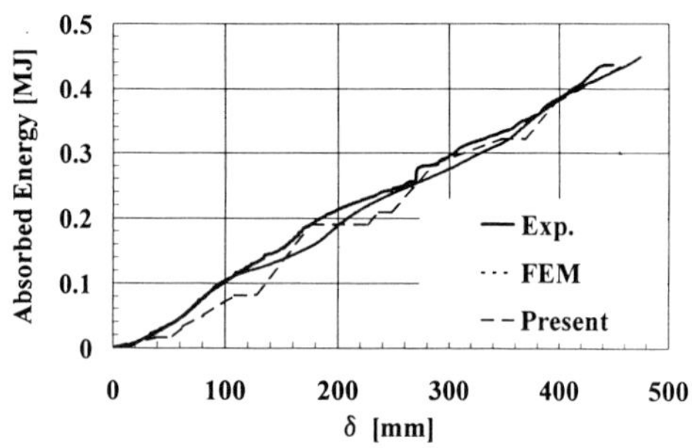

Figure 3: Absorbed energy for model BC-A

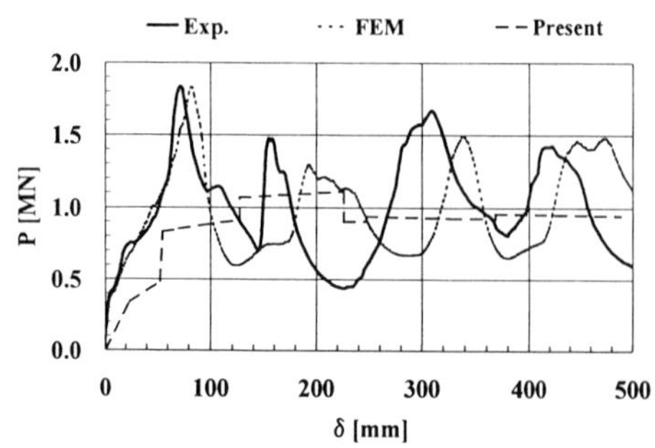

Figure 4: P/ δ relation for model BC-B

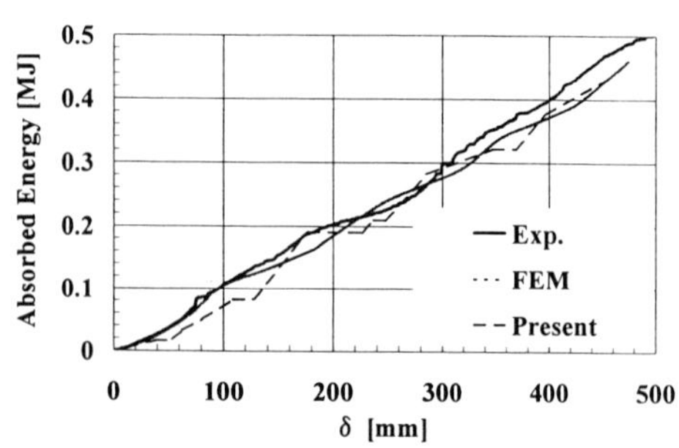

Figure 5: Absorbed energy for model BC-B

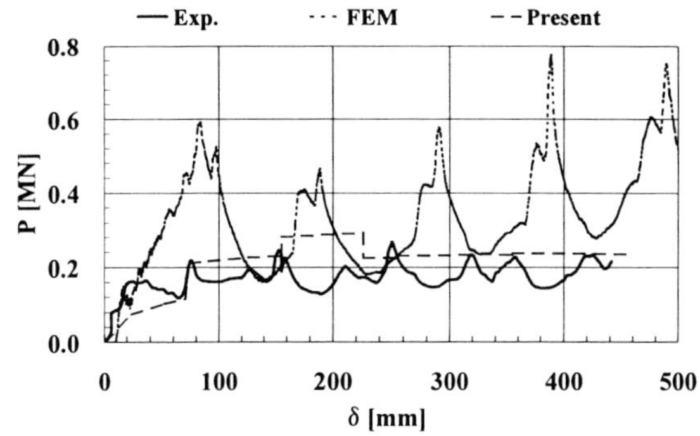

Figure 6: P/ δ relation for model BC-C

2.3.3 Energy absorption capacity

Relations between the absorbed energy and the displacement for BC-A and BC-B are shown in Figures 3 and 5 respectively. Energy absorption capacity for both models was found to be about the same.

3 COLLAPSE ANALYSIS

3.1 Finite Element Analysis

Elasto-plastic finite element analysis was carried out by using FEA program "LS-DYNA 3D" for the cases of the crush test.

3.1.1 FEA model

Overview of finite element model is illustrated with boundary condition and load application condition in Figure 7. The element size was so determined as to be able to simulate the folding mechanism of shell.

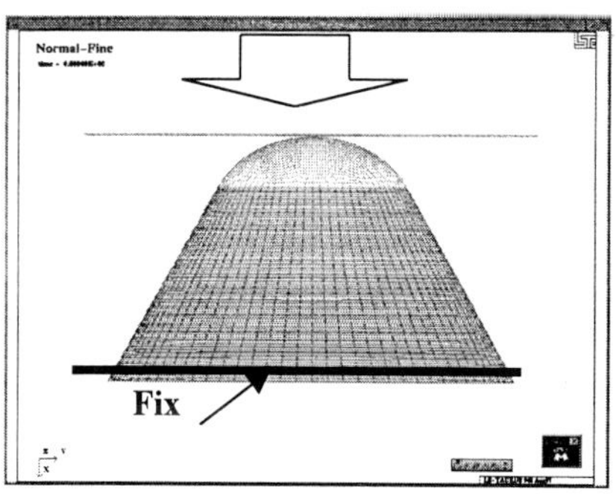

Figure 7: FEA model

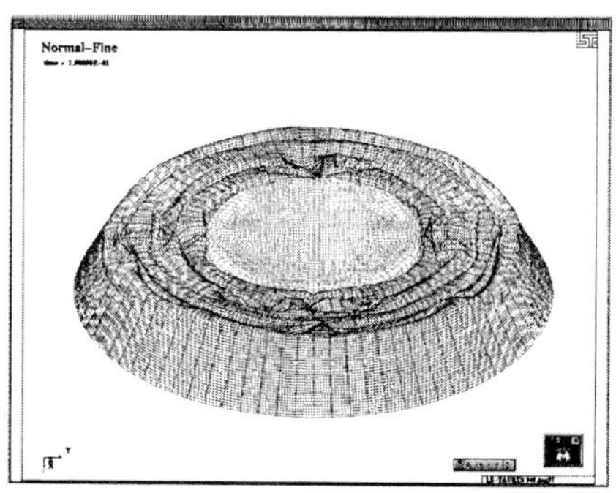

Figure 8: Crushed pattern in FEA model

3.1.2 Analysis condition

The model was fixed at the back end and was pressed from top of the bow with a rigid board with right angle. Colliding velocity of rigid board was set to 5 m/s considering the calculation time. The criteria for the breaking strain of elements were decided from the strain/gauge-length curves which were obtained from material tensile tests.

3.1.3 Analysis results

P- δ relation and E(absorbed energy)- δ relation obtained by FEA are shown in Figures 2 to 6. As for P- δ relation for model BC-A and BC-B, the magnitude of reaction force by FEA is found to agree with the test results though the intervals between peaks are slightly different. As for absorbed energy which is considered to be the most important design factor for buffer bow, FEA results are found to give satisfactory agreements with the test results.

3.2 Simplified Analysis

Endo & Yamada (2001) have developed a simplified analysis method to estimate P- δ relation for the buffer bow structure. The analysis method is summarized as follows:

1. The force deformation relation is obtained on the basis of plastic analysis.
2. The bow structure is modeled as the combined structure of a conical shell and segmented longitudinal girders.
3. The behavior of conical shell is estimated by Lehmann's method, Lehmann (1995).
4. The behavior of longitudinal girders is estimated by Ohtsubo's method, Ohtsubo (1994).

Please see Endo & Yamada (2001) for the detailed procedure of the analysis method.

The results obtained by the simplified analysis are shown in Figures 2 to 6 with a notation "present" comparing with the results obtained by test "Exp." and FEA "FEM". This analysis method is found to give a good estimation for the average tendency of reaction force and absorbed energy though it can tell us nothing about the fluctuation of reaction force changing from peak to trough in a short period.

4 GUIDELINES ON PRACTICAL DESIGN OF BUFFER BOW STRUCTURE

This guideline is presented on the premise that the design criteria for bow structures are entirely based on the rule of classification societies.

1. The reaction force per unit sectional area must be minimized to reduce the damage to the struck ship,
2. though the total force reacting on the whole sectional area should be maximized to enhance energy absorbing capacity.
3. A blunt shaped bow is advantageous when taking account of above-mentioned conditions. A sharp bow tip should be avoided.
4. The transverse stiffening system (ring frame) is advantageous in reducing thrust strength while keeping local out-of-plane resistance to a certain extent.
5. Decreasing the size of the section at the root of bulbous bow is another effective way as it is expected to be easily bent against horizontal and vertical bending induced by collision. Longer bulbous bow is also preferable in the same sense.
6. The stem structure should be designed as not so sharp and stiff considering that it might touch the side hull of struck ship at an early stage of collision.

5 CONCLUDING REMARKS

The bow structure stiffened by ring frame is expected to collapse progressively from bow tip with spherical concentric folding mode during collision. In this collapse mode the reaction force keeps a

certain degree of level while the peak value is not so high. This performance is quite advantageous for the buffer bow.
The results obtained by FEA and the simplified analysis have been confirmed to be reliable in estimating the performance of buffer bow structure.
In this research the bow was simplified as conical shaped and the side hull as a rigid board. The study needs to be extended to the next step where more realistic structure will be treated.

Acknowledgements

The authors are grateful to the Shipbuilding Research Association of Japan (JSRA) and the members of the Regulation Research panels No.76 for their valuable advice and discussions. The authors would like to extend their sincere gratitude to the Nippon Foundation's support to the activities of JSRA..

References

ASIS (The Association for Structural Improvement of Shipbuilding). (1997-1998). Report on Research in Environmental Preservation Technology, (in Japanese)

Cheung L. (1969). A Soft Bow for Ships. *European Shipbuilding* **No.3,** 52-53

Endo H. and Yamada Y. (2001). The Performance of Buffer Bow Structures against Collision (1st Report: Collapse Strength of the Simplified Structure Models). *Journal of the Society of Naval Architects of Japan.* **189**, (in Japanese)

JSRA (1971). Studies on Collision Protective Structures of Nuclear Powered Ships. *Report of NSR3* **71**, (in Japanese)

JSRA (1978-1980). Research on Collision Protective Performance of Nuclear Powered Ships. *Report of Nuclear Ship Research Committee 10*, (in Japanese)

JSRA (2000). Research on the Safety Sea-transport of Radioactive Materials. *Report of RR46*, (in Japanese)

JSRA (Shipbuilding Research Association of Japan) (2001). Research on the Reduction of Oil Spill from Tanker. *Report of RR76*, (in Japanese)

Kitamura O., Kuroiwa T., Kawamoto Y and Kaneko E. (1998). A Study on the Improved Tanker Structure against Collision and Grounding Damage. *Proceedings of the 7th PRADS,* 173-179.

Kitamura O. (2000). Buffer Bow Design for the Improved Safety of Ships. *Proc. of SSC/SNAME/ASNE Symposium*

Lee J. W. (1983). On the Optimization Design of Soft Bow Structure. *Proc. of the 2nd Internl. Symposium on Practical Design in Shipbuilding*, 429-435

Lehmann E. and Yu X. (1995). Progressive Folding of Bulbous Bows. *Jahrbuch der Schiffbautechnischen Gesellscaft* **89,** 88-97, (in German)

Ohnishi T., Kawakami H., Yasukawa W. and Nagasawa H. (1983). On the Ultimate Strength of Bow Construction (Bow in Collision). *Proc. of the 2nd Internl. Symposium on Practical Design in Shipbuilding*, 635-642

Ohtsubo H. and Suzuki K. (1994). The Crushing Mechanics of Bow Structure in Head on Collision (1st Report: The Derivation of Simplified Equations and Their Verification). *Journal of the Society of Naval Architects of Japan* **176,** 301-308, (in Japanese)

Practical Design of Ships and Other Floating Structures
You-Sheng Wu, Wei-Cheng Cui and Guo-Jun Zhou (Eds)

CALCULATION OF COLLISIONS WITH THE AID OF LINEAR FE MODELS

E. Lehmann, E.D. Egge, M. Scharrer, L. Zhang

Germanischer Lloyd AG, Vorsetzen 35, 20459 Hamburg, Germany

ABSTRACT

Today, state-of-the-art methods enable us to analyse collisions and groundings to a certain degree of accuracy by applying the nonlinear finite element technology. However, for the purpose of practical ship structural design, such calculations are too time-consuming and costly. Therefore, Germanischer Lloyd is developing a method that permits collision analyses with adequate accuracy by using the FE models generated for global structural investigations. The method is described in detail. Also, problems concerning the ultimate load of stiffened plates, box columns and other profiles, rational failure strain values and local contact forces are discussed.

KEYWORDS

Collision resistance, Linear FE model, Folding, Mean ultimate stress, Failure strain.

1 INTRODUCTION

Due to a number of serious tanker accidents in the past decade, the world-wide interest concerning ship collisions and grounding has increased considerably. Owing to the recurrence of major disasters, the safety regulations for tanker transport, tanker operation and newbuildings have been tightened considerably (MARPOL 73/76, 73/78, OPA 90, TOCA). For instance a the compulsory double hull design for tanker newbuildings was introduced. Also, methods for calculating the vessel's collision have evolved significantly. In the future, methods and programs, with which the ship collision resistance can be calculated, can also be applied for the proof of damage stability for cargo vessels subject to compliance with the SOLAS convention. In these regulations alternative design for the proofs are accepted, if the same degree of safety as required is verified.

Besides, the evidence of ecologically advantageous constructions has become an important sales argument. Also for this reason, Germanischer Lloyd established an additional class notation COLL classifying the ship's collision resistance in 1984 [6].

It is clear that quick and economically feasible methods will be needed for optimization of the collision resistance in the design phase. Germanischer Lloyd is therefore pursuing the strategy of first implementing the structural description in the design program POSEIDON and then using the FE model generated by this program for collision calculations.

2 THE ECONOMICAL CALCULATION METHOD FOR THE DESIGN PHASE

Previous investigations [1-5] have revealed that relatively simple approaches can provide useful solutions for practical applications. Investigations on the collision behaviour of RoRo ships as presented in [5] were carried out by using the substructure method developed by Germanischer Lloyd [3,4]. For verification, a reference calculation was performed using the FE program DYNA3D. Comparison of the deformation energies over the indentation depth (fig. 1) indicates that remarkably good solutions can be obtained. However, this pleasing agreement should not distract from the fact that some important influences remain unaccounted for when using the substructure method, such as coupled consideration of the inner and outer collision mechanics as well as complete interactions of all structural components including the global strength.

Accordingly, Germanischer Lloyd has commenced developing a new method and coding [7]. It is designed to cover oblique off-centre ship impacts and the interaction between the inner and outer collision mechanics. It also allows to consider the global strength of the hull. Since the interaction of the components needs to be accounted for, it is evident that this can only be achieved by using the finite element method. In this context it is obvious to use FE models established for the analysis of linear strength. These FE models usually consist of elements, the dimensions of which correspond to distances between structural components, such as decks, bulkheads and walls. For such element sizes, it is not possible to simulate detailed folding processes directly. As an alternative, the structural behaviour can be considered via appropriate material characteristics. Based on this idea, the elements were formulated with standard elastic-plastic characteristics under tension, while under compression, the initial ultimate load and the mean contact force in the range of the effective folding length were represented in the material characteristics. During the final reinforcement phase, no significant deformation energy is absorbed by the elements. The applied material characteristics are shown in fig. 2. For the formulation of the yield condition, a function $\eta(\varepsilon)$ was introduced to represent the relationship between the stress $\sigma(\varepsilon)$ and the dynamic yielding point R_{eHd}:

$$\sigma(\varepsilon) = \eta(\varepsilon) \cdot R_{eHd}$$

For differing plate field geometries, this function depends on the plate orientation and results in different stress characteristics for the longitudinal and transverse directions. Applying this description to the von Mises yield criterion, a modified formulation can be derived as follows:

$$\sigma_F(\varepsilon_1, \varepsilon_2) = \sqrt{\sigma_1^2/\eta_1^2 + \sigma_2^2/\eta_2^2 - \sigma_1 \cdot \sigma_2/(\eta_1 \cdot \eta_2)}$$

In the tensile zone, $\eta_1 = \eta_2 = 1$, and so the von Mises yield criterion is complied with. For compressive strains, the von Mises ellipse is distorted according to the relevant η values.

Figure 3 shows a typical folding force-deformation behaviour of a box column. The integration of the contact force $F(\delta)$ over the compression length δ is the deformation energy Π_p, from which a mean ultimate force F_m, stress σ_m and an associated linear energy curve Π_{pm} can be derived:

$$\Pi_p = \int F(\delta) \cdot d\delta = F_m \cdot \delta$$

Several authors [8-14] derived analytical solutions for fundamental structural components, such as L, T, X and box section profiles, as well as ring-stiffened bulbous bow constructions. For further confirmation of the analytical solutions, several parametric studies were performed by using the explicit FE program DYNA3D.

Figures 4 through 6 show comparisons of analytical solutions and FE analyses for the fundamental structural components. From these diagrams it becomes evident that box columns, X-profiles and profile combinations, which describe intersections of stiffened plates, feature a higher folding resistance than T-profiles. Since the number of stiffeners (i. e. T-profiles) in a ship structure is normally much higher than the number of intersections, the mean ultimate stresses should trend towards the stresses of the T-profiles. Accordingly, the following simple formula has been derived for σ_m(fig. 6):

$$\sigma_m / R_{eHd} = 3 \cdot (b/t)^{-2/3} \quad \text{where b = plate width and t = plate thickness}$$

Another important question of structural characteristics concerns the compression ε_u at the beginning of folding. The load-indentation curves of the profile combinations were evaluated and the following empirical relationship for ε_u was derived:

$$\varepsilon_u = 0.05 \cdot \ln\left(42 \cdot t/b\right)$$

In the final reinforcement phase, in which the folds lie on top of each other, a computational limiting stress σ_f is introduced. It considers the steep rise of the contact force in this phase ($\varepsilon \geq 0.73$).
In case of compression, there is practically no failure or fracture of materials. Therefore, only when the compression strain is greater than 73% will the correspondent elements be removed from the calculation. This assumption has repeatedly been justified by actual cases of collision damages. Figure 7 shows the typical folded car deck of a RoRo ferry after collision.

3 ASSUMPTION OF FAILURE STRAINS

Under tension, the load-carrying behaviour and the energy absorption is limited when the failure strain is attained. Evaluations of stress-strain curves from laboratory tests on tensile specimen show two ranges of differing strain behaviours. From the point at which the yield stress is reached, the specimen extends with plastic flow and the cross-section is reduced, whereby the volume remains constant. Owing to the strain hardening effect, the stress and strain increase uniformly over the length of the specimen until the nominal fracture stress is attained. This behaviour of the tensile specimen describes the zone of uniform strain. Once the nominal fracture stress is reached, the strain behaviour changes. With decreasing nominal stress, the necking of the tensile specimen is constricted only within a small region, where it extends rapidly until fracture takes place at high plastic strains.
Calculations using DYNA3D have shown that the deformation energies react sensitively with respect to the assumed failure criteria. Figure 9 contrasts the deformations and energy curves for two comparable collisions but different values for the failure strains (10% and 20%). It can be seen that, if an unrealistic failure strain is assumed, this can lead to considerable errors in estimating the energy absorption and indentation depth. Therefore, it is important to set realistic failure criteria. First of all, the element types and the fineness of the FE mesh play important roles [15]. Because the length of the necking zone is of the order of magnitude of the plate thickness, it is impossible to consider necking and fracture details with large element sizes. In order to obtain rational failure strain assumptions, it is necessary to consider and combine both the uniform strain and the necking strain. For different element sizes, the failure strain should occur for the same structural deformation energy.
But if the failure strains are derived from tensile tests only, fairly large failure strain values result for large elements because the uniform strains from tensile tests for ship structural steels are approximately 20%, while observations of actual collision damages have shown that there are no considerable deformations near the direct contact zone (fig. 8). At this point it is worthwhile recalling that, for the methods in which large and simple structural elements are used, low failure strains (approx. 5%) were assumed. In order to clarify which failure strains are to be assumed for the calculation procedure, failure strains were determined by thickness measurements at damaged structural details of actual collisions and of full-scale collision test models. The measured strains were divided into uniform and necking strain. Since the material volume does not change in the yielding process, it may be assumed that plastic strains occur whenever the thickness changes. However, information about strain directions cannot be obtained from measurements. It may be assumed that the prevailing strain component is perpendicular to the fracture line. This assumption can be verified by observing the fracture line flows on coatings after damage. But measurements also include other phenomena, e.g., influences of multiaxiality of the stresses (for a biaxial stress state, failure strain can be as low as half the value as results for a uniaxial stress state), of direction and kind of collision forces (tension, compression, bending and shear), of strain rate as well as of notch effects of the local structure. So, in contrast to laboratory tensile tests, there is the advantage that such influences are not neglected, but considered in a general sense.

The samples were flame-cut from plates of damaged structures which contained fracture lines. These were then devided into strips perpendicular to the fracture lines. The thickness near the fractures were measured with particularly fine resolution. Figure 10 shows a typical strain distribution curve resulting from such a thickness measurement. It can clearly be recognized that in a distance more than 20 mm from the rupture location, only uniform strain exists, while in a small region close to the rupture location, the necking strains are predominant and peak up to 60%.

With the aid of nonlinear regression analysis, the local strain $\varepsilon(x)$ was approximated as follows:

$$\varepsilon(x) = \varepsilon_g + \varepsilon_m \cdot \exp(-2 \cdot x/x_e)$$

$$\Delta l_e = 2 \cdot \int_0^{\infty} \varepsilon_m \cdot \exp(-2 \cdot x/x_e) \cdot dx = \varepsilon_m \cdot x_e$$

where ε_g = uniform strain
ε_m = max. necking strain
x_e = necking length
Δl_e = extension due to necking

For the evaluation, the measured x-coordinates were extrapolated back to the initial state.

Thickness measurements were carried out on specimen taken from three actual average losses and two full-scale collision test models. Structural deformations were caused by dynamic processes. Mean values of measured uniform and necking strains from all samples were used for the evaluation. The following table summarises the mean values as well as scatter ranges.

Measurement series	Uniform strain [%]		Necking strain [%]	
	Mean value	Scatter	Mean value	Scatter
Bottom damage by grounding	6.59	2.79	58.57	17.14
Side damage by collision	3.99	1.68	37.71	20.07
Damage by slamming	7.92	1.69	90.03	10.75
Collision test, outer shell, scale 1:1	3.71	2.26	85.47	24.47
Collision test, tanker, scale 1:3 [16]	8.65	1.98	70.43	18.04

From the measurements, mean values of 5.6% for uniform strain and of 54% for necking strain have been derived. From these strain parameters, the computational failure strains ε_{fl} for different element sizes can be determined as follows:

$$\varepsilon_{fl}(l_e) = 0.056 + \frac{0.54 \cdot t}{l_e}$$

Where t = plate thickness
l_e = element size in strain direction

From this relationship, failure strains are obtained which agree well with the typically low values given for the empirical-analytical calculation methods [2-6].

Since the size of plane stress elements typically differs in the longitudinal and transverse direction, different failure strains result. In order to consider the interaction of the failure strains, an elliptical approach is proposed which basically corresponds to the plastic equivalent strain. Here, the existing strains were referred to the corresponding failure strains, and a criterion was introduced to indicate rupture of an element (if $D \geq 1.0$). In the case of isotropic fracture behaviour, the actual plastic equivalent strain will be obtained.

$$D = \sqrt{\frac{2}{3} \cdot \left(e_I^2 + e_{II}^2 + e_{III}^2\right)}$$

e_I, e_{II}, e_{III} = Reference strains for the primary strain directions

$$e_I = \varepsilon_I/\varepsilon_{bI} \qquad e_{II} = \varepsilon_{II}/\varepsilon_{bII} \qquad e_{III} = [1-(1+e_I)\cdot(1+e_{II})] / [(1+e_I)\cdot(1+e_{II})]$$

4 CONTACT OF COLLISION PARTICIPANTS

Geometry changes of two collision participants due to nonlinear contact are considered by means of special contact elements. These elements only transfer compression forces and frictions, but no tensile forces. Two different types of contact are envisaged, with which most contact situations can be covered adequately:

- Contact between point and surface

Here, a nodal point comes in contact with an element surface. The forces are transferred between the nodal point and the corner nodes of the element. The contact force F is derived as follows:

$F = \alpha \cdot R_{eHd} \cdot t \cdot u$ Where $\alpha = 2.46$ [17] for quadratic plates; $\alpha = 4.00$ according to GL
u = indentation depth

This approach only considers the load-carrying behaviour of a plate without stiffeners.

- Contact between two crossed edges

Here, the edge of an element comes in contact with the edge of another plate element. Forces [13,17] are transferred between the nodes of the participating edges:

$$F = Min(F_{lin}, F_{nl})$$

$$F_{lin} = 0.34 \cdot 10^{-3} \cdot E \cdot t \cdot u$$

$$F_{nl} = 0.653 \cdot F_{ref} \cdot (u / H)^{1/4}$$

$$F_{ref} = 6.77 \cdot R_{eHd} \cdot t^{5/3} \cdot b^{1/3}$$

$$H = 0.333 \cdot (b^2 \cdot t)^{1/3}$$

Where
F = contact force
F_{lin} = force, linear-elastic
F_{nl} = force, nonlinear-plastic
F_{ref} = tearing force at the stiffeners (concertina tearing)
b = plate width
H = folding length

5 COLLISION CALCULATION

The collision calculations are carried out in the time domain. This involves solving a system of differential equations with initial conditions. Each degree of freedom of a node represents its own differential equation of second order, which couples to other equations nonlinearly. The acceleration is calculated as a function of the deformation, the speed and the load. By transformation, two differential equations of first order can be obtained, which are solved implicitly in a numerically stable manner using the Newmark integration method. The effective stiffness matrix contains the normal stiffness matrix, but also mass and damping components. A singularity of the stiffness matrix can therefore be removed; the shorter the time step chosen, the more effectively this is done. The effective load vector also contains terms dependent on mass and damping properties. Having performed a numerically expensive computation, an effective displacement vector is obtained, from which the actual acceleration, speed and displacement vectors at the end of the time step are determined. The deformation energy is obtained by integration of the internal forces.

6 SUMMARY

Germanischer Lloyd is pursuing the strategy of performing collision analyses by utilizing the linear FE models generated for elastic stress proofs. The basic concepts underlying the development of a procedure for collision calculations are explained within the scope of this presentation. The essential point is the use of large-size elements with particular characteristics accounting for tension and folding. The design of the procedure allows to simulate collisions entirely in the time domain so that interactions of outer and inner collision mechanics are accounted for. An empirical relationship between failure strain and element size is derived based on thickness measurements performed on damaged structural components from actual collisions. Using this procedure, only little additional input data is required in order to assess the collision resistance of the ship's hull by use of a previously established linear FE model.

References

[1] Minorsky V.U. (1959). An Analysis of Ship Collisions with Reference to Protection of Nuclear Power Plants. *Journal of Ship Research.* **3:2**.

[2] Mc Dermott J.F. (1974). Tanker Structural Analysis for Minor Collisions, *SNAME.* **82**.

[3] Reckling K.A. (1983). Mechanics of Minor Ship Collisions, *International Symposium on*

Structural Crashworthiness. University Liverpool.

[4] Egge E.D., Böckenhauer M. (1991). Calculation of the Collision Resistance of Ships and its Assessment for Classification Purposes, *Marine Structures*. **4**, 36-56.

[5] Scharrer M., Egge E.D. (1997). Untersuchungen zum Kollisionswiderstand von RoRo-Schiffen. *Jahrbuch der Schiffbautechnischen Gesellschaft*. Bd. 91.

[6] Germanischer Lloyd. (2000). *Rules for Classification and Construction*. I Ship Technology, 1 Seagoing Ships.

[7] Scharrer M., Zhang L., Egge E.D. (2000). Abschlußbericht zum Vorhaben MTK 05884; Versagensverhalten von Schiffen bei Kollision, *Germanischer Lloyd Bericht*.

[8] Abramovicz W. (1983). The Effective Crushing Distance in Axially Compressed Thin-Walled Metal Columns, *Int. J. of Impact Engineering*. **1:3**, 309-317.

[9] Amdahl J. (1983). Energy Absorption in Ship-Platform Impacts, *PhD. Thesis*, Department of Marine Technology, University of Trondheim, Norway, Report No. UR-83-34.

[10] Yang P. D. C., Caldwell J. B. (1988). Collision Energy Absorption of Ships' Bow Structures, *Int. J. of Impact Engineering*. **7:2**, 181-196.

[11] Paik J. K., Pederson P. T. (1995) Ultimate and Crushing Strength of Plated Structures. *Journal of Ship Research*. **39:3**, 259-261.

[12] Lee J.W.: On the Optimization Design of Soft Bow Structures, *PRADS 1983*, 429-435.

[13] Choi S.K., Wierzbicki T. Culbertson-Driscoll J. (1994). Crushing Strength of Web Girder, Joint MIT-Industry Program on Tanker Safety. Rep. **23**.

[14] Lehmann E., Yu X. (1995). Progressive Folding of Bulbous Bows. *Jahrbuch der Schiffbautechnischen Gesellschaft*. Bd. 89.

[15] Yu X. (1997). Strukturverhalten mit großer Verformung bis zum Brucheintritt und mit dynamischer Zusammenfaltung. *Dissertation*, Bericht Nr. 579, IfS der Uni. Hamburg.

[16] Peschmann J. (2000). Berechnung der Energieabsorption der Stahlstruktur von Schiffen bei Kollisionen und Grundberührungen. *Dissertation*, TU Hamburg-Harburg.

[17] Pedersen P.T., Servis D.P., Zhang S., Samuelides M. Design for Structural Safety under Extreme Loads Collision Mechanics, EU Research Project, No. BE97-4375.

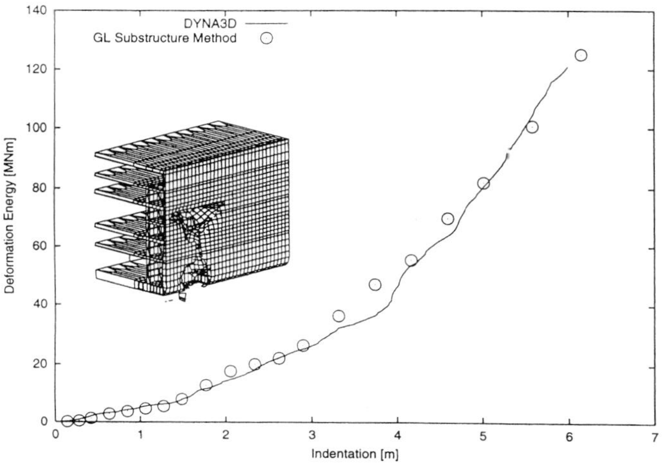

Fig. 1: Comparison of Deformation Energy Calculated by GL Substructure Methcd and DYNA3D

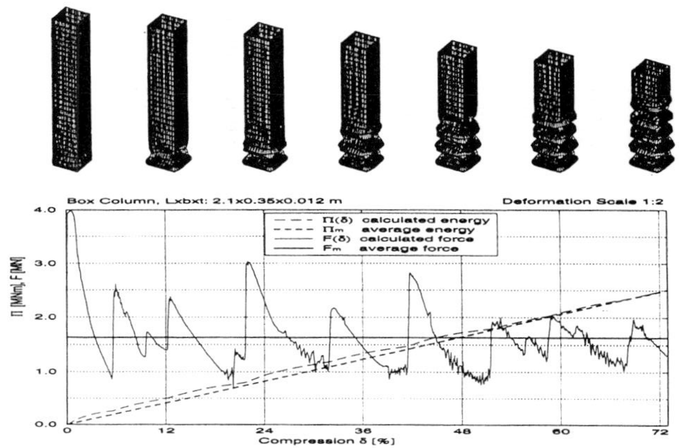

Fig. 3: Deformation, Collision Force and Deformation Energy of a Box Column

R_{eHd}: Dynamic yield stress.

σ_m: Mean ultimate stress within the effective folding length.

σ_u: Ultimate stress by compression.

σ_f: Limit stress at the complete compression.

E: Young's modulus.

ε_{ft}: Failure strain by tension dependent on element length and width.

ε_u: Compression at begin of folding.

ε_d: Strain at end of the effective folding length.

e: Range of final folding phase.

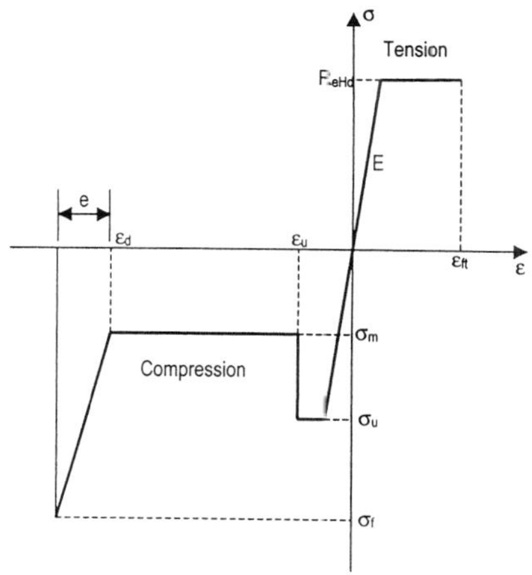

Fig. 2: Material Characteristics Considering Structural E'fects

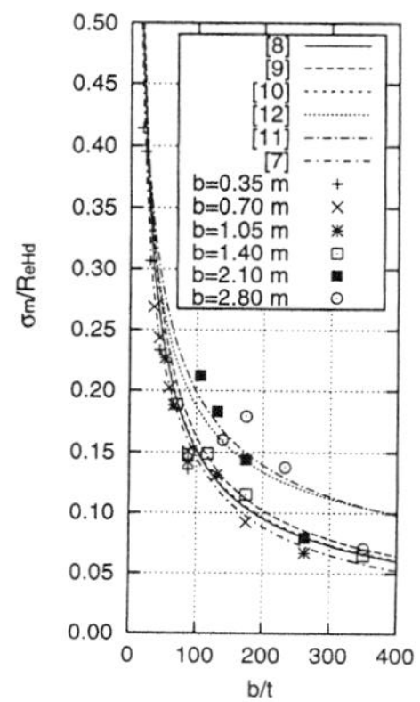

Fig. 4: Box Column

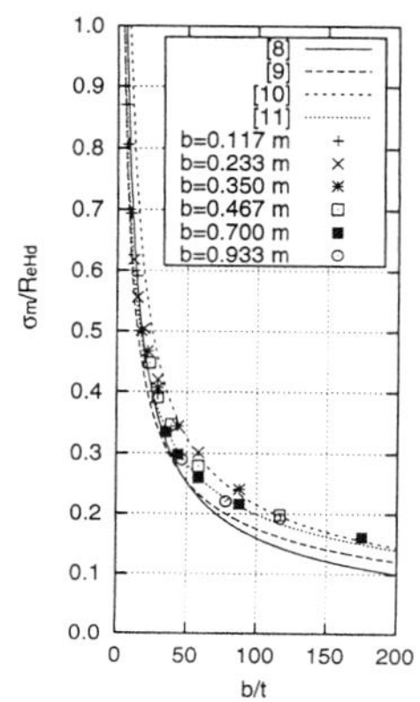

Fig. 5: Profile Combination

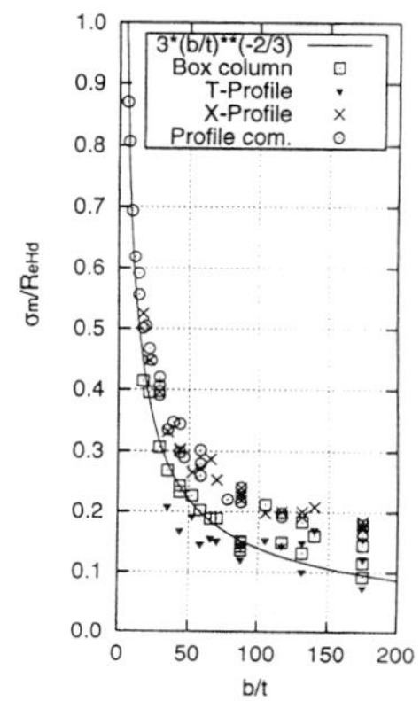

Fig. 6: All Profiles

Mean Ultimate Stresses Calculated by Analytical Solutions (Lines) and DYNA3D (Points)

Fig. 7: Folded Main Deck of a RoRo Ship caused by a Collision

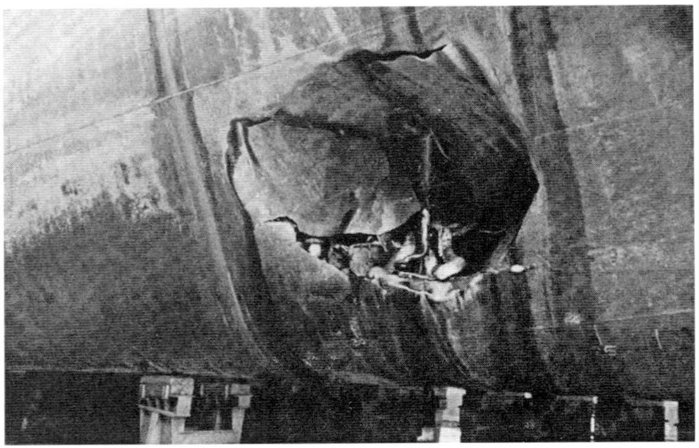

Fig. 8: Deformation and Crack Area of a Shell by a Bulbous Bow Collision

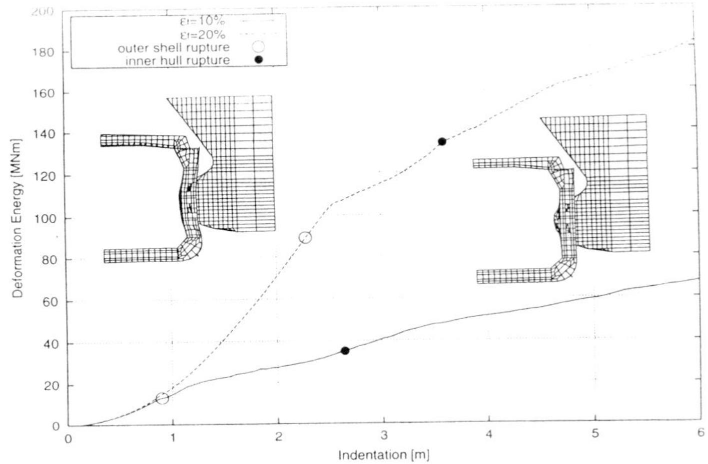

Fig. 9: Comparison of Deformation Energy under Different Failure Strain Assumptions

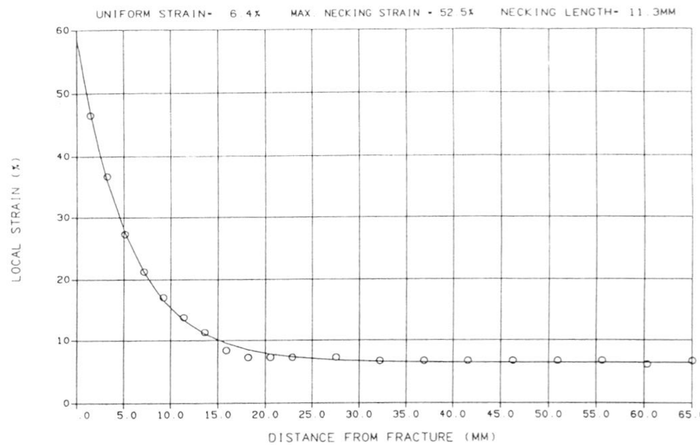

Fig. 10: Evaluation of a Typical Thickness Measurement

Practical Design of Ships and Other Floating Structures
You-Sheng Wu, Wei-Cheng Cui and Guo-Jun Zhou (Eds)

A SIMPLIFIED INTERNAL AND EXTERNAL MECHANICS MODEL FOR SHIPS' COLLISION

K. Suzuki[1], H. Ohtsubo[2] and K.S. Sajit[3]

[1] Institute of Environmental Study, The University of Tokyo
[2] Department of Environmental and Ocean Engineering, The University of Tokyo
7-3-1 Hongo Bunkyo-ku Tokyo Japan
[3] American Bureau of Shipping, Asia Pacific

ABSTRACT

The purpose of this paper is to present a simplified analytical method for the external and internal mechanics during a collision situation. The simplified external mechanics model can simulate the rigid body motion of the colliding ships. Added mass at infinite frequency is assumed to take into account the surrounding water effects. The derived mathematical model is applied to a collision between two ships. The results are compared with the results of FEM simulations and good agreement has been achieved. The simplified external mechanics model can be well integrated with the simplified internal mechanics model and the simultaneous interaction between the external and internal mechanics can be simulated.

KEYWORDS

CollisionAnalysis, Simplified method, Internal mechanics, External mechanics

1 INTRODUCTION

In the classical ship collision theory, the collision problem is separated into two independent problems; the first being the internal mechanics and the second is the external mechanics. Internal mechanics involves the evaluation of force-indentation responses of the striking and struck ships. The external mechanics deals with the motion of the colliding ships and their interaction with the surrounding water. This separation is possible because the colliding ships and the surrounding water are treated as an undamped system, using the theory of absolutely inelastic collisions between two bodies.

Minorsky (1959), in his famous paper, has equated the absorbed energy and the loss of kinetic energy of ships involved in the collision. He has also assumed that the hydrodynamic effect was represented by an increased inertia force due to added mass effect and estimated the added mass coefficient of the struck ship to be 0.4; added mass coefficient for lateral vibration. It has been observed that the ratio of external force to the acceleration of the struck ship varies with the time elapsed; the added mass is not constant but time dependent. There is always a simultaneous interaction between external and internal mechanics. Motora et al (1969) first introduced the concept of equivalent added mass which, dividing the external force by it, will give an acceleration equal to the exact value of acceleration at the end of collision. This equivalent added mass is also time dependent. If the duration of collision is very small, the added mass at infinite frequency is a good approximation.

In this paper, the simplified external mechanics model can be well integrated with the simplified internal mechanics model and the simultaneous interaction between the external and internal mechanics is simulated.

2 INTERNAL MECHANICS

Recent development of the explicit dynamic finite element codes enabled the simulation of the collision situation. However, the application of this method is prohibitively expensive in both model generation and computation. The objective of the simplified analysis is to derive theoretical models based on observed deformation patterns, which capture the major details without tremendous modeling effort. The procedure is to find the kinematically admissible deformation of the structure that yields the lowest force - least upper bound.

2.1 The Crushing Mechanics of the Side Structure

The deformation mode of the side structure is based on the Ohtsubo el. al. (1995). The intruding object is assumed to be a sharp diamond, and webs were modelled as plates subjected to lateral concentrated loads, thereby denting type of deformation occurs (Fig 1 left). The deformation of the side plate can be dominated by membrane behaviour. (Stretching mode Fig. 1 right) The deformation of the web plate is

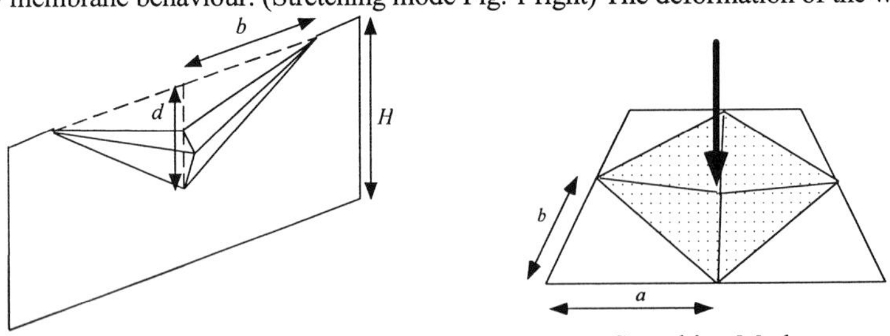

Denting Mode　　　　Stretching Mode

Figure 1: Damage of the side structure

The mean strength of the plate is found as:

$$F = 2N_0 \frac{a_1^2 + b_1^2}{a_1 b_1} \delta_1 \quad N_0 = \sigma_0 t_{out}^p \tag{1}$$

Resistance due to denting is given by:

$$F_d = 2.32\,\sigma_0 (2b)^{0.33} (t^w)^{1.67} \tag{2}$$

2.2 The Crushing Mechanics of Bow Structure

During collision, the axial crumpling behaviour of intersections of plate flanges plays a very important part in the energy absorption ability of a ship bow. The intersection can be categorized as L-shape, T-shape, and X-shape as shown in Figure 2 (Wierzbicki (1983)). Wang et al. (1995) proposed a simple one-term formula for predicting the crushing strength of a bow structure, after introducing two factors that represent energy absorption ability of structures and energy absorption reduction effect caused by inclination of plates with respect to collision load.

The mean crushing strength can be expressed as:

$$P_m = \alpha_1 \alpha_E \sigma_o A \tag{3}$$

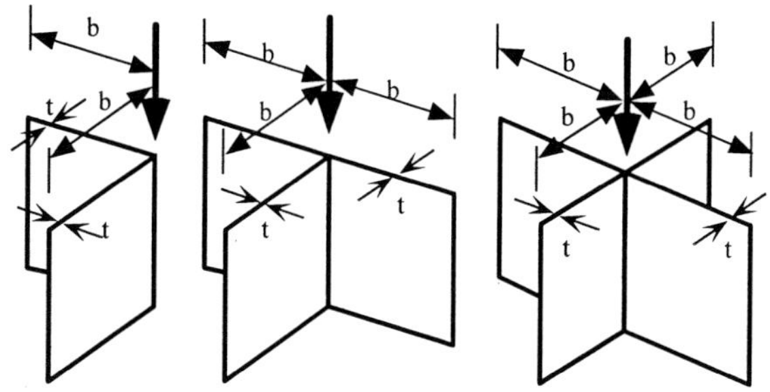

Figure 2: Basic Plate intersections

3 EXTERNAL MECHANICS

The external mechanics involves the simulation of the rigid body motion of the colliding ships including surrounding water effects.

The external mechanics involve the simulation of the rigid body motion of the colliding ships including surrounding water effects. It has been observed that the ratio of external force to the acceleration of the struck ship varies with the time elapsed; ie the added mass is not a constant but time dependent. If the duration of collision is small, the equivalent added mass can be approximated to the added mass at infinite frequency. There is always a simultaneous interaction between external and internal mechanics.

The following assumptions are made in the present external mechanics analysis:

(1) The ships are treated as stiff bodies.
(2) Interaction is only at the collision point.
(3) No coupling between surge, sway and yaw on one hand and heave, pitch and roll on the other hand.
(4) Gravitational and hydrostatic loads are ignored.
(5) No further structural damage once the relative velocity at the collision point reduces to zero

From the strip theory the equations (in the body - fixed coordinate system) of sway and yaw motions, after eliminating the roll components, can be found to be

$$\text{Sway:} \quad A_{SS}\dot{v} + B_{SS}v + A_{SY}\ddot{\psi} + B_{SY}\dot{\psi} = F_{ex,y} \tag{4}$$

$$\text{Yaw:} \quad A_{YS}\dot{v} + B_{YS}v + A_{YY}\ddot{\psi} + B_{YY}\dot{\psi} = N_{ex,z} \tag{5}$$

These are the equations of motion for the struck ship. The added mass at infinite frequency can be determined by Landweber's method (Landweber and Macagno (1957)) using two-parameter conformal mapping. Since the duration of collision is assumed to be very small (there by infinite frequency) the damping terms can be eliminated.

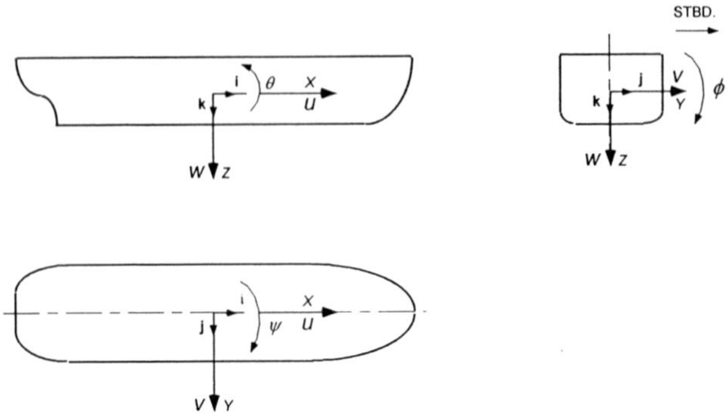

Figure3: The body-fixed coordinate system

For the striking ship only the surge motion need to be considered. The hydrodynamic forces related to the surge motion cannot be found by strip method. The force is however very small compared to sway force. According to Motora et al (1969), the added mass can be approximated to $(0.02 - 0.07)m$, where m is the mass of the ship.

The sway and yaw motions of the struck ship along with the surge motion of the striking ship can be written as follows. The suffixes '1' and '2' represent the striking as well as the struck ship respectively. In these equations there is no coupling between the motions of the two ships. The interaction of them comes from internal mechanics. The non-linear spring is introduced between them, and the spring relation between deformation and force is derived using internal mechanics, as described in the next section.

$$\begin{bmatrix} (A_{SS})_2 & (A_{SY})_2 & 0 \\ (A_{YS})_2 & (A_{YY})_2 & 0 \\ 0 & 0 & (M_{XX})_1 \end{bmatrix} \begin{bmatrix} (\dot{v})_2 \\ (\ddot{\psi})_2 \\ (\dot{u})_1 \end{bmatrix} = \begin{bmatrix} (Y_C)_2 \\ (N_C)_2 \\ (X_C)_1 \end{bmatrix} - \begin{bmatrix} (B_{SS})_2 & (B_{SY})_2 & 0 \\ (B_{YS})_2 & (B_{YY})_2 & 0 \\ 0 & 0 & 0 \end{bmatrix} \begin{bmatrix} (v)_2 \\ (\dot{\psi})_2 \\ (u)_1 \end{bmatrix} \tag{6}$$

For a center line collision as shown in Figure 4, if we assume the external force to a stepwise one of magnitude for a duration, the longitudinal motion of the striking ship and the sway motion of the struck ship can be written as:

$$\left[m_1 + \int_{x_a}^{x_f} m_x dx \right] (\dot{u})_1 = -Y_{C0} \tag{7}$$

$$\left[m_2 + \int_{x_a}^{x_f} m_y dx \right] (\dot{v}_P)_2 = Y_{C0} \tag{8}$$

Penetration of the first ship into the second one i.e the spring deformation can be written as follows.

$$\Delta S = \int_0^{\delta t} \left[(u)_1 - (v_P)_2 \right] dt \tag{9}$$

The force penetration curves for the colliding ships can be determined by the method described in the internal mechanics section. The resultant of the forces acting on the striking ship must be in equilibrium with the resultant of the forces acting on the struck ship at any time during the collision. The force deformation characteristics of the spring can be derived from the force balancing between the striking and the struck ships and the motions of the two ships after collision can be simulated. A non-central collision also can be simulated without difficulty.

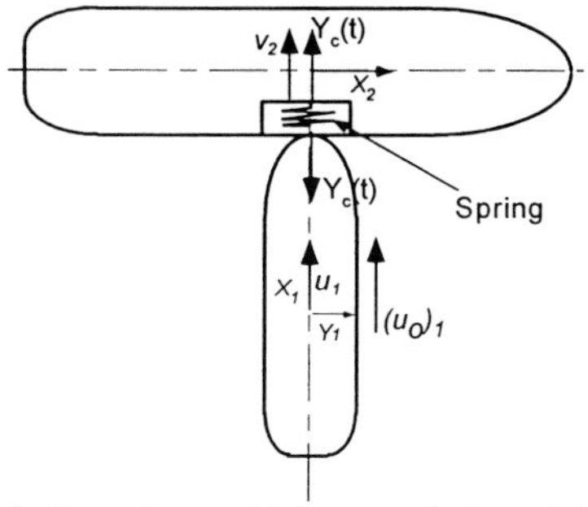

Figure 4: Centerline collision simulation with spring

4 SIMULATION OF COLLISION

The above method is used to evaluate the strength of a 7000 t ship with anti-collision structure side being hit by a Suezmax tanker (150,000 DWT) and a VLCC (280,000 DWT). The collisions are assumed to be central right-angled as this situation is more pertinent inducing greater damage to the struck side. The results are compared with the finite element analysis of the same situations by Mitsubishi Heavy Industries, Japan. Two conditions are analyzed each for the Suezmax and VLCC tankers. The first one is the ballast condition and the second is the fully loaded condition.

4.1 Suezmax Tanker VS 7000 ton ship

The Suezmax tanker is under ballast and the struck ship is assumed to be stationary. The ballast displacement of the tanker is 65,000 tons. The displacement of the struck ship is 7,000 tons. Considering the momentum equilibrium, the loss of kinetic energy during collision is given by

$$\mathrm{KE} = \frac{m_1}{2+1.43\mathrm{m}_1/\mathrm{m}_2} * (U_0)_1^2 = 253 \text{ MJ} \tag{10}$$

The center line cross section of the Suezmax tanker can be seen in Figure.5 left. Figure 5 right shows force-deformation curve for striking bow (left) and struck side (right). This relation gives non-linear spring that is introduced in the external mechanics.

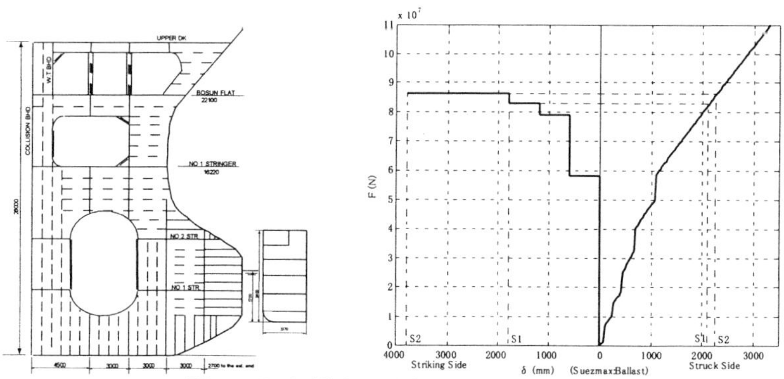

Figure 5: Collision with Suezmax Tanker

The proposed external mechanics simulation was applied to simulate the collision between the two ships. The numerical simulation (dt = 0.01 sec) with the above spring characteristics yields the following velocity curves, which is shown along with the FEM simulation.

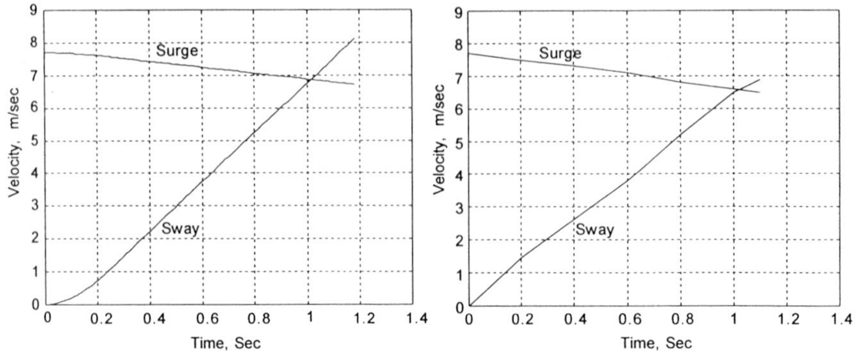

Simplified Method FEM Simulation

Figure 6: Comparison of Simplified and FEM simulations

The same final drift velocity can be seen in fig. 13. No further damage is assumed once the ships attain this common velocity. From the velocity curves the collision force as a function of time is derived and can be seen in Figure 7.

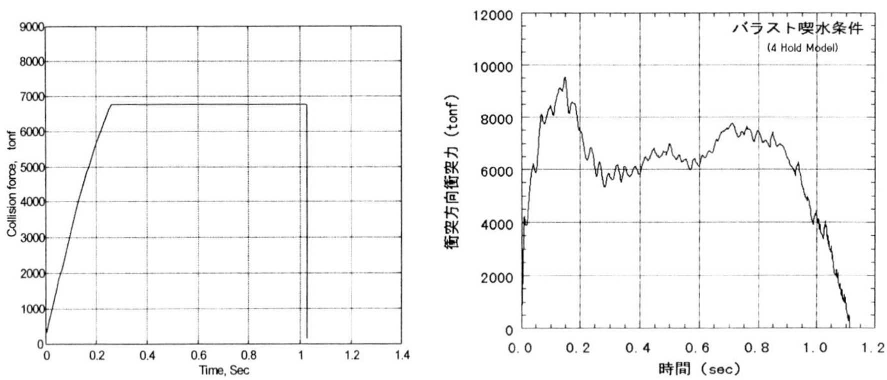

Simplified method FEM Simulation (MHI)

Figure 7: Collision Force vs Time

5 CONCLUSIONS

In this paper, a simplified external mechanics model was derived, and integrated with the simplified internal mechanics model to simulate a collision scenario. Two practical cases were analyzed and compared with FEM simulations. The results are quite encouraging. The greatest challenge for the designer will be arriving at realistic force-indentation curves for the colliding ships. This part of the analysis will be more difficult as it depends on the structural details of the colliding ships, the location of collision, the relative displacements of the ships, the loading conditions, the striking speed etc. A central right-angled collision is assumed in the analysis, as this situation is more pertinent inducing severe structural damage. In certain situations, the energy loss due to friction will be quite substantial and has to be taken into account. In the present simplified analysis, it is assumed that an increase in the flow stress takes into account the material dynamic effects. In certain practical situations a higher collision resistance for a side structure has been observed due to material dynamic effects. The authors strongly believe that the above simplified collision analysis can be used for a quick assessment of a collision situation within a day with the help of a personal computer.

References

Minorsky VU. (1959). An Analysis of Ship Collisions with Reference to Protection of Nuclear Power Plants. *Journal of Ship Research*, October: 1 - 4

Motora S, Fujino M, et al. (1969). Equivalent Added mass of Ships in Collisions. *Journal of the Society of Naval Architects of Japan*. **126**, 138-147

Landweber I, Macagno M. (1957). Added mass of two-dimensional forms oscillating in a free surface. *Journal of Ship Research*. November: 20-30

Ohtsubo H, Suzuki K, Yonesato N. (1995). Prediction of Collision Strength of Side Structures. *Journal of the Society of Naval Architects of Japan*. **178**, 421-427

Wang G., Suzuki K, Ohtsubo H. (1995). Predicting Collision Strength of Bow Structure. MARIENV'95 Proceedings. **1**, 40 - 43

Wierzbicki, T. (1983). Crushing Behaviour of Plate Intersections. *Structural Crashworthiness*. edited by N. Jones and T. Wierzbicki, Chapter 3, Butterworth & Co. Ltd.

Practical Design of Ships and Other Floating Structures
You-Sheng Wu, Wei-Cheng Cui and Guo-Jun Zhou (Eds)

NUMERICAL SIMULATION OF SHIP-SUBMARINE COLLISIONS

R. Donner F. Besnier H. Le Sourne

[1] Institut de Recherches de la Construction Navale,
Nantes, France

ABSTRACT

The continuously increase of maritime traffic makes the risk of collisions greater and several incidents have been reported during the last few years. Research work has been done for tanker collisions: this paper describes numerical methods, tools and methodology used in the simulation of a ship-submarine collision in various configurations. Though ship-submarine collisions are a particular case of ship-ship collisions, they exhibit specific difficulties, like the difference of displacements between the two ships, leading to large amplitude rotational motions for the submarine and the high local stiffness of the submarine hull compared to the impacting bow stiffness. The global collision dynamics, i.e. global motions of the two ships considered as rigid bodies are computed under collision forces and hydrodynamic forces. The influence of hydrodynamic effects will be highlighted in the ship-submarine collisions. The local collision mechanics, i.e. local crushes of the two ships are investigated using an explicit finite element approach. Collision forces are computed in this step. The importance of mesh size and material behaviour laws will be shown. Tools and methodology presented above have been applied to two LNG bow-submarine collisions and to the numerical simulation of a scaled ship-submarine collision test. Main difficulties and results will be presented and also on going developments of the tools which will allow to carry out industrial applications in submarine design.

KEYWORDS

Collisions, Submarines, Numerical methods, Finite elements, Hydrodynamic effects.

1 INTRODUCTION

Several ship-ship collisions and ship grounding with heavy consequences for passengers and marine environment have been reported for fifteen years. Extensive research has been done about prediction of structural damage and increased crashworthiness of ships, especially tankers. A seven-year research project on tanker structural failure and crashworthiness was carried out by Japan, Holland and Germany as described in Kitamura et al (1998). A numerical simulation system based on the explicit finite element code LS-DYNA was developed and verified by large-scale tanker collisions and grounding experiments.

Some cases of ship-submarine collisions near the sea-surface and head-on collisions of submarines with rocks occurred during the last few years, a head-on collision is analysed in Chauvet and Montfort (1996). A research program on submarine behaviour in ship-submarine collisions has been initiated by the French Navy, in 1995. The numerical simulation system with LS-DYNA described in Kuroiwa et al (1995) and in Kitamura et al (1998) has been used in this research program and is now under development. In this paper, applications of the tool to numerical simulations of LNG bow-submarine collisions and to the simulation of a scaled bow-submarine collision test are presented with further improvements of the methods.

2 NUMERICAL METHODS AND COMPUTATIONAL TOOLS

The ship-ship collision mechanics, described in Pedersen (1995) and Petersen (1982) can be subdivided into inner collision mechanics and outer collision dynamics.

The inner collision mechanics is a crash problem governed by buckling, yielding, rupture of materials and assemblies. It is investigated using the explicit finite element code LS-DYNA. Collision forces are computed at this step.

The outer collision dynamics, i.e. global motions of the two ships considered as rigid bodies under collision forces and hydrodynamic pressure forces, are computed with the program MCOL, coupled with LS-DYNA. Hydrodynamic loads are calculated with the hydrostatic restoring matrix, the added mass matrix and the frequency dependant added damping matrices computed for each ship by the 3D sea-keeping code AQUA+. A model of the collision analysis is shown on figure 1.

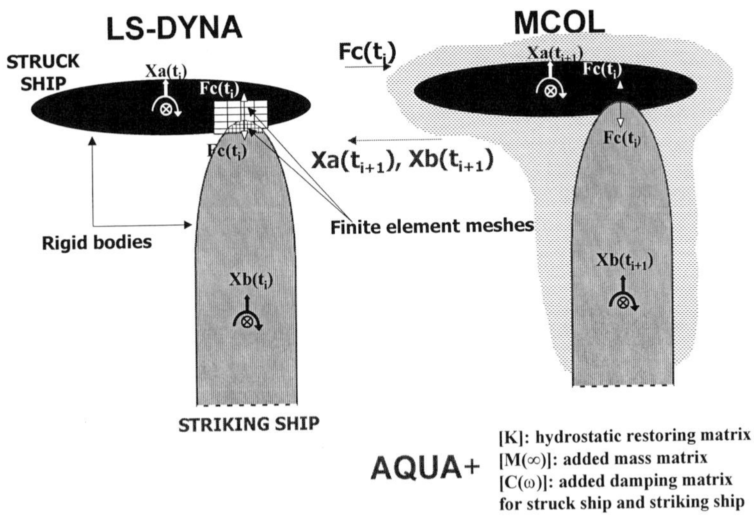

Figure 1: Collision simulation

3 STUDY OF COLLISION SCENARIOS

Two collision scenarios between a bulk carrier of 93 000 tons displacement with 10.6 m draught and 12 knots forward speed and a submarine of 3 000 tons DWT resting at periscopic immersion, or at free surface were investigated. A view of the two scenarios is shown in figure 2.

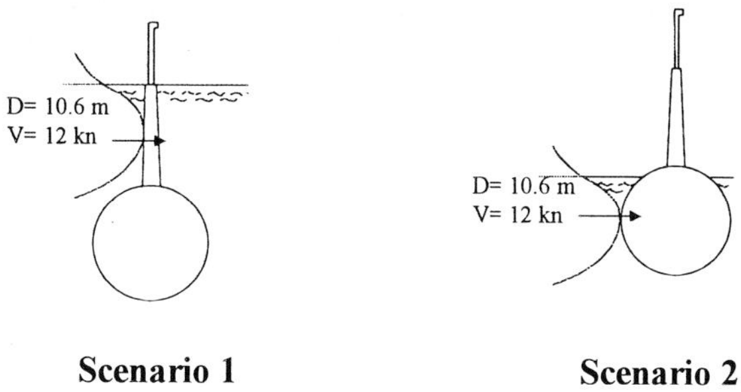

Figure 2: Collision scenarios

The deformations of the structures at the end of two numerical simulations of the scenario 1, using LS-DYNA and LS-DYNA + MCOL with simplified bi-dimensional models are shown on figure 3.

In the two analyses the submarine superstructures exhibit a significative bending. The global motion of the submarine is purely translational in the first simulation and contains a large rolling component in the second one, as can be expected from the displacement differences between the two ships. The importance of hydrodynamic effects in ship-submarine collisions is demonstrated and also the necessity of taking into account large rotations into MCOL, not accounted for in the original program.

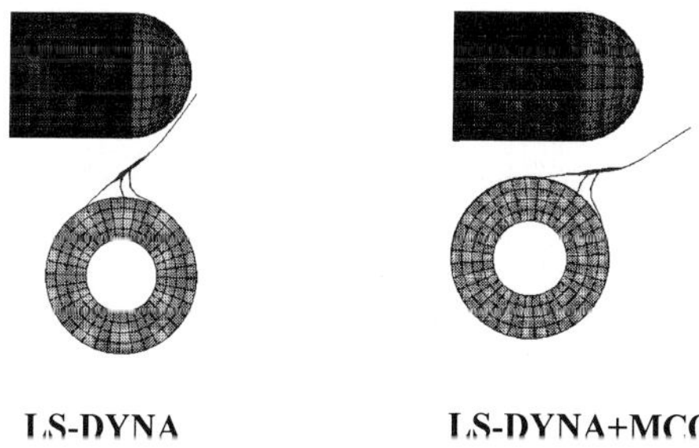

Figure 3: Bulb/superstructures collision

The deformations of the ship bow and of the submarine hull at the end of two numerical simulations of the scenario 2, using two mesh sizes of 40 cm and 10 cm for the bow and one mesh size of 10 cm for the submarine hull are shown on figure 4.

The bow is made of mild steel with plating thickness ranging from 10 mm to 15 mm. The submarine hull is made of high strength steel, with a thickness greater than 30 mm. So in this case the local stiffness of the struck hull is greater than the bow stiffness, contrary to the situation encountered in ship-ship collisions where the striking bow is often assumed rigid. In the second simulation with fine meshes, the bulb mainly crushes against the submarine hull, as expected. In the first simulation, the coarse meshing of the bow makes it too much stiff relatively to the submarine hull and a non-physical deformation of the hull is obtained. The importance of mesh size and material behaviour in ship-submarine collisions is highlighted.

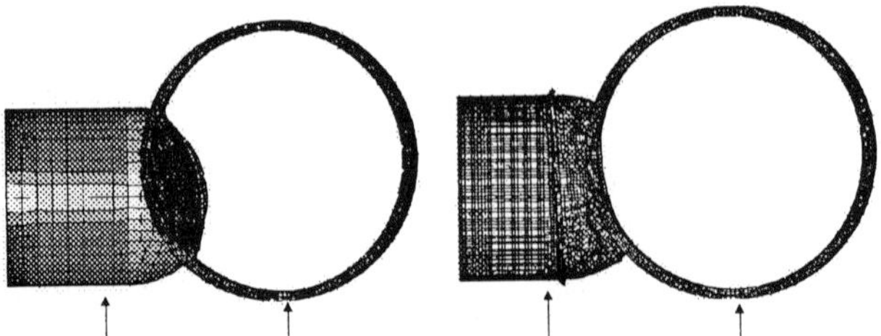

Figure 4: Bulb/hull collision

4 NUMERICAL SIMULATION OF A COLLISION TEST

A scaled collision test between a ship and a submarine was carried out by the French Navy in 1974. Two views of the collision experiment are shown on figure 5.

The submarine has a DWT of 1150 kg, a length of 6.40 m and a diameter of 0.53 m. The striking ship has a DWT of 1040 kg, a length of 5.51 m and a bow length of 1.52 m. The ship collides at an angle of 135° and with a forward speed of 15 knots with the submarine.

Figure 6 shows the experimental damage and the computed damage to the submarine hull. The damages are localised, the experimental indentation is 1.5 mm, the numerical indentation is 2.5 mm.

Figure 7 shows the damaged inside of the bow and the related computational result. The crushing of the plating is well-reproduced by the numerical simulation.

The plating and the hull are connected by spotwelds and a proper modelling of failure of the plating and rupture of the welding, based on cumulative plastic strain was used in LS-DYNA in order to obtain a correct crushing of the bow and a correct indentation of the submarine hull. The initial numerical simulations of the collision tests were carried out without taking into account the rupture of the spotwelds and led to an overstiffening of the bow with an indentation of 20 mm of the submarine hull. The importance of taking into account properly the failure phenomena in numerical simulation of ship-ship collisions is well known, as pointed out in Kuroiwa et al (1995).

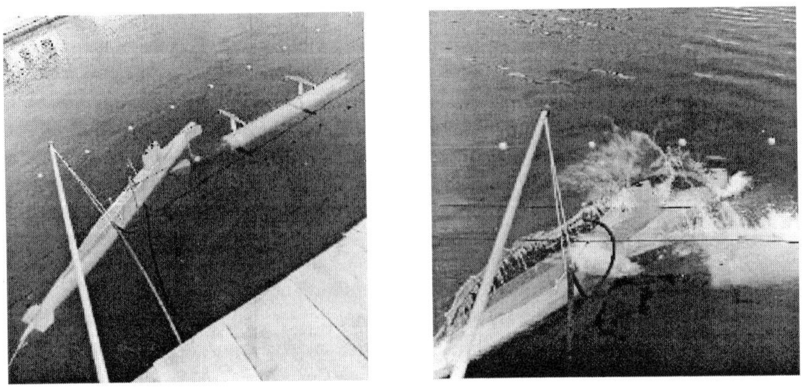

Figure 5: Collision experiment

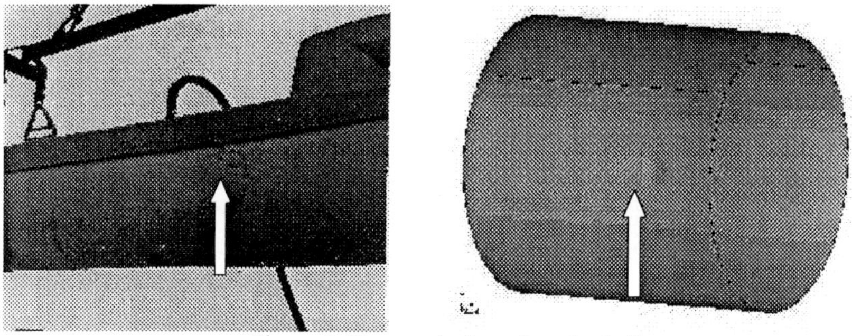

Figure 6: Damage to the submarine hull

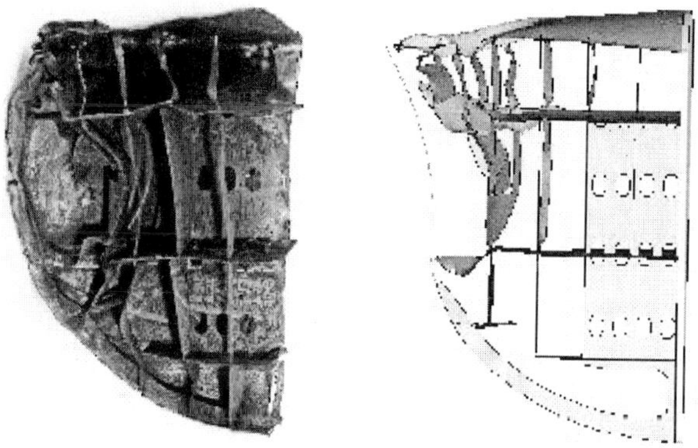

Figure 7: Damage to the ship bow

5 ON GOING DEVELOPMENTS OF MCOL

As shown in section 3, due to the great displacement difference between the striking ship and the struck submarine, the submarine will be subjected to large amplitude translational and rotational motions during the collision and after the collision.

In order to compute properly the submarine movements, the following developments are now being introduced in MCOL:

1) Introduction of large rotations and Coriolis and centrifugal couplings,
2) Introduction of non linear drag damping and uncoupling of MCOL from LS-DYNA, in order to simulate the submarine motions after loss of contact and during a long time (15-20 seconds).

Two numerical simulations of the scenario 1 described in section 3 during the collision and after loss of contact were carried out with the original MCOL program and by introducing drag damping. The resulting roll speeds for the submarine are shown in figure 8, as expected, the drag forces lower significantly the roll speed.

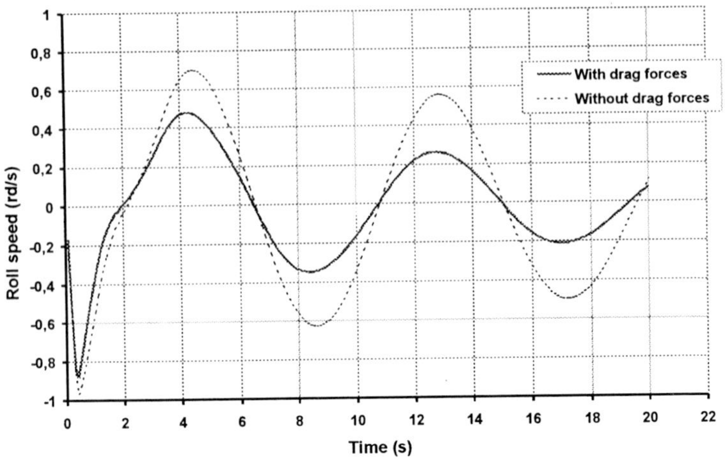

Figure 8: Influence of drag damping

6 CONCLUSIONS

Numerical simulations of ship-submarine collisions can be carried out using the special features for ship-ship collisions integrated to LS-DYNA.

However ship-submarine collisions exhibit specific difficulties, like the high stiffness of the submarine hull compared to the impacting bow stiffness and large amplitude motions of the submarine.

The on going development of the MCOL program, i.e. introduction of large rotations, drag damping and uncoupling of MCOL from LS-DYNA will allow to apply the resulting intregrated tool for industrial applications in submarine design and will improve the capabilities of surface ship collisions.

References

Chauvet M., Montfort C. (1996). Analyse de la collision d'un sous-marin. *Bulletin de l'Association Technique Maritime et Aéronautique 96,* 61-88

Kitamura O., Kuroiwa T., Kawamoto Y. and Kameko E. (1998). A Study on the Improved Tanker Structure against Collision and Grounding. *Proceedings of PRADS 98,* 173-179.

Kuroiwa T., Kawamoto Y., Kusuba S. and Stillman D. (1995). Numerical Simulation of Collision and Grounding of Ships. *Proceedings of MARIENV'95,* **1**, 66-70.

Pedersen P.T. (1995). Collision and Grounding Mechanics. *Proceedings of WEMT'95,* 125-157.

Petersen M.J. (1982). Dynamics of Ship Collisions. *Ocean Engineering* **9:2**, 295-329.

Practical Design of Ships and Other Floating Structures
You-Sheng Wu, Wei-Cheng Cui and Guo-Jun Zhou (Eds)

FLUID MESH MODELING ON SURFACE SHIP SHOCK RESPONSE UNDER UNDERWATER EXPLOSION

Sang-Gab Lee[1], Jeong-Il Kwon[1] and Jung-Hoon Chung[2]

[1]Division of Marine Systems Engineering, Korea Maritime University
#1, Dongsam-Dong Yeongdo-Gu, Pusan, 606-791, Korea
[2]Dept. of Structural Systems Eng., Korea Institute of Machinery and Materials,
Yusung P.O. Box 101, Daejeon, 305-600, Korea

ABSTRACT

In this study, for the investigation of the effects of several important parameters, such as fluid mesh size, its box or cylinder shape, its dimension of breadth, length and depth on free surface, on the reliable shock response of a finite element model under underwater explosion, analysis of a simplified surface ship was carried out using the LS-DYNA3D/USA code. The shock responses were not much affected by the fluid mesh parameters. The computational time was greatly dependent on the number of DAA boundary segments. It is desirable to reduce the DAA boundary segments, and it is not necessary to cover the fluid mesh boundary to or beyond the bulk cavitation zone just for the concerns about the initial shock wave response.

KEYWORDS

Underwater explosion, Surface ship, LS-DYNA3D/USA code, Shock response, DAA boundary segments, Bulk cavitation

1 INTRODUCTION

Surface ship structure and shipboard equipments must be designed to withstand severe shock excitations induced by underwater explosion. The ship shock test/trials identify the design and construction deficiencies giving a serious negative effect on the survivability of ship, equipment and crew, and also validate shock hardening criteria and performance. Unfortunately, the ship shock trials are very time consuming and expensive. With the advent and ongoing advances in simulation capabilities and sophisticated simulation tools, numerical modeling and simulation has become a viable, less costly alternative as well as more reliable aids to live fire testing (Mair 1997).

Surface ship shock simulation under underwater explosion is generally complicated by free surface effects, such as bulk cavitation resulting from surface reflection wave, in addition to local cavitation, gas bubble oscillation and migration toward free surface, and cavitation closure pulses (Santiago 1996).

Furthermore, complex fluid-structure interaction phenomena occur, as well as the complicated dynamic behavior of the ship and shipboard equipments. For a reliable and accurate shock response of surface ship, therefore, the surrounding fluid mesh model must be constructed to mate exactly with the finite element mesh of ship structure, and must be of sufficient size to capture a bulk cavitation zone. Since the bulk cavitation zone can become quite large, greater computational memory and time are required, depending on the size and depth of the charge.

Three-dimensional surface ship shock analyses were performed using a large scale finite element model of a coupled ship and surrounding fluid using LS-DYNA code (LSTC 2000) coupled with USA code (Shin 1996), and the predicted results were compared with ship shock test results (Shin 1999). The effects of reducing the box type fluid mesh size were investigated on the accuracy of the structural response of Navy's Floating Shock Platform (FSP) under underwater explosion, also using the LS-DYNA3D/USA coupled code (Smith 1996). LS-DYNA is a nonlinear dynamic analysis of structure in three dimensions based on the explicit integration, whereas the USA code, a boundary element code for the underwater shock fluid-structure interaction problem based on doubly asymptotic approximation (DAA) consisting of three main modules: FLUMAS, AUGMAT and TIMINT.

In this study, to investigate the effects of several important parameters, such as fluid mesh size, its box or cylinder shape, its dimension of breadth, length and depth on free surface, on the reliable shock response of a finite element model under underwater explosion, analysis of a simplified surface ship is carried out using the LS-DYNA3D/USA code.

2 FLUID MESH MODELING AND SIMULATION SCENARIOS

Two different charge location geometries were used in the shock simulation runs for this study, as shown in Figure 1, where both geometries consisted of the same TNT 100 lb charge and of the same Keel Shock Factor. One attack geometry placed the charge directly under ship model in midsection plane at depth of 30.0 ft with standoff distance of 29.0 ft, whereas the second one, from longitudinal center plane by 18.1 ft and at depth of 19.1 ft with standoff distance 25.6 ft. Figure 2 shows the dimensions of ship model, and TABLE 1, a list of material properties of mild steel plate and sea water. UNDEX parameters of the explosion are also summarized in TABLE 2.

The computed bulk cavitation zones for both geometries are shown in Figure 3 in their entirety, and five fluid mesh boundary sizes of cylinder type with respect to ship model, also in Figure 4(a) and (b) on a large scale. Their radius in midsection plane was set to 5.0, 7.5, 10.0, 12.5, and 15.0 ft for charge under ship model and offset charge, respectively. Figure 5 shows the configuration of finite element meshes of ship and fluid mesh models typically with radius 7.5 and 15.0 ft, respectively. The fluid mesh element size to the radial direction increases proportionally for both attack geometries. The ship model contains 1,464 4-noded shell elements, and the wetted surface, 464 elements. LS-DYNA's Material Type 90 (acoustic pressure 8-noded solid element) is used to model the pressure wave translation properties of water. The length of each fluid mesh model fore and aft ship is the same as its breadth at free surface. TABLE 3 shows a list of fluid mesh model such as total number of nodes, acoustic elements, and ADD boundary segments, and CPU time. An important aspect of fluid mesh model is the finite element size next to the structural mesh. For cavitation analysis using the USA code, critical element size is determined by the equation, $2T\rho / \rho_s t_s \leq 5$, where ρ =density of water, T=thickness of the fluid element in the direction normal to the wetted surface of the structure, ρ_s =density of the submerged structure, and t_s =thickness of the submerged structure. The elements adjacent to the structural model were set a value less than this value of thickness.

In addition, the effect of fluid mesh element size to the radial direction in midsection plane on the shock response was considered comparing proportional increasing element size (Case 1_2) with uniform one (Case 1_6), as shown in Figures 5(a) and 6(a). The following fluid mesh parameters were also examined: fluid mesh shape such as cylinder type (Case 1_2) and box one (Case 2_1) in Figure 6(b), and fluid mesh dimensions at free surface such as depth (Case 2_2), breadth (Case 2_3), and length (Case 2_4) compared with standard one (Case 2_1), respectively. L is referred to fluid mesh length fore or aft ship at free surface, B, its breadth to port or starboard, and D, its depth in TABLE 3.

3 SHOCK SIMULATION RESULTS

The shock simulation results for the scenarios in the previously mentioned chapter were compared to the kick off velocity of ship under free surface and the peak pressure under the keel for both attack geometries from the empirical formulation, as shown in TABLE 4.

The vertical velocity responses and fluid mesh pressure profiles were almost the same as the empirical calculations, and the results of five fluid mesh boundary models for each attack geometry, as shown in Figures. 7~10, except that cavitation closure pulses occurred a little bit late according to an increase of fluid mesh boundary, and a little bit of fluctuation phenomena on the response were found in the case of relatively small fluid mesh boundary, 5.0 ft (Case 1_1). There was not much difference in the effects of several fluid mesh parameters on the shock response, as shown in Figures. 11 and 12, such as element size of fluid mesh boundary to the radial direction, fluid mesh shape type, dimensions of fluid mesh boundary at free surface.

Total CPU computational time was affected greatly by the number of DAA boundary segments, as shown in TABLE 3. In addition to the restriction of number of DAA boundary segments of the USA code, it could be found again to reduce its segment number as small as possible. It might be desirable to model the fluid mesh boundary in a way to increase element size proportionally to the radial direction. The lower cavitation boundary must not be included to the fluid mesh boundary for the consideration of the bulk cavitation effects and the initial shock wave response. However, too small fluid mesh boundary is not desirable through this study..

4 CONCLUSIONS

In this study, the shock simulations have been carried out to investigate the effects of the several fluid mesh parameters on the shock response with respective to the simplified surface ship using LS-DYNA/USA code. The shock responses were not much affected by the fluid mesh parameters. The computational time was greatly dependent on the number of DAA boundary segments. It is desirable to reduce the DAA boundary segments, and it is not necessary to cover the fluid mesh boundary to or beyond the bulk cavitation zone just for the concerns about the initial shock wave response. As the next study, more reliable response of cavitation closure time and bubble characteristics will be examined using the test results.

References

LSTC, LS/DYNA3D User's Manual, Version 9.50. (2000). Livermore Software Technology Corp.

Mair, Hans U., Reese, Ronald M. and Hartsough, Kurt. (1997). Simulated ship shock Tests/Trials?, *Proceedings of the 68th Shock and Vibration Symposium*, 475~482.

Santiago, Leonard D. (1996). Fluid-interaction and cavitation effects on a surface ship model due to an underwater explosion. *Master Thesis*. Naval Postgraduate School, Monterey, CA.

Shin, Y. S., DeRunts, J. A. (1996). USA/LS-DYNA3D Software *Training Course*. **5**.

Shin, Y. S and Park, S. Y. (1999). Ship shock trial simulation of USS John Paul Jones (DGG53) using LS-DYNA3D/USA: Three dimensional analysis, *Proceedings of the 70th Shock and Vibration*

Symposium, Albuquerque, NM., USA.
Smith, James R. (1996). Effect of fluid mesh truncation on the response of a Floating Shock Platform subjected to an underwater explosion. *Master Thesis*. Naval Postgraduate School, Monterey, CA., USA.

TABLE 1
MATERIAL PROPERTIES OF STEEL PLATE AND SEA WATER

Shell Plate	Density	7.350E-04 lbs-sec^2 / in^4
	Modulus of Elasticity	3.00E+07 psi
	Poisson's Ratio	0.3
	Yield Stress	34.545 ksi
	Ultimate Stress	65.415 ksi
Sea Water	Density	9.345E-05 lbs-sec^2 / in^4
	Sound Speed	5.916E+04 in/sec

TABLE 2
UNDEX PARAMETERS FOR CHARGE

Parameters	Charge undership model	Offset charge
Standoff Distance	348.00 in	294.48 in
P_{max}	2589.63 psi	3153.66 psi
□	0.3778 msec	0.3663 msec
T	0.627 sec	0.747 sec
A_{max}	177.36 in	190. 20 in

TABLE 3
LIST OF FLUID MESH MODEL AND TOTAL CPU

Case			No. of Nodes	No. of Acoustic Elements	No. of DAA Boundary Segments	CPU Time (hr. min. sec) Charge Location	
						Under	Offset
Cylinder Type Fluid Model (ft)	1_1	Radius 5.0 (Rough)	16,621	13,920	2,040	2. 08. 41	2. 08. 48
	1_2	Radius 7.5 (Rough)	18,856	16,012	2,144	2. 14. 43	2. 14. 23
	1_3	Radius 10.0 (Rough)	31,651	28,032	2,664	2. 51. 03	2. 50. 44
	1_4	Radius 12.5 (Rough)	34,534	30,748	2,768	2. 59. 20	2. 59. 30
	1_5	Radius 15.0 (Rough)	50,569	45,888	3,288	3. 48. 02	3. 45. 38
	1_6	Radius 7.50 (Fine)	48,046	43,620	3,080	3. 23. 27	-
Block Type Fluid Model (L×B×D) (ft)	2_1	5.0×5.0×7.5 ft	29,914	26,424	3,772	5. 09. 14	-
	2_2	5.0×5.0×12.5 ft	40,267	36,336	4,640	7. 07. 36	-
	2_3	5.0×10.0×7.5 ft	44,794	40,560	4,516	6. 54. 16	-
	2_4	10.0×5.0×7.5 ft	51,034	46,336	5,124	8. 39. 11	-

TABLE 4
KICK OFF VELOCITY AND PRESSURE FROM EMPIRICAL FORMULATION

Items	Charge under ship model	Offset Charge
Kick off velocity (in/sec)	605.55	579.07
Peak pressure under the keel (psi)	2488.08	3000.16

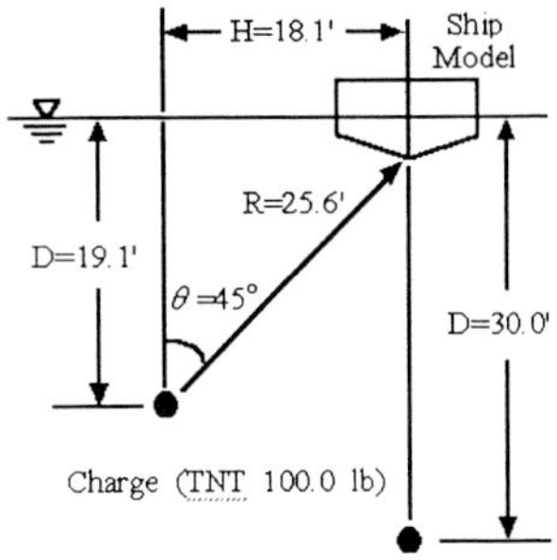

Figure 1: Geometry of surface ship model

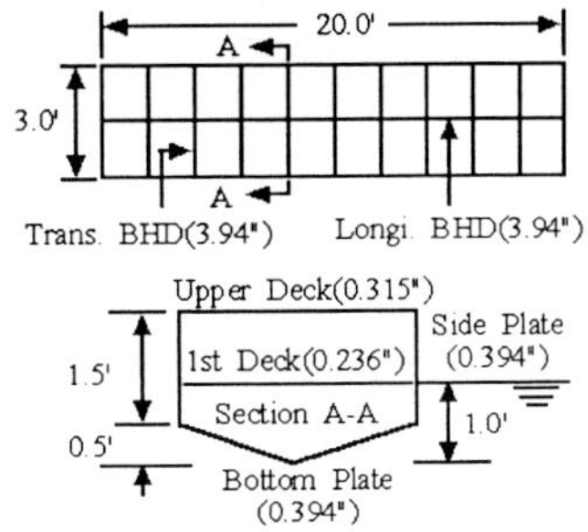

Figure 2: Dimensions of surface ship model

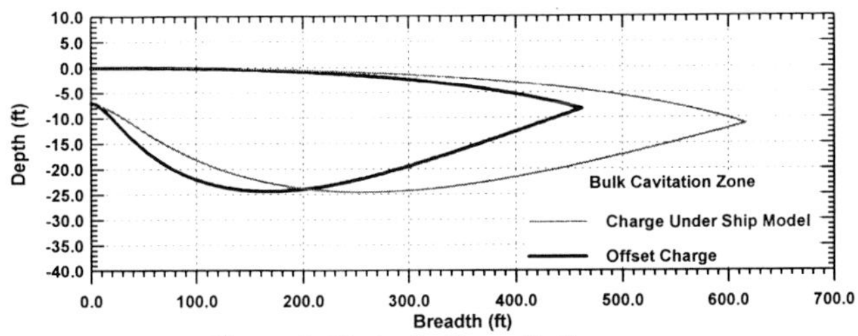

Figure 3: Entire bulk cavitation zones

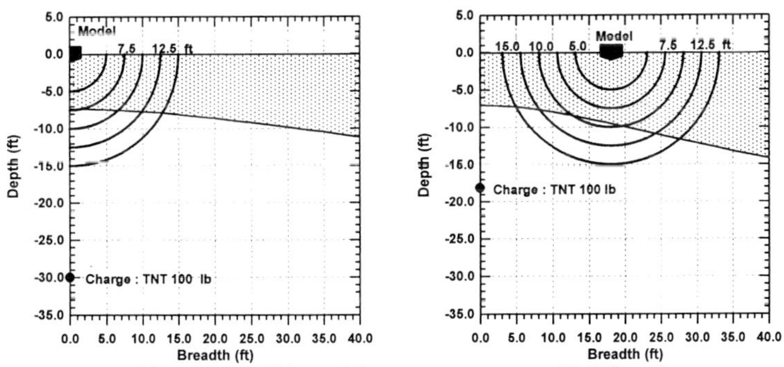

(a) Charge under ship model (b) Offset charge

Figure 4: Ship models and fluid mesh boundary locations in bulk cavitation zones

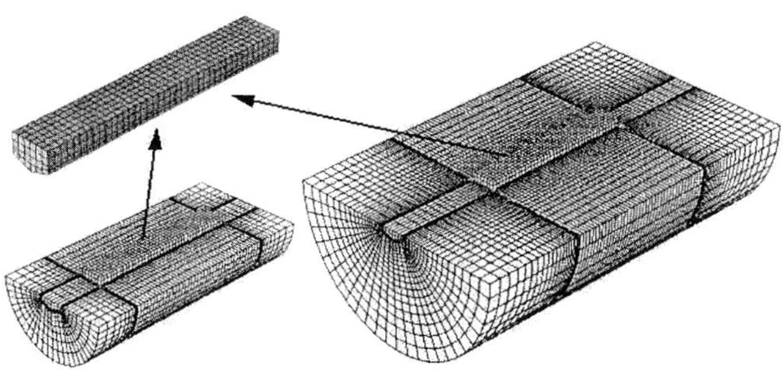

(a) Case 1_2 (Radius 7.5 ft) (b) Case 1_5 (Radius 15.0 ft)

Figure 5: Configuration of finite element meshes with radius 7.5 and 1.0 ft

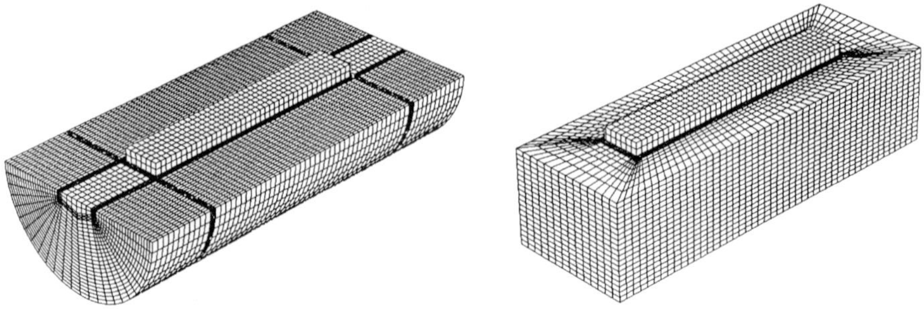

(a) Case 1_6 (5.0× 5.0× 7.5 ft) (b) Case 2_2 (5.0× 5.0× 7.5 ft)

Figure 6: Configuration of finite element meshes with radius 7.5 ft

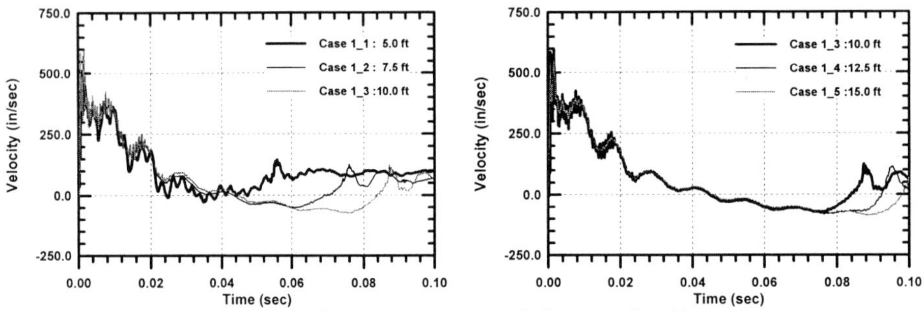

Figure 7: Vertical velocity response w/ charge under ship model

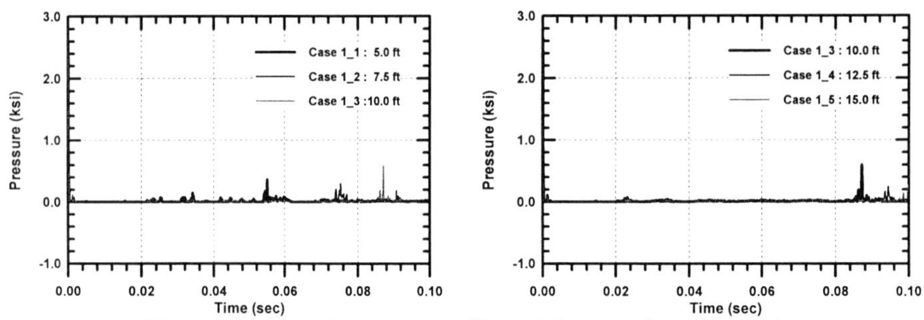

Figure 8: Fluid mesh pressure profiles w/ charge under ship model

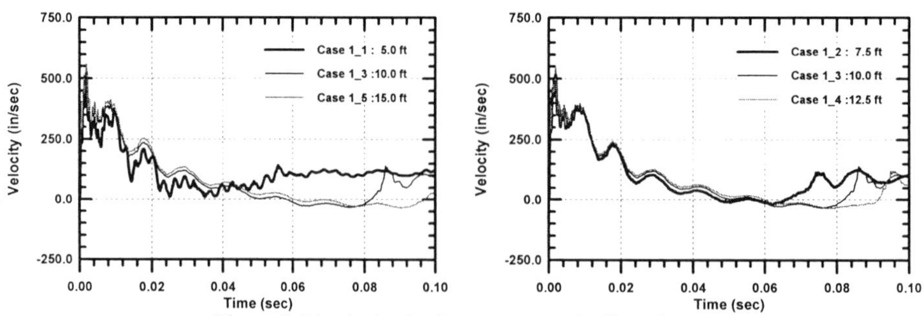

Figure 9: Vertical velocity response w/ offset charge

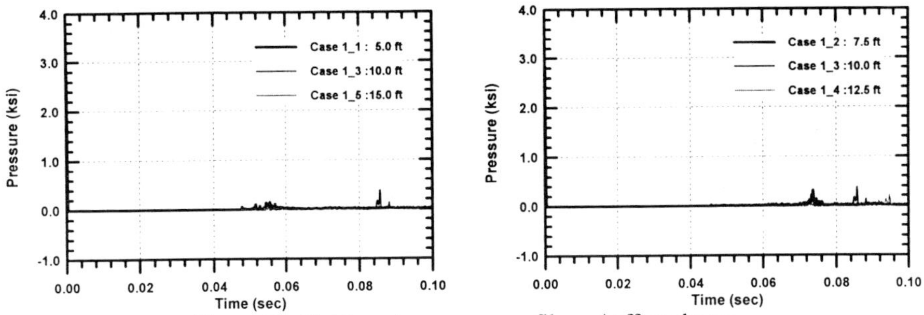

Figure 10: Fluid mesh pressure profiles w/ offset charge

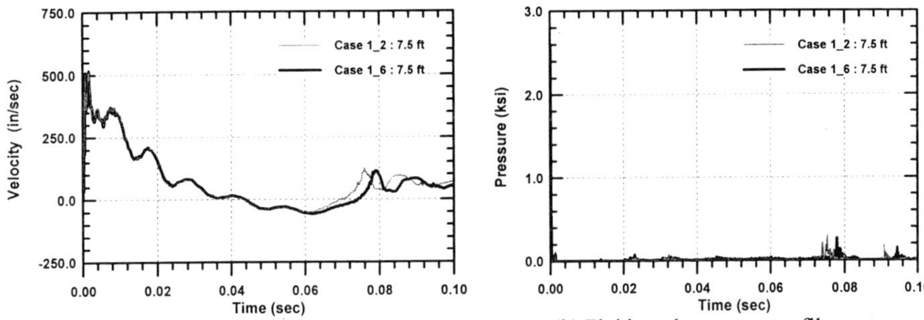

(a) Vertical velocity response (b) Fluid mesh pressure profiles

Figure 11: Effect of fluid mesh element size on shock response w/ charge under ship

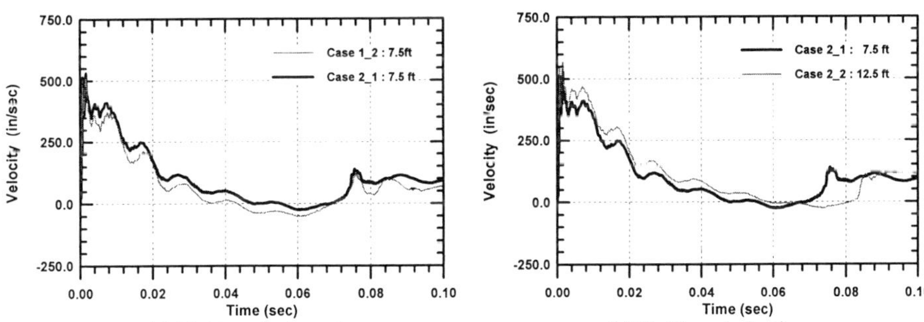

(a) Fluid mesh shape type (b) Fluid mesh depth

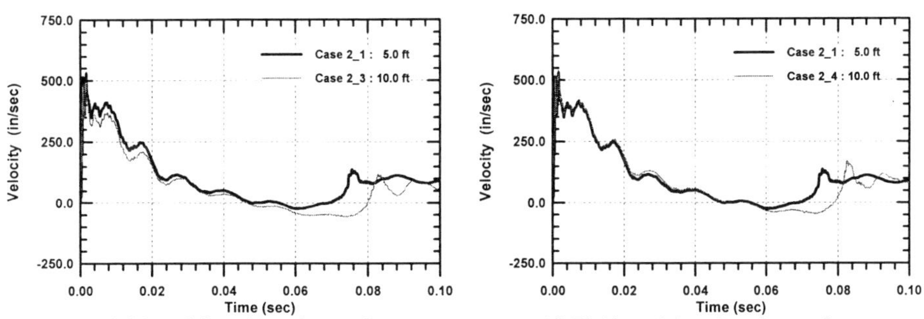

(c) Fluid mesh breadth at free surface (d) Fluid mesh length at free surface

Figure 12: Vertical velocity response w/ charge under ship model

Practical Design of Ships and Other Floating Structures
You-Sheng Wu, Wei-Cheng Cui and Guo-Jun Zhou (Eds)

WEIGHT REDUCTION IN SANDWICH STRUCTURES BY USE OF CURVED PANELS

C. Berggreen and B. C. Simonsen

Department of Mechanical Engineering, Maritime Engineering
Technical University of Denmark
DK-2800 Kgs. Lyngby, Denmark

ABSTRACT

The purpose of this paper is to investigate whether the membrane effects of laterally loaded, curved sandwich panels can save weight. Previous studies have indicated that significant weight reductions may be obtained, but the studies have been based on highly idealized boundary conditions and none of the studies have actually quantified the weight savings.

In this paper a representative bottom panel in a sandwich vessel is considered. The considered panel is 2000mm wide and 2700mm high, has a camber of 240mm and is subjected to uniform lateral pressure, of an amplitude varying between 37 kPa and 180 kPa.

In the first part of the paper the panel and the surrounding ship structure is modeled in a standard non-linear finite element program. In order to determine the effect of panel curvature a model with a plane panel with the same dimensions is also considered. The comparative analyses show that the minimum weight of the curved panel is 20% lower than that of the plane panel. Thus, the weight reduction potential is significant.

In order to utilize the membrane effect of the shell and obtain the weight reduction, it is essential, however, that the structure around the panel can hold the compressive membrane forces. In the second part of the paper, different surrounding structures are investigated, in order to determine under which conditions the advantages of the panel curvature can be obtained. First of all, it turns out that the real structure behaves very differently from the idealized, fully clamped boundary conditions considered in other studies. Secondly, it is shown how much structure is needed around a panel edge, in order to obtain the desired membrane effect. It turns out that the structure around the panel needs to be quite rigid. If the panel is at a weather deck, for example, the effect of the panel curvature on the weight will be insignificant.

KEYWORDS

Curved, Sandwich, Panel, Weight, Reduction, Optimization, Boundary, Condition, Surrounding, Structure, Stiffness

1 INTRODUCTION AND SCOPE OF WORK

In the last 30 years the demand from navies and civilian operators for lighter and faster vessels has been steadily increasing. Fibre-reinforced plastic (FRP) composites are among the most commonly used building materials for high speed light craft (HSLC) hulls. This is mainly because of the high strength-to-weight and stiffness-to-weight ratios but also due to advantageous properties with regards to thermal and acoustic insulation, fatigue and corrosion.

The key aspect in the design of weight-critical structures it to be able to take advantage of the building material and use it to the limits. This, in turn, leads to requirements to theoretical tools for accurate prediction of the loads and the structural response. This paper is focused on structural optimization and on the structural response, but it is recognized that in order to push the design further towards the true optimum, further investigation of the loads is necessary.

The understanding of geometrically non-linear behavior, due to large lateral deflections, may in some cases be essential in order to produce an optimal design. For initially flat plates it has long been known that even at relatively low levels of load, the effect of the geometrical non-linearity may be significant. The non-linearity arises because the membrane stresses in the plate are activated gradually as the maximum deflection of the plate has a finite magnitude. This effect has been documented experimentally by Shenoi, Moy and Allen (1996) for single skin plates and by Bau-Madsen, Svendsen and Kildegaard (1993) for sandwich plates. Furthermore, Riber (1997) presented a thorough theoretical analysis of flat sandwich plates and showed that it is possible to derive a closed-form expression for the non-linear, membrane effect of a sandwich plate, similar to the one DNV (1991) is using for single skin panels.

The same type of membrane effect will also come into play for a laterally loaded panel, which is initially curved outwards. For such geometry the membrane loads are activated immediately without requirements to a finite deflection from the initial geometry. Berggreen (2000), Bozhevolnaya (1998) and Smidt (1993) have presented analyses of this type of geometry. The two types of membrane effects are schematically illustrated in Figure 1.

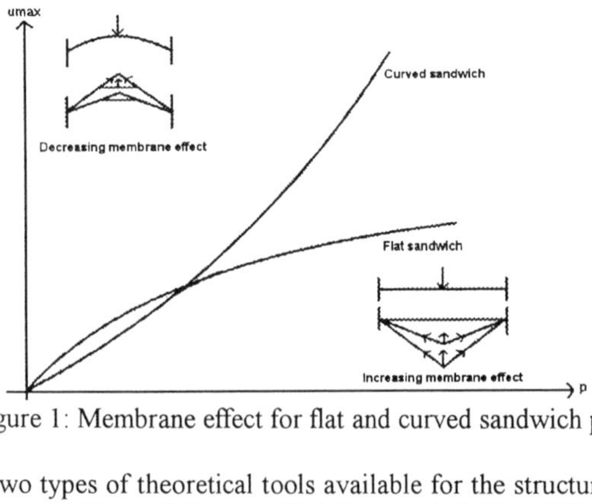

Figure 1: Membrane effect for flat and curved sandwich panel

The ship designer has two types of theoretical tools available for the structural analysis: the analytical rules of the classification societies and the so-called direct calculations. The analytical rules are quite straightforward to use and they do not require any computer-aided modeling. However, for all classification societies, the analytical rules for design of sandwich-panels are based on linear theory for a flat plate, so these rules do not take into account the membrane effects mentioned above. Therefore, some conservatism or non-optimality will be the result of their use, if the membrane effects can actually be utilized.

On the other hand, the so-called direct calculations, i.e. finite element modeling of the structure, may account for complex geometries, boundary conditions and membrane effects. The classification societies have allowed for direct calculations in the rules and specify requirements to the loads, to the modeling techniques and also to the allowable utilization of the nominal material strength and maximum deflections, see for example Det Norske Veritas (1991,1999).

To activate and take advantage of the membrane loads obviously requires that the structure around the considered panel provides the strength and stiffness necessary to hold the in-plane loads. In the work of Riber (1997), for example, it was assumed that the boundaries were fixed against in-plane movement. In the work done by Bozhevolnaya (1998) it was shown, that the magnitude of extra loading capacity in a curved sandwich panel is closely linked to the stiffness of the panel's boundaries.

The purpose of the first part of the paper is to investigate how much weight can be saved by taking advantage of the panel curvature. The purpose of the second part of the paper is to investigate what requirements should be put to the surrounding structure in order to hold the membrane loads.

2 THE VESSEL, MATERIALS AND DESIGN PARAMETERS

The analyses are based on one specific case: a bottom panel in a representative medium size vessel. The vessel length is 25m, breadth is 7.0m, draught is 1.8m, displacement is 125tons and speed is 12 knots.

One specific representative bottom panel is considered in the analyses. The panel is located between 16.0 and 18.0 m from the aft end of the vessel. The breadth of the panel is 2000mm and the average height (the chord length) is 2700mm. The panel is approximately cylindrical with a camber of 240mm, i.e. the radius of curvature of 3917mm. The geometry of the considered panel can be seen in .

The faces of the sandwich panel are built of quadri-axial glass fiber mats, with polyester resin. Thus, each laminate consists of a number of sets of layers of unidirectional (UD) plies angled in 0, 45, 90 and –45 degrees. On the top and bottom of each face laminate is a chopped stand mat (CSM). The mechanical properties of both laminates and the lay-up sequence can be seen on Table 1.

TABLE 1
UNI-DIRECTIONAL AND CHOPPED STRAND MAT GRP LAMINA

UD-lamina: E-glass/polyester				**CSM-lamina: E-glass/polyester, 300 g/m²**			
E_1	25.8 GPa	X_t	720 MPa	E_1	6.0 GPa	X_t	80 MPa
E_2	8.0 GPa	X_c	-351 MPa	E_2	6.0 GPa	X_c	-100 MPa
ν_{12}	0.26	Y_t	70 MPa	ν_{12}	0.35	Y_t	80 MPa
G_{12}	4.9 GPa	Y_c	-122 MPa	G_{12}	2.3 Gpa	Y_c	-100 MPa
t	varying	S	63 MPa	t	0.64 mm	S	20 MPa
Quad-mat with lay-up sequence: [0 45 90 –45]							

The core consists of cellular PVC foam with densities ranging from 45 to 250 kg/m^3 (Divinycell H-series). The mechanical properties for this material can be seen on Table 2. The maximum thickness of the core is set to 100mm.

TABLE 2
MECHANICAL PROPERTIES FOR THE CORE MATERIAL

Core Material (Divinycell H-series)						
Code	**ρ [kg/m^3]**	**E [MPa]**	**G [MPa]**	**X_t [MPa]**	**X_c [MPa]**	**S [MPa]**
H60	60	55	22	1.5	-0.8	0.7
H80	80	75	30	2.1	-1.25	1.05
H100	100	110	45	3.1	-1.7	1.4
H130	130	165	55	4.0	-2.8	2.15
H160	160	230	66	5.1	-3.4	2.6
H200	200	310	85	6.4	-4.4	3.3
H250	250	400	108	8.8	-5.8	4.5

3 PANEL DESIGN BY USE OF ANALYTICAL AND NUMERICAL CALCULATIONS

The purpose of this section is to investigate how much weight can be saved by taking advantage of the panel curvature. Furthermore, the analysis will show how much weight is saved by going from a linear analytical rule calculation (based on a flat plate) to a direct calculation for this particular configuration.
The designs are developed in agreement with the Rules of Det Norske Veritas (1991, 1996, 1999). The vessel is designed to Class Light Craft (LC), Type: Patrol, Service Restriction: R0. The design pressure is 36.4 kPa and is acting from the keel to the middle deck, see Figure 2.
Figure 2When the panel is designed by use of the analytical formulae given in the Rules, the optimum weight is found to be 80.8 kg, assuming fully clamped boundaries according to normal standard.
In order to design the panel by use of direct calculations, the finite element model shown in Figure 2 is used, and the commercial FEM code, **ANSYS**, is used for the analyses. The model includes the ship structure around the panel, from keel to deck and between the frames 13.5m and 19.5m from the aft section. The free boundaries of the shell plating are fully fixed and the nodes in the center plane are treated with a symmetry condition. The model contains 5116 elements and 15220 nodes. A non-linear 8 node layered composite shell element (SHELL91) has been used. The element has 4 integration point in the plane and 3 through the thickness for each layer. The element incorporates furthermore a sandwich option, which takes into account the out-of-plane shear deformation
The weight of the optimized curved panel becomes 72.8*kg*, when the rules for direct calculation are applied, see Configuration 3 in TABLE 3. The final design is determined by the required minimum face laminate thickness, the stress level in core and the maximum panel deflection. This design is 9.8% lighter than the basis design, Configuration 1 in TABLE 3. This saving of weight is caused partically by going from a linear plate theory to direct non-linear calculations and by going from a flat plate to a curved panel. In order to extract the effect of panel curvature, Configuration 2, was considered. Configuration 2 was
designed based on direct calculation with a modified finite element model with a plane panel. It is seen that in this case the minimum weight becomes 89.7 kg, which corresponds to a weight *increase* of 11%. The reason why this panel becomes heavier than Configuration 1 is that this design was based on the assumption of fully clamped boundaries. By comparison of Configurations 2 and 3 it can be seen that the
weight saved by use of the panel curvature is 18.8 % of the total panel weight. Thus, the reduction of weight is quite significant. The extra load-carrying capacity is obtained by taking advantage of membrane loads. In order to reach general conclusions regarding the effect of panel curvature, it is therefore necessary to investigate what in-plane stiffness and strength is required for the structure around the panel. That is the objective of the following section.

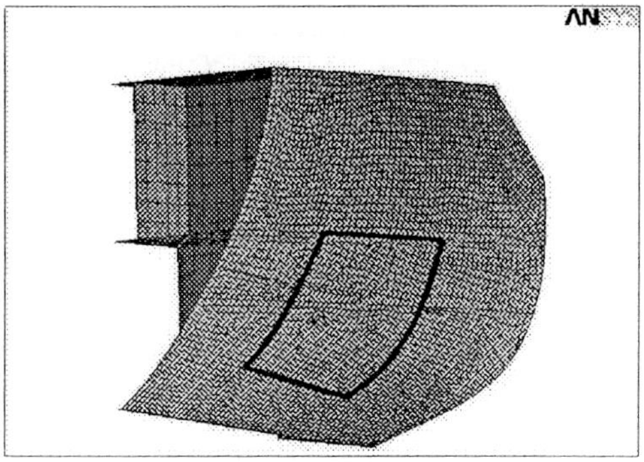

Figure 2: Finite element representation of the vessel

TABLE 3
RESULTS FROM THE WEIGHT OPTIMIZATION

Results of Weight Optimization						
Configuration	$t_{f,inner}$ [mm]	$t_{f,outer}$ [mm]	t_c [mm]	**Core density** [kg/m^3]	**Weight** [kg]	**Savings** %
1. Linear theory CL	3.64	3.64	60	80	80.8	0%
2. FEM flat	3.64	3.64	80	80	89.7	-11.0%
3. FEM curved	3.64	3.64	26	130	72.8	+9.8%
4. FEM curved (allow small $t_{f,inner}$)	2.46	3.64	37	100	59.5	+26.4%

4 EFFECT OF BOUNDARY IN-PLANE STIFFNESS ON MINIMUM PANEL WEIGHT

In order to investigate the effect of boundary stiffness on the weight of the curved panel, the model shown in Figure is considered. The model is an idealization of the real geometry using a perfectly cylindrical center part surrounded by two plane parts extending in the tangential direction to the keel and weather deck. The basic idea of this model is that strips may be removed from the upper boundary of the model, reducing the in-plane stiffness of the upper boundary. The plates are restricted against rotation around the q and r lines.

The results are shown in the two graphs in Figure 4 for two different pressures. For both configurations the internal structure is made sufficiently stiff that it does not deflect significantly. The figures show the weight of the optimized panel as a function of the strip-size, i.e. the amount of structure above the panel, see Figure. The weight is made non-dimensional by the weight of the panel if the boundaries had been fully fixed. From the left plot in Figure 4 it is seen that the weight ratio is gradually decreasing from 1.5 to 1.37 as the strip size is increased from 0 to 2000mm. This means that if the upper boundary is a free edge (for example at the weather deck) the weight is 13% higher than if the panel was embedded with a large amount of structure around it. On the other hand, if the in-plane deformations of the boundary could be fully held, 50% weight could be saved. For the panel with a very high load, the right plot in Figure 4, the numbers are different but the tendencies are the same.

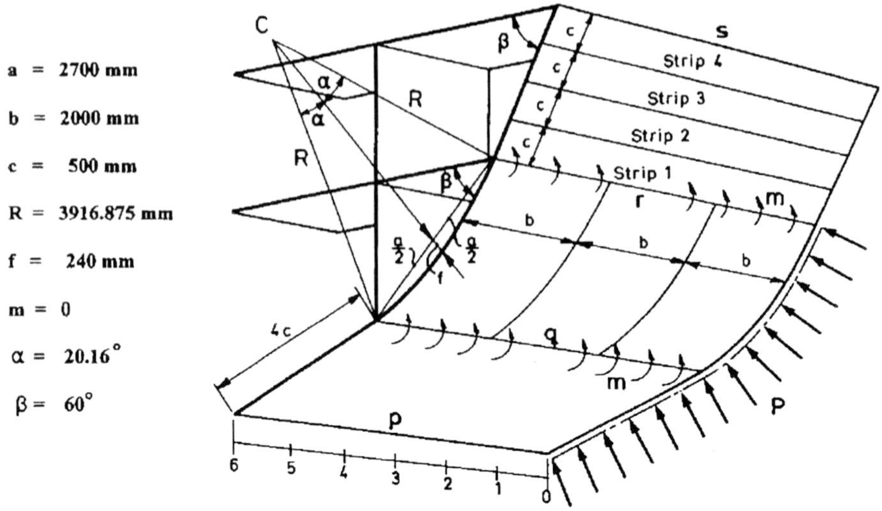

Figure3: Geometry of idealized section with removable plate strips

5 CONCLUSIONS

Previous studies have shown that the weight of curved sandwich panels may be reduced, if the membrane effect is taken properly into account in the structural analysis. It has, however, never been made clear how much weight can actually be saved and also what extra requirements the structure around the panel should fulfill, since it has to carry an additional in-plane load. The analyses presented here have considered a curved bottom panel in a representative medium size.
In the first part of the paper, the bottom panel is designed for minimum weight both by use of DNV's Rules for High Speed Light Craft and by direct calculations. The Rules are based on linear theory for a plane panel, with the usual idealized boundary conditions. The direct calculations are carried out for a large section of the hull in order to model boundary conditions correctly. Furthermore, the direct calculations are carried out both for the actual curved panel of the vessel and for a plane panel with the same dimensions embedded in the same structure. The analyses show that by taking into account the panel curvature in the structural analysis, approximately 20% weight can be saved. This weight reduction is sufficiently high that it should be utilized in truly weight critical ship structures.

The weight reduction is made possible by activation of compressive membrane forces in the panel. Therefore, in order to obtain this weight reduction, the structure around the panel must be able to hold those forces. The purpose of the second part of the paper is to investigate how close to a free boundary the panel can be placed without losing the potential of weight reduction. The analysis shows that there must be a significant in-plane stiffness of the structure around the panel in order to reach the weight reduction. For the considered panel, the width of the plate strip around the panel must be comparable in size to the panel itself in order to reach the 10-15% weight reduction. Remarkably, the analysis shows that the difference in weight between a panel that is fully clamped and one that has only a small strip around it, is 40-50% for certain loads. Similar results may apply to plane panels with large deflections. This means that previously published analyses, which are typically based an assumption of full fixation against in-plane movement of the boundaries, should be followed up by a critical assessment of the realism of this assumption.

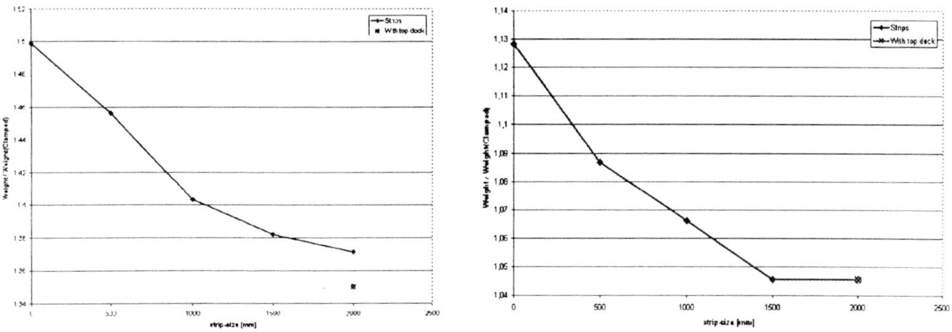

Figure 4: The effect of the surrounding structure on the optimum panel weight. The panel weight is made non-dimensional with the minimum weight of the fully clamped panel. This is shown as a function of the width of the plate strip above the panel. Left: p=60 kPa. Right: p=180kPa.

The reliability of the structure is closely linked to the uncertainty of the loads. At present the Rules prescribe uniform pressure loads. For a plane panel the, stresses and deflections are not very sensitive to the distribution of the lateral pressure. When the panel curvature is utilized, on the other hand, the pressure distribution may have a large effect on the response (cf. loading of an egg shell) so the load requirements should be revised to check if some non-uniform pressure distribution would be more reasonable.

References

Bau-Madsen N.K., Svendsen, K.H. and Kildegaard A. (1993), Large Deflections of Sandwich Plates - an Experimental Investigation, *Composite Structures* **23:1**, 47-52

Berggreen C. (2000), *Load-carrying Capacity of Curved Sandwich Panels (in Danish),* Master thesis, Department of Naval Architecture and Offshore Engineering, Technical University of Denmark, Lyngby.

Bozhevolnaya E. (1998), *A Theoretical and Experimental Study of Shallow Sandwich Panels,* Ph.D thesis, Department of Mechanical Engineering, Aalborg University, Aalborg, Denmark.

Det Norske Veritas (1991-1999), *Rules for Classification, High Speed Light Craft and Naval Surface Craft, Pt.3, Ch.1, 4 and 9, Pt.5, Ch.6,* Høvik, Norway

Rasmussen J. (1989), *Design of Sandwich Structures (in Danish)*, Ph.D thesis, Department of Ocean Engineering, Technical University of Denmark, Lyngby.

Riber, H.J. (1997), Non-linear analytical solutions for laterally loaded sandwich plates, *Composite Structures*, **39:1-2**, 63-83

Shenoi, R.A., Allen H.G. and Moy S.S.J. (1996), Strength and Stiffness of FRP Plates, *Proc. Inst. Civil Engrs., Structures and Buildings*, May

Skvortsov V. and Bozhevolnaya E. (2000), *Two-Dimensional Analysis of the Shallow Singly-Curved Sandwich Panels,* Report No.104, Department of Mechanical Engineering, Aalborg University, Aalborg, Denmark

Smidt S. (1993), *Curved Sandwich Beams and Panels: Theoretical and Experimental studies*, Report No.93-10, Department of Lightweight Structures, Royal Institute of Technology, Stockholm, Sweden

Practical Design of Ships and Other Floating Structures
You-Sheng Wu, Wei-Cheng Cui and Guo-Jun Zhou (Eds)

USE OF LARGE-DEFLECTION THEORIES FOR DESIGN OF FRP PANELS

Brian Hayman, Martin Jung Larsen, Dag McGeorge and Philippe Noury

Det Norske Veritas, N-1322 Høvik, Norway

ABSTRACT

Large-deflection plate theory forms the basis of the strength and deflection requirements for FRP single-skin laminates in the High Speed and Light Craft Rules of Det Norske Veritas (DNV). In the current version of the DNV Rules the restrictions set for allowable deflections limit the benefit to be gained from using this non-linear formulation. The paper examines the benefits of using a non-linear formulation within the current deflection limits, and the potential benefits of relaxing these limits. The current non-linear formulation is evaluated and compared with non-linear finite element calculations.

KEYWORDS

High speed craft, structural design, classification rules, fibre reinforced polymers, single-skin panels, strength, large deflections, geometrical non-linearity.

1 INTRODUCTION

Composite structures consisting of fibre reinforced plastics (FRP) are widely used in high speed and light craft and certain types of naval vessels. The applications range from sailing yachts to passenger ferries and high-performance marine vehicles such as patrol, rescue and fast attack craft.

The structural design of FRP hulls and superstructures is covered by Chapter 3 Part 4 of the DNV Rules for Classification of High Speed Light Craft and Naval Vessels (Det Norske Veritas, 1991). These Rules, which are currently under review, apply acceptance criteria that are concerned with both *strength* and *stiffness*. For simple structural geometries and FRP lay-ups the strength requirements are expressed in terms of allowable stresses. Formulae are provided for calculating the stresses for a given lateral pressure loading, for cases of both single-skin panels (i.e. single laminates) and sandwich panels. The stiffness requirements are expressed in terms of permissible deflections, which vary slightly according to the application and location of the panel. For single-skin panels they are expressed as a specified fraction of the laminate thickness. Either the strength or the stiffness/deflection requirement may be the governing criterion, depending on (amongst other parameters) the intensity of loading for which the panel is being designed.

For both single-skin and sandwich applications there has been some concern in the industry that the deflection limits being applied are more severe than they need to be. In particular there is a feeling that there are benefits to be gained from allowing greater non-linear membrane effects to develop so that the panel carries a significant proportion of the lateral loading through in-plane tensile stresses rather than by bending alone.

The present paper focuses on the case of single-skin panels. For such panels the DNV Rules already include a non-linear analytical formulation, although the imposed deflection limits are commonly believed to prevent significant non-linearity from developing. Studies are being conducted to explore the possibility of developing the Rules to allow larger deflections while ensuring a safe and reliable structural design. As a first step an assessment has been made of the benefits achieved by using the existing non-linear formulation within the existing deflection limits, and of the potential benefits of allowing a larger deflection limit. The range of validity of the current non-linear formulation has also been tested by comparison with non-linear finite element calculations. The results of these studies and their significance for future design are discussed in the sections that follow.

2 LARGE-DEFLECTION FORMULATION FOR SINGLE-SKIN PANELS

2.1 Formulation in Current DNV Rules

For single-skin FRP panels under uniformly distributed pressure loading, the current DNV Rules for High Speed and Light Craft (DNV, 1991) provide strength and stiffness requirements based on the following non-linear formulation. This formulation is restricted to rectangular panels for which the principal directions of reinforcement are parallel to the edges of the panel, and the difference between the moduli of elasticity in the two principal directions of reinforcement is not more than 20%.

First the location of the panel is established in relation to the entries in Table 1. This table specifies both the allowable value of stress σ_c and the allowable deflection parameter δ for the case in question. The deflection parameter is defined by

$$\delta = w/t \qquad (1)$$

where w is the mid-panel deflection and t is the thickness. The required panel thickness t is then given by substituting this value of δ into the equation

$$t = 178b\left(\frac{p}{\delta E\left(C_1 + \delta^2 C_2\right)}\right)^{\frac{1}{4}} \quad \text{(mm)} \qquad (2)$$

where p is the applied pressure loading in kPa, E the Young's modulus in MPa and b the smaller panel dimension in metres. The coefficients C_1 and C_2 are functions of the panel aspect ratio b/a (where a is the larger panel dimension) as given in Fig. 1.

Next the maximum in-plane tensile stress due to the combination of bending and membrane stresses is calculated from

$$\sigma_c = \left(\frac{t}{1000b}\right)^2 \delta E\left[C_1 C_3 + \delta C_4 C_2^{2/3}\right] \text{ (MPa)} \qquad (3)$$

where C_3 and C_4 are also given by Fig. 1. The maximum stress σ_c is now compared with the maximum allowable value in Table 1, in which σ_{nu} is the ultimate tensile strength of the laminate material. If the stress given by Equn. 3 does not exceed the allowable value, the previously calculated value of thickness t is satisfactory since both deflection and strength criteria are satisfied.

If, on the other hand, the stress given by Equn. 3 exceeds the allowable value from Table 1, the thickness must be increased until the strength criterion is satisfied. This can be done iteratively, by trying a smaller value of the deflection parameter δ in Equn. 2 to give an increased t. These revised values of δ and t are inserted in Equn. 3 to give the corresponding value of the maximum combined stress σ_c. This process is repeated until the calculated value of σ_c is less than or equal to the allowable value in Table 1.

It is convenient to re-write Equn. 2 and 3 such that the variables are all expressed in consistent units:

$$\frac{t}{b} = \left(\frac{p}{\delta E\left(C_1 + \delta^2 C_2\right)} \right)^{\frac{1}{4}} \tag{4}$$

$$\frac{\sigma_c}{E} = \left(\frac{t}{b}\right)^2 \delta \left[C_1 C_3 + \delta C_4 C_2^{\,2/3}\right] \tag{5}$$

TABLE 1.
MAXIMUM ALLOWABLE VALUES OF THE IN-PLANE COMBINED TENSILE STRESS σ_c AND DEFLECTION PARAMETER δ ACCORDING TO DNV RULES FOR HIGH SPEED AND LIGHT CRAFT

Structural member	σ_c	δ
Bottom panels exposed to slamming	0.3 σ_{nu}	1.0
Remaining bottom and inner bottom; side, deck and bulkhead structures; superstructures; deckhouses	0.3 σ_{nu}	0.9
All structures exposed to long time static loads	0.2 σ_{nu}	0.5

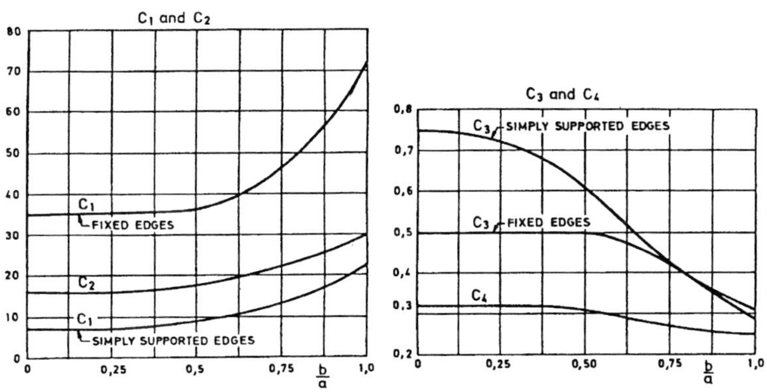

Figure 1: Coefficients C_1, C_2, C_3 and C_4 given as functions of the panel aspect ratio b/a, from DNV Rules for High Speed and Light Craft.

factor C_1 (which is associated with the bending deformation) is very much larger for fixed edges than clamped edges, whereas C_2 (which is associated with membrane deformation) is the same for both cases.

- The strength criterion curves for linear and non-linear analyses are very close for pressures above about 60 kPa.

Figure 2b shows also the effect of relaxing the deflection criterion by a factor of 2. This appears to allow lower thicknesses. However, as this only applies at pressures below the transition point between the deflection and strength criteria, i.e. at very low applied pressures, the benefit is limited.

4 ANALYSES TO CHECK THE VALIDITY OF THE NON-LINEAR FORMULATION

4.1 Approach

Figure 2b cast severe doubt on the validity of the non-linear formulation for the case of clamped edges, particularly with regard to the application of the strength criterion. To explore this in some detail, the results are compared with finite element analyses, and the theoretical basis of the non-linear analytical formulation as used in the DNV Rules is investigated. This is checked for the simply supported panel as well as the clamped one.

4.2 Non-linear finite element analyses

A series of non-linear finite element analyses has been performed using the ABAQUS/ADVANCE package. Comprehensive mesh refinement studies were performed but only the results of the final analyse after such refinement are reported here. The analyses were performed using a model of one-quarter of the panel, with symmetry conditions applied on two adjacent edges. Type S8R5 thin shell elements were used. An isotropic material model was assumed; the values of E and σ_c were the same as for the analytical solutions described previously, but now the in-plane Poisson's ratio $\nu_{12} = 0.17$ was inserted in addition, this being typical for a traditional marine GRP laminate with glass woven roving.

The procedure used was to assume a thickness value and then apply a series of increasing pressure loads until the maximum tensile stress exceeded the allowable stress. This was then repeated for a series of different thickness values so as to give the relation between applied pressure and the required thickness to satisfy the strength criterion.

The results for the case with clamped edges, shown in Fig. 3b, are discussed first. The curve is very different from that given by the non-linear analytical formulation; in fact it is quite close to that given by the linear formulation. However, all three curves are very close together for $p > 40$ kPa; below this the deflection limit applies in any case. The main conclusion is thus that the deflection limit $\delta < 1$ must be kept if the validity of the non-linear analytical formulation is to be maintained.

A corresponding finite element study on the panel with simply supported edges gave the results shown in Fig. 3a. For this case the form of the strength criterion curve is the same for both the finite element analysis and the non-linear analytical solution. However there is a horizontal shift between the curves. Closer study of this difference showed that this difference was due (at least partly) to differences in the material properties assumed in the two cases. Study of the origin of the analytical formulae in the DNV Rules revealed that they are based on an isotropic material model with $\nu_{12} = 0.3$, whereas in the FE analyses the Young's modulus is the same but ν_{12} was only 0.17. Some fresh FE analyses with $\nu_{12} = 0.3$ indicated a curve that was much closer to the non-linear analytical curve (see Fig. 3a).

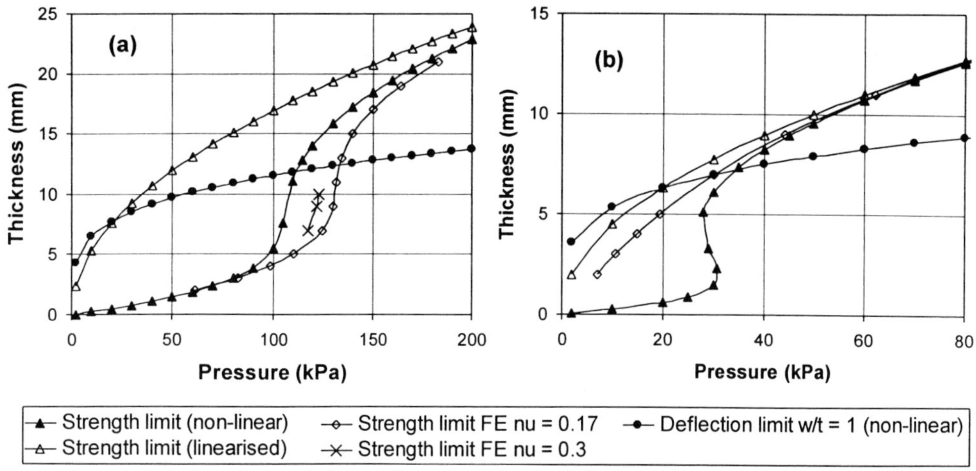

Figure 3: Required thicknesses according to strength and deflection criteria by non-linear finite element calculation, compared with analytical formulation, for (a) simply supported and (b) clamped edges.

4.3 Explanation of the differences between non-linear analytical and FE models

The non-linear analytical formulation in the DNV Rules is taken from a design manual for structural plastics published by the American Society of Civil Engineers (1984). The formulation is achieved by superposing the linear solution considering only bending deformations and a non-linear solution that considers only membrane deformations. For the simply supported panel the deformed shape for the membrane and bending cases are fairly similar until quite large out-of-plane displacements are reached, so this method of combination works quite well. However, this is not the case for the panel with clamped edges since the bending solution has a zero slope at the panel edges while the pure membrane solution (which is the same for both clamped and simply supported edges) has a non-zero slope. This is illustrated in Fig. 4, which shows the combined deformation predicted by the FE analysis for a fairly high load level together with the deformed shape for bending of a clamped beam. The non-linear FE solution shows a much sharper curvature at the panel edge caused by the panel trying to take up the membrane shape.

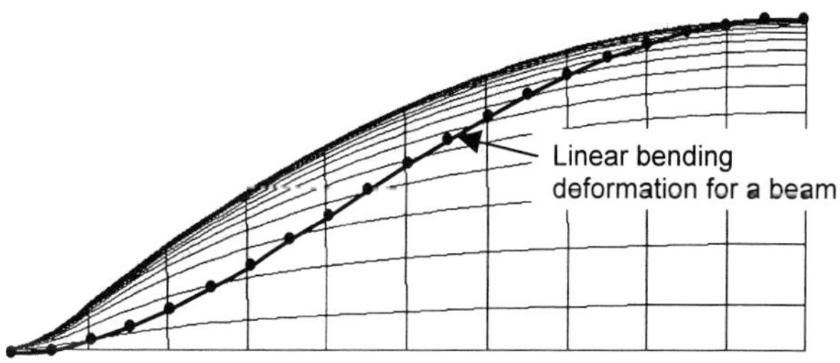

Figure 4: Deformed shape of panel with clamped edges (seen from end of panel) predicted by non-linear FE analysis compared with deformed shape from linear beam bending theory (both magnified and scaled to give the same maximum deflection).

The high bending stresses associated with this curvature cannot be reproduced by the linear bending part of the solution, so the analytical model greatly underestimates these stresses. Thus it is unsafe to use the non-linear analytical model when the deflections reach a level at which this happens. It turns out that applying the deflection limit $\delta = 1$ ensures that the approximation remains fairly accurate. However, in this range there is little difference between the linear and non-linear solutions for the clamped edge case.

5 MAIN CONCLUSIONS AND FINAL COMMENTS

The following are the main conclusions from this study of the application to singe-skin GRP laminates of the non-linear analytical formulation for stresses and deflections used in the DNV Rules:

- The non-linear formulation appears satisfactory for simply supported panels. Relaxing the currently applied deflection limit $\delta < 1$ for these panels may allow lower thicknesses for low design pressures, but at higher pressures the strength criterion governs in any case. In fact there is a substantial gain from using the non-linear formulation rather than a linear formulation even without relaxing the deflection limit.
- The non-linear analytical formulation is much less satisfactory for clamped panels. It is necessary to apply the deflection limit $\delta < 1$ for these panels to avoid serious errors and under-design due to the highly approximate formulation. Some benefit is gained from using the non-linear formulation rather than a linear formulation, but this is much less than for simply supported edges.
- If larger deflections are to be allowed than $\delta = 1$ for a clamped panel, an alternative non-linear formulation must be found; alternatively the linear formulation may be used, this being slightly conservative.

In practice there is usually some uncertainty about the boundary conditions that are appropriate for a given application, and it is not uncommon to analyse a panel assuming both sets of boundary conditions in the expectation that these represent opposite extremes. The studies show that the simply supported panel assumption leads to higher thicknesses to satisfy the deflection criterion (though only slightly higher when a non-linear formulation is used). However, for much of the range of design pressures considered, the simply supported panel requires lower thicknesses than the clamped one in order to satisfy the strength criterion.

The conclusions have been based on studies of panels with conventional glass reinforcements, and may have to be modified for laminates with appreciably different mechanical properties from these.

Acknowledgements

The work was carried out as part of the Nordsandwich project, which is a Nordic research collaboration on sandwich technology, co-ordinated by VTT Manufacturing Technology. Financial support was provided by the Nordic Industrial Fund and the Research Council of Norway.

References

Det Norske Veritas (1991). *Rules for the Classification of High Speed and Light Craft*, Part 3 Chapter 4, *Hull Structural Design, Fibre Composite and Sandwich Constructions*, Det Norske Veritas, Høvik, Norway.

American Society of Civil Engineers (1984). *Structural Plastics Design Manual*, ASCE Manuals and Reports on Engineering Practice, American Society of Civil Engineers, New York, USA.

Practical Design of Ships and Other Floating Structures
You-Sheng Wu, Wei-Cheng Cui and Guo-Jun Zhou (Eds)

DESIGN OF TEE CONNECTIONS IN FRP SHIPS USING AN ANALYTICAL APPROACH

R.A.Shenoi and W.Wang

School of Engineering Sciences, University of Southampton, Southampton, SO17 1BJ,UK

ABSTRACT

An elasticity-theory-based approach is developed for studying the design of the curved overlaminates forming the boundary angles in tee connections in FRP ship structures. The approach ensures an accurate description of the through-thickness and in-plane stresses in curved laminate beams. The solutions for various geometrical configurations are provided. The effects of various parameters, such as stacking sequence of the laminate, thickness of the skin are also studied in the applications.

KEYWORDS

FRP ships, Structural design, Through-thickness stress; Delamination, Curved composite laminate

1 INTRODUCTION

Fibre reinforced plastic (FRP) composite materials are finding increased usage in a wide variety of structural applications in the aerospace, civil construction, marine and offshore industries. A key feature of most such structures is the presence of bonded joints connecting two plate panels either butted together in-plane or placed in a tee fashion for out-of plane load transfer. In either case straps of laminated plates (or overlaminates) are bonded together to the base place to effect the load transfer. The strength of the joints then is dependent on correct selection of the adhesive and proper design of the overlaminate.

A typical ship section illustrating the boundary angles for top hat stiffeners and bulkhead to hull connections is shown in Figure 1. Most work on the strength characterisation of such joints has been done using experimental and/or numerical modelling techniques, for example Pei & Shenoi (1996). These are expensive and time consuming. This paper presents a simple approach which simulates the behaviour of the overlaminate as curved beam on an elastic foundation. The problem is solved using classical elasticity theory principles.

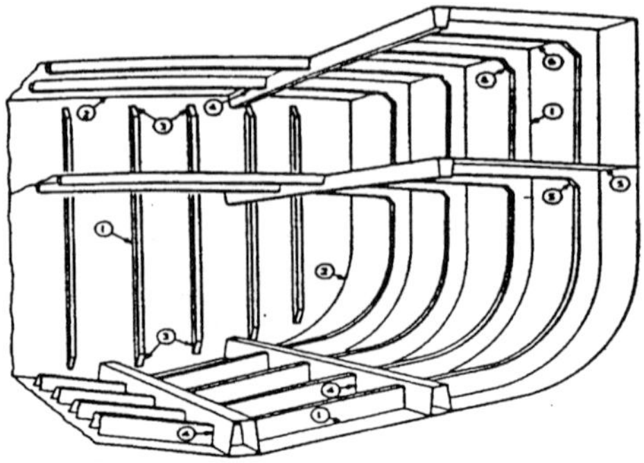

(1) Frame/shell (2) Bulkhead/shell
(3) Stiffener ending (4) Stiffener intersection
(5) Deck-edge (tee) (6) Deck-edge (knee)

Figure 1: Ship Hull Compartment showing typical tee connections

2 PROBLEM STATEMENT AND APPROACH FOR SOLUTION

The case of an overlaminate in a tee joint modelled as a cylindrical shell with large length (cylinder axial direction) is considered here, and the applied forces and boundary conditions are both uniform in cylinder axial direction (z-direction). The problem here could then be restricted to a curved layered beam of unit width, and it is assumed ε_z=0. A further assumption is made in that the transverse pressures load q and p are constants.

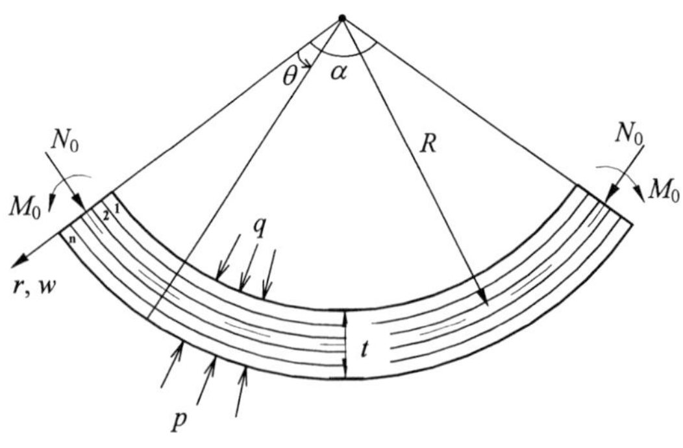

Figure 2: Typical curved composite beam under loads

Considering a general curved composite beam, as shown in Figure 2. Based on the theory of elasticity, the stress components can be written using an Airy stress function in polar coordinates $\Phi(r,\theta)$:

$$\begin{cases} \sigma_r = \dfrac{1}{r}\dfrac{\partial \Phi}{\partial r} + \dfrac{1}{r^2}\dfrac{\partial^2 \Phi}{\partial \theta^2} \\ \sigma_\theta = \dfrac{\partial^2 \Phi}{\partial r^2} \\ \tau_{r\theta} = -\dfrac{\partial}{\partial r}\left(\dfrac{1}{r}\dfrac{\partial \Phi}{\partial \theta}\right) \end{cases} \tag{1}$$

Under no-circumferential-dependence condition (e.g. p and q are both constants and this curved composite beam is in equilibrium state), the stress components do not depend on θ and $\tau_{r\theta}$ is zero. As far as a homogeneous curved beam (e.g. one layer of the composite beam) is concerned, considering the material with cylindrical anisotropy, a compatibility equation can be obtained as:

$$\frac{d^4\Phi}{dr^4}+\frac{2}{r}\cdot\frac{d^3\Phi}{dr^3}-\frac{\lambda}{r^2}\cdot\frac{d^2\Phi}{dr^2}+\frac{\lambda}{r^3}\cdot\frac{d\Phi}{dr}=0 \tag{2}$$

where $\lambda=E_1/E_2$. E_1 and E_2 are the moduli in circumferential and radial directions respectively. By solving this ordinary differential equation, stress components can be obtained:

$$\begin{cases}\sigma_r = 2C_2 + (1+\sqrt{\lambda})C_3\cdot r^{\sqrt{\lambda}-1} + (1-\sqrt{\lambda})C_4\cdot r^{-\sqrt{\lambda}-1}\\ \sigma_\theta = 2C_2 + (1+\sqrt{\lambda})\sqrt{\lambda}C_3\cdot r^{\sqrt{\lambda}-1} - (1-\sqrt{\lambda})\sqrt{\lambda}C_4\cdot r^{-\sqrt{\lambda}-1}\end{cases} \tag{3}$$

where C_2, C_3 and C_4 are constants which need to be determined.

The appropriate boundary conditions for any one of the three parts in a general curved sandwich beam--see Figure 2--then should be determined to calculate the coefficients in the above expressions. From the equilibrium of considered layer, the axial force and bending moment in every layer can be expressed as, see Shenoi & Wang (2001):

$$\begin{cases}N^{(i)} = \sigma_r^{(i)} r_i - \sigma_r^{(i-1)} r_{i-1}\\ M^{(i)} = -D_{11}^{(i)}\dfrac{4C_0^{(i)}}{\left(r^{(i)}+r^{(i-1)}\right)^2}\end{cases} \tag{4}$$

where $A_{11}^{(i)}, B_{11}^{(i)}$, and $D_{11}^{(i)}$ are respectively components of tension, coupling and bending stiffness of ith beam (the ith layer in the general laminate—see Figure 2), and $C_0^{(i)}$ is the radial displacement of ith beam. These can be taken as boundary conditions for every beam:

$$\begin{cases}r=r_{i-1}: \quad \sigma_r=\sigma_r^{(i-1)}\\ r=r_i: \quad \sigma_r=\sigma_r^{(i)}\\ \int_{r_{i-1}}^{r_i}\sigma_\theta dr = N^{(i)}\\ \int_{r_{i-1}}^{r_i}\sigma_\theta r dr = M^{(i)}\end{cases} \tag{5a}$$

For the first and third beam (inner face and outer face):

$$\begin{cases}r=R_i: \quad \sigma_r=0\\ r=R_o: \quad \sigma_r=0\end{cases} \tag{5b}$$

By substituting the expressions of stress components (3) into boundary conditions (5a), it can be noted that the third equation in (5a) is automatically satisfied when the first and second equations are satisfied. These equations about C_2, C_3 and C_4 in each layer can then be solved and the stresses in every layer can then be determined by the expression containing $C_0^{(i)}$ and $\sigma_r^{(i)}$.

Here it should also be noted that the above solution for anisotropy cannot be extended to the isotropic case (e.g. core in sandwich beam or 90° stacking laminae) only taking $\lambda=E_2/E_1=1$, although convergence to the isotropic solution can be achieved by putting λ close to 1. In isotropic materials, the stress distribution is, see Timoshenko (1934):

$$\begin{cases}\sigma_r = \dfrac{a}{r^2}+b(1+2\ln r)+2c\\ \sigma_\theta = -\dfrac{a}{r^2}+b(3+2\ln r)+2c\\ \tau_{r\theta}=0\end{cases} \tag{6}$$

If the same boundary conditions as in equations (5a) and (5b) are given, the constants a, b, c can then be determined by the expression containing $C_0^{(i)}$ and $\sigma_r^{(i)}$.

The displacement can then be obtained as a function of $C_0^{(i)}$ and $\sigma_r^{(i)}$ for either the anisotropic or the isotropic condition through a plane-strain analysis by invoking Hook's Law and the strain compatibility condition.

As can be seen, in the analysis of the whole sandwich beam, there exist five unknown variables $\sigma_r^{(i)}\big|_{i=1,2,\cdots,n-1}$ and $C_0^{(i)}\big|_{i=1,2,\cdots,n}$. To solve these unknowns, 2(*n*-1) equations can be obtained by the displacement compatibility condition on the interfaces:

$$\begin{cases} u_r^{(i+1)}(r_i) = u_r^{(i)}(r_i) & i = 1,2,\cdots,n-1 \\ u_\theta^{(i+1)}(r_i) = u_\theta^{(i)}(r_i) & i = 1,2,\cdots,n-1 \end{cases} \tag{7}$$

$u_r^{(i)}(r), u_\theta^{(i)}(r)$ are expressed differently depending on whether the layer under consideration is orthotropic or isotropic in nature (Figure 2). The (2*n*-1)th equation is obtained from the general boundary condition. For the case of curved beam on an elastic foundation under no-circumferential-dependence condition, there is the global force boundary condition from the equilibrium of the general curved beam:

$$\begin{cases} \sum_{i=1}^{n} N^{(i)} = N_0 = pR_o - qR_i \\ p = kC_0^{(n)} \end{cases} \tag{8a}$$

where k is the elastic modulus of foundation. And for the case of curved composite beam subjected to pure bending, there is another global force boundary condition from the equilibrium of the general curved beam and constitutive equation of composite laminate:

$$M_o = -\left(D_{11}^0 - \frac{B_{11}^{0\,2}}{A_{11}^0} \right) \frac{4C_0^{(n)}}{(R_i + R_o)^2} \tag{8b}$$

where A_{11}^0, B_{11}^0, and D_{11}^0 are respectively components of tension, coupling and bending stiffnesses of the global composite laminate, M_0 are global bending moment.

Thus the problem can be solved and the through-thickness stress through whole thickness of the curved sandwich beam can then be obtained. Actually note that $\sigma_r^{(i)}\big|_{i=1,2,\cdots,n}$ are just the interfacial tension stresses between every two adjacent layers.

3 APPLICATION TO TEE JOINT STRUCTURES

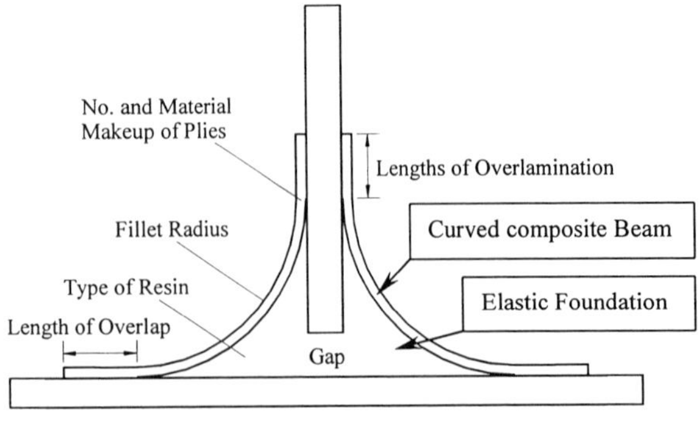

Figure 3: A typical tee-joint configuration

The developed approach could be applicable in practical contexts involving the design and characterisation of tee joints, for example. A typical tee joint, with variables influencing the design, is shown in Figure 3. Currently such structures are analysed using physical or numerical modelling, for example Shenoi and Hawkins (1995). Much of the experimental (and FEA) work relates to a 45° pull-off condition, where the tee piece is clamped on the two flanges and a load (at 45°) is applied to the tip of the web. In most of the work it has been shown that the performance of the joint is dependent on the strength of the boundary angle. A critical condition determining performance is the value of through-thickness inter-laminar tensile stresses, which induce the first delamination in the overlaminate or boundary angle and thus lead to eventual failure of the joint. Currently such conclusions can only be drawn after comprehensive and detailed numerical modelling.

The analytical approach developed in this paper is amenable for a relatively simple application to the boundary angle in the tee joint. Table 1 lists the design particulars of four different joint designs – all subjected to a 17.23kN pull-off load. The inter-laminar tensile stresses calculated from the curved beam model are listed in Table 2 and compared with values deduced from previous FEA analyses, see Shenoi & Hawkins (1995). As can be observed, there is close qualitative agreement of results in all cases. Importantly the trend is consistent. The reasons for the differences are two-fold. Firstly, deducing the boundary conditions at the edge of the boundary angles for purposes of analytical calculations proved difficult. Therefore some simplifications had to be made regarding the fixity and rigidity of the root region of the joint. These assumptions led to increased apparent stiffness in the boundary angle, leading to increased stresses. Secondly, the numerical analyses were carried out with a geometrical non-linearity option whereas the curved beam analytical model is based on inherent geometric and material linearities. This again would tend to increase (deflection, strains and thus) stresses. Such simple closed form, curved beam analyses can thus be used as a quick design reference to check for adequacy in a load-bearing context.

TABLE 1

DESIGN DETAILS OF THE TEE JOINTS

Sample	Boundary angle thickness (mm)	Fillet radius (mm)	Fillet overlay	Resin	Edge gap (mm)
B	15	50	---	CR1200	20
F	2	75	2WR*	CR1200	15
K	2	50	2WR	CR1200	15
L	2	75	2WR+CSM+	polyester	15

* WR – woven roving; + CSM – chopped strand mat

TABLE 2

COMPARISON OF FEA WITH CURVED BEAM MODEL RESULTS

Sample	Numerical Result	Theoretical Result	
	Maximum T-T* Stress (MPa)	Maximum T-T Stress (MPa)	Location
B	10.85	Very large (>50.00)	Near midplane
F	2.79	5.79	interface
K	9.10	14.28	interface
L	8.05	8.09	interface

* Through-thickness

4 CONCLUDING REMARKS

In conclusion, the paper has presented a new approach to the design of tee connections in FRP ships. An analytical formulation based on the flexural behaviour of a curved orthotropic beam on an elastic foundation has been applied to the overlaminate in a typical tee joints.

References

Pei J. and Shenoi R.A. (1996). Examination Of Key Aspects Defining The Performance Characteristics Of Out-Of-Plane Joints In FRP Marine Structures. *Composites* **27A:2**, 89-103.

Shenoi R.A. and Hawkins G.L. (1995). An Investigation into the Performance Characteristics Of Top-Hat Stiffener to Shell Plating Joints. *Composite Structures* **30**, 109-121.

Shenoi R.A. and Wang W. (2001). Flexural Behaviour of A Curved Orthotropic Beam on an Elastic Foundation. *Journal of Strain Analysis for Engineering Design* **36:1**, 1-16.

Shenoi R.A. and Wang W. Through-Thickness Stresses in Curved Composite Laminate and Sandwich, *Composite Science and Tecnology*, (accepted for publication, to appear).

Timoshenko S. (1934). *Theory of Elasticity*. McGraw-Hill Book Company, Inc., New York and London.

Practical Design of Ships and Other Floating Structures
You-Sheng Wu, Wei-Cheng Cui and Guo-Jun Zhou (Eds)

HISTORY OF PRADS

Masaki Mano

Consulting Naval Architect,
1988-57, Totsuka-Cho, Totsuka-Ku, Yokohama 244-0003, Japan

ABSTRACT

PRADS started in 1977 as a memorial event of the Society of Naval Architects Japan's 80 Years Anniversary with two main aims. One was to discuss the practical problems which are useful for actual ship design not to incline toward an academic theory. The other was to promote the friendship among people in shipbuilding industry and academic circles in mainly Asian countries. It has been a quarter century since PRADS started and at the beginning of the 21st century it will be meaningful to review the history of PRADS and to find the way to go on.
In this paper the Author reviewed statistically the tendency of PRADS, numbers of papers, numbers of participants, kinds of papers (practical or academic, hydrodynamic, structure etc.). The result of the review shows that the first aim of PRADS, to discuss practical problems which are useful for actual ship design, seems to be faded and the second aim, to promote the friendship among people in shipbuilding industry and academic circles, is developing not only in Asian region but also in world wide community.
The main reason why the first aim seems to be faded, the Author considers, will be small numbers of the participants from shipyards where practical problems are being generated every day. How to solve this problem is discussed in this paper.
The Author participated all PRADS since 1978 and has many memories of the important events which are to be kept in mind of PRADS participants. In this paper the important events in each PRADS are introduced.

KEYWORDS

Ship design, Practical design, International symposium, International friendship, Technical conference, Society of naval architects, Structure, hydrodynamics

1 OUTLINE OF PRADS

The idea of PRADS was born around 1975 in Ship Structure Committee, the Society of Naval Architects Japan. One of the main aims was to promote the friendship among the Asian naval architects and marine engineers. And the contacts with the presidents of the societies of naval architects in Asian countries show that they have positive opinions. In April 1976 PRADS Organizing Committee and PRADS Executive Committee started with a financial support of Japan Ship Promotion Association. And the covering area of the symposium was expanded not only ship structure but also to ship resistance, propulsion and ship motion etc.. PRADS symposium was to be held as an event celebrating 80 years anniversary of the Society of Naval Architects Japan.

The first aim of PRADS was to discuss the practical problems which are useful for the actual ship design not to incline toward academic theories. And the second was to promote the friendship among people in shipbuilding industry and academic circles in mainly Asian countries. However, as the announcement was distributed not only to Asian area but also to European and American areas, many European and American jointed.

The first PRADS was held under the auspices of the Society of Naval Architects Japan from 17th (Mon.) to 21st (Fri.) of October 1977 at Sasagawa Memorial Hall, Tokyo, Japan with 245 participants among which 90 persons came from 26 countries out of Japan. As the general lectures, the presidents of societies of naval architects of six Asian countries (Mr.Wu Ruen-ting of P.R.O.C., Mr.S.Hadisoemarto of Indonesia, Mr.S.J.Yim of Korea, Mr.V.Jovellanos of Philippines, Mr.Chua Chor Teck of Singapore and Mr.Y.Akita, Chairman of PRADS Organizing Committee) introduced the existing condition of the society of naval architects and ship building industry of each country. And 56 papers were presented. Almost of them were the paper on the practical ship design. On 21st Oct. a technical tour visiting IHI (Ishikawajima-Harima Heavy Industries Co. Ltd.) Yokohama Shipyard and Kamakura (famous old town) was held.

The second PRADS was held under the auspices of the Society of Naval Architects Japan and the Society of Naval Architects Korea on 17th (Mon.) and 18th (Tues.) of October 1983 at Sasagawa Memorial Hall, Tokyo, Japan and on 21st (Fri.) and 22nd (Sat.) of October 1983 at Hotel Lotte, Seoul, Korea. After first PRADS it was decided by the important persons of several countries that PRADS will be an international symposium and this time 366 participants came from all over the world. And 85 papers were presented. Between Tokyo and Seoul conferences on 19th (Wed.) and 20th (Thurs.) three technical tours were arranged as follows.
Course A : Tokyo - Pusan - Chungmu (stay one night) - Samsung Shipyard - Daewoo Shipyard - Pusan - Seoul.
Course B: Tokyo - Pusan - Korean Shipbuilding & Engineering Corp. - Ulsan (stay one night) - Hyundai Shipyard - Kyeonju (famous old capital) - Pusan - Seoul.
Course C : Tokyo - Sumitomo Oppama Shipyard - Kamakura - Tokyo Bay Sea Transport Control Centre - Tokyo

The third PRADS was held under the auspices of Norwegian Institute of Technology (NIT.) from 22nd (Mon.) to 26th (Fri.) of June 1987 at NIT., Trondheim, Norway with 235 participants. And 110 papers were presented. Every evening social event was held under the midnight sun until midnight as follows.
22nd, 22:00 pm. : Reception at Maritime Technical Centre with a demonstration of old ships in maneuvering tank
23rd, 19:00 pm. : After Nidelva river cruising, landed on Munkholmen Island off Trondheim and fishing around the island.
24th, 22:00 pm. : Midnight golf with "Midnight golf certificate"
25th, 19:00 pm. : Banquet

The fourth PRADS was held under the auspices of Bulgarian Ship Hydrodynamic Centre (BSHC) from 23rd (Mon.) to 28th (Sat.) of October 1989 at Chernomore Hotel, Varna, Bulgaria with 310 participants. And 141 papers were presented. Prof. S.Motora confirmed the characteristics of PRADS in his official speech at the banquet as follows.

1) Any one who has interest in the PRADS objective can join.
2) Any one can submit his paper. The paper submitted is to be selected by the committee.
3) The contents of the symposium is to be well balanced in basic and application, hydrodynamics, structure and design.
4) Any one other than the authors can join to the discussion.

The fifth PRADS was held under the auspices of British Maritime Technology Ltd. (BMT), British Shipbuilders, Kokums Computer Systems A.B., Lloyd's Register of Shipping, Marine Technology Directorate Ltd., North East Coast Institution of Engineers and Shipbuilders (NECIES), Press Offshore Ltd., Swan Hunter Shipbuilders and the University of Newcastle upon Tyne from 17th (Sun.) to 22nd (Fri.) of May 1992 at the University of Newcastle upon Tyne, Newcastle upon Tyne, UK. with 369 participants. And 107 papers were presented. After opening ceremony, two general lectures were given as follows.
"Engineering Design : Its Research and its Improvement" by Prof. Peter Hills of UK's Science and Engineering Research Council
"Marine Design : Advancing with Realism" by Dr. Marshal Meek, President of RINA

The sixth PRADS was held under the auspices of the Society of Naval Architects Korea from 17th (Sun.) to 22nd (Fri.) of September 1995 at Korea Exhibition Centre (KOEX), Seoul, Korea with 293 participants. And 125 papers were presented. After opening ceremony, two plenary lectures were given as follows.
"What about Double Hull for Bulk Carrier" by Dr. Frank S.B.Chao, President of Wahkwong Shipping Co. Ltd. Hongkong
"Korean's Role in the World Shipbuilding Industry" by Mr.Y.S.Song, President of Korean Shipbuilding Association
The cultural tour was visiting Korean Folk's Village in south part of Seoul, and the technical tour was to visit Samsung and Daewoo shipyards.

The seventh PRADS was held with Dr. M.W.C.Oosterveld, President of MARIN, in chair from 20th (Sun.) to 25th (Fri.) of September 1998 at the Netherlands Congress Centre, the Hague, the Netherlands with 235 participants. And 126 papers were presented. After opening ceremony, five plenary lectures were given as follows.
"Safety and Ship Design" and "Construction of Complicated Ship" by Prof. A.Aalbers of Delf University of Technology
"Model Test and Calculated Result of Ship and Structure in Waves" by Prof. J.A.Pinkster of Delf University of Technology
"Consideration on Hull Structure and Material Design" by Mr. G.T.M.Janssen of the Netherlands Applied Research Association
"Review on Ocean Technology" by Mr. J.H.Vugts of Royal Technical Society
The technical and cultural tour was held on 23rd afternoon visiting MARIN in Wageningen followed by a visit to the National Museum Kroller-Muller in the National Park the 'Hoge Veluwe'..

2 STATISTICAL ANALYSIS AND CONSIDERATIONS

The numbers of participants and the numbers of papers were statistically analyzed. Table 1 shows the numbers of the participants. Very roughly speaking, the foreign participants is about 70 % in the conferences held in Europe and 40 % in the conferences held in Asia because of many participants from the host country of Korea or Japan. The ratio of participants from the universities is about 40 %, that from the shipyards is about 20 % and that from others (research organizations, classification societies, governments etc.) is about 40 %. Naval architects and marine engineers in shipyards, the

Author considers, should attend PRADS conferences and absorb new knowledges to build better ships. PRADS started in 1977 aiming to discuss the practical problems which are useful for the actual ship design not to incline toward academic theories. This is a very important point, the Author considers, because a ship which brings happiness to people in the world carrying necessary materials can be designed and built by the naval architects and marine engineers in shipyards. From the other viewpoint PRADS should be more attractive to the naval architects and marine engineers in shipyards.

TABLE 1

NUMBERS AND PERCENTAGES OF PARTICIPANTS

	Total	Foreigners	from Shipyard	from Univ.	from Others
PRADS'77 Tokyo	245	90 (37%)			
PRADS'83 Tokyo & Seoul	366				
PRADS'87 Trondheim	235	163 (69%)	61 (26%)		
PRADS'89 Varna	310	204 (66%)			
PRADS'92 Newcastle	369	240 (65%)	54 (15%)	145 (39%)	170 (46%)
PRADS'95 Seoul	293	125 (43%)	81 (27%)	118 (40%)	94 (33%)
PRADS'98 The Hague	235	183 (78%)	47 (20%)	78 (33%)	110 (47%)

Table 2 shows the numbers of papers applied, presented, for design, for theoretical structure, for theoretical hydrodynamics and others. It seems that the capacity of the PRADS is limited up to about 120-140 papers. In PRADS'89 Varna, over 400 papers were applied among which 141 papers were selected for presentation. In PRADS'95 Seoul and in PRADS'98 The Hague over 200 papers were applied and 119 and 126 papers were presented respectively.

In PRADS, as its name shows, papers on practical design will be particularly welcome. From this viewpoint, PRADS'77 Tokyo and PRADS'92 Newcastle upon Tyne can be acceptable. PRADS'77 Tokyo was the first PRADS aiming to discuss the practical problems which are useful for the actual ship design not to incline toward academic theories. And PRADS'92 Newcastle upon Tyne was held focusing strictly on DESIGN as its central theme. These will be the reason of the success. Generally speaking, the ratio of the papers for design is 20 %, , that for theoretical structure is 30 %, that for theoretical hydrodynamics is 35 % and that for others is 15 %. The ratio of the papers for design of 20 % happened to coincide the ratio of the participants from the shipyards of 20 %. Less papers for design caused less participants from shipyards or less participants from shipyards caused less papers for design. This is a story of egg and hen. Anyway this is a point to be improved in future PRADS.

The ratio between numbers of papers of structure and hydrodynamics is well balanced except PRADS'89 Varna where Bulgarian Ship Hydrodynamics Centre - BSHC is famous and active. The ratio between numbers of papers of each category will be controlled by the committee of the host country, so it will be natural. The author considers that the papers for theoretical structure will be a little more effective in the practical ship design than those for theoretical hydrodynamics, because the structure of a ship sustains its life on the other hand the resistance and the propulsion improve the operational efficiency. On this viewpoint the ratio for theoretical structure of 30 % and that for theoretical hydrodynamics of 35 % will be reverse.

3 MEMORIAL EVENTS

PRADS'83 Tokyo & Seoul was successfully organized by the cooperation of Korean and Japanese Organizing Committees. Especially Korean Committee members, the Author appreciates, played their role wonderfully. The participants moved from Tokyo to Seoul enjoying two days technical tour in three groups. Everything was best prepared and all of the participants arrived at Seoul with full of satisfaction. Visiting and observing shipyards will be effective for the naval architects and marine engineers in shipyards. The modern and powerful four Korean shipyards, Samsung, Daewoo, KSEC and Hyundai, were open to the participants.

TABLE 2
NUMBERS AND PERCENTAGES OF PAPERS

	Applied	Presented	Design	Structure	Hydrodynamics	Others
PRADS'77 Tokyo		56	35 (63%)	4 (7%)	11 (20%)	6 (10%)
PRADS'83 Tko & Sel	119	85	21 (25%)	32 (38%)	20 (23%)	12 (14%)
PRADS'87 Trondheim		110	21 (19%)	33 (30%)	39 (35%)	17 (16%)
PRADS'89 Varna	over 400	141	39 (28%)	20 (14%)	53 (38%)	29 (20%)
PRADS'92 Newcastle		107	52 (48%)	17 (16%)	19 (18%)	19 (18%)
PRADS'95 Seoul	over 200	119	23 (19%)	35 (30%)	43 (36%)	18 (15%)
PRADS'98 The Hague	over 200	126	17 (13%)	35 (28%)	38 (30%)	36 (29%)

In PRADS'87 Trondheim, thanks to the midnight sun the participants enjoyed after conference time until midnight. The weather was not so good and the fishing tour was held in cold and rainy wind. Only three fishes were caught two of which were by Prof. Schmiechen of Berlin University..

The motto of PRADS'89 Varna was "The Shipbuilding Industry on the Threshold of 21st Century". Through the conference it had been fine weather. At the end of the conference Prof. J.B.Caldwell of the University of Newcastle upon Tyne asked the participants "Come to PRADS'92 Newcastle upon Tyne bringing your wife or girl friend, probably not both".

PRADS'92 Newcastle was held on perfectly fine days no cloud at all. The conference was prepared mainly by old men, Prof. J.B.Caldwell, Mr.R.C.Gray and Dr.G.Ward who retired or retiring. One evening the participants visited Beamish open air museum which is a reconstruction of life in the year 1913 as it was experienced in the North East of England.

In PRADS'95 Seoul the most impressive was the welcome address of Prof. K.C.Kim of Seoul National University, that was "---In the history of the Korean shipbuilding industry, the year 1981 was an epoch-making year in sense that the annual ship production exceeded 1,000,000 GT. The PRADS'83 gave a considerable impact on the promotion of the technical research and development activity of Korean shipbuilding industry. ---- The significance we host the PRADS'95 in Seoul is to recompense the community of the world maritime industry for their friendship with Korean shipbuilding industry. ---".

4 PROPOSAL AND CONCLUSIONS

From a viewpoint that a ship which brings happiness to people in the world carrying necessary materials can be designed and built by the naval architects and marine engineers in shipyards, they should more often attend and present papers on practical ship design and construction. In this respect, the Author would like propose following items.

1) In the call for papers, DESIGN is to be footlighted as PRADS'92 Newcastle upon Tyne.

2) More naval architects and marine engineers in shipyards are to be members of the standing committee or organizing committee.

3) The committee in the host country is to control the number of papers of each category to meet the PRADS original idea.

As a conclusion, the Author would like to emphasize that PRADS will develop successfully in future provided it is practical and attractive to shipyards.

References (in Japanese)

Mano M. (1984). Report of PRADS'83 (No.1). *Bulletin of the Society of Naval Architects Japan* **655**, 26-33

Mano M. (1984). Report of PRADS'83 (No.2). *Bulletin of the Society of Naval Architects Japan* **656**, 42-48

Mano M. (1987). Report of PRADS'87. *Bulletin of the Society of Naval Architects Japan* **698**, 41-44

Mano M. (1990). Report of PRADS'89 VARNA. *Science of Ship* **43-1**, 92-97

Mano M. (1992). Report of PRADS'92 NEWCASTLE UPON TYNE. *Science of Ship* **45-9**, 42-46

Mano M. (1995). Report of PRADS'95 SEOUL. *Science of Ship* **48-12**, 48-53

Mano M. (1998). Report of PRADS'98 (The Hague, The Netherlands). *Science of Ship* **52-3**, 43-51

PRADS Executive Committee (1977). PRADS (International Ship Design Symposium). *Bulletin of the Society of Naval Architects Japan* **581**, 10-15

AUTHOR INDEX

Volumes I and II

KEYWORD INDEX

Volumes I and II